STUDY GUIDE/SOLUTIONS MANUAL
TO ACCOMPANY

GENERAL CHEMISTRY

Table of Atomic Masses

Element	Symbol	Atomic Number	Atomic Mass
Actinium	Ac	89	(227)†
Aluminum	Al	13	26.98
Americium	Am	95	(243)
Antimony	Sb	51	121.8
Argon	Ar	18	39.95
Arsenic	As	33	74.92
Astatine	At	85	(210)
Barium	Ba	56	137.3
Berkelium	Bk	97	(247)
Beryllium	Be	4	9.012
Bismuth	Bi	83	209.0
Boron	B	5	10.81
Bromine	Br	35	79.90
Cadmium	Cd	48	112.4
Calcium	Ca	20	40.08
Californium	Cf	98	(251)
Carbon	C	6	12.01
Cerium	Ce	58	140.1
Cesium	Cs	55	132.9
Chlorine	Cl	17	35.45
Chromium	Cr	24	52.00
Cobalt	Co	27	58.93
Copper	Cu	29	63.55
Curium	Cm	96	(247)
Dysprosium	Dy	66	162.5
Einsteinium	Es	99	(252)
Erbium	Er	68	167.3
Europium	Eu	63	152.0
Fermium	Fm	100	(257)
Fluorine	F	9	19.00
Francium	Fr	87	(223)
Gadolinium	Gd	64	157.3
Gallium	Ga	31	69.72
Germanium	Ge	32	72.59
Gold	Au	79	197.0
Hafnium	Hf	72	178.5
Helium	He	2	4.003
Holmium	Ho	67	164.9
Hydrogen	H	1	1.008
Indium	In	49	114.8
Iodine	I	53	126.9
Iridium	Ir	77	192.2
Iron	Fe	26	55.85
Krypton	Kr	36	83.80
Lanthanum	La	57	138.9
Lawrencium	Lr	103	(260)
Lead	Pb	82	207.2
Lithium	Li	3	6.941
Lutetium	Lu	71	175.0
Magnesium	Mg	12	24.31
Manganese	Mn	25	54.94
Mendelevium	Md	101	(258)
Mercury	Hg	80	200.6
Molybdenum	Mo	42	95.94
Neodymium	Nd	60	144.2
Neon	Ne	10	20.18
Neptunium	Np	93	(237)
Nickel	Ni	28	58.70
Niobium	Nb	41	92.91
Nitrogen	N	7	14.01
Nobelium	No	102	(259)
Osmium	Os	76	190.2
Oxygen	O	8	16.00
Palladium	Pd	46	106.4
Phosphorus	P	15	30.97
Platinum	Pt	78	195.1
Plutonium	Pu	94	(244)
Polonium	Po	84	(209)
Potassium	K	19	39.10
Praseodymium	Pr	59	140.9
Promethium	Pm	61	(145)
Protactinium	Pa	91	(231)
Radium	Ra	88	226.0
Radon	Rn	86	(222)
Rhenium	Re	75	186.2
Rhodium	Rh	45	102.9
Rubidium	Rb	37	85.47
Ruthenium	Ru	44	101.1
Samarium	Sm	62	150.4
Scandium	Sc	21	44.96
Selenium	Se	34	78.96
Silicon	Si	14	28.09
Silver	Ag	47	107.9
Sodium	Na	11	22.99
Strontium	Sr	38	87.62
Sulfur	S	16	32.06
Tantalum	Ta	73	180.9
Technetium	Tc	43	(98)
Tellurium	Te	52	127.6
Terbium	Tb	65	158.9
Thallium	Tl	81	204.4
Thorium	Th	90	232.0
Thulium	Tm	69	168.9
Tin	Sn	50	118.7
Titanium	Ti	22	47.90
Tungsten	W	74	183.9
Uranium	U	92	238.0
Vanadium	V	23	50.94
Xenon	Xe	54	131.3
Ytterbium	Yb	70	173.0
Yttrium	Y	39	88.91
Zinc	Zn	30	65.38
Zirconium	Zr	40	91.22

*The values given here are to four significant figures. A table of more accurate atomic masses is given in Appendix E.

†A value given in parentheses denotes the mass of the longest-lived isotope.

STUDY GUIDE/SOLUTIONS MANUAL
TO ACCOMPANY

GENERAL CHEMISTRY

THIRD EDITION

Carole H. McQuarrie

Donald A. McQuarrie
University of California, Davis

Peter A. Rock
University of California, Davis

W. H. Freeman and Company
New York

ISBN 0-7167-2179-1

Printed in the United States of America

3 4 5 6 7 8 9 0   MB   9 9 8 7 6 5 4 3

CONTENTS

Acknowledgments   *vii*

To the Student   *ix*

1   Chemistry and the Scientific
    Method   *1*

2   Atoms and Molecules   *18*

3   The Periodic Table and Chemical
    Periodicity   *33*

4   Chemical Reactivity   *43*

5   Chemical Calculations   *63*

6   Chemical Calculations for
    Solutions   *91*

7   Properties of Gases   *113*

8   Thermochemistry   *149*

9   The Quantum Theory and Atomic
    Structure   *173*

10 Ionic Bonds and Compounds   *202*

11 Lewis Formulas   *218*

12 Prediction of Molecular
    Geometries   *240*

13 Covalent Bonding   *258*

14 Liquids and Solids   *278*

15 Colligative Properties of
    Solutions   *302*

16 Rates and Mechanisms of Chemical
    Reactions   *330*

17 Chemical Equilibrium   *365*

18 Acids and Bases I   *402*

19 Acids and Bases II   *436*

20 Solubility and Precipitation
    Reactions   *471*

21 Oxidation-Reduction Reactions   *503*

22 Entropy, Gibbs Free Energy, and
    Chemical Reactivity   *526*

23 Electrochemistry   *558*

24 Nuclear Chemistry and
    Radiochemistry   *586*

25 The Chemistry of the Main-Group
    Elements I   *612*

26 The Chemistry of the Main-Group
    Elements II   *632*

27 The Chemistry of the Transition
    Metals   *648*

28 Transition-Metal Complexes   *658*

29 Reactions of Organic
    Compounds   *676*

30 Synthetic and Natural Polymers   *698*

Glossary   *G-1*

ACKNOWLEDGMENTS

We thank Kay Ueno, Margot Getman, and Philip McCaffrey of W. H. Freeman and Company for coordinating and directing the publication of this Manual in an expeditious and professional manner. We also thank Dr. Joseph E. Ledbetter and Dr. Robert Kren for careful and detailed reviews of the manuscript and Martha Gleason for a first-rate copy-editing job.

TO THE
STUDENT

This Study Guide/Solutions Manual accompanies the text *General Chemistry* by Donald A. McQuarrie and Peter A. Rock. For each chapter in the text, this Manual has sections entitled

A.   Outline of the Chapter

B.   Self-Test

C.   Calculations You Should Know How to Do

D.   Solutions to the Odd-Numbered Problems

E.   Answers to the Self-Test

At the end, there is also a Glossary that is cross-referenced to the text.

The "Outline of the Chapter" (Section A) lists the headings for each section of the text, together with a few concise sentences that describe the key contents of each section.

The "Self-Test" (Section B) consists of about 40 short questions such as true/false or fill-in-the-blank questions. The Self-Test questions will give you a good indication of your understanding of the material in each chapter of the text, and we recommend that you answer these questions before you go on to do the numerical problems. The Answers to the Self-Test questions are given in Section E.

"Calculations You Should Know How to Do" (Section C) outlines each type of calculation that is presented in the chapter. It tells you what calculations you are expected to be able to do and keys these calculations to the worked Examples in the chapter and to the Problems at the end of the chapter. If you understand each type of calculation outlined in this section, then you are ready to go on to the Problems.

Occasionally, Section C also contains a detailed review of mathematical topics that appear in the chapter. For example, Chapter 1 requires a knowledge of exponents and writing numbers in scientific notation. Consequently, we treat these topics in this Manual and include a number of worked Examples and practice Exercises (with Answers). Other mathematical topics that are discussed are the equation of a straight line and its use to plot data (Chapter 7), natural logarithms (Chapter 16), common logarithms (Chapter 18), and the quadratic equation and its solutions, the quadratic formula (Chapter 17).

We think that "Solutions to the Odd-Numbered Problems" (Section D) is an especially valuable part of this manual. Most general chemistry courses tend to emphasize numerical problems, and so we have included detailed solutions to all the odd-numbered problems.

There is definitely a correct way and an incorrect way to use these problem solutions. When you are assigned a specific odd-numbered problem, you should do that problem first and compare your answer with the answer given in Appendix K of the text. If you are unable to get the correct answer, then you should refer to the detailed solution presented here to understand how to do the problem. Note that many problems in the text have been grouped and labeled according to topic, so if you have difficulty with a particular topic, you should work on additional problems, using this Manual as an aid. It is of little or no value for you to refer to the solutions given here before you make an honest, thoughtful attempt to do the problems yourself. Only by using the solutions as an aid to your understanding of how to do the problems will you benefit from having them.

Many of the solutions are presented in a step-by-step manner. Often the intermediate answers are given to one more significant figure than the final answer. If you solve the problem continuously on your calculator, your answer may differ slightly from the answer obtained in a step-by-step manner using the intermediate answers expressed to the proper number of significant figures. To avoid these discrepancies, do not round-off to the proper number of significant figures until the final answer.

Carole H. McQuarrie
Donald A. McQuarrie
Peter A. Rock

December 1990

1 CHEMISTRY AND THE SCIENTIFIC METHOD

A. OUTLINE OF CHAPTER 1

1-1 Why should you study chemistry?

1-2 Chemistry is an experimental science.

The scientific method is the use of controlled experiments to answer scientific questions.

Qualitative data consist of descriptive observations.

Quantitative data consist of numbers obtained by measurement.

A hypothesis is a possible explanation for an observation.

Experiments are performed to test the validity of a hypothesis.

A scientific law or natural law is a concise statement of a relationship among phenomena as determined by experiments.

Theory provides an explanation of scientific laws and aids in making predictions that lead to new knowledge.

1-3 Chemistry is based on quantitative measurements.

A quantitative measurement is one in which the result is expressed as a number.

A qualitative observation is a notation of a particular (nonnumerical) characteristic such as color or taste.

Antoine Lavoisier was the founder of modern chemistry.

The law of conservation of mass: In an ordinary chemical reaction, the total mass of the substances reacted is equal to the total mass of the products formed.

1-4 The metric system of units and standards is used in scientific work.

The numerical value of a physical quantity depends on the units.

The standard metric units are called SI units (for Système International).

Three basic SI units are of length, mass, and temperature.

Derived SI units are combinations of basic SI units.

The basic SI unit of length is the meter, whose symbol is m.

One meter is the distance light travels in a vacuum in $1/2.9979 \times 10^8$ seconds.

Prefixes are used to indicate factors-of-10 multiples and fractions of SI units.

Some of the common SI-unit prefixes are given in Table 1-2.

The SI unit of volume is derived from the meter.

The common measure of volume in the laboratory is the liter (L).

One liter is equal to 1000 cubic centimeters (cm^3).

One milliliter is equal to one cubic centimeter.

The basic SI unit of mass is the kilogram, whose symbol is kg.

Mass and weight are not synonomous.

Mass is the inherent amount of material in an object.

Weight is a measure of the force with which a body is attracted to a much larger body, such as the earth or the moon.

The mass of an object is determined by means of an analytical balance (Figure 1-9).

Temperature is a basic SI unit that gives a quantitative measure of the escaping tendency of heat from a body.

Numerical temperature scales are established by assigning temperatures to two reference systems.

Temperature is measured by means of a thermometer.

Three common temperature scales are the Celsius temperature scale, the Fahrenheit temperature scale, and the absolute Kelvin temperature scale.

The absolute Kelvin temperature scale is the most fundamental temperature scale.

The absolute Kelvin temperature scale is related to the Celsius temperature by the equation

$$T(\text{in K}) = t(\text{in °C}) + 273.15 \qquad (1\text{-}1)$$

Celsius temperatures are related to Fahrenheit temperatures by the equation

$$t(\text{in °C}) = \tfrac{5}{9}\,[t(\text{in °F}) - 32.0] \qquad (1\text{-}2)$$

Density is the amount of mass per unit volume of a substance or $d = m/V$.

The units of density are compound units such as $g \cdot cm^{-3}$ or $kg \cdot m^{-3}$.

Intensive properties of a substance are properties that are independent of the amount of the substance.

Density and temperature are intensive properties.

Extensive properties of a substance are properties that are dependent on the amount of the substance.

Volume and mass are extensive properties.

1-5   The precision of a measured quantity is indicated by the number of significant figures.

There are uncertainties associated with measured quantities.

The precision of a result conveys how well repeated measurements of a quantity agree with one another and how sensitive the measuring instrument is.

The accuracy of a result refers to how close the result is to the actual value.

High precision is no guarantee of high accuracy.

Rules for the determination of the number of significant figures in a measured quantity are given on page 15 of the text.

1-6   Calculated numerical results should show the correct number of significant figures.

The result of multiplication and division should not be expressed to more significant figures than the factor in the calculation with the fewest significant figures.

The result of addition and subtraction can be expressed to no more figures after the decimal point than the quantity in the calculation with the least number of figures after the decimal point.

1-7   Dimensional analysis is used to simplify many types of chemical calculations.

Units of the various quantities involved in calculations are treated as algebraic quantities.

The calculation is set up in a way that the undesired units cancel and the numerical answer is obtained in the desired units.

A physical quantity is converted from one unit to another by using a unit conversion factor.

1-8   The Guggenheim notation is used to label table headings and the axes of graphs.

Numbers are listed in tables of data without units.

The heading for tabulated data is given in the form: name, or symbol of the quantity/unit of the quantity.

The heading in a table of numerical values is treated as an algebraic quantity.

The axes of graphs in figures are labeled like the column headings for tabulated data.

B. SELF-TEST

1. The way to answer a scientific question is to _____

_____ .

2. The scientific method is _____

_____ .

3. A hypothesis precedes a theory. *True/False*

4. A scientific law is a concise explanation of experimental results. *True/False*

5. A theory can be used to make predictions that lead to new knowledge. *True/False*

6. Experiments may prove a theory incorrect. *True/False*

7. Scientific theories never change. *True/False*

8. A quantitative measurement is one in which the result is expressed as _____ _____ .

9. Explain how a qualitative result and a quantitative result differ.

10. State the law of conservation of mass.

11. In an ordinary chemical reaction, the total mass of substances that react is equal to _____ .

12. The preferred system of units in scientific work is the _____ system which uses _____ units.

13. The statement that the mass of a substance is 0.0289 is meaningless. *True/False*

14. The basic SI unit of length is the _____ .

15. A meter stick is *(longer/shorter)* than a yardstick.

16. The prefix kilo- indicates _____ of a SI unit.

17. The prefix centi- indicates _____ of a SI unit.

18. Measures of the SI unit of volume are derived from the SI unit for _____ .

19. The liter is the more usual unit of volume in laboratory work. *True/False*

20. One milliliter is equal to one cubic meter. *True/False*

21. The basic SI unit of mass is the _____ .

22. The terms mass and weight may be used interchangeably. *True/False*

23. The mass of an object is greater on earth than on the moon. *True/False*

24. An analytical balance is used in the laboratory to determine _____ _____ of an object.

25. Temperature is a basic SI unit that gives a quantitative measure of _____ _____ from a body.

26. The temperature of a body may be measured by means of a _____.

27. Three common temperature scales are the _____ temperature scale, the _____ temperature scale, and the _____ temperature scale.

28. The most fundamental temperature scale is the _____ temperature scale.

29. The Kelvin temperature scale is related to the Celsius temperature scale by the relation _____.

30. The Celsius temperature scale is not related to the Fahrenheit temperature scale. *True/False*

31. Density is defined as _____.

32. As an equation, the density $d =$, where _____ _____.

33. The units of a density depend upon the units of mass and volume used. *True/False*

34. An intensive property of a substance is *(independent/dependent)* on the amount of the substance.

35. An extensive property of a substance is *(independent/dependent)* on the amount of the substance.

36. The number of significant figures in a result indicates the *(precision/accuracy)* of the result.

37. A result of high precision is necessarily a highly accurate result. *True/False*

38. The zeros in the result 0.0289 m are significant figures. *True/False*

39. The zero in the result 45.70 mL is a significant figure. *True/False*

40. The result of the calculation 8.436×2.09 should be expressed to _____ significant figures.

41. The result of the calculation $8.436 + 2.09$ should be expressed to _____ digits after the decimal point.

42. Dimensional analysis is a method of calculation involving quantities with units. *True/False*

43. Units may be treated as algebraic quantities. *True/False*

44. A physical quantity is converted from one unit to another by the use of a

_____ .

45. If data for the speed of an object are tabulated in the form $speed/10^4 \, m \cdot s^{-1}$, then the value of the speed for the entry 1.46 is _____ .

46. In the Guggenheim notation, the numbers on figure axes are unitless. *True/False*

C. CALCULATIONS YOU SHOULD KNOW HOW TO DO

The SI units and conversion factors are listed on the inside back cover of the text.

1. Indicate the number of significant figures in quantities derived from calculations. See Examples 1-4, 1-5, and 1-6 and Problems 1-7 through 1-10.

2. Convert from one set of units to another set using a unit conversion factor and dimensional analysis. See Examples 1-7, 1-8, and 1-9 and Problems 1-11 through 1-16.

3. Use the Guggenheim notation to tabulate data or to plot data. See Examples 1-10 and 1-11 and Problems 1-17 through 1-20.

Scientific Notation and Calculations Using Exponents

The numbers encountered in chemistry are often extremely large (such as the number of atoms in a given quantity of substance) or extremely small (such as the mass of an electron in kilograms). To work with such numbers it is convenient to express them in scientific notation, whereby the number is written as a number between one and ten multiplied by 10 raised to the appropriate power. For example, the number 2831 is 2.831×1000, which is written 2.821×10^3 in scientific notation. Some other examples are

$$42500 = 4.25 \times 10^4$$

$$293100 = 2.931 \times 10^5$$

The zeros in these numbers are not regarded as significant figures and are dropped. Notice that in each case the power of 10 is the number of places that the decimal point is moved to the left.

$$42500_{\circlearrowleft} \qquad 293100_{\circlearrowleft}$$
$$\text{4 places} \qquad \text{5 places}$$

When numbers less than one are expressed in scientific notation, 10 is raised to a negative power. For example, 0.529 becomes 5.29×10^{-1}. Recall that a negative exponent is governed by the relation

$$10^{-n} = \frac{1}{10^n}$$

Some other examples are

$$0.006 = 6 \times 10^{-3}$$

$$0.000000742 = 7.42 \times 10^{-7}$$

Notice that the power of 10 in each case is the number of places that the decimal point is moved to the right.

$$0_{\odot}006 \qquad 0_{\odot}000000742$$

3 places   7 places

EXAMPLE 1 Express the following numbers in scientific notation.
(a) 0.000126   (b) 7380000000

Solution (a) We move the decimal point four places to the right to obtain 1.26×10^{-4}.
(b) We move the decimal point nine places to the left to obtain 7.38×10^9. Note that we do not retain the zeros following 8 because they are not significant figures.

It is necessary to be able to work with numbers in scientific notation. To add or subtract two or more numbers expressed in scientific notation, the power of 10 must be the same in each number. For example, consider the sum

$$1.711 \times 10^3 + 9.056 \times 10^2$$

We rewrite the first number to the power of 10^2:

$$1.711 \times 10^3 = 17.11 \times 10^2$$

Note that in having changed the 10^3 factor to 10^2, we have made the factor in front of 10^2 one power of 10 larger. Thus we have

$$1.711 \times 10^3 + 9.056 \times 10^2 = (17.11 + 9.056) \times 10^2$$
$$= 26.17 \times 10^2$$
$$= 2.617 \times 10^3$$

Similarly, we have

$$(6.287 \times 10^{-6}) - (1.562 \times 10^{-7}) = (6.287 - 0.1562) \times 10^{-6} = 6.131 \times 10^{-6}$$

When multiplying two numbers, we add the powers of 10 because of the relation

$$(10^x)(10^y) = 10^{x+y}$$

For example,

$$(2.00 \times 10^7)(6.00 \times 10^3) = (2.00)(6.00) \times 10^{10}$$
$$= 12.0 \times 10^{10}$$
$$= 1.20 \times 10^{11}$$

and

$$(5.014 \times 10^4)(7.143 \times 10^{-6}) = (5.014)(7.143) \times 10^{-2}$$
$$= 35.82 \times 10^{-2}$$
$$= 3.582 \times 10^{-1}$$

To divide, we subtract the power of 10 of the number in the denominator from the power of 10 of the number in the numerator because of the relation

$$\frac{10^x}{10^y} = 10^{x-y}$$

For example,

$$\frac{3.0 \times 10^{10}}{6.0 \times 10^{23}} = \left(\frac{3.0}{6.0}\right) \times 10^{10-23}$$
$$= 0.50 \times 10^{-13}$$
$$= 5.0 \times 10^{-14}$$

and

$$\frac{3.56 \times 10^{-6}}{8.73 \times 10^{-12}} = \left(\frac{3.56}{8.73}\right) \times 10^{-6+12}$$
$$= 0.408 \times 10^6$$
$$= 4.08 \times 10^5$$

EXAMPLE 2 Evaluate

$$x = \frac{(3.076 \times 10^{-4})(1.38 \times 10^{12})}{(6.67 \times 10^{-32})(7.110 \times 10^{21})}$$

Solution We rewrite x as

$$x = \frac{(3.076)(1.38)}{(6.67)(7.110)} \times 10^{-4+12+32-21}$$
$$= 0.0895 \times 10^{19}$$
$$= 8.95 \times 10^{17}$$

Note that we express our answer to three significant figures.

To raise a number to a power, we use the relation

$$(10^x)^n = 10^{nx}$$

For example,

$$\begin{aligned}
(3.141 \times 10^4)^3 &= (3.141)^3 \times 10^{12} \\
&= 30.99 \times 10^{12} \\
&= 3.099 \times 10^{13}
\end{aligned}$$

To take a root of a number, we use the relation

$$\sqrt[n]{10^x} = (10^x)^{1/n} = 10^{x/n}$$

Thus the power of 10 must be written so that it is divisible by the root. For example, the cube root of 6.40×10^7 is

$$\begin{aligned}
\sqrt[3]{6.40 \times 10^7} &= (6.40 \times 10^7)^{1/3} = (64.0 \times 10^6)^{1/3} \\
&= (64.0)^{1/3} \times 10^2 = 4.00 \times 10^2 \\
\sqrt{4.60 \times 10^5} &= (4.60 \times 10^5)^{1/2} = (46.0 \times 10^4)^{1/2} \\
&= (46.0)^{1/2} \times 10^2 = 6.78 \times 10^2
\end{aligned}$$

These calculations can be carried out directly on scientific hand calculators with a y^x function key. For example, to find $(6.40 \times 10^7)^{1/3}$, enter 6.40×10^7, press the y^x key, enter 0.33333 . . . , and press the equals sign (=) key to get 4.00×10^2.

EXAMPLE 3 Evaluate

$$x = \left(\frac{4.16 \times 10^{-3}}{9.723 \times 10^{12}} \right)^{1/3}$$

Solution
$$\begin{aligned}
x &= \left(\frac{4.16}{9.723} \times 10^{-15} \right)^{1/3} = (0.428 \times 10^{-15})^{1/3} \\
&= (428 \times 10^{-18})^{1/3} \\
&= (428)^{1/3} \times 10^{-6} \\
&= 7.54 \times 10^{-6}
\end{aligned}$$

You should realize that you can carry out all these calculations directly on your hand calculator. You can enter numbers in exponential notation, multiply and divide them, take roots and powers, and so on. It is well worth the effort to learn how to do this on your own hand calculator. Use your calculator to do the following exercises.

Exercises

Evaluate

1 $(4.164 \times 10^{-16})(9.275 \times 10^{12})$

2 $\dfrac{1.00 \times 10^4}{7.25 \times 10^8}$

3   $(6.176 \times 10^7)^{1/2}$

4   $(5.60 \times 10^{-3})^5$

5   $\dfrac{(2.14 \times 10^6)(7.813 \times 10^{-12})}{(8.89 \times 10^{16})}$

6   $\dfrac{(0.0929)(1728)}{(6.626 \times 10^{14})}$

7   $\left[\dfrac{(4.49 \times 10^5)(7.071 \times 10^{29})}{(1.019 \times 10^{-6})(6.88 \times 10^8)} \right]^{1/4}$

8   $\left[\dfrac{(5.716 \times 10^{-6})(4.28)}{(14.67 \times 10^2)} \right]^{1/3}$

Answers

1	3.862×10^{-3}	5	1.88×10^{-22}
2	1.38×10^{-5}	6	2.42×10^{-13}
3	7.859×10^3	7	1.46×10^8
4	5.51×10^{-12}	8	2.55×10^{-3}

D. SOLUTIONS TO THE ODD-NUMBERED PROBLEMS

1-1   The easiest way to order these quantities is to convert all the lengths to units of meters. We use the conversion factors according to the prefix as indicated in Table 1-2.

(a) 100×10^{-9} m $= 1.00 \times 10^{-7}$ m
(b) 1.0×10^3 m
(c) $1.0 \times 10^3 \times 10^{-2}$ m $= 1.0 \times 10^1$ m
(d) 100×10^{-12} m $= 1.00 \times 10^{-10}$ m
(e) $1.00 \times 10^3 \times 10^{-9}$ m $= 1.00 \times 10^{-6}$ m
(f) 1000 m $= 1.00 \times 10^3$ m

Thus, the order is d $<$ a $<$ e $<$ c $<$ b $=$ f.

1-3   The volume of the sphere is given by

$$V = \tfrac{4}{3}\pi r^3$$

If we substitute in the given value of r, then we have that

$$V = \tfrac{4}{3}\pi (100 \text{ pm})^3 = 4.19 \times 10^6 \text{ pm}^3$$
$$= \tfrac{4}{3}\pi (100 \times 10^{-12} \text{ m})^3 = 4.19 \times 10^{-30} \text{ m}^3$$

1-5   The density is given by

$$d = \frac{m}{V} = \frac{20.4 \text{ g}}{1.50 \text{ cm}^3} = 13.6 \text{ g} \cdot \text{cm}^{-3}$$

1-7   (a)   0.0390 has three significant figures: 3, 9, 0.
        (b) 6.022×10^{23} has four significant figures: 6, 0, 2, 2.

(c) 3.652×10^{-5} has four significant figures: 3, 6, 5, 2.

(d) 1,200,000 has two significant figures: 1, 2.

(e) The integer 16 is an exact number; there is no uncertainty associated with an integer.

1-9 (a) 2 (The result cannot be more accurate than zero digits past the decimal point.)

(b) 20800 or 2.08×10^4 (The result cannot be expressed to more than three significant figures.)

(c) 2.8 (The result cannot be expressed to more than two significant figures.)

(d) 3.4×10^{22} (The result cannot be expressed to more than two significant figures.)

1-11 (a) $(1.00 \ \cancel{L}) \left(\dfrac{1 \text{ qt}}{0.94633 \ \cancel{L}} \right) = 1.06 \text{ qt}$

(b) $(186{,}000 \ \cancel{\text{mile}} \cdot \text{s}^{-1}) \left(\dfrac{1.6093 \ \cancel{\text{km}}}{1 \ \cancel{\text{mile}}} \right) \left(\dfrac{10^3 \text{ m}}{1 \ \cancel{\text{km}}} \right) = 2.99 \times 10^8 \text{ m} \cdot \text{s}^{-1}$

(c) $(8.314 \ \cancel{J} \cdot \text{K}^{-1} \cdot \text{mol}^{-1}) \left(\dfrac{0.23901 \text{ cal}}{1 \ \cancel{J}} \right) = 1.987 \text{ cal} \cdot \text{K}^{-1} \cdot \text{mol}^{-1}$

1-13 The distance is given by

$$\text{distance} = \text{speed} \times \text{time}$$

$$= (3.00 \times 10^8 \text{ m} \cdot \cancel{\text{s}^{-1}}) (1 \ \cancel{\text{yr}}) \left(\frac{365 \ \cancel{d}}{1 \ \cancel{\text{yr}}} \right) \left(\frac{24 \ \cancel{h}}{1 \ \cancel{d}} \right) \left(\frac{60 \ \cancel{\text{min}}}{1 \ \cancel{h}} \right) \left(\frac{60 \ \cancel{s}}{1 \ \cancel{\text{min}}} \right)$$

$$= 9.46 \times 10^{15} \text{ m}$$

The distance in miles is

$$\text{distance} = (9.46 \times 10^{15} \ \cancel{\text{m}}) \left(\frac{1 \ \cancel{\text{km}}}{10^3 \ \cancel{\text{m}}} \right) \left(\frac{1 \text{ mile}}{1.6093 \ \cancel{\text{km}}} \right)$$

$$= 5.88 \times 10^{12} \text{ mile}$$

1-15 The total volume of soda in a six-pack is

$$\text{volume} = (6)(16 \ \cancel{\text{oz}}) \left(\frac{0.94633 \text{ L}}{32 \ \cancel{\text{oz}}} \right) = 2.84 \text{ L}$$

The cost per liter of soda in the six-pack is

$$\text{cost per L} = \frac{\$3.50}{2.84 \text{ L}} = \$1.23 \text{ per L}$$

The cost per liter of soda in the 2-L bottle is

$$\text{cost per L} = \frac{\$1.49}{2.00 \text{ L}} = \$0.75 \text{ per L}$$

The 2-L bottle is the better buy.

1-17   We shall use temperature/°C and volume/L as table headings. Thus the table is

temperature/°C	volume/L
0	1.000
100	1.37
200	1.73
300	2.10

1-19   The graph is

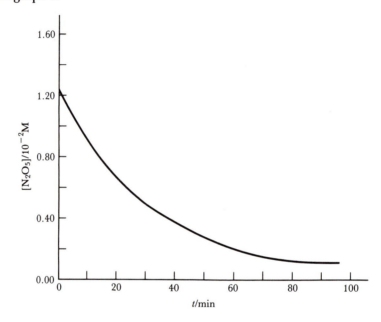

1-21   We use Equation (1-2) to convert from the Celsius temperature scale to the Fahrenheit temperature scale.

$$t(\text{in } ^\circ\text{C}) = \tfrac{5}{9}[t(\text{in } ^\circ\text{F}) - 32]$$
$$= \tfrac{5}{9}(68 - 32) = 20\,^\circ\text{C}$$

We use Equation (1-1) to convert from the Celsius temperature scale to the Kelvin temperature scale.

$$t(\text{in } \text{K}) = t(\text{in } ^\circ\text{C}) + 273.15$$
$$= 20 + 273.15 = 293 \text{ K}$$

1-23   The mass of the sulfuric acid solution is

$$\text{mass solution} = dV = \left(\frac{1.845 \text{ g}}{1 \text{ cm}^3}\right)(2.20 \text{ L})\left(\frac{10^3 \text{ cm}^3}{1 \text{ L}}\right)$$
$$= 4059 \text{ g}$$

The mass of the sulfuric acid is

$$\text{mass } H_2SO_4 = \left(\frac{96.7}{100}\right)(4059 \text{ g})\left(\frac{1 \text{ kg}}{10^3 \text{ g}}\right) = 3.925 \text{ kg} = 3.93 \text{ kg}$$

$$= (3.925 \text{ kg})\left(\frac{2.2046 \text{ lb}}{1 \text{ kg}}\right) = 8.65 \text{ lb}$$

1-25 We first calculate the smallest volume and largest volume that are allowed. The volume of a sphere is given by $V = \frac{4}{3}\pi r^3$. We can find the radius r of the baseball in each case from its circumference. The circumference of a circle C is equal to $2\pi r$. Thus, $r = C/2\pi$ and

$$\text{smallest radius} = \frac{9.00 \text{ in}}{2\pi} = 1.432 \text{ in}$$

$$\text{largest radius} = \frac{9.25 \text{ in}}{2\pi} = 1.472 \text{ in}$$

The volume of the baseball in each case is

$$V_{\text{smallest}} = \frac{4}{3}\pi(1.432 \text{ in})^3 = 12.30 \text{ in}^3$$

$$V_{\text{largest}} = \frac{4}{3}\pi(1.472 \text{ in})^3 = 13.36 \text{ in}^3$$

Thus, the range in densities of allowed baseballs is

$$d_{\text{least}} = \frac{m}{V} = \frac{5.00 \text{ oz}}{13.36 \text{ in}^3} = 0.374 \text{ oz} \cdot \text{in}^{-3}$$

$$d_{\text{greatest}} = \frac{m}{V} = \frac{5.25 \text{ oz}}{12.30 \text{ in}^3} = 0.427 \text{ oz} \cdot \text{in}^{-3}$$

1-27

$$\text{mass} = (135 \text{ lb})\left(\frac{1 \text{ kg}}{2.2046 \text{ lb}}\right) = 61.2 \text{ kg}$$

$$\text{height} = (5 \text{ ft } 7 \text{ in}) = \{[(5 \times 12) + 7] \text{ in}\}\left(\frac{2.54 \text{ cm}}{1 \text{ in}}\right)$$

$$= 170 \text{ cm or } 1.70 \times 10^2 \text{ cm}$$

1-29 The radius r of the container is one half the diameter of the container. Thus,

$$V = \pi r^2 h = \pi\left[\left(\frac{2.50 \text{ in}}{2}\right)\left(\frac{2.54 \text{ cm}}{1 \text{ in}}\right)\right]^2 (4.75 \text{ in})\left(\frac{2.54 \text{ cm}}{1 \text{ in}}\right)$$

$$= 382 \text{ cm}^3 = 382 \text{ mL}$$

1-31 The time it takes sound to travel one mile at sea level is

$$t = \frac{\text{distance}}{\text{speed}} = \left(\frac{1 \text{ mi}}{770 \text{ mi} \cdot \text{h}^{-1}}\right)\left(\frac{60 \text{ min}}{1 \text{ h}}\right)\left(\frac{60 \text{ s}}{1 \text{ min}}\right) = 4.675 \text{ s}$$

The time it takes light to travel one mile is

$$t = \frac{\text{distance}}{\text{speed}} = \left(\frac{1 \text{ mi}}{2.997925 \times 10^8 \text{ m·s}^{-1}}\right)\left(\frac{1.6093 \text{ km}}{1 \text{ mi}}\right)\left(\frac{10^3 \text{ km}}{1 \text{ km}}\right)$$
$$= 5.3680 \times 10^{-6} \text{ s}$$

The ratio of the two times is given by

$$\text{ratio} = \frac{t_{\text{sound}}}{t_{\text{light}}} = \frac{4.675 \text{ s}}{5.3680 \times 10^{-6} \text{ s}} = 8.71 \times 10^5$$

It takes sound almost 10^6 times as long to travel one mile as it does light.

1-33 What is meant by that statement is that no two snowflakes have yet been observed to be the same.

1-35 mass air $= dV = (1.20 \text{ g·L}^{-1})(6.0 \text{ L}) = 7.2 \text{ g}$

$$\text{mass oxygen} = \left(\frac{20}{100}\right)(7.2 \text{ g}) = 1.4 \text{ g}$$

1-37
$$\text{mass mercury} = dV = (13.56 \text{ g·cm}^{-3})(10.0 \text{ mL})\left(\frac{1 \text{ cm}^3}{1 \text{ mL}}\right)$$
$$= 136 \text{ g}$$
$$\text{volume of acetone} = \frac{m}{d} = \left(\frac{136 \text{ g}}{0.792 \text{ g·cm}^{-3}}\right)\left(\frac{1 \text{ mL}}{1 \text{ cm}^3}\right)$$
$$= 172 \text{ mL}$$

1-39 We must change the volumes of the ingredients to the metric system and the temperature to the Celsius scale.

$$\text{volume milk} = (2 \text{ cup})\left(\frac{8 \text{ fl oz}}{1 \text{ cup}}\right)\left(\frac{0.94633 \text{ L}}{32 \text{ fl oz}}\right)\left(\frac{10^3 \text{ mL}}{1 \text{ L}}\right)$$
$$= 473 \text{ mL}$$
$$\text{volume baking soda} = (1 \text{ tbs})\left(\frac{0.50 \text{ fl oz}}{1 \text{ tbs}}\right)\left(\frac{0.94633 \text{ L}}{32 \text{ fl oz}}\right)\left(\frac{10^3 \text{ mL}}{1 \text{ L}}\right)$$
$$= 15 \text{ mL}$$
$$t(\text{in °C}) = \tfrac{5}{9}[t(\text{in °F}) - 32] = \tfrac{5}{9}(350 - 32) = 180°C$$

1-41
$$t(\text{in °C}) = \tfrac{5}{9}[t(\text{in °F}) - 32] = \tfrac{5}{9}(104 - 32) = 40°C$$
$$T(\text{in K}) = t(\text{in °C}) + 273.15 = 40 + 273.15 = 313 \text{ K}$$

1-43
$$T(\text{in K}) = t(\text{in °C}) + 273.15$$
$$t(\text{in °C}) = \tfrac{5}{9}[t(\text{in °F}) - 32]$$

Substituting the expression for t(in °C) into the expression for T(in K) gives

$$T(\text{in K}) = \tfrac{5}{9}[t(\text{in °F}) - 32] + 273.15$$
$$= \tfrac{5}{9}\, t(\text{in °F}) - 17.78 + 273.15$$
$$= \tfrac{5}{9}\, t(\text{in °F}) + 255.37$$

1-45

$$\text{volume container} = \frac{m}{d} = \frac{6780 \text{ g}}{13.6 \text{ g} \cdot \text{mL}^{-1}} = 498.5 \text{ mL}$$

The density of carbon tetrachloride is given by

$$\text{density} = \frac{m}{V} = \frac{797 \text{ g}}{498.5 \text{ mL}} = 1.60 \text{ g} \cdot \text{mL}^{-1}$$

1-47

$$\text{distance} = (440 \text{ yd}) \left(\frac{1 \text{ m}}{1.0936 \text{ yd}} \right) = 402 \text{ m}$$

The 440-yd race is 2 m longer than a 400-m race.

1-49

$$\text{volume container} = \frac{m}{d} = \frac{250 \text{ g}}{1.00 \text{ g} \cdot \text{mL}^{-1}} = 250 \text{ mL}$$

$$\text{density gasoline} = \frac{m}{V} = \frac{175 \text{ g}}{250 \text{ mL}} = 0.700 \text{ g} \cdot \text{mL}^{-1}$$

1-51

$$\text{speed} = \left(\frac{100 \text{ m}}{9.80 \text{ s}} \right) \left(\frac{1 \text{ km}}{10^3 \text{ m}} \right) \left(\frac{1 \text{ mi}}{1.6093 \text{ km}} \right) \left(\frac{60 \text{ s}}{1 \text{ min}} \right) \left(\frac{60 \text{ min}}{1 \text{ h}} \right)$$
$$= 22.8 \text{ mi} \cdot \text{h}^{-1}$$

1-53

$$t(\text{in °F}) = \tfrac{9}{5} t(\text{in °C}) + 32 = \tfrac{9}{5}(3410) + 32 = 6170\text{°F}$$

1-55

$$\text{volume} = (40 \text{ ft})(20 \text{ ft})(6 \text{ ft}) = 4800 \text{ ft}^3$$
$$= (4800 \text{ ft}^3) \left(\frac{12 \text{ in}}{1 \text{ ft}} \right)^3 \left(\frac{2.54 \text{ cm}}{1 \text{ in}} \right)^3 \left(\frac{1 \text{ mL}}{1 \text{ cm}^3} \right) \left(\frac{1 \text{ L}}{10^3 \text{ mL}} \right)$$
$$= 1.36 \times 10^5 \text{ L}$$

$$\text{mass water} = dV = (1.0 \text{ g} \cdot \text{mL}^{-1}) \left(\frac{10^3 \text{ mL}}{1 \text{ L}} \right) (1.36 \times 10^5 \text{ L}) \left(\frac{1 \text{ kg}}{10^3 \text{ g}} \right)$$
$$= 1.36 \times 10^5 \text{ kg}$$

1-57 (a)

$$\text{volume room} = (30.58 \text{ ft})(41.0 \text{ ft})(9.00 \text{ ft})$$
$$= 11280 \text{ ft}^3$$

$$\text{mass air} = dV$$
$$= (1.184 \times 10^{-3} \text{ g} \cdot \text{cm}^{-3}) \left(\frac{1 \text{ lb}}{453.59 \text{ g}} \right) \left(\frac{2.54 \text{ cm}}{1 \text{ in}} \right)^3 \left(\frac{12 \text{ in}}{1 \text{ ft}} \right)^3 (11{,}280 \text{ ft}^3)$$
$$= 834 \text{ lb}$$

(b)

$$\text{volume} = \frac{m}{d} = \frac{1.0 \text{ g}}{1.184 \times 10^{-3} \text{ g} \cdot \text{cm}^{-3}} = 840 \text{ cm}^3$$

1-59
$$\text{area covered} = (350 \text{ ft}^2) \left(\frac{12 \text{ in}}{1 \text{ ft}}\right)^2 \left(\frac{2.54 \text{ cm}}{1 \text{ in}}\right)^2$$
$$= 3.252 \times 10^5 \text{ cm}^2$$

$$\text{volume paint} = (1 \text{ gal}) \left(\frac{3.7854 \text{ L}}{1 \text{ gal}}\right) \left(\frac{10^3 \text{ mL}}{1 \text{ L}}\right) \left(\frac{1 \text{ cm}^3}{1 \text{ mL}}\right)$$
$$= 3.785 \times 10^3 \text{ cm}^3$$

$$\text{thickness} = \frac{V}{A} = \frac{3.785 \times 10^3 \text{ cm}^3}{3.252 \times 10^5 \text{ cm}^2} = 0.0116 \text{ cm} = 0.116 \text{ mm}$$

1-61   The time it takes light to travel from the sun to the earth is given by

$$\text{time} = \frac{\text{distance}}{\text{speed}}$$
$$= \frac{(93 \times 10^6 \text{ mi})(1.61 \text{ km}/1 \text{ mi})(10^3 \text{ m}/1 \text{ km})}{(2.9979 \times 10^8 \text{ m} \cdot \text{s}^{-1})(60 \text{ s}/1 \text{ min})} = 8.3 \text{ min}$$

E. ANSWERS TO THE SELF-TEST

1.   to perform an experiment

2.   the use of experiments to answer scientific questions

3.   true

4.   true

5.   true

6.   true

7.   false

8.   a number

9.   See Section 1-3.

10.   In an ordinary chemical reaction, the total mass of the substances reacted is equal to the total mass of the products formed.

11.   the total mass of the products formed

12.   SI, metric

13.   true (The units must be given.)

14.   meter

15.   longer

16.   one thousand (10^3)

17.   one hundredth (10^{-2})

18.   length

19.   true

20.   false (It is equal to one cubic centimeter.)

21.   kilogram

22.   false

23.   false (The masses are the same; the weights are different.)

24.   the mass

25.   the escaping tendency of heat

26.   thermometer

27.   Celsius; Farhenheit; absolute Kelvin

28.   Kelvin

29. $T(\text{in K}) = t(\text{in } °C) + 273.15$

30.   false

31.   the mass per unit volume

32. m/V; m is the mass of the body and V is its volume.

33. true

34. independent

35. dependent

36. precision

37. false

38. false

39. true

40. three

41. two

42. true

43. true

44. unit conversion factor

45. $1.46 \times 10^4 \ \mathrm{m \cdot s^{-1}}$

46. true

2

ATOMS AND MOLECULES

A. OUTLINE OF CHAPTER 2

2-1   Elements are the simplest substances.

A mixture is composed of substances that are mixed together without altering their basic chemistry.

An element is a substance that cannot be broken down into simpler substances.

A compound is a pure substance that can be broken down into simpler substances.

2-2   About three-fourths of the elements are metals.

The elements can be classified into metals and nonmetals.

Some common metals and their chemical symbols are given in Table 2-3.

The nonmetals vary greatly in their appearance.

Some common nonmetals, their chemical symbols, and their appearances are given in Table 2-5.

Molecules are composed of two or more atoms joined together.

Some nonmetals exist as diatomic molecules (Figure 2-4).

2-3   The law of constant composition states that the relative amount of each element in a compound is always the same.

The law of constant composition is based on the chemical analysis of compounds.

The mass percentage composition of a compound can be calculated from the chemical analysis of the compound (Example 2-1).

2-4   Dalton's atomic theory explains the law of constant composition.

The postulates of Dalton's atomic theory are given on page 36 of the text.

Atoms are the small, indivisible particles of which matter is composed.

Atomic mass is the mass of one atom relative to that of another atom.

The atomic mass unit (amu) is a unit assigned to atomic masses.

2-5   Molecules are groups of atoms joined together.

A molecular picture of chemical reactions is introduced.

2-6   Compounds are named by an orderly system of chemical nomenclature.

If a compound is composed of a metal and a nonmetal, the metal is named first and then the nonmetal, with the ending of the name of the nonmetal changed to *-ide* (Table 2-6).

Subscripts in a chemical formula of a compound indicate the relative numbers of each kind of atom in the compound.

If a compound is composed of two nonmetals, then the number of each element is indicated in the name of the compound by Greek prefixes (Table 2-7).

2-7   Molecular mass is the sum of the atomic masses of the atoms in a molecule.

The molecular mass is also known as the formula mass.

The molecular mass of a compound can be calculated from its chemical formula and from the atomic masses of the elements in the compound.

Mass percentage compositions are calculated using atomic masses and molecular masses (Example 2-5).

2-8   Most of the mass of an atom is concentrated in its nucleus.

An atom is composed of subatomic particles.

The electron was discovered in 1897 by J. J. Thomson.

An electron is a negatively charged particle with a mass 1/1837 that of a hydrogen atom.

Radioactivity is the process in which certain nuclei spontaneously break apart.

α-particles, β-particles, and γ-rays are products of radioactive disintegrations (Table 2-8).

The nucleus was discovered by Ernest Rutherford and Ernest Marsden from the scattering of α-particles by a thin gold foil (Figure 2-15).

The proton was discovered by Ernest Rutherford.

A proton has a positive charge equal in magnitude to that of an electron but opposite in sign and has a mass that is almost the same as the mass of a hydrogen atom.

All the positive charge and essentially all the mass of an atom are concentrated in the very small volume in the center of the atom, called the nucleus.

The electrons in an atom are located throughout the space surrounding the nucleus.

2-9   Atoms consist of protons, neutrons, and electrons.

The neutron was postulated to account for the mass of a nucleus and was discovered by James Chadwick in 1932.

A neutron is an uncharged particle with essentially the same mass as a proton.

Atomic number is the number of protons in an atom and is denoted by Z.

In a neutral atom, the number of electrons is equal to the protons.

Mass number is the total number of protons and neutrons in an atom and is denoted by A.

The atomic number characterizes the element.

2-10   Most elements occur in nature as mixtures of isotopes.

Isotopes are atoms that contain the same number of protons but different numbers of neutrons.

An isotope is denoted by $_Z^A X$.

Atomic mass is the mass of an atom relative to the mass of the carbon-12 isotope.

The natural abundance of isotopes of an element, the naturally occurring percentages of the isotopes of an element, can be measured (Table 2-9).

Calculations involving natural abundances of isotopes and the atomic mass of an element are illustrated.

2-11   Isotopic masses are determined with a mass spectrometer.

Ions are atoms that have either more or fewer electrons than the neutral atom.

Positive ions have fewer electrons than the neutral atom and are called cations.

Negative ions have more electrons than the neutral atoms and are called anions.

The charge of an ion is indicated by a superscript following the symbol for the ion (page 53 of the text).

Species that contain the same number of electrons are called isoelectronic.

2-12   Heterogeneous mixtures are not uniform in composition from point to point.

The substances in a mixture exist together without combining chemically.

A heterogeneous mixture is not uniform from point to point in the mixture and may be separated by physical means.

A solution is a homogeneous mixture; that is, it is uniform from point to point.

A solid can be separated from a solution by filtration. The solution passes through the filter paper while the solid remains on the filter paper (Figure 2-21).

The solids in a water solution may be separated from the water by evaporation of the water.

A mixture containing a volatile component may be separated by distillation. One component is boiled away from the remaining components and may be converted back to a liquid or solid in a condenser (Figure 2-23).

2-13   Liquid or gas mixtures can be separated by chromatography.

Liquid-solid chromatography is used to separate the components of a solution.

In liquid-solid chromatography, the liquid is passed through a column packed with a solid that interacts with the components of the solution to differing degrees.

Gas-liquid chromatography is used to separate mixtures of gases of volatile liquids.

The process of gas-liquid chromatography is shown schematically in Figure 2-24.

A chromatogram is the record of the separation of a mixture by chromatography.

Paper chromatography and thin-layer chromatography are also used to separate compounds in solution.

In paper chromatography and thin-layer chromatography, the components of a solution interact to differing degrees with paper or the thin-layer chromatography absorbent.

B. SELF-TEST

1. A pure substance that cannot be decomposed into simpler substances is called a

_____.

2. A pure substance that can be decomposed into simpler substances is called a

_____.

3. Three properties of a metal are (a)

(b)   , and (c)

4. Most metals are solids at room temperature. *True/False*

5. Nonmetals are usually good conductors of electricity. *True/False*

6. Give the chemical symbol for the following metals: aluminum _____, calcium _____,

and potassium _____.

7. Name the elements whose symbol is O _____, C _____,

and Na _____.

8. All nonmetals are gases at room temperature. *True/False*

9. Two examples of elements that exist as diatomic molecules are _____

and _____.

10. The mass percentage of each element in a compound depends on how the compound is prepared. *True/False*

11. The law of constant composition states that the mass percentages of each element in a compound are equal. *True/False*

12. Chemical analysis of a compound is a determination of the _____

_____.

13. State in your own words the postulates of Dalton's atomic theory.

(a)

(b)

(c)

(d)

(e)

14. Explain the meaning of the scale of atomic masses.

15. The unit of the atomic mass scale is called _____ .

16. The atomic mass of an element is the mass in grams of one atom of the element. *True/False*

17. Explain the law of constant composition in terms of Dalton's atomic theory.

18. In a chemical reaction, the atoms of the reacting compounds are _____ _____ to form the products.

19. The rules for naming binary compounds composed of a metal and a nonmetal are

(a)

(b)

20. The molecular mass of a compound is the _____ _____ .

21. A beam of electrons passing between a voltage applied across electrodes will be deflected toward the _____ charged electrode.

22. A beam of protons passing between the same voltage as in Question 21 will be deflected in the same direction as the beam of electrons. *True/False*

23. A beam of protons passing between a voltage will be deflected to (*the same/a greater/ a lesser*) extent than a beam of electrons.

24. Compare the relative charges and masses of a proton, a neutron, and an electron.

25. A neutron has the same mass as an electron. *True/False*

26. Explain why some α-particles will be deflected through large angles when a beam of α-particles is directed at a thin gold foil.

27. Compare the mass of an atom to the mass of its nucleus.

28. The number of protons in a neutral atom is equal to the number of _____.

29. The atomic number of an atom indicates the number of neutrons in the atom. *True/False*

30. An element can be identified by its mass number. *True/False*

31. All isotopes of an element have the same atomic number. *True/False*

32. Isotopes of an element contain the same number of _____ but different numbers of _____.

33. The notation for one isotope of oxygen is $^{16}_{8}O$. The superscript designates the _____ of the isotope and the subscript designates the _____.

34. The isotopes chloride-35 and chlorine-37 differ by _____.

35. Deuterium is a naturally occurring isotope of which element?

36. Why does naturally occurring chlorine have an atomic mass that is not close to a whole number?

37. Atomic mass is the mass of an element relative to the mass of _____ isotope.

38. Isotopes of an element can be separated in a mass spectrometer. *True/False*

39. A positively charged ion is an atom that has _____

_____ .

40. A negatively charged ion is an atom that has _____

_____ .

41. The ion Ca^{2+} has a charge of _____ .

42. A heterogeneous mixture is not uniform from point to point. *True/False*

43. A homogeneous mixture is not uniform from point to point. *True/False*

44. Filtration may be used to separate a solid from a liquid solution. *True/False*

45. During the process of filtration, the insoluble solid passes through the filter paper. *True/False*

46. Distillation may be used to separate a volatile compound from the other components of a solution. *True/False*

47. A condenser is used to _____

_____ in a distillation process.

48. Gas-liquid chromatography is used to separate mixtures of _____ or

_____ .

49. In a gas-liquid chromatography process, all the volatile components travel down the column at the same rate. *True/False*

50. The vapors that interact least with the liquid coating on the column solid material exit the column last. *True/False*

51. In a liquid-solid chromatography process, all the components in the solution move through the column at the same rate. *True/False*

52. The components of the liquid that interact least with the solid of a column in liquid-solid chromatography exit the column first. *True/False*

C. CALCULATIONS YOU SHOULD KNOW HOW TO DO

1. Calculate the mass percentages of the elements that comprise a compound from chemical analysis of the compound. See Example 2-1 and Problems 2-5 through 2-10.

2. Calculate the molecular mass of a compound using the atomic masses of the elements that comprise the compound. See Example 2-5 and Problems 2-17 through 2-20.

3. Calculate the mass percentages of the elements that comprise a compound from the molecular mass of the compound and the atomic masses of the elements that comprise the compound. See Example 2-5 and Problems 2-21 through 2-26.

4. Calculate the atomic mass of an element using the isotopic masses and natural abundance of its isotopes. See Example 2-8 and Problems 2-33 through 2-36.

5. Calculate the natural abundance of one isotope of an element using the atomic mass of the element and the natural abundances of the remaining isotopes. See Problems 2-37 through 2-40.

D. SOLUTIONS TO THE ODD-NUMBERED PROBLEMS

2-1 (a) Se (b) In (c) Mn (d) Tm
(e) Hg (f) Kr (g) Pd (h) Tl
(i) U (j) W

See the alphabetical list of the elements on the inside front cover.

2-3 (a) germanium (b) scandium (c) iridium
(d) cesium (e) strontium (f) americium
(g) molybdenum (h) indium (i) plutonium
(j) xenon

See the alphabetical list of the elements on the inside front cover.

2-5 The mass percentage of sodium is given by

$$\text{mass \% of Na} = \frac{\text{mass of Na}}{\text{mass of compound}} \times 100$$

$$= \frac{0.978 \text{ g}}{1.659 \text{ g}} \times 100 = 59.0\%$$

The mass percentage of oxygen is given by

$$\text{mass \% of O} = \frac{\text{mass of O}}{\text{mass of compound}} \times 100$$

$$= \frac{0.681 \text{ g}}{1.659 \text{ g}} \times 100 = 41.0\%$$

2-7   The mass percentage of copper in the compound is given by

$$\text{mass \% of Cu} = \frac{\text{mass of Cu}}{\text{mass of compound}} \times 100$$

$$= \frac{1.28 \text{ g}}{1.60 \text{ g}} \times 100 = 80.0\%$$

The mass of sulfur in the compound is

$$\text{mass of S} = \text{mass of compound} - \text{mass of Cu}$$
$$= 1.60 \text{ g} - 1.28 \text{ g} = 0.32 \text{ g}$$

The mass percentage of sulfur in the compound is given by

$$\text{mass \% of S} = \frac{\text{mass of S}}{\text{mass of compound}} \times 100$$

$$= \frac{0.32 \text{ g}}{1.60 \text{ g}} \times 100 = 20\%$$

2-9   The respective mass percentages are

$$\text{mass \% of K} = \frac{\text{mass of K}}{\text{mass of potassium cyanide}} \times 100$$

$$= \frac{7.58 \text{ mg}}{12.63 \text{ mg}} \times 100 = 60.0\%$$

$$\text{mass \% of C} = \frac{\text{mass of C}}{\text{mass of potassium cyanide}} \times 100$$

$$= \frac{2.33 \text{ mg}}{12.63 \text{ mg}} \times 100 = 18.4\%$$

$$\text{mass \% of N} = \frac{\text{mass of N}}{\text{mass of potassium cyanide}} \times 100$$

$$= \frac{2.72 \text{ mg}}{12.63 \text{ mg}} \times 100 = 21.5\%$$

2-11 Using Table 2-6 and the list of the elements on the inside front cover, we have

(a)   lithium sulfide   (b)   barium oxide
(c)   magnesium phosphide   (d)   cesium bromide

2-13 (a)   silicon carbide   (b)   gallium phosphide
(c)   aluminum oxide   (d)   beryllium chloride

2-15 (a)   chlorine trifluoride and chlorine pentafluoride
(b)   sulfur tetrafluoride and sulfur hexafluoride
(c)   krypton difluoride and krypton tetrafluoride
(d)   bromine oxide and bromine dioxide

2-17   Refer to the inside front cover for the atomic masses.

(a)   The molecular mass of TiO_2 is

$$\text{molecular mass} = (\text{atomic mass of Ti}) + (2 \times \text{atomic mass of O})$$
$$= (47.90) + (2 \times 16.00) = 79.90$$

(b)   The molecular mass of Fe_2O_3 is

$$\text{molecular mass} = (2 \times \text{atomic mass of Fe}) + (3 \times \text{atomic mass of O})$$
$$= (2 \times 55.85) + (3 \times 16.00) = 159.70$$

(c)   The molecular mass of V_2O_5 is

$$\text{molecular mass} = (2 \times \text{atomic mass of V}) + (5 \times \text{atomic mass of O})$$
$$= (2 \times 50.94) + (5 \times 16.00) = 181.88$$

(d)   The molecular mass of P_4O_{10} is

$$\text{molecular mass} = (4 \times \text{atomic mass of P}) + (10 \times \text{atomic mass of O})$$
$$= (4 \times 30.97) + (10 \times 16.00) = 283.88$$

2-19   (a)   molecular mass of BrN_3 $= (\text{atomic mass of Br}) + (3 \times \text{atomic mass of N})$
$$= (79.90) + (3 \times 14.01)$$
$$= 121.93$$

(b)   molecular mass of $NaIO_3$ $= (\text{atomic mass of Na}) + (\text{atomic mass of I}) +$
$$(3 \times \text{atomic mass of O})$$
$$= (22.99) + (126.9) + (3 \times 16.00)$$
$$= 197.9$$

(c)   molecular mass of CCl_2F_2 $= (\text{atomic mass of C}) +$
$$(2 \times \text{atomic mass of Cl}) +$$
$$(2 \times \text{atomic mass of F})$$
$$= (12.01) + (2 \times 35.45) + (2 \times 19.00)$$
$$= 120.91$$

(d)   molecular mass of $C_{14}H_9Cl_6$ $= (14 \times \text{atomic mass of C}) +$
$$(9 \times \text{atomic mass of H}) +$$
$$(6 \times \text{atomic mass of Cl})$$
$$= (14 \times 12.01) + (9 \times 1.008) + (6 \times 35.45)$$
$$= 389.91$$

2-21   molecular mass of ClF_3 $= (\text{atomic mass of Cl}) + (3 \times \text{atomic mass of F})$
$$= (35.45) + (3 \times 19.00)$$
$$= 92.45$$

$$\text{mass \% of Cl} = \frac{\text{atomic mass of Cl}}{\text{molecular mass of ClF}_3} \times 100$$

$$= \frac{35.45}{92.45} \times 100 = 38.35\%$$

$$\text{mass \% of F} = \frac{3 \times \text{atomic mass of F}}{\text{molecular mass of ClF}_3} \times 100$$

$$= \frac{3 \times 19.00}{92.45} \times 100 = 61.65\%$$

2-23   molecular mass of $C_{12}H_{22}O_{11} = 342.30$

$$\text{mass \% of C} = \frac{12 \times \text{atomic mass of C}}{\text{molecular mass of C}_{12}H_{22}O_{11}} \times 100$$

$$= \frac{12 \times 12.01}{342.30} \times 100 = 42.10\%$$

$$\text{mass \% of H} = \frac{22 \times \text{atomic mass of H}}{\text{molecular mass of C}_{12}H_{22}O_{11}} \times 100$$

$$= \frac{22 \times 1.008}{342.30} \times 100 = 6.479\%$$

$$\text{mass \% of O} = \frac{11 \times \text{atomic mass of O}}{\text{molecular mass of C}_{12}H_{22}O_{11}} \times 100$$

$$= \frac{11 \times 16.00}{342.30} \times 100 = 51.42\%$$

2-25   molecular mass of $XeF_4 = 207.3$

$$\text{mass \% of Xe} = \frac{131.3}{207.3} \times 100 = 63.34\%$$

$$\text{mass of Xe} = \frac{\text{mass \% of Xe}}{100} \times \text{mass of XeF}_4$$

$$= \frac{63.34}{100} \times 2.000 \text{ g} = 1.267 \text{ g}$$

2-27   From the atomic numbers, we find that

	Protons	Electrons	Neutrons
(a) iodine-131	53	53	$131 - 53 = 78$
(b) cobalt-60	27	27	$60 - 27 = 33$
(c) potassium-43	19	19	$43 - 19 = 24$
(d) indium-113	49	49	$113 - 49 = 64$

2-29   The atomic number determines the element. The mass number is the sum of the atomic number and the number of neutrons.

Symbol	Atomic number	Number of neutrons	Mass number
$^{14}_{6}C$	6	8	14
$^{241}_{95}Am$	95	146	241
$^{123}_{53}I$	53	70	123
$^{18}_{8}O$	8	10	18

2-31

Symbol	Atomic number	Number of neutrons	Mass number
$^{67}_{31}Ga$	31	36	67
$^{15}_{7}N$	7	8	15
$^{58}_{27}Co$	27	31	58
$^{133}_{54}Xe$	54	79	133

2-33
$$\text{atomic mass of H} = (1.0078)\left(\frac{99.985}{100}\right) + (2.0141)\left(\frac{0.015}{100}\right)$$
$$= 1.0080$$

2-35
$$\text{atomic mass of Ne} = (19.99)\left(\frac{90.51}{100}\right) + (20.99)\left(\frac{0.27}{100}\right) + (21.99)\left(\frac{9.22}{100}\right)$$
$$= 20.18$$

2-37   Let x be the percentage of bromine-79 in naturally occurring bromine. The percentage of bromine-81 must be $100 - x$. Now set up the equation

$$\text{atomic mass of Br} = 79.904 = (78.9183)\left(\frac{x}{100}\right) + (80.9163)\left(\frac{100-x}{100}\right)$$

Multiply this equation through by 100 to obtain

$$7990.4 = 78.9183x + 8091.63 - 80.9163x$$

Collecting terms, we get

$$1.9980x = 101.2$$

or

$$x = 50.65\% = \% \text{ bromine-79}$$

The percentage of bromine-81 is

$$\% \text{ bromine-81} = 100 - x = 49.35\%$$

2-39 Let x be the percentage of nitrogen-15 in naturally occurring nitrogen. The percentage of nitrogen-14 must be $100 - x$. Now set up the equation

$$\text{atomic mass of N} = 14.0067 = (14.0031)\left(\frac{100 - x}{100}\right) + (15.0001)\left(\frac{x}{100}\right)$$

Multiply through by 100 to obtain

$$1400.67 = 1400.31 - 14.0031x + 15.0001x$$

Collecting terms, we get

$$0.9970x = 0.36$$

or

$$x = 0.36\% = \% \text{ nitrogen-15}$$

2-41 The number of electrons = atomic number − ionic charge
(a) 54   (b) 54   (c) 36   (d) 10

2-43 (a) 54   (b) 78   (c) 24   (d) 18

2-45 (a) Ca^{2+}, Cl^-, S^{2-} (b) Rb^+, Sr^{2+}, Br^-, Se^{2-}
(c) Na^+, Mg^{2+}, F^-, O^{2-} (d) Cs^+, Ba^{2+}, Te^{2-}

2-47 (a) molecular mass of $OH^- = 15.994 + 1.0079 = 17.002$
(b) molecular mass of $H_3O^+ = (3 \times 1.0079) + 15.994 = 19.018$
(c) molecular mass of $AlF_6^{3-} = 26.98154 + (6 \times 18.998403) = 140.97196$
(d) molecular mass of $PCl_4^+ = 30.97376 + (4 \times 35.453) = 172.786$

2-49 Use a magnet, which attracts only the iron filings, to remove the iron from the aluminum powder.

2-51 See Section 2-12 of the text for a description of filtration.

2-53 The role of the condenser in distillation is to convert the vapor from the distillation chamber into a liquid in order to collect the resulting liquid.

2-55 The role of the carrier gas in gas-liquid chromatography is to sweep the sample through the column.

2-57 The original separation using the technique of chromatography involved colored constituents. *Chroma-* comes from a Greek word meaning color.

2-59 See the description of evaporation in Section 2-12 of the text.

2-61 Gold is more dense than sand. The gold settles at the center of the swirled pan while the less dense sand particles are swirled out of the pan.

2-63 See Section 2-12 of the text for a description of sluicing for gold.

2-65 Atomic mass is the ratio of the mass of an atom to the mass of an atom of carbon-12. Thus, we have that

$$\frac{\text{atomic mass hydrogen}}{1} = \frac{1.008}{12}$$

$$\text{atomic mass hydrogen} = 0.0840$$

$$\frac{\text{atomic mass oxygen}}{1} = \frac{16.00}{12}$$

$$\text{atomic mass oxygen} = 1.333$$

2-67 Some examples of adsorption-based separations are baking soda in the refrigerator to remove odors (gaseous substances) and gas masks to remove noxious fumes.

2-69 (a) 53 protons in I; 53 protons in I^-
       (b) $131 - 53 = 78$ neutrons in I; 78 neutrons in I^-
       (c) 53 electrons in I; $53 + 1 = 54$ electrons in I^-

E. ANSWERS TO THE SELF-TEST

1. element

2. compound

3. luster, malleability, good conductivity of heat and electricity

4. true

5. false

6. Al, Ca, K

7. oxygen, carbon, sodium

8. false

9. oxygen, nitrogen, chlorine

10. false

11. false

12. mass percentages of each element that comprise the compound

13. See Section 2-4 of the text.

14. See Section 2-4 of the text.

15. atomic mass unit, amu

16. false

17. See Section 2-4 of the text.

18. rearranged

19. (a) name the metal; (b) name the nonmetal, changing the end of the name with -ide.

20. sum of the atomic masses of the atoms in the compound

21. positively

22. false

23. a lesser

24.

	Relative mass	Relative charge
Proton	1	+1
Neutron	1	0
Electron	1/1837	−1

25. false

26. See Section 2-8 of the text.

27. The proton and neutrons are located in the nucleus, and the electrons are located outside the nucleus. Consequently, the mass of an atom is essentially equal to the mass of its nucleus.

28. electrons

29. false

30. false

31. true

32. protons, neutrons

33. mass number, atomic number

34. two neutrons

35. hydrogen

36. Naturally occurring chlorine is composed of more than one isotope.

37. carbon-14

38. true

39. lost one or more electrons

40. gained one or more electrons

41. positive two

42. true

43. false

44. true

45. false

46. true

47. convert the volatile compound from a gas to a liquid

48. gases, volatile liquids

49. false

50. false

51. false

52. true

3

THE PERIODIC TABLE AND CHEMICAL PERIODICITY

A. OUTLINE OF CHAPTER 3

3-1   New substances are formed in chemical reactions.

Chemical reactions are represented by chemical equations in terms of the chemical formulas of the reactants and the products.

The reactants are the substances that react with each other.

The products are the substances that are formed in the reaction.

The symbols (s), (l), and (g) denote that a substance is a solid, liquid, or gas, respectively.

3-2   A chemical equation must be balanced.

The conservation of atoms in chemical reactions means that individual atoms of various types are neither created nor destroyed in a chemical reaction.

A chemical equation must have the same number of each type of atom on both sides of the equation.

Chemical equations are balanced by placing the appropriate numbers, called balancing coefficients, in front of the chemical formulas for the reactants and products.

The method of balancing chemical equations by inspection is summarized on pages 73 and 74 of the text.

3-3   Elements can be grouped according to their chemical properties.

Representative reactions of the alkali metals, the alkaline earth metals, and the halogens are presented.

The metals lithium, sodium, and potassium are called alkali metals.

The metals magnesium, calcium, strontium, and barium are called alkaline earth metals.

The nonmetals fluorine, chlorine, bromine, and iodine are called halogens.

The halogens react with the alkali metals to give products called halides.

The prediction of reaction products can be made using representative reactions.

3-4   The elements show a periodic pattern when listed in order of increasing atomic number.

Dmitri Mendeleev arranged the elements in order of increasing atomic mass and showed that the chemical properties of the elements exhibit repetitive behavior.

In a modern periodic table of the elements, the elements are ordered according to increasing atomic number (Figure 3-10).

3-5   Elements in the same column in the periodic table have similar chemical properties.

Groups or families of elements appear in the same column of the periodic table.

The extreme left-hand column of the periodic table contains the group of metals called the alkali metals or the Group 1 metals.

The second left-hand column contains the group of metals called the alkaline earth metals or the Group 2 metals.

The halogens occur in Group 7.

The extreme right-hand column, Group 8, contains the group of relatively unreactive elements called the noble gases.

Predictions of the products of reactions between elements or simple compounds can be made using the periodic table.

Mendeleev used periodicity to predict chemical and physical properties of undiscovered elements.

The more common versions of the periodic table have the lanthanide series and the actinide series placed at the bottom of the table (Figure 3-14).

The columns in the periodic table are called groups or families.

Elements in the same group or family have similar chemical properties.

The horizontal rows in the periodic table are called periods.

3-6   Elements are arranged as main-group elements, transition metals, and inner transition metals.

The periodic table organizes the elements into
    Groups (columns)
    Periods (rows)
    Semimetals
    Metals
    Nonmetals
    Main-group elements
    Transition metals
    Inner transition metals [the lanthanides ($Z = 57$ to $Z = 70$) and the actinides ($Z = 89$ to $Z = 102$)]

The main-group elements are the elements in the groups headed by the numbers 1 through 8 (Figures 3-14 and 3-15).

The transition metals are the elements in the groups not headed by numbers; they span the region between the main-group elements.

The lanthanides are also called the rare-earth elements.

The actinides are radioactive elements.

Elements can be classified based on their positions in the periodic table (Figure 3-19).

3-7   Periodic trends contain some irregularities.

Hydrogen does not fit readily into any group.

Hydrogen is not a metal like the Group 1 metals.

The first member of a group often reacts somewhat differently than the other members of the group.

The first member of a group often reacts similarly to the second member of the following group. These similarities are called diagonal relationships (Figure 3-21).

3-8   Many atoms form ions that have a noble-gas electron arrangement.

The nuclear model of the atom is reviewed.

The electron arrangement in an atom determines the chemical properties of that element.

Noble-gas electron arrangements are unusually stable.

Ions are formed by the loss or gain of electrons by atoms or molecules (Examples 3-7 and 3-8).

Some metals lose electrons to obtain a noble-gas electron configuration.

Some nonmetals gain electrons to obtain a noble-gas electron configuration.

3-9   Knowledge of ionic charges helps us write correct chemical formulas.

Metal atoms lose electrons and form positive ions.

An ionic compound has no net charge — the total positive charge equals the total negative charge.

A formula unit consists of the number of atoms of each type designated by the chemical formula.

Ionic charges are positive or negative numbers assigned to elements based on their position in the periodic table (Figure 3-22).

The prediction of chemical formulas of ionic compounds is made using ionic charges, such that the total positive charge is equal to the total negative charge.

Many transition-metal ions have more than one possible ionic charge (Table 3-5).

Roman numerals in parentheses after the name of the transition metal in the name of the compound are used to denote the charge on transition-metal ions.

B. SELF-TEST

1. Substances formed in chemical reactions have a combination of the properties of the substances from which they are produced. *True/False*

2. Sodium chloride has properties similar to those of sodium and chlorine. *True/False*

3. The chemical equation $H_2(g) + O_2(g) \rightarrow H_2O(l)$ is balanced as written. *True/False*

4. Balancing coefficients are placed _____ of the chemical formulas of the reactants and products of a chemical equation.

5. The total number of each kind of atom in all the reactants must _____ _____ in the products of a balanced chemical equation.

6. The chemical formulas of the products of a chemical reaction can be changed to balance the equation. *True/False*

7. The chemical properties of lithium are similar to those of _____ and _____ .

8. Lithium belongs to the group of elements called the _____ metals.

9. The alkali metals react with water to form _____ .

10. The chemical properties of magnesium are similar to those of _____ , _____ , and _____ .

11. Magnesium belongs to the group of elements called the _____ metals.

12. The alkaline earth metals burn in oxygen to form the metal oxide. *True/False*

13. The alkaline earth metals react with water to form _____ .

14. The halogens consist of the elements _____ , _____ , _____ , _____ , and _____ .

15. The halogens react with the alkali metals to form _____ .

16. Mendeleev arranged the elements in order of increasing _____ .

17. In the modern periodic table, the elements are arranged in order of increasing _____ .

18. Elements with similar chemical properties appear in the same row of the periodic table. *True/False*

19. Elements that have similar chemical properties are placed in the same column of the periodic table. *True/False*

20. Group 1 metals are also called ————————————.

21. Group 2 metals are also called ————————————.

22. The halogens appear in column ———————————— of the periodic table.

23. The noble gases occur in column ———————————— of the periodic table.

24. The noble gases used to be called ————————————.

25. Silicon is a nonmetal. *True/False*

26. Some properties of semimetals are ————————————, ————————————,

and ————————————.

27. There are almost as many nonmetals as metals. *True/False*

28. Where are the main-group elements located in the periodic table?

29. The main-group elements have many chemical properties in common. *True/False*

30. The transition metals are more similar in their properties than the main-group elements. *True/False*

31. The transition metals occur between ———————————— and ————————————
in the periodic table.

32. The inner transition metals occur in the series that begin with
———————————— and with ————————————.

33. The lanthanide series is also called ————————————.

34. Members of the lanthanide series have very similar chemical properties. *True/False*

35. One of the properties that the elements in the actinide series have in common is that
they are ————————————.

36. Explain why the inner transition metal series are placed at the bottom of many versions of the periodic table.

37. Hydrogen is sometimes placed in Group 1 because it behaves like an alkali metal.
True/False

38. The first member of a main-group family is typical of the group and behaves identically to the other members. *True/False*

39. A diagonal relationship refers to the chemical similarity between _____ _____ in the periodic table.

40. Lithium is somewhat similar in many of its properties to magnesium. *True/False*

41. The electron arrangements of the noble gases seem to be exceptionally stable. *True/False*

42. The noble gases are reactive nonmetals. *True/False*

43. An element in Group 1 (*gains/loses*) one electron to attain the electron arrangement of a noble gas.

44. An element in Group 7 (*gains/loses*) one electron to attain the electron arrangement of a noble gas.

45. An ionic compound consists of _____ and _____ _____.

46. The net charge on an ionic compound is _____.

47. The ionic charge of an element depends on its position in the periodic table. *True/False*

48. All transition metals have only one possible ionic charge. *True/False*

49. The designation tin(IV) indicates that tin has an ionic charge of _____.

C. CALCULATIONS YOU SHOULD KNOW HOW TO DO

There are no calculations in Chapter 3; however, you should know how to do the following.

1. Balance equations by inspection. See Example 3-1 and Problems 3-1 through 3-8.
2. Predict the products of chemical reactions using representative reactions. See Section 3-3, Examples 3-2 and 3-3, and Problems 3-9 and 3-10.
3. Use the periodic table to predict chemical and physical properties. See Section 3-5, Examples 3-4 and 3-5, and Problems 3-11, 3-12, 3-15, and 3-16.
4. Write chemical formulas from the names of compounds using Figure 3-22 and Table

3-5. See Section 3-9, Examples 3-9, 3-10, and 3-11, and Problems 3-25 through 3-28, 3-33, and 3-34.

D. SOLUTIONS TO THE ODD-NUMBERED PROBLEMS

3-1   The procedure for balancing equations of the type considered in this chapter is the balancing by inspection method outlined in Section 3-2.

(a) $2P(s) + 3Br_2(l) \rightarrow 2PBr_3(l)$
(b) $2H_2O_2(l) \rightarrow 2H_2O(l) + O_2(g)$
(c) $4CoO(s) + O_2(g) \rightarrow 2Co_2O_3(s)$
(d) $PCl_5(s) + 4H_2O(l) \rightarrow H_3PO_4(l) + 5HCl(g)$

3-3   (a) $CaH_2(s) + 2H_2O(l) \rightarrow Ca(OH)_2(aq) + 2H_2(g)$
(b) $CaCO_3(s) + 2HCl(aq) \rightarrow CaCl_2(aq) + CO_2(g) + H_2O(l)$
(c) $C_6H_{12}O_2(aq) + 8O_2(g) \rightarrow 6CO_2(g) + 6H_2O(l)$
(d) $2Li(s) + 2CO_2(g) + 2H_2O(g) \rightarrow 2LiHCO_3(s) + H_2(g)$

3-5   To work a problem like this you have to know how to write a chemical formula from the name. Once we have written the formulas for the reactants and the products, we then proceed to balance the equation.

(a) $2Na(s) + S(s) \rightarrow Na_2S(s)$
(b) $Ca(s) + Br_2(l) \rightarrow CaBr_2(s)$
(c) $2Ba(s) + O_2(g) \rightarrow 2BaO(s)$
(d) $2SO_2(g) + O_2(g) \rightarrow 2SO_3(g)$
(e) $3Mg(s) + N_2(g) \rightarrow Mg_3N_2(s)$

3-7   (a) $NaH(s) + H_2O(l) \rightarrow NaOH(aq) + H_2(g)$
         sodium hydride   water   sodium hydroxide   hydrogen

(b) $2SO_2(g) + O_2(g) \rightarrow 2SO_3(g)$
        sulfur dioxide   oxygen   sulfur trioxide

(c) $H_2S(g) + 2LiOH(aq) \rightarrow Li_2S(aq) + 2H_2O(l)$
        hydrogen sulfide   lithium hydroxide   lithium sulfide   water

(d) $ZnO(s) + CO(g) \rightarrow Zn(s) + CO_2(g)$
        zinc oxide   carbon monoxide   zinc   carbon dioxide

3-9   See Section 3-3 for the representative reactions.

(a) $2Na(s) + I_2(s) \rightarrow 2NaI(s)$
                             sodium iodide

(b) $Sr(s) + H_2(g) \rightarrow SrH_2(s)$
                        strontium hydride

(c) $3Ca(s) + N_2(g) \rightarrow Ca_3N_2(s)$
                        calcium nitride

(d) $2Mg(s) + O_2(g) \rightarrow 2MgO(s)$
                        magnesium oxide

3-11   By analogy with the properties of the other halogens, we predict the following:

(a)   solid   (b)   NaAt   (c)   white   (d) At_2 (e)   black

3-13   Tl, a main-group (5) metal; Eu, an inner transition metal; Xe, a main-group (8) nonmetal; Hf, a transition metal; Ru, a transition metal; Am, an inner transition metal; B, a main-group (3) semimetal

3-15   Radium is a Group 2 metal, and thus we predict the reactions of Ra by analogy with the other Group 2 metals.

    (a) $2Ra(s) + O_2(g) \rightarrow 2RaO(s)$
    (b) $Ra(s) + Cl_2(g) \rightarrow RaCl_2(s)$
    (c) $Ra(s) + 2HCl(g) \rightarrow RaCl_2(s) + H_2(g)$
    (d) $Ra(s) + H_2(g) \rightarrow RaH_2(s)$
    (e) $Ra(s) + S(s) \rightarrow RaS(s)$

3-17   Using the symbol for the element makes it easier to find the element in the periodic table. The main-group elements are tin (Sn), antimony (Sb), and argon (Ar).

3-19   (a)   Si — semimetal   (b)   Sn — metal
    (c)   Sc — metal   (d)   Sb — semimetal
    (e)   Sm — metal

3-21   We first determine the number of electrons in the ion and then compare the result with the Z values for the noble gases.

    (a)   yes; Xe   (b)   no   (c)   yes; Ar   (d)   yes; Ar   (e)   yes; Ne
    (f)   no

3-23   We determine the charges on the atomic ions in ionic compounds using Figure 3-22.

    (a) $Mg^{2+} S^{2-}$ magnesium sulfide
    (b) $Al^{3+} P^{3-}$ aluminum phosphide
    (c) $Ba^{2+} F^{-}$ barium fluoride
    (d) $Ga^{3+} O^{2-}$ gallium oxide

3-25   To determine the chemical formula of an ionic compound, we use the procedure outlined in Example 3-9. The ionic charges are given in Figure 3-22.

    (a) Ga_2Se_3 (b)   AlP   (c)   KI   (d) SrF_2

3-27   (a) Li_3N (b) Ga_2Te_3 (c) Ba_3N_2 (d) $MgBr_2$

3-29   To determine the chemical formula of an ionic compound formed from a positive ion and a negative ion, we use the fact that the compound has no net charge.

    (a) Fe_2O_3 (b)   CdS
    (c) RuF_3 (d) Tl_2S

3-31   (a)   copper(I) iodide   (b)   mercury(I) bromide
    (c)   cobalt(II) fluoride   (d)   iron(II) oxide

3-33   (a)   CoP   (b) MnO_2
    (c) V_2O_5 (d) $TiCl_4$

3-35   We first write down the chemical formulas of the reactants and products and then balance the equations by inspection.

    (a) $2K(s) + 2H_2O(l) \rightarrow 2KOH(s) + H_2(g)$
    (b) $KH(s) + H_2O(l) \rightarrow KOH(s) + H_2(g)$

(c) $SiO_2(s) + 3C(s) \rightarrow SiC(s) + 2CO(g)$
(d) $SiO_2(s) + 4HF(g) \rightarrow SiF_4(g) + 2H_2O(l)$
(e) $2P(s) + 3Cl_2(g) \rightarrow 2PCl_3(l)$

3-37 The equation for the reaction between lithium nitride and water is

$$Li_3N(s) + 3H_2O(l) \longrightarrow NH_3(g) + 3LiOH(aq)$$

By analogy,

$$Na_3P(s) + 3H_2O(l) \longrightarrow PH_3(g) + 3NaOH(aq)$$

3-39 We would predict that LiF would dissolve the least in water because the first member of a group does not behave chemically as the other members of the group.

3-41 (a) SnF_4 (b) HgS
      (c) Co_2O_3 (d) CrP

3-43 (a) $2Rb(s) + H_2(g) \rightarrow \quad 2RbH(s)$
              rubidium hydride

      (b) $6Li(s) + N_2(g) \rightarrow 2Li_3N(s)$
              lithium nitride

      (c) $Sr(s) + Cl_2(g) \rightarrow \quad SrCl_2(s)$
              strontium chloride

      (d) $2Sc(s) + 3S(s) \rightarrow \quad Sc_2S_3(s)$
              scandium sulfide

3-45 See Table 3-5 for the ionic charge indicated by the older system of nomenclature of transition-metal ions.

      (a) chromium(II) chloride   (b) iron(III) iodide
      (c) mercury(II) chloride   (d) gold(I) sulfide

3-47 (a) $Ba(s) + I_2(s) \rightarrow \quad BaI_2(s)$
              barium iodide

      (b) $4Al(s) + 3O_2(g) \rightarrow \quad 2Al_2O_3(s)$
              aluminum oxide

      (c) $2Ga(s) + 3F_2(g) \rightarrow \quad 2GaF_3(s)$
              gallium fluoride

      (d) $2K(s) + Br_2(l) \rightarrow \quad 2KBr(s)$
              potassium bromide

E. ANSWERS TO THE SELF-TEST

1. false
2. false
3. false
4. in front

5. be the same as
6. false
7. sodium, potassium (or other Group 1 metals)

8. alkali

9. hydroxides

10. calcium, strontium, barium

11. alkaline earth

12. true

13. hydroxides

14. fluorine, chlorine, bromine, iodine, and astatine

15. halides

16. atomic mass

17. atomic number

18. false

19. true

20. alkali metals

21. alkaline earth metals

22. 7

23. 8

24. the inert gases

25. false (Si is a semimetal.)

26. semiconducting, brittle, dull surface, nonductile

27. false

28. in groups that are headed by numbers (first two columns on the left and last six columns on the right in the table)

29. false

30. true

31. Group 2 and Group 3

32. lanthanum, actinium

33. the rare earths

34. true

35. radioactive

36. The members of each series have such similar chemical properties that they can be placed in one position in the periodic table. Also, this version of the periodic table is more compact.

37. false

38. false (The properties of the first member are not as typical as those of the other members.)

39. the first member of a group and the second member of the following group

40. true

41. true

42. false

43. loses

44. gains

45. positive ions and negative ions

46. zero

47. true

48. false

49. $+4$

4 CHEMICAL REACTIVITY

A. OUTLINE OF CHAPTER 4

4-1 Oxidation-reduction reactions involve the transfer of electrons between species.

An oxidation-reduction reaction is an electron-transfer reaction.

In the oxidation-reduction reaction between a metal and a nonmetal, electrons are transferred from a metal atom to the nonmetal atom, resulting in an ionic compound.

The total number of electrons lost in an oxidation-reduction reaction must equal the total number of electrons gained.

An oxidation state can be assigned to each atom in a chemical species.

The rules for assigning oxidation states are given on page 102.

For cases not covered by the rules, oxidation states are assigned by analogy with similar compounds derived from elements in the same group in the periodic table.

An atom is oxidized when the oxidation state of the atom increases.

Oxidation denotes a loss of electrons.

An atom is reduced when the oxidation state of the atom decreases.

Reduction denotes a gain of electrons.

The species that loses electrons in an oxidation-reduction reaction is called the reducing agent.

The species that gains electrons in an oxidation-reduction reaction is called the oxidizing agent.

4-2 Oxidation states can be used to balance oxidation-reduction reactions.

The oxidation-state method of balancing oxidation-reduction equations is based on changes in oxidation states and on the conservation of electrons. (See Example 4-4.)

4-3 A combination reaction is the reaction of two substances to form a single product.

Combination reactions occur between two elements, between an element and a compound, or between two compounds.

Combination reactions involve two (occasionally three) reactants and a single product.

All combination reactions between a metal and a nonmetal are also oxidation-reduction reactions.

When the reactants in a combination reaction are a metal and a nonmetal, the product is usually an ionic compound.

Ionic compounds are generally solids at room temperature.

When an ionic compound dissolves in water, the species in solution are the individual ions surrounded by water molecules; the ions are solvated in solution.

The process of dissolution can be represented by a chemical equation.

A species in solution is designated by the symbol (aq).

Combustion reactions involve the burning of a compound in oxygen.

Molecular compounds are composed of molecules.

All combustion reactions are oxidation-reduction reactions.

Combustion reactions involve a fuel, which acts as the reducing agent, and an oxidizer, which acts as the oxidizing agent $[O_2(g)]$.

Combination reactions between compounds may not be oxidation-reduction reactions.

A polyatomic ion is an ion that contains more than one atom.

The nomenclature of compounds containing polyatomic ions is similar to that for binary compounds.

A compound containing a polyatomic ion is named using the information in Table 4-2.

The formula of a compound containing a polyatomic ion can be written from its name using Table 4-2.

When water-soluble salts containing polyatomic ions are dissolved in water, the polyatomic ions persist as solvated units unless some additional reaction occurs.

4-4 Soluble metal oxides yield bases and some soluble nonmetal oxides yield acids when dissolved in water.

Bases yield hydroxide ions, $OH^-(aq)$, when dissolved in water.

Basic anhydrides are oxides that yield bases when dissolved in water.

Water-soluble metal oxides are basic anhydrides (Figure 4-9).

Acids yield hydrogen ions, $H^+(aq)$, when dissolved in water.

The hydronium ion $H_3O^+(aq)$ is the dominant species of $H^+(aq)$.

Acidic anhydrides are nonmetal oxides that yield acids when dissolved in water.

The formula of the acid anhydride that corresponds to a given acid can be determined by subtracting a water molecule from the formula unit for the acid.

Hydrogen atoms that yield $H^+(aq)$ when an acid is dissolved in water are called acidic hydrogen atoms or acidic protons.

Oxyacids are acids that contain oxygen atoms.

Binary acids consist of two elements, one of which is hydrogen.

Some common acids are listed in Tables 4-3 and 4-4.

4-5 In a decomposition reaction, a substance is broken down into two or more less complex substances.

There is usually only one reactant in a decomposition reaction.

Some compounds require heat in order to undergo a decomposition reaction.

The designation "high T" over the arrow in the equation of a reaction indicates that the reaction must be run at a high temperature.

Some decomposition reactions involve metal oxides, carbonates, sulfites, and sulfates.

Many metal azides decompose violently.

4-6 In a single-replacement reaction, one element in a compound is replaced by another.

A single-replacement reaction is also called a substitution reaction.

One metal may replace another metal in a compound.

A reactive metal replaces hydrogen in a dilute acid to produce hydrogen gas.

4-7 Metals can be ordered in terms of relative reactivity based on single-replacement reactions.

A more reactive metal will replace a less reactive metal in a compound.

The metals can be ranked in order of their reactivity in a reactivity series of the metals (Table 4-5).

A metal will displace from a compound any metal that lies below it in the reactivity series.

Single-replacement reactions do not have to take place in solution.

The reaction of carbon and a metal oxide to produce the free metal is an important type of single-replacement reaction.

4-8 The reactivity order of the halogens is $F_2 > Cl_2 > Br_2 > I_2$.

4-9 In a double-replacement reaction, the cations and anions of two ionic compounds exchange to form new compounds.

A precipitate is an insoluble product of a reaction that occurs in solution.

A precipitation reaction is a double-replacement reaction involving the formation of an insoluble product.

Precipitate formation is one type of driving force for chemical reactions that occur in solution.

Net ionic equations are used to describe double-replacement reactions.

The net ionic equation for a precipitation reaction shows only the ions that form the insoluble product and the precipitate product.

Spectator ions are not directly involved in double-replacement reactions.

The driving force for a double-replacement reaction can be the formation of a molecular compound from ionic reactants.

A neutralization reaction is the reaction between an acid and a base.

The formation of water is the driving force for neutralization reactions.

The net ionic equation of a neutralization reaction is

$$H^+(aq) + OH^-(aq) \longrightarrow H_2O(l)$$

A salt is an ionic compound formed in the reaction between an acid and a base.

Some properties of acids and bases are listed in Table 4-6.

Litmus paper is used to test whether a solution is acidic or basic.

A third type of driving force for double-replacement reactions is the formation of a gaseous product.

4-10   The names of oxyacids and their anions are based on oxidation states.

Table 4-2 contains many of the oxyanions, the anions of oxyacids.

Some elements, such as chlorine, nitrogen, phosphorus, and sulfur, can form more than one oxyacid.

One of the oxyacids containing a certain element is designated as the key acid and the other names are derived from the name of the key acid (Table 4-7).

The key acids to know are

Chloric acid	$HClO_3$
Nitric acid	HNO_3
Phosphoric acid	H_3PO_4
Sulfuric acid	H_2SO_4

The anions of the key acids are named by dropping the -ic ending and adding -ate.

If the oxidation state of the central atom in the key acid is reduced by 2, then the resulting acid is named by changing the -ic ending to -ous, and the corresponding anion is named by changing the -ate to -ite.

If the oxidation state of the central atom in the key acid is increased by 2, then the resulting acid and its corresponding anion are named by adding the prefix *per-* to the name of the key acid and its corresponding anion.

If the oxidation state of the central atom in the key acid is reduced by 4, then the resulting acid and its corresponding anion are named by adding the prefix *hypo-* to the name of the -ous acid and its corresponding anion.

B. SELF-TEST

1. An oxidation-reduction reaction involves _____

_____ .

2. An oxidation-reduction reaction is also called an _____ reaction.

3. In an oxidation-reduction reaction, the number of electrons lost by the element that is oxidized is equal to _____.

4. Free elements are assigned an oxidation state of _____.

5. The sum of the oxidation states of the atoms in a species must be equal to _____.

6. The alkali metals in compounds are always assigned an oxidation state of _____.

7. Fluorine in compounds is always assigned an oxidation state of _____.

8. The alkaline earth metals, Zn, and Cd are always assigned an oxidation state of _____.

9. Hydrogen in compounds is assigned an oxidation state of _____.

10. Oxygen in compounds is assigned an oxidation state of _____.

11. For cases not covered by the rules, oxidation states are assigned by analogy using the periodic table as a guide. *True/False*

12. An atom is oxidized when the oxidation state of the atom _____.

13. An atom is _____ when the oxidation state of the atom decreases.

14. In an oxidation-reduction reaction, the reducing agent is the species that _____.

15. In an oxidation-reduction reaction, the _____ agent is the species that gains electrons.

16. The oxidation-state method of balancing oxidation-reduction equations is based on changes in _____ and the conservation of _____.

17. The reaction $2Rb(s) + Br_2(l) \rightarrow 2RbBr(s)$ is an example of a _____ reaction.

18. The reaction between a reactive metal and a reactive nonmetal results in an (*ionic/molecular*) compound.

19. Ionic compounds are usually solids at room temperature. *True/False*

20. The reaction between sulfur and oxygen to produce SO_2 is an example of a combustion reaction. *True/False*

21. All combustion reactions are oxidation-reduction reactions. *True/False*

22. In the reaction between sulfur and oxygen to produce sulfur dioxide, _____ is the fuel and _____ is the oxidizer.

23. The carbonate ion is an example of a _____ ion.

24. Magnesium sulfate, $MgSO_4$, dissolved in water consists of _____ ions and _____ ions.

25. All metal oxides are soluble in water. *True/False*

26. Metal oxides that dissolve in water yield _____ ions in solution.

27. A base is a compound that _____ when dissolved in water.

28. The compound $Na_2O(s)$ is an example of a _____ anhydride.

29. An acid is a compound that _____ when dissolved in water.

30. The hydrogen ion exists as H^+ in aqueous solution. *True/False*

31. An acidic anhydride yields _____ when dissolved in water.

32. The compound SO_3 is an example of an _____ anhydride.

33. The formula of the acid anhydride that corresponds to a given acid by subtracting one or more _____ molecules from the formula unit for the acid.

34. Binary acids contain only two elements. *True/False*

35. An example of a sulfur oxyacid is _____ .

36. Decomposition reactions involve (*one/two/three*) reactant(s).

37. Oxygen can be produced in the laboratory by heating _____ _____ .

38. Many metal carbonates decompose upon heating to produce _____ _____ and _____ gas.

39. In a single-replacement reaction, one element in a compound is replaced by _____

_____ .

40. When a reactive metal is added to an aqueous solution of an acid, _____
gas is produced.

41. A metal will replace a less reactive metal from a compound. *True/False*

42. The metals may be ordered with respect to their reactivity. *True/False*

43. Any metal will replace any other metal that lies (*above/below*) it in the reactivity series
of the metals.

44. Fluorine F_2 will replace iodine in the compound sodium iodide, NaI. *True/False*

45. The most reactive halogen is _____ .

46. A double-replacement reaction is a reaction in which _____

_____ .

47. A precipitate is an _____ product of a reaction that takes place in solution.

48. Consider the equation

$$AgNO_3(aq) + NaCl(aq) \longrightarrow NaNO_3(aq) + AgCl(s)$$

   (a)   The spectator ions are _____ and _____ .
   (b)   The precipitate is _____ .
   (c)   The net ionic equation is

49. The driving force for the chemical reaction between NaCl(*aq*) and $AgNO_3$(*aq*) is

_____ .

50. Consider the equation

$$HNO_3(aq) + KOH(aq) \longrightarrow KNO_3(aq) + H_2O(l)$$

   (a)   The spectator ions are _____ and _____ .
   (b)   The net ionic equation is

51. The driving force for the chemical reaction between HNO_3(*aq*) and KOH(*aq*) is

_____ .

52. The reaction between an acid and a base is called a _____ reaction.

53. An ionic compound formed in the reaction between an acid and a base is called a

_____ .

54. An acidic solution tastes _____ .

55. A basic solution tastes _____ .

56. The color of litmus is _____ in an acidic solution and _____ in a basic solution.

57. A driving force for a double-replacement reaction is the formation of a (*ionic/molecular*) compound.

58. Nitrogen forms more than one oxyacid. *True/False*

59. The name of an oxyacid depends upon the oxidation state of the element that occurs with oxygen. *True/False*

60. The key acid in a series of oxyacids is given an *-ic* name. *True/False*

61. The anion corresponding to a key acid in a series of oxyacids is named by dropping the *-ic* ending on the acid and adding _____ .

62. If the oxidation state of the central atom in a key acid is reduced by 2, then the resulting acid is named by changing the *-ic* ending to _____ .

63. The anion corresponding to the acid given in Question 62 is named by changing the *-ate* ending to _____ .

64. If the oxidation state of the central atom in a key acid is increased by 2, then the resulting acid is named by adding _____ to _____ .

65. If the oxidation state of the central atom in a key acid is reduced by 4, then the resulting acid is named by adding _____ to _____ .

C. CALCULATIONS YOU SHOULD KNOW HOW TO DO

There are no calculations in this chapter; however, you should be able to do the following.

1. Assign oxidation states to atoms in compounds using the rules given on page 102 of the text. See Examples 4-1 and 4-2 and Problems 4-1 through 4-14.

2. Determine the atom that is oxidized, the atom that is reduced, the oxidizing agent, and the reducing agent in an oxidation-reduction reaction. See Example 4-3 and Problems 4-15 through 4-20.

3. Balance oxidation-reduction equations by the oxidation-state method. See Example 4-4 and Problems 4-15 through 4-20.

4. Predict reaction products for (a) combination reactions, (b) decomposition reactions, (c) single-replacement reactions, and (d) double-replacement reactions. See Examples 4-11 through 4-15 and 4-17 and Problems 4-25 through 4-32.

5. Write net ionic equations. See Examples 4-16 through 4-18 and Problems 4-33 through 4-36.

6. Using Table 4-2, name compounds containing polyions. See Example 4-6 and Problems 4-41 through 4-44 and 4-49 and 4-50.

7. Write the formulas of compounds containing polyions from their names. See Example 4-7 and Problems 4-45 through 4-48 and 4-51 and 4-52.

8. Using the key acids given in Table 4-7, name the oxyacids and their corresponding anions. See Examples 4-19 and 4-20 and Problems 4-53 through 4-56.

D. SOLUTIONS TO THE ODD-NUMBERED PROBLEMS

4-1 (a) CaC_2. We assign calcium an oxidation state of $+2$ (rule 5).
     (b) Al_2O_3. We assign oxygen an oxidation state of -2 (rule 7). The oxidation state x of aluminum is (rule 2)

$$2x + 3(-2) = 0 \quad \text{or} \quad x = +3$$

     (c) VO_2^+. We assign oxygen an oxidation state of -2 (rule 7). The oxidation state x of vanadium is (rule 2)

$$x + 2(-2) = +1 \quad \text{or} \quad x = +5$$

     (d) Co_3O_4. We assign oxygen an oxidation state of -2 (rule 7). The oxidation state x of cobalt is (rule 2)

$$3x + 4(-2) = 0 \quad \text{or} \quad x = \frac{8}{3}$$

4-3 (a) $LiAlH_4$. We assign lithium an oxidation state of $+1$ (rule 3), and aluminum is assigned an oxidation state of $+3$. (See Problem 4-1b or Table 4-1.) The oxidation state x of hydrogen is (rule 2)

$$+1 + (+3) + 4x = 0 \quad \text{or} \quad x = -1$$

     (b) ClO_2. We assign oxygen an oxidation state of -2 (rule 7). The oxidation state x of chlorine is (rule 2)

$$x + 2(-2) = 0 \quad \text{or} \quad x = +4$$

(c) $NaBrO_3$. We assign sodium an oxidation state of $+1$ (rule 3) and oxygen an oxidation state of -2 (rule 7). The oxidation state x of bromine is (rule 2)

$$+1 + x + 3(-2) = 0 \quad \text{or} \quad x = +5$$

(d) $HAsO_2$. We assign hydrogen an oxidation state of $+1$ (rule 6) and oxygen an oxidation state of -2 (rule 7). The oxidation state x of arsenic is (rule 2)

$$+1 + x + 2(-2) = 0 \quad \text{or} \quad x = +3$$

4-5   In each case we assign oxygen an oxidation state of -2 (rule 7) and calculate the oxidation state of nitrogen by rule 2. Thus

(a) NO_2; $x + 2(-2) = 0$, or $x = +4$.
(b) N_2O; $2x + (-2) = 0$, or $x = +1$.
(c) N_2O_5; $2x + 5(-2) = 0$, or $x = +5$.
(d) N_2O_3; $2x + 3(-2) = 0$, or $x = +3$.

4-7   (a) H_2CO. We assign oxygen an oxidation state of -2 (rule 7) and hydrogen an oxidation state of $+1$ (rule 6). The oxidation state x of carbon is (rule 2)

$$2(+1) + x + 1(-2) = 0 \quad \text{or} \quad x = 0$$

(b) CH_4. We assign hydrogen an oxidation state of $+1$ (rule 6). The oxidation state x of carbon is (rule 2)

$$x + 4(+1) = 0 \quad \text{or} \quad x = -4$$

(c) CH_3OH. We assign oxygen an oxidation state of -2 (rule 7) and hydrogen an oxidation state of $+1$ (rule 6). The oxidation state x of carbon is (rule 2)

$$x + 3(+1) + 1(-2) + 1(+1) = 0 \quad \text{or} \quad x = -2$$

(d)   HCOOH. We assign oxygen an oxidation state of -2 (rule 7) and hydrogen an oxidation state of $+1$ (rule 6). The oxidation state x of carbon is (rule 2)

$$2(+1) + x + 2(-2) = 0 \quad \text{or} \quad x = +2$$

4-9   (a) $SbCl_3$. The name of $SbCl_3$ is antimony(III) chloride, and so we assign chlorine an oxidation state of -1, by analogy with fluorine (rule 4). Thus the oxidation state of antimony is (rule 2)

$$x + 3(-1) = 0 \quad \text{or} \quad x = +3$$

(b) Sb_4O_6. We assign oxygen an oxidation state of -2 (rule 7), and so the oxidation state of antimony is (rule 2)

$$4x + 6(-2) = 0 \quad \text{or} \quad x = +3$$

(c) SbF_5^{2-}. We assign fluorine an oxidation state of -1 (rule 4), and so the oxidation state of antimony is (rule 2)

$$x + 5(-1) = -2 \quad \text{or} \quad x = +3$$

(d) $SbCl_6^{3-}$. We assign chlorine an oxidation state of -1 [part (a)], and so the oxidation state of antimony is (rule 2)

$$x + 6(-1) = -3 \quad \text{or} \quad x = +3$$

4-11 (a) N_2O_4. We assign oxygen an oxidation state of -2 (rule 7). The oxidation state x of nitrogen is (rule 2)

$$2x + 4(-2) = 0 \quad \text{or} \quad x = +4$$

(b) NO. We assign oxygen an oxidation state of -2 (rule 7). The oxidation state x of nitrogen is (rule 2)

$$x + 1(-2) = 0 \quad \text{or} \quad x = +2$$

(c) NO_3^-. We assign oxygen an oxidation state of -2 (rule 7). The oxidation state x of nitrogen is (rule 2)

$$x + 3(-2) = -1 \quad \text{or} \quad x = +5$$

(d) NO_2^-. We assign oxygen an oxidation state of -2 (rule 7). The oxidation state x of nitrogen is (rule 2)

$$x + 2(-2) = -1 \quad \text{or} \quad x = +3$$

4-13 (a) $Cr_2O_7^{2-}$. We assign oxygen an oxidation state of -2 (rule 7). The oxidation state of chromium x is (rule 2)

$$2x + 7(-2) = -2 \quad \text{or} \quad x = +6$$

(b) MoO_4^{2-}. We assign oxygen an oxidation state of -2 (rule 7). The oxidation state of molybdenum x is (rule 2)

$$x + 4(-2) = -2 \quad \text{or} \quad x = +6$$

(c) $Cr(OH)_4^-$. We assign oxygen an oxidation state of -2 (rule 7) and hydrogen an oxidation state of $+1$ (rule 6). The oxidation state of chromium x is (rule 2)

$$x + 4(-2) + 4(+1) = -1 \quad \text{or} \quad x = +3$$

(d) Mn_2O_7. We assign oxygen an oxidation state of -2 (rule 7). The oxidation state of manganese x is (rule 2)

$$2x + 7(-2) = 0 \quad \text{or} \quad x = +7$$

4-15 (a) We first note that the oxidation state of sulfur in Na_2SO_4 is $+6$ and that in Na_2S is -2. Thus, the change in the oxidation state of sulfur is -8, which corresponds to a gain of eight electrons. The oxidation state of carbon in C is zero and that in CO is $+2$. Thus, the change in the oxidation state of carbon is $+2$, which corresponds to a loss of two electrons. We can diagram the changes in oxidation states and the transfer of electrons as

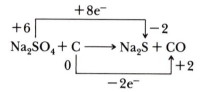

The number of electrons that are gained must equal the number of electrons that are lost. Each Na_2SO_4 gains eight electrons, but each C loses only two electrons. Thus, four ($4 \times 2 = 8$ electrons) carbons are required. The balanced equation is

$$Na_2SO_4(s) + 4C(s) \longrightarrow Na_2S(s) + 4CO(g)$$

The oxidizing agent is $Na_2SO_4(s)$ and the reducing agent is $C(s)$.

(b) We can diagram the oxidation states of iodine and carbon and the transfer of electrons as

$$
\begin{array}{c}
\overset{+10e^-}{\overbrace{}} \\
+5 \overset{}{\big|} \qquad\qquad\qquad \downarrow 0 \\
I_2O_5 + CO \longrightarrow CO_2 + I_2 \\
+2 \big|\underset{}{} \uparrow +4 \\
\underset{-2e^-}{\underbrace{}}
\end{array}
$$

Each I_2O_5 that reacts gains 10 (2×5) electrons and each CO that reacts loses 2 electrons. Thus, five ($5 \times 2 = 10$ electrons) CO molecules are required. The balanced equation is

$$I_2O_5(s) + 5CO(g) \longrightarrow 5CO_2(g) + I_2(s)$$

The oxidizing agent is $I_2O_5(s)$ and the reducing agent is $CO(g)$.

4-17 (a) In this reaction, NO_2 acts as both an oxidizing agent *and* a reducing agent. The oxidation state of some nitrogen atoms is increased from $+4$ to $+5$, while that for others is reduced from $+4$ to $+2$. Because the NO_2 acts as both an oxidizing agent and a reducing agent, it is convenient to write it twice on the left-hand side of the chemical equation:

$$NO_2 + NO_2 + H_2O \longrightarrow HNO_3 + NO$$

We can diagram the oxidation states of nitrogen and the transfer of electrons as

$$
\begin{array}{c}
\overset{\displaystyle +2e^-}{\overset{\displaystyle +4\,\boxed{}\,\downarrow +2}{}} \\
NO_2 + NO_2 + H_2O \longrightarrow HNO_3 + NO \\
+4\,\boxed{}\,\uparrow +5 \\
-e^-
\end{array}
$$

Each NO_2 that is reduced gains two electrons and each NO_2 that is oxidized loses one electron. Thus, two ($2 \times 1 = 2$ electrons) NO_2 that are oxidized are required for each NO_2 that is reduced. The balanced equation is

$$3NO_2(g) + H_2O(l) \longrightarrow 2HNO_3(aq) + NO(g)$$

The electron donor is NO_2 and the electron acceptor is NO_2.

(b) We can diagram the oxidation states of hydrogen in CaH_2 and H_2O and the transfer of electrons as

$$
\begin{array}{c}
+e^- \\
+1\,\boxed{}\,\downarrow 0 \\
CaH_2 + H_2O \longrightarrow H_2 + Ca(OH)_2 \\
-1\,\boxed{}\,\uparrow 0 \\
-2e^-
\end{array}
$$

Each CaH_2 that is oxidized loses two electrons, and each water that is reduced gains one electron. Note that only one of the hydrogen atoms in H_2O gains an electron. The oxidation state of the other hydrogen atom remains $+1$ (see $Ca(OH_2)$). Thus, two ($2 \times 1 = 2$ electrons) H_2O that are reduced are required for each CaH_2 that is oxidized. The balanced equation is

$$CaH_2(s) + 2H_2O(l) \longrightarrow 2H_2(g) + Ca(OH)_2(aq)$$

The electron donor is CaH_2 and the electron acceptor is H_2O.

4-19 (a) We can diagram the oxidation states of carbon and chlorine and the transfer of electrons as

$$
\begin{array}{c}
+1e^- \\
+4\,\boxed{}\,\downarrow +3 \\
NaOH + Ca(OH)_2 + C + ClO_2 \longrightarrow NaClO_2 + CaCO_3 + H_2O \\
0\,\boxed{}\,\uparrow +4 \\
-4e^-
\end{array}
$$

Each ClO_2 that reacts gains one electron and each C that reacts loses four electrons. Thus, four ($4 \times 1 = 4$ electrons) ClO_2 are required for each C.

The equation balanced with respect to chlorine and carbon is

$$NaOH(aq) + Ca(OH)_2(aq) + C(s) + 4ClO_2(g) \longrightarrow$$
$$4NaClO_2(aq) + CaCO_3(s) + H_2O(l)$$

We now balance the other elements to give the balanced equation

$$4NaOH(aq) + Ca(OH)_2(aq) + C(s) + 4ClO_2(g) \longrightarrow$$
$$4NaClO_2(aq) + CaCO_3(s) + 3H_2O(l)$$

The species reduced is ClO_2 and the species oxidized is C.

(b) We can diagram the oxidation states of iodine and nitrogen and the transfer of electrons as

$$
\begin{array}{c}
\overset{+2e^-}{0 \boxed{} -1} \\
I_2 + N_2H_4 \longrightarrow HI + N_2 \\
-2 \boxed{} 0 \\
-4e^-
\end{array}
$$

Each I_2 that reacts gains two (2×1) electrons and each N_2H_4 that reacts loses four (2×2) electrons. Thus, two $(2 \times 2 = 4$ electrons) I_2 react for each N_2H_4. The balanced equation is

$$2I_2(s) + N_2H_4(aq) \longrightarrow 4HI(aq) + N_2(g)$$

The species reduced is I_2 and the species oxidized is N_2H_4.

4-21 (a) Calcium atoms are neutral and so have an oxidation state of zero. A chlorine molecule is neutral; each chlorine atom has an oxidation state of zero. In $CaCl_2$ the oxidation state of calcium is $+2$ and the oxidation state of chlorine is -1. Thus calcium is oxidized and chlorine is reduced.

(b) Aluminum atoms have an oxidation state of zero, and oxygen atoms in O_2 have an oxidation state of zero. In Al_2O_3 the oxidation states of aluminum and oxygen are $+3$ and -2, respectively. Thus aluminum is oxidized and oxygen is reduced.

(c) Rubidium is oxidized and bromine is reduced.

(d) Sodium is oxidized and sulfur is reduced.

4-23 (a) Each calcium atom loses two electrons, and each of the two chlorine atoms gains one electron. Thus a total of two electrons is transferred.

(b) Each aluminum atom loses three electrons, and each oxygen atom gains two electrons. Because the formula unit of the product Al_2O_3 involves two aluminum atoms, a total of 2×3 or 6 electrons is transferred per formula unit of product. The overall balanced equation shows that two formula units $(2Al_2O_3)$ are produced.

(c) One electron. However, two formula units are produced.

(d) Two electrons

4-25 (a) decomposition   (b) combination

       (c) single-replacement   (d) double-replacement

4-27   (a)   decomposition; already balanced
           (b)   combination

$$4Fe(s) + 3O_2(g) \longrightarrow 2Fe_2O_3(s)$$

           (c)   single-replacement

$$2Al(s) + Mn_2O_3(s) \longrightarrow 2Mn(s) + Al_2O_3(s)$$

           (d)   double-replacement

$$2AgNO_3(aq) + H_2SO_4(aq) \longrightarrow Ag_2SO_4(s) + 2HNO_3(aq)$$

           (e)   double-replacement

$$Ca(OH)_2(aq) + 2HBr(aq) \longrightarrow CaBr_2(aq) + 2H_2O(l)$$

           (f)   single-replacement

$$Cd(s) + 2HCl(aq) \longrightarrow CdCl_2(aq) + H_2(g)$$

4-29   (a) $3Mg(s) + N_2(g) \rightarrow Mg_3N_2(s)$
           (b) $H_2(g) + S(s) \rightarrow H_2S(g)$
           (c) $2K(s) + Br_2(l) \rightarrow 2KBr(s)$
           (d) $4Al(s) + 3O_2(g) \rightarrow 2Al_2O_3(s)$
           (e) $MgO(s) + SO_2(g) \rightarrow MgSO_3(s)$

4-31   In order for a metal-metal replacement reaction to occur, the free metal must be a more reactive metal than the metal in the compound (see Table 4-5). Also, a more reactive free halogen will replace a less reactive halogen in compounds.

           (a) $Zn(s) + 2HBr(aq) \rightarrow ZnBr_2(aq) + H_2(g)$
           (b) $2Al(s) + Fe_2O_3(s) \rightarrow 2Fe(s) + Al_2O_3(s)$
           (c) $Pb(s) + Cu(NO_3)_2(aq) \rightarrow Cu(s) + Pb(NO_3)_2(aq)$
           (d) $Br_2(l) + 2NaI(aq) \rightarrow 2NaBr(aq) + I_2(s)$

4-33   To determine the net ionic equation that corresponds to the complete ionic equation, we determine the precipitate or molecular product formed. The ions that are used to form this species are the reactants in the net ionic equation.

           (a) $2H^+(aq) + S^{2-}(aq) \rightarrow H_2S(g)$
           (b) $Pb^{2+}(aq) + S^{2-}(aq) \rightarrow PbS(s)$
           (c) $H^+(aq) + OH^-(aq) \rightarrow H_2O(l)$
           (d) $Na_2O(s) + 2H^+(aq) \rightarrow 2Na^+(aq) + H_2O(l)$
           (e) $NH_3(aq) + H^+(aq) \rightarrow NH_4^+(aq)$

4-35   (a) $Fe(NO_3)_3(aq) + 3NaOH(aq) \rightarrow Fe(OH)_3(s) + 3NaNO_3(aq)$
                 $Fe^{3+}(aq) + 3OH^-(aq) \rightarrow Fe(OH)_3(s)$
           (b) $Zn(ClO_4)_2(aq) + K_2S(aq) \rightarrow ZnS(s) + 2KClO_4(aq)$
                 $Zn^{2+}(aq) + S^{2-}(aq) \rightarrow ZnS(s)$
           (c) $Pb(NO_3)_2(aq) + 2KOH(aq) \rightarrow Pb(OH)_2(s) + 2KNO_3(aq)$
                 $Pb^{2+}(aq) + 2OH^-(aq) \rightarrow Pb(OH)_2(s)$

(d) $Zn(NO_3)_2(aq) + Na_2CO_3(aq) \rightarrow ZnCO_3(s) + 2NaNO_3(aq)$
$Zn^{2+}(aq) + CO_3^{2-}(aq) \rightarrow ZnCO_3(s)$

(e) $Cu(ClO_4)_2(aq) + Na_2CO_3(aq) \rightarrow CuCO_3(s) + 2NaClO_4(aq)$
$Cu^{2+}(aq) + CO_3^{2-}(aq) \rightarrow CuCO_3(s)$

4-37 (a) acidic   (b) acidic   (c) basic   (d) acidic   (e) basic

4-39 (a) $2HClO_3(aq) + Ba(OH)_2(aq) \rightarrow \underset{\text{barium chlorate}}{Ba(ClO_3)_2(aq)} + 2H_2O(l)$

$$2H^+(aq) + 2OH^-(aq) \longrightarrow 2H_2O(l)$$

(b) $HC_2H_3O_2(aq) + KOH(aq) \rightarrow \underset{\text{potassium acetate}}{KC_2H_3O_2(aq)} + H_2O(l)$

$$HC_2H_3O_2(aq) + OH^-(aq) \longrightarrow H_2O(l) + C_2H_3O_2^-(aq)$$

(c) $2HI(aq) + Mg(OH)_2(s) \rightarrow \underset{\text{magnesium iodide}}{MgI_2(aq)} + 2H_2O(l)$

$$2H^+(aq) + Mg(OH)_2(s) \longrightarrow Mg^{2+}(g) + 2H_2O(l)$$

(d) $H_2SO_4(aq) + 2RbOH(aq) \rightarrow \underset{\text{rubidium sulfate}}{Rb_2SO_4(aq)} + 2H_2O(l)$

$$2H^+(aq) + 2OH^-(aq) \longrightarrow 2H_2O(l)$$

4-41 In order to name the following compounds, we need to know the names and formulas of the polyatomic ions given in Table 4-2.

(a) calcium cyanide   (b) silver perchlorate
(c) potassium permanganate   (d) strontium chromate

4-43 (a) ammonium sulfate   (b) ammonium phosphate
(c) calcium phosphate   (d) potassium phosphate

4-45 In order to write the chemical formula for each of the following compounds, we need to know the formula and ionic charge of the polyatomic ions given in Table 4-2. Recall that the net charge on the ionic compound must be zero.

(a) $Na_2S_2O_3$ (b) $KHCO_3$ (c) $NaClO$ (d) $CaSO_3$

4-47 See the solution to Problem 4-45.

(a) Na_2SO_3 (b) K_3PO_4 (c) Ag_2SO_4 (d) NH_4NO_3

4-49 The following compounds contain transition-metal ions. The ionic charge of the metal ion is indicated by a Roman numeral. See Section 3-9 of the text.

(a) mercury(I) chloride   (b) chromium(III) nitrate
(c) cobalt(II) bromide   (d) copper(II) carbonate

4-51 The Roman numeral following the name of the transition-metal ion indicates the ionic charge of the metal ion.

(a) Cr_2O_3 (b) $Sn(OH)_2$ (c) $Cu(C_2H_3O_2)_2$ (d) $Co_2(SO_4)_3$

4-53 (a) The key acid is sulfuric acid. The oxidation state of sulfur in this acid is reduced by 2 from the key acid. Thus, this acid is sulfurous acid.
(b) By analogy with the chlorine oxyacids, this acid is bromic acid.
(c) The key acid is phosphoric acid. The oxidation state of phosphorus in this acid is reduced by 4 from the key acid. Thus, this acid is hypophosphorous acid.
(d) By analogy with the chlorine oxyacids, the key acid is iodic acid. The oxidation state of iodine in this acid is increased by 2 from the key acid. Thus, this acid is periodic acid.

4-55 (a) potassium hypobromite   (b) calcium hydrogen phosphite
(c) lead(II) chlorite   (d) nickel(II) perchlorate

4-57 The diagram showing the oxidation states and transfer of electrons is

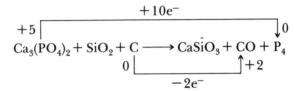

Each $Ca_3(PO_4)_2$ that reacts gains 10 (2×5) electrons and each C that reacts loses 2 electrons. Thus, five ($5 \times 2 = 10$ electrons) C react for each $Ca_3(PO_4)_2$. The equation balanced with respect to phosphorus and carbon is

$$2Ca_3(PO_4)_2(s) + SiO_2(s) + 10C(s) \longrightarrow CaSiO_3(l) + 10CO(g) + P_4(g)$$

Balancing the other elements gives the balanced equation

$$2Ca_3(PO_4)_2(s) + 6SiO_2(s) + 10C(s) \longrightarrow 6CaSiO_3(l) + 10CO(g) + P_4(g)$$

4-59 (a) $+4$ (b) $+2$ (c) $+4$ (d) $+2$
(e) 0 (f) -1 (g) -2 (h) $-\frac{4}{3}$

4-61 (a) $HCl(aq) + KCN(aq) \rightarrow HCN(g) + KCl(aq)$
(b) $2K(s) + 2H_2O(l) \rightarrow H_2(g) + 2KOH(aq)$
(c) $2H_2O_2(aq) \rightarrow O_2(g) + 2H_2O(l)$
(d) $H_2(g) + Br_2(l) \rightarrow 2HBr(g)$

4-63 $2Pb(l) + O_2(g) \rightarrow 2PbO(s)$
$Ag(l) + O_2(g) \rightarrow$ no reaction

4-65 $HgS(s) + O_2(g) \xrightarrow{heat} Hg(g) + SO_2(g) \xrightarrow{cool} Hg(l) + SO_2(g)$

4-67 See Table 4-5. Na, Fe, Sn, Au

4-69 (a) $2Na(s) + H_2(g) \rightarrow 2NaH(s)$
(b) $2Al(s) + 3S(s) \rightarrow Al_2S_3(s)$
(c) $H_2O(g) + C(s) \rightarrow CO(g) + H_2(g)$
(d) $C(s) + 2H_2(g) \rightarrow CH_4(g)$
(e) $PCl_3(l) + Cl_2(g) \rightarrow PCl_5(s)$

4-71 $2CuO(s) + C(s) \xrightarrow{\text{high T}} 2Cu(s) + CO_2(g)$

$SnO_2(s) + C(s) \xrightarrow{\text{high T}} Sn(s) + CO_2(g)$

$2Fe_2O_3(s) + 3C(s) \xrightarrow{\text{high T}} 4Fe(s) + 3CO_2(g)$

4-73 (a) The diagram showing the oxidation states of phosphorus and oxygen and the transfer of electrons is

$$
\begin{array}{c}
\overset{+4e^-}{\overbrace{}} \\
\overset{0}{|} \qquad \downarrow^{-2} \quad \downarrow^{-2} \\
PH_3 + O_2 \longrightarrow P_4O_{10} + H_2O \\
\underset{-3}{|} \qquad \qquad \uparrow^{+5} \\
\underset{-8e^-}{\underbrace{}}
\end{array}
$$

Each O_2 that reacts gains four (2 × 2) electrons and each PH_3 that reacts loses eight electrons. Thus, two (2 × 4 = 8 electrons) O_2 react with each PH_3. The balanced equation is

$$4PH_3(g) + 8O_2(g) \longrightarrow P_4O_{10}(s) + 6H_2O(l)$$

(b)

$$
\begin{array}{c}
\overset{+2e^-}{\overbrace{}} \\
\overset{-1}{|} \qquad \qquad \downarrow^{-2} \\
Cr^{3+} + OH^- + H_2O_2 \longrightarrow CrO_4^{2-} + H_2O \\
\underset{+3}{|} \qquad \qquad \uparrow^{+6} \\
\underset{-3e^-}{\underbrace{}}
\end{array}
$$

Each H_2O_2 that reacts gains two (2 × 1) electrons and each Cr^{3+} that reacts loses three electrons. Thus, three (3 × 2 = 6 electrons) H_2O_2 react with two (2 × 3 = 6 electrons) Cr^{3+}. The balanced equation is

$$2Cr^{3+}(aq) + 10OH^-(aq) + 3H_2O_2(aq) \longrightarrow 2CrO_4^{2-}(aq) + 8H_2O(l)$$

(c)

$$
\begin{array}{c}
\overset{+4e^-}{\overbrace{}} \\
\overset{+5}{|} \qquad \qquad \downarrow^{+1} \\
Sn^{2+} + HNO_3 + H^+ \longrightarrow N_2O + Sn^{4+} + H_2O \\
\underset{+2}{|} \qquad \qquad \uparrow^{+4} \\
\underset{-2e^-}{\underbrace{}}
\end{array}
$$

Each HNO_3 that reacts gains four electrons and each Sn^{2+} that reacts loses two electrons. Thus, two Sn^{2+} react with one HNO_3. The balanced equation is

$$4Sn^{2+}(aq) + 2HNO_3(aq) + 8H^+(aq) \longrightarrow N_2O(g) + 4Sn^{4+}(aq) + 5H_2O(l)$$

(d)

$$
\begin{array}{c}
\overset{+2e^-}{\overbrace{}} \\
0\,|\downarrow\!-1 \\
F_2 + H_2O \longrightarrow HF + O_3 \\
-2\,|\uparrow 0 \\
\underset{-2e^-}{\underbrace{}}
\end{array}
$$

Each F_2 that reacts gains two (2×1) electrons and each H_2O that reacts loses two electrons. Thus, one F_2 reacts with one H_2O. The balanced equation is

$$3F_2(g) + 3H_2O(l) \longrightarrow 6HF(g) + O_3(g)$$

4-75   (a) $+5$ (b) $+4$ (c) $+7$ (d) $+6$

E. ANSWERS TO THE SELF-TEST

1. the transfer of electrons between reactants
2. electron-transfer
3. the number of electrons gained by the element that is reduced
4. zero
5. the charge on the species
6. $+1$
7. -1
8. $+2$
9. $+1$
10. -2
11. true
12. increases
13. reduced
14. loses electrons
15. oxidizing
16. oxidation states . . . electrons
17. oxidation-reduction reaction or combination
18. ionic
19. true
20. true
21. true
22. sulfur . . . oxygen
23. polyatomic
24. $Mg^{2+}(aq)$, $SO_4^{2-}(aq)$
25. false (For example, Al_2O_3 is insoluble in water.)
26. hydroxide ions, $OH^-(aq)$
27. yields hydroxide ions
28. basic
29. yields hydrogen ions, $H^+(aq)$
30. false [The hydrogen ion in water is associated with one or more water molecules and is represented as $H^+(aq)$.]
31. an acid
32. acidic
33. water
34. true
35. sulfuric acid, H_2SO_4, or sulfurous acid, H_2SO_3

36. one

37. $KClO_3$, potassium chlorate, or HgO, mercury(II) oxide

38. the metal oxide and carbon dioxide, CO_2

39. another element

40. hydrogen

41. true

42. true

43. below

44. true

45. fluorine

46. the cations in two different compounds exchange anions

47. insoluble

48. (a) $Na^+(aq)$ and $NO_3^-(aq)$
    (b) $AgCl(s)$
    (c) $Ag^+(aq) + Cl^-(aq) \rightarrow AgCl(s)$

49. the formation of the precipitate, $AgCl(s)$

50. (a) $NO_3^-(aq)$ and $K^+(aq)$
    (b) $H^+(aq) + OH^-(aq) \rightarrow H_2O(l)$

51. the formation of water molecules

52. neutralization

53. salt

54. sour

55. bitter

56. red, blue

57. molecular

58. true

59. true

60. true

61. *-ate*

62. *-ous*

63. *-ite*

64. the prefix *per-* . . . the name of the key acid

65. the prefix *hypo-* . . . the name of the *-ous* acid

5 CHEMICAL CALCULATIONS

A. OUTLINE OF CHAPTER 5

5-1   The quantity of a substance that is equal to its formula mass in grams is called a mole.

Formula mass refers to both atomic mass and molecular mass.

Formula unit refers to an atom, molecule, ion, or a group of ions.

A mole is the quantity of a substance whose mass in grams is numerically equal to its formula mass.

One formula mass in grams of any substance contains the same number of formula units as one formula mass of any other substance.

The molar mass of a compound is the number of grams necessary to make one mole of the compound and has the units grams per mole ($g \cdot mol^{-1}$).

The number of moles in a given mass of a substance is calculated using the unit conversion factor

$$1 = \frac{1 \text{ mol}}{\text{formula mass in grams}}$$

In order to calculate the number of moles in a given mass of a chemical compound, it is necessary to know the chemical formula of the compound.

5-2   One mole of any substance contains Avogadro's number of formula units.

Avogadro's number, 6.022×10^{23}, is the number of formula units in one mole of a substance.

The mass of a formula unit is calculated by dividing the formula mass in grams by Avogadro's number (Example 5-2).

The number of atoms and molecules in a given mass of a substance is found using Avogadro's number and the formula mass of the substance (Example 5-3).

5-3   Simplest formulas can be determined by chemical analysis.

Stoichiometry is the calculation of the quantities of elements or compounds involved in chemical equations.

The symbol $\approx$ means "is stoichiometrically equivalent to" or "combines with."

The simplest chemical formula of a substance is given as the number of atoms of each type in the substance expressed as smallest whole numbers (Examples 5-4 and 5-5).

The empirical formula is the simplest chemical formula of a substance.

5-4   Empirical formulas can be used to determine an unknown atomic mass.

The atomic mass of an element can be determined from the empirical formula of one of its compounds (Example 5-6).

5-5   An empirical formula along with the molecular mass determines the molecular formula.

The molecular formula of a substance is found from the empirical formula and the molecular mass of the substance (Example 5-7).

5-6   The coefficients in chemical equations can be interpreted as numbers of moles.

The balancing coefficients in a chemical equation are the relative numbers of moles of each reactant and product in the balanced equation.

Balancing coefficients are also called stoichiometric coefficients.

The various interpretations of chemical equations in terms of molecules, moles, and grams are summarized in Table 5-2.

Two substances involved in a chemical reaction are related by a stoichiometric unit conversion factor.

Some calculations involving chemical reactions are discussed.

5-7   Calculations involving chemical reactions are carried out in terms of moles.

The procedure for calculations involving chemical equations is given in Figure 5-5.

5-8   It is not always necessary to know the chemical equation to carry out stoichiometric calculations.

Some calculations involving reactions whose chemical equation is not given are discussed.

Some calculations involving mixtures are discussed.

5-9   When two or more substances react, the mass of the product is determined by the limiting reactant.

The limiting reactant is the reactant that is consumed completely in a chemical reaction.

An excess reactant is a reactant that is not consumed completely in a chemical reaction.

When the quantities of two or more reactants are given, it is necessary to determine which is the limiting reactant (Example 5-14).

5-10 For many chemical reactions the amount of the desired product obtained is less than the theoretical amount.

The theoretical yield of a reaction is the mass of a particular product that is calculated from the limiting reactant.

The actual yield is the mass of the product that is actually obtained.

The percent yield is defined as

$$\text{percent yield} = \left(\frac{\text{actual yield}}{\text{theoretical yield}}\right) \times 100 \qquad (5\text{-}2)$$

B. SELF-TEST

1. The atomic mass of an element is the mass of an atom of the element relative to

_____ .

2. The atomic mass of helium is 4.033 g. *True/False*

3. The formula mass of a substance is the _____

_____ .

4. A sample of 4.0 g of helium contains the same number of atoms as 4.0 g of carbon. *True/False*

5. One mole of a substance is the quantity of the substance that is equal to

_____

_____ .

6. One mole of ammonia, NH_3, has the same mass as one mole of nitrogen dioxide, NO_2. *True/False*

7. The value of Avogadro's number is _____ .

8. _____ of any substance contains Avogadro's number of formula units.

9. Two moles of hydrogen chloride, HCl, contain _____ molecules.

10. One mole of carbon dioxide, CO_2, contains the same number of molecules as one mole of carbon monoxide, CO. *True/False*

11. One mole of CO_2 contains Avogadro's number of oxygen atoms. *True/False*

12. Two moles of NaCl contain _____ chloride ions.

13. The symbol _____ means "is stoichiometrically equivalent to."

14. The simplest chemical formula of a compound in which two atoms of Cu combine with one atom of O is _____.

15. Chemical analysis of a compound provides us with the number of each kind of atom in the compound. *True/False*

16. The expression 88.82 g Cu $\leftrightarrows$ 11.18 g O means that _____

_____

_____.

17. The number of moles of Cu and O that combine can be found from the data in Question 16 by _____

_____.

18. The simplest chemical formula of the compound of Cu and O given in Question 16 is Cu_4O_2. *True/False*

19. The empirical formula of a compound is always the actual molecular formula of the compound. *True/False*

20. The molecular formula of a compound can be determined from the empirical formula and its _____.

21. A substance has the empirical formula CH_2 and its molecular mass is 42. The molecular formula is _____.

22. The balancing coefficients in a chemical equation can be interpreted as _____

_____.

Questions 23 through 26 refer to the balanced chemical equation

$$2C_2H_6(g) + 7O_2(g) \longrightarrow 4CO_2(g) + 6H_2O(l)$$

23. How many moles of C_2H_6 react with 14 mol of O_2?

24. In order to calculate the number of grams of CO_2 produced from 14 mol of O_2, we first calculate the _____ of CO_2 produced.

25. We can calculate the mass of CO_2 produced from 14 mol of O_2 by _____

_____.

26. The procedure to calculate the mass of O_2 that is necessary for the complete combustion of a given mass of C_2H_6 is

(1)

(2)

(3)

27. The chemical equation of a reaction must be known in order to calculate the amount of product obtained from a given quantity of a reactant. *True/False*

28. In a reaction between 10 g of hydrogen, H_2, and 52 g of oxygen, O_2, 6 g of hydrogen remain after the consumption of all the oxygen. The limiting reactant is

_____ and the excess reactant is _____.

29. The *(limiting/excess)* reactant is used to calculate the number of grams of product that can be obtained from the reaction.

30. The actual yield of a product in a reaction is always equal to the theoretical yield. *True/False*

C. CALCULATIONS YOU SHOULD KNOW HOW TO USE

Chapter 5 contains many different types of calculations. You should learn how all the calculations presented are unified by the concept of a mole. The procedure for any calculation involving chemical equations is summarized in Figure 5-5. You should be able to do the following types of calculations.
    You will find that it will be helpful to memorize the values of the atomic masses of a few elements, in particular, H, C, O, N, Na, Cl.

1. Calculate the number of moles in a given mass of a substance. You must know the chemical formula of the substance to be able to do this type of calculation. See Example 5-1 and Problems 5-1 through 5-4.

2. Calculate the mass of a given number of formula units of a substance. See Example 5-2 and Problems 5-7 through 5-10.

3. Calculate the number of molecules in a given mass of a substance. See Example 5-3 and Problems 5-11 and 5-12.

4. Determine the simplest formula from chemical analysis. See Examples 5-4 and 5-5 and Problems 5-13 through 5-20.

5. Determine the atomic mass of an element from the empirical formula of one of its compounds. See Example 5-6 and Problems 5-21 through 5-24.

6. Determine the molecular formula from the empirical formula and the molecular mass. See Example 5-7 and Problems 5-25 through 5-28.

7. Calculate the quantities of reactants and products that are involved in chemical reactions. See Examples 5-8, 5-9, 5-10, and 5-11 and Problems 5-29 through 5-36.

8. Calculate the quantities of reactants and products that are involved in chemical reactions whose chemical equation is not given. See Examples 5-12 and 5-13 and Problems 5-37 through 5-42.

9. Calculate the quantities involving limiting reactants. See Example 5-14 and Problems 5-43 through 5-46.

10. Calculate the percent yield of a product. See Example 5-15 and Problems 5-47 through 5-50.

D. SOLUTIONS TO THE ODD-NUMBERED PROBLEMS

5-1 (a) formula mass of $H_2O = 18.02$

$$\text{number of moles} = (28.0 \text{ g}) \left(\frac{1 \text{ mol } H_2O}{18.02 \text{ g } H_2O} \right) = 1.55 \text{ mol}$$

(b) formula mass of $C = 12.01$

$$\text{number of moles} = (200 \text{ mg}) \left(\frac{1 \text{ g}}{10^3 \text{ mg}} \right) \left(\frac{1 \text{ mol } C}{12.01 \text{ g } C} \right) = 0.0167 \text{ mol}$$

(c) formula mass of $NaCl = 58.44$

$$\text{number of moles} = (454 \text{ g}) \left(\frac{1 \text{ mol } NaCl}{58.44 \text{ g } NaCl} \right) = 7.77 \text{ mol}$$

(d) formula mass of $CaO = 56.08$

$$\text{number of moles} = (1000 \text{ kg}) \left(\frac{1000 \text{ g}}{1 \text{ kg}} \right) \left(\frac{1 \text{ mol } CaO}{56.08 \text{ g } CaO} \right) = 1.78 \times 10^4 \text{ mol}$$

5-3 (a) formula mass of malathion $= 330.34$

$$\text{number of moles} = (1.00 \text{ kg}) \left(\frac{10^3 \text{ g}}{1 \text{ kg}} \right) \left(\frac{1 \text{ mol } C_{10}H_{19}O_6PS_2}{330.34 \text{ g } C_{10}H_{19}O_6PS_2} \right)$$

$$= 3.03 \text{ mol}$$

(b) formula mass of aluminum sulfate = 342.14

$$\text{number of moles} = (75.0 \text{ g}) \left(\frac{1 \text{ mol } Al_2(SO_4)_3}{342.14 \text{ g } Al_2(SO_4)_3} \right) = 0.219 \text{ mol}$$

(c) formula mass of oil of peppermint = 156.26

$$\text{number of moles} = (50.0 \text{ mg}) \left(\frac{1 \text{ g}}{10^3 \text{ mg}} \right) \left(\frac{1 \text{ mol } C_{10}H_{20}O}{156.26 \text{ g } C_{10}H_{20}O} \right)$$

$$= 3.20 \times 10^{-4} \text{ mol}$$

(d) formula mass of potassium dichromate = 294.20

$$\text{number of moles} = (2.756 \text{ g}) \left(\frac{1 \text{ mol } K_2Cr_2O_7}{294.20 \text{ g } K_2Cr_2O_7} \right) = 9.368 \times 10^{-3} \text{ mol}$$

5-5 mass of Avogadro's number of baseballs $= (142 \text{ g})(6.022 \times 10^{23})$
$$= 8.55 \times 10^{25} \text{ g}$$

The mass of the earth is 6.0×10^{24} kg $= 6.0 \times 10^{27}$ g. The ratio of the mass of 1 mol of baseballs to the mass of the earth is

$$\frac{8.55 \times 10^{25} \text{ g}}{6.0 \times 10^{27} \text{ g}} = 1.4 \times 10^{-2}$$

The mass of 1 mol of baseballs is 1.4×10^{-2} that of the mass of the earth.

5-7 (a) formula mass of $CO_2 = 44.01$

$$\text{mass of one } CO_2 \text{ molecule} = \left(\frac{44.01 \text{ g } CO_2}{1 \text{ mol } CO_2} \right) \left(\frac{1 \text{ mol}}{6.022 \times 10^{23} \text{ molecules}} \right)$$

$$= 7.308 \times 10^{-23} \text{ g}$$

(b) formula mass of $C_6H_{12}O_6 = 180.16$

$$\text{mass of one } C_6H_{12}O_6 \text{ molecule} = \left(\frac{180.16 \text{ g } C_6H_{12}O_6}{1 \text{ mol } C_6H_{12}O_6} \right)$$

$$\times \left(\frac{1 \text{ mol}}{6.022 \times 10^{23} \text{ molecules}} \right)$$

$$= 2.992 \times 10^{-22} \text{ g}$$

(c) formula mass of $CaCl_2 = 110.98$

$$\text{mass of one } CaCl_2 \text{ formula unit} = \left(\frac{110.98 \text{ g } CaCl_2}{1 \text{ mol } CaCl_2} \right)$$

$$\times \left(\frac{1 \text{ mol}}{6.022 \times 10^{23} \text{ formula units}} \right)$$

$$= 1.843 \times 10^{-22} \text{ g}$$

5-9   (a)   formula mass of Fe $= 55.85$

$$\text{mass of 200 Fe atoms} = \left(\frac{55.85 \text{ g Fe}}{1 \text{ mol Fe}}\right)\left(\frac{1 \text{ mol}}{6.022 \times 10^{23} \text{ atoms}}\right)(200 \text{ atoms})$$
$$= 1.855 \times 10^{-20} \text{ g}$$

(b)   formula mass of $H_2O = 18.02$

$$\left(\begin{array}{c}\text{mass of } 1.0 \times 10^{16} \text{ H}_2\text{O} \\ \text{molecules}\end{array}\right) = \left(\frac{18.02 \text{ g H}_2\text{O}}{1 \text{ mol H}_2\text{O}}\right)\left(\frac{1 \text{ mol}}{6.022 \times 10^{23} \text{ molecules}}\right)$$
$$\times (1.0 \times 10^{16} \text{ molecules})$$
$$= 3.0 \times 10^{-7} \text{ g}$$

(c)   formula mass of O $= 16.00$

$$\text{mass of } 1.0 \times 10^{6} \text{ O atoms} = \left(\frac{16.00 \text{ g O}}{1 \text{ mol O}}\right)\left(\frac{1 \text{ mol}}{6.022 \times 10^{23} \text{ atoms}}\right)$$
$$\times (1.0 \times 10^{6} \text{ atoms})$$
$$= 2.7 \times 10^{-17} \text{ g}$$

(d)   formula mass of $O_2 = 32.00$

$$\left(\begin{array}{c}\text{mass of } 1.0 \times 10^{6} \text{ O}_2 \\ \text{molecules}\end{array}\right) = \left(\frac{32.00 \text{ g O}_2}{1 \text{ mol O}_2}\right)\left(\frac{1 \text{ mol}}{6.022 \times 10^{23} \text{ molecules}}\right)$$
$$\times (1.0 \times 10^{6} \text{ molecules})$$
$$= 5.3 \times 10^{-17} \text{ g}$$

5-11   The number of moles that corresponds to 50.0 g of H_2O is

$$\text{number of moles} = (50.0 \text{ g})\left(\frac{1 \text{ mol H}_2\text{O}}{18.02 \text{ g H}_2\text{O}}\right) = 2.77 \text{ mol}$$

The number of H_2O molecules in 2.77 mol is

$$\text{number of molecules} = (2.77 \text{ mol})\left(\frac{6.022 \times 10^{23} \text{ molecules}}{1 \text{ mol}}\right)$$
$$= 1.67 \times 10^{24} \text{ molecules}$$

There are three atoms in each H_2O molecule, and so the total number of atoms is

$$\text{number of atoms} = (1.67 \times 10^{24} \text{ molecules})\left(\frac{3 \text{ atoms}}{1 \text{ H}_2\text{O molecule}}\right)$$
$$= 5.01 \times 10^{24} \text{ atoms}$$

5-13   Take a 100-g sample and write

$$62.5 \text{ g Ca} \approx 37.5 \text{ g C}$$

Divide each quantity by its corresponding atomic mass to get

$$(62.5 \text{ g})\left(\frac{1 \text{ mol Ca}}{40.08 \text{ g Ca}}\right) = 1.56 \text{ mol Ca} \approx (37.5 \text{ g})\left(\frac{1 \text{ mol C}}{12.01 \text{ g C}}\right) = 3.12 \text{ mol C}$$

Divide by the smaller quantity (1.56) to obtain

$$1.00 \text{ mol Ca} \approx 2.00 \text{ mol C}$$

The empirical formula is CaC_2.

5-15   The mass percentage of Cu in the compound is

$$\text{mass \% Cu} = \left(\frac{2.46 \text{ g}}{5.22 \text{ g}}\right) \times 100 = 47.1\%$$

Therefore, mass % Cl = 52.9%. Assume a 100-g sample and write

$$47.1 \text{ g Cu} \approx 52.9 \text{ g Cl}$$

Divide each quantity by the corresponding atomic mass to get

$$(47.1 \text{ g})\left(\frac{1 \text{ mol Cu}}{63.55 \text{ g Cu}}\right) = 0.741 \text{ mol Cu} \approx (52.9 \text{ g})\left(\frac{1 \text{ mol Cl}}{35.45 \text{ g Cl}}\right) = 1.49 \text{ mol Cl}$$

Divide both quantities by the smaller number (0.741) to get

$$1.00 \text{ mol Cu} \approx 2.01 \text{ mol Cl}$$

Now recognize that 2.01 is essentially 2 and write the empirical formula as $CuCl_2$.

5-17   We are given that 28.1 g of cobalt react with chlorine to yield 61.9 g of the compound. Therefore, the mass of chlorine that reacted is 61.9 g − 28.1 g = 33.8 g. Thus the mass of Cl in the compound is 33.8 g, and so we have the stoichiometric correspondence

$$28.1 \text{ g Co} \approx 33.8 \text{ g Cl}$$

Divide each quantity by its corresponding atomic mass to get

$$0.4768 \text{ mol Co} \approx 0.9535 \text{ mol Cl}$$

or

$$1.00 \text{ mol Co} \approx 2.00 \text{ mol Cl}$$

The empirical formula is $CoCl_2$.

5-19   (a)   Take a 100-g sample and write

$$46.45 \text{ g Li} \backsimeq 53.55 \text{ g O}$$

$$\frac{46.45 \text{ g}}{6.941 \text{ g}\cdot\text{mol}^{-1}} = 6.692 \text{ mol Li} \backsimeq \frac{53.55 \text{ g}}{16.00 \text{ g}\cdot\text{mol}^{-1}} = 3.347 \text{ mol O}$$

$$2.00 \text{ mol Li} \backsimeq 1.00 \text{ mol O}$$

The empirical formula is Li_2N.

(b)   Take a 100-g sample and write

$$59.78 \text{ g Li} \backsimeq 40.22 \text{ g N}$$

$$\frac{59.78 \text{ g}}{6.941 \text{ g}\cdot\text{mol}^{-1}} = 8.613 \text{ mol Li} \backsimeq \frac{40.22 \text{ g}}{14.01 \text{ g}\cdot\text{mol}^{-1}} = 2.871 \text{ mol N}$$

$$3.00 \text{ mol Li} \backsimeq 1.00 \text{ mol N}$$

The empirical formula is Li_3N.

(c)   Take a 100-g sample and write

$$14.17 \text{ g Li} \backsimeq 85.83 \text{ g N}$$

$$\frac{14.17 \text{ g}}{6.941 \text{ g}\cdot\text{mol}^{-1}} = 2.041 \text{ mol Li} \backsimeq \frac{85.83 \text{ g}}{14.01 \text{ g}\cdot\text{mol}^{-1}} = 6.126 \text{ mol N}$$

$$1.00 \text{ mol Li} \backsimeq 3.00 \text{ mol N}$$

The empirical formula is LiN_3.

(d)   Take a 100-g sample and write

$$36.11 \text{ g Ca} \backsimeq 63.89 \text{ g Cl}$$

$$\frac{36.11 \text{ g}}{40.08 \text{ g}\cdot\text{mol}^{-1}} = 0.9009 \text{ mol Ca} \backsimeq \frac{63.89 \text{ g}}{35.45 \text{ g}\cdot\text{mol}^{-1}} = 1.802 \text{ mol Cl}$$

$$1.00 \text{ mol Ca} \backsimeq 2.00 \text{ mol Cl}$$

The empirical formula is $CaCl_2$.

5-21   The mass percentage of the metal is

$$\text{mass \% M} = \left(\frac{1.443 \text{ g}}{1.683 \text{ g}}\right) \times 100 = 85.74\%$$

Assuming a 100-g sample, we have

$$85.74 \text{ g M} \backsimeq 100.00 \text{ g} - 85.74 \text{ g} = 14.26 \text{ g O}$$

We do not know the atomic mass of M, but we can divide 14.26 g O by the atomic mass of O (16.00) to obtain

$$85.74 \text{ g M} \approx 0.8913 \text{ mol O}$$

We know from the given empirical formula that

$$2 \text{ mol M} \approx 3 \text{ mol O}$$

Thus we have

$$85.74 \text{ g M} \approx 0.8913 \text{ mol O} \left(\frac{2 \text{ mol M}}{3 \text{ mol O}} \right)$$

or

$$85.74 \text{ g M} \approx 0.5942 \text{ mol M}$$

Divide by 0.5942 to get

$$144.3 \text{ g M} \approx 1.000 \text{ mol M}$$

The atomic mass of M is 144.3, which corresponds to neodymium (Nd).

5-23
$$\text{mass \% of } H_2O = \left(\frac{0.0949 \text{ g}}{0.642 \text{ g}} \right) \times 100 = 14.8\%$$

Therefore, mass % MCl_2 = 85.2%. Assuming a 100-g sample, we have

$$14.8 \text{ g } H_2O \approx 85.2 \text{ g } MCl_2$$

Divide by the formula mass of H_2O (we do not know the formula mass of MCl_2 because we do not know the atomic mass of M) to obtain

$$0.821 \text{ mol } H_2O \approx 85.2 \text{ g } MCl_2$$

We know from the chemical formula of the compound that

$$2 \text{ mol } H_2O \approx 1 \text{ mol } MCl_2$$

Thus we have

$$0.821 \text{ mol } H_2O \left(\frac{1 \text{ mol } MCl_2}{2 \text{ mol } H_2O} \right) \approx 85.2 \text{ g } MCl_2$$

or

$$0.4105 \text{ mol } MCl_2 \approx 85.2 \text{ g } MCl_2$$

Divide both sides of the correspondence by 0.4105 to get

$$1.00 \text{ mol MCl}_2 \backsimeq 208 \text{ g MCl}_2$$

There are two chlorine atoms in MCl_2, and so the atomic mass of M is

$$\text{atomic mass of M} = 208 - (2 \times 35.45) = 137$$

The metal is barium.

5-25   Take a 100-g sample and write

$$62.0 \text{ g C} \backsimeq 10.4 \text{ g H} \backsimeq 27.5 \text{ g O}$$

Divide by the corresponding atomic masses to get

$$5.16 \text{ mol C} \backsimeq 10.3 \text{ mol H} \backsimeq 1.72 \text{ mol O}$$

Now divide by the smallest quantity (1.72) to get

$$3.00 \text{ mol C} \backsimeq 5.99 \text{ mol H} \backsimeq 1.00 \text{ mol O}$$

The simplest formula is C_3H_6O, whose formula mass is 58.1. Given that the formula mass is 58.1, the molecular formula is C_3H_6O.

5-27   Assume a 100-g sample and write

$$22.5 \text{ g Na} \backsimeq 30.4 \text{ g P} \backsimeq 47.1 \text{ g O}$$

Divide by the corresponding atomic masses to get

$$0.979 \text{ mol Na} \backsimeq 0.982 \text{ mol P} \backsimeq 2.94 \text{ mol O}$$

Divide through by 0.979 to obtain

$$1.00 \text{ mol Na} \backsimeq 1.00 \text{ mol P} \backsimeq 3.00 \text{ mol O}$$

The empirical formula is $NaPO_3$. The formula mass corresponding to this empirical formula is 102, and this divides into the observed molecular mass (612) six times. Thus the molecular formula is $Na_6P_6O_{18}$.

5-29   The number of moles that corresponds to 10.0 g of propane, C_3H_8, is

$$\text{moles of } C_3H_8 = (10.0 \text{ g}) \left(\frac{1 \text{ mol } C_3H_8}{44.09 \text{ g } C_3H_8} \right) = 0.2268 \text{ mol}$$

We see from the reaction that 5 mol of O_2 are required for each mol of C_3H_8 that reacts

$$\text{moles of } O_2 = (0.2268 \text{ mol } C_3H_8) \left(\frac{5 \text{ mol } O_2}{1 \text{ mol } C_3H_8} \right)$$

$$= 1.134 \text{ mol}$$

The mass of O_2 required is

$$\text{mass of } O_2 = (1.134 \text{ mol}) \left(\frac{32.00 \text{ g } O_2}{1 \text{ mol } O_2} \right) = 36.3 \text{ g}$$

5-31 The number of moles of MnO_2 reacted is

$$\text{moles of } MnO_2 = (100 \text{ g}) \left(\frac{1 \text{ mol } MnO_2}{86.94 \text{ g } MnO_2} \right) = 1.150 \text{ mol}$$

According to the equation, 1 mol of Cl_2 is produced from 1 mol of MnO_2. Therefore,

$$\text{moles of } Cl_2 = (1.150 \text{ mol } MnO_2) \left(\frac{1 \text{ mol } Cl_2}{1 \text{ mol } MnO_2} \right)$$
$$= 1.150 \text{ mol}$$

The mass of Cl_2 prepared is

$$\text{mass of } Cl_2 = (1.150 \text{ mol}) \left(\frac{70.90 \text{ g } Cl_2}{1 \text{ mol } Cl_2} \right) = 81.5 \text{ g}$$

5-33 The number of moles of ND_3 produced is

$$\text{moles of } ND_3 = (200 \text{ mg}) \left(\frac{1 \text{ g}}{10^3 \text{ mg}} \right) \left(\frac{1 \text{ mol } ND_3}{20.05 \text{ g } ND_3} \right) = 0.009975 \text{ mol}$$

According to the equation, 3 mol of D_2O are required to produce 1 mol of ND_3. Therefore,

$$\text{moles of } D_2O = (0.009975 \text{ mol } ND_3) \left(\frac{3 \text{ mol } D_2O}{1 \text{ mol } ND_3} \right) = 0.02993 \text{ mol}$$

The mass of heavy water required is

$$\text{mass of } D_2O = (0.02993 \text{ mol}) \left(\frac{20.03 \text{ g } D_2O}{1 \text{ mol } D_2O} \right) = 0.599 \text{ g} = 599 \text{ mg}$$

5-35 Note that 1 mol of Zn results from each mole of ZnS. The number of moles of ZnS is

$$\text{moles of ZnS} = (2.00 \times 10^5 \text{ kg}) \left(\frac{10^3 \text{ g}}{1 \text{ kg}} \right) \left(\frac{1 \text{ mol ZnS}}{97.44 \text{ g ZnS}} \right) = 2.05 \times 10^6 \text{ mol}$$

This is the number of moles of Zn produced. The mass of Zn produced is

$$\text{mass of Zn} = (2.05 \times 10^6 \text{ mol}) \left(\frac{65.38 \text{ g Zn}}{1 \text{ mol Zn}} \right) = 1.34 \times 10^8 \text{ g}$$
$$= 1.34 \times 10^5 \text{ kg}$$

5-37 Because all the nitrogen in NH_3 ends up in N_2H_4, we have

$$1 \text{ mol } N_2H_4 \backsimeq 2 \text{ mol } NH_3$$

The number of moles of NH_3 is given by

$$\text{moles of } NH_3 = (10.0 \text{ metric ton}) \left(\frac{10^3 \text{ kg}}{1 \text{ metric ton}} \right) \left(\frac{10^3 \text{ g}}{1 \text{ kg}} \right) \left(\frac{1 \text{ mol } NH_3}{17.03 \text{ g } NH_3} \right)$$
$$= 5.872 \times 10^5 \text{ mol}$$

The number of moles of N_2H_4 is given by

$$\text{moles of } N_2H_4 = (5.872 \times 10^5 \text{ mol } NH_3) \left(\frac{1 \text{ mol } N_2H_4}{2 \text{ mol } NH_3} \right)$$
$$= 2.936 \times 10^5 \text{ mol}$$

The number of metric tons of N_2H_4 is given by

$$\text{mass of } N_2H_4 = (2.936 \times 10^5 \text{ mol}) \left(\frac{32.05 \text{ g } N_2H_4}{1 \text{ mol } N_2H_4} \right) \left(\frac{1 \text{ kg}}{10^3 \text{ g}} \right) \left(\frac{1 \text{ metric ton}}{10^3 \text{ kg}} \right)$$
$$= 9.41 \text{ metric ton}$$

5-39 Because all the thallium in the sample ends up in TlI, we have

$$1 \text{ mol } Tl_2SO_4 \backsimeq 2 \text{ mol } TlI$$

The number of moles of TlI that was precipitated is

$$\text{moles of TlI} = (0.6112 \text{ g}) \left(\frac{1 \text{ mol TlI}}{331.3 \text{ g TlI}} \right) = 1.8449 \times 10^{-3} \text{ mol}$$

The number of moles of Tl_2SO_4 is given by

$$\text{moles of } Tl_2SO_4 = (1.8449 \times 10^{-3} \text{ mol TlI}) \left(\frac{1 \text{ mol } Tl_2SO_4}{2 \text{ mol TlI}} \right)$$
$$= 9.2245 \times 10^{-4} \text{ mol}$$

The mass of Tl_2SO_4 in the sample is given by

$$\text{mass of } Tl_2SO_4 = (9.2245 \times 10^{-4} \text{ mol}) \left(\frac{504.9 \text{ g } Tl_2SO_4}{1 \text{ mol } Tl_2SO_4} \right)$$
$$= 0.4657 \text{ g}$$

The percentage of Tl_2SO_4 in the sample is

$$\% \ Tl_2SO_4 = \left(\frac{\text{mass } Tl_2SO_4}{\text{mass sample}} \right) \times 100 = \left(\frac{0.4657 \text{ g}}{12.76 \text{ g}} \right) \times 100$$
$$= 3.650 \%$$

5-41 Because all the aluminum in bauxite ends up as aluminum metal, we have

$$1 \text{ mol } Al_2O_3 \cdot 2H_2O \rightleftharpoons 2 \text{ mol } Al$$

The number of moles of bauxite is given by

$$\text{moles of } Al_2O_3 \cdot 2H_2O = (100 \times 10^6 \text{ g}) \left(\frac{1 \text{ mol } Al_2O_3 \cdot 2H_2O}{137.99 \text{ g } Al_2O_3 \cdot 2H_2O} \right)$$
$$= 7.247 \times 10^5 \text{ mol}$$

The number of moles of aluminum produced is given by

$$\text{moles of } Al = (7.247 \times 10^5 \text{ mol } Al_2O_3 \cdot 2H_2O) \left(\frac{2 \text{ mol } Al}{1 \text{ mol } Al_2O_3 \cdot 2H_2O} \right)$$
$$= 1.449 \times 10^6 \text{ mol}$$

The number of metric tons of aluminum produced is

$$\text{mass of } Al = (1.449 \times 10^6 \text{ mol}) \left(\frac{26.98 \text{ g } Al}{1 \text{ mol } Al} \right) \left(\frac{1 \text{ kg}}{10^3 \text{ g}} \right) \left(\frac{1 \text{ metric ton}}{10^3 \text{ kg}} \right)$$
$$= 39.1 \text{ metric tons}$$

5-43 Because we are given the quantities of two reactants, we must check to see if one of them is a limiting reactant. The number of moles of KCl is

$$\text{moles of KCl} = (50.0 \text{ kg}) \left(\frac{10^3 \text{ g}}{1 \text{ kg}} \right) \left(\frac{1 \text{ mol KCl}}{74.55 \text{ g KCl}} \right) = 670.7 \text{ mol}$$

and the number of moles of HNO_3 is

$$\text{moles of } HNO_3 = (50.0 \text{ kg}) \left(\frac{10^3 \text{ g}}{1 \text{ kg}} \right) \left(\frac{1 \text{ mol } HNO_3}{63.02 \text{ g } HNO_3} \right) = 793.4 \text{ mol}$$

Because 1 mol of KCl reacts with 1 mol of HNO_3, we see that HNO_3 is in excess and KCl is the limiting reactant. The mass of KNO_3 produced is

$$\text{mass of } KNO_3 = (670.7 \text{ mol KCl}) \left(\frac{4 \text{ mol } KNO_3}{4 \text{ mol KCl}} \right) \left(\frac{101.11 \text{ g } KNO_3}{1 \text{ mol } KNO_3} \right)$$
$$= 6.78 \times 10^4 \text{ g} = 67.8 \text{ kg}$$

The mass of Cl_2 produced is

$$\text{mass of } Cl_2 = (670.7 \text{ mol KCl}) \left(\frac{2 \text{ mol } Cl_2}{4 \text{ mol KCl}} \right) \left(\frac{70.90 \text{ g } Cl_2}{1 \text{ mol } Cl_2} \right)$$
$$= 2.38 \times 10^4 \text{ g} = 23.8 \text{ kg}$$

5-45   The number of moles of NaOH is

$$\text{moles of NaOH} = (60.0 \text{ g}) \left(\frac{1 \text{ mol NaOH}}{40.00 \text{ g NaOH}} \right) = 1.50 \text{ mol}$$

and the number of moles of H_2SO_4 is

$$\text{moles of } H_2SO_4 = (20.0 \text{ g}) \left(\frac{1 \text{ mol } H_2SO_4}{98.08 \text{ g } H_2SO_4} \right) = 0.204 \text{ mol}$$

Each mole of H_2SO_4 requires 2 mol of NaOH, or 0.204 mol of H_2SO_4 requires 0.408 mol of NaOH. Thus the NaOH is in excess and H_2SO_4 is the limiting reactant. The mass of Na_2SO_4 that will be produced is

$$\text{mass of } Na_2SO_4 = (0.204 \text{ mol } H_2SO_4) \left(\frac{1 \text{ mol } Na_2SO_4}{1 \text{ mol } H_2SO_4} \right) \left(\frac{142.04 \text{ g } Na_2SO_4}{1 \text{ mol } Na_2SO_4} \right)$$
$$= 29.0 \text{ g}$$

5-47   The theoretical yield of $TiCl_4$ is

$$\binom{\text{theoretical}}{\text{yield}} = (50.0 \text{ g } TiO_2) \left(\frac{1 \text{ mol } TiO_2}{79.90 \text{ g } TiO_2} \right) \left(\frac{1 \text{ mol } TiCl_4}{1 \text{ mol } TiO_2} \right) \left(\frac{189.70 \text{ g } TiCl_4}{1 \text{ mol } TiCl_4} \right)$$
$$= 118.7 \text{ g}$$

The percent yield is

$$\% \text{ yield} = \left(\frac{\text{actual yield}}{\text{theoretical yield}} \right) \times 100$$
$$= \frac{55.0 \text{ g}}{118.7 \text{ g}} \times 100 = 46.3\%$$

5-49   The theoretical yield of $C_2H_5O_2CC_2H_5$ is

$$\binom{\text{theoretical}}{\text{yield}} = (250 \text{ g } C_2H_5OH) \left(\frac{1 \text{ mol } C_2H_5OH}{46.07 \text{ g } C_2H_5OH} \right) \left(\frac{1 \text{ mol } C_2H_5O_2CC_2H_5}{1 \text{ mol } C_2H_5OH} \right)$$
$$\times \left(\frac{102.13 \text{ g } C_2H_5O_2CC_2H_5}{1 \text{ mol } C_2H_5O_2CC_2H_5} \right)$$
$$= 554 \text{ g}$$

The percent yield is

$$\% \text{ yield} = \left(\frac{\text{actual yield}}{\text{theoretical yield}} \right) \times 100$$
$$= \frac{349 \text{ g}}{554 \text{ g}} \times 100 = 63.0\%$$

5-51 The numbers of moles of the reactants are

$$\text{moles of } CS_2 = (1000 \text{ g}) \left(\frac{1 \text{ mol } CS_2}{76.13 \text{ g } CS_2} \right) = 13.1 \text{ mol}$$

$$\text{moles of NaOH} = (1000 \text{ g}) \left(\frac{1 \text{ mol NaOH}}{40.00 \text{ g NaOH}} \right) = 25.0 \text{ mol}$$

Because 25.0 mol NaOH requires only 12.5 mol CS_2, the CS_2 is in excess and NaOH is the limiting reagent. The mass of each product that is produced is

$$\text{mass of } Na_2CS_3 = (25.0 \text{ mol NaOH}) \left(\frac{2 \text{ mol } Na_2CS_3}{6 \text{ mol NaOH}} \right) \left(\frac{154.17 \text{ g } Na_2CS_3}{1 \text{ mol } Na_2CS_3} \right)$$

$$= 1280 \text{ g}$$

$$\text{mass of } Na_2CO_3 = (25.0 \text{ mol NaOH}) \left(\frac{1 \text{ mol } Na_2CO_3}{6 \text{ mol NaOH}} \right) \left(\frac{105.99 \text{ g } Na_2CO_3}{1 \text{ mol } Na_2CO_3} \right)$$

$$= 442 \text{ g}$$

$$\text{mass of } H_2O = (25.0 \text{ mol NaOH}) \left(\frac{3 \text{ mol } H_2O}{6 \text{ mol NaOH}} \right) \left(\frac{18.02 \text{ g } H_2O}{1 \text{ mol } H_2O} \right)$$

$$= 225 \text{ g}$$

5-53 The number of moles of $KMnO_4$, H_2O_2, and HCl are

$$\text{moles of } KMnO_4 = (20.0 \text{ g}) \left(\frac{1 \text{ mol } KMnO_4}{158.04 \text{ g } KMnO_4} \right) = 0.1266 \text{ mol}$$

$$\text{moles of } H_2O_2 = (10.0 \text{ g}) \left(\frac{1 \text{ mol } H_2O_2}{34.02 \text{ g } H_2O_2} \right) = 0.2939 \text{ mol}$$

$$\text{moles of HCl} = (1.00 \times 10^2 \text{ g}) \left(\frac{1 \text{ mol HCl}}{36.46 \text{ g HCl}} \right) = 2.743 \text{ mol}$$

We can find the limiting reactant by dividing the number of available moles of each reactant by the corresponding balancing coefficient in the chemical equation. Thus

$$\frac{0.1266 \text{ mol } KMnO_4}{2 \text{ mol } KMnO_4} = 0.06330$$

$$\frac{0.2939 \text{ mol } H_2O_2}{5 \text{ mol } H_2O_2} = 0.05878$$

$$\frac{2.743 \text{ mol HCl}}{6 \text{ mol HCl}} = 0.4572$$

The smallest (limiting) value is that for H_2O_2, and so we see that H_2O_2 is the limiting reactant. The number of grams of $MnCl_2$ that can be produced is

$$\text{mass of } MnCl_2 = (0.2939 \text{ mol } H_2O_2)\left(\frac{2 \text{ mol } MnCl_2}{5 \text{ mol } H_2O_2}\right)\left(\frac{125.84 \text{ g } MnCl_2}{1 \text{ mol } MnCl_2}\right)$$

$$= 14.8 \text{ g}$$

The percent yield is

$$\% \text{ yield} = \left(\frac{\text{actual yield}}{\text{theoretical yield}}\right) \times 100$$

$$= \frac{9.82 \text{ g}}{14.8 \text{ g}} \times 100 = 66.4\%$$

5-55   The mass of As in the ore is equal to the mass of As in Ag_3AsO_4.

$$\text{mass of As} = (3.09 \text{ g } Ag_3AsO_4)\left(\frac{1 \text{ mol } Ag_3AsO_4}{462.6 \text{ g } Ag_3AsO_4}\right)\left(\frac{1 \text{ mol As}}{1 \text{ mol } Ag_3AsO_4}\right)$$
$$\times \left(\frac{74.92 \text{ g As}}{1 \text{ mol As}}\right)$$

$$= 0.5004 \text{ g}$$

The percent arsenic in the ore sample is

$$\% \text{ As} = \frac{0.5004 \text{ g As}}{5.00 \text{ g ore}} \times 100 = 10.0\%$$

5-57   The equation for the combustion of sugar is

$$C_{12}H_{22}O_{11}(s) + 12O_2(g) \longrightarrow 12CO_2(g) + 11H_2O(g)$$

The mass of sugar that produces 2.20 g of CO_2 is

$$\text{mass of } C_{12}H_{22}O_{11} = (2.20 \text{ g } CO_2)\left(\frac{1 \text{ mol } CO_2}{44.01 \text{ g } CO_2}\right)\left(\frac{1 \text{ mol } C_{12}H_{22}O_{11}}{12 \text{ mol } CO_2}\right)$$
$$\times \left(\frac{342.30 \text{ g } C_{12}H_{22}O_{11}}{1 \text{ mol } C_{12}H_{22}O_{11}}\right)$$

$$= 1.43 \text{ g}$$

The mass of NaCl in the sample is

$$\text{mass of NaCl} = 5.00 \text{ g} - 1.43 \text{ g} = 3.57 \text{ g}$$

The percent NaCl in the sample is

$$\% \text{ NaCl} = \frac{3.57 \text{ g NaCl}}{5.00 \text{ g sample}} \times 100 = 71.4\%$$

5-59   The equations for the reactions that produce hydrogen gas are

$$Zn(s) + 2HCl(aq) \longrightarrow ZnCl_2(aq) + H_2(g)$$
$$Mg(s) + 2HCl(aq) \longrightarrow MgCl_2(aq) + H_2(g)$$

The number of moles of hydrogen produced is

$$\text{moles of } H_2 = (0.204 \text{ g}) \left(\frac{1 \text{ mol } H_2}{2.016 \text{ g } H_2} \right) = 0.1012 \text{ mol}$$

From the stoichiometry of the two equations, we have that

$$1 \text{ mol } H_2 \backsimeq 1 \text{ mol metal}$$

The number of moles of metal (both zinc and magnesium) present is given by

$$\text{moles of metal} = (0.1012 \text{ mol } H_2) \left(\frac{1 \text{ mol metal}}{1 \text{ mol } H_2} \right) = 0.1012 \text{ mol}$$

Let

$$x \text{ g} = \text{number of grams of Zn}$$

Because the mass of the original sample is 4.07 g,

$$4.07 \text{ g} - x \text{ g} = \text{number of grams of Mg}$$

The number of moles of Zn is given by

$$\text{moles of Zn} = (x \text{ g}) \left(\frac{1 \text{ mol Zn}}{65.38 \text{ g Zn}} \right) = \frac{x \text{ mol}}{65.38}$$

and the number of moles of Mg is given by

$$\text{moles of Mg} = (4.07 \text{ g} - x \text{ g}) \left(\frac{1 \text{ mol Mg}}{24.31 \text{ g Mg}} \right) = \frac{(4.07 - x) \text{ mol}}{24.31}$$

We have that

$$\text{moles of Zn} + \text{moles of Mg} = 0.1012 \text{ mol}$$

Substituting the expressions for moles of Zn and moles of Mg, we have

$$\frac{x \text{ mol}}{65.38} + \frac{(4.07 - x) \text{ mol}}{24.31} = 0.1012 \text{ mol}$$

Solving for x, we have

$$0.015295x + 0.1674 - 0.041135x = 0.1012$$
$$0.02584x = 0.0662$$
$$x = 2.56$$

The number of grams of zinc in the sample is

$$\text{mass of Zn} = 2.56 \text{ g}$$

The mass percentage of Zn in the sample is

$$\% \text{ Zn} = \left(\frac{2.56 \text{ g}}{4.07 \text{ g}}\right) \times 100 = 62.9\%$$
$$\% \text{ Mg} = 100\% - 62.9\% = 37.1\%$$

5-61   The number of moles of $BaSO_4$ precipitated is given by

$$\text{moles of BaSO}_4 = (4.37 \text{ g}) \left(\frac{1 \text{ mol BaSO}_4}{233.4 \text{ g BaSO}_4}\right) = 0.01872 \text{ mol}$$

Let

$$x \text{ g} = \text{number of grams of } K_2SO_4$$

and

$$3.00 \text{ g} - x \text{ g} = \text{number of grams of MnSO}_4$$

Then

$$\text{moles of K}_2SO_4 = (x \text{ g}) \left(\frac{1 \text{ mol K}_2SO_4}{174.26 \text{ g K}_2SO_4}\right) = \frac{x \text{ mol}}{174.26}$$

and

$$\text{moles of MnSO}_4 = (3.00 \text{ g} - x \text{ g}) \left(\frac{1 \text{ mol MnSO}_4}{151.00 \text{ g MnSO}_4}\right) = \frac{(3.00 - x) \text{ mol}}{151.00}$$

Because

$$K_2SO_4(aq) + Ba(NO_3)_2(aq) \longrightarrow 2KNO_3(aq) + BaSO_4(aq)$$
$$MnSO_4(aq) + Ba(NO_3)_2(aq) \longrightarrow 2Mn(NO_3)_2(aq) + BaSO_4(aq)$$

the total number of moles of K_2SO_4 and $MnSO_4$ in the sample is equal to the number of moles of $BaSO_4$ precipitated or

$$\text{moles of K}_2SO_4 + \text{moles of MnSO}_4 = \text{moles of BaSO}_4$$

Thus, we have that

$$\frac{x \text{ mol}}{174.26} + \frac{(3.00 - x) \text{ mol}}{151.00} = 0.01872 \text{ mol}$$

Solving for x, we have

$$5.7386 \times 10^{-3}\, x + 0.01987 - 6.6225 \times 10^{-3}\, x = 0.01872$$
$$0.884 \times 10^{-3}\, x = 0.00115$$
$$x = 1.30$$

The mass of K_2SO_4 is 1.30 g.

The mass percentage of K_2SO_4 in the sample is

$$\% \ K_2SO_4 = \left(\frac{1.30 \text{ g}}{3.00 \text{ g}}\right) \times 100 = 43.3\%$$

The mass percentage of $MnSO_4$ in the sample is

$$\% \ MnSO_4 = 100\% - 43.3\% = 56.7\%$$

5-63 Let

$$x \text{ g} = \text{number of grams of Al}$$

Therefore, we have that

$$9.87 \text{ g} - x \text{ g} = \text{number of grams of Mg}$$

The number of moles of aluminum and of magnesium are given by

$$\text{moles of Al} = (x \text{ g})\left(\frac{1 \text{ mol Al}}{26.98 \text{ g Al}}\right) = \left(\frac{x}{26.98}\right) \text{mol}$$
$$\text{moles of Mg} = (9.87 \text{ g} - x \text{ g})\left(\frac{1 \text{ mol Mg}}{24.31 \text{ g Mg}}\right) = \left(0.4060 - \frac{x}{24.31}\right) \text{mol}$$

The equations for the reactions are

$$2Al(s) + 6HCl(aq) \longrightarrow 2AlCl_3(aq) + 3H_2(g)$$
$$Mg(s) + 2HCl(aq) \longrightarrow MgCl_2(aq) + H_2(g)$$

The number of moles of hydrogen produced is given by

$$\text{moles of } H_2 = (0.998 \text{ g})\left(\frac{1 \text{ mol } H_2}{2.016 \text{ g } H_2}\right) = 0.4950 \text{ mol}$$

We have that the total number of moles of hydrogen produced is equal to the number of moles of hydrogen produced by the reaction of aluminum plus the number of moles of hydrogen produced by the reaction of magnesium.

$$0.4950 \text{ mol H}_2 = \left(\frac{x}{26.98}\right) \text{mol Al} \left(\frac{3 \text{ mol H}_2}{2 \text{ mol Al}}\right)$$
$$+ \left(0.4060 - \frac{x}{24.31}\right) \text{mol Mg} \left(\frac{1 \text{ mol H}_2}{1 \text{ mol Mg}}\right)$$
$$0.4950 = 0.05560 \, x + 0.4060 - 0.04114 \, x$$

Solving for x, we have that

$$x = 6.15$$

Thus,

$$\text{mass of Al} = 6.15 \text{ g}$$
$$\text{mass of Mg} = 9.87 \text{ g} - 6.15 \text{ g} = 3.72 \text{ g}$$

5-65   The number of carbon atoms in 12.01 g of carbon is 6.022×10^{23}. The number of carbon atoms in 12.01 lb of carbon is

$$N = \left(\frac{6.022 \times 10^{23} \text{ atom}}{12.01 \text{ g C}}\right)\left(\frac{10^3 \text{ g}}{1 \text{ kg}}\right)\left(\frac{1 \text{ kg}}{2.205 \text{ lb}}\right)(12.01 \text{ lb C})$$
$$= 2.731 \times 10^{26} \text{ atom}$$

Thus, if Avogadro's number were defined as the number of carbon atoms in 12.01 lb of carbon, then Avogadro's number would equal 2.731×10^{26}.

5-67   The mass of $CaCO_3$ in the sample of limestone is

$$\text{mass of } CaCO_3 = \left(\frac{86.4}{100}\right)(3.24 \text{ kg})\left(\frac{10^3 \text{ g}}{1 \text{ kg}}\right) = 2.799 \times 10^3 \text{ g}$$
$$\text{moles of } CaCO_3 = (2.799 \times 10^3 \text{ g})\left(\frac{1 \text{ mol } CaCO_3}{100.09 \text{ g } CaCO_3}\right) = 27.96 \text{ mol}$$
$$\text{moles of } CaO = (27.97 \text{ mol } CaCO_3)\left(\frac{1 \text{ mol } CaO}{1 \text{ mol } CaCO_3}\right) = 27.96 \text{ mol}$$

The mass of CaO produced is

$$\text{mass of } CaO = (27.96 \text{ mol})\left(\frac{56.08 \text{ g } CaO}{1 \text{ mol } CaO}\right) = 1570 \text{ g} = 1.57 \text{ kg}$$

5-69   We must first check to see which reactant is the limiting reactant.

$$\text{moles of KCN} = (50.0 \text{ g}) \left(\frac{1 \text{ mol KCN}}{65.12 \text{ g KCN}} \right) = 0.7678 \text{ mol}$$

$$\text{moles of CuSO}_4 \cdot 5\text{H}_2\text{O} = (75.0 \text{ g}) \left(\frac{1 \text{ mol CuSO}_4 \cdot 5\text{H}_2\text{O}}{249.69 \text{ g CuSO}_4 \cdot 5\text{H}_2\text{O}} \right) = 0.3004 \text{ mol}$$

If we divide the number of moles of each reactant by its respective stoichiometric coefficient, then we see that (moles of $\text{KCN}/2$) = 0.3839 mol is greater than (moles of $\text{CuSO}_4 \cdot \text{H}_2\text{O}/1$) = 0.3004 mol. Therefore, the limiting reactant is $\text{CuSO}_4 \cdot 5\text{H}_2\text{O}$. We have 1 mol of CuSO_4 in each mole of $\text{CuSO}_4 \cdot 5\text{H}_2\text{O}$, and so

$$\text{moles of C}_2\text{N}_2 = (0.3004 \text{ mol CuSO}_4) \left(\frac{1 \text{ mol C}_2\text{N}_2}{2 \text{ mol CuSO}_4} \right) = 0.1502 \text{ mol}$$

$$\text{mass of C}_2\text{N}_2 = (0.1502 \text{ mol}) \left(\frac{52.04 \text{ g C}_2\text{N}_2}{1 \text{ mol C}_2\text{N}_2} \right) = 7.82 \text{ g}$$

5-71   Because all the silicon in SiO_2 ends up in SiC, we have that

$$1 \text{ mol SiO}_2 \leftrightarrows 1 \text{ mol SiC}$$

$$\text{moles of SiO}_2 = (8.65 \times 10^3 \text{ g}) \left(\frac{1 \text{ mol SiO}_2}{60.09 \text{ g SiO}_2} \right) = 144.0 \text{ mol}$$

$$\text{moles of SiC} = \text{moles of SiO}_2 = 144.0 \text{ mol}$$

$$\text{mass of SiC} = (144.0 \text{ mol}) \left(\frac{40.10 \text{ g SiC}}{1 \text{ mol SiC}} \right) = 5770 \text{ g} = 5.77 \text{ kg}$$

5-73   We must first check to see which reactant is the limiting reactant.

$$\text{moles of SiO}_2 = (300 \text{ g}) \left(\frac{1 \text{ mol SiO}_2}{60.09 \text{ g SiO}_2} \right) = 4.993 \text{ mol}$$

$$\text{moles of C} = (50.0 \text{ g}) \left(\frac{1 \text{ mol C}}{12.01 \text{ g C}} \right) = 4.163 \text{ mol}$$

If we divide the number of moles of each reactant by its respective stoichiometric coefficient, then we see that (moles of $\text{SiO}_2/1$) = 4.993 mol is greater than (moles of $\text{C}/2$) = 2.082 mol. Therefore, the limiting reactant is C.

$$\text{moles of SiCl}_4 = (4.163 \text{ mol C}) \left(\frac{1 \text{ mol SiCl}_4}{2 \text{ mol C}} \right) = 2.082 \text{ mol}$$

$$\text{theoretical yield of SiCl}_4 = (2.082 \text{ mol}) \left(\frac{169.89 \text{ g SiCl}_4}{1 \text{ mol SiCl}_4} \right) = 353.7 \text{ g}$$

From the definition of percent yield (Equation 5-2), we have that

$$\text{actual yield} = \frac{(\% \text{ yield})(\text{theoretical yield})}{100}$$

$$\text{actual yield SiCl}_4 = \frac{(90.7)(353.7 \text{ g})}{100} = 321 \text{ g}$$

5-75 Take a 100-g sample of the ore. The mass of chromium in the sample is 42.7 g. Thus,

$$\text{moles of Cr} = (42.7 \text{ g}) \left(\frac{1 \text{ mol Cr}}{52.00 \text{ g Cr}} \right) = 0.8212 \text{ mol}$$

We have that

$$2 \text{ mol Cr} \backsimeq 1 \text{ mol Cr}_2\text{O}_3$$

$$0.8212 \text{ mol Cr} \backsimeq 0.4106 \text{ mol Cr}_2\text{O}_3$$

$$\text{mass of Cr}_2\text{O}_3 = (0.4106 \text{ mol}) \left(\frac{152.00 \text{ g Cr}_2\text{O}_3}{1 \text{ mol Cr}_2\text{O}_3} \right) = 62.41 \text{ g}$$

The mass percentage of Cr_2O_3 in the ore is

$$\% \text{ Cr}_2\text{O}_3 = \frac{62.41 \text{ g Cr}_2\text{O}_3}{100 \text{ g sample}} \times 100 = 62.4\%$$

5-77 Let

$$x \text{ g} = \text{mass of Na}_2\text{SO}_4$$

$$2.606 \text{ g} - x \text{ g} = \text{mass of NaHSO}_4$$

$$\text{moles of Na}_2\text{SO}_4 = (x \text{ g}) \left(\frac{1 \text{ mol Na}_2\text{SO}_4}{142.04 \text{ g Na}_2\text{SO}_4} \right) = \frac{x \text{ mol}}{142.04} = 0.0070403x \text{ mol}$$

$$\text{moles of NaHSO}_4 = (2.606 \text{ g} - x \text{ g}) \left(\frac{1 \text{ mol NaHSO}_4}{120.06 \text{ g NaHSO}_4} \right)$$

$$= 0.021706 \text{ mol} - 0.0083292x \text{ mol}$$

$$\text{moles of BaSO}_4 = (4.688 \text{ g}) \left(\frac{1 \text{ mol BaSO}_4}{233.4 \text{ g BaSO}_4} \right) = 0.020086 \text{ mol}$$

Now, from the given chemical equations, we have that

$$\text{moles of Na}_2\text{SO}_4 + \text{moles of NaHSO}_4 = \text{moles of BaSO}_4$$

$$0.0070403x \text{ mol} + 0.021706 \text{ mol} - 0.0083292x \text{ mol} = 0.020086 \text{ mol}$$

$$0.0012889x = 0.00162$$

$$x = 1.257$$

$$\text{mass of Na}_2\text{SO}_4 = 1.257 \text{ g}$$

The mass percentages of Na_2SO_4 and $NaHSO_4$ are

$$\% \ Na_2SO_4 = \frac{1.257 \text{ g } Na_2SO_4}{2.606 \text{ g sample}} \times 100 = 48.2\%$$

$$\% \ NaHSO_4 = 100\% - 48.2\% = 51.8\%$$

5-79

$$\text{moles of } Mo_2O_3 = (12.64 \text{ g}) \left(\frac{1 \text{ mol } Mo_2O_3}{239.88 \text{ g } Mo_2O_3} \right) = 0.052693 \text{ mol}$$

$$\text{moles of } Mo = (0.052693 \text{ mol } Mo_2O_3) \left(\frac{2 \text{ mol } Mo}{1 \text{ mol } Mo_2O_3} \right) = 0.10539 \text{ mol}$$

$$\text{mass of } Mo = (0.10539 \text{ mol}) \left(\frac{95.94 \text{ g } Mo}{1 \text{ mol } Mo} \right) = 10.111 \text{ g}$$

The mass of Mo in the new oxide is 10.111 g and the mass of oxygen in the oxide is 13.48 g $-$ 10.111 g = 3.37 g. Thus, we have that

$$10.11 \text{ g } Mo \backsimeq 3.37 \text{ g } O$$

$$0.1054 \text{ mol } Mo \backsimeq 0.2106 \text{ mol } O$$

or, upon dividing by 0.1054, we have

$$1 \text{ mol } Mo \backsimeq 2.00 \text{ mol } O$$

The empirical formula of the new oxide is MoO_2.

5-81 The number of moles of titanium produced is

$$\text{moles of } Ti = (10.0 \text{ metric ton}) \left(\frac{10^3 \text{ kg}}{1 \text{ metric ton}} \right) \left(\frac{10^3 \text{ g}}{1 \text{ kg}} \right) \left(\frac{1 \text{ mol } Ti}{47.90 \text{ g } Ti} \right)$$

$$= 2.088 \times 10^5 \text{ mol}$$

Because all the titanium in TiO_2 ends as the free metal, we have that

$$1 \text{ mol } Ti \backsimeq 1 \text{ mol } TiO_2$$

$$\text{mass of } TiO_2 = (2.088 \times 10^5 \text{ mol}) \left(\frac{79.90 \text{ g } TiO_2}{1 \text{ mol } TiO_2} \right)$$

$$= 16.68 \times 10^6 \text{ g} = 16.68 \text{ metric tons}$$

The mass of the ore required is given by

$$\text{mass of ore} = \frac{16.68 \text{ metric tons}}{0.652} = 25.6 \text{ metric tons}$$

5-83 The number of moles of Sb_2S_3 is

$$\text{moles of } Sb_2S_3 = (500 \text{ g}) \left(\frac{1 \text{ mol } Sb_2S_3}{339.8 \text{ g } Sb_2S_3} \right) = 1.471 \text{ mol}$$

From the two equations, we see that 4 mol of Sb are produced from 2 mol of Sb_2S_3. Thus

$$\text{moles of Sb} = (1.471 \text{ mol } Sb_2S_3)\left(\frac{4 \text{ mol Sb}}{2 \text{ mol } Sb_2S_3}\right) = 2.942 \text{ mol}$$

The mass of Sb produced is

$$\text{mass of Sb} = (2.942 \text{ mol})\left(\frac{121.8 \text{ g Sb}}{1 \text{ mol Sb}}\right) = 358 \text{ g}$$

5-85
$$\text{moles of C} = (1.7 \text{ g})\left(\frac{1 \text{ mol C}}{12.01 \text{ g C}}\right) = 0.142 \text{ mol}$$

number of C atoms $= (0.142 \text{ mol})(6.022 \times 10^{23} \text{ atom} \cdot \text{mol}^{-1}) = 8.6 \times 10^{22} \text{ atom}$

5-87 Using the definition of density, we have that the volume of 1 mol of any substance is given by

$$V = \frac{M}{d}$$

where d is the density of the substance and M is its molar mass.

(a) $\quad V = \dfrac{63.55 \text{ g} \cdot \text{mol}^{-1}}{8.92 \text{ g} \cdot \text{cm}^{-3}} = 7.12 \text{ cm}^3 \cdot \text{mol}^{-1}$

(b) $\quad V = \dfrac{18.02 \text{ g} \cdot \text{mol}^{-1}}{1.00 \text{ g} \cdot \text{cm}^{-3}} = 18.02 \text{ cm}^3 \cdot \text{mol}^{-1}$

(c) $\quad V = \dfrac{12.01 \text{ g} \cdot \text{mol}^{-1}}{3.52 \text{ g} \cdot \text{cm}^{-3}} = 3.41 \text{ cm}^3 \cdot \text{mol}^{-1}$

(d) $\quad V = \dfrac{2.016 \text{ g} \cdot \text{mol}^{-1}}{8.99 \times 10^{-2} \text{ g} \cdot \text{L}^{-1}} = (22.4 \text{ L} \cdot \text{mol}^{-1})\left(\dfrac{10^3 \text{ mL}}{1 \text{ L}}\right)\left(\dfrac{1 \text{ cm}^3}{1 \text{ mL}}\right)$

$\qquad\qquad = 2.24 \times 10^4 \text{ cm}^3 \cdot \text{mol}^{-1}$

5-89 We must first check to see which reactant is the limiting reactant.

$$\text{moles of } P_4O_6 = (10.0 \text{ g})\left(\frac{1 \text{ mol } P_4O_6}{219.88 \text{ g } P_4O_6}\right) = 0.04548 \text{ mol}$$

$$\text{moles of } I_2 = (10.0 \text{ g})\left(\frac{1 \text{ mol } I_2}{253.8 \text{ g } I_2}\right) = 0.03940 \text{ mol}$$

If we divide the number of moles of each reactant by its respective stoichiometric coefficient, then we see that (moles of $P_4O_6/5) = 0.0091$ mol is greater than (moles of $I_2/8) = 0.0049$ mol. Therefore, I_2 is the limiting reactant.

$$\text{moles of } P_2I_4 = (0.03940 \text{ mol } I_2)\left(\frac{4 \text{ mol } P_2I_4}{8 \text{ mol } I_2}\right) = 0.01970 \text{ mol}$$

$$\text{mass of } P_2I_4 = (0.01970 \text{ mol } P_2I_4)\left(\frac{569.5 \text{ g } P_2I_4}{1 \text{ mol } P_2I_4}\right) = 11.2 \text{ g}$$

5-91
$$\text{moles of FeSAs} = (10.0 \text{ g As}) \left(\frac{1 \text{ mol As}}{74.92 \text{ g As}} \right) \left(\frac{1 \text{ mol FeSAs}}{1 \text{ mol As}} \right)$$
$$= 0.1335 \text{ mol}$$

$$\text{mass of FeSAs} = (0.1335 \text{ mol FeSAs}) \left(\frac{162.83 \text{ g FeSAs}}{1 \text{ mol FeSAs}} \right) = 21.7 \text{ g}$$

5-93 (a)
$$\text{moles of Na}_2\text{CO}_3 = (25.0 \text{ g}) \left(\frac{1 \text{ mol Na}_2\text{CO}_3}{105.99 \text{ g Na}_2\text{CO}_3} \right) = 0.2359 \text{ mol}$$

$$\text{number of formula units in Na}_2\text{CO}_3 = (0.2359 \text{ mol})$$
$$\times \left(\frac{6.022 \times 10^{23} \text{ formula units}}{1 \text{ mol}} \right)$$
$$= 1.42 \times 10^{23} \text{ formula units}$$

Because there is one carbon atom per formula unit of Na_2CO_3, the number of carbon atoms in 25.0 g of Na_2CO_3 is 1.42×10^{23} atoms.

(b)
$$\text{number of C atoms in C}_{27}\text{H}_{46}\text{O} = (50 \times 10^{-3} \text{ g C}_{27}\text{H}_{46}\text{O})$$
$$\times \left(\frac{1 \text{ mol C}_{27}\text{H}_{46}\text{O}}{386.64 \text{ g C}_{27}\text{H}_{46}\text{O}} \right)$$
$$\times \left(\frac{6.022 \times 10^{23} \text{ formula units}}{1 \text{ mol}} \right)$$
$$\times \left(\frac{27 \text{ C atoms}}{1 \text{ C}_{27}\text{H}_{46}\text{O molecule}} \right)$$
$$= 2.1 \times 10^{21} \text{ atoms}$$

(c)
$$\text{number of C atoms in (C}_2\text{H}_5)_4\text{Pb} = (2.50 \text{ g})(\text{C}_2\text{H}_5)_4\text{Pb})$$
$$\times \left(\frac{1 \text{ mol (C}_2\text{H}_5)_4\text{Pb}}{323.4 \text{ g (C}_2\text{H}_5)_4\text{Pb}} \right)$$
$$\times \left(\frac{6.022 \times 10^{23} \text{ formula units}}{1 \text{ mol}} \right)$$
$$\times \left(\frac{8 \text{ C atoms}}{1 \text{ (C}_2\text{H}_5)_4\text{Pb molecule}} \right)$$
$$= 3.72 \times 10^{22} \text{ atoms}$$

(d)
$$\text{number of C atoms in C}_{14}\text{H}_{20}\text{N}_2\text{O}_8\text{S}_2 = (2.50 \times 10^3 \text{ g C}_{14}\text{H}_{20}\text{N}_2\text{O}_8\text{S}_2)$$
$$\times \left(\frac{1 \text{ mol C}_{14}\text{H}_{20}\text{N}_2\text{O}_8\text{S}_2}{408.44 \text{ g C}_{14}\text{H}_{20}\text{N}_2\text{O}_8\text{S}_2} \right)$$
$$\times \left(\frac{6.022 \times 10^{22} \text{ formula units}}{1 \text{ mol}} \right)$$
$$\times \left(\frac{14 \text{ C atoms}}{1 \text{ C}_{14}\text{H}_{20}\text{N}_2\text{O}_8\text{S}_2 \text{ molecule}} \right)$$
$$= 5.16 \times 10^{25} \text{ atoms}$$

E. ANSWERS TO THE SELF-TEST

1. carbon-12 being assigned a mass of exactly 12

2. false (Atomic masses have no units because they are relative quantities.)

3. mass of the formula unit of the substance on the atomic mass scale

4. false

5. its formula mass in grams or contains Avogadro's number of formula units

6. false

7. 6.022×10^{23}

8. one mole

9. $2 \times 6.022 \times 10^{23} = 1.204 \times 10^{24}$

10. true

11. false

12. $2 \times 6.022 \times 10^{23} = 1.204 \times 10^{24}$

13. ⇌

14. Cu_2O

15. false (only relative numbers)

16. 88.82 g of Cu combines with or is stoichiometrically equivalent to 11.18 g of O

17. dividing the mass of each by its atomic mass

18. false (The simplest formula is Cu_2O.)

19. false

20. molecular mass

21. C_3H_6

22. See Table 5-2.

23. 4 mol

24. number of moles

25. multiplying the number of moles of CO_2 (4 mol) by the unit conversion factor 44.01 g CO_2/1 mol CO_2

26. (1) Calculate the number of moles of C_2H_6.
    (2) Calculate the number of moles of O_2 that is required to react with the number of moles of C_2H_6 in step (1).
    (3) Calculate the mass of the number of moles of O_2 in step (2).

27. false

28. oxygen, hydrogen

29. limiting

30. false

6 CHEMICAL CALCULATIONS FOR SOLUTIONS

A. OUTLINE OF CHAPTER 6

6-1 A solution is a homogeneous mixture of two or more substances.

The components of a solution are the pure substances that are mixed to form the solution.

The solvent is the substance in which other substances are dissolved.

The solute is the dissolved substance.

The various types of solutions are described in Table 6-1.

The process of dissolving a substance in water can be represented by a chemical equation.

Ions in aqueous solutions are solvated by water molecules; that is, they are surrounded by a loosely bound shell of water molecules.

A saturated solution contains the maximum quantity of solute that can be dissolved in the solvent at equilibrium.

An unsaturated solution is able to dissolve additional solute.

The solubility of a solute is the maximum quantity of the solute that can be dissolved in a given quantity of solvent.

Solubility can be expressed in a variety of units.

The solubilities of most salts in liquids increase with increasing temperature (Figure 6-2).

6-2 Molarity is the most common unit of concentration.

Molarity is defined as the number of moles of solute per liter of solution; the equation is

$$\text{molarity} = \frac{\text{moles of solute}}{\text{liters of solution}} \qquad (6\text{-}1)$$

or in symbols,

$$M = \frac{n}{V} \tag{6-2}$$

The molarity of a solution can be calculated from the mass of solute in a given volume of solution.

The procedure for preparing a solution of known molarity is described (Figure 6-3).

A volumetric flask is a precision piece of glassware used to prepare precise volumes.

The concentration of a solution may be given as the mass percentage of the solute (Example 6-2).

A more dilute solution can be prepared from a more concentrated solution by dilution with solvent.

The number of moles of solute does not change upon dilution:

$$\text{moles of solute} = n = M_1 V_1 = M_2 V_2 \tag{6-4}$$

6-3 Molarity is used in stoichiometric calculations for reactions that take place in solution.

Calculations involving reactions in solutions are discussed.

6-4 Molarity can be used to calculate quantities in precipitation reactions.

Calculations involving reactions between two solutions to form an insoluble product are discussed.

6-5 Solubility rules can be used to predict whether a compound is soluble or insoluble.

If the solubility is less than 0.01 M, then we say that the salt is insoluble.

If the solubility is greater than 0.1 M, then we say that the salt is soluble.

The solubility rules for salts in water are given on page 181 of the text.

The solubility rules must be applied in the order given.

The rule with the lower number supersedes in case of a conflict.

The solubility rules can be used to predict whether a precipitation reaction occurs.

6-6 The concentration of an acid or a base can be determined by titration.

The neutralization reaction between an acid and a base is a double-replacement reaction in which a molecular compound is formed from ionic reactants.

A titration is a process in which a given volume of a base or acid is neutralized by a measured volume of an acid or base of known concentration (Figure 6-8).

A buret is used to measure the volume of the solution added in a titration.

Titration calculations involve the equation $n = MV$ where M usually is expressed as $mmol \cdot mL^{-1}$ and V as mL.

The formula mass of an acid or base can be determined by a titration experiment.

6-7 Many oxidation-reduction reactions are used in chemical analysis.

Calculations involving oxidation-reduction reactions where one or more of the reactants are in solution are discussed.

B. SELF-TEST

1. A solution is prepared by dissolving 1.0 g of sodium chloride in 100 mL of water. The solute is _____ and the solvent is _____ .

2. The process of dissolving sodium chloride in water is represented by the equation

$$NaCl(s) \longrightarrow$$

3. A species in aqueous solution is designated by writing _____ after the species.

4. The components of a solution are uniformly dispersed throughout the solution. *True/ False*

5. The ions in an aqueous solution are surrounded by a loosely bound shell of _____ molecules.

6. A solution is said to be saturated when no further solid can be dissolved in the solution at equilibrium. *True/False*

7. The solubility of a solute is the quantity of solute that can be dissolved in a given volume of a saturated solution at equilibrium. *True/False*

8. More solute can be dissolved in a (*saturated/unsaturated*) solution.

9. The solubility of most salts in water (*increases/decreases*) as the temperature increases.

10. Molarity is defined as _____
_____ .

11. A 1.0 M solution of sodium chloride is prepared by dissolving 1.0 mol of sodium chloride in 1.0 L of water. *True/False*

12. The procedure to prepare 500 mL of a 0.50 M solution of sodium chloride is as follows:

13. The equation that expresses the number of moles of solute in a given volume of solution whose molarity is known is $n =$ _____.

14. A more dilute solution can be prepared from a more concentrated solution. *True/False*

15. The number of moles of solute decreases upon dilution with solvent. *True/False*

16. The driving force for some double-replacement reactions is the formation of an insoluble product. *True/False*

17. A net ionic equation involves only the spectator ions. *True/False*

18. The equation $n = MV$ cannot be used with calculations involving double-replacement reactions. *True/False*

19. When given amounts of two solutions are mixed, one of the solutions may be a limiting reactant. *True/False*

20. Solubility rules can be used to predict if a compound is _____.

21. A solute is said to be insoluble if its solubility is less than _____.

22. A solute is said to be soluble if its solubility is greater than _____.

23. The solubility rules given in the text can be applied in any order. *True/False*

24. All sodium, potassium, and ammonium salts are (*soluble/insoluble*) in water.

25. All silver, lead, and mercury(I) salts are (*soluble/insoluble*).

26. All nitrates, acetates, and perchlorates are (*soluble/insoluble*).

27. Calcium sulfate is (*soluble/insoluble*).

28. The solubility rules can be used to predict whether a double-replacement reaction will occur. *True/False*

29. In a titration experiment, an acid solution is added to a given volume of a base solution until _____

_____.

30. The concentration of a solution of a base can be determined by a titration experiment using a solution of NaOH(*aq*) of known concentration. *True/False*

31. In a titration experiment, the volume is often expressed as mL. *True/False*

32. The formula mass of an acid can be determined by a titration experiment. *True/False*

33. Oxidation-reduction reactions may be used in chemical analysis. *True/False*

C. CALCULATIONS YOU SHOULD KNOW HOW TO DO

1. Calculate the molarity of a solution using Equation (6-2). See Problems 6-1 through 6-4.

2. Calculate quantities in order to prepare a solution of known molarity. See Example 6-1 and Problems 6-9 and 6-10.

3. Calculate the molarity of a solution given the mass percentage of solute. See Example 6-2 and Problems 6-11 and 6-12.

4. Prepare a solution by diluting a more concentrated solution using Equation (6-4). See Example 6-3 and Problems 6-7 and 6-8.

5. Calculate quantities of reactants and products of reactions that take place in solution. See Examples 6-4, 6-5, and 6-6 and Problems 6-13 through 6-21.

6. Calculate quantities of reactants and products of acid-base reactions. See Example 6-9 and Problems 6-29 through 6-36.

7. Calculate the formula mass of an acid or a base in a titration experiment. See Example 6-10.

8. Calculate the quantity of a species in a mixture or solution using oxidation-reduction reactions. See Examples 6-11 and 6-12 and Problems 6-37 through 6-44.

D. SOLUTIONS TO THE ODD-NUMBERED PROBLEMS

Many of the calculations in this chapter require the volume of a solution to be expressed in liters. Often the volume is given in milliliters. The unit conversion factor is

$$1 = \frac{10^3 \text{ mL}}{1 \text{ L}}$$

You should become proficient in converting from milliliters to liters and from liters to milliliters. The conversion from one unit to the other will not be shown in many of the solutions.

6-1 The number of moles of $Ca(OH)_2$ is

$$\text{moles of Ca(OH)}_2 = (0.185 \text{ g}) \left(\frac{1 \text{ mol Ca(OH)}_2}{74.10 \text{ g Ca(OH)}_2} \right) = 2.50 \times 10^{-3} \text{ mol}$$

The molarity is calculated by using Equation (6-2):

$$\text{molarity} = \frac{n}{V} = \left(\frac{2.50 \times 10^{-3} \text{ mol}}{100 \text{ mL}} \right) \left(\frac{10^3 \text{ mL}}{1 \text{ L}} \right) = 0.0250 \text{ M}$$

6-3 The number of moles of NaOH is

$$\text{moles of NaOH} = (572 \text{ g}) \left(\frac{1 \text{ mol NaOH}}{40.00 \text{ g NaOH}} \right) = 14.3 \text{ mol}$$

Because the 14.3 mol are dissolved in 1 L of solution, the molarity of the solution is 14.3 M.

6-5   (a)   The number of moles of $K_2Cr_2O_7$ in the solution is

$$\text{moles of } K_2Cr_2O_7 = MV$$
$$= (0.1255 \text{ mol} \cdot L^{-1})(25.46 \text{ mL}) \left(\frac{1 \text{ L}}{10^3 \text{ mL}} \right)$$
$$= 3.195 \times 10^{-3} \text{ mol}$$

(b)   The number of moles of $C_6H_{12}O_6$ in the solution is

$$\text{moles of } C_6H_{12}O_6 = MV$$
$$= (0.020 \text{ mol} \cdot L^{-1})(50 \text{ } \mu L) \left(\frac{1 \text{ L}}{10^6 \text{ } \mu L} \right)$$
$$= 1.0 \times 10^{-6} \text{ mol}$$

6-7   From Equation (6-4) we have

$$M_1 V_1 = M_2 V_2$$

Thus

$$(18.0 \text{ mmol} \cdot \text{mL}^{-1})V_1 = (0.30 \text{ mmol}^{-1} \cdot \text{mL}^{-1})(500 \text{ mL})$$

and

$$V_1 = \frac{(0.30)(500 \text{ mL})}{18.0} = 8.3 \text{ mL}$$

6-9   The number of moles of $C_{12}H_{22}O_{11}$ in 500 mL of a 0.250 M solution is

$$\text{moles of } C_{12}H_{22}O_{11} = MV = (0.250 \text{ mol} \cdot L^{-1})(0.500 \text{ L})$$
$$= 0.125 \text{ mol}$$

The mass of $C_{12}H_{22}O_{11}$ required is

$$\text{mass of } C_{12}H_{22}O_{11} = (0.125 \text{ mol}) \left(\frac{342.30 \text{ g } C_{12}H_{22}O_{11}}{1 \text{ mol } C_{12}H_{22}O_{11}} \right) = 42.8 \text{ g}$$

Dissolve 42.8 g of sucrose in about 250 mL of water in a 500-mL volumetric flask and then dilute the solution to the 500-mL mark, and shake to make the solution homogeneous.

6-11   Take a 100.0-g sample of concentrated sulfuric acid. We have 97.0 g of H_2SO_4 and 3.0 g of H_2O. The volume of the 100.0-g sample is

$$V = \frac{m}{d} = \frac{100.0 \text{ g}}{1.84 \text{ g} \cdot \text{mL}^{-1}} = 54.35 \text{ mL} = 0.05435 \text{ L}$$

The number of moles of H_2SO_4 in the sample is

$$\text{moles of } H_2SO_4 = (97.0 \text{ g}) \left(\frac{1 \text{ mol } H_2SO_4}{98.08 \text{ } H_2SO_4} \right) = 0.9890 \text{ mol}$$

The molarity of the concentrated sulfuric acid is

$$M = \frac{n}{V} = \frac{0.9890 \text{ mol}}{0.05435 \text{ L}} = 18.2 \text{ M}$$

6-13 The number of moles of silicon reacted is

$$\text{moles of Si} = (12.5 \text{ g}) \left(\frac{1 \text{ mol Si}}{28.09 \text{ g Si}} \right) = 0.4450 \text{ mol}$$

The number of moles of sodium hydroxide required is

$$\text{moles of NaOH} = (0.4450 \text{ mol Si}) \left(\frac{2 \text{ mol NaOH}}{1 \text{ mol Si}} \right) = 0.8900 \text{ mol}$$

The volume of sodium hydroxide required is given by

$$V = \frac{n}{M} = \frac{0.8900 \text{ mol}}{6.00 \text{ mol} \cdot L^{-1}} = 0.148 \text{ L} = 148 \text{ mL}$$

The number of moles of hydrogen produced is given by

$$\text{moles of } H_2 = (0.4450 \text{ mol Si}) \left(\frac{2 \text{ mol } H_2}{1 \text{ mol Si}} \right) = 0.8900 \text{ mol}$$

The number of grams of hydrogen produced is given by

$$\text{mass of } H_2 = (0.8900 \text{ mol } H_2) \left(\frac{2.016 \text{ g } H_2}{1 \text{ mol } H_2} \right) = 1.79 \text{ g}$$

6-15 The number of moles of BaO_2 reacted is

$$\text{mass of } BaO_2 = (17.6 \text{ g}) \left(\frac{1 \text{ mol } BaO_2}{169.3 \text{ g } BaO_2} \right) = 0.1040 \text{ mol}$$

The number of moles of H_2SO_4 required to react with the BaO_2 is given by

$$\text{moles of } H_2SO_4 = (0.1040 \text{ mol } BaO_2) \left(\frac{1 \text{ mol } H_2SO_4}{1 \text{ mol } BaO_2} \right) = 0.1040 \text{ mol}$$

The volume of sulfuric acid required is given by

$$V = \frac{n}{M} = \frac{0.1040 \text{ mol}}{3.75 \text{ mol} \cdot L^{-1}} = 0.0277 \text{ L} = 27.7 \text{ mL}$$

6-17 We must first check to see which reactant is the limiting reactant. The number of moles of each reactant is given by

$$\text{moles of } H_2SO_4 = MV = (1.75 \text{ mol} \cdot \text{L}^{-1})(0.250 \text{ L}) = 0.4375 \text{ mol}$$

$$\text{moles of } MnO_2 = (20.0 \text{ g}) \left(\frac{1 \text{ mol } MnO_2}{86.94 \text{ g } MnO_2} \right) = 0.2300 \text{ mol}$$

$$\text{moles of KI} = (75.0 \text{ g}) \left(\frac{1 \text{ mol KI}}{166.0 \text{ g KI}} \right) = 0.4518 \text{ mol}$$

If we divide the number of moles of each reactant by its respective stoichiometric coefficient, then we see that (moles of $H_2SO_4/2$) = 0.2188 mol is less than (moles of KI/2) = 0.2259 mol or (moles of $MnO_2/1$) = 0.2300 mol. Therefore, H_2SO_4 is the limiting reactant. The number of moles of I_2 produced is

$$\text{moles of } I_2 = (0.4375 \text{ mol } H_2SO_4) \left(\frac{1 \text{ mol } I_2}{2 \text{ mol } H_2SO_4} \right) = 0.2188 \text{ mol}$$

The number of grams of I_2 produced is

$$\text{mass of } I_2 = (0.2188 \text{ mol}) \left(\frac{253.8 \text{ g } I_2}{1 \text{ mol } I_2} \right) = 55.5 \text{ g}$$

6-19 The number of moles of Zn that react is

$$\text{moles of Zn} = (2.55 \text{ g}) \left(\frac{1 \text{ mol Zn}}{65.38 \text{ g Zn}} \right) = 3.90 \times 10^{-2} \text{ mol}$$

The number of moles of HCl required is

$$\text{moles of HCl} = (3.90 \times 10^{-2} \text{ mol Zn}) \left(\frac{2 \text{ mol HCl}}{1 \text{ mol Zn}} \right) = 7.80 \times 10^{-2} \text{ mol}$$

The volume of a 2.00 M solution to use can be found by using Equation (6-3).

$$V = \frac{n}{M} = \frac{7.80 \times 10^{-2} \text{ mol}}{2.00 \text{ mol} \cdot \text{L}^{-1}} = 0.0390 \text{ L} = 39.0 \text{ mL}$$

6-21 We must first determine the number of moles of NaOH. Using Equation (6-3), we have

$$\text{moles of NaOH} = MV = (6.00 \text{ mol} \cdot \text{L}^{-1})(5.00 \text{ L}) = 30.0 \text{ mol}$$

The number of moles of Cl_2 required is

$$\text{moles of } Cl_2 = (30.0 \text{ mol NaOH}) \left(\frac{1 \text{ mol } Cl_2}{2 \text{ mol NaOH}} \right) = 15.0 \text{ mol}$$

and the number of grams of Cl_2 is

$$\text{mass of } Cl_2 = (15.0 \text{ mol}) \left(\frac{70.90 \text{ g } Cl_2}{1 \text{ mol } Cl_2} \right) = 1060 \text{ g}$$

6-23 (a) insoluble, rule 3   (b) soluble, rule 2
     (c) soluble, rule 1   (d) soluble, rule 1
     (e) insoluble, rule 5

6-25 (a) insoluble, rule 5   (b) soluble, rule 2
     (c) soluble, rule 4   (d) insoluble, rule 5
     (e) insoluble, rule 6

6-27 (a) CuS is insoluble by rule 5.

$$CuCl_2(aq) + Na_2S(aq) \longrightarrow CuS(s) + 2NaCl(aq)$$
$$Cu^{2+}(aq) + S^{2-}(aq) \longrightarrow CuS(s)$$

     (b) $MgCO_3$ is insoluble by rule 5.

$$MgBr_2(aq) + K_2CO_3(aq) \longrightarrow MgCO_3(s) + 2KBr(aq)$$
$$Mg^{2+}(aq) + CO_3^{2-}(aq) \longrightarrow MgCO_3(s)$$

     (c) $BaSO_4$ is insoluble by rule 6.

$$BaCl_2(aq) + K_2SO_4(aq) \longrightarrow BaSO_4(s) + 2KCl(aq)$$
$$Ba^{2+}(aq) + SO_4^{2-}(aq) \longrightarrow BaSO_4(s)$$

     (d) Hg_2Cl_2 is insoluble by rule 3.

$$Hg_2(NO_3)_2(aq) + 2KCl(aq) \longrightarrow Hg_2Cl_2(s) + 2KNO_3(aq)$$
$$Hg_2^{2+}(aq) + 2Cl^-(aq) \longrightarrow Hg_2Cl_2(s)$$

6-29 The number of millimoles of NaOH required to neutralize the HCl solution is

$$\text{millimoles of NaOH} = MV = (0.155 \text{ mmol} \cdot mL^{-1})(27.5 \text{ mL})$$
$$= 4.263 \text{ mmol}$$

We see from the neutralization reaction

$$NaOH(aq) + HCl(aq) \longrightarrow NaCl(aq) + H_2O(l)$$

that it requires 1 mol of NaOH to neutralize 1 mol of HCl. Thus we have

$$\text{millimoles of HCl} = \text{millimoles of NaOH} = 4.263 \text{ mmol}$$

The concentration of the HCl solution is

$$\text{molarity} = \frac{n}{V} = \frac{4.263 \text{ mmol}}{25.0 \text{ mL}} = 0.171 \text{ M}$$

6-31 (a) The number of millimoles of $Ca(OH)_2$ that is neutralized is

$$\text{millimoles of } Ca(OH)_2 = MV = (0.010 \text{ mmol} \cdot mL^{-1})(15.0 \text{ } \mu L)\left(\frac{1 \text{ mL}}{10^3 \text{ } \mu L}\right)$$

$$= 1.5 \times 10^{-4} \text{ mmol}$$

We see from the neutralization equation

$$2HNO_3(aq) + Ca(OH)_2(aq) \longrightarrow Ca(NO_3)_2(aq) + 2H_2O(l)$$

that it requires 2 mol of HNO_3 to neutralize 1 mol of $Ca(OH)_2$. Thus we have

$$\text{moles of } HNO_3 = [1.5 \times 10^{-4} \text{ mmol } Ca(OH)_2]\left[\frac{2 \text{ mmol } HNO_3}{1 \text{ mmol } Ca(OH)_2}\right]$$

$$= 3.0 \times 10^{-4} \text{ mmol}$$

The volume of HNO_3 required is

$$V = \frac{n}{M} = \frac{3.0 \times 10^{-4} \text{ mmol}}{0.108 \text{ mmol} \cdot mL^{-1}} = 0.0028 \text{ mL} = 2.8 \text{ } \mu L$$

(b) The number of millimoles of NaOH that is neutralized is

$$\text{millimoles of NaOH} = MV = (0.200 \text{ mmol} \cdot mL^{-1})(25.0 \text{ mL})$$

$$= 5.00 \text{ mmol}$$

From the neutralization equation

$$2NaOH(aq) + H_2SO_4(aq) \longrightarrow Na_2SO_4(aq) + 2H_2O(l)$$

we have

$$\text{millimoles of } H_2SO_4 = (5.00 \text{ mmol NaOH})\left(\frac{1 \text{ mmol } H_2SO_4}{2 \text{ mmol NaOH}}\right)$$

$$= 2.50 \text{ mmol}$$

The volume of the H_2SO_4 solution required is

$$V = \frac{n}{M} = \frac{2.50 \text{ mmol}}{0.300 \text{ mmol} \cdot mL^{-1}} = 8.33 \text{ mL}$$

6-33 The number of moles of $OH^-(aq)$ added to the $HCl(aq)$ solution is

$$\text{moles of } OH^- = \text{moles of KOH} = (40.0 \text{ g}) \left(\frac{1 \text{ mol KOH}}{56.11 \text{ g KOH}} \right) = 0.713 \text{ mol}$$

The number of moles of $HCl(aq)$ initially present is

$$\text{moles of HCl} = MV = (0.125 \text{ mol} \cdot \text{L}^{-1})(2.00 \text{ L}) = 0.250 \text{ mol}$$

Thus an amount of KOH *in excess* of that required to neutralize all of the HCl was added. The neutralization reaction is

$$HCl(aq) + KOH(aq) \longrightarrow H_2O(l) + KCl(aq)$$

The number of moles of KCl produced is equal to the number of moles of HCl initially present, which is 0.250 mol. The volume of the final solution is

$$\text{final volume} = \text{initial volume} + \text{volume added}$$
$$= 2.00 \text{ L} + 0.20 \text{ L} = 2.20 \text{ L}$$

Thus,

$$\text{molarity of KCl} = \frac{n}{V} = \frac{0.250 \text{ mol}}{2.20 \text{ L}} = 0.114 \text{ M}$$

6-35 The number of moles of NaOH present in the sample is equal to the number of moles of HCl required for neutralization. Thus,

$$\text{moles of NaOH} = (0.0317 \text{ L})(0.150 \text{ mol} \cdot \text{L}^{-1} \text{ HCl}) \left(\frac{1 \text{ mol NaOH}}{1 \text{ mol HCl}} \right)$$
$$= 0.00476 \text{ mol}$$
$$\text{mass of NaOH} = (0.00476 \text{ mol}) \left(\frac{40.00 \text{ g NaOH}}{1 \text{ mol NaOH}} \right)$$
$$= 0.190 \text{ g}$$

The mass percentage of NaOH in the mixture equals the mass of NaOH divided by the total mass times 100:

$$\% \text{ NaOH} = \frac{0.190 \text{ g NaOH}}{0.365 \text{ g mixture}} \times 100$$
$$= 52.1\%$$

6-37 The number of millimoles of $BrO_3^- (aq)$ is

$$n = MV = (0.125 \text{ mmol} \cdot \text{mL}^{-1})(43.7 \text{ mL}) = 5.46 \text{ mmol}$$

The number of millimoles of antimony that reacts with 5.46 mmol of BrO_3^- is

$$\text{millimoles of Sb}^{3+} = (5.46 \text{ mmol BrO}_3^-)\left(\frac{3 \text{ mmol Sb}^{3+}}{1 \text{ mmol BrO}_3^-}\right)$$

$$= 16.38 \text{ mmol}$$

The number of grams of antimony that reacts is

$$\text{mass of Sb} = (1.638 \times 10^{-2} \text{ mol})\left(\frac{121.8 \text{ g Sb}}{1 \text{ mol Sb}}\right) = 1.995 \text{ g}$$

$$\% \text{ antimony} = \frac{1.995 \text{ g Sb}}{9.62 \text{ g ore}} \times 100 = 20.7\%$$

6-39 The number of millimoles of $Na_2S_2O_3$ used to titrate the I_3^- is

$$\text{millimoles of Na}_2\text{S}_2\text{O}_3 = (0.330 \text{ M})(36.4 \text{ mL}) = 12.0 \text{ mmol}$$

The number of millimoles of I_3^- is

$$\text{millimoles of I}_3^- = (12.0 \text{ mmol Na}_2\text{S}_2\text{O}_3)\left(\frac{1 \text{ mmol I}_3^-}{2 \text{ mmol Na}_2\text{S}_2\text{O}_3}\right) = 6.00 \text{ mmol}$$

The concentration of I_3^- (aq) is

$$M = \frac{6.00 \text{ mmol}}{15.0 \text{ mL}} = 0.400 \text{ M}$$

6-41 The balanced equation for the reaction is

$$2P_4 \, (s) + 5BaSO_4(s) \longrightarrow 2P_4O_{10}(s) + 5BaS(s)$$

The amount of phosphorus required to react with 2.16 g of $BaSO_4(s)$ is

$$\text{mass of P}_4 = (2.16 \text{ g BaSO}_4)\left(\frac{1 \text{ mol BaSO}_4}{233.4 \text{ g BaSO}_4}\right)\left(\frac{2 \text{ mol P}_4}{5 \text{ mol BaSO}_4}\right)\left(\frac{123.88 \text{ g P}_4}{1 \text{ mol P}_4}\right)$$

$$= 0.459 \text{ g}$$

6-43 The number of millimoles of $K_2C_2O_4$ that reacted is

$$\text{millimoles of K}_2\text{C}_2\text{O}_4 = MV = (0.336 \text{ mmol} \cdot \text{mL}^{-1})(50.0 \text{ mL})$$

$$= 16.80 \text{ mmol}$$

The number of millimoles of $KMnO_4$ required is

$$\text{millimoles of KMnO}_4 = (16.80 \text{ mmol K}_2\text{C}_2\text{O}_4)\left(\frac{2 \text{ mmol KMnO}_4}{5 \text{ mmol K}_2\text{C}_2\text{O}_4}\right)$$

$$= 6.720 \text{ mmol}$$

The number of milliliters of $KMnO_4$ required is given by

$$V = \frac{n}{M} = \frac{6.720 \text{ mmol}}{0.475 \text{ mmol} \cdot mL^{-1}} = 14.1 \text{ mL}$$

6-45 The number of moles of $Sc(NO_3)_3$ is

$$\text{moles of } Sc(NO_3)_3 = (2.86 \text{ g}) \left(\frac{1 \text{ mol } Sc(NO_3)_3}{230.99 \text{ g } Sc(NO_3)_3} \right) = 0.01238 \text{ mol}$$

The concentration of $Sc(NO_3)_3$ (aq) is

$$M = \frac{n}{V} = \frac{0.01238 \text{ mol}}{0.100 \text{ L}} = 0.124 \text{ M}$$

Thus, we have that

$$\text{concentration of } Sc^{3+}(aq) = 0.124 \text{ M}$$
$$\text{concentration of } NO_3^-(aq) = (3)(0.124 \text{ M}) = 0.372 \text{ M}$$

6-47 Take a 1-L (1000-mL) sample of the stock solution. The mass of the sample is

$$\text{mass of sample} = dV = (1.67 \text{ g} \cdot mL^{-1})(1000 \text{ mL}) = 1670 \text{ g}$$

The mass of perchloric acid in the sample is

$$\text{mass of } HClO_4 = \left(\frac{70.5}{100} \right) (1670) = 1177 \text{ g}$$

The number of moles of perchloric acid in the sample is

$$\text{moles of } HClO_4 = (1177 \text{ g}) \left(\frac{1 \text{ mol } HClO_4}{100.46 \text{ g } HClO_4} \right) = 11.72 \text{ mol}$$

The concentration of perchloric acid is

$$M = \frac{n}{V} = \frac{11.72 \text{ mol}}{1.00 \text{ L}} = 11.7 \text{ M}$$

6-49 The number of moles of $Cr^{3+}(aq)$ in 250 mL (0.250 L) of the solution is

$$\text{moles of } Cr^{3+} = MV = (0.500 \text{ mol} \cdot L^{-1})(0.250 \text{ L}) = 0.125 \text{ mol}$$

The number of moles of $CrCl_3 \cdot 6H_2O(s)$ required is

$$\text{moles of } CrCl_3 \cdot 6H_2O = (0.125 \text{ mol } Cr^{3+}) \left(\frac{1 \text{ mol } CrCl_3 \cdot 6H_2O}{1 \text{ mol } Cr^{3+}} \right)$$
$$= 0.125 \text{ mol}$$

The number of grams of $CrCl_3 \cdot 6H_2O(s)$ required is

$$\text{mass of } CrCl_3 \cdot 6H_2O = (0.125 \text{ mol}) \left(\frac{266.45 \text{ g } CrCl_3 \cdot 6H_2O}{1 \text{ mol } CrCl_3 \cdot 6H_2O} \right)$$
$$= 33.3 \text{ g}$$

6-51   We use Equation (6-4) in the form

$$V_2 = \frac{M_1 V_1}{M_2}$$

The number of milliliters of commercial phosphoric acid required is

$$V = \frac{(0.650 \text{ mol} \cdot L^{-1})(1.00 \text{ L})}{14.6 \text{ mol} \cdot L^{-1}} = 0.0445 \text{ L} = 44.5 \text{ mL}$$

6-53   The equation for the reaction is

$$2AgNO_3(aq) + CaCl_2(aq) \longrightarrow 2AgCl(s) + Ca(NO_3)_2(aq)$$

The number of millimoles of $Ag^+(aq)$ precipitated is

$$\text{millimoles of } AgNO_3 = MV = (0.500 \text{ mmol} \cdot mL^{-1})(35.0 \text{ mL})$$
$$= 17.5 \text{ mmol}$$

The number of millimoles of $CaCl_2(aq)$ required is

$$\text{millimoles of } CaCl_2 = (17.5 \text{ mmol } AgNO_3) \left(\frac{1 \text{ mmol } CaCl_2}{2 \text{ mmol } AgNO_3} \right)$$
$$= 8.75 \text{ mmol}$$

The number of milliliters of $CaCl_2(aq)$ required is

$$V = \frac{n}{M} = \frac{8.75 \text{ mmol}}{0.865 \text{ mmol} \cdot mL^{-1}} = 10.1 \text{ mL}$$

6-55   The equation for the reaction is

$$Zn(s) + CuSO_4(aq) \longrightarrow ZnSO_4(aq) + Cu(s)$$

We must first check to see which reactant is the limiting reactant.

$$\text{moles of } Zn = (30.0 \text{ g}) \left(\frac{1 \text{ mol } Zn}{65.38 \text{ g } Zn} \right) = 0.4589 \text{ mol}$$
$$\text{moles of } CuSO_4 = MV = (0.165 \text{ mol} \cdot L^{-1})(0.500 \text{ L}) = 0.08250 \text{ mol}$$

If we divide the number of moles of each reactant by its respective stoichiometric coefficient, then we see that (moles of Zn/1) = 0.4589 mol is greater than (moles of CuSO₄/1) = 0.0825 mol. Therefore, $CuSO_4$ is the limiting reactant.

$$\text{mass of Cu} = (0.0825 \text{ mol CuSO}_4)\left(\frac{1 \text{ mol Cu}}{1 \text{ mol CuSO}_4}\right)\left(\frac{63.55 \text{ g Cu}}{1 \text{ mol Cu}}\right)$$

$$= 5.24 \text{ g}$$

6-57 We have that

$$\text{mass \%} = \frac{\text{mass CaCl}_2}{\text{mass solution}} \times 100$$

$$\text{mass solution} = \frac{\text{mass CaCl}_2}{\text{mass \%}} \times 100$$

$$= \frac{3.25 \text{ g}}{14.0} \times 100 = 23.2 \text{ g}$$

6-59 The number of moles of $Al_2(SO_4)_3$ in 500 mL of the solution is

$$\text{moles of Al}_2(SO_4)_3 = MV = (3.00 \text{ mol} \cdot L^{-1})(0.500 \text{ L}) = 1.50 \text{ mol}$$

$$\text{mass of Al}_2(SO_4)_3 \cdot 18H_2O = [1.50 \text{ mol Al}_2(SO_4)_3]\left[\frac{1 \text{ mol Al}_2(SO_4)_3 \cdot 18H_2O}{1 \text{ mol Al}_2(SO_4)_3}\right]$$

$$\times \left[\frac{666.43 \text{ g Al}_2(SO_4)_3 \cdot 18H_2O}{1 \text{ mol Al}_2(SO_4)_3 \cdot 18H_2O}\right]$$

$$= 1000 \text{ g} = 1.00 \text{ kg}$$

6-61 Let

$$x \text{ g} = \text{mass of NaCl}$$

and so

$$0.3146 \text{ g} - x \text{ g} = \text{mass of KBr}$$

The number of moles of each is

$$\text{moles of NaCl} = (x \text{ g})\left(\frac{1 \text{ mol NaCl}}{58.44 \text{ g NaCl}}\right) = 0.017112 \, x \text{ mol}$$

$$\text{moles of KBr} = (0.3146 \text{ g} - x \text{ g})\left(\frac{1 \text{ mol KBr}}{119.00 \text{ g KBr}}\right)$$

$$= 2.6437 \times 10^{-3} \text{ mol} - 8.4034 \times 10^{-3} \, x \text{ mol}$$

The number of moles of $Ag^+(aq)$ required is

$$\text{moles of Ag}^+ = MV = (0.08765 \text{ mol} \cdot L^{-1})(0.03760 \text{ L})$$

$$= 3.2956 \times 10^{-3} \text{ mol}$$

From the stoichiometry of the preciptitation reaction, we have that

$$\text{moles of Ag}^+ = \text{moles of AgCl} + \text{moles of AgBr}$$
$$= \text{moles of NaCl} + \text{moles of KBr}$$

and so we can write

$$3.2956 \times 10^{-3} \text{ mol} = 0.017112 \, x \text{ mol} + 2.6437 \times 10^{-3} \text{ mol} -$$
$$8.4034 \times 10^{-3} \, x \text{ mol}$$

or

$$6.519 \times 10^{-4} = 8.709 \times 10^{-3} \, x$$
$$x = 0.07485$$

The mass of NaCl is 0.07485 g. The mass percentage of NaCl in the mixture is

$$\text{mass \% NaCl} = \frac{0.07485 \text{ g}}{0.3146 \text{ g}} \times 100 = 23.8\%$$
$$\text{mass \% KBr} = 100\% - 23.8\% = 76.2\%$$

6-63   The number of moles of KOH in 500 mL (0.500 L) of a 6.00 M solution is

$$\text{moles of KOH} = MV = (6.00 \text{ mol} \cdot \text{L}^{-1})(0.500 \text{ L}) = 3.00 \text{ mol}$$
$$\text{mass of KOH} = (3.00 \text{ mol}) \left(\frac{56.11 \text{ g KOH}}{1 \text{ mol KOH}} \right) = 168.3 \text{ g}$$

The mass percentage of KOH in the KOH(s) sample is $(100 - 8.75)\% = 91.25\%$. Using the formula

$$\text{mass \% KOH} = \frac{\text{mass of KOH}}{\text{mass of KOH sample}} \times 100$$

and solving for the mass of the KOH(s) sample gives

$$\text{mass of KOH sample} = \frac{\text{mass of KOH}}{\text{mass \% KOH}} \times 100$$
$$= \frac{168.3 \text{ g}}{91.25} \times 100 = 184 \text{ g}$$

6-65   We have that

$$\text{millimoles of Fe}^{2+} = MV = (0.633 \text{ mmol} \cdot \text{mL}^{-1})(45.0 \text{ mL})$$
$$= 28.49 \text{ mmol}$$
$$\text{millimoles of KMnO}_4 = MV = (0.200 \text{ mmol} \cdot \text{mL}^{-1})(28.5 \text{ mL})$$
$$= 5.70 \text{ mmol}$$

Let x equal the number of moles of Fe^{2+} that reacts with 1 mol of MnO_4^-. We have that

$$\text{millimoles of } MnO_4^- = (28.49 \text{ mmol } Fe^{2+})\left(\frac{1 \text{ mmol } MnO_4^-}{x \text{ mmol } Fe^{2+}}\right)$$

$$= 5.70 \text{ mmol}$$

Solving for x gives

$$x = \frac{28.49}{5.70} = 5$$

The oxidation state of iron increases by 1. Because 5 mol of Fe^{2+} reacts with 1 mol of $KMnO_4$, the oxidation state of manganese must decrease by 5. The oxidation state of manganese in $KMnO_4$ is $+7$, and so, the final oxidation state of manganese is $+7 - 5 = +2$.

6-67

$$\text{millimoles of HCl} = MV = (0.150 \text{ mmol}\cdot\text{mL}^{-1})(27.5 \text{ mL})$$
$$= 4.125 \text{ mmol}$$

$$\text{millimoles of } NH_3 = (4.125 \text{ mmol HCl})\left(\frac{1 \text{ mmol } NH_3}{1 \text{ mmol HCl}}\right)$$

$$= 4.125 \text{ mmol}$$

$$\text{mass of nitrogen} = (4.125 \times 10^{-3} \text{ mol } NH_3)\left(\frac{1 \text{ mol N}}{1 \text{ mol } NH_3}\right)\left(\frac{14.01 \text{ g N}}{1 \text{ mol N}}\right)$$

$$= 0.05779 \text{ g}$$

$$\% \text{ nitrogen} = \frac{0.05779 \text{ g}}{2.25 \text{ g}} \times 100 = 2.57\%$$

6-69 The equation for the reaction is

$$BaCl_2(aq) + K_2SO_4(aq) \longrightarrow BaSO_4(s) + 2KCl(aq)$$

We must first check to see which reactant is the limiting reactant.

$$\text{millimoles of } BaCl_2 = MV = (0.450 \text{ mmol}\cdot\text{mL}^{-1})(20.0 \text{ mL})$$
$$= 9.00 \text{ mmol}$$

$$\text{moles of } K_2SO_4 = MV = (0.250 \text{ mmol}\cdot\text{mL}^{-1})(36.0 \text{ mL})$$
$$= 9.00 \text{ mmol}$$

If we divide the number of millimoles of each reactant by its respective stoichiometric coefficient, then we see that (millimoles of $BaCl_2/1$) $= 9.00$ mmol is equal to (millimoles of $K_2SO_4/1$) = 9.00 mmol. Therefore, neither reactant is a limiting reactant. The number of moles of $BaSO_4$ precipitated is

$$\text{moles of } BaSO_4 = (9.00 \times 10^{-3} \text{ mol } BaCl_2) \left(\frac{1 \text{ mol } BaSO_4}{1 \text{ mol } BaCl_2} \right)$$

$$= 9.00 \times 10^{-3} \text{ mol}$$

$$\text{mass of } BaSO_4 = (9.00 \times 10^{-3} \text{ mol}) \left(\frac{233.4 \text{ g } BaSO_4}{1 \text{ mol } BaSO_4} \right) = 2.10 \text{ g}$$

6-71 The equation for the reaction is

$$Zn(NO_3)_2(aq) + Na_2S(aq) \longrightarrow ZnS(s) + 2NaNO_3(aq)$$

We must first check to see which reactant is the limiting reactant.

$$\text{millimoles of } Zn(NO_3)_2 = MV = (1.76 \text{ mmol} \cdot mL^{-1})(30.0 \text{ mL})$$
$$= 52.8 \text{ mmol}$$

$$\text{millimoles of } Na_2S = MV = (2.18 \text{ mmol} \cdot mL^{-1})(30.0 \text{ mL}) = 65.4 \text{ mmol}$$

If we divide the number of millimoles of each reactant by its respective stoichiometric coefficient, then we see that (millimoles of $Zn(NO_3)_2/1$) = 52.8 mmol is less than (millimoles of $Na_2S/1$) = 65.4 mmol. Therefore, $Zn(NO_3)_2$ is the limiting reactant.

$$\text{moles of } ZnS = [0.0528 \text{ mol } Zn(NO_3)_2] \left[\frac{1 \text{ mol } ZnS}{1 \text{ mol } Zn(NO_3)_2} \right]$$

$$= 0.0528 \text{ mol}$$

$$\text{mass of } ZnS = (0.0528 \text{ mol}) \left(\frac{97.44 \text{ g } ZnS}{1 \text{ mol } ZnS} \right) = 5.14 \text{ g}$$

6-73 The number of moles of $PbCl_2$ precipitated is

$$\text{moles of } PbCl_2 = (12.79 \text{ g}) \left(\frac{1 \text{ mol } PbCl_2}{278.1 \text{ g } PbCl_2} \right) = 0.04599 \text{ mol}$$

$$\text{moles of } Pb(NO_3)_2 = (0.04599 \text{ mol } PbCl_2) \left(\frac{1 \text{ mol } Pb(NO_3)_2}{1 \text{ mol } PbCl_2} \right)$$

$$= 0.04599 \text{ mol}$$

The molarity of the solution is

$$M = \frac{n}{V} = \frac{0.04599 \text{ mol}}{0.2000 \text{ L}} = 0.2300 \text{ M}$$

6-75 The equation for the precipitation of AsO_4^{3-} is

$$AsO_4^{3-}(aq) + 3AgNO_3(aq) \longrightarrow Ag_3AsO_4(s) + 3NO_3^-(aq)$$

$$\text{millimoles of AgNO}_3 = MV = (0.655 \text{ mmol} \cdot \text{mL}^{-1})(37.5 \text{ mL})$$
$$= 24.56 \text{ mmol}$$

$$\text{millimoles of AsO}_4^{3-} = (24.56 \text{ mmol AgNO}_3)\left(\frac{1 \text{ mmol AsO}_4^{3-}}{3 \text{ mmol AgNO}_3}\right)$$
$$= 8.1881 \text{ mmol}$$

All the arsenic in As_4O_6 ends up in AsO_4^{3-}; thus

$$1 \text{ mol As}_4O_6 \backsimeq 4 \text{ mol AsO}_4^{3-}$$

$$\text{moles of As}_4O_6 = (8.188 \times 10^{-3} \text{ mol AsO}_4^{3-})\left(\frac{1 \text{ mol As}_4O_6}{4 \text{ mol AsO}_4^{3-}}\right)$$
$$= 2.047 \times 10^{-3} \text{ mol}$$

$$\text{mass As}_4O_6 = (2.047 \times 10^{-3} \text{ mol})\left(\frac{395.68 \text{ g As}_4O_6}{1 \text{ mol As}_4O_6}\right) = 0.8099 \text{ g}$$

The mass percentage of As_4O_6 in the pesticide is

$$\text{mass \% As}_4O_6 = \left(\frac{0.8099 \text{ g}}{11.75 \text{ g}}\right) \times 100 = 6.89\%$$

6-77 $$\text{millimoles of NH}_3 = (3.52 \text{ mg})\left(\frac{1 \text{ mmol NH}_3}{17.03 \text{ mg NH}_3}\right) = 0.2067 \text{ mmol}$$

$$\text{millimoles of OBr}^- = (0.2067 \text{ mmol NH}_3)\left(\frac{3 \text{ mmol OBr}^-}{2 \text{ mmol NH}_3}\right)$$
$$= 0.3100 \text{ mmol}$$

The molarity of the solution is

$$M = \frac{n}{V} = \frac{0.3100 \text{ mmol}}{10.00 \text{ mL}} = 0.0310 \text{ M}$$

6-79 $$\text{moles of NaOH} = (0.100 \text{ mol} \cdot \text{L}^{-1})(0.02135 \text{ L})$$
$$= 2.135 \times 10^{-3} \text{ mol}$$

$$\text{moles of H}_2SO_4 = (2.135 \times 10^{-3} \text{ mol NaOH})\left(\frac{1 \text{ mol H}_2SO_4}{2 \text{ mol NaOH}}\right)$$
$$= 1.0675 \times 10^{-3} \text{ mol}$$

All the sulfur in the fuel ends up in sulfuric acid; thus

$$1 \text{ mol S} \approxeq 1 \text{ mol H}_2\text{SO}_4$$

$$\text{mass of S} = (1.0675 \times 10^{-3} \text{ mol H}_2\text{SO}_4)\left(\frac{1 \text{ mol S}}{1 \text{ mol H}_2\text{SO}_4}\right)\left(\frac{32.06 \text{ g S}}{1 \text{ mol S}}\right)$$

$$= 0.03422 \text{ g}$$

The mass percentage of sulfur in the fuel is

$$\text{mass \%} = \frac{0.03422 \text{ g}}{5.63 \text{ g}} \times 100 = 0.608\%$$

6-81 $$\text{millimoles of NaAl(OH)}_2\text{CO}_3 = (0.33 \text{ g})\left[\frac{1 \text{ mol NaAl(OH)}_2\text{CO}_3}{144.00 \text{ g NaAl(OH)}_2\text{CO}_3}\right]$$

$$= 2.29 \times 10^{-3} \text{ mol} = 2.29 \text{ mmol}$$

$$\text{millimoles of HCl} = [2.29 \text{ mmol NaAl(OH)}_2\text{CO}_3]\left(\frac{4 \text{ mmol HCl}}{1 \text{ mmol NaAl(OH)}_2\text{CO}_3}\right)$$

$$= 9.17 \text{ mmol}$$

The volume of HCl that can be neutralized is

$$V = \frac{n}{M} = \frac{9.17 \text{ mmol}}{0.14 \text{ mmol} \cdot \text{mL}^{-1}} = 66 \text{ mL}$$

6-83   Take 1000 mL of solution. The mass of HCl in 1.00 L of solution is

$$\text{mass of HCl} = (1.20 \text{ g} \cdot \text{mL}^{-1})(1000 \text{ mL})(0.40)$$

$$= 480 \text{ g}$$

The number of moles of HCl in 1000 mL of solution is

$$\text{moles of HCl} = (480 \text{ g})\left(\frac{1 \text{ mol HCl}}{36.46 \text{ g HCl}}\right) = 13.2 \text{ mol}$$

The molarity of the solution is

$$M = \frac{n}{V} = \frac{13.2 \text{ mol}}{1.00 \text{ L}} = 13.2 \text{ M}$$

6-85   We must first check to see which reactant is the limiting reactant.

$$\text{moles of Cr}_2\text{O}_7^{2-} = MV = (0.560 \text{ mol} \cdot \text{L}^{-1})(0.0550 \text{ L})$$

$$= 0.0308 \text{ mol}$$

$$\text{moles of C}_2\text{H}_5\text{OH} = MV = (0.963 \text{ mol} \cdot \text{L}^{-1})(0.100 \text{ L})$$

$$= 0.0963 \text{ mol}$$

If we divide the number of moles of each by its respective stoichiometric coefficient, then we see that (moles of $Cr_2O_7^{2-}/2) = 0.0154$ mol is less than (moles of $C_2H_5OH/3) = 0.0321$ mol. Therefore, the limiting reactant is $Cr_2O_7^{2-}(aq)$. Thus the number of grams of $HC_2H_3O_2$ produced is

$$\text{mass of } HC_2H_3O_2 = (0.0308 \text{ mol } Cr_2O_7^{2-}) \left(\frac{3 \text{ mol } HC_2H_3O_2}{2 \text{ mol } Cr_2O_7^{2-}} \right)$$
$$\times \left(\frac{60.05 \text{ g } HC_2H_3O_2}{1 \text{ mol } HC_2H_3O_2} \right)$$
$$= 2.77 \text{ g}$$

6-87   From Equation (6-4) we have

$$M_1V_1 = M_2V_2$$

Thus

$$V_1 = \frac{(0.050 \text{ mol} \cdot L^{-1})(500 \text{ mL})}{1.00 \text{ mol} \cdot L^{-1}} = 25 \text{ mL}$$

Thus we add 25 mL of 1.0 M $NaH_2PO_4(aq)$ to a 500-mL volumetric flask that is about half-filled with water, swirl the solution, and dilute with water to the 500-mL mark on the flask.

6-89   The number of moles of base required to neutralize the acid is

$$\text{moles of NaOH} = MV = (0.250 \text{ mol} \cdot L^{-1})(0.0666 \text{ L}) = 0.01665 \text{ mol}$$

Therefore, the number of moles of acid present in the original 100-mL solution was 0.01665 mol. Thus we have

$$1.00 \text{ g acid} \backsimeq 0.01665 \text{ mol acid}$$

Dividing by 0.01665, we obtain

$$60.1 \text{ g} \backsimeq 1.00 \text{ mol}$$

The formula mass of the acid is 60.1.

E. ANSWERS TO THE SELF-TEST

1. $NaCl; H_2O$

2. $NaCl(s) \xrightarrow[H_2O(l)]{} Na^+(aq) + Cl^-(aq)$

3. (aq)

4. true

5. water

6. true

7. true

8. unsaturated

9. increases

10. the number of moles of solute per liter of solution

11. false

12. Dissolve 0.50 mol (29.22 g) of NaCl in about 400 mL of water and then dilute to 500 mL.

13. MV

14. true

15. false

16. true

17. false

18. false

19. true

20. soluble in water

21. 0.01 M

22. 0.1 M

23. false

24. soluble

25. insoluble

26. soluble

27. insoluble

28. true

29. the base is just completely neutralized

30. false (It can be determined using an acid of known concentration.)

31. true

32. true

33. true

7 PROPERTIES OF GASES

A. OUTLINE OF CHAPTER 7

7-1   Most of the volume of a gas is empty space.

> The physical states of matter are solid, liquid, and gas.
>
> A solid has a fixed volume and shape.
>
> A solid is an ordered array of closely packed particles.
>
> A liquid has a fixed volume but assumes the shape of its container.
>
> A liquid is a dense, random array of particles.
>
> A gas occupies the entire volume of its container.
>
> The molecules of a gas are widely separated and in constant chaotic motion.

7-2   A manometer is used to measure the pressure of a gas.

> Gas molecules are in constant motion.
>
> The pressure exerted by a gas is due to the collisions of the gas molecules with the walls of the vessel that contains the gas.
>
> The pressure of a gas can be expressed as the height of a column of mercury supported by the gas (Figure 7-3).
>
> The pressure unit mmHg, millimeters of mercury, is called a torr.
>
> The height of a column of liquid supported by a gas depends on the density of the liquid.

7-3   A standard atmosphere is 760 torr.

> Barometric pressure is the pressure due to the atmosphere.
>
> A barometer is used to measure the pressure due to the atmosphere.
>
> Pressure is defined as a force per unit area.
>
> The pascal is the SI unit of pressure.

One standard atmosphere equals 1.013×10^5 Pa.

The various units of pressure are given in Table 7-1.

7-4   The volume of a gas is inversely proportional to its pressure and directly proportional to its Kelvin temperature.

Boyle's law can be expressed as $V \propto 1/P$ or as $V = c/P$ (constant temperature), where c is a proportionality constant (Equation 7-1).

The absolute temperature scale, or Kelvin temperature scale, is related to the Celsius scale by the expression

$$T(\text{K}) = t(°\text{C}) + 273.15 \qquad (7\text{-}2)$$

The absolute (Kelvin) temperature scale has the units of kelvin, K.

Absolute zero, 0 K, is the lowest possible temperature.

The equation for a straight line is $y = mx + b$, where b is the intercept of the straight line with the y axis and m is the slope of the line, which is a measure of its steepness.

The equation expressing Charles's law is $V = mT$ (constant pressure), where m is a proportionality constant (Equation 7-4).

A gas thermometer is based on Charles's law (Figure 7-11).

7-5   Equal volumes of gases at the same pressure and temperature contain equal numbers of molecules.

Gay-Lussac's law of combining volumes states that when all volumes are measured at the same pressure and temperature, the volumes in which gases combine in chemical reactions are related by simple, whole numbers (Figure 7-13).

7-6   The ideal-gas equation is a combination of Boyle's, Charles's, and Avogadro's laws.

The ideal-gas law is expressed by the ideal-gas equation

$$PV = nRT \qquad (7\text{-}6)$$

where P is the pressure, V is the volume, n is the number of moles of gas molecules, T is the Kelvin temperature, and R is the molar gas constant.

One value of the gas constant R is 0.0821 L·atm·mol^{-1}·K^{-1}.

An ideal gas is a gas that obeys the ideal-gas law.

The molar volume of an ideal gas is 22.4 L at 0°C and 1.00 atm.

Applications of the ideal-gas equation are discussed.

7-7   The ideal-gas equation can be used to calculate the molecular masses of gases.

The molar mass of a gas can be determined using the ideal-gas equation (Example 7-8).

The density ρ of an ideal gas in grams per liter is given by

$$\rho = \frac{MP}{RT} \qquad (7\text{-}9)$$

Gas density increases as gas pressure increases and as gas temperature decreases.

The molecular mass of a gas can be determined from the density of the gas (Example 7-9).

7-8 The total pressure of a mixture of ideal gases is the sum of the partial pressures of all the gases in the mixture.

Dalton's law of partial pressures states that for a mixture of two ideal gases

$$P_{total} = P_1 + P_2 \qquad (7\text{-}10)$$

The total pressure of a mixture of gases is determined by the total number of moles of gas in the mixture, or

$$P_{total} = n_{total} \frac{RT}{V} \qquad (7\text{-}12)$$

The mole fraction of species i in a two-component mixture, denoted by X_i, is given by

$$X_i = \frac{n_i}{n_1 + n_2} \qquad i = 1,2 \qquad (7\text{-}14)$$

The partial pressure of species i is given by

$$P_i = X_i P_{total} \qquad i = 1,2 \qquad (7\text{-}15)$$

The molecular fraction of species i, denoted by X_i, is given by

$$X_i = \frac{N_i}{N_1 + N_2} \qquad i = 1,2 \qquad (7\text{-}16)$$

where N_i is the number of molecules of species i and $N_1 + N_2$ is the total number of molecules.

The partial pressure of each gas in a mixture is proportional only to the number of molecules of that gas.

When a gas is collected over water, Dalton's law of partial pressures

$$P_{total} = P_{gas} + P_{H_2O}$$

is used to find the pressure of the dry gas.

7-9 The molecules of a gas have a distribution of speeds.

Collisions of gas molecules with the container walls give rise to the gas pressure.

Kinetic energy is the energy associated with the motion of a body.

The kinetic energy of a particle of mass m and speed v is given by $E_k = \frac{1}{2}mv^2$ (Equation 7-18).

The SI unit of energy is the joule, J, where $1\,J = 1\;kg \cdot m^2 \cdot s^{-2}$.

The molecules in a gas do not all travel at the same speed.

The distribution of molecular speeds in a gas is called a Maxwell-Boltzmann distribution and is shown in Figure 7-19.

The distribution of molecular kinetic energies is shown in Figure 7-20.

7-10   The average kinetic energy of a gas is proportional to its Kelvin temperature.

The postulates of the kinetic theory of gases are listed on page 220 in the text.

The average kinetic energy per mole of a gas is given by

$$\overline{E}_k = \tfrac{1}{2} M_{kg} \overline{v^2} \qquad (7\text{-}20)$$

where $\overline{v^2}$ is the average of the square of the speeds of the gas molecules.

The average of the square of the speeds of the gas molecules is not equal to the square of the average speed of the gas molecules.

The fourth postulate of the kinetic theory of gases states that

$$\overline{E_k} \propto T \qquad (7\text{-}21)$$

where T is the absolute temperature of the gas.

7-11   The kinetic theory of gases allows us to calculate the root-mean-square speed of a molecule.

The pressure exerted by a gas depends on the three factors given on page 222 of the text.

The fundamental equation of the kinetic theory of gases relates the pressure exerted by a gas to the average of the square of the molecular speeds $\overline{v^2}$ and is given by

$$PV = \left(\frac{1}{3}\right) N_0 m \overline{v^2} \qquad (7\text{-}24)$$

where N_0 is Avogadro's number and m is the mass of the gas molecule.

A fundamental result of the kinetic theory of gases is

$$\overline{E_k} = \frac{3}{2} RT \qquad (7\text{-}26)$$

where R is the gas constant and T is the absolute temperature.

The root-mean-square speed v_{rms} is the square root of the mean (average) of the square of the molecular speeds and is given by

$$v_{rms} = \left(\frac{3RT}{M_{kg}}\right)^{1/2} \qquad (7\text{-}28)$$

where M_{kg} is the molar mass in kilograms per mole.

The root-mean-square speed is a good measure of the average molecular speed in a gas.

Values of v_{rms} for several gases are given in Table 7-2.

7-12   Gases can be separated by effusion.

Effusion is the process of a gas leaking through a small hole.

The rate of effusion is directly proportional to the root-mean-square speed of the gas molecules.

Graham's law of effusion is expressed by the relation

$$\frac{rate_A}{rate_B} = \left(\frac{M_B}{M_A}\right)^{1/2} \tag{7-29}$$

The isotope uranium-235 is separated from naturally occurring uranium by effusion.

7-13   The average distance a molecule travels between collisions is called the mean free path.

The mean free path is inversely proportional to the number of molecules per unit volume.

The mean free path is inversely proportional to the cross section of the molecules.

The mean free path is given by

$$l = \frac{3.7 \times 10^8 \ pm^3 \cdot mol \cdot L^{-1}}{\sigma^2(n/V)} \tag{7-31}$$

where σ is the molecular diameter of the gas molecule (see Table 7-3).

The number of collisions a gas molecule undergoes per second is called the collision frequency and is given by

$$z = \frac{v_{rms}}{l} \tag{7-32}$$

7-14   The van der Waals equation accounts for deviations from gas ideality.

Deviations from ideal-gas behavior are due to the volume of the gas molecules and the attraction between gas molecules.

The ideal-gas equation is not valid at high pressure (Figure 7-23).

One equation for non-ideal-gas behavior is the van der Waals equation

$$\left(P + \frac{n^2 a}{V^2}\right)(V - nb) = nRT \tag{7-33}$$

where a and b are van der Waals constants, whose values depend upon the particular gas (Table 7-4).

B. SELF-TEST

1. The particles in a solid move throughout the solid. *True/False*

2. The particles in a liquid move throughout the liquid. *True/False*

3. When a solid melts, the liquid has a much lower density. *True/False*

4. When a liquid is vaporized, there is a large increase in (*volume/density*).

5. Most of the volume of a gas is ——————————————.

6. A gas has a large compressibility because ————————————————
——————————————————————————————————.

7. A manometer is a device used to measure ————————————————.

8. The height of mercury in a manometer depends on ——————————————
——————————————————————————————.

9. The pressure of a gas is given as millimeters of Hg, which is also called ————
——————.

10. The height of the column of liquid that can be supported by a gas is the same for all liquids. *True/False*

11. A barometer is a device used to measure ——————————————.

12. The pressure due to the atmosphere is a constant. *True/False*

13. At sea level atmospheric pressure is always 760 torr. *True/False*

14. One standard atmosphere is defined as ——————————.

15. The SI unit of pressure is ——————————————.

16. Boyle's law states that the volume of a gas is —————————— proportional to the —————————— at constant ——————————.

17. A volume of 2.4 L of gas whose pressure is 1.5 atm is compressed to 1.2 L with no change in temperature. The pressure of the gas is now (*greater/less*) than 1.5 atm.

18. The absolute temperature scale is also called the ————————————
scale and has the unit ——————————.

19. The relationship between the absolute temperature scale and the Celsius temperature scale is ——————————————————————.

20. The lowest possible temperature on the absolute temperature scale is _____ ; on the Celsius scale it is _____ .

21. Charles's law states that the volume of a gas is _____ proportional to the _____ at constant _____ .

22. A volume of 3.4 L of gas at $100\,^\circ C$ is heated to $200\,^\circ C$ with no change in pressure. The volume of the gas is now 6.8 L. *True/False*

23. A constant-pressure gas thermometer measures _____ through its direct proportionality to the _____ of a gas at _____
_____ .

24. At the same temperature and pressure, 3 volumes of hydrogen, H_2, and 1 volume of nitrogen, N_2, produce _____ volumes of ammonia, NH_3.

25. The ideal-gas law can be expressed as _____ = _____ .

26. An ideal gas is a gas that _____ .

27. The value of the molar gas constant depends on the units used. *True/False*

28. One value of the molar gas constant is _____ .

29. When $R = 0.821$ $L \cdot atm \cdot mol^{-1} \cdot K^{-1}$ is used in the ideal-gas equation, P must be expressed in _____ V in _____ n in _____ , and T in _____ .

30. One mole of an ideal gas at $0\,^\circ C$ and 1 atm occupies a volume of _____
_____ .

31. The volume of a gas (*increases/decreases*) when the temperature of the gas decreases and the pressure of the gas remains constant.

32. The pressure of a gas (*increases/decreases*) when the temperature of the gas increases and the volume of the gas remains constant.

33. A reaction between a gas and a solid to produce a solid product takes place at constant temperature and volume. After the consumption of part of the gas, the pressure due to the gas (*increases/decreases*).

34. The gas density (*increases/decreases*) when the pressure of the gas increases at constant temperature.

35. The gas density (*increases/decreases*) when the temperature of the gas decreases at constant pressure.

36. The molar mass of a gas may be determined from the density of the gas. *True/False*

37. The partial pressure of one gas in a mixture of gases depends on the pressure of the other gases. *True/False*

38. The mole fraction of oxygen in a mixture of O_2 and CO_2 is given by X_{O_2} = _____.

39. The partial pressure of O_2 in a mixture of O_2 and CO_2 is related to the mole fraction of O_2 by P_{O_2} = _____.

40. When a gas such as nitrogen is collected over water, the pressure of the nitrogen gas is the total pressure (*plus/minus*) the pressure due to water vapor.

41. The pressure exerted by a gas is caused by _____

_____.

42. The kinetic energy of a body in motion is given by _____,

where _____.

43. The kinetic energy of 1 mol of gas molecules is given by _____,

where _____

_____.

44. All the molecules of a gas travel at the same speed. *True/False*

45. More molecules travel at (*higher/lower*) speeds at higher temperatures.

46. A Maxwell-Boltzmann distribution is a distribution of molecular _____in a gas.

47. All the molecules in a gas have the same kinetic energy. *True/False*

48. State the postulates of the kinetic theory of gases.

(1)

(2)

(3)

(4)

49. The average kinetic energy is denoted by ———— .

50. The average kinetic energy per mole of a gas is given by ————————
———————— , where ————————————————————————
—————————————————————————————————————— .

51. The average kinetic energy of a gas depends only upon its (*temperature/pressure/ volume*).

52. The average kinetic energy per mole of all gases is the same at the same temperature. *True/False*

53. The pressure that a gas exerts on the walls of its container (*increases/decreases*) when the number density increases.

54. The pressure that a gas exerts on the walls of its container (*increases/decreases*) when the average speed of the molecules increases.

55. The pressure that a gas exerts on the wall of its container (*increases/decreases*) when the frequency of collisions with the walls increases.

56. The fundamental equation of the kinetic theory of gases is $PV =$ ———————— ,
where ————————————————————————————————————
—————————————————————————————————————— .

57. The average kinetic energy of a gas is related to the temperature of the gas by
———————————————————— , where ————————————————
—————————————————————————————— .

58. The root-mean-square speed is the square of the average of the molecular speeds. *True/False*

59. The root-mean-square speed depends on the (*temperature/pressure/volume*) of the gas.

60. All gases effuse at the same rate. *True/False*

61. Propane (C_3H_8) will effuse at a slower rate than methane (CH_4). *True/False*

62. The average distance traveled between collisions by a molecule in a gas whose pressure is 1 atm is quite small when compared to the distance between the walls of the container. *True/False*

63. The mean free path is the average distance ————————————————
—————————————————————————————————— .

64. The mean free path of a molecule depends on the _____

and _____ .

65. The number of collisions that a gas molecule undergoes in 1 s can be calculated from

the _____ and the _____ of the molecule.

66. The ideal-gas equation is valid at high pressures and low temperatures. *True/False*

67. The van der Waals equation describes (*ideal/nonideal*) gas behavior.

68. The van der Waals constant b is proportional to the _____ of a gas molecule.

69. The van der Waals constant a is related to the _____ the gas molecules.

C. CALCULATIONS YOU SHOULD KNOW HOW TO DO

1. Convert from one unit of pressure to another using Table 7-1. See Example 7-2, and Problems 7-1 and 7-2.

2. Convert between Kelvin and Celsius temperature scales, using the relation $T(K) = t(°C) + 273.15$. See Example 7-3 and Problems 7-5 and 7-6.

3. Use Boyle's law in the form $P_iV_i = P_fV_f$. See Problems 7-3 and 7-4.

4. Use Charles's law in the form $V_i/T_i = V_f/T_f$. See Example 7-3 and Problems 7-7 and 7-8.

5. Use Gay-Lussac's law of combining volumes. See Problems 7-9 and 7-10.

6. Use the ideal-gas law, $PV = nRT$. See Examples 7-4 through 7-7 and Problems 7-11 through 7-18 and 7-53 through 7-58, in which the pressure is expressed in pascals.

7. Use the ideal-gas law in stoichiometric calculations. See Examples 7-5 and 7-6 and Problems 7-19 through 7-24.

8. Use the ideal-gas law in the form $\rho = MP/RT$ to calculate gas density ρ and to determine molecular mass. See Examples 7-8 through 7-10 and Problems 7-25 through 7-30.

9. Use Dalton's law of partial pressures: $P_{total} = P_1 + P_2$, with $P_1 = X_1P_{total}$ and $P_2 = X_2P_{total}$. See Examples 7-11 through 7-13 and Problems 7-31 through 7-36.

10. Calculate the root-mean-square speed of a gas molecule using $v_{rms} = (3RT/M_{kg})^{1/2}$. See Example 7-15 and Problems 7-37 through 7-42.

11. Calculate the mean free path of a gas molecule by using the equation $l = (3.7 \times 10^8 \text{ pm}^3 \cdot \text{mol} \cdot \text{L}^{-1})/\sigma^2(n/V)$ and Table 7-3. See Problems 7-43 through 7-46.

12. Calculate the number of collisions (collision frequency) that a gas molecule under-goes per second using $z = v_{rms}/l$. See Problems 7-45 and 7-46.

13. Use Graham's law of effusion: $rate_A/rate_B = t_B/t_A = (M_B/M_A)^{1/2}$. See Example 7-16 and Problems 7-47 through 7-50.

14. Calculate the pressure of a gas by using the van der Waals equation (Equation 7-37) and Table 7-4. See Problems 7-51 and 7-52.

Plotting Data

It is usually desirable to plot equations or experimental data such that a straight line is obtained. The mathematical equation of a straight line is of the form

$$y = mx + b \tag{1}$$

In this equation, m and b are constants: m is the *slope* of the line and b is its *intercept* with the y axis. The slope of a straight line is a measure of its steepness; it is defined as the ratio of its vertical rise to the corresponding horizontal distance.

Let's plot the two straight lines

(I) $y = x + 1$ and   (II) $y = 2x - 2$

We first make a table of values of x and y:

Equation I		Equation II	
x	y	x	y
-3	-2	-3	-8
-2	-1	-2	-6
-1	0	-1	-4
0	1	0	-2
1	2	1	0
2	3	2	2
3	4	3	4
4	5	4	6
5	6	5	8

These results are plotted in Figure 1. Note that curve I intersects the y axis at $y = 1$ ($b = 1$) and has a slope of 1 ($m = 1$). Curve II intersects the y axis at $y = -2$ ($b = -2$) and has a slope of 2 ($m = 2$).

Usually the equation to be plotted will not appear to be of the form of Equation (I) at first. For example, consider Boyle's law

$$V = \frac{c}{P} \quad \text{(constant temperature)} \tag{2}$$

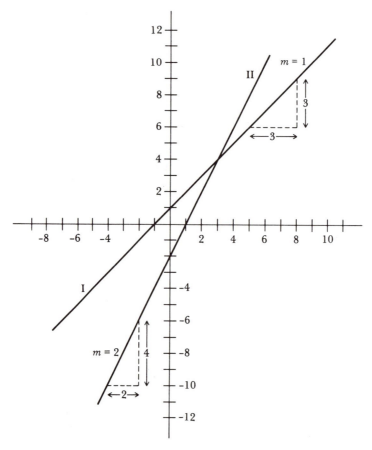

Figure 1 Plot of Equation (I): $y = x + 1$ Plot of Equation (II): $y = 2x - 2$

where c is a proportionality constant. The value of the proportionality constant depends on the temperature for a given sample. For example, for a 0.29-g sample of air at 25°C, $c = 0.244$ L·atm. Some results for such a sample are presented in Table 1. The data in Table 1 are plotted as volume versus pressure in Figure 2.

Table 1   Pressure-volume data for a sample of 0.29 g of air at 25°C

P/atm	V/L	$\frac{1}{P}$/atm^{-1}	P/atm	V/L	$\frac{1}{P}$/atm^{-1}
0.26	0.938	3.85	2.10	0.116	0.48
0.41	0.595	2.44	2.63	0.093	0.38
0.83	0.294	1.20	3.14	0.078	0.32
1.20	0.203	0.83			

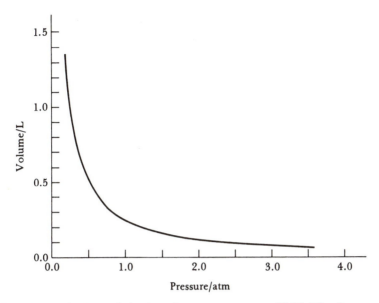

Figure 2 The volume of 0.29 g of air plotted versus pressure at 25°C. The data are given in Table 1. The curve in this figure obeys the equation $V = 0.244$ L·atm/P, which is an expression of Boyle's law.

It may appear at first sight that Equation (2) is not of the form $y = mx + b$. However, if we let $V = y$ and $1/P = x$, then Equation (2) becomes $y = cx$ which is the equation of a straight line with $m = c$ and $b = 0$. Thus, if we plot V versus $1/P$ instead of P, a straight line will result. The data in Table 1 are plotted as V versus $1/P$ in Figure 3. Note that a straight line is obtained.

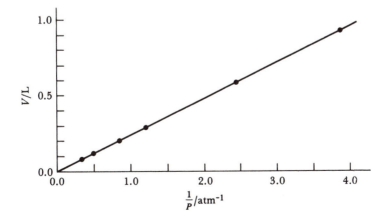

Figure 3 The volume of 0.29 g of air plotted versus the reciprocal of the pressure ($1/P$) at 25°C. If we compare this curve to that of Figure 2, we see that a straight line results by plotting V versus $1/P$ instead of versus P.

D. SOLUTIONS TO THE ODD-NUMBERED PROBLEMS

7-1   The conversion factors for the units of pressure are given in Table 7-1.

(a)   The atmospheric pressure at the surface of Venus is

$$(75\ \text{atm})\left(\frac{760\ \text{torr}}{1\ \text{atm}}\right) = 5.7 \times 10^4\ \text{torr}$$

and

$$(75\ \text{atm})\left(\frac{1.013\ \text{bar}}{1\ \text{atm}}\right) = 76\ \text{bar}$$

(b)   The atmospheric pressure in Mexico City is

$$(580\ \text{torr})\left(\frac{1\ \text{atm}}{760\ \text{torr}}\right) = 0.76\ \text{atm}$$

$$(580\ \text{torr})\left(\frac{1\ \text{atm}}{760\ \text{torr}}\right)\left(\frac{1013\ \text{mbar}}{1\ \text{atm}}\right) = 770\ \text{mbar}$$

(c)   The pressure of CO_2 in the gas cylinder is

$$(5.2\ \text{atm})\left(\frac{1.013 \times 10^5\ \text{Pa}}{1\ \text{atm}}\right) = 5.3 \times 10^5\ \text{Pa}$$

$$(5.2\ \text{atm})\left(\frac{101.3\ \text{kPa}}{1\ \text{atm}}\right) = 530\ \text{kPa}$$

(d)   The pressure of the N_2 gas is

$$(920\ \text{torr})\left(\frac{1.013 \times 10^5\ \text{Pa}}{760\ \text{torr}}\right) = 1.23 \times 10^5\ \text{Pa}$$

$$(920\ \text{torr})\left(\frac{1\ \text{atm}}{760\ \text{torr}}\right) = 1.21\ \text{atm}$$

7-3 Boyle's law problems are worked by using Boyle's law in the form

$$P_i V_i = P_f V_f \quad \text{(constant } T\text{)}$$

where i stands for initial and f for final. Thus we have for the volume V_f

$$V_f = \frac{P_i V_i}{P_f} = \frac{(3.46 \text{ atm})(0.650 \text{ mL})}{(1.00 \text{ atm})} = 2.25 \text{ mL}$$

The relation between the volume and the radius of a sphere is

$$V = \tfrac{4}{3}\pi r^3$$

where r is the radius. The diameter is twice the radius; thus

$$V = \tfrac{4}{3}\pi \left(\frac{d}{2}\right)^3 = \tfrac{1}{6}\pi d^3$$

and

$$d = \left(\frac{6V}{\pi}\right)^{1/3}$$

For the 0.650-cm^3 (1 mL = 1 cm^3) bubble we have

$$d = \left(\frac{6 \times 0.650 \text{ cm}^3}{3.14}\right)^{1/3} = 1.07 \text{ cm}$$

At the surface the volume of the bubble is 2.25 cm^3, thus

$$d = \left(\frac{6 \times 2.25 \text{ cm}^3}{3.14}\right)^{1/3} = 1.63 \text{ cm}$$

7-5 The relationship between the Celsius scale and the Kelvin scale is $T \text{ (K)} = t \text{ (°C)} + 273.15$ (Equation (7-4).

(a) $T = 37 + 273.15 = 310 \text{ K}$ (b) $T = 20 + 273.15 = 293 \text{ K}$
(c) $T = -259 + 273.15 = 14 \text{ K}$ (d) $T = 199 + 273.15 = 472 \text{ K}$

7-7 Charles's law problems are worked using Charles's law in the form of Equation (7-5). Celsius temperatures must be converted to Kelvin temperatures before substituting into Charles's law.

$$T_i = 0 + 273 = 273 \text{ K}$$
$$T_f = 100 + 273 = 373 \text{ K}$$

Thus, the volume of the gas at 100°C is given by

$$V_f = \frac{T_f V_i}{T_i} = \frac{(373 \text{ K})(14.7 \text{ mL})}{273 \text{ K}} = 20.1 \text{ mL}$$

7-9   We see from Gay-Lussac's law and the reaction stoichiometry that one volume of CH_4 reacts with two volumes of O_2, or 5.0 L of CH_4 reacts with 10.0 L of O_2. Because air is 20% O_2, the volume of air needed is

$$(0.20)V = 10.0 \text{ L}$$

$$V = \frac{10.0 \text{ L}}{0.20} = 50 \text{ L}$$

7-11   The first step in the solution of an ideal-gas problem is to write the ideal-gas law

$$PV = nRT$$

We solve the ideal-gas equation for V to obtain

$$V = \frac{nRT}{P}$$

Recall that using the gas constant R in the units 0.0821 L·atm·mol^{-1}·K^{-1} requires P in atm and T in kelvin. We must convert P and T to the proper units:

$$P = (600 \text{ torr}) \left(\frac{1 \text{ atm}}{760 \text{ torr}} \right) = 0.789 \text{ atm}$$

$$T = 37 + 273 = 310 \text{ K}$$

Substituting these values into the ideal-gas equation yields

$$V = \frac{(0.65 \text{ mol})(0.0821 \text{ L·atm·mol}^{-1}\text{·K}^{-1})(310 \text{ K})}{(0.789 \text{ atm})}$$

$$= 21 \text{ L}$$

It is a good idea to check the cancellation of units to make sure that the answer is obtained in the desired units and that the right units were used for the various quantities involved in the calculation.

7-13   In order to use the ideal-gas law we must first convert the mass of water to moles of water:

$$n = (18 \text{ g}) \left(\frac{1 \text{ mol H}_2\text{O}}{18.02 \text{ g H}_2\text{O}} \right) = 1.00 \text{ mol}$$

Solving the ideal-gas equation for P, we have

$$P = \frac{nRT}{V}$$

Converting the temperature to kelvin, we obtain

$$T = 100 + 273 = 373 \text{ K}$$

and substituting the values for n, R, T, and V into the expression for P, we have

$$P = \frac{(1.00 \text{ mol})(0.0821 \text{ L} \cdot \text{atm} \cdot \text{mol}^{-1} \cdot \text{K}^{-1})(373 \text{ K})}{(18 \text{ L})} = 1.7 \text{ atm}$$

The volume of water is calculated from the density of water, $d = m/V$, or

$$\text{volume of water} = \frac{m}{d} = \frac{18 \text{ g}}{1.00 \text{ g} \cdot \text{mL}^{-1}} = 18 \text{ mL}$$

7-15 We first use the ideal-gas equation to compute the number of moles n of helium. Then we compute the number of molecules by multiplying n by Avogadro's number. The expression for n is

$$n = \frac{PV}{RT}$$

The temperature in kelvin is

$$T = -200 + 273 = 73 \text{ K}$$

Substituting the values for P, V, R, and T into the expression for n, we have

$$n = \frac{(0.0010 \text{ atm})(1.0 \text{ L})}{(0.0821 \text{ L} \cdot \text{mol}^{-1} \cdot \text{K}^{-1})(73 \text{ K})} = 1.67 \times 10^{-4} \text{ mol}$$

The number of molecules of helium is given by

$$\text{number of molecules} = (1.67 \times 10^{-4} \text{ mol})(6.022 \times 10^{23} \text{ molecules} \cdot \text{mol}^{-1})$$
$$= 1.0 \times 10^{20} \text{ molecules}$$

Recall that at 1.0 atm and 0°C, 1 mol of an ideal gas occupies 22.4 L. Thus the number of moles in 1.0 L at 1.0 atm and 0°C is

$$n = \frac{1.0 \text{ L}}{22.4 \text{ L} \cdot \text{mol}^{-1}} = 4.46 \times 10^{-2} \text{ mol}$$

and the number of molecules is

$$\text{number of molecules} = (4.46 \times 10^{-2} \text{ mol})(6.022 \times 10^{23} \text{ molecules} \cdot \text{mol}^{-1})$$
$$= 2.7 \times 10^{22} \text{ molecules}$$

We could also have calculated n at 0°C, 1.0 atm, and 1.0 L from the ideal-gas equation. There are $(2.7 \times 10^{22} \text{ molecules}/1.0 \times 10^{20} \text{ molecules} = 2.7 \times 10^2) = 270$ times as many molecules at 0°C and 1.0 atm.

7-17 We use the ideal-gas equation to calculate the number of moles of ozone in 1.0 mL and then multiply n by Avogadro's number to obtain the number of molecules.

$$n = \frac{PV}{RT}$$

The temperature in kelvin is

$$T = -23 + 273 = 250 \text{ K}$$

Substituting the values for P, V, R, and T into the expression for n, we have

$$n = \frac{(1.4 \times 10^{-7} \text{ atm})(1.0 \times 10^{-3} \text{ L})}{(0.0821 \text{ L} \cdot \text{atm} \cdot \text{mol}^{-1} \cdot \text{K}^{-1})(250 \text{ K})} = 6.82 \times 10^{-12} \text{ mol}$$

$$\text{number of molecules} = (6.82 \times 10^{-12} \text{ mol})(6.022 \times 10^{23} \text{ molecules} \cdot \text{mol}^{-1})$$
$$= 4.1 \times 10^{12} \text{ molecules}$$

7-19 We must first check to see which reactant is the limiting reactant.

$$\text{moles of CaC}_2 = (100 \text{ g}) \left(\frac{1 \text{ mol CaC}_2}{64.10 \text{ g CaC}_2} \right) = 1.56 \text{ mol}$$

and the number of moles of water is

$$\text{moles of H}_2\text{O} = (100 \text{ g}) \left(\frac{1 \text{ mol H}_2\text{O}}{18.02 \text{ g H}_2\text{O}} \right) = 5.55 \text{ mol}$$

Thus we see that CaC_2 is the limiting reactant. The number of moles of C_2H_2 produced is

$$\text{moles of C}_2\text{H}_2 = (1.56 \text{ mol CaC}_2) \left(\frac{1 \text{ mol C}_2\text{H}_2}{1 \text{ mol CaC}_2} \right) = 1.56 \text{ mol}$$

We now use the ideal-gas equation to compute the volume V occupied by 1.56 mol of C_2H_2 at the given values of T and P:

$$V = \frac{nRT}{P}$$

At 0°C and 1.00 atm, we have

$$V = \frac{(1.56 \text{ mol})(0.0821 \text{ L} \cdot \text{atm} \cdot \text{mol}^{-1} \cdot \text{K}^{-1})(273 \text{ K})}{(1.00 \text{ atm})} = 35.0 \text{ L}$$

At 120°C and 1.00 atm, we have

$$V = \frac{(1.56 \text{ mol})(0.0821 \text{ L} \cdot \text{atm} \cdot \text{mol}^{-1} \cdot \text{K}^{-1})(393 \text{ K})}{(1.00 \text{ atm})} = 50.3 \text{ L}$$

7-21 The number of moles of glucose in 1.00 g is

$$n = (1.00 \text{ g}) \left(\frac{1 \text{ mol C}_6\text{H}_{12}\text{O}_6}{180.16 \text{ g C}_6\text{H}_{12}\text{O}_6} \right) = 0.00555 \text{ mol}$$

From the reaction stoichiometry, the number of moles of CO_2 produced is

$$n = (0.00555 \text{ mol } C_6H_{12}O_6) \left(\frac{6 \text{ mol } CO_2}{1 \text{ mol } C_6H_{12}O_6} \right) = 0.0333 \text{ mol}$$

We now compute the volume by using the ideal-gas equation.

$$V = \frac{nRT}{P} = \frac{(0.0333 \text{ mol})(0.0821 \text{ L} \cdot \text{atm} \cdot \text{mol}^{-1} \cdot \text{K}^{-1})(310 \text{ K})}{(1.00 \text{ atm})}$$
$$= 0.848 \text{ L} = 848 \text{ mL}$$

7-23 The number of moles of chlorine desired is

$$n = \frac{PV}{RT} = \frac{(750 \text{ torr})(1 \text{ atm}/760 \text{ torr})(0.500 \text{ L})}{(0.0821 \text{ L} \cdot \text{atm} \cdot \text{mol}^{-1} \cdot \text{K}^{-1})(298 \text{ K})}$$
$$= 0.02017 \text{ mol}$$

The mass of MnO_2 required is

$$\text{mass of } MnO_2 = (0.02017 \text{ mol } Cl_2) \left(\frac{1 \text{ mol } MnO_2}{1 \text{ mol } Cl_2} \right) \left(\frac{86.94 \text{ g } MnO_2}{1 \text{ mol } MnO_2} \right)$$
$$= 1.75 \text{ g}$$

7-25 We shall use Equation (7-9) to calculate the density of $H_2O(g)$.

$$\rho = \frac{MP}{RT}$$

Thus for $H_2O(g)$ at 100°C and 1.00 atm, we have

$$\rho = \frac{(18.02 \text{ g} \cdot \text{mol}^{-1})(1.00 \text{ atm})}{(0.0821 \text{ L} \cdot \text{atm} \cdot \text{mol}^{-1} \cdot \text{K}^{-1})(373 \text{ K})} = 0.588 \text{ g} \cdot \text{L}^{-1}$$

The density of liquid water at 100°C is 0.958 $\text{g} \cdot \text{mL}^{-1}$, or

$$d = (0.958 \text{ g} \cdot \text{mL}^{-1}) \left(\frac{1000 \text{ mL}}{1 \text{ L}} \right) = 958 \text{ g} \cdot \text{L}^{-1}$$

The ratio of the densities of $H_2O(l)$ and $H_2O(g)$ at 100°C is

$$\frac{958 \text{ g} \cdot \text{L}^{-1}}{0.588 \text{ g} \cdot \text{L}^{-1}} = 1630$$

7-27 We first calculate the molecular mass of the gas. We can use Equation (7-6) to calculate n:

$$n = \frac{PV}{RT} = \frac{(765 \text{ torr})(1 \text{ atm}/760 \text{ torr})(0.294 \text{ L})}{(0.0821 \text{ L} \cdot \text{atm} \cdot \text{mol}^{-1} \cdot \text{K}^{-1})(373 \text{ K})}$$
$$= 0.00966 \text{ mol}$$

Thus, we have the stoichiometric correspondence

$$0.271 \text{ g} \approx 0.00966 \text{ mol}$$
$$28.1 \text{ g} \approx 1.00 \text{ mol}$$

The formula mass of CH_2 is 14.03. Thus the molecular formula of the compound is C_2H_4.

7-29 We first determine the empirical formula of the compound. Taking a 100-g sample, we have

$$85.60 \text{ g C} \approx 14.40 \text{ g H}$$

Dividing the mass of each by its atomic mass, we have

$$7.127 \text{ mol C} \approx 14.29 \text{ mol H}$$

Dividing by 7.127, we have

$$1.00 \text{ mol C} \approx 2.00 \text{ mol H}$$

Thus the empirical formula is CH_2. The molar mass is computed from the gas density at a known temperature and pressure.

$$M = \frac{\rho RT}{P}$$
$$= \frac{(0.9588 \text{ g} \cdot \text{L}^{-1})(0.0821 \text{ L} \cdot \text{atm} \cdot \text{mol}^{-1} \cdot \text{K}^{-1})(298 \text{ K})}{(635 \text{ torr})(1 \text{ atm}/760 \text{ torr})}$$
$$= 28.1 \text{ g} \cdot \text{mol}^{-1}$$

The formula mass of CH_2 is 14.03. Thus the molecular formula of ethylene is C_2H_4.

7-31 The number of moles of H_2 is

$$\text{moles of } H_2 = (0.513 \text{ g}) \left(\frac{1 \text{ mol } H_2}{2.016 \text{ g } H_2} \right) = 0.254 \text{ mol}$$

and the number of moles of N_2 is

$$\text{moles of } N_2 = (16.1 \text{ g}) \left(\frac{1 \text{ mol } N_2}{28.02 \text{ g } N_2} \right) = 0.575 \text{ mol}$$

The total number of moles is 0.829 mol. We can calculate the total pressure by using the ideal-gas equation, Equation (7-6).

$$P = \frac{nRT}{V} = \frac{(0.829 \text{ mol})(0.0821 \text{ L} \cdot \text{atm} \cdot \text{mol}^{-1} \cdot \text{K}^{-1})(293 \text{ K})}{10.0 \text{ L}}$$
$$= 1.99 \text{ atm}$$

The partial pressure of each gas is

$$P_{H_2} = X_{H_2} P_{total} = \left(\frac{0.254 \text{ mol}}{0.829 \text{ mol}}\right)(1.99 \text{ atm}) = 0.610 \text{ atm}$$

$$P_{N_2} = X_{N_2} P_{total} = \left(\frac{0.575 \text{ mol}}{0.829 \text{ mol}}\right)(1.99 \text{ atm}) = 1.38 \text{ atm}$$

7-33 The pressure of nitrogen when the volume is 35.0 mL is

$$P_{N_2} = 740 \text{ torr}$$

If we now increase the volume available to the N_2 from 35.0 mL to 50.0 mL, then the partial pressure of nitrogen decreases. Because T and n_{N_2} are constant we have

$$P_i V_i = P_f V_f$$

and thus

$$P_f \text{ (of } N_2) = \frac{P_i V_i}{V_f} = \frac{(740 \text{ torr})(35.0 \text{ mL})}{(50.0 \text{ mL})}$$
$$= 518 \text{ torr}$$

The mixture of N_2 and O_2 in the 50.0-mL volume has a total pressure of 740 torr:

$$P_{total} = 740 \text{ torr} = P_{O_2} + P_{N_2}$$

thus

$$P_{O_2} = 740 \text{ torr} - 518 \text{ torr} = 222 \text{ torr}$$

7-35 Because the gases are collected at 25°C, $H_2O(g)$ will condense to $H_2O(l)$ and so 4 mol of nitroglycerin yield $12 + 6 + 1 = 19$ mol of gas. Thus 10 g of nitroglycerin yields the following number of moles of gas at 25°C:

$$n = (10 \text{ g nitro}) \left(\frac{1 \text{ mol nitro}}{227.10 \text{ g nitro}}\right)\left(\frac{19 \text{ mol gas}}{4 \text{ mol nitro}}\right) = 0.209 \text{ mol}$$

The volume is calculated from the ideal-gas equation:

$$V = \frac{nRT}{P} = \frac{(0.209 \text{ mol})(0.0821 \text{ L} \cdot \text{atm} \cdot \text{mol}^{-1} \cdot \text{K}^{-1})(298 \text{ K})}{(1.0 \text{ atm})} = 5.1 \text{ L}$$

The pressure produced when the reaction is confined to 0.50 L is given by

$$P = \frac{nRT}{V} = \frac{(0.209 \text{ mol})(0.0821 \text{ L}\cdot\text{atm}\cdot\text{mol}^{-1}\cdot\text{K}^{-1})(298 \text{ K})}{0.50 \text{ L}} = 10 \text{ atm}$$

7-37   Recall that in working kinetic theory problems we use the value of $R = 8.314 \text{ J}\cdot\text{mol}^{-1}\cdot\text{K}^{-1}$. The root-mean-square speed in $\text{m}\cdot\text{s}^{-1}$ of a gas molecule is calculated from the equation

$$v_{\text{rms}} = \left(\frac{3RT}{M_{\text{kg}}}\right)^{1/2}$$

where M_{kg} is the molar mass in kilograms per mole. For F_2 we have

$$M_{\text{kg}} = \frac{38.00 \text{ g}\cdot\text{mol}^{-1}}{1000 \text{ g}\cdot\text{kg}^{-1}} = 0.03800 \text{ kg}\cdot\text{mol}^{-1}$$

Thus at 298 K we have for the root-mean-square speed of a F_2 molecule

$$v_{\text{rms}} = \left[\frac{(3)(8.314 \text{ J}\cdot\text{mol}^{-1}\cdot\text{K}^{-1})(298 \text{ K})}{0.03800 \text{ kg}\cdot\text{mol}^{-1}}\right]^{1/2} = 442 \text{ m}\cdot\text{s}^{-1}$$

7-39   Application of the equation for the root-mean-square speed

$$v_{\text{rms}} = \left(\frac{3RT}{M_{\text{kg}}}\right)^{1/2}$$

to the temperatures T_f and T_i yields

$$\frac{v_{\text{rms}_f}}{v_{\text{rms}_i}} = \frac{(3RT_f/M_{\text{kg}})^{1/2}}{(3RT_i/M_{\text{kg}})^{1/2}} = \left(\frac{3RT_f/M_{\text{kg}}}{3RT_i/M_{\text{kg}}}\right)^{1/2} = \left(\frac{T_f}{T_i}\right)^{1/2}$$

But $T_f = 2T_i$. Thus,

$$\frac{v_{\text{rms}_f}}{v_{\text{rms}_i}} = \left(\frac{2T_i}{T_i}\right)^{1/2} = 2^{1/2} = \sqrt{2}$$

Solving for v_{rms_f}, we have

$$v_{\text{rms}_f} = \sqrt{2}\, v_{\text{rms}_i}$$

The root-mean-square speed of a molecule is increased by a factor of $\sqrt{2}$ when the temperature T is doubled.

7-41   The root-mean-square speed of a gas molecule decreases as the molar mass increases at the same temperature. Thus we have

$$^{238}\text{UF}_6 < {}^{235}\text{UF}_6 < \text{NO}_2 < \text{CO}_2 < \text{O}_2 < \text{N}_2 < \text{H}_2\text{O}$$

7-43 In order to calculate the mean free path by using the equation

$$l = \frac{3.7 \times 10^8 \text{ pm}^3 \cdot \text{mol} \cdot \text{L}^{-1}}{\sigma^2(n/V)}$$

we need to convert n/V to the unit, $\text{mol} \cdot \text{L}^{-1}$.

$$\frac{n}{V} = \left(\frac{1 \text{ atom}}{1 \text{ m}^3}\right)\left(\frac{1 \text{ mol}}{6.022 \times 10^{23} \text{ atoms}}\right)\left(\frac{1 \text{ m}}{100 \text{ cm}}\right)^3\left(\frac{1000 \text{ cm}^3}{1 \text{ L}}\right)$$
$$= 1.7 \times 10^{-27} \text{ mol} \cdot \text{L}^{-1}$$

Thus, the mean free path is

$$l = \frac{3.7 \times 10^8 \text{ pm}^3 \cdot \text{mol} \cdot \text{L}^{-1}}{(100 \text{ pm})^2 (1.7 \times 10^{-27} \text{ mol} \cdot \text{L}^{-1})} = 2.2 \times 10^{31} \text{ pm}$$
$$= (2.2 \times 10^{31} \text{ pm})\left(\frac{1 \text{ m}}{10^{12} \text{ pm}}\right) = 2.2 \times 10^{19} \text{ m}$$

7-45 The number of collisions per second (collision frequency) is given by

$$z = \frac{v_{\text{rms}}}{l}$$

The root-mean-square speed of H_2 molecules at $20°C$ is

$$v_{\text{rms}} = \left(\frac{3RT}{M_{\text{kg}}}\right)^{1/2} = \left[\frac{3(8.314 \text{ J} \cdot \text{mol}^{-1} \cdot \text{K}^{-1})(293 \text{ K})}{2.016 \times 10^{-3} \text{ kg} \cdot \text{mol}^{-1}}\right]^{1/2}$$
$$= 1904 \text{ m} \cdot \text{s}^{-1}$$

To calculate the mean free path, we must first calculate the density n/V using the ideal-gas equation:

$$\frac{n}{V} = \frac{P}{RT} = \frac{1 \text{ atm}}{(0.0821 \text{ L} \cdot \text{atm} \cdot \text{mol}^{-1} \cdot \text{K}^{-1})(293 \text{ K})} = 0.0416 \text{ mol} \cdot \text{L}^{-1}$$

Thus, using the value $\sigma_{H_2} = 280$ pm from Table 7-4,

$$l = \frac{3.7 \times 10^8 \text{ pm}^3 \cdot \text{mol} \cdot \text{L}^{-1}}{\sigma^2(n/V)}$$
$$= \frac{3.7 \times 10^8 \text{ pm}^3 \cdot \text{mol} \cdot \text{L}^{-1}}{(280 \text{ pm})^2 (0.0416 \text{ mol} \cdot \text{L}^{-1})} = 1.13 \times 10^5 \text{ pm}$$

Thus,

$$z = \frac{1904 \text{ m} \cdot \text{s}^{-1}}{1.13 \times 10^{-7} \text{ m}} = 1.7 \times 10^{10} \text{ collisions} \cdot \text{s}^{-1}$$

7-47   The ratio of the rates of effusion of the two gases is given by Graham's law:

$$\frac{\text{rate}_A}{\text{rate}_B} = \left(\frac{M_B}{M_A}\right)^{1/2}$$

If we take A = helium and B = nitrogen, then we have

$$\text{rate}_{He} = (\text{rate}_{N_2})\left(\frac{M_{N_2}}{M_{He}}\right)^{1/2}$$

Substituting in the values for rate_{N_2}, M_{N_2}, and M_{He}, we have

$$\text{rate}_{He} = (75 \text{ mL}\cdot\text{h}^{-1})\left(\frac{28.02}{4.003}\right)^{1/2} = 200 \text{ mL}\cdot\text{h}^{-1}$$

7-49   Graham's law gives

$$\frac{\text{rate}_A}{\text{rate}_B} = \left(\frac{M_B}{M_A}\right)^{1/2}$$

or

$$\frac{M_A}{M_B} = \left(\frac{\text{rate}_A}{\text{rate}_B}\right)^2$$

If we let B = unknown gas and A = nitrogen, then we have

$$M_{\text{unknown}} = (M_{N_2})\left(\frac{\text{rate}_{N_2}}{\text{rate}_{\text{unknown}}}\right)^2$$

The rate of effusion of N_2 is

$$\text{rate}_{N_2} = \frac{1.00 \text{ mL}}{145 \text{ s}} = 6.90 \times 10^{-3} \text{ mL}\cdot\text{s}^{-1}$$

The rate of effusion of the unknown gas is

$$\text{rate}_{\text{unknown}} = \frac{1.00 \text{ mL}}{230 \text{ s}} = 4.35 \times 10^{-3} \text{ mL}\cdot\text{s}^{-1}$$

Thus,

$$M_{\text{unknown}} = (28.02 \text{ g}\cdot\text{mol}^{-1})\left(\frac{6.90 \times 10^{-3} \text{ mL}\cdot\text{s}^{-1}}{4.35 \times 10^{-3} \text{ mL}\cdot\text{s}^{-1}}\right)^2 = 70.5 \text{ g}\cdot\text{mol}^{-1}$$

7-51   We shall use Equation (7-37) to calculate the pressure

$$P = \frac{nRT}{V - nb} - \frac{n^2a}{V^2}$$

The number of moles of NH_3 is

$$n = (24.5 \text{ g}) \left(\frac{1 \text{ mol NH}_3}{17.03 \text{ g NH}_3} \right) = 1.439 \text{ mol}$$

We obtain the values of a and b for NH_3 from Table 7-4. Thus we have

$$P = \frac{(1.439 \text{ mol})(0.0821 \text{ L} \cdot \text{atm} \cdot \text{mol}^{-1} \cdot \text{K}^{-1})(300 \text{ K})}{2.15 \text{ L} - (1.439 \text{ mol})(0.0371 \text{ L} \cdot \text{mol}^{-1})}$$
$$- \frac{(1.439 \text{ mol})^2(4.170 \text{ L}^2 \cdot \text{atm} \cdot \text{mol}^{-2})}{(2.15 \text{ L})^2}$$
$$= 16.90 \text{ atm} - 1.87 \text{ atm} = 15.0 \text{ atm}$$

The pressure calculated by using the ideal-gas equation is

$$P = \frac{nRT}{V} = \frac{(1.439 \text{ mol})(0.0821 \text{ L} \cdot \text{atm} \cdot \text{mol}^{-1} \cdot \text{K}^{-1})(300 \text{ K})}{2.15 \text{ L}}$$
$$= 16.5 \text{ atm}$$

7-53 From the ideal-gas equation we have for the number of moles of Cl_2,

$$n = \frac{PV}{RT}$$

Note that $1 \text{ Pa} = 1 \text{ N} \cdot \text{m}^{-2}$ and $1 \text{ J} = 1 \text{ N} \cdot \text{m}$. When we use R in the units $J \cdot K^{-1} \cdot \text{mol}^{-1}$, we must express the volume in the units m^3. Thus

$$V = (5.00 \text{ mL}) \left(\frac{1 \text{ cm}^3}{1 \text{ mL}} \right) \left(\frac{1 \text{ m}}{100 \text{ cm}} \right)^3 = 5.00 \times 10^{-6} \text{ m}^3$$
$$n = \frac{(2.15 \times 10^4 \text{ N} \cdot \text{m}^{-2})(5.00 \times 10^{-6} \text{ m}^3)}{(8.314 \text{ N} \cdot \text{m} \cdot \text{mol}^{-1} \cdot \text{K}^{-1})(313 \text{ K})} = 4.13 \times 10^{-5} \text{ mol}$$

The number of molecules is obtained by multiplying n by Avogadro's number:

$$\text{number of molecules} = (4.13 \times 10^{-5} \text{ mol})(6.022 \times 10^{23} \text{ molecules} \cdot \text{mol}^{-1})$$
$$= 2.49 \times 10^{19} \text{ molecules}$$

7-55 Using the ideal-gas law, we have

$$\frac{P_i V_i}{T_i} = \frac{P_f V_f}{T_f}$$

Solving for the final volume, we have

$$V_f = V_i \left(\frac{T_f}{T_i} \right) \left(\frac{P_i}{P_f} \right)$$

Thus,

$$V_f = (7.12 \ \mu L) \left(\frac{273 \ K}{295 \ K} \right) \left(\frac{8.72 \times 10^4 \ Pa}{1.013 \times 10^5 \ Pa} \right)$$

$$= 5.67 \ \mu L$$

The number of moles of radon is

$$n = \frac{PV}{RT}$$

We must first convert the volume to cubic meters, m^3:

$$V = (7.12 \ \mu L) \left(\frac{1 \ mL}{10^3 \ \mu L} \right) \left(\frac{1 \ cm^3}{1 \ mL} \right) \left(\frac{1 \ m}{100 \ cm} \right)^3 = 7.12 \times 10^{-9} \ m^3$$

Thus

$$n = \frac{(8.72 \times 10^4 \ N \cdot m^{-2})(7.12 \times 10^{-9} \ m^3)}{(8.314 \ N \cdot m \cdot K^{-1} \cdot mol^{-1})(295 \ K)} = 2.53 \times 10^{-7} \ mol$$

The mass of radon is

$$m = (2.53 \times 10^{-7} \ mol) \left(\frac{222 \ g \ Rn}{1 \ mol \ Rn} \right) = 5.62 \times 10^{-5} \ g$$

$$= 56.2 \ \mu g$$

Note that either set of conditions can be used in the ideal-gas equation to calculate n.

7-57 From the ideal-gas equation we have

$$\rho = \frac{MP}{RT}$$

$$= \frac{(20.06 \ g \cdot mol^{-1})(2.00 \times 10^3 \ N \cdot m^{-2})}{(8.314 \ N \cdot m \cdot K^{-1} \cdot mol^{-1})(273 \ K)}$$

$$= 17.7 \ g \cdot m^{-3}$$

7-59 We must calculate the force exerted by a column of mercury 760.0 mm high and 1 m^2 in cross-sectional area. The volume of the column of mercury in cm^3 is

$$V = (760.0 \ mm) \left(\frac{1 \ cm}{10 \ mm} \right) (1 \ m^2) \left(\frac{10^2 \ cm}{1 \ m} \right)^2$$

$$= 7.600 \times 10^5 \ cm^3$$

The mass of the mercury column is

$$\text{mass of Hg} = (13.59 \text{ g} \cdot \text{cm}^{-3})(7.600 \times 10^5 \text{ cm}^3)$$
$$= 1.033 \times 10^7 \text{ g}$$

The force exerted by the column of mercury is

$$F = mg$$
$$= (1.033 \times 10^7 \text{ g}) \left(\frac{1 \text{ kg}}{10^3 \text{ g}} \right) (9.806 \text{ m} \cdot \text{s}^{-2})$$
$$= 1.013 \times 10^5 \text{ kg} \cdot \text{m} \cdot \text{s}^{-2}$$
$$= 1.013 \times 10^5 \text{ N}$$

and so the pressure is $1.013 \times 10^5 \text{ N} \cdot \text{m}^{-2}$.

7-61

V/L	$\frac{1}{P}/\text{atm}^{-1}$	$PV/L \cdot \text{atm}$
0.938	3.8	0.24
0.595	2.4	0.24
0.294	1.2	0.24
0.203	0.83	0.244
0.116	0.48	0.244
0.093	0.38	0.244
0.078	0.32	0.245

Boyle's law in an equation states that

$$V = \frac{c}{P}$$

Thus

$$PV = \text{constant}$$

as the data show. A plot of V versus $1/P$ should be a straight line of the form $y = ax$, where $y = V$ and $x = 1/P$.

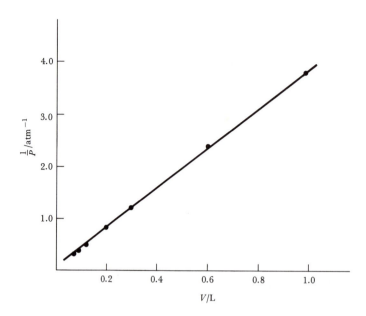

7-63 We use the ideal-gas equation to calculate the volume.

$$V = \frac{nRT}{P} = \frac{(1.00 \text{ mol})(0.0821 \text{ L·atm·mol}^{-1}\text{·K}^{-1})(1073 \text{ K})}{75 \text{ atm}}$$
$$= 1.2 \text{ L}$$

7-65 The balanced equation for the combustion of octane is

$$2C_8H_{18}(g) + 25O_2(g) \longrightarrow 16CO_2(g) + 18H_2O(l)$$

The mass of 1 gal of gasoline is

$$\text{mass} = (1.00 \text{ gal})\left(\frac{4 \text{ qt}}{1 \text{ gal}}\right)\left(\frac{0.946 \text{ L}}{1 \text{ qt}}\right)\left(\frac{10^3 \text{ mL}}{1 \text{ L}}\right)\left(\frac{0.70 \text{ g}}{1 \text{ mL}}\right) = 2.65 \times 10^3 \text{ g}$$

The number of moles of oxygen required to react with 2.65×10^3 g of C_8H_{18} is

$$\text{moles of } O_2 = (2.65 \times 10^3 \text{ g } C_3H_8)\left(\frac{1 \text{ mol } C_8H_{18}}{114.22 \text{ g } C_8H_{18}}\right)\left(\frac{25 \text{ mol } O_2}{2 \text{ mol } C_8H_{18}}\right)$$
$$= 2.90 \times 10^2 \text{ mol}$$

The volume of O_2 required at 0°C and 1 atm is

$$V = \frac{nRT}{P} = \frac{(2.90 \times 10^2 \text{ mol})(0.0821 \text{ L·atm·mol}^{-1}\text{·K}^{-1})(273 \text{ K})}{1.00 \text{ atm}}$$
$$= 6.50 \times 10^3 \text{ L}$$

The volume of air required is

$$V \text{ (of air)} = \frac{6.50 \times 10^3 \text{ L}}{0.20} = 3.25 \times 10^4 \text{ L}$$
$$= 8600 \text{ gal}$$

7-67   The equation for the reaction that takes place when the mixture is heated is

$$2KClO_3(s) \longrightarrow 2KCl(s) + 3O_2(g)$$

The pressure of O_2 is

$$P = 756 \text{ torr} - 15.5 \text{ torr} = 740 \text{ torr}$$

The number of moles of O_2 produced is

$$n = \frac{PV}{RT} = \frac{(740 \text{ torr})(1 \text{ atm}/760 \text{ torr})(0.0807 \text{ L})}{(0.0821 \text{ L} \cdot \text{atm} \cdot \text{mol}^{-1} \cdot \text{K}^{-1})(291 \text{ K})}$$
$$= 3.289 \times 10^{-3} \text{ mol}$$

The mass of $KClO_3$ that produced 3.289×10^{-3} mol of O_2 is

$$\text{mass of } KClO_3 = (3.289 \times 10^{-3} \text{ mol } O_2) \left(\frac{2 \text{ mol } KClO_3}{3 \text{ mol } O_2} \right) \left(\frac{122.55 \text{ g } KClO_3}{1 \text{ mol } KClO_3} \right)$$
$$= 0.2687 \text{ g}$$

The mass percentage of $KClO_3$ in the mixture is

$$\text{mass \% of } KClO_3 = \frac{0.2687 \text{ g}}{0.428 \text{ g}} \times 100 = 62.8\%$$

7-69   The number of moles of H_2 produced is

$$n = \frac{PV}{RT} = \frac{(750 \text{ torr})(1 \text{ atm}/760 \text{ torr})(0.150 \text{ L})}{(0.0821 \text{ L} \cdot \text{atm} \cdot \text{mol}^{-1} \cdot \text{K}^{-1})(283 \text{ K})}$$
$$= 6.371 \times 10^{-3} \text{ mol}$$

The number of moles of NaOH required is

$$n = (6.371 \times 10^{-3} \text{ mol } H_2) \left(\frac{2 \text{ mol NaOH}}{3 \text{ mol } H_2} \right)$$
$$= 4.247 \times 10^{-3} \text{ mol}$$

The volume of NaOH(aq) required is

$$V = \frac{n}{M} = \frac{4.247 \times 10^{-3} \text{ mol}}{0.200 \text{ mol} \cdot \text{L}^{-1}}$$
$$= 0.0212 \text{ L} = 21.2 \text{ mL}$$

7-71   The number of moles of NH_3 added is

$$\text{moles of } NH_3 = (5.0 \text{ g}) \left(\frac{1 \text{ mol } NH_3}{17.03 \text{ g } NH_3} \right) = 0.294 \text{ mol}$$

and the number of moles of HCl added is

$$\text{moles of HCl} = (10.0 \text{ g}) \left(\frac{1 \text{ mol HCl}}{36.46 \text{ g HCl}} \right) = 0.2743 \text{ mol}$$

We see that HCl is the limiting reactant. The number of moles of NH_3 remaining is

$$\text{moles of } NH_3 = 0.294 \text{ mol} - 0.2743 \text{ mol} = 0.020 \text{ mol}$$

The pressure of NH_3 is

$$P = \frac{nRT}{V} = \frac{(0.020 \text{ mol})(0.0821 \text{ L} \cdot \text{atm} \cdot \text{mol}^{-1} \cdot \text{K}^{-1})(348 \text{ K})}{1.00 \text{ L}}$$
$$= 0.57 \text{ atm}$$

7-73   The volume of air breathed in per day is

$$V \text{ of air} = \left(\frac{0.5 \text{ L}}{1 \text{ breath}} \right) \left(\frac{14 \text{ breaths}}{1 \text{ min}} \right) \left(\frac{60 \text{ min}}{1 \text{ hr}} \right) \left(\frac{24 \text{ hr}}{1 \text{ day}} \right)$$
$$= 1.0 \times 10^4 \text{ L}$$

The number of moles of O_2 in 1.0×10^4 L of air is [recall that the partial pressure of O_2 in air is $(0.20)(1.00 \text{ atm})$]

$$n = \frac{PV}{RT} = \frac{(0.20 \text{ atm})(1.0 \times 10^4 \text{ L})}{(0.0821 \text{ L} \cdot \text{atm} \cdot \text{mol}^{-1} \cdot \text{K}^{-1})(310 \text{ K})}$$
$$= 79 \text{ mol}$$

where the temperature is taken to be body temperature $(37 \degree C)$. The number of moles of O_2 utilized per day is

$$n = (0.25)(79 \text{ mol}) = 20 \text{ mol}$$

The mass of O_2 utilized per day is

$$\text{mass of } O_2 = (20 \text{ mol}) \left(\frac{32.00 \text{ g } O_2}{1 \text{ mol } O_2} \right) = 640 \text{ g}$$
$$= (640 \text{ g}) \left(\frac{1 \text{ lb}}{453.5 \text{ g}} \right) = 1.4 \text{ lb}$$

The mass of O_2 used per day is $(1.4 \text{ lb}/20 \text{ lb}$ times $100)$ 70% of the mass of solid food eaten per day.

7-75 The relationship between the root-mean-square speed and the temperature is

$$v_{\text{rms}} = \left(\frac{3RT}{M_{\text{kg}}}\right)^{1/2}$$

Squaring both sides of this equation and solving for T yields

$$T = \frac{M_{\text{kg}}v_{\text{rms}}^2}{3R}$$

Thus we have

$$T = \frac{(0.04401 \text{ kg} \cdot \text{mol}^{-1})(1000 \text{ m} \cdot \text{s}^{-1})^2}{(3)(8.314 \text{ J} \cdot \text{K}^{-1} \cdot \text{mol}^{-1})} = 1764 \text{ K}$$

7-77 The equation for the reaction is

$$Zn(s) + 2HCl(aq) \longrightarrow ZnCl_2(aq) + H_2(g)$$

The number of moles of hydrogen produced is

$$\text{moles of H}_2 = n = \frac{PV}{RT} = \frac{(745 \text{ torr})(1 \text{ torr}/760 \text{ atm})(0.250 \text{ L})}{(0.0821 \text{ L} \cdot \text{atm} \cdot \text{mol}^{-1} \cdot \text{K}^{-1})(293 \text{ K})}$$
$$= 0.01019 \text{ mol}$$

$$\text{moles of HCl} = (0.01019 \text{ mol H}_2)\left(\frac{2 \text{ mol HCl}}{1 \text{ mol H}_2}\right) = 0.02038 \text{ mol}$$

The number of milliliters of hydrochloric acid required is

$$V = \frac{n}{M} = \frac{0.02038 \text{ mol}}{0.620 \text{ mol} \cdot \text{L}^{-1}} = 0.0329 \text{ L} = 32.9 \text{ mL}$$

7-79 The number of moles of sodium peroxide present is

$$\text{moles of Na}_2\text{O}_2 = (1000 \text{ g})\left(\frac{1 \text{ mol Na}_2\text{O}_2}{77.98 \text{ g Na}_2\text{O}_2}\right) = 12.82 \text{ mol}$$

$$\text{moles of CO}_2 = (12.82 \text{ mol Na}_2\text{O}_2)\left(\frac{2 \text{ mol CO}_2}{2 \text{ mol Na}_2\text{O}_2}\right) = 12.82 \text{ mol}$$

The number of liters of carbon dioxide absorbed is

$$V = \frac{nRT}{P} = \frac{(12.82 \text{ mol})(0.0821 \text{ L} \cdot \text{atm} \cdot \text{mol}^{-1} \cdot \text{K}^{-1})(273 \text{ K})}{1 \text{ atm}}$$
$$= 287 \text{ L}$$

The number of moles of O_2 produced is

$$\text{moles of } O_2 = (12.82 \text{ mol Na}_2O_2)\left(\frac{1 \text{ mol } O_2}{2 \text{ mol Na}_2O_2}\right) = 6.410 \text{ mol}$$

The number of liters of oxygen produced is

$$V = \frac{nRT}{P} = \frac{(6.410 \text{ mol})(0.0821 \text{ L·atm·mol}^{-1}\text{·K}^{-1})(273 \text{ K})}{1 \text{ atm}}$$

$$= 144 \text{ L}$$

7-81 The number of moles of acetylene in the cannister is

$$\text{moles of } C_2H_2 = \frac{PV}{RT} = \frac{(3.00 \text{ atm})(0.500 \text{ L})}{(0.0821 \text{ L·atm·mol}^{-1}\text{·K}^{-1})(291 \text{ K})}$$

$$= 0.06278 \text{ mol}$$

$$\text{moles of } CaC_2 = (0.06278 \text{ mol } C_2H_2)\left(\frac{1 \text{ mol CaC}_2}{1 \text{ mol C}_2H_2}\right) = 0.06278 \text{ mol}$$

$$\text{mass of } CaC_2 = (0.06278 \text{ mol})\left(\frac{64.10 \text{ g CaC}_2}{1 \text{ mol CaC}_2}\right) = 4.02 \text{ g}$$

7-83 We use Graham's law to find the molecular mass of the unknown gas. The rate of effusion of nitrogen is

$$\text{rate of } N_2 = \frac{1850 \text{ torr} - 915 \text{ torr}}{30.0 \text{ min}} = 31.17 \text{ torr·min}^{-1}$$

The rate of effusion of the unknown gas is

$$\text{rate of unknown} = \frac{1850 \text{ torr} - 915 \text{ torr}}{54.3 \text{ min}} = 17.22 \text{ torr·min}^{-1}$$

From Graham's law, we have that

$$\frac{\text{rate}_{N_2}}{\text{rate}_X} = \left(\frac{M_x}{M_{N_2}}\right)^{1/2}$$

where X stands for the unknown gas. Substituting in the values of rate_{N_2}, rate_X, and M_{N_2}, we have that

$$\frac{31.17 \text{ torr·min}^{-1}}{17.22 \text{ torr·min}^{-1}} = 1.81 = \left(\frac{M_x}{28.02}\right)^{1/2}$$

$$M_x = (1.81)^2(28.02) = 91.8$$

The molecular mass of the unknown gas is 91.8.

7-85 The volume of the room is

$$V = (3.0 \text{ m})(5.0 \text{ m})(6.0 \text{ m}) \left(\frac{100 \text{ cm}}{1 \text{ m}} \right)^3$$
$$= 9.0 \times 10^7 \text{ cm}^3 = 9.0 \times 10^7 \text{ mL} = 9.0 \times 10^4 \text{ L}$$

The number of moles of water vapor in the room is

$$\text{moles of } H_2O = \frac{PV}{RT} = \frac{(35 \text{ torr})(1 \text{ atm}/760 \text{ torr})(9.0 \times 10^4 \text{ L})}{(0.0821 \text{ L} \cdot \text{atm} \cdot \text{mol}^{-1} \cdot \text{K}^{-1})(308 \text{ K})}$$
$$= 164 \text{ mol}$$
$$\text{mass of } H_2O = (164 \text{ mol}) \left(\frac{18.02 \text{ g } H_2O}{1 \text{ mol } H_2O} \right) = 2960 \text{ g}$$

Using the fact that the density of water is $1.00 \text{ g} \cdot \text{mL}^{-1}$, we have that the volume of water obtained would be

$$V = \frac{m}{d} = \frac{2960 \text{ g}}{1.00 \text{ g} \cdot \text{mL}^{-1}} = 3000 \text{ mL} = 3.0 \text{ L}$$

7-87 We must first check to see which reactant is the limiting reactant.

$$\text{moles of HCl} = MV = (0.1150 \text{ mol} \cdot \text{L}^{-1})(0.300 \text{ L}) = 0.03450 \text{ mol}$$
$$\text{moles of } KMnO_4 = (6.75 \text{ g}) \left(\frac{1 \text{ mol } KMnO_4}{158.04 \text{ g } KMnO_4} \right) = 0.04271 \text{ mol}$$

If we divide the number of moles of each reactant by its respective stoichiometric coefficient, then we see that (moles of HCl/16) = 0.0022 mol is less than (moles of $KMnO_4/2$) = 0.02136 mol. Therefore, the limiting reactant is HCl.

$$\text{moles of } Cl_2 = (0.03450 \text{ mol HCl}) \left(\frac{5 \text{ mol } Cl_2}{16 \text{ mol HCl}} \right) = 0.01078 \text{ mol}$$

The volume of chlorine is

$$V = \frac{nRT}{P} = \frac{(0.01078 \text{ mol})(0.0821 \text{ L} \cdot \text{atm} \cdot \text{mol}^{-1} \cdot \text{K}^{-1})(288 \text{ K})}{(815 \text{ torr})(1 \text{ atm}/760 \text{ torr})}$$
$$= 0.238 \text{ L} = 238 \text{ mL}$$

7-89 After the valve has been opened, the volume of each of the gases is

$$\text{final volume} = 650 \text{ mL} + 500 \text{ mL} = 1150 \text{ mL}$$

We can use Boyle's law in the form $P_f V_f = P_i V_i$ to find the partial pressure of each gas (P_f).

$$\text{partial pressure of } N_2 = P_f = \frac{P_i V_i}{V_f} = \frac{(825 \text{ torr})(650 \text{ mL})}{1150 \text{ mL}}$$
$$= 466 \text{ torr}$$

$$\text{partial pressure of } O_2 = P_f = \frac{P_i V_i}{V_f} = \frac{(730 \text{ torr})(500 \text{ mL})}{1150 \text{ mL}}$$
$$= 317 \text{ torr}$$

$$\text{total pressure} = 466 \text{ torr} + 317 \text{ torr} = 783 \text{ torr}$$

7-91 The equation for the reaction is

$$Zn(s) + 2HCl(aq) \longrightarrow ZnCl_2(aq) + H_2(g)$$

$$\text{moles of } H_2 = \frac{PV}{RT} = \frac{(752 \text{ torr})(1 \text{ atm}/760 \text{ torr})(1.65 \text{ L})}{(0.0821 \text{ L} \cdot \text{atm} \cdot \text{mol}^{-1} \cdot \text{K}^{-1})(293 \text{ K})}$$
$$= 0.06787 \text{ mol}$$

$$\text{moles of HCl} = (0.06787 \text{ mol } H_2)\left(\frac{2 \text{ mol HCl}}{1 \text{ mol } H_2}\right) = 0.1357 \text{ mol}$$

The molarity of hydrochloric acid is

$$M = \frac{n}{V} = \frac{0.1357 \text{ mol}}{0.300 \text{ L}} = 0.452 \text{ M}$$

7-93 Let

$$x \text{ g} = \text{mass of NaH}$$
$$3.75 \text{ g} - x \text{ g} = \text{mass of } CaH_2$$

$$\text{moles of NaH} = (x \text{ g})\left(\frac{1 \text{ mol NaH}}{24.00 \text{ g NaH}}\right) = 0.04167x \text{ mol}$$

$$\text{moles of } CaH_2 = (3.75 \text{ g} - x \text{ g})\left(\frac{1 \text{ mol } CaH_2}{42.10 \text{ g } CaH_2}\right)$$
$$= 0.08907 \text{ mol} - 0.02375x \text{ mol}$$

$$\text{moles of } H_2 = \frac{PV}{RT} = \frac{(742 \text{ torr})(1 \text{ atm}/760 \text{ torr})(4.12 \text{ L})}{(0.0821 \text{ L} \cdot \text{atm} \cdot \text{mol}^{-1} \cdot \text{K}^{-1})(290 \text{ K})}$$
$$= 0.1689 \text{ mol}$$

The equations for the reactions are

$$NaH(s) + H_2O(l) \longrightarrow NaOH(aq) + H_2(g)$$
$$CaH_2(s) + 2H_2O(l) \longrightarrow Ca(OH)_2(aq) + 2H_2(g)$$

Thus, we have that 1 mol of $H_2(g)$ is produced for each mole of NaH(s) that reacts and 2 mol for each mole of CaH_2 that reacts. Thus,

$$\text{moles of } H_2 = \text{moles of NaH} + 2(\text{moles of } CaH_2)$$

$$0.1689 \text{ mol} = 0.04167x \text{ mol} + 2 \,(0.08907 \text{ mol} - 0.02375x \text{ mol})$$

$$0.1689 = 0.04167x + 0.1781 - 0.04750x$$

$$5.83 \times 10^{-3} \, x = 9.2 \times 10^{-3}$$

$$x = 1.6$$

The mass of NaH is 1.6 g. The mass percentage of each in the mixture is

$$\text{mass \% NaH} = \frac{1.6 \text{ g}}{3.75 \text{ g}} \times 100 = 43\%$$

$$\text{mass \% CaH}_2 = 100\% - 42\% = 57\%$$

7-95   From the expressions for v_{rms} and $\bar{v}$, we have

$$\frac{v_{rms}}{\bar{v}} = \frac{(3RT/M_{kg})^{1/2}}{(8RT/\pi M_{kg})^{1/2}} = \left(\frac{3RT/M_{kg}}{8RT/\pi M_{kg}} \right)^{1/2}$$

$$= \left(\frac{(3RT)\,(\pi M_{kg})}{(M_{kg})\,(8RT)} \right)^{1/2} = \left(\frac{3\pi}{8} \right)^{1/2} = 1.085$$

E. ANSWERS TO THE SELF-TEST

1.   false

2.   true

3.   false (The density usually does not change much.)

4.   volume

5.   empty space

6.   the gas molecules occupy only a small fraction of the available space

7.   gas pressure

8.   the difference in pressures on the two mercury surfaces

9.   torr

10.   false

11.   the pressure due to the atmosphere

12.   false (It depends on weather conditions and altitude.)

13.   false

14.   760 torr or 1.013×10^5 Pa

15.   the pascal

16.   inversely . . . pressure . . . temperature

17.   greater

18.   Kelvin temperature . . . kelvin, K

19. $T \, (\text{K}) = t(°\text{C}) + 273.15$

20.   0 K . . . -273.15 °C

21.   inversely . . . absolute temperature . . . pressure

22.   false

23. temperature . . . volume . . . constant pressure

24. 2

25. $PV = nRT$

26. obeys the ideal-gas equation $PV = nRT$

27. true

28. 0.0821 L·atm·mol^{-1}·K^{-1} or 8.314 J·mol^{-1}·K^{-1}

29. atmospheres . . . liters . . . moles . . . kelvin

30. 22.4 L

31. decreases

32. increases

33. decreases

34. increases

35. increases

36. true

37. false

38. $n_{O_2}/(n_{O_2} + n_{CO_2})$

39. $X_{O_2}P_{total}$

40. minus

41. collisions of the gas molecules with the walls of the container

42. $E_k = \frac{1}{2}mv^2$ . . . m is the mass of the body and v is its speed

43. $E_k = \frac{1}{2}M_{kg}\overline{v^2}$ . . . M_{kg} is the molar mass of the gas in units of kilograms and v is the speed of the molecules

44. false

45. higher

46. speeds

47. false

48. See Section 7-10 of the text.

49. $\overline{E_k}$

50. $\overline{E_k} = \frac{1}{2}M_{kg}\overline{v^2}$ . . . M_{kg} is the molar mass of the gas in units of kilograms and $\overline{v^2}$ is the average of the square of the speeds of the gas molecules

51. temperature

52. true

53. increases

54. increases

55. increases

56. $\frac{1}{3}Nm\overline{v^2}$ . . . N is the number of gas molecules, m is the mass of one gas molecule, and $\overline{v^2}$ is the average of the square of the speeds of the molecules

57. $\overline{E_k} = \frac{3}{2}RT$ . . . R is the gas constant and T is the absolute temperature

58. false

59. temperature

60. false

61. true

62. true

63. traveled by a molecule between collisions

64. number of molecules per unit volume and the cross section of the molecule

65. mean free path and the root-mean-square speed

66. false

67. nonideal

68. volume

69. attraction between

8 THERMOCHEMISTRY

A. OUTLINE OF CHAPTER 8

8-1   Energy is conserved.

The law of conservation of energy states that, during any process, energy is neither created nor destroyed but is converted from one form to another.

A body in motion has kinetic energy given by

$$E_k = \tfrac{1}{2}\, mv^2 \tag{8-1}$$

where m is the mass of the body and v is its speed.

A body can have potential energy due to its position, which is given by

$$E_p = gmh \tag{8-2}$$

where g is a proportionality constant whose value is 9.81 m·s^{-2}, m is its mass, and h is its height above the ground.

The sum of the kinetic energy and the potential energy of a body is a constant.

Power is the rate at which energy is produced or utilized.

The SI unit of power is a watt, which is defined as one joule per second.

A kilowatt-hour is the energy used by a one-kilowatt device operating for one hour.

8-2   The transfer of energy between a reaction system and its surroundings occurs as heat or work.

Chemical reactions almost always involve a change in energy.

Thermodynamics is the study of the transfer of energy between a reaction system and its surroundings.

The first law of thermodynamics states that energy is neither created nor destroyed but is simply converted from one form to another.

Energy can be transferred as work or as heat.

Work is the magnitude of a force times the displacement caused by the force.

The amount of energy transferred as work is denoted by w.

The work done on a system is given by

$$w = -F \, \Delta X \tag{8-4}$$

where F is the force and ΔX is the displacement.

The work done on a system that is compressed or expanded at a constant pressure P is given by

$$w = -P \, \Delta V = -P(V_f - V_i) \tag{8-6}$$

where V_f is the final volume and V_i is the initial volume.

The work done on a system due to compression is a positive quantity.

The conversion factor between joules and liter-atmospheres is $1 \, \text{L} \cdot \text{atm} = 101.3 \, \text{J}$.

Energy as heat flows spontaneously from higher to lower temperature regions.

The energy transferred as heat is denoted by q.

The energy change of a system is denoted by ΔU.

The energy change of a system is given by

$$\Delta U = q + w \tag{8-7}$$

where q is the energy transferred as heat to the system and w is the energy transferred to the system as work.

The energy change ΔU depends upon only the initial and final states and is independent of the path taken from the initial state to the final state.

A state function in thermodynamics is a function that depends only upon the state of the system and not upon how that state was achieved.

Energy is a state function.

Work and heat are not state functions.

The energy change of a chemical reaction is denoted by ΔU_{rxn}.

When a reaction occurs at constant pressure, the energy change is equal to

$$\Delta U_{rxn} = q - P \, \Delta V \tag{8-9}$$

where q is the heat involved, P is the pressure, and ΔV is the change in volume of the system.

When a reaction takes place at constant volume, the energy change is equal to the heat evolved or absorbed.

$$\Delta U_{rxn} = q_V \qquad \text{at constant volume} \tag{8-10}$$

The enthalpy H is defined as $H = U + PV$ (Equation 8-11).

When a reaction takes place at constant pressure, the enthalpy change is equal to the heat evolved or absorbed.

$$\Delta H_{rxn} = q_P \qquad \text{at constant pressure} \qquad (8\text{-}14)$$

ΔH_{rxn} is also called the heat of reaction.

8-3 Chemical reactions evolve or absorb energy as heat.

An exothermic reaction is a reaction that evolves energy as heat.

An endothermic reaction is a reaction that absorbs energy as heat.

The heat of combustion is the heat evolved in a combustion reaction.

The enthalpy change for a chemical reaction is

$$\Delta H_{rxn} = H_{prod} - H_{react} \qquad (8\text{-}15)$$

For an exothermic reaction, $\Delta H_{rxn} < 0$.

For an endothermic reaction, $\Delta H_{rxn} > 0$ (Figure 8-6).

The standard enthalpy change for a reaction, for which the reactants and products are at 1 atm pressure, is denoted by ΔH_{rxn}°.

The enthalpy change for a chemical reaction is approximately equal to the energy change.

8-4 Enthalpy changes for chemical equations are additive.

Hess's law states that if two or more chemical equations are added together, then the value of ΔH_{rxn}° for the resulting equation is equal to the sum of the ΔH_{rxn}° values for the separate equations.

$$\Delta H_{rxn}^{\circ}(\text{reverse}) = -\Delta H_{rxn}^{\circ}(\text{forward}) \qquad (8\text{-}20)$$

The rules for Hess's law calculations are summarized on page 259 of the text.

8-5 Heats of reaction can be calculated from tabulated heats of formation.

The standard molar enthalpy of formation ΔH_f° is the value of ΔH_{rxn}° of the reaction in which one mole of a substance is formed from its constituent elements in their most stable form at 25°C and 1 atm.

The standard molar enthalpy of formation is also known as the heat of formation.

The value of ΔH_f° is zero for an element in its normal state at 25°C and 1 atm.

For a chemical equation, the value of ΔH_{rxn}° is given by

$$\Delta H_{rxn}^{\circ} = \Delta H_f^{\circ}(\text{all products}) - \Delta H_f^{\circ}(\text{all reactants}) \qquad (8\text{-}21)$$

Values of ΔH_f° of some substances are given in Table 8-2.

8-6 Heat capacity measures the ability of a substance to take up energy as heat.

The heat capacity of a substance is the heat required to raise the temperature of the substance by one degree Celsius, or one kelvin.

At constant pressure, the heat capacity c_P is given by

$$c_P = \frac{q_P}{\Delta T} \tag{8-22}$$

where q_P is the heat input and ΔT is the change in temperature.

The molar heat capacity C_P is the heat capacity per mole of a substance at constant pressure.

The molar heat capacities at constant pressure of some substances are given in Table 8-3.

The specific heat C_{sp} is the heat capacity per gram of a substance.

If we combine two nonreactive samples at different temperatures, then the higher temperature sample will transfer energy to the lower temperature sample until the combined system attains a uniform temperature.

If we combine two nonreactive samples at different temperatures, then

$$c_{P,h}(T_f - T_h) + c_{P,l}(T_f - T_l) = 0 \tag{8-27}$$

where $c_{P,h}$ is the heat capacity of the higher temperature sample, $c_{P,l}$ is the heat capacity of the lower temperature sample, T_h is the temperature of the higher temperature sample, T_l is the temperature of the lower temperature sample, and T_f is the final temperature.

8-7   The magnitudes of heat capacities have a molecular interpretation.

Monatomic gases can increase their energy only by traveling at a faster speed.

Polyatomic molecules can take up energy as translational energy, rotational energy, and vibrational energy.

Translational motion is motion in a straight line.

The larger a molecule is, the more ways it can vibrate and rotate, and so its heat capacity is greater.

The rule of Dulong and Petit is the experimental observation that the values of C_P for solid metals are approximately equal to $3R$.

Water has an unusually large heat capacity due to strong intermolecular interactions.

8-8   A calorimeter is a device used to measure the amount of heat evolved or absorbed in a reaction.

For a reaction that takes place in a calorimeter (Figure 8-11)

$$\Delta H_{rxn} = -c_{P,calorimeter} \, \Delta T \tag{8-30}$$

where ΔT is the measured temperature change and $c_{p,calorimeter}$ is the heat capacity of the calorimeter.

The heat of combustion of a substance can be measured in a bomb calorimeter (Figure 8-12).

In a bomb calorimeter, $\Delta U_{rxn} = q_V$.

In most cases, $\Delta U_{rxn} \approx \Delta H_{rxn}$.

8-9 Food is fuel.

The popular term calorie is actually a kilocalorie.

One calorie is equal to 4.184 joules.

The approximate energy values of some common foods are given in Table 8-6.

8-10 The United States utilizes about 80 quadrillion kilojoules of energy per year.

Ninety-three percent of the energy utilized in the United States and 70 percent of that in Canada is generated from the combustion of oil, coal, and natural gas.

The total U.S. energy supply is given in Table 8-7.

8-11 Energy is obtained from fossil fuels by combustion reactions.

Fossil fuels are essentially mixtures of hydrocarbons.

The heats of combustion for a variety of fuels are given in Table 8-8.

Gasohol is a mixture of gasoline and ethyl alcohol.

8-12 Rockets are powered by highly exothermic reactions with gaseous products.

8-13 Sunlight is a major energy source.

An energy storage device is necessary so that the solar energy can be used when the sun is not shining.

Some storage devices are described.

B. SELF-TEST

1. The law of conservation of energy states that _____
_____.

2. A body in motion has kinetic energy. *True/False*

3. A body can also have kinetic energy due to its position. *True/False*

4. The potential energy of a body can be converted to kinetic energy. *True/False*

5. The potential energy of a body is always equal to its kinetic energy. *True/False*

6. The sum of the potential energy and kinetic energy of a body is a constant. *True/False*

7. Power is defined as _____.

8. The SI unit of power is the _____.

9. A kilowatt-hour is the energy used by a kilowatt device operating for 1 hour. *True/False*

10. The first law of thermodynamics states that _____
_____.

11. Chemical reactions involve energy transfers in the form of _____
and _____.

12. The transfer of energy as work involves the action of a _____
_____.

13. The work done on a system is a (*positive/negative*) quantity.

14. The work done on a system that expands is a (*positive/negative*) quantity.

15. The transfer of energy as heat requires a _____
_____.

16. A state function depends only upon the initial and final states of the system. *True/ False*

17. The value of ΔU_{rxn} depends on the path taken from the initial state to the final state. *True/False*

18. For a chemical reaction, the energy change ΔU_{rxn} is equal to the sum of _____
_____.

19. For a chemical reaction that takes place at constant volume, the energy change is equal to _____.

20. The thermodynamic function, enthalpy H, is defined as _____.

21. For a chemical reaction that takes place at constant pressure, the enthalpy change ΔH_{rxn} is equal to _____.

22. When a chemical reaction is run in a reaction vessel open to the atmosphere, the reaction takes place at constant _____.

23. The enthalpy change for a chemical reaction ΔH_{rxn} is also called the heat of reaction. *True/False*

24. Reactions that give off energy as heat are called _____.

25. Reactions that take up energy as heat are called _____.

26. If the value of ΔH_{rxn} is negative, then the reaction is _____ thermic.

27. If the value of ΔH_{rxn} is positive, then heat is absorbed by a reaction. *True/False*

28. The standard enthalpy change ΔH_{rxn}° is the enthalpy change of a reaction that takes place at constant pressure. *True/False*

29. The standard state of a substance is its state at one atmosphere pressure. *True/False*

30. Enthalpy changes for chemical equations are additive. *True/False*

31. The enthalpy change for the reverse equation for a chemical equation is related to the original equation by the relationship _____ .

32. State Hess's law in your own words.

_____

_____

33. The standard molar enthalpy of formation ΔH_f° of a substance is defined as _____

_____

_____

_____ .

34. The heat of formation of an element in its standard state is zero. *True/False*

35. Hess's law can be used to calculate the standard enthalpy change for a reaction from the heats of formation of the reactants and products. *True/False*

36. For the equation

$$CH_4(g) + 2O_2(g) \longrightarrow CO_2(g) + 2H_2O(g)$$

the values of ΔH_f° of the reactants and products are known. The value of ΔH_{rxn}° can be found by the relation $\Delta H_{rxn}^\circ = $ _____

_____ .

37. The amount of heat required to raise the temperature of one mole of a substance by one degree Celsius is the _____ of the substance.

38. The heat capacity per gram of water is $4.18 \ J \cdot K^{-1} \cdot g^{-1}$ and the heat capacity per gram of sodium is $1.34 \ J \cdot K^{-1} \cdot g^{-1}$. It will require (*more/less*) heat to raise the temperature of one gram of water by one degree than one gram of sodium.

39. Suggest a reason why the heat capacity of a substance is given as the molar heat capacity instead of simply as the heat capacity.

_____

_____ .

40. The specific heat of a substance is the heat capacity per _____ of the substance.

41. If we combine two nonreactive samples at different temperatures, then the (*higher/ lower*) temperature sample will transfer energy to the (*higher/lower*) temperature sample.

42. If we combine two nonreactive samples at different temperatures, then the combined system will come to a uniform temperature. *True/False*

43. Monatomic gases can take up energy by _____ motion.

44. The heat capacity of a polyatomic molecule is (*greater/the same/less*) than that of a monatomic gas of similar molecular mass.

45. The larger a molecule, the (*greater/less*) is its heat capacity.

46. The more ways that a molecule can rotate and vibrate, the greater is its heat capacity. *True/False*

47. The rule of Dulong and Petit is the experimental observation that _____

_____.

48. A calorimeter is a device used to measure _____

_____.

49. Only heats of combustion can be measured in a calorimeter. *True/False*

50. Why is the heat absorbed or given off by a calorimeter equal to the heat of the reaction taking place in the calorimeter?

_____

_____

51. In a calorimeter, the change in _____ is the physical change measured.

52. The value of ΔH_{rxn} is determined from the _____ of the calorimeter and the change in _____.

53. Combustion is an example of an _____ thermic reaction.

54. An apparatus in which ΔU_{rxn} can be measured is called a _____ calorimeter.

55. The enthalpy change for a reaction often is almost equal to the energy change for the reaction. *True/False*

56. Why is the heat of combustion of a substance important in deciding whether a substance can be used as a fuel? _____

_____

57. The common unit for the energy content of food is a _____.

58. A calorie is defined as the heat required _____

_____.

59. The main source of the energy utilized in the United States is the combustion of fossil fuels. *True/False*

60. Fossil fuels are mixtures of _____.

61. A fuel has a high heat of combustion. *True/False*

62. Gasohol is a mixture of _____ and _____.

63. A rocket fuel must be highly endothermic. *True/False*

64. Solar energy can be stored for use as energy. *True/False*

65. Glauber's salt can be used to store solar energy. *True/False*

C. CALCULATIONS YOU SHOULD KNOW HOW TO DO

1. Use kinetic energy and potential energy relationships. See Example 8-1 and Problems 8-1 through 8-6.

2. Calculate the work required to compress a gas by exerting a constant pressure using Equation (8-6). See Example 8-3 and Problems 8-11 through 8-14.

3. Use Hess's law to calculate ΔH°_{rxn} for an equation by adding or subtracting ΔH°_{rxn}'s for two or more equations. See Examples 8-5 and 8-6 and Problems 8-15 through 8-24.

4. Calculate ΔH°_{rxn} by using values of ΔH°_f given in Table 8-2 and Equation (8-21), $\Delta H^\circ_{rxn} = \Delta H^\circ_f$ (all products) $- \Delta H^\circ_f$ (all reactants). See Examples 8-8 and 8-9 and Problems 8-25 through 8-36.

5. Calculate the heat capacity of a substance from the increases in temperature when a known amount of heat is added to the substance. See Examples 8-10 and 8-12 and Problems 8-37 and 8-38.

6. Use heat capacity to calculate the heat absorbed or evolved by a temperature change. See Example 8-11 and Problems 8-39 and 8-40.

7. Calculate the final temperature when two substances at different temperatures are brought into contact. See Example 8-13 and Problems 8-41 through 8-43.

8. Calculate ΔH_{rxn}° by using data that are obtained from running the reaction in a calorimeter. See Examples 8-14 and 8-15 and Problems 8-45 through 8-54.

D. SOLUTIONS TO THE ODD-NUMBERED PROBLEMS

8-1 All the initial energy is kinetic energy. Therefore,

$$E(\text{initial}) = E(\text{total}) = \tfrac{1}{2} mv^2$$
$$= \tfrac{1}{2}(0.500 \text{ kg})(50 \text{ m}\cdot\text{s}^{-1})^2 = 625 \text{ J}$$

At the apex, all the energy is potential energy. Therefore,

$$E(\text{total}) = E(\text{apex}) = mgh$$

Solving for h, the maximum height, we have that

$$h = \frac{E(\text{total})}{mg} = \frac{625 \text{ J}}{(0.500 \text{ kg})(9.81 \text{ m}\cdot\text{s}^{-2})} = 130 \text{ m}$$

8-3 All of the initial energy is kinetic energy. Therefore,

$$E(\text{initial}) = E(\text{total}) = \tfrac{1}{2} mv^2$$

At the apex, all the energy is potential energy. Therefore,

$$E(\text{total}) = E(\text{apex}) = mgh$$

Setting the two equations for the total energy equal, we have that

$$\tfrac{1}{2} mv^2 = mgh$$

Solving for h, the maximum height, we have that

$$h = \frac{\tfrac{1}{2} v^2}{g} = \frac{\tfrac{1}{2}(9.2 \text{ m}\cdot\text{s}^{-1})^2}{9.81 \text{ m}\cdot\text{s}^{-2}} = 4.3 \text{ m}$$

8-5 The power of the sunlight incident on the collector is

$$\text{power} = (0.65 \text{ kW}\cdot\text{m}^{-2})(\text{area})$$

The number of kilowatt-hours is given by

$$\text{kW}\cdot\text{h} = (\text{power})(10 \text{ h}) = (0.65 \text{ kW}\cdot\text{m}^{-2})(\text{area})(10 \text{ h})$$
$$= (6.5 \text{ kW}\cdot\text{h}\cdot\text{m}^{-2})(\text{area})$$

The available energy is given by

$$\text{available energy} = (0.20)(6.5 \text{ kW} \cdot \text{h} \cdot \text{m}^{-2})(\text{area})$$
$$= (1.3 \text{ kW} \cdot \text{h} \cdot \text{m}^{-2})(\text{area})$$

The energy needed is $100 \text{ kW} \cdot \text{h}$; thus

$$100 \text{ kW} \cdot \text{h} = (1.3 \text{ kW} \cdot \text{h} \cdot \text{m}^{-2})(\text{area})$$

Solving for the area, we have that

$$\text{area} = \frac{100 \text{ kW} \cdot \text{h}}{1.3 \text{ kW} \cdot \text{h} \cdot \text{m}^{-2}} = 77 \text{ m}^2$$

8-7   We are given the amount of heat evolved (1503 kJ) when 30.0 g of methane is burned. The amount of heat that is evolved when 1 mol of methane is burned is

$$q = \left(\frac{1503 \text{ kJ}}{30.0 \text{ g}}\right)\left(\frac{16.04 \text{ g CH}_4}{1 \text{ mol CH}_4}\right) = 804 \text{ kJ} \cdot \text{mol}^{-1}$$

8-9   The reaction is

$$C(s) + 2S(s) \longrightarrow CS_2(l)$$

One mole of CS_2 is formed when 1 mol of carbon reacts. Therefore,

$$\text{heat evolved per mole of } CS_2 = \left(\frac{9.52 \text{ kJ}}{1.280 \text{ g}}\right)\left(\frac{12.01 \text{ g C}}{1 \text{ mol C}}\right) = 89.3 \text{ kJ} \cdot \text{mol}^{-1}$$

8-11   The work done by the system in expanding at constant pressure is given by

$$w = -P \, \Delta V = -(3.5 \text{ atm})(25 \text{ L} - 10 \text{ L}) = -52.5 \text{ L} \cdot \text{atm}$$

Work is generally expressed in joules; therefore, we must convert from $L \cdot atm$ to J.

$$w = -(52.5 \text{ L} \cdot \text{atm})\left(\frac{101.3 \text{ J}}{1 \text{ L} \cdot \text{atm}}\right) = -5300 \text{ J} = -5.3 \text{ kJ}$$

8-13   We use Equation (8-13).

$$\Delta H^\circ_{\text{rxn}} = \Delta U^\circ_{\text{rxn}} + P \, \Delta V$$

Solving for $\Delta U^\circ_{\text{rxn}}$, we have that

$$\Delta U^\circ_{\text{rxn}} = \Delta H^\circ_{\text{rxn}} - P \, \Delta V$$

For this system,

$$P \, \Delta V = \Delta n R T$$

with $\Delta n = -1$ mol. Thus

$$\Delta U_{rxn}^{\circ} = \Delta H_{rxn}^{\circ} - \Delta nRT$$

$$= -206 \text{ kJ} - (-1 \text{ mol})(8.314 \text{ J} \cdot \text{mol}^{-1} \cdot \text{K}^{-1}) \left(\frac{1 \text{ kJ}}{10^3 \text{ J}} \right) (298 \text{ K})$$

$$= -206 \text{ kJ} + 2.48 \text{ kJ} = -204 \text{ kJ}$$

8-15 We can obtain the second equation from the first by reversing the first equation and dividing the balancing coefficients by 2. Thus

$$\Delta H_{rxn}^{\circ}(2) = \frac{[-\Delta H_{rxn}^{\circ}(1)]}{2} = \frac{290.8 \text{ kJ}}{2} = 145.4 \text{ kJ}$$

8-17 To obtain the third equation, reverse the second equation and add it to the first equation:

$$C_2H_5OH(l) + 3O_2(g) \longrightarrow 2CO_2(g) + 3H_2O(g) \qquad \Delta H_{rxn}^{\circ} = -1234.7 \text{ kJ}$$

$$\underline{2CO_2(g) + 3H_2O(g) \longrightarrow CH_3OCH_3(l) + 3O_2(g) \qquad \Delta H_{rxn}^{\circ} = -(-1328.3 \text{ kJ})}$$

$$C_2H_5OH(l) \longrightarrow CH_3OCH_3(l) \qquad\qquad\qquad \Delta H_{rxn}^{\circ} = 1328.3 \text{ kJ} - 1234.7 \text{ kJ}$$
$$= 93.6 \text{ kJ}$$

8-19 The equation that we want can be obtained from the two given equations by reversing the first equation, multiplying it by 3, and then adding it to 2 times the second equation:

$$3Fe_2O_3(s) \longrightarrow 6Fe(s) + \tfrac{9}{2}O_2(g) \qquad\qquad \Delta H_{rxn}^{\circ} = -(3)(-823.41 \text{ kJ})$$
$$= 2470.23 \text{ kJ}$$

$$6Fe(s) + 4O_2(g) \longrightarrow 2Fe_3O_4(s) \qquad\qquad \Delta H_{rxn}^{\circ} = (2)(-1120.48 \text{ kJ})$$
$$= -2240.96 \text{ kJ}$$

$$\overline{3Fe_2O_3(s) \longrightarrow 2Fe_3O_4(s) + \tfrac{1}{2}O_2(g) \qquad \Delta H_{rxn}^{\circ} = 2470.23 \text{ kJ} - 2240.96 \text{ kJ}}$$
$$= +229.27 \text{ kJ}$$

8-21 The equations that correspond to the combustion reactions are

(1) $C_{12}H_{22}O_{11}(s) + 12O_2(g) \longrightarrow 12CO_2(g) + 11H_2O(l)$ $\Delta H_{rxn}^{\circ}(1) = -5646.7 \text{ kJ}$
    sucrose

(2) $C_6H_{12}O_6(s) + 6O_2(g) \longrightarrow 6CO_2(g) + 6H_2O(l)$ $\qquad\qquad \Delta H_{rxn}^{\circ}(2) = -2815.8 \text{ kJ}$
    glucose

(3) $C_6H_{12}O_6(s) + 6O_2(g) \longrightarrow 6CO_2(g) + 6H_2O(l)$ $\qquad\qquad \Delta H_{rxn}^{\circ}(3) = -2826.7 \text{ kJ}$
    fructose

To obtain the desired equation, reverse Equations (2) and (3) and add them to Equation (1):

$$C_{12}H_{22}O_{11}(s) + 12O_2(g) \longrightarrow 12CO_2(g) + 11H_2O(l) \quad \Delta H_{rxn}^\circ = -5646.7 \text{ kJ}$$
sucrose

$$6CO_2(g) + 6H_2O(l) \longrightarrow C_6H_{12}O_6(s) + 6O_2(g) \qquad \Delta H_{rxn}^\circ = +2815.8 \text{ kJ}$$
glucose

$$6CO_2(g) + 6H_2O(l) \longrightarrow C_6H_{12}O_6(s) + 6O_2(g) \qquad \Delta H_{rxn}^\circ = +2826.7 \text{ kJ}$$
fructose

(4) $C_{12}H_{22}O_{11}(s) + H_2O(l) \longrightarrow C_6H_{12}O_6(s) + C_6H_{12}O_6(s)$
sucrose   glucose   fructose

$$\Delta H_{rxn}^\circ(4) = -5646.7 \text{ kJ} + 2815.8 \text{ kJ} + 2826.7 \text{ kJ} = -4.2 \text{ kJ}$$

8-23 If we let the first two equations be 1 and 2, then the equation

$$N_2(g) + O_2(g) \longrightarrow 2NO(g)$$

(equation 3) can be obtained from the first two equations by

$$\text{equation (3)} = \frac{\text{equation (1)}}{2} - \frac{\text{equation (2)}}{2}$$

According to Hess's law then, we write

$$\Delta H_{rxn}^\circ(3) = \frac{\Delta H_{rxn}^\circ(1)}{2} - \frac{\Delta H_{rxn}^\circ(2)}{2}$$

$$= \frac{-1170 \text{ kJ}}{2} + \frac{1530 \text{ kJ}}{2} = 180 \text{ kJ}$$

8-25 Using Equation (8-21) and the data given, we have

$$\Delta H_{rxn}^\circ = \Delta H_f^\circ[O_3(g)] - \Delta H_f^\circ[O_2(g)] - \Delta H_f^\circ[O(g)]$$
$$= (1 \text{ mol})(142 \text{ kJ} \cdot \text{mol}^{-1}) - (0) - (1 \text{ mol})(247.5 \text{ kJ} \cdot \text{mol}^{-1})$$
$$= -106 \text{ kJ}$$

8-27 (a) $\Delta H_{rxn}^\circ = \Delta H_f^\circ[N_2(g)] + 2 \, \Delta H_f^\circ[H_2O(g)] - \Delta H_f^\circ[N_2H_4(l)] - \Delta H_f^\circ[O_2(g)]$

Using the data in Table 8-2, we have

$$\Delta H_{rxn}^\circ = (1 \text{ mol})(0 \text{ kJ} \cdot \text{mol}^{-1}) + (2 \text{ mol})(-241.8 \text{ kJ} \cdot \text{mol}^{-1})$$
$$- (1 \text{ mol})(50.6 \text{ kJ} \cdot \text{mol}^{-1}) - (1 \text{ mol})(0 \text{ kJ} \cdot \text{mol}^{-1})$$
$$= -534.2 \text{ kJ} \qquad \text{exothermic}$$

(b) $\Delta H_{rxn}^\circ = \Delta H_f^\circ[C_2H_5OH(l)] - \Delta H_f^\circ[C_2H_4(g)] - \Delta H_f^\circ[H_2O(l)]$
$$= (1 \text{ mol})(-277.7 \text{ kJ} \cdot \text{mol}^{-1}) - (1 \text{ mol})(52.28 \text{ kJ} \cdot \text{mol}^{-1})$$
$$- (1 \text{ mol})(-285.8 \text{ kJ} \cdot \text{mol}^{-1})$$
$$= -44.2 \text{ kJ} \qquad \text{exothermic}$$

(c) $\Delta H_{rxn}^{\circ} = \Delta H_f^{\circ}[CCl_4(l)] + 4\,\Delta H_f^{\circ}[HCl(g)] - \Delta H_f^{\circ}[CH_4(g)] - 4\,\Delta H_f^{\circ}[Cl_2(g)]$

$\quad\quad = (1\text{ mol})(-135.4\text{ kJ}\cdot\text{mol}^{-1}) + (4\text{ mol})(-92.31\text{ kJ}\cdot\text{mol}^{-1})$
$\quad\quad\quad\quad\quad\quad - (1\text{ mol})(-74.86\text{ kJ}\cdot\text{mol}^{-1}) - (4\text{ mol})(0\text{ kJ}\cdot\text{mol}^{-1})$

$\quad\quad = -429.8\text{ kJ}\quad\quad\text{exothermic}$

8-29 (a) $\Delta H_{rxn}^{\circ} = 2\,\Delta H_f^{\circ}[CO_2(g)] + 3\,\Delta H_f^{\circ}[H_2O(l)] - \Delta H_f^{\circ}[C_2H_5OH(l)]$
$\quad\quad\quad\quad\quad\quad\quad\quad\quad\quad\quad\quad\quad\quad\quad\quad\quad\quad\quad\quad\quad\quad - 3\,\Delta H_f^{\circ}[O_2(g)]$

$\quad\quad = (2\text{ mol})(-393.5\text{ kJ}\cdot\text{mol}^{-1}) + (3\text{ mol})(-285.8\text{ kJ}\cdot\text{mol}^{-1})$
$\quad\quad\quad\quad\quad\quad\quad\quad - (1\text{ mol})(-277.7\text{ kJ}\cdot\text{mol}^{-1}) - (0)$

$\quad\quad = -1366.7\text{ kJ}$

The heat of combustion of $C_2H_5OH(l)$ per gram is

$$\left(\frac{-1366.7\text{ kJ}}{1\text{ mol}}\right)\left(\frac{1\text{ mol }C_2H_5OH}{46.07\text{ g }C_2H_5OH}\right) = -29.67\text{ kJ}\cdot\text{g}^{-1}$$

(b) $\Delta H_{rxn}^{\circ} = 2\,\Delta H_f^{\circ}[CO_2(g)] + 3\,\Delta H_f^{\circ}[H_2O(l)] - \Delta H_f^{\circ}[C_2H_6(g)]$
$\quad\quad\quad\quad\quad\quad\quad\quad\quad\quad\quad\quad\quad\quad\quad\quad\quad\quad\quad\quad\quad\quad - \tfrac{7}{2}\,\Delta H_f^{\circ}[O_2(g)]$

$\quad\quad = (2\text{ mol})(-393.5\text{ kJ}\cdot\text{mol}^{-1}) + (3\text{ mol})(-285.8\text{ kJ}\cdot\text{mol}^{-1})$
$\quad\quad\quad\quad\quad\quad\quad\quad - (1\text{ mol})(-84.68\text{ kJ}\cdot\text{mol}^{-1}) - (0)$

$\quad\quad = -1559.7\text{ kJ}$

The heat of combustion of $C_2H_6(g)$ per gram is

$$\left(\frac{-1559.7\text{ kJ}}{1\text{ mol}}\right)\left(\frac{1\text{ mol }C_2H_6}{30.07\text{ g }C_2H_6}\right) = -51.87\text{ kJ}\cdot\text{g}^{-1}$$

The combustion of $C_2H_6(g)$ produces almost twice as much heat per gram as does the combustion of $C_2H_5OH(l)$.

8-31 $\Delta H_{rxn}^{\circ} = 6\,\Delta H_f^{\circ}[CO_2(g)] + 6\,\Delta H_f^{\circ}[H_2O(l)] - \Delta H_f^{\circ}[C_6H_{12}O_6(s)] - 6\,\Delta H_f^{\circ}[O_2(g)]$

In this case we are given ΔH_{rxn}° and must determine $\Delta H_f^{\circ}[C_6H_{12}O_6(s)]$. Using the data in Table 8-2, we have

$\quad\quad -2826.7\text{ kJ} = (6\text{ mol})(-393.5\text{ kJ}\cdot\text{mol}^{-1}) + (6\text{ mol})(-285.8\text{ kJ}\cdot\text{mol}^{-1})$
$\quad\quad\quad\quad\quad\quad\quad\quad\quad\quad\quad\quad - (1\text{ mol})\,\Delta H_f^{\circ}[C_6H_{12}O_6(s)] - (0)$

Solving for $\Delta H_f^{\circ}[C_6H_{12}O_6(s)]$, we have

$\quad\quad (1\text{ mol})\,\Delta H_f^{\circ}[C_6H_{12}O_6(s)] = (6\text{ mol})(-393.5\text{ kJ}\cdot\text{mol}^{-1})$
$\quad\quad\quad\quad\quad\quad\quad\quad\quad\quad + (6\text{ mol})(-285.8\text{ kJ}\cdot\text{mol}^{-1}) + 2826.7\text{ kJ}$

$\quad\quad\quad\quad \Delta H_f^{\circ}[C_6H_{12}O_6(s)] = -1249.1\text{ kJ}\cdot\text{mol}^{-1}$

8-33 (a) $\Delta H^\circ_{rxn} = 2\,\Delta H^\circ_f[N(g)] - \Delta H^\circ_f[N_2(g)]$

945.2 kJ $= (2$ mol$)\,\Delta H^\circ_f[N(g)] - (0)$

or

$$\Delta H^\circ_f[N(g)] = \frac{945.2\text{ kJ}}{2\text{ mol}} = 472.6\text{ kJ}\cdot\text{mol}^{-1}$$

Similarly,

(b) $\Delta H^\circ_f[F(g)] = \dfrac{158.0\text{ kJ}}{2\text{ mol}} = 79.0\text{ kJ}\cdot\text{mol}^{-1}$

(c) $\Delta H^\circ_f[H(g)] = \dfrac{436.0\text{ kJ}}{2\text{ mol}} = 218.0\text{ kJ}\cdot\text{mol}^{-1}$

(d) $\Delta H^\circ_f[Cl(g)] = \dfrac{243.4\text{ kJ}}{2\text{ mol}} = 121.7\text{ kJ}\cdot\text{mol}^{-1}$

The bond strength of each diatomic molecule given is equal to the value of the corresponding ΔH°_{rxn}. Therefore, $N_2(g)$ has the greatest bond strength.

8-35 The reaction is

$$CCl_4(l) \longrightarrow CCl_4(g)$$

The heat required to vaporize 1 mol of CCl_4 at $25\,^\circ$C is equal to ΔH°_{rxn}. Thus

$$\text{heat} = \Delta H^\circ_{rxn} = \Delta H^\circ_f[CCl_4(g)] - \Delta H^\circ_f[CCl_4(l)]$$

From Table 8-2 we have

$\Delta H^\circ_{rxn} = (1\text{ mol})(-103.0\text{ kJ}\cdot\text{mol}^{-1}) - (1\text{ mol})(-135.4\text{ kJ}\cdot\text{mol}^{-1})$
$= 32.4$ kJ

8-37 Use Equation (8-22), with $q_P = 1105$ J and $\Delta T = 12.3\,^\circ$C $= 12.3$ K. The total heat capacity is

$$c_P = \frac{q_P}{\Delta T} = \frac{1105\text{ J}}{12.3\text{ K}} = 89.84\text{ J}\cdot\text{K}^{-1}$$

for the 36.5-g sample of C_2H_5OH. The molar heat capacity is

$$c_P = \left(\frac{89.84\text{ J}\cdot\text{K}^{-1}}{36.5\text{ g}}\right)\left(\frac{46.07\text{ g }C_2H_5OH}{1\text{ mol }C_2H_5OH}\right) = 113\text{ J}\cdot\text{K}^{-1}\cdot\text{mol}^{-1}$$

8-39   The heat capacity of 10.0 kg of water is

$$c_P = (75.2\ \mathrm{J\cdot K^{-1}\cdot mol^{-1}}) \left(\frac{1\ \mathrm{mol\ H_2O}}{18.02\ \mathrm{g\ H_2O}}\right)(10.0 \times 10^3\ \mathrm{g})$$
$$= 4.17 \times 10^4\ \mathrm{J\cdot K^{-1}}$$

Using Equation (8-22), we have for $\Delta T = (100.0 - 20.0)°\mathrm{C} = 80.0°\mathrm{C} = 80.0\ \mathrm{K}$

$$q_P = c_P\,\Delta T = (4.17 \times 10^4\ \mathrm{J\cdot K^{-1}})(80.0\ \mathrm{K}) = 3.34 \times 10^6\ \mathrm{J}$$

8-41   The sample of copper, being at a higher temperature than the water, will give up heat to the water. Using Equation (8-27), we have that

$$c_{P,Cu}\,\Delta T_{Cu} = -c_{P,H_2O}\,\Delta T_{H_2O}$$

For the sample of copper

$$\Delta T_{Cu} = t_f - 90.0°\mathrm{C}$$

and for the sample of water

$$\Delta T_{H_2O} = t_f - 20.0°\mathrm{C}$$

where t_f is the final Celsius temperature. The value of $c_{P,Cu}$ is

$$c_{P,Cu} = (24.5\ \mathrm{J\cdot K^{-1}\cdot mol^{-1}}) \left(\frac{1\ \mathrm{mol\ Cu}}{63.55\ \mathrm{g\ Cu}}\right)(25.0\ \mathrm{g})$$
$$= 9.638\ \mathrm{J\cdot K^{-1}}$$

The value of c_{P,H_2O} is

$$c_{P,H_2O} = (75.2\ \mathrm{J\cdot K^{-1}\cdot mol^{-1}}) \left(\frac{1\ \mathrm{mol\ H_2O}}{18.02\ \mathrm{g\ H_2O}}\right)(100.0\ \mathrm{g})$$
$$= 417.3\ \mathrm{J\cdot K^{-1}}$$

Therefore

$$(9.638\ \mathrm{J\cdot K^{-1}})(t_f - 90.0°\mathrm{C}) = -(417.3\ \mathrm{J\cdot K^{-1}})(t_f - 20.0°\mathrm{C})$$
$$9.638\ \mathrm{J\cdot K^{-1}}t_f - 867.4\ \mathrm{J\cdot K^{-1}\cdot °C} = -417.3\ \mathrm{J\cdot K^{-1}}t_f + 8346\ \mathrm{J\cdot K^{-1}\cdot °C}$$
$$426.9\ \mathrm{J\cdot K^{-1}}t_f = 9213\ \mathrm{J\cdot K^{-1}\cdot °C}$$

Solving for t_f

$$t_f = \frac{9213\ \mathrm{J\cdot K^{-1}\cdot °C}}{426.9\ \mathrm{J\cdot K^{-1}}} = 21.6°\mathrm{C}$$

8-43   Using Equation (8-27), we have that

$$c_{P,Al} \Delta T_{Al} = -c_{P,Cu} \Delta T_{Cu}$$

For the aluminum

$$c_{P,Al} = (24.2 \text{ J} \cdot \text{K}^{-1} \cdot \text{mol}^{-1})(1.00 \text{ kg}) \left(\frac{10^3 \text{ g}}{1 \text{ kg}}\right)\left(\frac{1 \text{ mol Al}}{26.98 \text{ g Al}}\right)$$

$$= 897 \text{ J} \cdot \text{K}^{-1}$$

$$\Delta T_{Al} = t_f - 500°\text{C}$$

For the copper

$$c_{P,Cu} = (24.5 \text{ J} \cdot \text{K}^{-1} \cdot \text{mol}^{-1})(1.00 \text{ kg}) \left(\frac{10^3 \text{ g}}{1 \text{ kg}}\right)\left(\frac{1 \text{ mol Cu}}{63.55 \text{ g Cu}}\right)$$

$$= 386 \text{ J} \cdot \text{K}^{-1}$$

$$\Delta T_{Cu} = t_f - 10°\text{C}$$

Putting all this into Equation (8-27) gives

$$(897 \text{ J} \cdot \text{K}^{-1})(t_f - 500°\text{C}) = -(386 \text{ J} \cdot \text{K}^{-1})(t_f - 10°\text{C})$$

or

$$1283 \, t_f = 4.53 \times 10^5 °\text{C}$$

or

$$t_f = 350°\text{C}$$

8-45   The heat evolved by the reaction is given by Equation (8-30):

$$\Delta H_{rxn} = -c_{P,\text{calorimeter}} \Delta T$$
$$= -(480 \text{ J} \cdot \text{K}^{-1})(2.34 \text{ K}) = -1123 \text{ J}$$

This amount of heat is evolved when 0.100 L of 0.200 M solutions react. The number of moles that react is given by

$$n = MV = (0.200 \text{ mol} \cdot \text{L}^{-1})(0.100 \text{ L}) = 0.0200 \text{ mol}$$

The heat of reaction for 1 mol of reactants is

$$\Delta H_{rxn}^{\circ} = \frac{-1123 \text{ J}}{0.0200 \text{ mol}} = -56.2 \text{ kJ} \cdot \text{mol}^{-1}$$

8-47 The value of ΔH_{rxn} is

$$\Delta H_{rxn} = -(4.92 \text{ kJ} \cdot \text{K}^{-1})(0.300 \text{ K}) = -1.48 \text{ kJ}$$

This is the heat evolved when 1.00 g of NH_4NO_3 reacts. The heat of reaction for 1.00 kg of NH_4NO_3 reacting is

$$\Delta H_{rxn} = \left(\frac{-1.48 \text{ kJ}}{1.00 \text{ g}}\right)(1.00 \times 10^3 \text{ g}) = -1.48 \times 10^3 \text{ kJ}$$

8-49 The temperature of the calorimeter *decreases*, and so heat is absorbed in the process of dissolving KCl in water. The heat absorbed is

$$\Delta H_{rxn} = -(4.51 \text{ kJ} \cdot \text{K}^{-1})(-0.256 \text{ K}) = +1.155 \text{ kJ}$$

A 5.00-g sample of KCl corresponds to

$$(5.00 \text{ g})\left(\frac{1 \text{ mol KCl}}{74.55 \text{ g KCl}}\right) = 0.06707 \text{ mol}$$

The molar heat of solution of KCl in water is

$$\Delta H^{\circ}_{soln} = \frac{1.155 \text{ kJ}}{0.06707 \text{ mol}} = 17.2 \text{ kJ} \cdot \text{mol}^{-1}$$

8-51 $\Delta H_{rxn} = -(32.7 \text{ kJ} \cdot \text{K}^{-1})(42.5 \text{ K}) = -1390 \text{ kJ}$

$$\Delta H^{\circ}_{comb} = \frac{-1390 \text{ kJ}}{30.0 \text{ g}} = -46.3 \text{ kJ} \cdot \text{g}^{-1}$$

$$= (-46.3 \text{ kJ} \cdot \text{g}^{-1})\left(\frac{44.09 \text{ g C}_3\text{H}_8}{1 \text{ mol C}_3\text{H}_8}\right) = -2040 \text{ kJ} \cdot \text{mol}^{-1}$$

8-53 $\Delta H_{rxn} = -(8.75 \text{ kJ} \cdot \text{K}^{-1})(0.780 \text{ K}) = -6.825 \text{ kJ}$

$$\Delta H^{\circ}_{comb} = \left(\frac{-6.825 \text{ kJ}}{2.50 \text{ g}}\right)\left(\frac{90.04 \text{ g H}_2\text{C}_2\text{O}_4}{1 \text{ mol H}_2\text{C}_2\text{O}_4}\right) = -246 \text{ kJ} \cdot \text{mol}^{-1}$$

The equation for the combustion of oxalic acid is

$$H_2C_2O_4(s) + \tfrac{1}{2} O_2(g) \longrightarrow 2CO_2(g) + H_2O(l)$$

The value of ΔH°_{rxn} for the combustion of oxalic acid is given by

$$\Delta H^{\circ}_{rxn} = 2 \Delta H^{\circ}_f[CO_2(g)] + \Delta H^{\circ}_f[H_2O(l)] - \Delta H^{\circ}_f[H_2C_2O_4(s)] - \tfrac{1}{2} \Delta H^{\circ}_f[O_2(g)]$$

Using the above value for ΔH°_{rxn} and the data in Table 8-2, we have

$$-246 \text{ kJ} = (2 \text{ mol})(-393.5 \text{ kJ} \cdot \text{mol}^{-1}) + (1 \text{ mol})(-285.8 \text{ kJ} \cdot \text{mol}^{-1})$$
$$- (1 \text{ mol}) \, \Delta H^\circ_f[\text{H}_2\text{C}_2\text{O}_4(s)] - (\tfrac{1}{2} \text{ mol})(0)$$
$$= -1072.8 \text{ kJ} - (1 \text{ mol}) \, \Delta H^\circ_f[\text{H}_2\text{C}_2\text{O}_4(s)]$$

Solving for $\Delta H^\circ_f[\text{H}_2\text{C}_2\text{O}_4(s)]$, we get

$$\Delta H^\circ_f[\text{H}_2\text{C}_2\text{O}_4(s)] = -827 \text{ kJ} \cdot \text{mol}^{-1}$$

8-55 Using Table 8-7, the annual energy production from oil is $39 \times 10^{15} \text{ kJ} \cdot \text{y}^{-1}$. From Section 8-10 we find that 50 percent of the oil used is imported; thus the energy derived from oil per day is

$$\text{imported oil energy per day} = \left(\frac{50}{100}\right)(39 \times 10^{15} \text{ kJ} \cdot \text{y}^{-1})\left(\frac{1 \text{ y}}{365 \text{ d}}\right)$$
$$= 5.34 \times 10^{13} \text{ kJ} \cdot \text{d}^{-1}$$

We now need to find a conversion factor between kilojoules and barrels of oil. We are given in Section 8-10 that 10 kW per person corresponds to 51 barrels of oil per person per year. First we calculate the number of kilojoules per year in 10 kilowatts.

$$10 \text{ kW} = (10 \text{ kJ} \cdot \text{s}^{-1})\left(\frac{60 \text{ s}}{1 \text{ min}}\right)\left(\frac{60 \text{ min}}{1 \text{ h}}\right)\left(\frac{24 \text{ h}}{1 \text{ d}}\right)\left(\frac{365 \text{ d}}{1 \text{ y}}\right)$$
$$= 3.15 \times 10^8 \text{ kJ} \cdot \text{y}^{-1}$$

Thus the conversion factor is

$$1 = \frac{51 \text{ barrels} \cdot \text{person}^{-1} \cdot \text{y}^{-1}}{3.15 \times 10^8 \text{ kJ} \cdot \text{person}^{-1} \cdot \text{y}^{-1}}$$

The number of barrels of oil imported per day is

$$\text{number of barrels} = (5 \cdot 34 \times 10^{13} \text{ kJ} \cdot \text{d}^{-1})\left(\frac{51 \text{ barrels} \cdot \text{person}^{-1} \cdot \text{y}^{-1}}{3.15 \times 10^8 \text{ kJ} \cdot \text{person}^{-1} \cdot \text{y}^{-1}}\right)$$
$$= 8.6 \times 10^6 \text{ barrels} \cdot \text{d}^{-1}$$

Almost 10 million barrels of oil are imported per day.

8-57 In Section 8-13 of the text we are given that $\text{Na}_2\text{SO}_4 \cdot 10\text{H}_2\text{O}(l)$ absorbs $354 \text{ kJ} \cdot \text{L}^{-1}$. The volume of 100 kg of $\text{Na}_2\text{SO}_4 \cdot 10\text{H}_2\text{O}(l)$ is

$$V = \left(\frac{100 \times 10^3 \text{ g}}{1.5 \text{ g} \cdot \text{mL}^{-1}}\right)\left(\frac{1 \text{ L}}{10^3 \text{ mL}}\right) = 66.7 \text{ L}$$

The number of kilojoules stored is given by

$$\text{kilojoules} = (354 \text{ kJ} \cdot \text{L}^{-1})(66.7 \text{ L}) = 24,000 \text{ kJ}$$

8-59 The equation for the partial oxidation of ethane is given in Section 8-11 of the text as

(1) $C_2H_6(g) + \frac{1}{2}O_2(g) \longrightarrow C_2H_5OH(l)$ $\Delta H^\circ_{rxn} = -193 \text{ kJ}$

The equation for the combustion of ethane is

(2) $C_2H_6(g) + \frac{7}{2}O_2(g) \longrightarrow 2CO_2(g) + 3H_2O(l)$ $\Delta H^\circ_{rxn} = -1427 \text{ kJ}$

The equation for the combustion of ethyl alcohol is

(3) $C_2H_5OH(l) + 3O_2(g) \longrightarrow 2CO_2(g) + 3H_2O(l)$

The value of the heat of combustion of ethyl alcohol is given by

$$\Delta H^\circ_{rxn}(3) = -\Delta H^\circ_{rxn}(1) + \Delta H^\circ_{rxn}(2)$$
$$= -(-193 \text{ kJ}) + (-1427 \text{ kJ}) = -1234 \text{ kJ}$$

8-61 Petroleum liquids are easier to distribute to the site at which they are used because they can be pumped from the source in underground pipes. Petroleum liquids do not have to be stored on the site of use as does coal. Petroleum liquids can be pumped directly to the device used to burn the fuel rather than transported by the more cumbersome means that coal requires.

8-63 The heat required to raise the temperature of 1.0 L of water by 37°C is

$$\Delta H = (75.2 \text{ J} \cdot \text{K}^{-1} \cdot \text{mol}^{-1}) \left(\frac{1 \text{ mol H}_2\text{O}}{18.02 \text{ g H}_2\text{O}} \right) (1.00 \text{ g} \cdot \text{mL}^{-1})(1000 \text{ mL})(37 \text{ K})$$
$$= 150 \text{ kJ}$$

Given that 1 g of body fat yields 39 kJ, we see that

$$\frac{150 \text{ kJ}}{39 \text{ kJ} \cdot \text{g}^{-1}} = 3.8 \text{ g}$$

of body fat must be burned. The number of moles of ice required to produce the same effect is

$$150 \text{ kJ} = (6.0 \text{ kJ} \cdot \text{mol}^{-1})n + (75.2 \text{ J} \cdot \text{K}^{-1} \cdot \text{mol}^{-1})(37 \text{ K})n$$
$$= (8.8 \text{ kJ} \cdot \text{mol}^{-1})n$$
$$n = \frac{150 \text{ kJ}}{8.8 \text{ kJ} \cdot \text{mol}^{-1}} = 17 \text{ mol}$$

The mass of ice required is

$$\text{mass of ice} = (17 \text{ mol}) \left(\frac{18.02 \text{ g } H_2O}{1 \text{ mol } H_2O} \right) = 310 \text{ g}$$

8-65 We set up the following table for the proposed formula of the compound of thallium and chlorine.

Proposed formula	Value of N in Dulong and Petit's rule	Predicted value of $C_P/J \cdot K^{-1} \cdot mol^{-1}$	Observed value of C_P from specific heat and proposed formula
TlCl	2	50	$(0.208 \text{ J} \cdot K^{-1} \cdot g^{-1}) \left(\frac{239.9 \text{ g}}{1 \text{ mol}} \right)$ $= 49.9 \text{ J} \cdot K^{-1} \cdot mol^{-1}$
TlCl$_2$	3	75	$(0.208 \text{ J} \cdot K^{-1} \cdot g^{-1}) \left(\frac{275.3 \text{ g}}{1 \text{ mol}} \right)$ $= 57.3 \text{ J} \cdot K^{-1} \cdot mol^{-1}$
TlCl$_3$	4	100	$(0.208 \text{ J} \cdot K^{-1} \cdot g^{-1}) \left(\frac{310.8 \text{ g}}{1 \text{ mol}} \right)$ $= 64.6 \text{ J} \cdot K^{-1} \cdot mol^{-1}$

Because of the agreement between the predicted and observed values of C_P, we conclude that the formula of the compound is TlCl.

8-67 The molar heat capacity of stilleite, using Dulong and Petit's rule, is

$$C_P = (2)(25 \text{ J} \cdot K^{-1} \cdot mol^{-1}) = 50 \text{ J} \cdot K^{-1} \cdot mol^{-1}$$

if stilleite is ZnSe. The observed molar heat capacity of stilleite is

$$C_P = (0.348 \text{ J} \cdot K^{-1} \cdot g^{-1}) \left(\frac{144.34 \text{ g ZnSe}}{1 \text{ mol ZnSe}} \right) = 50.2 \text{ J} \cdot K^{-1} \cdot mol^{-1}$$

Thus we determine the formula of stilleite to be ZnSe.

8-69 We assume that the energy consumed in riding is due to the combustion of body fat. Given that one gram of body fat yields 39 kJ, the energy produced by 1 lb of body fat is

$$\text{energy} = (39 \text{ kJ} \cdot g^{-1})(1 \text{ lb}) \left(\frac{454 \text{ g}}{1 \text{ lb}} \right) = 17,700 \text{ kJ}$$

The number of hours of bicycle riding that is required to consume 17,700 kJ is

$$\text{number of hours of riding} = \frac{17,700 \text{ kJ}}{2000 \text{ kJ} \cdot \text{h}^{-1}} = 9.0 \text{ h}$$

The distance traveled in 9.0 h is

$$\text{distance} = (13 \text{ mi} \cdot \text{h}^{-1})(9.0 \text{ h}) = 120 \text{ mi}$$

8-71   The relation between ΔH_{rxn}° and ΔU_{rxn}° is given by Equation (8-13):

$$\Delta H_{rxn}^\circ = \Delta U_{rxn}^\circ + P \, \Delta V$$

(a) $\Delta U_{rxn}^\circ = \Delta H_{rxn}^\circ - P \, \Delta V$

$$= (-572 \text{ kJ}) - (1.00 \text{ atm})(-67.2 \text{ L})\left(\frac{101.3 \text{ J}}{1 \text{ L} \cdot \text{atm}}\right)$$

$$= -565 \text{ kJ}$$

(b) $\Delta U_{rxn}^\circ = \Delta H_{rxn}^\circ - P \, \Delta V$

$$= (-545 \text{ kJ}) - (1.00 \text{ atm})(-44.8 \text{ L})\left(\frac{101.3 \text{ J}}{1 \text{ L} \cdot \text{atm}}\right)$$

$$= -540 \text{ kJ}$$

(c) $\Delta U_{rxn}^\circ = \Delta H_{rxn}^\circ - P \, \Delta V$

$$= (-180 \text{ kJ}) - (1.00 \text{ atm})(0 \text{ L})$$

$$= -180 \text{ kJ}$$

Notice that $\Delta U_{rxn}^\circ \approx \Delta H_{rxn}^\circ$ in each case.

8-73 $\Delta H_{rxn}^\circ = 3 \, \Delta H_f^\circ[N_2(g)] + 4 \, \Delta H_f^\circ[H_2O(g)] - 2 \, \Delta H_f^\circ[N_2H_4(l)] - \Delta H_f^\circ[N_2O_4(l)]$

$$= (3 \text{ mol})(0 \text{ kJ} \cdot \text{mol}^{-1}) + (4 \text{ mol})(-241.8 \text{ kJ} \cdot \text{mol}^{-1})$$
$$- (2 \text{ mol})(50.6 \text{ kJ} \cdot \text{mol}^{-1}) - (1 \text{ mol})(-19.5 \text{ kJ} \cdot \text{mol}^{-1})$$

$$= -1049 \text{ kJ}$$

8-75   The rate of consumption of fuel in terms of milliliters per minute is

$$\text{rate of consumption} = (80 \text{ gal} \cdot \text{min}^{-1})\left(\frac{4 \text{ qt}}{1 \text{ gal}}\right)\left(\frac{1 \text{ L}}{1.06 \text{ qt}}\right)\left(\frac{10^3 \text{ mL}}{1 \text{ L}}\right)$$

$$= 3.02 \times 10^5 \text{ mL} \cdot \text{min}^{-1}$$

The rate of consumption of fuel in terms of grams per minute is

$$\text{rate of consumption} = (3.02 \times 10^5 \text{ mL} \cdot \text{min}^{-1})(0.72 \text{ g} \cdot \text{mL}^{-1})$$

$$= 2.17 \times 10^5 \text{ g} \cdot \text{min}^{-1}$$

The power produced by the aircraft is

$$\text{power} = (2.17 \times 10^5 \text{ g} \cdot \text{min}^{-1})(50.0 \text{ kJ} \cdot \text{g}^{-1}) \left(\frac{1 \text{ min}}{60 \text{ s}} \right)$$
$$= 1.8 \times 10^5 \text{ kJ} \cdot \text{s}^{-1} = 1.8 \times 10^5 \text{ kW}$$

8-77   We have that

$$\Delta U = c^2 \, \Delta m = (3.00 \times 10^8 \text{ m} \cdot \text{s}^{-1})^2 (1.0 \text{ g}) \left(\frac{1 \text{ kg}}{10^3 \text{ g}} \right)$$
$$= 9.0 \times 10^{13} \text{ J} = 9.0 \times 10^{10} \text{ kJ}$$

Most of the energies encountered in this chapter are about 500 kJ·mol^{-1}. If we assume a molecular mass of 50, then the energy per gram produced by a chemical reaction is about 10 kJ. Thus, we see that the conversion of mass to energy is around 10^{10} times greater than the energies of ordinary chemical processes.

8-79   The equation for the reaction is

$$C_2H_4(g) + H_2O(g) \longrightarrow C_2H_5OH(l)$$

Using Equation (8-21) and Table 8-2, we have

$$\Delta H^\circ_{\text{rxn}} = \Delta H^\circ_f[C_2H_5OH(l)] - \Delta H^\circ_f[C_2H_4(g)] - \Delta H^\circ_f[H_2O(g)]$$
$$= (1 \text{ mol})(-277.7 \text{ kJ} \cdot \text{mol}^{-1}) - (1 \text{ mol})(+52.28 \text{ kJ} \cdot \text{mol}^{-1})$$
$$- (1 \text{ mol})(-241.8 \text{ kJ} \cdot \text{mol}^{-1})$$
$$= -88.2 \text{ kJ}$$

E. ANSWERS TO THE SELF-TEST

1. energy is neither created nor destroyed during any process

2. true

3. false (energy due to position is potential energy)

4. true

5. false

6. true

7. the rate at which energy is used or produced

8. watt

9. true

10. energy is neither created nor destroyed but is transformed from one form to another; energy is conserved

11. heat . . . work

12. force through a distance

13. positive

14. negative

15. difference in temperature

16. true

17. false

18. the energy transferred as heat plus the energy transferred as work, $\Delta U_{rxn} = q + w$

19. heat evolved or absorbed, q_V

20. $U + PV$

21. heat evolved or absorbed, q_P

22. pressure

23. true

24. exothermic

25. endothermic

26. exo-

27. true

28. false (at 1 atm pressure)

29. true

30. true

31. ΔH_{rxn}(reverse reaction) $= -\Delta H_{rxn}$(forward reaction)

32. See Section 8-4.

33. the energy that is evolved or absorbed as heat when one mole of the substance is formed directly from its elements in their standard forms at one atm and 25°C

34. true

35. true

36. $\Delta H_f^\circ[CO_2(g)] + 2\,\Delta H_f^\circ[H_2O(g)] - \Delta H_f^\circ[CH_4(g)] - 2\,\Delta H_f^\circ[O_2(g)]$

37. molar heat capacity

38. more

39. The value of the heat capacity of a substance depends on its mass.

40. gram

41. higher . . . lower

42. true

43. translational

44. greater than

45. greater

46. true

47. the value of C_p for solid metals is approximately equal to $3R$

48. the heat evolved or absorbed in a process by measuring the change in temperature

49. false

50. All the heat that is evolved or absorbed by the reaction is absorbed or supplied by the calorimeter.

51. temperature

52. heat capacity . . . temperature

53. exo-

54. bomb

55. true

56. The heat of combustion is the quantity of heat that is evolved when the substance burns.

57. calorie

58. to raise the temperature of one gram of water one degree Celsius

59. true

60. hydrocarbons

61. true

62. gasoline . . . ethyl alcohol

63. false

64. true

65. true

9 THE QUANTUM THEORY AND ATOMIC STRUCTURE

A. OUTLINE OF CHAPTER 9

9-1 First ionization energy is one of many periodic properties of the elements.

The ionization energy is the minimum energy required to remove an electron from an atom or ion.

The first ionization energy is the minimum energy required to remove an electron from an atom A to produce an A^+ ion.

The second ionization energy is the minimum energy required to remove an electron from an A^+ ion to produce an A^{2+} ion.

A plot of the first ionization energy I_1 against atomic number (Figure 9-1) displays the periodic nature of I_1.

I_1 increases across each row of the periodic table (Figure 9-2).

The electronic structure of the noble gases is relatively stable.

9-2 The values of successive ionization energies of atoms suggest a shell structure.

Table 9-1 lists successive ionization energies of the elements hydrogen through argon.

The inner-core electronic structure of an atom is that of the preceding noble gas.

A plot of the ionization energies versus number of electrons removed suggests a shell structure.

The electrons in the outermost occupied shell of an atom are also called valence electrons.

A Lewis electron-dot formula (Table 9-2) is a representation of an atom with the noble-gas-like inner electrons represented by the symbol of the element and the outer, or valence, electrons represented by dots around the symbol.

9-3   The electromagnetic spectrum is characterized by radiation of different wavelengths.

The electromagnetic theory of radiation states that all forms of radiation are propagated through space as oscillating electric and magnetic fields.

The wavelength λ and frequency ν of electromagnetic radiation are related by

$$\lambda\nu = c \qquad (9\text{-}1)$$

where c is the speed of light, 3.00×10^8 m·s^{-1}.

The complete range of wavelengths or frequencies of electromagnetic radiation is called the electromagnetic spectrum.

9-4   The emission spectra of atoms consist of series of lines.

Electromagnetic radiation can be separated into its components.

The visible region consists of electromagnetic radiation in the range 400 nm to 700 nm.

A continuous spectrum consists of all wavelengths.

A line spectrum consists of only certain wavelengths.

An atomic spectrum is a line spectrum; part of the spectrum of the hydrogen atom is shown in Figure 9-7.

The Rydberg-Balmer equation is an empirical equation for the wavelengths of the lines in the visible spectrum of hydrogen and is given as

$$\frac{1}{\lambda} = (1.097 \times 10^7 \text{ m}^{-1}) \left(\frac{1}{4} - \frac{1}{n^2} \right) \qquad (9\text{-}2)$$

where n is an integer greater than 2.

The constant 1.097×10^7 m^{-1} is known as the Rydberg constant.

Each type of atom has a characteristic atomic spectrum (Figure 9-9).

Atomic spectra can be used to determine the atomic composition of a substance.

Atomic spectroscopy is the study of the interaction of electromagnetic radiation and atoms.

9-5   Electromagnetic radiation can be viewed as a beam of photons.

Blackbody radiation is continuous radiation emitted by solid bodies when they are heated to high temperatures.

Planck assumed that radiation could be emitted only in energy packets called quanta.

The energy associated with these quanta is related to the frequency ν by

$$E = h\nu \qquad (9\text{-}3)$$

where h is Planck's constant, which is equal to 6.626×10^{-34} J·s.

The photoelectric effect is the ejection of electrons from the surface of a metal when it is irradiated with ultraviolet radiation.

The threshold frequency ν_0 of a metal is the minimum frequency required to eject electrons from the metal.

Each metal has a characteristic value of ν_0.

The graph of the kinetic energy of the ejected electrons versus the frequency of the radiation is a straight line when the frequency of the radiation exceeds the threshold frequency (Figure 9-11).

Einstein postulated that the radiation consists of little packets of energy, $E = h\nu$ or

$$E = \frac{hc}{\lambda} \tag{9-4}$$

The packets of energy of electromagnetic radiation are called photons.

9-6 Einstein applied the principle of conservation of energy to explain the photoelectric effect.

The work function of a metal is the minimum energy required to remove an electron from the surface of the metal.

The work function is denoted by Φ.

The work function is related to ν_0 by

$$\Phi = h\nu_0 \tag{9-5}$$

The kinetic energy of the ejected electrons is given by

$$\text{K.E.} = h\nu - \Phi = h\nu - h\nu_0 \tag{9-6}$$

The quantization of energy is the idea that energy exists in discrete little packets.

In the quantum theory, only certain discrete values of the energy are allowed.

9-7 De Broglie was the first to propose that matter has wavelike properties.

The wave-particle duality of light suggested to de Broglie that matter may appear wavelike.

De Broglie proposed that a moving mass has a wavelength associated with it.

The wavelength of a particle of mass m and speed v is given by

$$\lambda = \frac{h}{mv} \tag{9-7b}$$

The wavelength associated with a moving body is called its de Broglie wavelength.

9-8 The electron microscope utilizes the wavelike properties of electrons.

A beam of electrons can behave similarly to a beam of X-rays (Figure 9-14).

Wave-particle duality of matter is the concept that electromagnetic radiation and matter can exhibit either wavelike properties or particlelike properties.

The wavelike nature of matter is most evident when the mass is very small.

9-9   The energy of the electron in a hydrogen atom is quantized.

Bohr postulated that the electron in a hydrogen atom is restricted to only certain circular orbits about the nucleus.

The radius r of a stable orbit is subject to the quantum condition

$$2\pi r = n\lambda \qquad n = 1,2,3 \ldots \tag{9-8}$$

Bohr showed that the energies of the electron in a hydrogen atom are given by

$$E_\text{n} = \frac{-2.18 \times 10^{-18}\,\text{J}}{n^2} \qquad n = 1,2,3 \ldots \tag{9-9}$$

The energy of the electron is quantized or restricted to only certain values.

A stationary state is an allowed energy state.

The ground state is the stationary state of lowest energy, $n = 1$.

An excited state is a stationary state of higher energy than the ground state, $n = 2,3. \ldots$

9-10   Atoms emit or absorb electromagnetic radiation when they undergo transitions from one stationary state to another.

The energy states of the electron in a hydrogen atom are shown in Figure 9-19.

When an electron in a hydrogen atom goes from a higher energy state to a lower energy state, the energy emitted is given by

$$\Delta E = E_2 - E_1 \tag{9-10}$$

The energy is emitted as electromagnetic radiation.

When the electron in a hydrogen atom goes from a higher energy state to the ground state, the frequency of the emitted radiation is given by

$$\nu_{n1} = (3.29 \times 10^{15}\,\text{s}^{-1}) \left(\frac{1}{1^2} - \frac{1}{n^2} \right) \qquad n = 2,3,4. \ldots \tag{9-13}$$

These frequencies correspond to the series of lines in the hydrogen atom emission spectrum called the Lyman series (Table 9-5).

When the electron goes from a higher energy state to the $n = 2$ state, the frequency of the emitted radiation is given by

$$\nu_{n2} = (3.29 \times 10^{15}\,\text{s}^{-1}) \left(\frac{1}{2^2} - \frac{1}{n^2} \right) \qquad n = 3,4,5. \ldots \tag{9-14}$$

These frequencies correspond to the Balmer series in the hydrogen atom emission spectrum (Figure 9-19).

The Bohr theory provided a theoretical explanation of the Rydberg-Balmer equation.

An emission spectrum occurs when excited atoms return to the ground state.

An absorption spectrum occurs when atoms in the ground state absorb energy and are excited to higher energy states (Figure 9-21).

The frequency of absorption is equal to the frequency of emission between the same two states.

9-11   The Heisenberg uncertainty principle limits the accuracy of the simultaneous measurement of the position and momentum of a particle.

The Bohr theory cannot be extended to atoms other than hydrogen.

The Heisenberg uncertainty principle states that it is not possible to measure accurately both the position and the momentum of a particle simultaneously.

The uncertainty in the position Δx and the momentum Δp are related by

$$(\Delta x)(\Delta p) \simeq h \tag{9-17}$$

where h is Planck's constant.

The uncertainty principle is a result of the act of measurement itself.

9-12   The Schrödinger equation is the central equation of the quantum theory.

The Schrödinger equation is consistent with the wave nature of particles and the Heisenberg uncertainty principle.

Solution of the Schrödinger equation for a hydrogen atom gives the same set of energy levels as that predicted by the Bohr theory.

The electron in the hydrogen atom is not restricted to certain sharp orbits.

The Schrödinger equation also provides an associated set of functions called wave functions (orbitals), which are designated by ψ.

The value of the square of the wave function is a probability density, which is the probability of the electron being located in a certain region.

The electron cannot be located precisely, but can be assigned only a probability of being located in a certain region.

Quantum numbers arise from the solution of the Schrödinger equation.

The principal quantum number n specifies the energy of the electron in a hydrogen atom.

The quantum numbers n, l, and m_l are needed to specify the electron wave functions in a atom.

The ground state of a hydrogen atom occurs when $n = 1$.

The ground state of a hydrogen atom is denoted by ψ_{1s} (Figure 9-24).

The ψ_{1s}, or $1s$, orbital is spherically symmetric.

The $1s$ orbital can be represented by a stippled diagram or a 99% probability sphere as shown in Figure 9-25.

9-13   The shape of an orbital depends on the value of the azimuthal quantum number.

The principal quantum number n specifies the extent of an orbit.

The azimuthal quantum number l is restricted to the following values: $l = 0, 1, \ldots, n - 1$.

s, p, d, and f orbitals are those for which $l = 0, 1, 2$, and 3, respectively.

Orbitals are usually designated by the values of n and l (Table 9-6).

All s orbitals are spherically symmetric (Figure 9-28).

The three $2p$ orbitals are cylindrically symmetric.

The $2p$ orbitals can be represented by a stippled diagram or a 99% probability surface as shown in Figure 9-29.

A surface on which the probability density is zero is a nodal surface.

The number of planar nodes in an orbital is equal to the angular quantum number l.

The number of spherical nodes in an orbital is equal to $n - 1 - l$.

9-14 The spatial orientation of an orbital depends on the value of the magnetic quantum number.

The magnetic quantum number m_l is restricted to the following values: $m_l = -l, -l + 1, \ldots, -1, 0, 1, 2, \ldots, l$.

The allowed values of the quantum numbers l and m_l for $n = 1$ to $n = 4$ are given in Table 9-7.

The three $2p$ orbitals are designated p_x, p_y, and p_z; the subscript indicates the axis along which the orbital is directed (Figure 9-30).

9-15 An electron has an intrinsic spin.

The spin quantum number m_s is restricted to the values $m_s = +\frac{1}{2}$ or $-\frac{1}{2}$.

The allowed combinations of the four quantum numbers for $n = 1$ to $n = 3$ are given in Table 9-8.

9-16 The energy states of atoms with two or more electrons depend on the values of both n and l.

Multielectron atoms involve electron-nucleus and electron-electron interactions.

The relative energies of the orbitals of atoms with two or more electrons are given in Figures 9-32 and 9-33.

9-17 The Pauli exclusion principle states that no two electrons in the same atom can have the same set of four quantum numbers.

The Pauli exclusion principle is used to assign electrons to orbitals.

Electrons with $m = +\frac{1}{2}$ are called spin up; electrons with $m = -\frac{1}{2}$ are called spin down.

When two electrons occupy an orbital, they have opposite spins and are said to have their electron spins paired.

The sets of allowed quantum numbers are given in Table 9-9.

Shells are the levels designated by n.

Subshells are the groups of orbitals designated by l within the shells.

9-18 Electron configurations designate the occupancy of electrons in atomic orbitals.

The assignment of electrons to orbitals of lowest energies gives the ground-state electron configuration of an atom or an ion.

The value of the first ionization energy of an atom depends on the electron configuration of the atom.

9-19 Hund's rule is used to predict ground-state electron configurations.

Hund's rule states that for any set of orbitals of the same energy, the ground-state electron configuration is obtained by placing the electrons into different orbitals of this set with parallel spins until each of the orbitals has one electron before any electrons are paired.

Hund's rule is used to predict the ground-state electron configuration of an atom with a partially filled subshell.

An atom may absorb electromagnetic radiation to promote an electron to an orbital of higher energy state.

When an electron has been promoted to an orbital of higher energy than the ground state, the atom is in an excited state.

9-20 Elements in the same column of the periodic table have similar valence-electron configurations.

The ground-state electron configurations of the elements correlate with their Lewis electron-dot formulas.

The number of dots displayed in a Lewis electron-dot formula is the same as the number of electrons in the outer shell.

The electron configurations of atoms are correlated to their positions in the periodic table (Figure 9-34).

The number of the row in the periodic table is equal to the principal quantum number of the outer s electrons.

The alkali metals have the electron configuration [noble gas]ns^1.

The alkaline earth metals have the electron configuration [noble gas]ns^2.

9-21 The occupied orbitals of highest energy are d orbitals for neutral transition metals and f orbitals for lanthanides and actinides.

In the first set of transition-metals-the five $3d$ orbitals are filled sequentially.

The $3d$ transition-metal series consists of the elements Sc through Zn (Figure 9-34).

The $4d$ transition-metal series consists of the elements Y through Cd (Figure 9-34).

Half-filled and completely filled d subshells are relatively stable.

The inner transition metals involve the filling of f orbitals.

The lanthanides La through Yb involve the filling of the $4f$ orbitals (Figure 9-34).

The lanthanides have similar chemical properties.

The 4f electrons tend to lie in the interior of the atom and have little effect on chemical activity.

The 5d transition metals follow the lanthanides and involve the filling of the 5d orbitals (Figure 9-34).

The actinide series, Ac through No, involves the filling of the 5f orbitals (Figure 9-34).

The transuranium elements are all radioactive.

The regions of the periodic table are designated as the s-block elements (Groups 1 and 2), the p-block elements (Groups 3 through 8), the d-block elements (the transition metals), and the f-block elements (the inner transition metals).

For the main group elements, the number of valence electrons is equal to the group number.

For the transition metals, the number of valence electrons varies and is equal to the oxidation state of the transition metal in the compound or species.

9-22 Atomic radius is a periodic property.

Atomic radii determined by X-ray analysis of crystal structures are called crystallographic radii (Figure 9-38).

Atomic radii usually decrease from left to right across the periodic table.

Atomic radii increase going down a column of the periodic table (Figure 9-39).

First ionization energies decrease going down a column of the periodic table.

B. SELF-TEST

1. The first ionization energy of an atom is always greater than the second ionization energy of that atom. *True/False*

2. The atoms of group _____ in the periodic table have the largest first ionization energies for a given row.

3. A small first ionization energy indicates a stable electronic structure. *True/False*

4. The alkali metals have relatively (*small/large*) values of the first ionization energy.

5. The chemically active electrons that are most responsible for the chemical activity of an atom are located _____ .

6. A plot of successive ionization energies versus the number of electrons removed suggests that the electrons in atoms are arranged in shells. *True/False*

7. A lithium atom consists of a _____ -like inner shell and _____ .

8. The Lewis electron-dot formula of a lithium atom is _____ .

9. The wavelength and frequency of electromagnetic radiation are related by the equation _____.

10. The speed of light is denoted by the symbol _____ and has the value _____ $m \cdot s^{-1}$.

11. The spectrum of white light is an example of a (*continuous/line*) spectrum.

12. A line spectrum consists of _____
_____.

13. An atomic spectrum is a (*continuous/line*) spectrum.

14. The Rydberg-Balmer equation gives an empirical relationship for the _____ of the lines in the visible spectrum of hydrogen.

15. The atomic spectrum of an element is unique to that element. *True/False*

16. Blackbody radiation is radiation that is emitted by _____
_____.

17. Planck assumed that radiation from a blackbody could be emitted _____
_____.

18. The energy associated with quanta of electromagnetic radiation is given by the equation $E =$ _____ where _____
_____.

19. Planck's constant is denoted by _____ and is equal to _____.

20. The photoelectric effect is the _____
_____.

21. Electrons are ejected from the surface of a metal when irradiated with radiation of any frequency. *True/False*

22. In order to explain the photoelectric effect, Einstein proposed that the incident radiation _____
_____.

23. The plot of the kinetic energy of the ejected electrons versus the frequency of the incident radiation is a straight line when the frequency is greater than v_0. *True/False*

24. A photon is a quantum of _____.

25. A moving particle may behave like a wave. *True/False*

26. The de Broglie wavelength of a moving particle is given by the expression $\lambda =$ _____ where _____.

27. The de Broglie wavelength of a particle is significant when the mass of the particle is (*small/large*).

28. The (*wave/particle*) property of electrons is exploited in electron microscopes.

29. The Bohr theory for an electron in a hydrogen atom predicts that the energies the electron may have are restricted to discrete values. *True/False*

30. The energies of the electron in a hydrogen atom are given by $E_n =$ _____ where _____.

31. The interaction energy between an electron and a proton is very small when they are (*close together/far apart*).

32. A negative energy state is (*more/less*) stable relative to a zero energy state.

33. The ground state in the quantum theory is _____.

34. Excited states in the quantum theory are _____.

35. When an atom undergoes a transition from a higher energy state to a lower energy state, the atom (*emits/absorbs*) electromagnetic radiation.

36. When the electron in a hydrogen atom goes from the $n = 3$ state to the $n = 1$ state, energy is (*emitted/absorbed*).

37. The atomic absorption spectrum of an element is due to _____ _____.

38. The frequency of absorption is the same as the frequency of emission between the same two energy states. *True/False*

39. The Bohr theory provides a theoretical explanation of the Rydberg-Balmer equation. *True/False*

40. The Bohr theory cannot be extended to atoms other than hydrogen. *True/False*

41. The Heisenberg uncertainty principle states that _____ _____ _____.

42. The uncertainty principle is due to poor measurement techniques. *True/False*

43. The Heisenberg uncertainty principle is important mainly for (*small/large/all*) particles.

44. The Bohr theory for the hydrogen atom is not consistent with the Heisenberg uncertainty principle. *True/False*

45. The central equation of the quantum theory is the _____.

46. Solutions of the Schrödinger equation for a hydrogen atom give the same set of values of the energy levels as does the Bohr theory. *True/False*

47. The Schrödinger equation can be applied to atoms other than hydrogen. *True/False*

48. The Schrödinger equation restricts the electron in a hydrogen atom to certain sharp orbits. *True/False*

49. Wave functions are a set of functions that are solutions of _____.

50. The value of the square of a wave function is the probability that an electron is found in a small region surrounding a point. *True/False*

51. The electron can be assigned a precise location. *True/False*

52. The principal quantum number n determines the _____ of the electron in a hydrogen atom.

53. The orbital that the electron occupies in the ground state of the hydrogen atom is designated _____.

54. The wave function that describes the ground state of a hydrogen atom depends only on the distance of the electron from the nucleus. *True/False*

55. The value of ψ_{1s}^2 (*increases/decreases*) with the distance of the electron from the nucleus.

56. The orbital ψ_{1s} is (*spherically/cylindrically*) symmetric.

57. A stippled diagram of an orbital represents _____
_____.

58. The azimuthal quantum number ____ determines the _____ of an orbital.

59. The azimuthal quantum number l may have the values _____ when $n = 2$.

60. The 2s orbital is (*spherically/cylindrically*) symmetric.

61. A 2p orbital is (*spherically/cylindrically*) symmetric.

62. For a $2p$ orbital, n equals _____ and l equals _____.

63. When $n = 3$ and $l = 2$, the orbital is designated _____.

64. The magnetic quantum number _____ determines the _____ _____ of an orbital.

65. The magnetic quantum number may have the values _____ when $n = 2$ and $l = 1$.

66. A $2p$ hydrogen atomic orbital for which $m_l = 0$ and one for which $m_l = 1$ differ in energy. *True/False*

67. The three $2p$ orbitals are directed along the _____ axes.

68. The spin quantum number m_s designates _____.

69. The values of m_s are _____.

70. The electronic energy of the hydrogen atom depends on the quantum numbers n and l. *True/False*

71. The electronic energy of a multielectron atom depends on the three quantum numbers n, l, and m_l. *True/False*

72. The $3p$ orbitals and the $3d$ orbitals of an iron atom have the same energy. *True/False*

73. The Pauli exclusion principle states that _____ _____.

74. The term "spin up" refers to electrons with _____.

75. The set of $2p$ orbitals may hold a maximum of _____ electrons.

76. The $1s$ orbital may hold two electrons with parallel spins. *True/False*

77. The L shell is the level for which n equals _____.

78. The L shell can contain a maximum of _____ electrons.

79. The L shell contains _____ subshells.

80. The symbol $2p^5$ signifies that _____.

81. The symbol _____ signifies that there are six electrons in the $3d$ orbitals.

82. In neutral atoms the $4s$ orbital is of higher energy than the $3d$ orbital. *True/False*

83. The 5*d* orbital is of higher energy than the 4*f* orbital in neutral atoms. *True/False*

84. Hund's rule states that _____

_____

_____.

85. The outer electron configuration of the ground state of the Group 2 metals is___.

86. The outer electron configuration of the ground state of the halogens is_____.

87. The first member of each row of the periodic table is a metal. *True/False*

88. The last member of each row of the periodic table is a_____.

89. The number of dots displayed in the Lewis electron-dot formula of a main-group element is the same as the total number of electrons in the element. *True/False*

90. Valence electrons are the electrons _____.

91. The transition metals occur because of the sequential filling of the _____ orbitals.

92. A half-filled subshell has an extra stability relative to other partially filled subshells. *True/False*

93. The lanthanides occur because of the sequential filling of the _____ orbitals.

94. The rare earths are difficult to separate because _____

_____.

95. The actinide series occurs because of the sequential filling of the _____ orbitals.

96. The actinides are all radioactive. *True/False*

97. The *s*-block elements belong to Groups _____.

98. The *p*-block elements belong to Groups _____.

99. The *d*-block elements belong to the _____.

100. The *f*-block elements belong to the _____.

101. Atomic radii can be determined from _____

_____.

102. Atomic radii (*increase/decrease*) in going from left to right across a row in the periodic table.

103. Atomic radii (*increase/decrease*) in going down a column in the periodic table.

104. The first ionization energy decreases as the radius increases because _____

_____ .

C. CALCULATIONS YOU SHOULD KNOW HOW TO DO

1. Convert between the frequency and wavelength of electromagnetic radiation by using Equation (9-1). See Example 9-2 and Problems 9-9 and 9-10.

2. Calculate the energy associated with electromagnetic radiation by using Equation (9-3) or (9-4). See Example 9-3 and Problems 9-11 and 9-12.

3. Calculate the number of photons in a given amount of electromagnetic radiation of a specified frequency or wavelength. To do this type of problem, first calculate the energy of one photon by using Equation (9-3) and then divide this result into the energy of the electromagnetic radiation. See Problems 9-13 and 9-14.

4. Calculate the work function or the threshold frequency of a metal by using Equation (9-5). See Example 9-4 and Problems 9-15 and 9-16.

5. Calculate the kinetic energy of electrons ejected from a metal by using Equation (9-6). See Example 9-5 and Problems 9-17 and 9-18.

6. Calculate the de Broglie wavelengths of moving bodies by using Equation (9-7b). See Example 9-6 and Problems 9-19 through 9-22.

7. Calculate the frequencies and wavelengths in the hydrogen atomic spectrum. The starting point for calculations like these is Equation (9-9), which gives the energy states of the electron in a hydrogen atom. Using Equation (9-9), the energy and the frequency associated with a transition from state n to state m are obtained by using conservation of energy. For a transition from state n to state m, we write

$$\nu_{nm} = \frac{E_m - E_n}{h} \qquad m > n \qquad \text{(absorption)}$$

$$\nu_{mn} = \frac{E_m - E_n}{h} \qquad m > n \qquad \text{(emission)}$$

See Examples 9-7 and 9-8 and Problems 9-23 through 9-26.

8. Calculate the uncertainty in a measurement of a moving body using the Heisenberg uncertainty principle. See Example 9-9.

9. Determine the allowed sets of quantum numbers (n, l, m_l, m_s). See Examples 9-11 and 9-12 and Problems 9-29 through 9-40.

10. Know the mnemonic for the order of the orbital energies in neutral gas atoms.

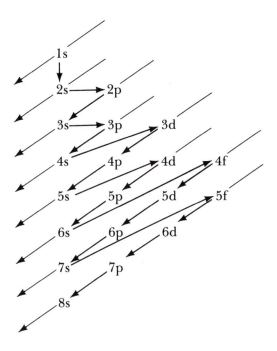

11. Assign electrons to orbitals in atoms using the Pauli exclusion principle and Hund's rule and write out ground-state electron configurations. See Examples 9-14 and 9-17 and Problems 9-41 through 9-46.

12. Correlate ground-state electron configurations and the periodic table. See Problems 9-47 through 9-54.

13. Write the ground-state electron configurations of ions. See Example 9-13 and Problems 9-55 through 9-60.

14. Write excited-state electron configurations. See Example 9-15 and Problems 9-61 and 9-62.

15. Use the periodic table to predict relative values of atomic and ionic radii. See Problems 9-63 through 9-66.

D. SOLUTIONS TO THE ODD-NUMBERED PROBLEMS

9-1   Of the four species listed, Be has the smallest ionization energy because it is a Group 2 metal. The ionization energies of the remaining three species decrease in the order He, Ne, and Kr because ionization energies decrease with increasing atomic size within a family. The farther away an electron is from the nucleus, the easier it is to remove the electron from the atom. Thus we write Be, Kr, Ne, and He.

9-3   Using the data in Table 9-1, we have for a boron atom

n	$I_n/\text{MJ}\cdot\text{mol}^{-1}$	$\log[I_n/\text{MJ}\cdot\text{mol}^{-1}]$
1	0.80	-0.097
2	2.42	0.384
3	3.66	0.563
4	25.02	1.400
5	32.82	1.516

The following plot suggests that the five electrons are arranged in two shells, with two electrons in an inner, tightly held shell and three in an outer shell.

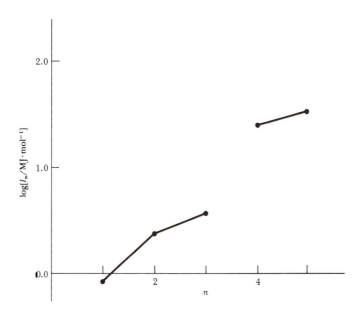

9-5   The alkali metals are in Group 1 and have one outer electron. The Lewis electron-dot formulas of the alkali metals are

$$\text{Li}\cdot \quad \text{Na}\cdot \quad \text{K}\cdot \quad \text{Rb}\cdot \quad \text{Cs}\cdot \quad \text{Fr}\cdot$$

The halogens are in Group 7 and have seven outer electrons. The Lewis electron-dot formulas of the halogen atoms are

$$:\!\overset{\displaystyle ..}{\underset{\displaystyle ..}{\text{F}}}\!\cdot \quad :\!\overset{\displaystyle ..}{\underset{\displaystyle ..}{\text{Cl}}}\!\cdot \quad :\!\overset{\displaystyle ..}{\underset{\displaystyle ..}{\text{Br}}}\!\cdot \quad :\!\overset{\displaystyle ..}{\underset{\displaystyle ..}{\text{I}}}\!\cdot \quad :\!\overset{\displaystyle ..}{\underset{\displaystyle ..}{\text{At}}}\!\cdot$$

9-7 Argon (Group 8) has eight outer electrons and so we write $:\overset{\cdot\cdot}{\underset{\cdot\cdot}{Ar}}:$, or simply Ar.

Sulfur (Group 6) has six outer electrons and so we write $\cdot\overset{\cdot\cdot}{S}\cdot$.

S^{2-} has eight outer electrons and so we write $:\overset{\cdot\cdot}{\underset{\cdot\cdot}{S}}:^{2-}$.

Aluminum (Group 3) has three outer electrons, Al^{3+} has zero outer electrons, and so we write Al^{3+}.

Chlorine (Group 7) has seven outer electrons, Cl^- has eight outer electrons, and so we write $:\overset{\cdot\cdot}{\underset{\cdot\cdot}{Cl}}:^-$.

9-9 Solve Equation (9-1) for v.

$$v = \frac{c}{\lambda} = \frac{3.00 \times 10^8 \text{ m} \cdot \text{s}^{-1}}{633 \times 10^{-9} \text{ m}} = 4.74 \times 10^{14} \text{ s}^{-1}$$

9-11 We first convert the value of the first ionization energy of potassium from units of $kJ \cdot mol^{-1}$ to units of $J \cdot atom^{-1}$:

$$E = (419 \text{ kJ} \cdot \text{mol}^{-1}) \left(\frac{10^3 \text{ J}}{1 \text{ kJ}}\right)\left(\frac{1 \text{ mol}}{6.022 \times 10^{23} \text{ atom}}\right) = 6.96 \times 10^{-19} \text{ J} \cdot \text{atom}^{-1}$$

We then use Equation (9-4) to calculate the wavelength of the radiation that corresponds to this energy:

$$E = \frac{hc}{\lambda}$$

Solving for λ, we have

$$\lambda = \frac{hc}{E} = \frac{(6.626 \times 10^{-34} \text{ J} \cdot \text{s})(3.00 \times 10^8 \text{ m} \cdot \text{s}^{-1})}{6.96 \times 10^{-19} \text{ J}}$$
$$= 2.86 \times 10^{-7} \text{ m} = 286 \text{ nm}$$

9-13 The energy per photon of green light is given by Equation (9-4)

$$E = \frac{hc}{\lambda} = \frac{(6.626 \times 10^{-34} \text{ J} \cdot \text{s})(3.00 \times 10^8 \text{ m} \cdot \text{s}^{-1})}{510 \times 10^{-9} \text{ m}}$$
$$= 3.90 \times 10^{-19} \text{ J} \cdot \text{photon}^{-1}$$

The number of photons in 2.35×10^{-18} J of light is

$$\frac{2.35 \times 10^{-18} \text{ J}}{3.90 \times 10^{-19} \text{ J} \cdot \text{photon}^{-1}} = 6 \text{ photons}$$

9-15 The energy of 200-nm radiation is given by Equation (9-4)

$$E = \frac{hc}{\lambda} = \frac{(6.626 \times 10^{-34} \text{ J} \cdot \text{s})(3.00 \times 10^8 \text{ m} \cdot \text{s}^{-1})}{200 \times 10^{-9} \text{ m}}$$
$$= 9.94 \times 10^{-19} \text{ J}$$

This is larger than the work function of gold, and so electrons will be ejected from the surface of the gold.

9-17 The energy of 400-nm radiation is given by Equation (9-4).

$$E = \frac{hc}{\lambda} = \frac{(6.626 \times 10^{-34} \, \text{J} \cdot \text{s})(3.00 \times 10^8 \, \text{m} \cdot \text{s}^{-1})}{400 \times 10^{-9} \, \text{m}}$$
$$= 4.97 \times 10^{-19} \, \text{J}$$

We calculate the kinetic energy of an ejected electron by using Equation (9-6).

$$\text{K.E.} = h\nu - \Phi$$
$$= 4.97 \times 10^{-19} \, \text{J} - 2.90 \times 10^{-19} \, \text{J}$$
$$= 2.07 \times 10^{-19} \, \text{J}$$

9-19 We use Equation (9-7b),

$$\lambda = \frac{h}{m\nu}$$

The mass and the speed of the proton are given, and so

$$\lambda = \frac{6.626 \times 10^{-34} \, \text{J} \cdot \text{s}}{(1.67 \times 10^{-27} \, \text{kg})(1.00 \times 10^5 \, \text{m} \cdot \text{s}^{-1})}$$
$$= 3.97 \times 10^{-12} \, \text{m} = 3.97 \, \text{pm}$$

Recall that a joule is equal to a $\text{kg} \cdot \text{m}^2 \cdot \text{s}^{-2}$.

9-21 The mass of a hydrogen molecule is

$$m = \frac{2.016 \, \text{g} \cdot \text{mol}^{-1}}{6.022 \times 10^{23} \, \text{molecule} \cdot \text{mol}^{-1}} = 3.348 \times 10^{-24} \, \text{g}$$
$$= 3.348 \times 10^{-27} \, \text{kg}$$

Therefore,

$$\lambda = \frac{h}{m\nu} = \frac{6.626 \times 10^{-34} \, \text{J} \cdot \text{s}}{(3.348 \times 10^{-27} \, \text{kg})(2000 \, \text{m} \cdot \text{s}^{-1})} = 9.90 \times 10^{-11} \, \text{m}$$
$$= 99.0 \, \text{pm}$$

We have used the fact that a joule is equal to a $\text{kg} \cdot \text{m}^2 \cdot \text{s}^{-2}$.

9-23 The energy required for the electron in a hydrogen atom to make a transition from the $n = 2$ state to the $n = 3$ state is given by

$$\Delta E = E_3 - E_2 = -\frac{2.18 \times 10^{-18} \, \text{J}}{3^2} - \left(-\frac{2.18 \times 10^{-18} \, \text{J}}{2^2}\right)$$
$$= 3.03 \times 10^{-19} \text{J}$$

The wavelength of a photon with this energy is obtained from Equation (9-4).

$$\lambda = \frac{hc}{\Delta E} = \frac{(6.626 \times 10^{-34}\,\text{J} \cdot \text{s})(3.00 \times 10^8\,\text{m} \cdot \text{s}^{-1})}{3.03 \times 10^{-19}\,\text{J}} = 6.56 \times 10^{-7}\,\text{m}$$
$$= 656\,\text{nm}$$

9-25   We use Equation (9-13) to determine the state that results when a 97.2-nm photon is absorbed by a ground-state hydrogen atom.

$$\nu_{1n} = (3.29 \times 10^{15}\,\text{s}^{-1})\left(1 - \frac{1}{n^2}\right)$$

The frequency of the 97.2-nm photon is

$$\nu = \frac{c}{\lambda} = \frac{3.00 \times 10^8\,\text{m} \cdot \text{s}^{-1}}{97.2 \times 10^{-9}\,\text{m}} = 3.086 \times 10^{15}\,\text{s}^{-1}$$

Substituting in the value of ν_{1n} and solving for $(1 - 1/n^2)$, we have

$$\left(1 - \frac{1}{n^2}\right) = \frac{3.086 \times 10^{15}\,\text{s}^{-1}}{3.29 \times 10^{15}\,\text{s}^{-1}} = 0.938$$

Thus

$$\frac{1}{n^2} = 1 - 0.938 = 0.062$$

or

$$n^2 = 16$$

Therefore

$$n = 4$$

The electron then makes a transition from the $n = 4$ state to some lower state and emits a 486-nm photon. The final state can be determined from

$$\nu_{4n} = (3.29 \times 10^{15}\,\text{s}^{-1})\left(\frac{1}{n^2} - \frac{1}{4^2}\right)$$

The frequency of the 486-nm photon is

$$\nu = \frac{3.00 \times 10^8\,\text{m} \cdot \text{s}^{-1}}{486 \times 10^{-9}\,\text{m}} = 6.173 \times 10^{14}\,\text{s}^{-1}$$

Thus, we have

$$\frac{1}{n^2} - \frac{1}{16} = \frac{6.173 \times 10^{14}\,\mathrm{s}^{-1}}{3.29 \times 10^{15}\,\mathrm{s}^{-1}} = 0.188$$

or

$$\frac{1}{n^2} = 0.188 + \frac{1}{16} = 0.250$$

Thus

$$n^2 = 4$$

and

$$n = 2$$

9-27 To calculate the ionization energies, we let $Z = 2$ for He^+, $Z = 3$ for Li^{2+}, and $Z = 4$ for Be^{3+}. The results are

$$IE(He^+) = E_\infty - E_1 = \frac{(2.18 \times 10^{-18}\,\mathrm{J})2^2}{1^2} = 8.72 \times 10^{-18}\,\mathrm{J}$$

$$IE(Li^{2+}) = \frac{(2.18 \times 10^{-18}\,\mathrm{J})3^2}{1^2} = 1.96 \times 10^{-17}\,\mathrm{J}$$

$$IE(Be^{3+}) = \frac{(2.18 \times 10^{-18}\,\mathrm{J})4^2}{1^2} = 3.49 \times 10^{-17}\,\mathrm{J}$$

The values given in Table 9-1 are per mole, and so we must multiply each of these results by Avogadro's number in order to compare with Table 9-1. The results are

$$IE(He^+) = 5.25\ \mathrm{MJ \cdot mol^{-1}} = I_2$$
$$IE(Li^{2+}) = 11.8\ \mathrm{MJ \cdot mol^{-1}} = I_3$$
$$IE(Be^{3+}) = 21.0\ \mathrm{MJ \cdot mol^{-1}} = I_4$$

in excellent agreement with the values in Table 9-1.

9-29 (a) $n = 7, l = 0$ (possible)
 (b) $n = 1, l = 1$ (not possible, because $l = n$)
 (c) $n = 5, l = 2$ (possible)
 (d) $n = 2, l = 2$ (not possible, because $l = n$)
 (e) $n = 4, l = 3$ (possible)

9-31 (a)   If $n = 4$ and $l = 1$, then the orbital is a $4p$ orbital.
 (b)   If $n = 3$ and $l = 2$, then the orbital is a $3d$ orbital.
 (c)   If $n = 4$ and $l = 2$, then the orbital is a $4d$ orbital.
 (d)   If $n = 2$ and $l = 0$, then the orbital is a $2s$ orbital.

9-33 If $l = 2$, then n must be at least 3 because $l = 0, 1, 2, \ldots, n - 1$.

If $m_l = 3$, then l must be at least 3 because $m_l = -l, -l+1, \ldots, -1, 0, +1, \ldots, l$.

9-35 For a $3d$ orbital $n = 3$ and $l = 2$; thus m_l can be $-2, -1, 0, 1,$ or 2. For each value of m_l, m_s can be $+\frac{1}{2}$ or $-\frac{1}{2}$. Therefore, the 10 possible sets of four quantum numbers are

n	l	m_l	m_s
3	2	-2	$+\frac{1}{2}$
			$-\frac{1}{2}$
		-1	$+\frac{1}{2}$
			$-\frac{1}{2}$
		0	$+\frac{1}{2}$
			$-\frac{1}{2}$
		$+1$	$+\frac{1}{2}$
			$-\frac{1}{2}$
		$+2$	$+\frac{1}{2}$
			$-\frac{1}{2}$

9-37 Note that for each value of m_l, m_s can have the values $+\frac{1}{2}$ or $-\frac{1}{2}$. Thus the maximum number of electrons in a subshell is equal to two times the number of possible values of m_l.

				Maximum number of electrons
s orbital	$l = 0$	$m_l = 0$	$m_s = +\frac{1}{2}, -\frac{1}{2}$	$2 \times 1 = 2$ electrons
p orbital	$l = 1$	$m_l = -1, 0, 1$	$m_s = +\frac{1}{2}, -\frac{1}{2}$	$2 \times 3 = 6$ electrons
d orbital	$l = 2$	$m_l = -2, -1, 0, 1, 2$	$m_s = +\frac{1}{2}, -\frac{1}{2}$	$2 \times 5 = 10$ electrons
f orbital	$l = 3$	$m_l = -3, -2, -1, 0, 1, 2, 3$	$m_s = +\frac{1}{2}, -\frac{1}{2}$	$2 \times 7 = 14$ electrons

9-39 In a d transition series the five d orbitals are being filled. A set of five d orbitals can hold up to 10 electrons.

9-41 (a) Ruled out; the $2p$ orbitals cannot hold seven electrons.
(b) Ruled out; a $3s$ orbital cannot hold three electrons.
(c) Ruled out; the $3d$ orbitals hold only 10 electrons.
(d) Allowed

9-43 (a) $1s^22s^22p^63s^23p^2$ 14 electrons, silicon
      (b) $1s^22s^22p^63s^23p^64s^13d^5$ 24 electrons, chromium
      (c) $1s^22s^22p^63s^23p^64s^23d^{10}4p^2$ 32 electrons, germanium
      (d) $1s^22s^22p^63s^23p^64s^23d^{10}4p^5$ 35 electrons, bromine
      (e) $1s^22s^22p^1$ 5 electrons, boron

9-45 (a) Ti: $[Ar]4s^23d^2$ (b) K: $[Ar]4s^1$
      (c) Fe: $[Ar]4s^23d^6$ (d) As: $[Ar]4s^23d^{10}4p^3$

9-47 (a) Ca: $[Ar]4s^2$ (b) Br: $[Ar]4s^23d^{10}4p^5$
      (c) Ag: $[Kr]5s^14d^{10}$ (d) Zn: $[Ar]4s^23d^{10}$

9-49 (a) Groups 1 and 2   (b) Groups 3, 4, 5, 6, 7, 8
      (c) the transition metals   (d) the lanthanides and the actinides

9-51 We write the ground-state electron configuration and apply Hund's rule to each case.

      (a) Ge: $[Ar]4s^23d^{10}4p^2$; two unpaired electrons
      (b) Se: $[Ar]4s^23d^{10}4p^4$; two unpaired electrons
      (c) V: $[Ar]4s^23d^3$; three unpaired electrons
      (d) Fe: $[Ar]4s^23d^6$; four unpaired electrons

9-53 The number of electrons gained is equal to the number of electrons that must be added to attain the noble-gas configuration at the end of the row in which the element is located in the periodic table.

      (a) 1 electron $\cdot\text{H}\cdot^-$ helium

      (b) 2 electrons $:\ddot{\text{O}}:^{2-}$ neon

      (c) 4 electrons $:\ddot{\text{C}}:^{4-}$ neon

      (d) 2 electrons $:\ddot{\text{S}}:^{2-}$ argon

9-55 First we determine the number of electrons in the ion, and then we write the ground-state electron configuration.

      (a) $15 + 3 = 18$ electrons: $1s^22s^22p^63s^23p^6$; [Ar]

      (b) $35 + 1 = 36$ electrons: $1s^22s^22p^63s^23p^64s^23d^{10}4p^6$; [Kr]

      (c) $34 + 2 = 36$ electrons: $1s^22s^22p^63s^23p^64s^23d^{10}4p^6$; [Kr]

      (d) $56 - 2 = 54$ electrons: $1s^22s^22p^63s^23p^64s^23d^{10}4p^65s^24d^{10}5p^6$; [Xe]
      They all have attained a noble-gas ground-state electron configuration.

9.57 We first determine the total number of electrons in the ion, then we determine the ground-state electron configuration, and finally we count the number of unpaired electrons using Hund's rule.

	Species	Ground-state configuration	Number of unpaired electrons
(a)	F^+	$[He]2s^22p^4$	2
(b)	Sn^{2+}	$[Kr]5s^24d^{10}$	0
(c)	Bi^{3+}	$[Xe]6s^24f^{14}5d^{10}$	0
(d)	Ar^+	$[Ne]3s^23p^5$	1

9-59   In this problem we work out the electronic configurations of the reactants and the products.

    (a) $O(g) + 2e^- \rightarrow O^{2-}(g)$
          $[He]2s^2 2p^4 + 2e^- \rightarrow [He]2s^2 2p^6$ or $[Ne]$
    (b) $Ca(g) + Sr^{2+}(g) \rightarrow Sr(g) + Ca^{2+}(g)$
          $[Ar]4s^2 + [Kr] \rightarrow [Kr]5s^2 + [Ar]$

9-61   First we determine the electron configuration of the ground state; then we promote one electron from the highest energy subshell that is occupied to the lowest energy subshell that is unoccupied.

    (a)   The ground state of Be^{2+} ($4 - 2 = 2$ electrons) is $1s^2$. The highest energy occupied subshell is the $1s$, and the lowest energy unoccupied subshell is the $2s$; thus the first excited state is obtained by promoting an electron from the $1s$ subshell to the $2s$ subshell to yield $1s^1 2s^1$.
    (b)   The gound state of He^+ (1 electron) is $1s^1$. The first excited state is $2s^1$.
    (c)   The ground state of F^- (10 electrons) is $1s^2 2s^2 2p^6$. The first excited state is $1s^2 2s^2 2p^5 3s^1$.
    (d)   The ground state of O^{2-} is $1s^2 2s^2 2p^6$. The first excited state is $1s^2 2s^2 2p^5 3s^1$.

9-63   Recall that atomic radii increase as we move down a column of the periodic table, whereas atomic radii decrease as we move from left to right across a row of the periodic table.

    (a) $P > N$ (b) $P > S$
    (c) $S > Ar$ (d) $Kr > Ar$

9-65   (a) $Li < Na < Rb < Cs$
    (b) $P < Al < Mg < Na$
    (c) $Ca < Sr < Ba$

9-67   The electron configurations of the alkaline earth metals are classified as the [noble gas] ns^2 type. The energy of attraction between the nucleus and the outer electrons is greater the higher the value of the nuclear charge Z and is less the higher the value of n, the principal quantum number. These two effects oppose one another as we move down a group. However, the underlying electrons in the noble-gas-like core partially screen the nuclear charge, and the farther an electron is from the nucleus (larger n) the lower is the ionization energy.

9-69   The energy of one photon is given by

$$E = \frac{hc}{\lambda} = \frac{(6.626 \times 10^{-34} \text{ J} \cdot \text{s})(3.00 \times 10^8 \text{ m} \cdot \text{s}^{-1})}{10.6 \times 10^{-6} \text{ m}} = 1.88 \times 10^{-20} \text{ J}$$

The number of photons in a pulse of 1 J is

$$\frac{1 \text{ J}}{1.88 \times 10^{-20} \text{ J} \cdot \text{photon}^{-1}} = 5 \times 10^{19} \text{ photons}$$

9-71   The speed of the electrons is given by Equation (9-7):

$$v = \frac{h}{m\lambda} = \frac{(6.626 \times 10^{-34} \text{ J} \cdot \text{s})}{(9.11 \times 10^{-31} \text{ kg})(2.00 \times 10^{-9} \text{ m})} = 3.637 \times 10^5 \text{ m} \cdot \text{s}^{-1}$$

The kinetic energy of an electron with this speed is given by

$$E = \tfrac{1}{2}mv^2 = \frac{(9.11 \times 10^{-31} \text{ kg})(3.637 \times 10^5 \text{ m}\cdot\text{s}^{-1})^2}{2}$$

$$= 6.02 \times 10^{-20} \text{ J}$$

9-73 The frequencies of the lines in the Lyman series of the hydrogen atom are given by Equation (9-13):

$$\nu_{n1} = (3.29 \times 10^{15} \text{ s}^{-1}) \left(\frac{1}{1^2} - \frac{1}{n^2} \right) \quad n = 2, 3 \ldots$$

The values of ν and $1/n^2$ are given by

Transition	$\nu_{n \to 1}/10^{15} \text{ s}^{-1}$	$1/n^2$
$2 \to 1$	2.47	0.250
$3 \to 1$	2.92	0.111
$4 \to 1$	3.08	0.0625
$5 \to 1$	3.16	0.040

These data are plotted below. Note that a plot of $\nu_{n \to 1}$ versus $1/n^2$ is a straight line.

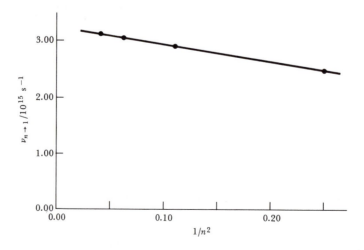

9-75 We must use Hess's law to do this problem. From Table 9-1, we find that

(a)
$$Li(g) \rightarrow Li^+(g) + e^-(g) \qquad \Delta H^\circ_{rxn} = I_1 = 0.52 \text{ MJ}$$
$$\underline{Na^+(g) + e^-(g) \rightarrow Na(g) \qquad \Delta H^\circ_{rxn} = -I_1 = -0.50 \text{ MJ}}$$
$$Li(g) + Na^+(g) \rightarrow Li^+(g) + Na(g) \qquad \Delta H^\circ_{rxn} = 0.020 \text{ MJ}$$
$$= 20 \text{ kJ}$$

(b)
$$Mg^{2+}(g) + e^-(g) \rightarrow Mg^+(g) \qquad \Delta H^\circ_{rxn} = -I_2 = -1.45 \text{ MJ}$$
$$\underline{Mg(g) \rightarrow Mg^+(g) + e^-(g) \qquad \Delta H^\circ_{rxn} = I_1 = 0.74 \text{ MJ}}$$
$$Mg^{2+}(g) + Mg(g) \rightarrow 2Mg^+(g) \qquad \Delta H^\circ_{rxn} = -0.71 \text{ MJ}$$
$$= -710 \text{ kJ}$$

(c)
$$Al^{3+}(g) + e^-(g) \rightarrow Al^{2+}(g) \qquad \Delta H^\circ_{rxn} = -I_3 = -2.74 \text{ MJ}$$
$$Al^{2+}(g) + e^-(g) \rightarrow Al^+(g) \qquad \Delta H^\circ_{rxn} = -I_2 = -1.82 \text{ MJ}$$
$$\underline{Al^+(g) + e^-(g) \rightarrow Al(g) \qquad \Delta H^\circ_{rxn} = -I_1 = -0.58 \text{ MJ}}$$
$$Al^{3+}(g) + 3e^-(g) \rightarrow Al(g) \qquad \Delta H^\circ_{rxn} = -5.14 \text{ MJ}$$

9-77 Once we determine the outer electron configuration of the element with $Z = 16$, then we proceed to find the next element with the same outer electron configuration.

(a) The electron configuration corresponding to $Z = 16$ electrons is $1s^2 2s^2 2p^6 3s^2 3p^4$. The outer electron configuration is $3p^4$. The next element with an np^4 configuration is $[Ar]4s^2 3d^{10} 4p^4$ and $Z = 34$. The next element with np^4 is $[Kr]5s^2 4d^{10} 5p^4$ and $Z = 52$. The next element with np^4 is $[Xe]6s^2 4f^{14} 5d^{10} 6p^4$ and $Z = 84$. The first element with np^4 is $[He]2s^2 2p^4$ and $Z = 8$.

(b) Eleven electrons correspond to the electron configuration $1s^2 2s^2 2p^6 3s^1$. The succeeding elements with ns^1 outer electron configurations are $[Ar]4s^1$, $Z = 19$; $[Kr]5s^1$, $Z = 37$; $[Xe]6s^1$, $Z = 55$, and $[Rn]7s^1$, $Z = 87$. The first elements are $[He]2s^1$, $Z = 3$, and $1s^1$, $Z = 1$.

9-79 (a) 15 electrons: $1s^2 2s^2 2p^6 3s^2 3p^3$
no d electrons
nine electrons with $l = 1$ (p electrons)
three unpaired electrons (Hund's rule)

(b) 26 electrons: $1s^2 2s^2 2p^6 3s^2 3p^6 4s^2 3d^6$
six d electrons
12 electrons with $l = 1$ (p electrons)
four unpaired electrons (Hund's rule)

(c) 32 electrons: $1s^2 2s^2 2p^6 3s^2 3p^6 4s^2 3d^{10} 4p^2$
ten d electrons
14 electrons with $l = 1$ (p electrons)
two unpaired electrons (Hund's rule)

9-81 If the $2s$ and $2p$ orbitals had the same energy, the ground-state outer electron configurations of the second-row elements would be

$$2s^1, \; 2s^1 2p_x^1, \; 2s^1 2p_x^1 2p_y^1, \; 2s^1 2p_x^1 2p_y^1 2p_z^1, \; 2s^2 2p_x^1 2p_y^1 2p_z^1, \; 2s^2 2p_x^2 2p_y^1 2p_z^1, \; 2s^2 2p_x^2 2p_y^2 2p_z^1, \; 2s^2 2p_x^2 2p_y^2 2p_z^2$$

9-83 The Heisenberg uncertainty principle tells us that the product of the uncertainties in the momentum p and the position x is given by

$$(\Delta p)(\Delta x) = h$$

where h is Planck's constant. The momentum p is equal to mv, where m is the mass and v is the velocity, and $\Delta p = m \, \Delta v$. The mass of a hydrogen atom is

$$\text{mass} = \left(\frac{1.008 \text{ g H}}{1 \text{ mol H}}\right)\left(\frac{1 \text{ kg}}{10^3 \text{ g}}\right)\left(\frac{1 \text{ mol}}{6.022 \times 10^{23} \text{ atom}}\right) = 1.67 \times 10^{-27} \text{ kg}$$

Thus

$$\Delta v = \frac{h}{m \, \Delta x} = \frac{6.626 \times 10^{-34} \text{ J} \cdot \text{s}}{(1.67 \times 10^{-27} \text{ kg})(1.00 \times 10^{-12} \text{ m})}$$

$$= 3.97 \times 10^5 \text{ m} \cdot \text{s}^{-1}$$

9-85   The number of valence electrons in a transition metal is equal to the oxidation state of the metal in its compound.

(a)   2   (b)   7   (c)   6   (d)   1   (e)   2

9-87   We are given that $1 \text{ J} = 1 \text{ C} \cdot \text{V}$; therefore,

$$1 \text{ V} = \frac{1 \text{ J}}{1 \text{ C}}$$

We multiply by the number of coulombs in one mole of electrons to get

$$1 \text{ eV} = \left(\frac{1 \text{ J}}{1 \text{ C}}\right)(1.61 \times 10^{-19} \text{ C} \cdot \text{electron}^{-1})(6.022 \times 10^{23} \text{ electrons} \cdot \text{mol}^{-1})$$

$$= 96950 \text{ J} \cdot \text{mol}^{-1} = 96.95 \text{ kJ} \cdot \text{mol}^{-1}$$

9-89   Equation (9-2) can be written as

$$\bar{v} = (1.097 \times 10^7 \text{ m}^{-1})\left(\frac{1}{4} - \frac{1}{n^2}\right)$$

The Rydberg constant becomes

$$(1.097 \times 10^7 \text{ m}^{-1})\left(\frac{1 \text{ m}}{10^2 \text{ cm}}\right) = 1.097 \times 10^5 \text{ cm}^{-1}$$

9-91   The energy of a 900-nm photon is given by Equation (9-4):

$$E = \frac{hc}{\lambda} = \frac{(6.626 \times 10^{-34} \text{ J} \cdot \text{s})(3.00 \times 10^8 \text{ m} \cdot \text{s}^{-1})}{900 \times 10^{-9} \text{ m}}$$

$$= 2.21 \times 10^{-19} \text{ J} \cdot \text{photon}^{-1}$$

The amount of energy required to raise the temperature of 1.00 L of water by 1.00°C is

$$q = (4.18 \text{ J} \cdot \text{K}^{-1} \cdot \text{g}^{-1})(1.00 \text{ L}) \left(\frac{1000 \text{ mL}}{1 \text{ L}} \right) (1.00 \text{ g} \cdot \text{mL}^{-1})(1.00 \text{ K})$$

$$= 4180 \text{ J}$$

The number of 900-nm photons required to supply this energy is given by

$$\text{number of photons} = \frac{4180 \text{ J}}{2.21 \times 10^{-19} \text{ J} \cdot \text{photon}^{-1}} = 1.89 \times 10^{22} \text{ photons}$$

9-93 The data are plotted below. Equation (9-6), K.E. $= h\nu - \Phi = h\nu - h\nu_0$, tells us that a plot of K.E. versus ν is a straight line whose slope is Planck's constant. The intercept of the straight line with the horizontal axis (where K.E. $= 0$) gives the threshold frequency. From the graph we find that $\nu_0 = 1.1 \times 10^{15} \text{ s}^{-1}$ and that the slope is given by (for example, if we take the second and fourth sets of data)

$$\text{slope} = h = \frac{(15.84 - 9.21) \times 10^{-19} \text{ J}}{(3.50 - 2.50) \times 10^{15} \text{ s}^{-1}} = 6.63 \times 10^{-34} \text{ J} \cdot \text{s}$$

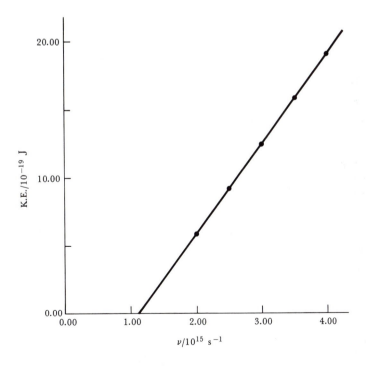

E. ANSWERS TO THE SELF-TEST

1. false

2. 8, or the noble gases

3. false

4. small

5. in the outer shell

6. true

7. helium . . . a lone outer electron

8. Li·

9. $\lambda v = c$

10. c . . . 3.00×10^8

11. continuous

12. a set of lines with definite wavelengths

13. line

14. wavelength

15. true

16. a solid body when heated to high temperatures

17. only in quantized packets

18. $E = hv$ . . . h is Planck's constant and v is the frequency.

19. h . . . 6.626×10^{-34} J·s

20. ejection of electrons from the surface of a metal when it is irradiated with ultraviolet radiation

21. false

22. was composed of little packets of energy, which he called photons

23. true

24. electromagnetic radiation

25. true

26. h/mv . . . h is Planck's constant, m is the mass of the particle, and v is its speed

27. small

28. wave

29. true

30. $-(2.18 \times 10^{-18} \text{ J})/n^2$ . . . $n = 1, 2, 3, \ldots$

31. far apart

32. more

33. the state of lowest energy

34. states with energies greater than the ground state

35. emits

36. emitted

37. electrons making transitions from states of lower energy to states of higher energies

38. true

39. true

40. true

41. it is not possible to measure accurately both the position and the momentum of a particle simultaneously

42. false

43. small

44. true

45. Schrödinger equation

46. true

47. true

48. false

49. the Schrödinger equation

50. true

51. false

52. energy

53. $1s$

54. true

55. decreases

56. spherically

57. the probability density of finding an electron

58. l . . . the shape

59. 0 and 1

60. spherically

61. cylindrically

62. $n = 2$ and $l = 1$

63. $3d$

64. m_l . . . the spatial orientation

65. $+1, 0, -1$

66. false (Energy levels of a hydrogen atom depend upon only n.)

67. x, y, and z

68. the spin of an electron

69. $+\frac{1}{2}$ and $-\frac{1}{2}$

70. false

71. false (It depends only on n and l.)

72. false

73. no two electrons in the same atom can have the same four quantum numbers

74. positive spin, or $m_s = +\frac{1}{2}$

75. six

76. false (The spins must be opposed.)

77. 2

78. eight

79. two (s and p)

80. there are five electrons in the $2p$ orbitals

81. $3d^6$

82. false

83. true

84. for any set of orbitals of the same energy, the ground-state electron configuration is obtained by placing the electrons into different orbitals of this set with parallel spins until each of the orbitals has one electron before any electrons are paired

85. ns^2

86. ns^2np^5

87. true

88. noble gas

89. false

90. that are in the outer shell of an atom

91. d

92. true

93. $4f$

94. they have similar chemical properties

95. $5f$

96. true

97. 1 and 2

98. 3 through 8

99. transition metals

100. inner transition metals

101. crystallographic data (X-ray analysis of crystal structure)

102. decrease

103. increase

104. the outer electron is farther from the nucleus

10 IONIC BONDS AND COMPOUNDS

A. OUTLINE OF CHAPTER 10

10-1 Solutions that contain ions conduct an electric current.

Ionic compounds yield ions when they dissolve in water.

The movement of ions through a solution constitutes an electric current.

Electrolytes in aqueous solution conduct an electric current.

Nonelectrolytes in aqueous solution do not conduct an electric current.

A solution of a weak electrolyte conducts electricity to a lesser extent than a solution of a strong electrolyte of the same concentration.

Some guidelines for predicting whether a substance is a strong, weak, or non-electrolyte are given on pages 356 and 357 of the text.

A salt that is a weak electrolyte does not dissociate completely in aqueous solution.

The degree of dissociation of a salt in aqueous solution can be determined by measuring the electrical conductance of the aqueous solution.

The molar conductance is the electrical conductance of a one molar aqueous solution.

The greater the molar conductance, the greater is the degree of dissociation of the dissolved salt.

10-2 The electrostatic force that binds oppositely charged ions together is called an ionic bond.

Electrostatic force binds ions together in an ionic bond.

Some metals lose electrons and some nonmetals gain electrons to attain noble-gas electron configurations (Figure 10-3).

A noble-gas electron configuration is a particularly stable arrangement.

The electron-transfer reaction between a metal and a nonmetal can be described using electron configurations or Lewis electron-dot formulas.

A cation is a positively charged ion.

An anion is a negatively charged ion.

An ionic compound is composed of cations and anions.

Reactive metals combine with reactive nonmetals to produce ionic compounds.

10-3 The common oxidation states of transition-metal ions can be understood in terms of electron configurations.

Some metals lose electrons to form ions with an 18-electron outer configuration, $ns^2np^6nd^{10}$ (Figure 10-4).

Some transition metals form ions with the outer electron configuration, [noble gas] $nd^{10} (n + 1)s^2$ (Figure 10-5).

In the d transition-metal ions, the nd orbital is of lower energy than the $(n + 1)s$ orbital.

The order of the orbital energies for transition-metal ions is $1s < 2s < 2p < 3s < 3p < 3d < 4s < 4p < 4d < 4f. \ldots$

10-4 Cations are smaller and anions are larger than their parent neutral atoms.

For positive ions in the same row of the periodic table, the higher the positive charge, the smaller is the ion (Figure 10-6).

Positive atomic ions are smaller than their parent neutral atoms because the excess positive charge draws the remaining electrons closer to the nucleus.

Negative atomic ions are larger than their parent neutral atoms because the extra electron(s) increases the electron-electron repulsion and causes the electron cloud to expand.

Isoelectronic species contain the same number of electrons.

The values of some ionic radii are given in Table 10-2.

10-5 Coulomb's law is used to calculate the energy of an ion pair.

The ionization energy is the energy required to remove an electron from an atom or ion.

The first electron affinity EA_1 of an atom is the energy change per mole for the process

$$A(g) + e^- \longrightarrow A^-(g)$$

The second electron affinity EA_2 is the energy change per mole for the process

$$A^-(g) + e^- \longrightarrow A^{2-}(g)$$

The values of the electron affinities of some nonmetals are given in Table 10-3.

The distance between the centers of the ions in an ion pair is the sum of the radii of the ions.

Coulomb's law states that the energy of interaction of two ions is directly proportional to the product of their electrical charges and is inversely proportional to the distance between them.

Coulomb's law is given by

$$E = \frac{(2.31 \times 10^{-16} \, \text{J} \cdot \text{pm})Z_1Z_2}{d} \qquad (10\text{-}1)$$

where E is the coulombic ion pairing energy, d is the sum of the ionic radii in units of picometers, and Z_1 and Z_2 are the charges on the ions.

The energy change for the reaction

$$M(g) + X(g) \longrightarrow M^+X^-(g)$$

is equal to the sum of the ionization energy of $M(g)$, the electron affinity of $X(g)$, and the coulombic energy of the ion pair (Figure 10-11).

An ionic bond is the result of the electrostatic (Coulomb's law) attraction between two oppositely charged ions.

The energy of the ionic bond between two ions is the energy that must be supplied to separate the ions.

10-6 The formation of ionic solids from the elements is an exothermic process.

The energy change for the reaction

$$M(s) + \tfrac{1}{2}X_2(s, l, \text{ or } g) \longrightarrow M^+X^-(s)$$

is equal to the sum of the vaporization energy of $M(s)$, one half of the vaporization energy and dissociation energy of X_2, the ionization energy of $M(g)$, the electron affinity of $X(g)$, and the lattice energy of $M^+X^-(s)$ (see Figure 10-13).

The lattice energy of an ionic solid is the energy released when the isolated ions are brought together to form the ionic solid.

The lattice energies of some ionic solids are given in Table 10-4.

The set of steps used to calculate the energy of formation of an ionic solid from its elements is called a Born-Haber cycle.

B. SELF-TEST

1. An aqueous solution of sodium chloride contains _____ and

_____ .

2. A substance that exists as ions when dissolved in water will conduct an electric current. *True/False*

3. All substances that dissolve in water produce solutions that conduct electricity. *True/False*

4. All electrolytes are good conductors of electricity. *True/False*

5. Weak electrolytes are substances that dissociate completely in water. *True/False*

6. All acids are strong electrolytes. *True/False*

7. The soluble hydroxides are (*strong/weak/non*) electrolytes.

8. Most soluble salts are (*strong/weak/non*) electrolytes.

9. Most organic compounds are (*strong/weak/non*) electrolytes.

10. A salt that is a weak electrolyte dissociates completely in aqueous solution. *True/False*

11. The greater the molar conductance of a dissolved salt, the (*greater/less*) is the degree of dissociation of the dissolved salt.

12. A Group 1 metal (*gains/loses*) an electron to obtain a noble-gas electron configuration.

13. A halogen atom (*gains/loses*) an electron to obtain a noble-gas electron configuration.

14. An ionic compound consists of _____ and _____.

15. The bond in an ionic compound is called _____ bond.

16. The Group 2 metals (*gain/lose*) electrons to attain a noble-gas electron configuration.

17. The Group 6 nonmetals (*gain/lose*) electrons to attain a noble-gas electron configuration.

18. The group of transition metals headed by Cu lose _____ electron(s) to attain a _____ configuration.

19. The energy of the $3d$ orbitals is lower than the $4s$ orbital in a transition-metal ion. *True/False*

20. The filling of the orbitals for transition-metal ions is as follows: _____ _____.

21. The radius of an ion is the same as the radius of the parent neutral atom. *True/False*

22. The radius of a chloride ion is (*larger/smaller*) than the radius of a chlorine atom because _____ _____.

23. The radius of a potassium ion is (*larger/smaller*) than the radius of a potassium atom because _____

_____.

24. Isoelectronic species always have the same size. *True/False*

25. Energy is required to remove an electron from a metal atom. *True/False*

26. Energy is required to add an electron to a fluorine atom. *True/False*

27. Electron affinity is the energy _____

_____.

28. The electron affinity of a chlorine atom has a (*positive/negative*) value.

29. The energy required to bring two ions together in the gas phase is found from _____ law.

30. The energy required to bring a positive ion and a negative ion together is negative. The ion pair is (*more/less*) stable than the separated ions.

31. Coulomb's law states that $E = $ _____

where _____

_____.

32. The coulombic energy for the process

$$M^+(g) + X^-(g) \longrightarrow MX(g)$$

is greater the (*greater/less*) the separation between the ions.

33. The energies used to calculate the energy evolved in the process

$$Na(g) + Cl(g) \longrightarrow Na^+Cl^-(g)$$

are

(a) _____

(b) _____

(c) _____

34. The energy of an ionic bond is the energy required to separate the ions. *True/False*

35. Energy is (*absorbed/released*) when isolated fluoride ions and isolated lithium ions are brought together to form crystalline lithium fluoride.

36. The lattice energy of an ionic crystal has a (*positive/negative*) value relative to the energy of the gaseous ions.

37. The reaction between a metal and a nonmetal to form an ionic crystal is an (*exothermic/endothermic*) reaction.

38. Write out the five steps used to analyze the energetics of the formation of LiF(*s*) from Li(*s*) and $F_2(g)$.

(a) _____

(b) _____

(c) _____

(d) _____

(e) _____

39. The sum of the energy terms in the five steps involved in Question 38 is equal to the energy change for the reaction _____.

40. The lattice energy of NaCl(*s*) is equal to the energy released in the process _____

_____.

41. The Born-Haber cycle for NaCl(*s*) is the process in which _____

_____.

C. CALCULATIONS YOU SHOULD KNOW HOW TO DO

1. Use electron configurations or Lewis electron-dot formulas to predict reaction products and chemical formulas. See Example 10-2 and Problems 10-1 through 10-4.

2. Determine the electron configurations of transition-metal ions. See Examples 10-3 through 10-5 and Problems 10-5 through 10-12.

3. Use electron configurations and the charges on the ions to predict the relative sizes of ions. See Example 10-6 and Problems 10-15 and 10-16.

4. Calculate the energy released in the formation of ion pairs from their gaseous ions as shown in Figure 10-11. See Example 10-7 and Problems 10-25 through 10-30.

5. Calculate the energy released in the formation of ionic crystals from the elements. See Section 10-6, Figure 10-13, and Example 10-8, and Problems 10-31 through 10-36.

D. SOLUTIONS TO THE ODD-NUMBERED PROBLEMS

10-1   In order to emphasize the attainment of noble-gas configurations for ions, we write the electron configurations in terms of noble-gas configurations.

(a) $Ca([Ar]4s^2) + 2F([He]2s^22p^5) \rightarrow Ca^{2+}([Ar]) + 2F^-([Ne]) \rightarrow CaF_2(g)$

(b) $Sr([Kr]5s^2) + 2Br([Ar]4s^23d^{10}4p^5) \rightarrow Sr^{2+}([Kr]) + 2Br^-([Kr]) \rightarrow SrBr_2(g)$

(c) $2Al([Ne]3s^23p^1) + 3O([He]2s^22p^4) \rightarrow 2Al^{3+}([Ne]) + 3O^{2-}([Ne]) \rightarrow Al_2O_3(g)$

10-3   (a) $3Li\cdot + \cdot \overset{\cdot\cdot}{\underset{\cdot}{N}}\cdot \rightarrow \underbrace{3Li^+ + :\overset{\cdot\cdot}{\underset{\cdot\cdot}{N}}:^{3-}}_{Li_3N}$

(b) $Na\cdot + H\cdot \rightarrow \underbrace{Na^+ + H:^-}_{NaH}$

(c) $\cdot\overset{\cdot}{Al}\cdot + 3:\overset{\cdot\cdot}{\underset{\cdot\cdot}{I}}\cdot \rightarrow \underbrace{Al^{3+} + 3:\overset{\cdot\cdot}{\underset{\cdot\cdot}{I}}:^-}_{AlI_3}$

10-5   We first determine the number of electrons in the ion, and then we write the electron configuration filling the orbitals in the order $1s$, $2s$, $2p$, $3s$, $3p$, $3d$, $4s$, and so forth. For Cr^{2+} we have $Z - 2 = 24 - 2 = 22$ electrons; thus we have

(a) $Cr^{2+}([Ar]3d^4)$ (b) $Cu^{2+}([Ar]3d^9)$

(c) $Co^{3+}([Ar]3d^6)$ (d) $Mn^{2+}([Ar]3d^5)$

10-7   For $+2$ ions the nth member of a d transition series has nd electrons. Thus the sixth member has six d electrons, and so forth.

(a)   Fe, Ru, Os   (b)   Zn, Cd, Hg

(c)   Sc, Y, Lu   (d)   Mn, Tc, Re

10-9   (a)   For $n = 3$ or 4, an 18-electron outer configuration is one of the type $ns^2np^6nd^{10}$. For $n = 4$ the corresponding noble gas is Kr (the fourth noble gas); thus for Cd^{2+} we have $Cd^{2+}([Kr]4d^{10})$.

(b) $In^{3+}([Kr]4d^{10})$

(c)   In this case, $n = 5$ and the corresponding 18-electron outer configuration is of the type $ns^2np^6(n-1)f^{14}nd^{10}$ or $5s^25p^64f^{14}5d^{10}$; thus we have $Tl^{3+}([Xe]4f^{14}5d^{10})$.

(d) $Zn^{2+}([Ar]3d^{10})$

10-11   We simply determine the number of electrons in the positive ion and in the negative ion. If the numbers are the same, then the ions are isoelectronic.

(a) $Li^+(2)$, $F^-(10)$, not isoelectronic

(b) $Na^+(10)$, $F^-(10)$, isoelectronic

(c) $K^+(18)$, $Br^-(36)$, not isoelectronic

(d)   KCl, isoelectronic ions, each has 18 electrons

(e) BaI_2, isoelectronic ions, each has 54 electrons

(f) AlF_3, isoelectronic ions, each has 10 electrons

10-13 Figure 10-3 gives the ionic charges of various ions that correspond to noble-gas electron configurations. We use these charges, together with the requirement that the ionic compound must be electrically neutral, to determine the formula.

(a) Y_2S_3 (b) $LaBr_3$ (c) $MgTe$

(d) Rb_3N (e) Al_2Se_3 (f) CaO

10-15 (a) Cl^- A negative ion is larger than a positive ion with the same number of electrons.

(b) Ag^+ The higher charged ion is smaller because of the larger nuclear attraction for the same number of electrons.

(c) Cu^+ Same reason as (b).

(d) O^{2-} The higher the negative charge for the same number of electrons, the larger the ion is, because the electrons repel each other.

10-17 See Table 10-3 for electron affinities. The larger the magnitude of the electron affinity of an atom, the easier it is to add an electron. Thus we have

$$Cl > Br > I > H$$

10-19 In going from Cl to Br and from Br to I, the electron affinity decreases in magnitude by about 25 kJ·mol^{-1} at each step. Thus we estimate the electron affinity of At to be about -270 kJ·mol^{-1}.

10-21 In each case we add the electron affinity of the second reactant to the ionization energy of the first reactant.

(a) $Li(g) + Br(g) \rightarrow Li^+(g) + Br^-(g)$

$$\Delta H^\circ_{rxn} = 520 \text{ kJ} \cdot \text{mol}^{-1} + (-324 \text{ kJ} \cdot \text{mol}^{-1}) = 196 \text{ kJ} \cdot \text{mol}^{-1}$$

(b) $I^-(g) + Cl(g) \rightarrow I(g) + Cl^-(g)$

The ionization energy of $I^-(g)$ is the negative of the electron affinity of $I(g)$. Thus

$$\Delta H^\circ_{rxn} = +295 \text{ kJ} \cdot \text{mol}^{-1} + (-348 \text{ kJ} \cdot \text{mol}^{-1}) = -53 \text{ kJ} \cdot \text{mol}^{-1}$$

(c) $Na(g) + H(g) \rightarrow Na^+(g) + H^-(g)$

$$\Delta H^\circ_{rxn} = 500 \text{ kJ} \cdot \text{mol}^{-1} + (-72 \text{ kJ} \cdot \text{mol}^{-1}) = 428 \text{ kJ} \cdot \text{mol}^{-1}$$

10-23 The equation for Coulomb's law is

$$E = \frac{(2.31 \times 10^{-16} \text{ J} \cdot \text{pm})(Z_1 Z_2)}{d}$$

From Table 10-2, the radius of $Zn^{2+} = 74$ pm and the radius of $O^{2-} = 140$ pm, and, $Z_1 = +2$, and $Z_2 = -2$. Thus we have

$$E = \frac{(2.31 \times 10^{-16} \text{ J} \cdot \text{pm})(2)(-2)}{(74 \text{ pm} + 140 \text{ pm})} = -4.32 \times 10^{-18} \text{ J}$$

10-25 (a) For the ionization of K

$$K(g) \longrightarrow K^+(g) + e^- \qquad I_1 = 419 \text{ kJ} \cdot \text{mol}^{-1}$$

(b) For the addition of an electron to Br (Table 10-3)

$$Br(g) + e^- \longrightarrow Br^-(g) \qquad EA_1 = -324 \text{ kJ} \cdot \text{mol}^{-1}$$

If we add steps (a) and (b), then we have

$$K(g) + Br(g) \longrightarrow K^+(g) + Br^-(g) \qquad \Delta H^{\circ}_{rxn} = 95 \text{ kJ} \cdot \text{mol}^{-1}$$

(c) We now bring together K^+ and Br^- to their ion-pair separation. According to Table 10-2, the radius of K^+ is 133 pm and the radius of Br^- is 195 pm. Their separation as an ion pair is 328 pm. We now use Coulomb's law to calculate the energy released by the formation of one ion pair from the separated ions:

$$K^+(g) + Br^-(g) \longrightarrow K^+Br^-(g)$$

$$E = \frac{(2.31 \times 10^{-16} \text{ J} \cdot \text{pm})(Z_1 Z_2)}{d}$$

$$= \frac{(2.31 \times 10^{-16} \text{ J} \cdot \text{pm})(+1)(-1)}{328 \text{ pm}}$$

$$= -7.04 \times 10^{-19} \text{ J}$$

For the energy released on the formation of one mole of ion pairs, we multiply this result by Avogadro's number:

$$E = \left(-\frac{7.04 \times 10^{-10} \text{ J}}{1 \text{ ion pair}} \right) \left(\frac{6.022 \times 10^{23} \text{ ion pair}}{1 \text{ mol}} \right)$$

$$= -424 \text{ kJ} \cdot \text{mol}^{-1}$$

Thus, we have

$$K^+(g) + Br^-(g) \longrightarrow K^+Br^-(g) \qquad \Delta H^{\circ}_{rxn} = -424 \text{ kJ} \cdot \text{mol}^{-1}$$
$$d = 328 \text{ pm}$$

We combine this result with the result

$$K(g) + Br(g) \longrightarrow K^+(g) + Br^-(g) \qquad \Delta H^{\circ}_{rxn} = 95 \text{ kJ} \cdot \text{mol}^{-1}$$

to obtain the final result

$$K(g) + Br(g) \longrightarrow K^+Br^-(g) \qquad \Delta H^{\circ}_{rxn} = -329 \text{ kJ} \cdot \text{mol}^{-1}$$
$$d = 328 \text{ pm}$$

10-27 (a)   For the ionization of Na (Table 9-1)

$$Na(g) \longrightarrow Na^+(g) + e^- \qquad I_1 = 500 \ kJ \cdot mol^{-1}$$

(b)   For the addition of an electron to H (Table 10-3),

$$H(g) + e^- \longrightarrow H^-(g) \qquad EA_1 = -72 \ kJ \cdot mol^{-1}$$

Adding steps (a) and (b), we have

$$Na(g) + H(g) \longrightarrow Na^+(g) + H^-(g) \qquad \Delta H^{\circ}_{rxn} = 428 \ kJ \cdot mol^{-1}$$

(c)   We now bring together Na^+ and H^- to their ion-pair separation. We use Coulomb's law to calculate the energy released by the formation of one ion pair:

$$
\begin{aligned}
E &= \frac{(2.31 \times 10^{-16} \ J \cdot pm)(Z_1 Z_2)}{d} \\
&= \frac{(2.31 \times 10^{-16} \ J \cdot pm)(+1)(-1)}{(95 \ pm + 154 \ pm)} \\
&= -9.28 \times 10^{-19} \ J
\end{aligned}
$$

The energy released on formation of one mole of ion pairs is

$$
\begin{aligned}
E &= \left(-\frac{9.28 \times 10^{-19} \ J}{1 \ \text{ion pair}} \right) \left(\frac{6.022 \times 10^{23} \ \text{ion pair}}{1 \ mol} \right) \\
&= -558 \ kJ \cdot mol^{-1}
\end{aligned}
$$

Thus, we have for the reaction

$$Na^+(g) + H^-(g) \longrightarrow \underset{d \ = \ 249 \ pm}{NaH(g)} \qquad \Delta H^{\circ}_{rxn} = -558 \ kJ \cdot mol^{-1}$$

Combining the above result with

$$Na(g) + H(g) \longrightarrow Na^+(g) + H^-(g) \qquad \Delta H^{\circ}_{rxn} = 428 \ kJ \cdot mol^{-1}$$

we obtain

$$Na(g) + H(g) \longrightarrow \underset{d \ = \ 249 \ pm}{NaH(g)} \qquad \Delta H^{\circ}_{rxn} = -130 \ kJ \cdot mol^{-1}$$

10-29

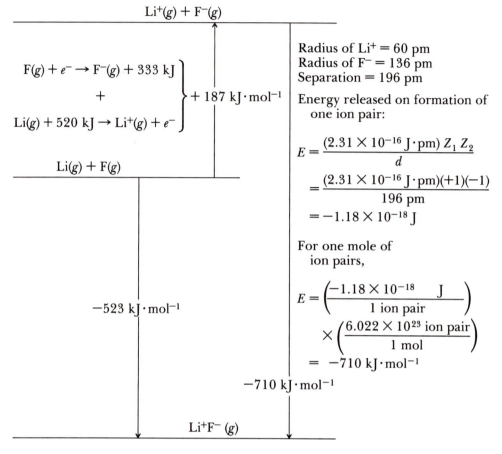

Radius of Li^+ = 60 pm
Radius of F^- = 136 pm
Separation = 196 pm

Energy released on formation of one ion pair:

$$E = \frac{(2.31 \times 10^{-16} \text{ J} \cdot \text{pm}) \, Z_1 \, Z_2}{d}$$

$$= \frac{(2.31 \times 10^{-16} \text{ J} \cdot \text{pm})(+1)(-1)}{196 \text{ pm}}$$

$$= -1.18 \times 10^{-18} \text{ J}$$

For one mole of ion pairs,

$$E = \left(\frac{-1.18 \times 10^{-18} \quad \text{J}}{1 \text{ ion pair}} \right)$$

$$\times \left(\frac{6.022 \times 10^{23} \text{ ion pair}}{1 \text{ mol}} \right)$$

$$= -710 \text{ kJ} \cdot \text{mol}^{-1}$$

10-31   Following the steps outlined in Section 10-6, we have

(a)   Vaporize one mole of sodium metal:

$$Na(s) \longrightarrow Na(g) \qquad \Delta H_{vap}^\circ = 93 \text{ kJ} \cdot \text{mol}^{-1}$$

(b)   Dissociate one-half mole of $F_2(g)$:

$$\tfrac{1}{2} F_2(g) \longrightarrow F(g) \qquad \Delta H_{rxn}^\circ = \tfrac{1}{2}(155 \text{ kJ} \cdot \text{mol}^{-1}) = 78 \text{ kJ} \cdot \text{mol}^{-1}$$

(c)   Ionize one mole of $Na(g)$ (Table 9-1):

$$Na(g) \longrightarrow Na^+(g) + e^- \qquad I_1 = 500 \text{ kJ} \cdot \text{mol}^{-1}$$

(d)   Attach one mole of electrons to one mole of fluorine atoms (Table 10-3):

$$F(g) + e^- \longrightarrow F^-(g) \qquad EA_1 = -333 \text{ kJ} \cdot \text{mol}^{-1}$$

(e)   Bring one mole of $Na^+(g)$ and one mole of $F^-(g)$ together to form crystalline $NaF(s)$ (lattice energy) (Table 10-4):

$$Na^+(g) + F^-(g) \longrightarrow NaF(s) \qquad \Delta H^\circ_{LE} = -919 \text{ kJ} \cdot \text{mol}^{-1}$$

(f)   Add the equations in steps (a) through (e) to obtain

$$Na(s) + \tfrac{1}{2}F_2(g) \longrightarrow NaF(s) \qquad \Delta H^\circ_{rxn} = -581 \text{ kJ} \cdot \text{mol}^{-1}$$

where $93 + 78 + 500 - 333 - 919 = -581$

10-33   (a)   Vaporize one mole of sodium metal:

$$Na(s) \longrightarrow Na(g) \qquad \Delta H^\circ_{vap} = 93 \text{ kJ} \cdot \text{mol}^{-1}$$

(b)   Sublime and dissociate one-half mole of I_2

$$\tfrac{1}{2}I_2(s) \longrightarrow I(g) \qquad \Delta H^\circ_{rxn} = \tfrac{1}{2}(214 \text{ kJ} \cdot \text{mol}^{-1})$$
$$= 107 \text{ kJ} \cdot \text{mol}^{-1}$$

(c)   Ionize one mole of $Na(g)$ (Table 9-1):

$$Na(g) \longrightarrow Na^+(g) + e^- \qquad I_1 = 500 \text{ kJ} \cdot \text{mol}^{-1}$$

(d)   Attach one mole of electrons to one mole of $I(g)$ (Table 10-3):

$$I(g) + e^- \longrightarrow I^-(g) \qquad EA_1 = -295 \text{ kJ} \cdot \text{mol}^{-1}$$

(e)   Bring one mole of $Na^+(g)$ and one mole of $I^-(g)$ together to form crystalline NaI (lattice energy) (Table 10-4):

$$Na^+(g) + I^-(g) \longrightarrow NaI(s) \qquad \Delta H^\circ_{LE} = -704 \text{ kJ} \cdot \text{mol}^{-1}$$

(f)   Add the equations in steps (a) through (e) to obtain

$$Na(s) + \tfrac{1}{2}I_2(s) \longrightarrow NaI(s) \qquad \Delta H^\circ_{rxn} = -299 \text{ kJ} \cdot \text{mol}^{-1}$$

where $93 + 107 + 500 - 295 - 704 = -299$

10-35   (a)   Vaporize one mole of calcium metal:

$$Ca(s) \longrightarrow Ca(g) \qquad \Delta H^\circ_{vap} = 193 \text{ kJ} \cdot \text{mol}^{-1}$$

(b)   Dissociate one mole of $Cl_2(g)$:

$$Cl_2(g) \longrightarrow 2Cl(g) \qquad \Delta H^\circ_{diss} = 244 \text{ kJ} \cdot \text{mol}^{-1}$$

(c)   Ionize one mole of $Ca(g)$ to $Ca^+(g)$:

$$Ca(g) \longrightarrow Ca^+(g) + e^- \qquad I_1 = 590 \text{ kJ} \cdot \text{mol}^{-1}$$

Ionize one mole of $Ca^+(g)$ to $Ca^{2+}(g)$:

$$Ca^+(g) \longrightarrow Ca^{2+}(g) + e^- \qquad I_2 = 1140 \text{ kJ} \cdot \text{mol}^{-1}$$

Thus for the production of $Ca^{2+}(g)$ from $Ca(g)$, we have

$$Ca(g) \longrightarrow Ca^{2+}(g) + 2e^- \qquad I = 1730 \text{ kJ} \cdot \text{mol}^{-1}$$

(d)   Attach two moles of electrons to two moles of $Cl(g)$:

$$2Cl(g) + 2e^- \longrightarrow 2Cl^-(g) \qquad \begin{aligned} \Delta H_{rxn}^{\circ} &= 2EA_1 = (2)(-348 \text{ kJ} \cdot \text{mol}^{-1}) \\ &= -696 \text{ kJ} \cdot \text{mol}^{-1} \end{aligned}$$

(e)   Bring one mole of $Ca^{2+}(g)$ and two moles of Cl^- together to form crystalline $CaCl_2$ (lattice energy):

$$Ca^{2+}(g) + 2Cl^-(g) \longrightarrow CaCl_2(s) \qquad \Delta H_{LE}^{\circ} = -2266 \text{ kJ} \cdot \text{mol}^{-1}$$

(f)   Add the equations in steps (a) through (e) to obtain

$$Ca(s) + Cl_2(g) \longrightarrow CaCl_2(s) \qquad \Delta H_{rxn}^{\circ} = -795 \text{ kJ} \cdot \text{mol}^{-1}$$

where $193 + 244 + 1730 - 696 - 2266 = -795$

10-37   The ions N^{3-}, O^{2-}, Na^+, Mg^{2+}, and Al^{3+} are isoelectronic with F^-.

10-39   The ground-state electron configuration of K^+ is $1s^2 2s^2 2p^6 3s^2 3p^6$ and that of Cu^+ is $1s^2 2s^2 2p^6 3s^2 3p^6 3d^{10}$. The outer electrons are in the same shell, but the nuclear charge of Cu is greater than that of K and so the electrons are attracted more strongly to the nucleus.

10-41   We use a Born-Haber cycle (Figure 10-13) and write

$$\begin{aligned} \Delta H_f^{\circ}[CaCl_2(s)] = {} & \text{enthalpy of vaporization of } Ca(s) \\ & + (Cl\!-\!Cl \text{ bond energy}) \\ & + (\text{the sum of the first two ionization} \\ & \quad \text{energies of } Ca) \\ & + 2[\text{electron affinity of } Cl(g)] \\ & + \text{lattice energy of } CaCl_2(s) \end{aligned}$$

If we substitute the given numerical data into this equation, then we obtain

$$\begin{aligned} -795.0 \text{ kJ} \cdot \text{mol}^{-1} = {} & (193 \text{ kJ} \cdot \text{mol}^{-1}) + (244 \text{ kJ} \cdot \text{mol}^{-1}) \\ & + (590 \text{ kJ} \cdot \text{mol}^{-1} + 1140 \text{ kJ} \cdot \text{mol}^{-1}) \\ & + 2[\text{electron affinity of } Cl(g)] \\ & + (-2266 \text{ kJ} \cdot \text{mol}^{-1}) \end{aligned}$$

Solving for the electron affinity of $Cl(g)$ gives

$$\text{electron affinity} = -348 \text{ kJ} \cdot \text{mol}^{-1}$$

10-43   The ground-state electron configuration of Fe^{2+} is $[Ar]3d^6$. According to Hund's rule, we fill all five $3d$ orbitals with one electron before we pair them up. Thus there are four unpaired electrons in Fe^{2+}.

The ground-state electron configuration of Zn^{2+} is $[Ar]3d^{10}$. Thus there are no unpaired electrons in Zn^{2+}.

10-45   There are two reasons: the magnitude of the electron affinity decreases and the ionic size of X^- increases, thus decreasing the coulombic energy of the ion pair.

10-47   The $+1$ charged d^3 transition-metal ions are Ti^+, Zr^+, and Hf^+.

10-49   (a)   strong electrolyte, rule 3
(b)   weak electrolyte, rule 4
(c)   nonelectrolyte, rule 5
(d)   nonelectrolyte, rule 5

10-51   From Problem 10-35, we have

(1) $Ca(g) \longrightarrow Ca^+(g) + e^-$ $\Delta H^\circ_{rxn} = I_1 = 590 \text{ kJ}$

(2) $Ca^+(g) \longrightarrow Ca^{2+}(g) + e^-$ $\Delta H^\circ_{rxn} = I_2 = 1140 \text{ kJ}$

We can compute the the value of ΔH°_{rxn} for the reaction

(3) $Ca(g) + Ca^{2+}(g) \longrightarrow 2Ca^+(g)$

using Hess's law because Equation (3) is obtained by subtracting Equation (2) from Equation (1). Thus,

$$\Delta H^\circ_{rxn} = \Delta H^\circ_{rxn}(1) - \Delta H^\circ_{rxn}(2)$$
$$= 590 \text{ kJ} - 1140 \text{ kJ} = -550 \text{ kJ}$$

Thus Equation (3) is highly exothermic and gaseous calcium atoms will readily reduce $Ca^{2+}(g)$ to $Ca^+(g)$. However, in aqueous solution Ca^{2+} binds much more strongly to water molecules because of its $+2$ charge than does Ca^+. The high solvation energy of Ca^{2+} stabilizes it relative to Ca^+ in aqueous solution and $Ca^+(aq)$ is unstable relative to $Ca^{2+}(aq)$ and $Ca(s)$.

10-53   A neutral iron atom has 26 electrons. Fe(II) is the Fe^{2+} ion, which has $26 - 2 = 24$ electrons. Fe(III) is the Fe^{3+} ion, which has $26 - 3 = 23$ electrons. The ground-state electron configurations are

$$\text{Fe(II): } 1s^2 2s^2 2p^6 3s^2 3p^6 3d^6$$

$$\text{Fe(III): } 1s^2 2s^2 2p^6 3s^2 3p^6 3d^5$$

10-55   The ions are P^{3-}, S^{2-}, and Cl^-. For the same row of the periodic table, the ion with the greatest negative charge will have the largest radius because the extra electrons will cause both a greater electron-electron repulsion and a greater electron cloud expansion. Thus, the smallest ion is the chloride ion because it has the smallest negative charge.

10-57 $HgCl_2$: weak electrolyte (rule 4); KCl: strong electrolyte (rule 3); $BaCl_2$: strong electrolyte (rule 3); and $CrCl_3$: strong electrolyte (rule 3).

10-59 The fluoride ion is so small that the coulombic force between the fluoride ion and a metal ion is greater than between any other anion and the metal ion.

10-61 Using the ion size data in Table 10-2, we have

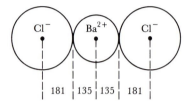
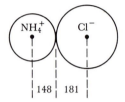

where the distances are in picometers. We can use Equation (10-1) to compute the ionic bond energy of these two species. For an ion pair

$$E = \frac{(2.31 \times 10^{-16}\,\text{J}\cdot\text{pm})(Z_1 Z_2)}{d}$$

For one mole of NH_4Cl ions pairs, we have

$$E = \left(\frac{(2.31 \times 10^{-16}\,\text{J}\cdot\text{pm})(1)(-1)}{148\,\text{pm} + 181\,\text{pm}}\right)(6.022 \times 10^{23}\,\text{mol}^{-1})$$
$$= -4.23 \times 10^5\,\text{J}\cdot\text{mol}^{-1} = -423\,\text{kJ}\cdot\text{mol}^{-1}$$

For one mole of $BaCl_2$ ion pairs, we have to consider the coulombic attraction between each Cl^- ion and the Ba^{2+} ion (separated by 181 pm + 135 pm = 316 pm) and the coulombic repulsion between the two Cl^- ions [separated by 2(181 pm + 135 pm = 632 pm)].

$$E = (2.31 \times 10^{-16}\,\text{J}\cdot\text{pm})\left[\frac{(2)(+2)(-1)}{316\,\text{pm}} + \frac{(-1)(-1)}{632\,\text{pm}}\right](6.022 \times 10^{23}\,\text{mol}^{-1})$$
$$= -1.54 \times 10^6\,\text{J}\cdot\text{mol}^{-1} = -1540\,\text{kJ}\cdot\text{mol}^{-1}$$

10-63 The Tl^+ ion and the K^+ ion have identical ionic charges and similar ionic radii (144 pm and 133 pm, respectively, as given in Table 10-2) and thus $Tl^+(aq)$ moves around in human tissue much as $K^+(aq)$ does.

E. ANSWERS TO THE SELF-TEST

1. $Na^+(aq)$ and $Cl^-(aq)$

2. true

3. false

4. false (Weak electrolytes are only poor conductors.)

5. false (Weak electrolytes are only partially dissociated.)

6. false

7. strong

8. strong

9. non

10. false

11. greater

12. loses

13. gains

14. positively charged ions (cations) and negatively charged ions (anions)

15. an ionic

16. lose

17. gain

18. one . . . 18-electron outer

19. true

20. $1s < 2s < 2p < 3s < 3p < 3d < 4s < 4p < 4d < 4f$ . . .

21. false

22. larger . . . interelectronic repulsion due to the additional electron

23. smaller . . . the net positive charge draws the electrons in toward the nucleus

24. false

25. true

26. false [Energy is released in the process $F(g) + e^- \rightarrow F^-(g)$.]

27. associated with the process of adding an electron to an atom or an ion

28. negative

29. Coulomb's

30. more

31. $(2.31 \times 10^{-16}$ J·pm$)Z_1Z_2/d$; . . . Z_1 and Z_2 are the ionic charges and d is the distance between the ion centers in picometers

32. less

33. (a) the ionization energy of Na(g)
    (b) the electron affinity of Cl(g)
    (c) the coulombic energy of the Na$^+$ Cl$^-$ ion pair

34. true

35. released

36. negative (The lattice is more stable than the gaseous ions.)

37. exothermic

38. (a) the sublimation of Li(s):

$$Li(s) \longrightarrow Li(g)$$

(b) the ionization of Li(g):

$$Li(g) \longrightarrow Li^+(g) + e^-$$

(c) the dissociation of F$_2$(g):

$$\tfrac{1}{2}F_2(g) \longrightarrow F(g)$$

(d) the formation of F$^-$(g) from F(g):

$$F(g) + e^- \longrightarrow F^-(g)$$

(e) the formation of LiF(s) from Li$^+$(g) and F$^-$(g):

$$Li^+(g) + F^-(g) \longrightarrow LiF(s)$$

39. $Li(s) + \tfrac{1}{2}F_2(g) \longrightarrow LiF(s)$

40. $Na^+(g) + Cl^-(g) \longrightarrow NaCl(s)$

41. the ionic crystal lattice of NaCl is formed from the elements sodium metal and chlorine (Cl$_2$) gas

11 LEWIS FORMULAS

A. OUTLINE OF CHAPTER 11

11-1   A covalent bond can be described as a pair of electrons shared by two atoms.

    The Lewis formula of a molecule indicates the covalent bond between two atoms as a line joining the atoms.

    The Lewis formula of a molecule indicates lone electron pairs on an atom as pairs of dots surrounding the atom.

    Molecular compounds form molecular crystals in the solid state (Figure 11-1).

    Bond length is the average distance between the nuclei of the two atoms joined by a covalent bond.

11-2   We always try to satisfy the octet rule when writing Lewis formulas.

    The octet rule states that when writing Lewis formulas, arrange the valence electrons about the atoms so that there are eight valence electrons about each atom other than hydrogen (most useful for carbon, nitrogen, oxygen, and fluorine).

    The unique atom in a molecule is often the central atom.

    The atom with the smallest number of valence electrons is a good choice for the central atom.

    A systematic method for writing Lewis formulas is given on pages 378 and 379 of the text.

11-3   Hydrogen atoms are almost always terminal atoms in Lewis formulas.

    A hydrogen atom achieves a noble-gas-like electron configuration by having two electrons around it.

    A hydrogen atom forms only one covalent bond.

11-4   Formal charges can be assigned to atoms in Lewis formulas.

Formal charge is a positive or negative number that is assigned to each atom in a Lewis formula.

Formal charges are determined by a set of rules and do not necessarily represent the actual charges on the atoms.

The formal charge on an atom is defined by Equation (11-1):

$$\begin{pmatrix} \text{formal charge} \\ \text{on an atom in a} \\ \text{Lewis formula} \end{pmatrix} = \begin{pmatrix} \text{total number of} \\ \text{valence electrons} \\ \text{in the free atom} \end{pmatrix} - \begin{pmatrix} \text{total number} \\ \text{of lone-pair} \\ \text{electrons} \end{pmatrix} - \frac{1}{2}\begin{pmatrix} \text{total number} \\ \text{of shared} \\ \text{electrons} \end{pmatrix}$$

Formal charge can be used to choose a preferred Lewis formula.

The preferred Lewis formula generally has the lowest formal charges.

11-5 It is not always possible to satisfy the octet rule by using only single bonds.

A covalent double bond consists of two pairs of electrons shared by two atoms.

A double bond is shorter and stronger than a single bond.

A covalent triple bond consists of three pairs of electrons shared by two atoms.

Many molecules contain double or triple covalent bonds.

11-6 A resonance hybrid is a superposition of Lewis formulas.

When it is possible to write two or more satisfactory Lewis formulas without altering the positions of the nuclei, the actual formula is viewed as an average or a superposition of the individual formulas.

Each of the individual Lewis formulas is said to be a resonance form and the use of resonance forms to describe the electron distribution is called resonance.

The charge on an ion that is represented by a resonance hybrid is delocalized; the charge is not on only one of the atoms that comprise the ion.

The energy of the resonance hybrid is less than that of any of the individual Lewis formulas.

The difference in energy between the resonance hybrid and an individual Lewis formula is called the resonance energy.

Benzene has two major resonance forms.

An abbreviated formula for benzene is .

All the carbon-carbon bonds in benzene are equivalent.

The extraordinary stability of benzene is called resonance stabilization.

11-7 A species with one or more unpaired electrons is called a free radical.

An electron-deficient compound does not have enough valence electrons to satisfy the octet rule for each atom.

A coordinate-covalent bond is a covalent bond that is formed when one atom contributes both electrons to the covalent bond.

A donor-acceptor complex results from a coordinate-covalent bond.

11-8   Atoms of elements below carbon through neon in the periodic table can expand their valence shells.

Some atoms expand their valence shell by using their d orbitals to accommodate more than eight electrons.

Some atoms can bond to more than four other atoms.

The Lewis formulas of a species containing an atom that can expand its valence shell must include the resonance forms containing the atom with its expanded valence shell.

11-9   The number of valence electrons for a transition metal is taken to be equal to its oxidation state.

Many compounds of transition metals are covalent or have covalent character.

Transition metals can expand their valence shells in forming covalent compounds.

11-10   Electronegativity is a periodic property.

Electronegativity is a measure of the force with which an atom attracts the electrons in its covalent bonds.

The larger the electronegativity, the greater is the attraction by the atom for the electrons in its covalent bonds.

Electronegativity is a derived quantity in that it cannot be directly measured.

The electronegativity scale devised by Linus Pauling is based upon the differences in the bond energies of a heteronuclear molecule AB and the homonuclear molecules A_2 and B_2.

The most electronegative element, fluorine, is assigned an electronegativity of 4.0.

The Pauling electronegativities are plotted against atomic number in Figure 11-4.

Electronegativities tend to decrease going down a column and to increase going across from left to right in a row in the periodic table (Figure 11-5).

The greater the electronegativity, the more nonmetallic is the element.

Only differences in electronegativities are meaningful.

11-11   We can use electronegativity to predict the polarity of chemical bonds.

If the electronegativities of the two atoms joined by a covalent bond differ, then the electrons in the bond are not shared equally.

The electrons in a nonpolar bond are shared equally by the two atoms.

A polar bond occurs when the electrons in the covalent bond are attracted to the more electronegative atom.

The more electronegative atom has a partial negative charge when joined to a less electronegative atom by a polar covalent bond.

11-12   Polyatomic molecules with polar bonds may be polar or nonpolar.

The dipole moment of a bond is a measure of the polarity of a bond.

A dipole moment has both magnitude and direction.

CO$_2$ is a linear molecule and has no net dipole moment (Figure 11-7).
H$_2$O is a bent molecule and has a net dipole moment (Figure 11-8).

B. SELF-TEST

1. A covalent bond results when electrons are _____ between two atoms.

2. The covalent bond in a Lewis formula of a molecule is represented as a (*line/pair of dots*) between two atoms.

3. A pair of electrons that is not shared between two atoms is called _____

_____ .

4. The constituent units of a molecular cyrstal are _____ .

5. A molecular crystal has a higher melting point than an ionic crystal. *True/False*

6. The attraction between the molecules of a molecular crystal is stronger than the attraction between the ions in an ionic crystal. *True/False*

7. Bond length is defined as _____

_____ .

8. A nitrogen atom usually forms (*one/two/three*) covalent bonds.

9. The octet rule states that _____

_____ .

10. The octet rule works best for carbon, oxygen, nitrogen, and fluorine. *True/False*

11. A hydrogen atom does not obey the octet rule because _____

_____ .

12. A hydrogen atom may form (*one/two/three*) covalent bond(s).

13. A double bond occurs when two atoms share (*two/four/six*) electrons.

14. A double bond can occur only when the two atoms are the same. *True/False*

15. A triple bond occurs when two atoms share (*two/four/six*) electrons.

16. A triple bond can occur only when the two atoms are the same. *True/False*

17. The atoms in a polyatomic ion are joined by ionic bonds. *True/False*

18. The formal charge of an atom is a charge that is assigned by using the equation

_____ .

19. The formal charge of an atom in a polyatomic species represents the actual charge on the atom. *True/False*

20. When it is possible to write two or more Lewis formulas for a molecule or ion without altering the positions of the nuclei, the actual structure is _____

_____ .

21. Each of the individual Lewis formulas referred to in Question 20 is said to be a

_____ .

22. Because of resonance in the nitrite ion NO_2^-, the bond lengths of the two N—O bonds are (*equal/different*).

23. The N—O bonds in the nitrite ion NO_2^- are both single bonds. *True/False*

24. The −1 charge on the nitrite ion NO_2^- is shared equally by the two oxygen atoms. *True/False*

25. A delocalized charge on an ion means that the charge is shared by two or more atoms. *True/False*

26. The energy of a resonance hybrid is (*less/greater*) than the energy of an individual resonance form.

27. Resonance energy is the energy of the resonance hybrid. *True/False*

28. A benzene molecule contains three double bonds. *True/False*

29. All the bonds in a benzene molecule are equivalent. *True/False*

30. The stability of a benzene molecule is due to _____ stabilization.

31. A free radical is a species in which _____

_____ .

32. A free radical must contain an odd number of electrons. *True/False*

33. An electron-deficient compound is a compound that _____

_____ .

34. An atom in a compound may have more than eight electrons in its valence shell. *True/False*

35. An atom may expand its valence shell by using its _____ orbitals.

36. All the resonance forms of a species containing an atom that can expand its valence shell contribute equally to the resonance hybrid of the species. *True/False*

37. The resonance forms with the (*lowest/highest*) formal charge of a species containing an atom that can expand its valence shell contribute to the resonance hybrid of the species.

38. Transition metals form only ionic compounds. *True/False*

39. The number of valence electrons for a transition metal is taken to be equal to

_____ .

40. A transition metal may expand its valence shell when forming a covalent compound. *True/False*

41. Electronegativity is a measure of _____

_____ .

42. The larger the electronegativity, the (*less/greater*) is the attraction by the atom for the electrons in its covalent bond.

43. Electronegativity is a measurable quantity. *True/False*

44. The Pauling electronegativity scale is based upon the differences in _____

_____ .

45. The most electronegative element is _____ .

46. Fluorine is assigned an electronegativity of _____ .

47. Electronegativity is a measure of the (*metallic/nonmetallic*) character of an element.

48. Electronegativities increase from _____ to _____
across the second and third rows of the periodic table.

49. The electrons in all covalent bonds are shared equally by the two atoms. *True/False*

50. The oxygen atom is (*more/less*) electronegative than the carbon atom.

51. The electrons in a covalent bond are shared equally by the two atoms when the electronegativities of the two atoms are (*similar/different*).

52. A polar bond occurs when the electrons are shared (*equally/unequally*) by the two atoms.

53. When the bond between two atoms is a polar bond, one of the atoms has a

_____ charge and the other has a _____

charge.

54. A partial negative charge is indicated by the symbol _____ .

55. The dipole moment of a bond is a measure of _____ .

56. A dipole moment has both _____ and

_____ .

57. A nonpolar molecule has no net dipole moment. *True/False*

C. CALCULATIONS YOU SHOULD KNOW HOW TO DO

There are not many calculations in this chapter. The determination of formal charge by means of Equation (11-1) is the only type of numerical problem in the chapter. See Examples 11-6, 11-7, and 11-10.

D. SOLUTIONS TO THE ODD-NUMBERED PROBLEMS

11-1 (a) Because sulfur is the unique atom in this molecule, we shall assume that it is central and that the two chlorine atoms are attached to it.

$$Cl \quad S \quad Cl$$

The total number of valence electrons is $(1 \times 6) + (2 \times 7) = 20$. We use four of the valence electrons to form sulfur-chlorine bonds. We now place valence electrons as lone pairs on the two chlorine atoms (accounting for 12 of the 16 valence electrons) and the remaining four on the sulfur atom. The completed Lewis formula is

$$:\ddot{Cl}—\ddot{S}—\ddot{Cl}:$$

Notice that all three atoms satisfy the octet rule.

(b) Because germanium is the unique atom in this molecule, we shall assume that it is central and that each chlorine atom is attached to it. The total number of valence electrons is $(1 \times 4) + (4 \times 7) = 32$. We use eight of the valence electrons to form Ge—Cl bonds, which satisfies the octet rule about the Ge atom. We place the remaining 24 as lone pairs on the Cl atoms. The completed Lewis formula is

$$
\begin{array}{c}
:\ddot{Cl}: \\
| \\
:\ddot{Cl}—Ge—\ddot{Cl}: \\
| \\
:\ddot{Cl}:
\end{array}
$$

(c) We shall assume that the arsenic atom is central and that each bromine atom is attached to it. The total number of valence electrons is $(1 \times 5) + (3 \times 7) = 26$. We use six of the valence electrons to form the As—Br bonds. We place 18 of the valence electrons as lone pairs on the three bromine atoms and the remaining two as a lone pair on the As atom. The Lewis formula is

$$: \overset{..}{\underset{..}{Br}} — \overset{..}{\underset{|}{As}} — \overset{..}{\underset{..}{Br}} :$$
$$: \overset{}{\underset{..}{Br}} :$$

(d) The total number of valence electrons is $(1 \times 5) + (3 \times 1) = 8$. We use six of the valence electrons to form the P—H bonds. We place the remaining two valence electrons as a lone pair on the P atom. The Lewis formula is

$$H — \overset{..}{\underset{|}{P}} — H$$
$$H$$

11-3 Since there is no unique atom in the molecule, we shall assume that the two oxygen atoms are bonded together and that the hydrogen atoms are attached to the two oxygen atoms.

$$H — O — O — H$$

The total number of valence electrons is $(2 \times 1) + (2 \times 6) = 14$. We use two valence electrons to form the O—O bond and four valence electrons to form the H—O bonds. We place the eight remaining valence electrons as lone pairs on the O atoms. The Lewis formula is

$$H — \overset{..}{\underset{..}{O}} — \overset{..}{\underset{..}{O}} — H$$

11-5 (a) The eight valence electrons are used to form the four C—H bonds.

$$H$$
$$H — \overset{|}{\underset{|}{C}} — H$$
$$H$$

(b) Six of the 14 valence electrons are used to form the C—H bonds, two to form the C—F bond and the six remaining valence electrons are placed as lone pairs on the F atom.

$$H$$
$$H — \overset{|}{\underset{|}{C}} — \overset{..}{\underset{..}{F}} :$$
$$H$$

(c)   The hydrogen atoms must be terminal atoms, and so we write

$$
\begin{array}{c}
\quad\quad\ \ \text{H} \\
\quad\quad\ \ | \\
\text{H}-\overset{\textstyle |}{\underset{\textstyle |}{\text{C}}}-\overset{\cdot\cdot}{\underset{\textstyle |}{\text{N}}}-\text{H} \\
\quad\quad\text{H}\ \ \ \text{H}
\end{array}
$$

11-7 (a)   The hydrogen atoms must be in terminal positions, and so we write

$$\text{H}\qquad\text{C}\qquad\text{C}\qquad\text{H}$$

There is a total of $(2 \times 1) + (2 \times 4) = 10$ valence electrons. If we add one bond between each atom and then try to satisfy the octet rule about the carbon atoms, we find that we are four electrons short. Thus we add two more bonds between the carbon atoms and obtain

$$\text{H}-\text{C}\equiv\text{C}-\text{H}$$

(b)   Arrange the atoms as

$$\text{H}\qquad\text{N}\qquad\text{N}\qquad\text{H}$$

There is a total of $(2 \times 1) + (2 \times 5) = 12$ valence electrons. If we add one bond between each H atom and each N atom and one bond between the two N atoms, we cannot satisfy the octet rule around each N atom. Thus we add one more bond between the two N atoms. We place the four remaining valence electrons as lone pairs on each N atom. The Lewis formula is

$$\text{H}-\overset{\cdot\cdot}{\text{N}}=\overset{\cdot\cdot}{\text{N}}-\text{H}$$

(c)   Arrange the atoms as

$$
\begin{array}{ccc}
 & \text{Cl} & \\
\text{Cl} & \text{C} & \text{O}
\end{array}
$$

There is a total of 24 valence electrons. If we add one bond between each Cl atom and the C atom and one between the O atom and the C atom, we cannot satisfy the octet rule around each atom. Thus we add one more bond between the C atom and the O atom. We place the remaining 16 valence electrons as lone pairs to satisfy the octet rule about each atom. The Lewis formula is

$$
\begin{array}{c}
\quad\ \ \overset{\cdot\cdot}{\text{Cl}}\!: \\
\quad\ \ | \\
:\!\overset{\cdot\cdot}{\underset{\cdot\cdot}{\text{Cl}}}-\text{C}=\overset{\cdot\cdot}{\underset{\cdot\cdot}{\text{O}}}
\end{array}
$$

11-9   There are 18 valence electrons, or nine electron pairs, in HCOOH. The only way to satisfy the octet rule for the carbon and oxygen atoms by using nine electron pairs is to write

$$\overset{\displaystyle \overset{\cdot\cdot}{O}\overset{\cdot\cdot}{}}{\underset{}{\underset{\parallel}{}}}$$

$$H-\overset{\parallel}{C}-\overset{\cdot\cdot}{\underset{\cdot\cdot}{O}}-H$$

11-11 The hydrogen atoms are terminal atoms. There are 18 valence electrons, or nine electrons pairs. In order to satisfy the octet rule about the carbon and chlorine atoms, we write

$$\overset{\displaystyle H \quad H}{\underset{}{\underset{|}{} \quad \underset{|}{}}}$$

$$H-C=C-\overset{\cdot\cdot}{\underset{\cdot\cdot}{Cl}}:$$

11-13 The Lewis formula of NF_3 in which the nitrogen atom and the fluorine atoms are connected in a row is

$$:\overset{\cdot\cdot}{\underset{\cdot\cdot}{F}}-\overset{\cdot\cdot}{\underset{\cdot\cdot}{N}}-\overset{\cdot\cdot}{\underset{\cdot\cdot}{F}}-\overset{\cdot\cdot}{\underset{\cdot\cdot}{F}}:$$

We now assign a formal charge to each atom according to Equation (11-1).

formal charge on the left $F = 7 - 6 - \frac{1}{2}(2) = 0$

formal charge on $N = 5 - 4 - \frac{1}{2}(4) = -1$

formal charge on the middle $F = 7 - 4 - \frac{1}{2}(4) = +1$

formal charge on the right $F = 7 - 6 - \frac{1}{2}(2) = 0$

Thus we write

$$:\overset{\cdot\cdot}{\underset{\cdot\cdot}{F}}-\overset{\ominus}{\overset{\cdot\cdot}{\underset{\cdot\cdot}{N}}}-\overset{\oplus}{\overset{\cdot\cdot}{\underset{\cdot\cdot}{F}}}-\overset{\cdot\cdot}{\underset{\cdot\cdot}{F}}:$$

A Lewis formula for NF_3 with lower formal charges (all zeros) is

$$:\overset{\cdot\cdot}{\underset{\cdot\cdot}{F}}-\overset{\cdot\cdot}{\underset{|}{N}}-\overset{\cdot\cdot}{\underset{\cdot\cdot}{F}}:$$

$$:\overset{\cdot\cdot}{\underset{\cdot\cdot}{F}}:$$

which represents the preferred Lewis formula.

11-15 The Lewis formula for the arrangement NNO is

$$\overset{\ominus}{\underset{\cdot\cdot}{:}}N=\overset{\oplus}{N}=O\overset{\cdot\cdot}{\underset{\cdot\cdot}{:}}$$

where we have assigned formal charges according to Equation (11-1).

$$\text{formal charge on N} = 5 - 4 - \tfrac{1}{2}(4) = -1$$
$$\text{formal charge on N} = 5 - 0 - \tfrac{1}{2}(8) = +1$$
$$\text{formal charge on O} = 6 - 4 - \tfrac{1}{2}(4) = 0$$

The Lewis formula for the arrangement NON is

$$\overset{\ominus}{:}\!N\!=\!\overset{2+}{O}\!=\!N\overset{\ominus}{:}$$

where we have assigned formal charges according to Equation (11-1).

$$\text{formal charge on each N} = 5 - 4 - \tfrac{1}{2}(4) = -1$$
$$\text{formal charge on O} = 6 - 0 - \tfrac{1}{2}(8) = +2$$

Thus we predict that the arrangement NNO is the more likely.

11-17 There are 18 valence electrons, or nine electron pairs, in $HCOO^-$. Two resonance forms are

$$
\begin{array}{cc}
\overset{\textstyle \cdot \cdot}{O} & \overset{\ominus}{:}\!\overset{\cdot\cdot}{O}\!: \\
\| & | \\
H\!-\!C\!-\!\overset{\cdot\cdot}{O}\!:^{\ominus} & \longleftrightarrow \quad H\!-\!C\!=\!O\!: \\
\end{array}
$$

The superposition of these two resonance forms gives the resonance hybrid

$$
\left[\begin{array}{c} O \\ \| \\ H\!-\!C\!\cdots\!O \end{array} \right]^{-}
$$

The two carbon-oxygen bonds in $HCOO^-$ are equivalent; they have the same bond length and the same bond energy.

11-19 The three resonance forms are

$$
\overset{\textstyle \cdot \cdot}{O} \qquad \qquad :\!\overset{\cdot\cdot}{O}\!:^{\ominus} \qquad\qquad :\!\overset{\cdot\cdot}{O}\!:^{\ominus}
$$
$$
^{\ominus}\!:\!\overset{\cdot\cdot}{O}\!-\!C\!-\!\overset{\cdot\cdot}{O}\!:^{\ominus} \longleftrightarrow\; :\!O\!=\!C\!-\!\overset{\cdot\cdot}{O}\!:^{\ominus} \longleftrightarrow\; ^{\ominus}\!:\!\overset{\cdot\cdot}{O}\!-\!C\!=\!O\!:
$$

The superposition of these three resonance forms is

$$
\left[\begin{array}{c} O \\ \| \\ O\!\cdots\!C\!\cdots\!O \end{array} \right]^{2-}
$$

The three bonds in CO_3^{2-} are equivalent; they have the same bond length and the same bond energy.

11-21 (a) NO_2 contains a total of 23 electrons of which 17 are valence electrons. Lewis formulas for NO_2 are

$$\cdot \ddot{\text{O}}-\text{N}=\ddot{\text{O}}\colon \longleftrightarrow \colon\!\ddot{\text{O}}=\text{N}-\ddot{\text{O}}\cdot \qquad \text{odd electron}$$

(b) CO contains 14 electrons and 10 valence electrons. The Lewis formula for CO is

$$\colon\!\overset{\ominus}{\text{C}}\equiv\overset{\oplus}{\text{O}}\colon$$

(c) O_3^- contains 25 electrons and 19 valence electrons. Lewis formuals for O_3^- are

$$\cdot\ddot{\text{O}}-\ddot{\text{O}}-\overset{\ominus}{\text{O}}\colon \longleftrightarrow \colon\!\overset{\ominus}{\text{O}}-\ddot{\text{O}}-\ddot{\text{O}}\cdot \qquad \text{odd electron}$$

(d) O_2^- contains 17 electrons and 13 valence electrons. Lewis formulas for O_2^- are

$$\colon\!\overset{\ominus}{\text{O}}-\ddot{\text{O}}\cdot \longleftrightarrow \cdot\ddot{\text{O}}-\overset{\ominus}{\text{O}}\colon \qquad \text{odd electron}$$

11-23 Write the C, N, N, O atoms in a row as suggested by the chemical formula and add the hydrogen atoms to the carbon atom. There is an odd number of valence electrons (23), and a Lewis formula is

$$\begin{array}{c} \text{H} \\ | \\ \text{H}-\overset{\displaystyle |}{\underset{\displaystyle |}{\text{C}}}-\ddot{\text{N}}-\ddot{\text{N}}=\ddot{\text{O}}\colon \\ | \\ \text{H} \end{array}$$

Methylnitrosamine is a free radical.

11-25 (a) There are 48 valence electrons, or 24 electron pairs, in PCl_6^-. Putting the phosphorus atom as the central atom, we have

$$\begin{array}{ccc} \ddot{\text{Cl}}\colon & & \ddot{\text{Cl}}\colon \\ \diagdown\!\overset{\ominus}{}\!\diagup & & \\ \colon\!\ddot{\text{Cl}}-\text{P}-\ddot{\text{Cl}}\colon & & \\ \diagup & \diagdown & \\ \ddot{\text{Cl}}\colon & & \ddot{\text{Cl}}\colon \end{array}$$

Notice that phosphorus does not obey the octet rule in this case.

(b) There are 22 valence electrons, or 11 electron pairs, in I_3^-.

$$\colon\!\ddot{\text{I}}-\overset{\ominus}{\ddot{\text{I}}}-\ddot{\text{I}}\colon$$

(c)   There are 24 electron pairs in SiF_6^{2-}. Placing the silicon atom as the central atom, we have

$$
\begin{array}{c}
:\ddot{F}: \quad :\ddot{F}: \\
\backslash \overset{\fbox{$2-$}}{} / \\
:\ddot{F}-Si-\ddot{F}: \\
/ \quad \backslash \\
:\ddot{F}: \quad :\ddot{F}:
\end{array}
$$

11-27   There are 28 valence electrons or 14 electron pairs in IF_3. Placing the iodine atom as the central atom, we have

$$
:\ddot{F}-\ddot{I}-\ddot{F}: \\
\phantom{:\ddot{F}}|\phantom{\ddot{F}} \\
:\ddot{F}:
$$

There are 42 valence electrons or 21 electron pairs in IF_5. Placing the iodine atom as the central atom, we have

$$
\begin{array}{c}
:\ddot{F}: \quad :\ddot{F}: \\
\backslash \quad / \\
:\ddot{F}-\ddot{I}-\ddot{F}: \\
| \\
:\ddot{F}:
\end{array}
$$

11-29   There are 34 valence electrons or 17 electron pairs in SF_4. Placing the sulfur atom as the central atom, we have

$$
:\ddot{F}: \\
| \\
:\ddot{F}-\ddot{S}-\ddot{F}: \\
| \\
:\ddot{F}:
$$

There are 48 valence electrons or 24 electron pairs in SF_6. Placing the sulfur atom as the central atom, we have

$$
\begin{array}{c}
:\ddot{F}: \quad :\ddot{F}: \\
\backslash \quad / \\
:\ddot{F}-S-\ddot{F}: \\
/ \quad \backslash \\
:\ddot{F}: \quad :\ddot{F}:
\end{array}
$$

11-31   The oxidation state of vanadium in VO_2^+ is $+5$ $[x + 2(-2) = +1$ or $x = +5]$; thus vanadium has five valence electrons giving $5 + (2 \times 6) - 1 = 16$ valence electrons. Placing the vanadium atom as the central atom, we have

$$
:\ddot{O}=\overset{\oplus}{V}=\ddot{O}:
$$

The oxidation state of vanadium in VO_3^{2-} is $+4$ $[x + 3(-2) = -2$ or $x = +4]$, thus vanadium has four valence electrons giving a total of $4 + (3 \times 6) + 2 = 24$ valence electrons. Placing the vanadium atom as the central atom, we have

$$:\overset{..}{\overset{\ominus}{\underset{..}{O}}}-V=\overset{..}{\underset{.}{O}}: \longleftrightarrow :\overset{.}{\underset{.}{O}}=V-\overset{..}{\underset{..}{O}}:\overset{\ominus}{} \longleftrightarrow \overset{\ominus}{}:\overset{..}{\underset{..}{O}}-V-\overset{..}{\underset{..}{O}}:\overset{\ominus}{}$$

$$\overset{|}{\underset{\ominus}{:\overset{}{\underset{..}{O}}:}} \qquad \overset{|}{\underset{\ominus}{:\overset{}{\underset{..}{O}}:}} \qquad \overset{\|}{.\overset{}{\underset{.}{O}}.}$$

11-33   The oxidation state of titanium in TiF_6^{2-} is $+4$ $[x + 6(-1) = -2$ or $x = +4]$; thus titanium has four valence electrons giving a total of $4 + (6 \times 7) + 2 = 48$ valence electrons. Placing the titanium atom as the central atom, we have

$$\begin{array}{ccc} \overset{..}{\underset{..}{F}} & & \overset{..}{\underset{..}{F}} \\ \diagdown & \overset{\textcircled{\scriptsize 2-}}{} & \diagup \\ :F- & Ti & -F: \\ \diagup & & \diagdown \\ \overset{..}{\underset{.}{F}} & & \overset{..}{\underset{.}{F}} \end{array}$$

The oxidation state of titanium in $TiBr_5^-$ is $+4$ $[x + 5(-1) = -1$ or $x = +4]$; thus titanium has four valence electrons giving a total of $4 + (5 \times 7) + 1 = 40$ valence electrons. Placing the titanium atom as the central atom, we have

$$\begin{array}{ccc} & \overset{..}{:\underset{}{Br}:} & \\ & \overset{|}{}\ominus & \\ :\overset{..}{\underset{..}{Br}}- & Ti & -\overset{..}{\underset{}{Br}}: \\ \diagup & & \diagdown \\ \overset{..}{\underset{.}{Br}} & & \overset{..}{\underset{.}{Br}} \end{array}$$

11-35   The Lewis formula of BrCl is

$$:\overset{..}{\underset{..}{Br}}-\overset{..}{\underset{..}{Cl}}:$$

The electronegativity of chlorine is greater than that of bromine, and so we have

$$\overset{\delta+}{:\overset{..}{\underset{..}{Br}}}-\overset{\delta-}{\overset{..}{\underset{..}{Cl}}}:$$

or

$$\overset{\longmapsto}{:\overset{..}{\underset{..}{Br}}-\overset{..}{\underset{..}{Cl}}:}$$

11-37   (a)   Fluorine is more electronegative than nitrogen, and so we have

$$\overset{3\delta+}{} \\ \delta-:\overset{..}{\underset{..}{F}}-\overset{..}{\underset{|}{N}}-\overset{..}{\underset{..}{F}}:\delta- \\ :\overset{..}{\underset{..}{F}}:\,\delta-$$

(b)   Fluorine is more electronegative than oxygen, and so we have

$$\delta- : \overset{\cdot\cdot}{\underset{\cdot\cdot}{F}} - \overset{2\delta+}{\underset{\cdot\cdot}{O}} - \overset{\cdot\cdot}{\underset{\cdot\cdot}{F}} : \delta-$$

(c)   Oxygen is more electronegative than bromine, and so we have

$$: \overset{\delta+}{\underset{\cdot\cdot}{Br}} - \overset{2\delta-}{\underset{\cdot\cdot}{O}} - \overset{\delta+}{\underset{\cdot\cdot}{Br}} :$$

11-39   (a)   Some resonance forms are

The superposition of these resonance forms gives

All four bonds are equivalent; they have the same bond lengths and the same bond energy.

(b)   Some resonance forms are

The superposition of these resonance forms is

$$
\left[\begin{array}{c} O \quad\quad O \\ P \\ O \quad\quad O \end{array}\right]^{3-}
$$

All four bonds are equivalent; they have the same bond lengths and the same bond energy.

(c)   Two resonance forms are

$$
\begin{array}{cc}
H \quad \overset{\cdot\cdot}{\overset{\cdot\cdot}{O}} \\
| \quad\; \| \\
H-\overset{|}{\underset{|}{C}}-\overset{\cdot\cdot}{\underset{\cdot\cdot}{C}}-\overset{\cdot\cdot}{\underset{\cdot\cdot}{O}}:^{\ominus} \\
H
\end{array}
\longleftrightarrow
\begin{array}{cc}
H \quad :\overset{\cdot\cdot}{O}:^{\ominus} \\
| \quad\; | \\
H-\overset{|}{\underset{|}{C}}-C=\overset{\cdot\cdot}{\underset{\cdot\cdot}{O}}: \\
H
\end{array}
$$

The superposition of these resonance forms is

$$
\left[\begin{array}{c}
H \quad O \\
| \quad\; \| \\
H-\overset{|}{\underset{|}{C}}-C\text{---}O \\
H
\end{array}\right]^{\ominus}
$$

The two C—O bonds are equivalent; they have the same bond lengths and the same bond energy.

11-41   (a)   32 valence electrons, 16 electron pairs, nitrogen atom central

$$
\begin{array}{c}
:\overset{\cdot\cdot}{F}: \\
| \\
:\overset{\cdot\cdot}{\underset{\cdot\cdot}{F}}-\overset{\oplus}{N}-\overset{\cdot\cdot}{\underset{\cdot\cdot}{F}}: \\
| \\
:\overset{}{\underset{\cdot\cdot}{F}}:
\end{array}
$$

(b)   34 valence electrons, 17 electron pairs, chlorine atom central

$$
\begin{array}{c}
:\overset{\cdot\cdot}{F}: \\
| \\
:\overset{\cdot\cdot}{\underset{\cdot\cdot}{F}}-\overset{\oplus}{\underset{\cdot}{Cl}}-\overset{\cdot\cdot}{\underset{\cdot\cdot}{F}}: \\
| \\
:\overset{}{\underset{\cdot\cdot}{F}}:
\end{array}
$$

(c)   8 valence electrons, 4 electron pairs, phosphorus atom central

$$
\begin{array}{c}
H \\
| \\
H-\overset{\oplus}{P}-H \\
| \\
H
\end{array}
$$

(d)   48 valence electrons, 24 electron pairs, arsenic atom central

$$
\begin{array}{ccc}
:\!\ddot{F}\!: & & :\!\ddot{F}\!: \\
\diagdown & \ominus & \diagup \\
:\!\ddot{F}\!-\!\!\!\! & \!\!As\!\! & \!\!\!\!-\ddot{F}\!: \\
\diagup & & \diagdown \\
:\!\ddot{F}\!: & & :\!\ddot{F}\!:
\end{array}
$$

(e)   36 valence electrons, 18 electron pairs, bromine atom central

$$
\begin{array}{c}
:\!\ddot{F}\!: \\
| \\
:\!\ddot{F}\!-\!\!\overset{\ominus}{Br}\!\!-\!\ddot{F}\!: \\
| \\
:\!\ddot{F}\!:
\end{array}
$$

11-43 (a)   26 valence electrons, 13 electron pairs, chlorine atom central

$$\overset{\ominus}{:}\!\ddot{O}\!-\!\!\overset{\overset{\ominus:\ddot{O}:}{|}}{\underset{}{\overset{2+}{Cl}}}\!\!-\ddot{O}\!-\!H \longleftrightarrow \overset{\cdot\dot{O}\cdot}{:\!O\!=\!\overset{\|}{Cl}\!-\!\ddot{O}\!-\!H} + \text{other resonance forms with an expanded valence shell}$$

(b)   18 valence electrons, 9 electron pairs, nitrogen atom central

$$:\!O\!=\!\ddot{N}\!-\!\ddot{O}\!-\!H$$

(c)   32 valence electrons, 16 electrons pairs, iodine atom central

$$\overset{\ominus}{:}\!\ddot{O}\!-\!\!\overset{\overset{:\ddot{O}:^{\ominus}}{|}}{\underset{\underset{:O:^{\ominus}}{|}}{\overset{3+}{I}}}\!\!-\ddot{O}\!-\!H \longleftrightarrow \overset{\cdot\dot{O}\cdot}{:\!O\!=\!\!\underset{\cdot\dot{O}\cdot}{\overset{\|}{I}}\!\!-\!\ddot{O}\!-\!H} + \text{other resonance forms with an expanded valence shell}$$

(d)   20 valence electrons, 10 electron pairs, bromine atom central

$$\overset{\ominus}{:}\!\ddot{O}\!-\!\overset{\oplus}{Br}\!-\!\ddot{O}\!-\!H \longleftrightarrow :\!O\!=\!\ddot{Br}\!-\!\ddot{O}\!-\!H$$

11-45 (a)   20 valence electrons, 10 electron pairs, sulfur atom central

$$:\!\ddot{F}\!-\!\ddot{S}\!-\!\ddot{F}\!:$$

(b)   34 valence electrons, 17 electron pairs, sulfur atom central

$$
\begin{array}{ccc}
\cdot\ddot{F}\cdot & & \cdot\ddot{F}\cdot \\
\diagdown & :\!\ddot{\;} & \diagup \\
 & S & \\
\diagup & & \diagdown \\
:\!\ddot{F}\cdot & & \cdot\ddot{F}\!:
\end{array}
$$

(c)   82 valence electrons, 41 electron pairs, sulfur atoms central

$$
\begin{array}{c}
\ddot{F}\cdot \quad \cdot\ddot{F}\cdot \quad \cdot\ddot{F}\cdot \quad \cdot\ddot{F}\cdot \\
\diagdown \quad \diagup \quad \diagdown \quad \diagup \\
:\ddot{F}{-}S{=\!=\!=\!=}S{-}\ddot{F}: \\
\diagup \quad \diagdown \quad \diagup \quad \diagdown \\
\cdot\ddot{F}\cdot \quad \cdot\ddot{F}\cdot \quad \cdot\ddot{F}\cdot \quad \cdot\ddot{F}\cdot
\end{array}
$$

(d)   26 valence electrons, 13 electron pairs, sulfur atoms central

$$:\ddot{F}{-}\ddot{S}{-}\ddot{S}{-}\ddot{F}:$$

11-47   There are 56 valence electrons, or 28 electron pairs, in $Cr_2O_7^{2-}$. The Lewis formulas are

$$
\ominus:\ddot{O}\overset{\textcircled{2+}}{\text{Cr}}{-}\ddot{O}{-}\overset{\textcircled{2+}}{\text{Cr}}\ddot{O}:\ominus \longleftrightarrow \ominus:\ddot{O}{-}\text{Cr}{-}\ddot{O}{-}\text{Cr}{-}\ddot{O}:\ominus
$$

+ other resonance forms with expanded valence shells

11-49   The Lewis formulas of each species are

(a)
$$
\begin{array}{cc}
\text{H} & \text{H} \\
| & | \\
\text{H}{-}\text{C}{=}\text{C}{-}\text{C}{\equiv}\text{N}:
\end{array}
$$

(b)
$$
\begin{array}{ccc}
& \cdot\ddot{O} & \cdot\ddot{O} \\
& \| & \| \\
\text{H}{-}\ddot{O}{-}\text{C}{-}\text{C}{-}\ddot{O}{-}\text{H}
\end{array}
$$

(c)
$$\ominus:\text{C}{\equiv}\text{C}:\ominus$$

(d)
$$
\begin{array}{ccc}
\text{H} & \text{H} & \text{H} \\
| & | & | \\
\text{H}{-}\text{C}{-}\text{C}{=}\text{C}{-}\text{H} \\
| \\
\text{H}
\end{array}
$$

The species with a triple bond are (a) and (c).

11-51   The Lewis formula of N_2F_2 with the arrangement FNNF is

$$:\ddot{F}{-}\ddot{N}{=}\ddot{N}{-}\ddot{F}:$$

The Lewis formula of N_2F_2 with the arrangement FFNN is

$$:\ddot{F}{-}\overset{\oplus}{\ddot{F}}{-}\ddot{N}{=}\overset{\ominus}{N}:$$

where the formal charges are assigned according to Equation (11-1). The arrangement FNNF has the lower formal charges, and so we predict that the structure of N_2F_2 is FNNF.

11-53 Line up the carbon and oxygen atoms as they appear in the chemical formula and place the hydrogen atoms on them in terminal positions to get

(a)

$$\begin{array}{c} \text{H}\quad\text{H} \\ |\quad\ | \\ \text{H}-\overset{|}{\underset{|}{\text{C}}}-\overset{|}{\underset{|}{\text{C}}}-\overset{..}{\underset{..}{\text{O}}}-\text{H} \\ |\quad\ | \\ \text{H}\quad\text{H} \end{array}$$

ethyl alcohol

Similarly, we have

(b)

$$\begin{array}{c} \text{H}\quad\text{H}\quad\text{H} \\ |\quad\ |\quad\ | \\ \text{H}-\overset{|}{\underset{|}{\text{C}}}-\overset{|}{\underset{|}{\text{C}}}-\overset{|}{\underset{|}{\text{C}}}-\overset{..}{\underset{..}{\text{O}}}-\text{H} \\ |\quad\ |\quad\ | \\ \text{H}\quad\text{H}\quad\text{H} \end{array}$$

n-propyl alcohol

and

(c)

$$\begin{array}{c} \text{H}\quad\text{H}\quad\text{H} \\ |\quad\ |\quad\ | \\ \text{H}-\overset{|}{\underset{|}{\text{C}}}-\overset{|}{\underset{|}{\text{C}}}-\overset{|}{\underset{|}{\text{C}}}-\text{H} \\ |\quad\ |\quad\ | \\ \text{H}\quad:\overset{}{\text{O}}:\quad\text{H} \\ | \\ \text{H} \end{array}$$

isopropyl alcohol

11-55 The Lewis formula for H_2O_2 is

$$\text{H}-\overset{..}{\underset{..}{\text{O}}}-\overset{..}{\underset{..}{\text{O}}}-\text{H}$$

When a proton is removed, the electron remains with the oxygen atom; thus the Lewis formula for the hydroperoxide ion is

$$\text{H}-\overset{..}{\underset{..}{\text{O}}}-\overset{..}{\underset{..}{\text{O}}}:^{\ominus}$$

The names are

$NaHO_2$	sodium hydroperoxide
$Ba(HO_2)_2$	barium hydroperoxide

11-57 (a) The thiocarbonate ion CS_3^{2-} is similar to the carbonate ion. Three Lewis formulas are

$$^{\ominus}:\overset{..}{\underset{..}{\text{S}}}-\overset{|}{\underset{\underset{\ominus}{:\overset{..}{\underset{..}{\text{S}}}:}}{\text{C}}}=\overset{..}{\underset{..}{\text{S}}}:\longleftrightarrow:\overset{..}{\underset{..}{\text{S}}}=\overset{|}{\underset{\underset{\ominus}{:\overset{..}{\underset{..}{\text{S}}}:}}{\text{C}}}-\overset{..}{\underset{..}{\text{S}}}:^{\ominus}\longleftrightarrow^{\ominus}:\overset{..}{\underset{..}{\text{S}}}-\overset{\overset{..}{\underset{..}{\text{S}}}}{\underset{\overset{..}{\underset{..}{\text{S}}}}{\overset{||}{\text{C}}}}-\overset{..}{\underset{..}{\text{S}}}:^{\ominus}$$

and their superposition is

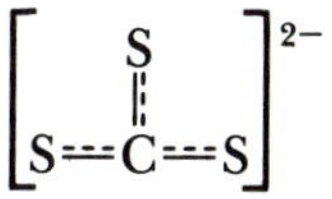

$$\left[\begin{matrix} & S \\ & \| \\ S \text{---} & C \text{---} S \end{matrix}\right]^{2-}$$

(b) 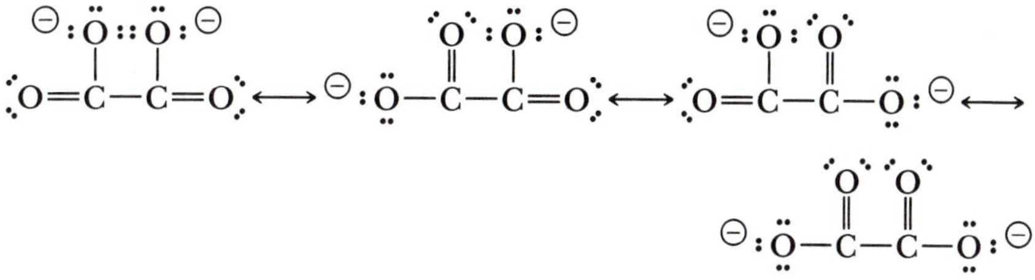 $\quad :O=C-C=O: \longleftrightarrow \ominus:\ddot{O}-C-C=O: \longleftrightarrow :O=C-C-\ddot{O}:\ominus \longleftrightarrow$

$$\ominus:\ddot{O}-C-C-\ddot{O}:\ominus$$

Their superposition is

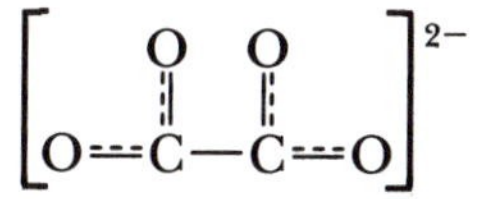

$$\left[\begin{matrix} & O & O \\ & \| & \| \\ O\text{---}C & - & C\text{---}O \end{matrix}\right]^{2-}$$

(c) $\quad :N\equiv C-\ddot{S}:\ominus \longleftrightarrow \ominus:\ddot{N}=C=\ddot{S}: \longleftrightarrow \textcircled{2-}:\ddot{N}-C\equiv S:\oplus$

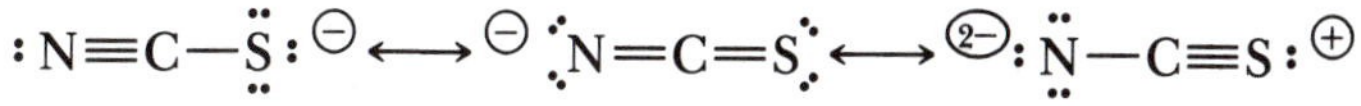

Their superposition is

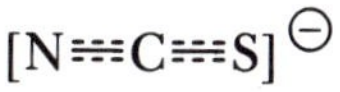

$$[N\text{≡≡}C\text{≡≡}S]^{\ominus}$$

(d)  $\quad \ominus:\ddot{O}-\ddot{O}\cdot \longleftrightarrow \cdot\ddot{O}-\ddot{O}:\ominus$

11-59 We shall use the abbreviation for the superimposed formula of benzene in writing the Lewis formula for the following benezene derivatives.

(a) 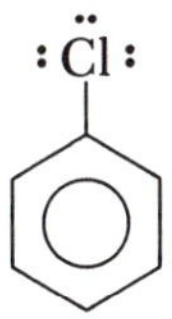 $\quad :\ddot{C}l:$

(b) $\quad H-\ddot{N}-H$

(c) $\quad :O=C-\ddot{O}-H$

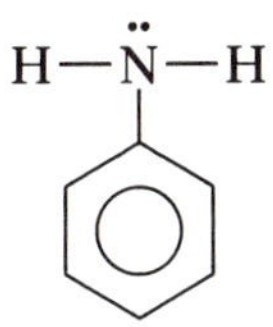

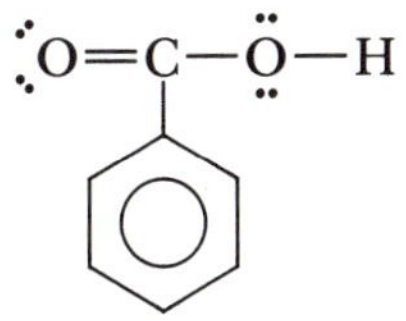

(d) 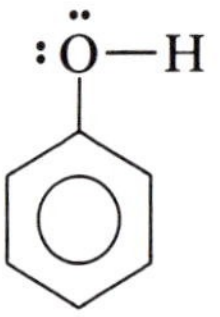 $\overset{\cdot\cdot}{:}\overset{\cdot\cdot}{O}$—H

11-61 Each sulfur atom is covalently bonded to two other sulfur atoms to form an eight-membered ring. The Lewis formula for solid sulfur is

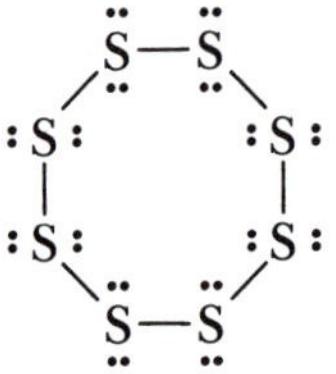

11-63 (a) 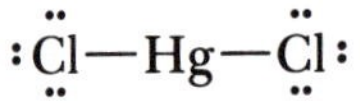 $:\overset{\cdot\cdot}{\underset{\cdot\cdot}{Cl}}$—Hg—$\overset{\cdot\cdot}{\underset{\cdot\cdot}{Cl}}:$

(b) 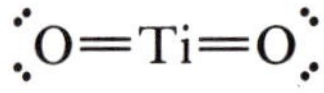 $\overset{\cdot\cdot}{:}O{=}Ti{=}O\overset{\cdot\cdot}{:}$

(c)

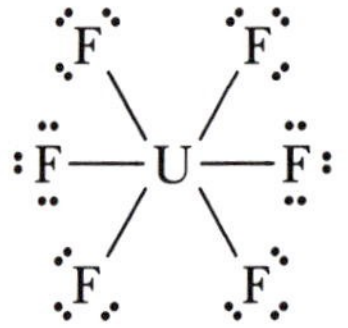

(d) 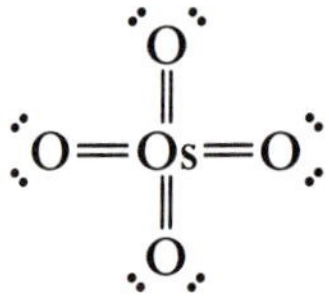 $\overset{\cdot\cdot}{:}O{=}Os{=}O\overset{\cdot\cdot}{:}$ plus other resonance forms

11-65 Because electronegativities increase as you go from left to right across a row in the periodic table and decrease as you go down a column in the periodic table, we predict that the order of electronegativities is

$$Cl > S > Se > Sb > In$$

E. ANSWERS TO THE SELF-TEST

1. shared

2. line

3. a lone pair

4. molecules

5. false

6. false

7. the distance between the nuclei of two atoms that are joined by a bond

8. three

9. many elements form covalent bonds so as to end up with eight electrons in their valence shells

10. true

11. it can achieve a noble-gas-like electron configuration by having two electrons in its outer shell

12. one

13. four (two pairs)

14. false

15. six (three pairs)

16. false

17. false

18. See Equation (11-1).

19. false

20. a superposition of the various Lewis formulas

21. resonance form

22. equal

23. false

24. true

25. true

26. less

27. false (It is the difference in energy between the resonance hybrid and an individual Lewis formula.)

28. false

29. true

30. resonance

31. there is one or more unpaired electrons

32. false

33. does not contain enough electrons to satisfy the octet rule for each atom in the compound

34. true

35. *d*

36. false

37. lowest

38. false

39. the oxidation state of the transition metal

40. true

41. the force with which an atom attracts the electrons in its covalent bonds

42. greater

43. false

44. the bond energies of a heteronuclear molecule AB and the homonuclear molecules A_2 and B_2

45. fluorine

46. 4.0

47. nonmetallic

48. left (to) right

49. false

50. more

51. similar

52. unequally

53. small positive . . . small negative

54. $\delta-$

55. the polarity of the bond

56. magnitude . . . direction

57. true

12 PREDICTION OF MOLECULAR GEOMETRIES

A. OUTLINE OF CHAPTER 12

12-1 Lewis formulas do not give us the shapes of molecules.

Lewis formulas do not show the geometrical arrangement of the nuclei in a molecule.

Geometrical isomers are molecules that have the same chemical formula but different spatial arrangements of their atoms.

A regular tetrahedron has four equivalent vertices and four identical faces, each of which is an equilateral triangle (Figure 12-2).

12-2 All four vertices of a regular tetrahedron are equivalent.

When four atoms are bonded to a carbon atom, they are arranged in a tetrahedral array around the carbon atom, which is in the center of the tetrahedron (Figure 12-1).

A space-filling model of a molecule illustrates the angles between bonds and the relative sizes of the atoms (Figure 12-3).

A ball-and-stick model of a molecule illustrates the angles between bonds (Figure 12-1).

The regular tetrahedron bond angle is 109.5° (for example, the H—C—H bond angle in CH_4 is 109.5°).

Structural chemistry is the study of the shapes and sizes of molecules.

12-3 Valence-shell electron-pair repulsion theory is used to predict the shapes of molecules.

VSEPR theory postulates that the shape of a molecule is determined by minimizing the mutual repulsion of the electron pairs in the valence shell of the central atom.

The electron-deficient molecule $BeCl_2$ is linear (Figure 12-6a).

The electron-deficient molecule BF_3 is trigonal planar (Figure 12-6b).
The bond angles of BF_3 are 120°.

12-4   The number of valence-shell electron pairs determines the shape of a molecule.

The arrangements of sets of electron pairs on the surface of spheres that minimize mutual repulsion between electron pairs are shown in Figure 12-6.

Some molecules with four covalent bonds about a central atom are tetrahedral.

The arrangement of five electron pairs in five covalent bonds about a central atom is a trigonal bipyramid (Figures 12-6d and 12-7).

The five vertices of a trigonal bipyramid are not equivalent; the three lying on the equator are called equatorial vertices, and the two lying at the poles are called axial vertices (Figures 12-6d and 12-7).

Molecules with six covalent bonds about a central atom are octahedral.

All six vertices of an octahedron are equivalent (Figure 12-8).

The bond angles for various molecular shapes are given in Table 12-1.

12-5   Lone electron pairs in the valence shell affect the shapes of molecules.

The total number of electron pairs in the valence shell of the central atom determines the shape of a molecule.

The geometry of a molecule is described by the positions of the nuclei in the molecule.

Lone pairs are more spread out than bond pairs.

The repulsion between a lone pair and the electron pair in a covalent bond is greater than the repulsion between the electron pairs in two covalent bonds.

Lone-pair – lone-pair repulsion > lone-pair – bond-pair repulsion > bond-pair – bond-pair repulsion [Equation (12-1)].

Many molecules can be classified by using the formula AX_mE_n, where A represents a central atom, X represents an attached ligand, and E is a lone pair of electrons on the central atom.

The numbers of bonds and lone pairs of electrons around the central atom result in the molecular shapes shown in Figure 12-11.

12-6   VSEPR theory is applicable to molecules that contain multiple bonds.

A double or triple bond is counted simply as one bond connecting the ligand X to the central atom A.

Multiple bonds repel single bonds more strongly than single bonds repel other single bonds.

VSEPR theory can be applied to molecules that are described by resonance.

12-7   Lone-pair electrons occupy the equatorial vertices of a trigonal bipyramid.

An equatorial lone pair has only two nearest neighbors at 90° (Figure 12-13a).

The electron repulsions due to a lone pair are minimized by placing the lone-pair electrons at an equatorial vertex of a trigonal bipyramid.

AX_4E molecules are seesaw-shaped (Figure 12-14a).

AX_3E_2 molecules are T-shaped (Figure 12-14b).

AX_2E_3 molecules are linear (Figure 12-14c).

12-8 Two lone electron pairs occupy opposite vertices of an octahedron.

The lone-pair–lone-pair repulsion is minimized when the two lone pairs occupy opposite vertices of an octahedron.

AX_5E molecules are square pyramidal (Figure 12-18b).

AX_4E_2 molecules are square planar (Figure 12-18c).

A summary of the results of VSEPR theory is given in Table 12-3.

At most, three possible geometries can arise for different numbers of atoms attached to a central atom in an octahedral skeleton.

12-9 Molecular geometry determines whether or not a molecule has a net dipole moment.

A symmetric molecule with polar bonds may have no net dipole moment.

A nonsymmetric molecule with polar bonds has a net dipole moment.

B. SELF-TEST

1. The four single bonds about a central carbon atom are directed toward _____
_____ .

2. All four vertices of a regular tetrahedron are equivalent. *True/False*

3. Dichloromethane, CH_2Cl_2, has (*zero/one/two*) geometrical isomers.

4. The shape of a molecule can be determined experimentally. *True/False*

5. The basis of the VSEPR theory is the postulate that the shape of a molecule is determined by _____
_____
_____ .

6. The shape of a molecule in which the central atom forms two bonds and has no lone electron pairs is _____ .

7. The VSEPR theory predicts that in a molecule with only two valence-shell electron pairs about the central atom, the two valence-shell electron pairs lie _____
_____ .

8. The VSEPR theory predicts that in a molecule with only three valence-shell electron pairs about the central atom, the three valence-shell electron pairs lie _____
_____ .

9. The VSEPR theory predicts that in a molecule with only four valence-shell electron pairs about the central atom, the four valence-shell electron pairs lie _____

_____.

10. The VSEPR theory predicts that in a molecule with only five valence-shell electron pairs about the central atom, the five valence-shell electron pairs lie _____

_____.

11. The VSEPR theory predicts that in a molecule with only six valence-shell electron pairs about the central atom, the six valence-shell electron pairs lie _____

_____.

12. The five vertices of a trigonal bipyramid are equivalent. *True/False*

13. The six vertices of an octahedron are equivalent. *True/False*

14. Axial vertices are the vertices of a _____ that lie

_____.

15. Lone-electron pairs on the central atom in a molecule or ion do not affect the shape of the molecule or ion. *True/False*

16. The H—N—H bond angle in ammonia is less than 109.5° because _____

_____.

17. The H—O—H bond angle in water is 109.5°. *True/False*

18. The shape of a molecule is described by the positions of the nuclei of the central atom and the ligands. *True/False*

19. The symbol AX_3E indicates a molecule or ion that has _____

_____ around a central atom.

20. The shape of a molecule of class AX_3E is _____ .

21. The symbol AX_4E indicates a molecule or ion that has _____

_____ around a central atom.

22. The lone electron pair in an AX_4E class of molecule is located at an (*axial/equatorial*) position on a trigonal bipyramid.

23. The shape of an AX_4E molecule is _____ .

24. The shape of an AX_3E_2 molecule is _____ .

25. The shape of an AX_2E_3 molecule is _____ .

26. In the molecule PCl_5, _____ is the central atom and _____ are ligands.

27. The F—Cl—F bond angle in the molecule ClF_3 is less than 120° because _____ _____ .

28. The symbol AX_5E indicates a molecule or ion that has _____ _____ around the central atom.

29. The shape of an AX_5E molecule is _____ .

30. It makes no difference at which vertex the lone electron pair in an AX_5E molecule is located. *True/False*

31. The shape of an AX_4E_2 molecule is _____ .

32. The double bond in the formaldehyde molecule H_2CO is considered as two bonds in applying VSEPR theory. *True/False*

33. The VSEPR theory cannot be applied to molecules that are described by resonance. *True/False*

34. The bond angles in a trigonal planar molecule are close to _____ .

35. The bond angles in a trigonal bipyramidal molecule are close to _____ , _____ , and _____ .

36. The bond angles in a square pyramidal molecule are all exactly 90°. *True/False*

37. A symmetric molecule with polar bonds may have no net dipole moment. *True/False*

C. CALCULATIONS YOU SHOULD KNOW HOW TO DO

There are no numerical calculations in this chapter. You should be able to use VSEPR theory to predict molecular shapes of a variety of molecules.

D. SOLUTIONS TO THE ODD-NUMBERED PROBLEMS

12-1   In this problem we first use VSEPR theory to predict the molecular shape and from the shape we determine if there are any 90° bond angles. (See Table 12-1.)

(a) TeF_6 is an AX_6 octahedral molecule and thus has 90° bond angles.
(b) $AsBr_5$ is an AX_5 trigonal bipyramidal molecule and thus has some 90° bond angles.

(c) GaI_3 is an AX_3 trigonal planar molecule and thus has no 90° bond angles.

(d) XeF_4 is an AX_4E_2 square planar molecule and thus has 90° bond angles.

12-3   (a) ClF_3 is an AX_3E_2 T-shaped molecule and thus has no 120° bond angles.

(b) $SbBr_6^-$ is an AX_6 octahedral ion and thus has no 120° bond angles.

(c) $SbCl_5$ is an AX_5 trigonal bipyramidal molecule and thus has some 120° bond angles.

(d) $InCl_3$ is an AX_3 trigonal planar molecule and thus has 120° bond angles.

12-5   See Figure 12-11 or Table 12-3.

(a)   :F̈—T̈e—F̈: AX_2E_2 bent

(b)   :B̈r—S̈n—B̈r: AX_2E bent

(c)   :F̈—K̈r—F̈: AX_2E_3 linear

(d)   :F̈—Ö—F̈: AX_2E_2 bent

12-7   (a)
$$
\begin{array}{c}
:\ddot{F}: \\
| \\
:\ddot{F}—\ddot{X}e—\ddot{F}: \\
| \\
:\ddot{F}:
\end{array}
$$
AX_4E_2 square planar

(b)
$$
\begin{array}{c}
:\ddot{O}:^{\ominus} \\
|(2+) \\
:\ddot{F}—\ddot{X}e—\ddot{F}: \\
| \\
:O:_{\ominus}
\end{array}
$$
AX_4E seesaw-shaped

(c)
$$
\begin{array}{c}
:\ddot{O}:^{\ominus} \\
|(+) \\
:\ddot{F}—N—\ddot{F}: \\
| \\
:\ddot{F}:
\end{array}
$$
AX_4 tetrahedral

(d)
$$
\begin{array}{c}
F \quad\quad F \\
\backslash \quad / \\
Se \\
/ \quad \backslash \\
F \quad\quad F
\end{array}
$$
AX_4E seesaw-shaped

Only NF_3O is tetrahedral.

12-9   (a)
$$
\begin{array}{c}
F \quad\quad F \\
\backslash \quad / \\
Br \\
/ \quad | \quad \backslash \\
F \quad :F: \quad F
\end{array}
$$
AX_5E square pyramidal

(b) $\ddot{C}l$ ... Sb ... $\ddot{C}l$ (structure) AX_5 trigonal bipyramidal

(c) $\ddot{F}$ ... Ge ... $\ddot{F}$ (structure, with $\ominus$ charge) AX_5 trigonal bipyramidal

(d) $\ddot{F}$ ... S ... $\ddot{F}$ (structure, with $\ominus$ charge) AX_5E square pyramidal

$SbCl_5$ and GeF_5 are trigonal bipyramidal.

12-11 (a) $:\!\ddot{F}\!-\!Te\!-\!\ddot{F}\!:$ (octahedral structure) AX_6 octahedral   90°
(1)

(b) $:\!\ddot{C}l\!-\!Sb\!-\!\ddot{C}l\!:$ (structure) AX_5 trigonal bipyramidal   90°, 120° (1, 3)

(c) $:\!\ddot{C}l\!-\!I\!-\!\ddot{C}l\!:$ (structure, with $\ominus$ charge) AX_4E_2 square planar   90° (1)

(d) $:\!\ddot{B}r\!-\!In\!-\!\ddot{B}r\!:$ (structure) AX_3 trigonal planar   120° (3)

12-13 (a) $H\!-\!Al\!-\!H$ (structure, with $\ominus$ charge) AX_4 tetrahedral   109.5° (2)

(b)

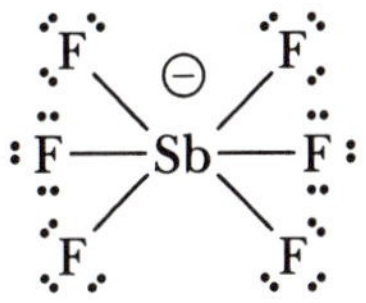

AX_6 octahedral   90°
(1)

(c)

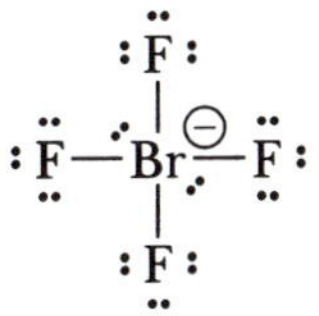

AX_4E_2 square planar   90°
(1)

(d)

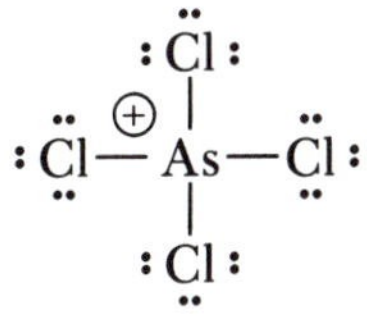

AX_4 tetrahedral   109.5°
(2)

12-15   AB compounds; for example

IF

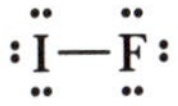

  linear

All AB compounds are linear.

AB_3 compounds; for example

IF$_3$

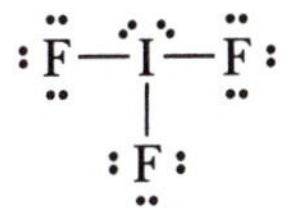

  AX_3E_2 T-shaped

Thus, BrF$_3$ and ClF$_3$ are T-shaped.

AB_5 compounds; for example

IF$_5$

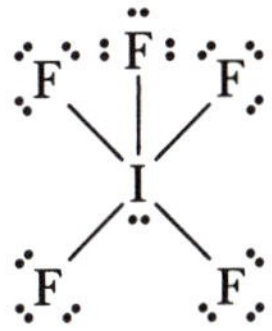

  AX_5E square pyramidal

Thus, BrF$_5$ and ClF$_5$ are square pyramidal.

12-17   (a)

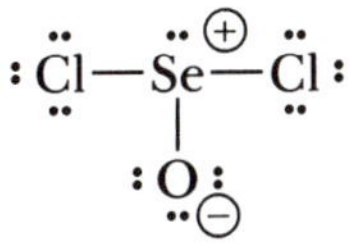

  or

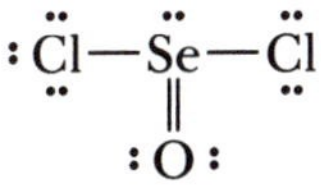

  AX_3E trigonal pyramidal

(b) $:\overset{..}{\underset{..}{Cl}}:$ ... $\overset{\ominus}{:}\overset{..}{\underset{..}{O}}-\overset{(2+)}{\underset{}{S}}\overset{..}{\underset{..}{O}}:\overset{\ominus}{}$ ... $:\overset{..}{\underset{..}{Cl}}:$ or $\overset{..}{\underset{..}{Cl}}:$ ... $:\overset{..}{O}=S=\overset{..}{O}:$ ... $:\overset{..}{\underset{..}{Cl}}:$ AX_4 tetrahedral

(c)   or AX_5 trigonal bipyramidal

(d)   or AX_4 tetrahedral

12-19 (a) AX_3 trigonal planar

(b) AX_4 tetrahedral

(c) $\overset{\ominus}{:}\overset{..}{N}=\overset{\oplus}{N}=\overset{..}{N}:\overset{\ominus}{}$ AX_2 linear

(d) $:\overset{..}{\underset{..}{Cl}}-\overset{..}{Sb}=\overset{..}{O}:$ AX_2E bent

12-21 (a) $\overset{\ominus}{:}\overset{..}{\underset{..}{O}}-\overset{..}{Br}=\overset{..}{O}:$ AX_2E_2 bent

(b) AX_5E square pyramidal

(c) $:\overset{..}{O}=\overset{}{S}=\overset{..}{O}:$ with Cl above and O below AX_4 tetrahedral

(d) :F̈—S̈⁺—F̈: AX_3E trigonal pyramidal

:F̈:

12-23 (a) AX_6 octahedral

(b) AX_5 trigonal bipyramidal

(c) ⊖:Ö—I—Ö:⊖ or :Ö=I=Ö: AX_3E trigonal pyramidal

(d) ⊖:Ö—I=Ö:⊖ or :Ö=I=Ö: AX_4 tetrahedral

12-25 The Lewis formula of the species NO_2^+ is

$$\overset{\oplus}{O=N=O}$$

Thus, NO_2^+ is a linear molecule with an O—N—O bond angle of 180°. The Lewis formula of NO_2^- is

$$[O{\cdots}\overset{\cdot\cdot}{N}{\cdots}O]^-$$

Thus, NO_2^- is a bent molecule with a predicted O—N—O bond angle of slightly less than 120°.

12-27 (a) :F̈—Xe—F̈: AX_2E_3 linear
                                            (no dipole moment)

(b) $\overset{..}{\underset{..}{F}}$ ... As structure ... AX_5 trigonal bipyramidal
(no dipole moment)

(c) Te with four Cl ... AX_4E seesaw-shaped
(dipole moment)

(d) $:\overset{..}{\underset{..}{Cl}}-\overset{..}{\underset{..}{O}}-\overset{..}{\underset{..}{Cl}}:$ AX_2E_2 bent
(dipole moment)

12-29 (a) Ga with three Cl ... AX_3 trigonal planar
(no dipole moment)

(b) $:\overset{..}{\underset{..}{Cl}}-\overset{..}{Te}-\overset{..}{\underset{..}{Cl}}:$ AX_2E_2 bent
(dipole moment)

(c) Te with four F ... AX_4E seesaw-shaped
(dipole moment)

(d) Sb with five Cl ... AX_5 trigonal bipyramidal
(no dipole moment)

12-31 (a) CF_4, AX_4, tetrahedral, nonpolar
(b) AsF_3, AX_3E, trigonal pyramidal, polar
(c) XeF_4, AX_4E_2, square planar, nonpolar
(d) SeF_4, AX_4E, seesaw-shaped, polar

12-33 (a) Fluorine is more electronegative than nitrogen, and so we have

$$\delta-:\overset{..}{\underset{..}{F}}-\overset{3\delta+}{\overset{..}{N}}-\overset{..}{\underset{..}{F}}:\delta-$$
$$\underset{\delta-}{\overset{|}{:\underset{..}{F}:}}$$

(b) Fluorine is more electronegative than oxygen, and so we have

$$\delta-:\overset{..}{\underset{..}{F}}-\overset{2\delta+}{\overset{..}{\underset{..}{O}}}-\overset{..}{\underset{..}{F}}:\delta-$$

(c)   Oxygen is more electronegative than bromine, and so we have

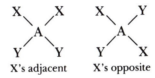

12-35 (a)   All the vertices of a tetrahedron are equivalent, and thus it makes no difference where the ligand Y is placed. Therefore, there is only one possible arrangement of the ligands for tetrahedral AX_3Y molecules.

(b)   There is only one possible arrangement of the ligands in tetrahedral AX_2YZ molecules.

(c)   All the vertices of a square are equivalent, and thus it makes no difference where ligand Y is placed. Therefore, there is only one possible arrangement of the ligands in a square planar AX_3Y molecule.

(d)   Two isomers with identical ligands either adjacent or opposite to one another:

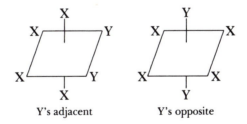

12-37 (a)   All octahedral vertices are equivalent. Thus there is only one possible arrangement for an octahedral AX_5Y molecule.

(b)   Two isomers, one with the two Y's opposite and one with the two Y's adjacent:

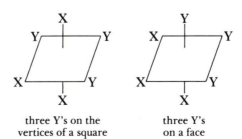

(c)   Two isomers, one with the three Y's on an octahedral face and one with the three Y's along three of the four vertices of a square:

12-39   The possible isomers of $C_2H_2F_2Cl_2$ are

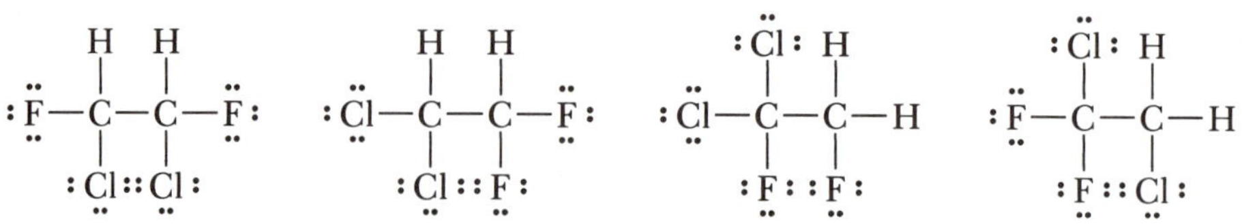

There are four isomers of $C_2H_2F_2Cl_2$.

12-41   (a)

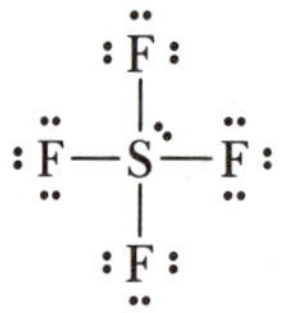

AX_4E seesaw-shaped

(b)

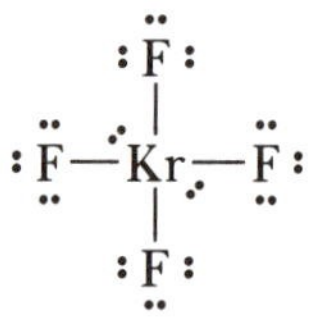

AX_4E_2 square planar

(c)

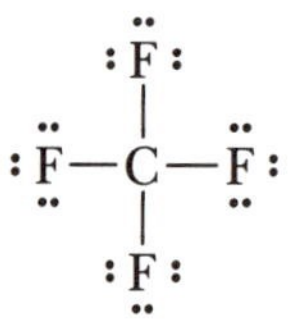

AX_4 tetrahedral

(d)

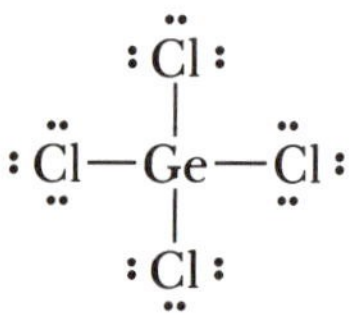

AX_4 tetrahedral

12-43   (a) H_2O (AX_2E_2)
          (b) NH_2^- (AX_2E_2)
          (c) AlH_4^- (AX_4)
          (d) SF_6 (AX_6)

See Table 12-3 for other examples of each case.

12-45   (a)

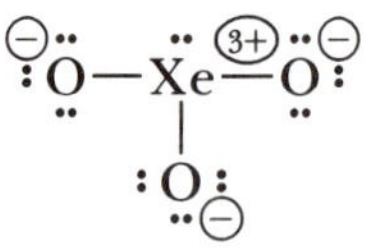

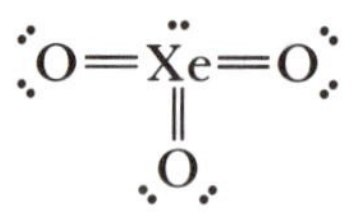

AX_3E trigonal pyramidal

(b)

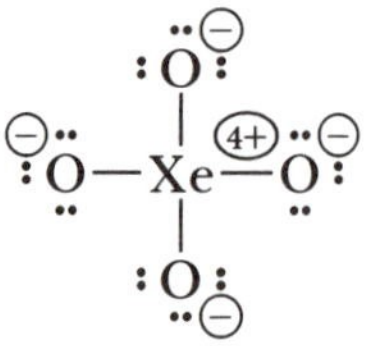

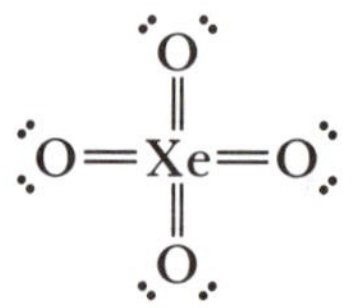

AX_4 tetrahedral

(c)   [Lewis structure of XeO_2F_2 with Xe^{2+}] or [Lewis structure with Xe double-bonded to two O and single-bonded to two F] AX_4E seesaw-shaped

(d)   [Lewis structure of XeO_6 with Xe^{2+}] or [Lewis structure with Xe double-bonded to O] AX_6 octahedral

12-47   (a)   [Lewis structure $^-O-S-O^-$] AX_2E_2 bent

(b)   [Lewis structure of S with double-bonded O and two single-bonded O^-]   + resonance forms AX_3E trigonal pyramidal

(c)   [Lewis structure of S^{2+} bonded to four O^- groups]   + resonance forms AX_4 tetrahedral

12-49   (a)   The electronegativities of the halogens increase as F > Cl > Br > I, and so the dipole moments of HX go as

$$HF > HCl > HBr > HI$$

(b)   Electronegativity goes as N > P > As, and so the XH_3 dipole moments go as

$$NH_3 > PH_3 > AsH_3$$

There is a small difference in the shapes of these molecules, but the electronegativity difference is more important.

(c)   Referring to part (a), we write

$$IF_3 > BrF_3 > ClF_3$$

(d)   Electronegativity increases as O > S > Se > Te, and so the dipole moments go as

$$H_2O > H_2S > H_2Se > H_2Te$$

There are small differences in the shapes of these molecules, but the electronegativity difference is more important.

12-51 Using the AX_mE_n notation, we find that (a) XeF_4 is AX_4E_2, (b) CF_4 is AX_4, (c) SF_2 is AX_2E_2, and (d) XeF_2 is AX_2E_3. The corresponding shapes are (a) square planar, (b) tetrahedral, (c) bent, and (d) linear. Thus, only XeF_4 has 90° bond angles.

12-53 (a)

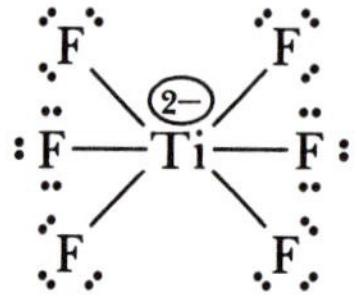

AX_6 octahedral

(b)

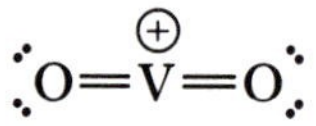

AX_2 linear

(c)

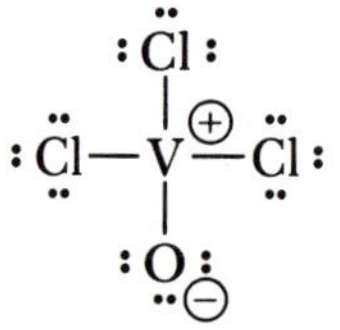

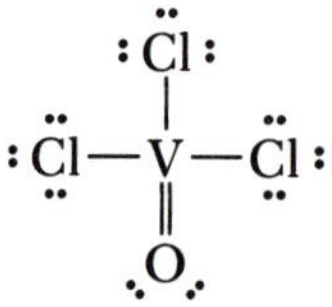

AX_4 tetrahedral

(d)

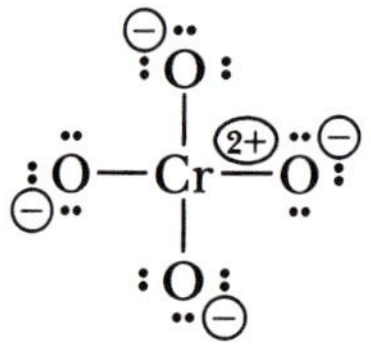

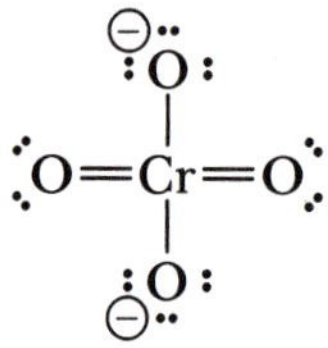

AX_4 tetrahedral

12-55 (a)

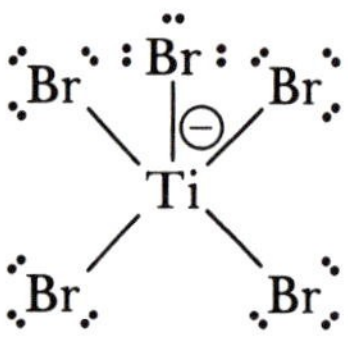

AX_5 trigonal bipyramidal

(b)

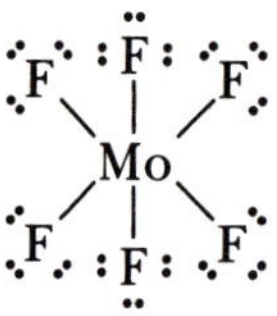

AX_6 octahedral

(c)

AX_6 octahedral

(d)

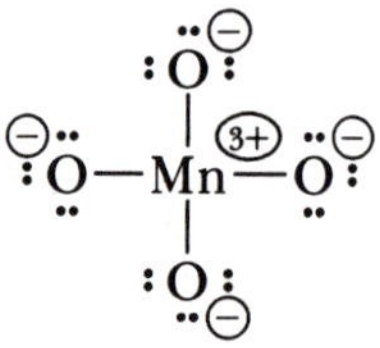

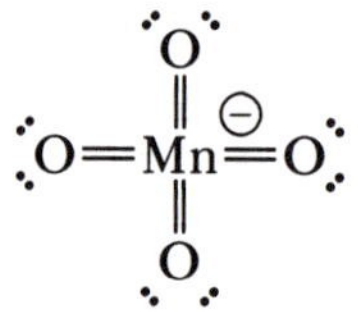

AX_4 tetrahedral

12-57 (a) AX_5 trigonal bipyramidal

(b) :F̈—S̈⁺—F̈:   or   :F̈—S̈—F̈: AX_3E trigonal pyramidal
                 | ‖
              :Ö:⊖   :O:

(c) :F̈—S²⁺—Ö:⊖   or   :F̈—S̈=Ö: AX_4 tetrahedral
          | |
        :F̈:   :F̈:

12-59 (a)   :Ö=Ös=Ö: AX_4 tetrahedral
              ‖
             .O.

(b) :B̈r—Ni—B̈r: AX_4 tetrahedral
         |²⊖
       :B̈r:

(c) :C̈l—Fe⊖—C̈l: AX_4 tetrahedral
         |
       :C̈l:

12-61   The triangle formed by the center sphere and any two others in Figure 12-2b is given by θ in

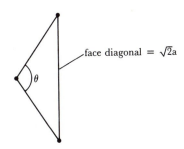

face diagonal $= \sqrt{2}a$

If we now draw a line from the center dot perpendicular to the face diagonal, then we bisect θ and the face diagonal:

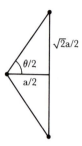

The length of this new line is $a/2$, because it is one-half the length of an edge (see Figure 12-2b).

Using trigonometry, we write

$$\tan\frac{\theta}{2} = \frac{\sqrt{2}a/2}{a/2} = \sqrt{2}$$

or

$$\frac{\theta}{2} = \tan^{-1}\sqrt{2}$$

Using the $\tan^{-1}$ key on your calculator, you find that

$$\frac{\theta}{2} = 54.74°$$

or that $\theta = 109.5°$.

E. ANSWERS TO THE SELF-TEST

1. the vertices of a tetrahedron

2. true

3. zero

4. true

5. the mutual repulsion of the electron pairs in the valence shell of the central atom

6. linear

7. on opposite sides of the central atom, at 180° from each other

8. at the vertices of an equilateral triangle

9. at the vertices of a tetrahedron

10. at the vertices of a trigonal bipyramid

11. at the vertices of an octahedron

12. false

13. true

14. trigonal bipyramid that lie at the poles (there are two axial vertices)

15. false

16. lone-pair – bond-pair repulsions are greater than bond-pair – bond-pair repulsions

17. false (It is slightly less than 109.5°.)

18. true

19. three ligands (X) and a lone pair (E)

20. trigonal pyramidal

21. four ligands (X) and a lone pair (E)

22. equatorial

23. seesaw-shaped

24. T-shaped

25. linear

26. P . . . Cl

27. of lone-pair – bond-pair repulsions

28. five ligands (X) and one lone pair (E)

29. square pyramidal

30. true

31. square planar

32. false

33. false

34. 120°

35. 120°, 90°, and 180°

36. false (Angles are somewhat less than 90° because of lone-pair – bond-pair repulsions.)

37. true

13 COVALENT BONDING

A. OUTLINE OF CHAPTER 13

13-1   A molecular orbital is a combination of atomic orbitals on different atoms.

The solution of the Schrödinger equation for H_2 gives the energies and wave functions that describe the two electrons in H_2.

The ground-state wave function of H_2 can be used to compute a contour map that shows the distribution of the electron density around the two nuclei in H_2.

A molecular orbital that encompasses both nuclei of H_2 is shown in Figure 13-1.

The buildup of electron density between nuclei is responsible for bond formation.

13-2   The hydrogen molecular ion H_2^+ is the simplest diatomic species.

Molecular orbital theory is a theory of chemical bonding.

A molecular orbital extends over two or more nuclei in a molecule.

The shapes of some of the molecular orbitals of H_2^+ are shown in Figure 13-2.

An internuclear axis is a line joining two nuclei.

Molecular orbitals that are cylindrically symmetric when viewed along the internuclear axis are called σ orbitals.

A molecular orbital that is concentrated in the region between two nuclei is called a bonding orbital.

A molecular orbital that has one or more nodal surfaces in the region between two nuclei and concentrates the electron density on the far sides of the nuclei is called an antibonding orbital.

An antibonding orbital is designated by a * superscript; for example, $\sigma*$ and $\pi*$.

The molecular orbitals of H_2^+ are listed in order of increasing energy in Figure 13-2.

A molecular orbital that has a cross section that is similar to an atomic p orbital when viewed along the internuclear axis is called a π orbital.

The electron configuration of homonuclear diatomic molecules can be constructed by placing electrons in the molecular orbitals shown in Figure 13-2 according to the Pauli exclusion principle (Table 13-1).

Antibonding electrons cancel the effect of bonding electrons.

Bond order is defined as

$$\text{bond order} = \frac{\left(\begin{array}{c}\text{number of electrons} \\ \text{in bonding orbitals}\end{array}\right) - \left(\begin{array}{c}\text{number of electrons} \\ \text{in antibonding orbitals}\end{array}\right)}{2} \quad (13\text{-}1)$$

Bond lengths decrease and bond energies increase with increasing bond order.

A bond order of zero implies that the diatomic molecule is not stable relative to the two separated atoms.

13-3   Molecular orbital theory predicts molecular electron configurations.

The order of the homonuclear diatomic molecular orbitals in terms of energy is given in Figure 13-5.

Electrons are placed in 1π and $1\pi^*$ orbitals according to Hund's rule.

O_2 contains two unpaired electrons, one in each of the $1\pi^*$ orbitals.

Oxygen molecules are paramagnetic because the electrons in each of the $1\pi^*$ orbital have spins that are unpaired.

The ground-state electron configurations for the homonuclear diatomic molecules Li_2 through Ne_2 are given in Table 13-2.

Molecular orbital theory can be applied to heteronuclear diatomic molecules if the atomic numbers of the two atoms in the molecule differ by one or two atomic numbers.

13-4   The bonding in polyatomic molecules can be described in terms of bond orbitals.

Localized bond orbitals are orbitals that are concentrated primarily in the region between the two atoms that form the covalent bond.

A localized covalent bond consists of two electrons that occupy a localized bond orbital.

The localized bond orbital approach is a simplified version of valence-bond theory.

13-5   Hybrid orbitals are combinations of atomic orbitals on the same atom.

A hybrid orbital is a linear combination of atomic orbitals on the same atom.

Atomic orbitals have positive and negative values.

An s orbital has a positive sign everywhere.

A $2p$ orbital has a positive sign in one lobe and a negative sign in the other lobe.

When atomic orbitals overlap, they generally reinforce in regions where they both have similar values and partially cancel in regions where they have opposite values.

The *sp* hybrid orbitals are the two equivalent hybrid orbitals that result from combining a 2*s* orbital and a 2*p* orbital on the same atom (Figure 13-11).

The *sp* hybrid orbitals are directed 180° from each other.

The bonding in BeH_2 can be described by two localized bond orbitals, each of which is a combination of an *sp* hybrid orbital on the Be atom and a 1*s* atomic orbital on the H atom (Figure 13-12).

A σ bond is a σ orbital occupied by two electrons of opposite spin.

There are two σ bonds in BeH_2.

13-6 sp^2 hybrid orbitals have trigonal planar symmetry.

The sp^2 hybrid orbitals are the three equivalent hybrid orbitals that are formed by combining a 2*s* orbital and two 2*p* orbitals on the same atom (Figure 13-14).

The sp^2 hybrid obitals lie in a plane and are directed toward the vertices of an equilateral triangle (Figure 13-14).

The bonding in BF_3 can be described by three localized bond orbitals, each of which is a combination of an sp^2 hybrid orbital on the central B atom and a 2*p* atomic orbital on the F atom (Figure 13-15).

There are three σ bonds in BF_3.

The prinicple of conservation of orbitals states that (1) if we combine atomic orbitals on the same atom to form hybrid orbitals, then the number of resulting hybrid orbitals is equal to the number of atomic orbitals combined, and (2) if we combine atomic orbitals from different atoms, then the number of resulting molecular orbitals is equal to the number of atomic orbitals combined.

13-7 sp^3 hybrid orbitals point toward the vertices of a tetrahedron.

The sp^3 hybrid orbitals are the four equivalent hybrid orbitals that result from combining the 2*s* orbital and the three 2*p* orbitals on the same atom. (Figure 13-16).

The sp^3 hybrid orbitals are directed toward the vertices of a tetrahedron. (Figure 13-16).

The bonding in CH_4 can be described by four localized bond orbitals, each of which is a combination of an sp^3 hybrid orbital on the central carbon atom and a 1*s* atomic orbital on the hydrogen atom (Figure 13-17).

The bonding in ethane, C_2H_6, is shown in Figure 13-19.

The geometry of the molecule determines which type of hybrid orbitals are appropriate to describe the bonding in that molecule.

13-8   We can use sp^3 orbitals to describe the bonding in molecules that have four electron pairs about the central atom.

The bonding in H_2O can be described in terms of sp^3 hybrid orbitals on the oxygen atom. Two of the sp^3 orbitals are used to form bonds with the hydrogen atoms, and the other two are occupied by the lone electron pairs on the oxygen atom (Figure 13-20).

The bonding in NH_3 can be described in terms of sp^3 orbitals on the nitrogen atom (Figure 13-21).

We can use sp^3 orbitals to describe the bonding in molecules that have no unique central atom (Figures 13-22 and 13-23).

13-9   Hybrid atomic orbitals can involve d orbitals.

The sp^3d hybrid orbitals are the five hybrid orbitals that result from combining the $3s$ orbital, the $3p$ orbitals, and one $3d$ orbital on the same atom.

The sp^3d orbitals are directed toward the vertices of a trigonal bipyramid.

The five sp^3d orbitals are not equivalent.

The sp^3d^2 hybrid orbitals are the six equivalent hybrid orbitals that result from combining the $3s$ orbital, the $3p$ orbitals, and two $3d$ orbitals on the same atom.

The sp^3d^2 orbitals are directed toward the vertices of a regular octahedron.

Only atomic orbitals of similar energy combine effectively to form hybrid orbitals.

The various hybrid orbitals discussed in this chapter are summarized in Table 13-3.

13-10   A double bond can be represented by a σ bond and a π bond.

The bonding in ethene, C_2H_4, is described in terms of sp^2 orbitals on the carbon atoms.

The σ-bond framework of a molecule indicates all the σ bonds in the molecule (Figure 13-25).

A π orbital is formed by the combination of two parallel $2p$ orbitals on two neighboring atoms (Figure 13-26).

A π orbital is similar to an atomic p orbital in cross section.

A π bond is formed when two electrons of opposite spins occupy a π orbital.

The double bond in ethylene consists of a σ bond (overlap of an sp^2 orbital from each carbon atom) and a π bond (overlap of a $2p$ orbital from each carbon atom).

The bond energy of a double bond is greater than that of a single bond, but not twice as great as that of a single bond.

13-11   There is limited rotation about a double bond.

There is no free rotation about carbon-carbon double bonds.

Stereoisomers are molecules with the same atom-to-atom bonding but different spatial arrangements.

Cis and trans isomers are stereoisomers.

The chemistry of vision involves cis-trans isomerization, which is the conversion of one isomer into the other.

In the visual process, light converts 11-*cis*-retinal to 11-*trans*-retinal.

The change in shape of the 11-retinal vision molecule triggers a response in the optic nerve cells that is eventually perceived as vision.

13-12   A triple bond can be represented by one σ bond and two π bonds.

The σ bonding in ethyne, C_2H_2, is described in terms of sp orbitals on the carbon atoms (Figure 13-28).

The carbon-carbon triple bond consists of one σ bond (overlap of an sp orbital from each carbon atom) and two π bonds (overlap of two $2p$ orbitals from each carbon atom) (Figure 13-29).

13-13   The π electrons in benzene are delocalized.

The bonding in benzene can be described in terms of σ bonds and π bonds.

It is appropriate to use sp^2 orbitals on the carbon atoms in benzene, which leads to the σ-bond framework shown in Figure 13-31.

The three π orbitals in benzene are spread uniformly over the entire ring (Figure 13-34).

The bonding π orbitals in benzene are not associated with any particular pair of carbon atoms and are called delocalized orbitals.

The delocalization of the π electrons in benzene is an example of charge delocalization.

13-14   Some π orbitals are nonbonding orbitals.

It is appropriate to use sp^2 hybrid orbitals to describe the bonding in ozone, O_3.

The σ-bond framework is shown in Figure 13-35.

The three $2p$ orbitals in O_3 combine to give three π molecular orbitals whose energy-level diagram is given in Figure 13-36.

The middle π orbital in the π-energy level diagram of O_3 is a nonbonding orbital.

Electrons in nonbonding orbitals neither contribute nor detract from bonding.

The bond order of ozone is three, two due to the localized oxygen-oxygen bonds and one due to the occupied delocalized π bond.

B. SELF-TEST

1. The solution of the Schrödinger equation for H_2 can be used to show the distribution of electron density around the two hydrogen nuclei. *True/False*

2. The buildup of electron density between the two nuclei in H_2 is responsible for the formation of a covalent bond between the two nuclei. *True/False*

3. The solution of the Schrödinger equation for H_2^+ results in a set of (*atomic/molecular*) orbitals.

4. All molecular orbitals have the same energy. *True/False*

5. A molecular orbital that has a circular cross section along the internuclear axis is called a _____ orbital.

6. A bonding orbital is a molecular orbital that is _____
_____.

7. An antibonding orbital is a molecular orbital that is _____
_____.

8. The antibonding orbital of lowest energy is designated _____.

9. The molecular orbital that has a cross section that is similar to a p orbital when viewed along the internuclear axis is called a _____ orbital.

10. The effect of electrons in antibonding orbitals cancels the effect of electrons in
_____.

11. The Pauli exclusion principle does not apply to the placement of electrons into molecular orbitals. *True/False*

12. The bond order of a molecule is defind as _____
_____.

13. The bond order of a molecule must always be a whole number. *True/False*

14. A bond order of zero for a molecule suggests that the molecule is not stable relative to the two separated atoms. *True/False*

15. Hund's rule is used to place electrons in the molecular orbitals of the oxygen molecule. *True/False*

16. The two electrons in the $1\pi^*$ orbital of O_2 have (*paired/unpaired*) spins.

17. The Lewis formula for O_2 predicts that oxygen is paramagnetic. *True/False*

18. A $1s$ orbital has regions of positive and negative values. *True/False*

19. A $2p$ orbital has regions of positive and negative values. *True/False*

20. Localized bond orbitals are concentrated in the region between the two atoms that are joined by the bond. *True/False*

21. Hybrid orbitals are combinations of _____
_____.

22. The sp hybrid orbitals are obtained by combining _____
_____.

23. There are (*one/two/three/four*) *sp* hybrid orbitals.

24. The *sp* hybrid orbitals are directed _____.

25. The two covalent bond orbitals in BeH_2 are formed by combining _____

_____.

26. A covalent bond is a bond orbital occupied by two electrons. *True/False*

27. The beryllium-hydrogen bonds in BeH_2 are σ bonds. *True/False*

28. The sp^2 hybrid orbitals are obtained by combining _____

_____.

29. The sp^2 hybrid orbitals point to _____.

30. There are (*one/two/three/four*) sp^2 hybrid orbitals.

31. The boron-fluorine bond in BF_3 is a σ bond. *True/False*

32. If we combine atomic orbitals on the same atom to form hybrid orbitals, then the number of resulting hybrid orbitals is equal to the number of _____

_____.

33. If we combine atomic orbitals on the different atoms, then the number of resulting molecular orbitals is equal to the number of _____.

34. The sp^3 hybrid orbitals are obtained by combining _____

_____.

35. The sp^3 hybrid orbitals are directed toward _____.

36. There are (*one/two/three/four*) sp^3 hybrid orbitals.

37. A carbon-hydrogen bond in CH_4 is a σ bond. *True/False*

38. _____ orbitals can be used to describe the bonding in molecules that have four electron pairs about the central atom.

39. The sp^3d hybrid orbitals are obtained by combining _____

_____.

40. There are _____ sp^3d hybrid orbitals.

41. The sp^3d hybrid orbitals are directed toward _____

_____.

42. The sp^3d^2 hybrid orbitals are obtained by combining _____ _____.

43. There are _____ sp^3d^2 hybrid orbitals.

44. The sp^3d^2 hybrid orbitals are directed toward _____.

45. The double bond in ethene, C_2H_4, consists of a _____ bond and a _____ bond.

46. The C—H bond in ethene, C_2H_4, is a combination of a(n) _____ orbital on the carbon atom and the _____ orbital of the hydrogen atom.

47. The σ-bond framework in a molecule indicates _____ _____.

48. The π bond in ethene is formed by combining the _____ _____.

49. The energy of a double bond is twice that of a single bond. *True/False*

50. Rotation cannot easily occur about carbon-carbon double bonds. *True/False*

51. In *trans*-1,2-dichloroethene, $ClCH_2CH_2Cl$, the chlorine atoms lie _____ _____.

52. The physical properties of cis-trans isomers are identical. *True/False*

53. Light causes the isomerization of 11-*cis*-retinal in the eye to give _____.

54. The triple bond in ethyne, C_2H_2, is described by a _____ bond and two _____ bonds.

55. A C—H bond in ethyne, C_2H_2, is a combination of a(n) _____ orbital on the carbon atom and a _____ orbital on the hydrogen atom.

56. The two π bonds in ethyne are formed by combining _____ _____.

57. All the carbon-carbon bond lengths in benzene are the same. *True/False*

58. The shape of the benzene molecule is _____ and _____.

59. The σ-bond framework in benzene is formed by overlapping a(n) _____ orbital from each carbon atom.

60. The π bonds in benzene are formed from the _____ _____.

61. Each π orbital in benzene is located between two adjacent carbon atoms. *True/False*

62. The π electrons in benzene are said to be _____.

63. The π electrons in ozone, O_3, are delocalized. *True/False*

64. Electrons in (*bonding/nonbonding/antibonding*) orbitals neither contribute nor detract from bonding.

C. CALCULATIONS YOU SHOULD KNOW HOW TO DO

The only numerical calculations in Chapter 13 involve Equation (13-1). You should be able to discuss the bonding of certain diatomic molecules in terms of molecular orbitals and the bonding of polyatomic molecules in terms of hybrid atomic orbitals.

D. ANSWERS TO THE ODD-NUMBERED PROBLEMS

13-1   There are eight electrons in diatomic beryllium. Using Figure 13-5, we place electrons into the molecular orbitals according to the Pauli exclusion principle. $(1\sigma)^2(1\sigma*)^2 (2\sigma)^2(2\sigma*)^2$ is the ground-state electron configuration of diatomic beryllium. There are four electrons in bonding orbitals and four electrons in antibonding orbitals, and so Be_2 has no net bonding. We predict that Be_2 does not exist.

13-3   Using Figure 13-5 and Equation (13-1) we find that the ground-state electron configurations and bond orders of N_2, N_2^+, O_2 and O_2^+ are as follows:

	Ground-state electron configuration	Bond order
N_2	$(1\sigma)^2(1\sigma*)^2(2\sigma)^2(2\sigma*)^2(1\pi)^4(3\sigma)^2$	3
N_2^+	$(1\sigma)^2(1\sigma*)^2(2\sigma)^2(2\sigma*)^2(1\pi)^4(3\sigma)^1$	$2\frac{1}{2}$
O_2	$(1\sigma)^2(1\sigma*)^2(2\sigma)^2(2\sigma*)^2(1\pi)^4(3\sigma)^2(1\pi*)^2$	2
O_2^+	$(1\sigma)^2(1\sigma*)^2(2\sigma)^2(2\sigma*)^2(1\pi)^4(3\sigma)^2(1\pi*)^1$	$2\frac{1}{2}$

We find that the bond order of N_2 is 3 while the bond order of N_2^+ is $2\frac{1}{2}$. The bond energy increases as the bond order increases; therefore, the bond energy of N_2 is greater than that of N_2^+. However, we find that the bond order of O_2 is 2, while the bond order of O_2^+ is $2\frac{1}{2}$. Therefore, the bond energy of O_2 is less than that of O_2^+.

13-5   Using Figure 13-5 and Equation (13-1), we find that the ground-state electron configurations and bond orders of C_2 and C_2^{2-} are as follows:

	Ground-state electron configuration	**Bond order**
C_2	$(1\sigma)^2(1\sigma^*)^2(2\sigma)^2(2\sigma^*)^2(1\pi)^4$	2
C_2^{2-}	$(1\sigma)^2(1\sigma^*)^2(2\sigma)^2(2\sigma^*)^2(1\pi)^4(3\sigma)^2$	3

Because the bond order of C_2^{2-} is greater than that of C_2, we predict that C_2^{2-} has a larger bond energy and a shorter bond length than C_2.

13-7   Determine the total number of electrons and use Figure 13-5 and Equation (13-1).

(a)   18 electrons
$(1\sigma)^2(1\sigma^*)^2(2\sigma)^2(2\sigma^*)^2(1\pi)^2(1\pi)^2(3\sigma)^2(1\pi^*)^2(1\pi^*)^2$
bond order = 1

(b)   11 electrons
$(1\sigma)^2(1\sigma^*)^2(2\sigma)^2(2\sigma^*)^2(1\pi)^3$
bond order = $1\frac{1}{2}$

(c)   7 electrons
$(1\sigma)^2(1\sigma^*)^2(2\sigma)^2(2\sigma^*)^1$
bond order = $\frac{1}{2}$

(d)   19 electrons
$(1\sigma)^2(1\sigma^*)^2(2\sigma)^2(2\sigma^*)^2(1\pi)^2(1\pi)^2(3\sigma)^2(1\pi^*)^2(1\pi^*)^2(3\sigma^*)^1$
bond order = $\frac{1}{2}$

13-9   If the additional electron occupies a bonding orbital, then a stronger net bonding will result. For example, the ground-state electron configuration of B_2 (10 electrons) is $(1\sigma)^2(1\sigma^*)^2(2\sigma)^2(2\sigma^*)^2(1\pi)^1(1\pi)^1$. The addition of an electron will produce $(1\sigma)^2(1\sigma^*)^2(2\sigma)^2(2\sigma^*)^2(1\pi)^2(1\pi)^1$ with one additional bonding electron.

13-11   Write the ground-state electron configuration and determine if there are unpaired electrons.

(a) $(1\sigma)^2(1\sigma^*)^2(2\sigma)^2(2\sigma^*)^2(1\pi)^2(1\pi)^2$
(b) $(1\sigma)^2(1\sigma^*)^2(2\sigma)^2(2\sigma^*)^2(1\pi)^1(1\pi)^1$
(c) $(1\sigma)^2(1\sigma^*)^2(2\sigma)^2(2\sigma^*)^2(1\pi)^1(1\pi)^1$
(d) $(1\sigma)^2(1\sigma^*)^2(2\sigma)^2(2\sigma^*)^2(1\pi)^2(1\pi)^2(3\sigma)^2(1\pi^*)^1(1\pi^*)^1$

The species B_2, C_2^{2+} and F_2^{2+} have unpaired electrons and are paramagnetic.

13-13   Boron has three valence electrons and each hydrogen has one, for a total of six valence electrons. The Lewis formula of BH_3 is

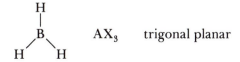

Because BH_3 is a trigonal-planar molecule, it is appropriate to use sp^2 orbitals on the boron atom. We can form the three localized bond orbitals by combining each boron sp^2 orbital with a hydrogen $1s$ orbital. The six valence electrons occupy the three localized bond orbitals pairwise to form the three localized covalent bonds.

13-15   There are four valence electrons from the carbon atom and seven from each fluorine atom, for a total of 32 valence electrons in CF_4. The Lewis formula of CF_4 is

$$:\!\overset{\displaystyle ..}{\underset{\displaystyle ..}{F}}\!: $$

$$:\!\overset{..}{\underset{..}{F}}\!-\!\underset{|}{\overset{|}{C}}\!-\!\overset{..}{\underset{..}{F}}\!:\qquad AX_4\qquad \text{tetrahedral}$$

$$:\!\overset{\displaystyle ..}{\underset{\displaystyle ..}{F}}\!: $$

Because CF_4 is tetrahedral, it is appropriate to use sp^3 orbitals on the carbon atom. We can form the four localized bond orbitals that point toward the vertices of a regular tetrahedron by combining each carbon sp^3 orbital with a fluorine $2p$ orbital. Eight of the valence electrons occupy these four localized bond orbitals pairwise to form the four localized bonds. The remaining 24 valence electrons are lone pairs on the fluorine atoms.

13-17   The Lewis formula of CH_3Cl is

$$H\!-\!\underset{\underset{\displaystyle H}{|}}{\overset{\overset{\displaystyle H}{|}}{C}}\!-\!\overset{..}{\underset{..}{Cl}}\!:\qquad AX_4\qquad \text{tetrahedral}$$

Because CH_3Cl is tetrahedral, it is appropriate to use sp^3 hybrid orbitals on the carbon atom. We can form three localized σ-bond orbitals by combining an sp^3 orbital on the carbon atom with a $1s$ orbital on one of the three hydrogen atoms and one localized σ-bond orbital by combining an sp^3 orbital on the carbon atom with a $3p$ orbital on the chlorine atom. Eight of the 14 valence electrons occupy the four bond orbitals pairwise to form four σ bonds and the remaining 6 valence electrons are lone-pair electrons on the chlorine atom.

13-19   The structure of H_3O^+ can be represented by

$$\overset{\displaystyle \overset{..}{O}{}^{\oplus}}{\underset{\displaystyle H^{110°}}{H\diagup\big|\nwarrow H}}$$

Because H_3O^+ is trigonal pyramidal, it is appropriate to use sp^3 orbitals on the oxygen atom. Each of the three σ-bond orbitals is formed by combining an sp^3 orbital on the oxygen atom with a $1s$ hydrogen orbital. Six of the eight valence electrons occupy these three σ-bond orbitals pairwise. The lone electron pair occupies the remaining sp^3 orbital on the oxygen atom.

13-21   The Lewis formula for NF_3 is

$$:\ddot{F}-\ddot{N}-\ddot{F}:\qquad AX_3E\qquad\text{trigonal pyramidal}$$
$$\underset{\displaystyle :\ddot{F}:}{\big|}$$

Because NF_3 is trigonal pyramidal, it is appropriate to use sp^3 orbitals on the nitrogen atom. The three localized bond orbitals are formed by combining an sp^3 orbital on the nitrogen atom with a $2p$ orbital on a fluorine atom. The three σ-bond orbitals are occupied pairwise by 6 of the 26 valence electrons. The lone electron pair on the nitrogen atom occupies the remaining sp^3 orbital and the remaining 18 valence electrons are lone pairs on the fluorine atoms.

13-23   There are five valence electrons from the phosphorus atom and seven from each of the chlorine atoms for a total of 26 valence electrons in PCl_3. The Lewis formula of PCl_3 is

$$:\ddot{Cl}-\ddot{P}-\ddot{Cl}:\qquad AX_3E\qquad\text{trigonal pyramidal}$$
$$\underset{\displaystyle :\ddot{Cl}:}{\big|}$$

Because PCl_3 is trigonal pyramidal, it is appropriate to use sp^3 orbitals on the phosphorus atom. We can form the three localized bond orbitals by combining three of the sp^3 orbitals on the phosphorus atom with a $3p$ orbital on each of the three chlorine atoms. Six of the valence electrons occupy these localized bond orbitals to form three localized bonds. Two of the remaining valence electrons occupy the fourth sp^3 orbital on the phosphorus atom as a lone pair, and 18 form three lone pairs on each chlorine atom.

13-25   The Lewis formula for hydroxylamine is

$$\underset{\displaystyle H-\ddot{O}-\underset{\displaystyle \ddot{}}{N}-H}{\overset{\displaystyle H}{\big|}}$$

Because there are four pairs of electrons about the nitrogen atom and about the oxygen atom, it is appropriate to use sp^3 orbitals on the nitrogen atom and the oxygen atom. The three σ-bond orbitals on the nitrogen atom are formed by combining two of the sp^3 orbitals on the nitrogen atom with two $1s$ hydrogen orbitals and one of the sp^3 nitrogen orbitals with an sp^3 orbital on the oxygen atom. The other σ-bond orbital on the oxygen atom is formed by combining one of the sp^3 orbitals on the oxygen atom with a $1s$ hydrogen orbital. Eight of the 14 valence electrons occupy the four σ-bond orbitals pairwise to form the four σ bonds. One electron pair occupies the remaining sp^3 orbital on the nitrogen atom, and two electron pairs occupy the remaining two sp^3 orbitals on the oxygen atom.

13-27   Because there are four pairs of electrons about the carbon atom and the nitrogen atom, it is appropriate to use sp^3 orbitals on both the carbon atom and the nitrogen atom. The formation of the localized bond orbitals can be illustrated by

$$
\begin{array}{c}
\text{H}\quad\text{H} \\
|\quad\ | \\
\text{H}-\text{C}-\text{N} \\
|\quad\ | \\
\text{H}\quad\text{H}
\end{array}
\quad sp^3(\text{N}) + 1s(\text{H})
$$

$$sp^3(\text{C}) + 1s(\text{H}) \qquad sp^3(\text{C}) + sp^3(\text{N})$$

Twelve of the 14 valence electrons occupy the six σ-bond orbitals and the remaining two occupy an sp^3 orbital on the nitrogen atom as a lone pair. The shape around the carbon atom is tetrahedral and the shape around the nitrogen atom is trigonal pyramidal.

13-29   Because there are four pairs of electrons about each of the two carbon atoms and the oxygen atom, it is appropriate to use sp^3 orbitals on the oxygen atom and the two carbon atoms. The formation of the localized bond orbitals can be illustrated by

$$
\begin{array}{c}
\text{H}\qquad\ \text{H} \\
|\qquad\ \ | \\
\text{H}-\text{C}-\text{O}-\text{C}-\text{H} \\
|\qquad\ \ | \\
\text{H}\qquad\ \text{H}
\end{array}
\quad sp^3(\text{C}) + 1s(\text{H})
$$

$$sp^3(\text{O}) + sp^3(\text{C})$$

Sixteen of the 20 valence electrons occupy the eight σ-bond orbitals and the remaining four occupy two of the oxygen sp^3 orbitals as lone pairs.

13-31   (a)   The Lewis formula for $Cl_2C{=}CH_2$ is

$$
\begin{array}{c}
:\!\ddot{\text{Cl}}\!:\qquad\qquad\text{H} \\
\diagdown\qquad\diagup \\
\text{C}=\text{C} \\
\diagup\qquad\diagdown \\
:\!\ddot{\text{Cl}}\!:\qquad\qquad\text{H}
\end{array}
$$

There are five σ bonds and one π bond.

(b)   The Lewis formula for $H_2C{=}CHCH{=}CH_2$ is

$$
\begin{array}{c}
\text{H}-\text{C}=\text{C}-\text{C}=\text{C}-\text{H} \\
|\quad\ |\quad\ |\quad\ | \\
\text{H}\quad\text{H}\quad\text{H}\quad\text{H}
\end{array}
$$

There are nine σ bonds and two π bonds.

(c)   The Lewis formula of CH_3COOH is

$$
\begin{array}{c}
\quad\;\; H \qquad\qquad \ddot{O}: \\
\quad\;\; | \qquad\qquad \parallel \\
H-C-C \\
\quad\;\; | \qquad\quad\; \diagdown \\
\quad\;\; H \qquad\qquad\; \ddot{O}\!-\!H
\end{array}
$$

There are seven σ bonds and one π bond.

(d)   The Lewis formula is

$$
\begin{array}{c}
H \diagdown \quad \diagup H \\
C \\
H-C \diagup \quad \diagdown C-H \\
\parallel \qquad\qquad \parallel \\
H-C \diagdown \quad \diagup C-H \\
C \\
H \diagup \quad \diagdown H
\end{array}
$$

There are 14 σ bonds and two π bonds.

13-33   The Lewis formula for ethylacetylene is

$$
\begin{array}{c}
\;\;\; H \;\; H \\
\;\;\; | \quad\; | \\
H-C-C-C\equiv C-H \\
\;\;\; | \quad\; | \\
\;\;\; H \;\; H
\end{array}
$$

There are nine σ bonds and two π bonds. There are $(6 \times 1) + (4 \times 4) = 22$ valence electrons, which occupy the 11 bond orbitals.

13-35   The Lewis formula of carbon monoxide is

$$
\ominus : C \equiv O : \oplus
$$

Because the molecule is linear, it is appropriate to use sp orbitals on the carbon atom and on the oxygen atom. The σ-bond orbital between the carbon and oxygen atoms is formed by combining an sp orbital on the carbon atom and an sp orbital on the oxygen atom. Each of the two π-bond orbitals is formed by combining a $2p$ orbital on the carbon atom and a $2p$ orbital on the oxygen atom. Two of the 10 valence electrons occupy the σ-bond orbital, four occupy the two π-bond orbitals, and the remaining four valence electrons are lone-pair electrons. One lone pair occupies an sp orbital on the carbon atom; the other pair occupies an sp orbital on the oxygen atom.

13-37   The Lewis formula for naphthalene is

Because the geometry about each carbon atom is trigonal planar, it is appropriate to use use sp^2 hybrid orbitals on each of the carbon atoms. The σ-bond framework is

The remaining $2p$ orbitals on each of the carbon atoms combine to form ten π-bond orbitals that are delocalized over the entire two rings as indicated by the circles in the Lewis formula. Thirty-eight of the 48 valence electrons occupy the nineteen σ-bond orbitals and the remaining 10 valence electrons occupy five of the delocalized π-bond orbitals.

13-39   Because NO_2^- is bent, it is appropriate to use sp^2 hybrid orbitals on each of the three atoms. The σ-bond framework is formed by combining two of the sp^2 orbitals on the nitrogen atom with one on each of the oxygen atoms. There are $5 + (2 \times 6) + 1 = 18$ valence electrons in NO_2^-. Four of the valence electrons occupy the two σ-bond orbitals to form two σ bonds, and 10 of the valence electrons occupy the remaining five sp^2 orbitals. The remaining four valence electrons occupy the π and the π^{nb} orbitals; two occupy the bonding orbital and two occupy the nonbonding orbital. The total bond order is 3; 2 from the σ bonds and 1 from the π bond. The bond order of 3 is shared between the two bonds.

13-41   Because H_2NCHO is known to be planar, it is appropriate to use sp^2 orbitals on the nitrogen, carbon, and oxygen atoms. The σ-bond framework is

The remaining $2p$ orbitals on the nitrogen, carbon, and oxygen atoms combine to form three delocalized π-bond orbitals. Ten of the $(3 \times 1) + 5 + 4 + 6 = 18$

valence electrons occupy the five localized σ-bond orbitals, four of them occupy the remaining two oxygen sp^2 orbitals as two lone pairs, and the remaining four valence electrons occupy two of the delocalized π-bond orbitals.

13-43   The Lewis formula of acetaldehyde is

$$H_3C \underset{H}{\overset{}{\diagdown}} C = \overset{\cdot\cdot}{\underset{\cdot\cdot}{O}}\!\!\cdot$$

Because there are three sets of electrons about the oxygen atom and its neighboring carbon atom, it is appropriate to use sp^2 hybrid orbitals about these two atoms. Because the geometry about the other ($-CH_3$) carbon atom is tetrahedral, it is appropriate to use sp^3 hybrid orbitals on the other carbon atom. The σ-bond framework in acetaldehyde is

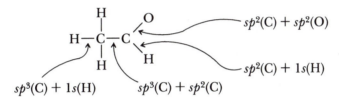

The π-bond orbital is formed by combining the remaining $2p$ orbital on the central carbon atom with the remaining $2p$ orbital on the oxygen atom. Twelve of the $(4 \times 1) + (2 \times 4) + 6 = 18$ valence electrons occupy the six σ-bond orbitals and two of them occupy the π-bond orbital. The two lone electron pairs on the oxygen atom occupy the other two sp^2 orbitals. The shape of acetaldehyde is trigonal planar around the central carbon atom.

13-45   The three carbon atoms are colinear, and so it is appropriate to use sp hybrid orbitals on the two carbon atoms connected by the triple bond. The geometry about the methyl ($-CH_3$) carbon atom is tetrahedral, and so it is appropriate to use sp^3 hybrid orbitals on the methyl carbon atom. The σ-bond framework of methylacetylene is

$$sp^3(C) + sp(C)$$
$$sp(C) + sp(C)$$
$$1s(H) + sp^3(C) \rightleftharpoons \overset{H}{\underset{H}{\underset{|}{\overset{|}{H-C-C-C-H}}}} \quad sp(C) + 1s(H)$$

The remaining bond orbitals between the two carbon atoms on the right are two π-bond orbitals formed from the two $2p$ orbitals on each atom. There are six σ-bond orbitals and two π-bond orbitals. There are $(4 \times 1) + (3 \times 4) = 16$ valence electrons, which occupy the eight bond orbitals.

13-47   The ground-state electron configuration of a tellurium atom is as follows: $[Kr](5s)^2(4d)^{10}(5p_x)^2(5p_y)^1(5p_z)^1$, indicating that two of the $5p$ orbitals are occupied by only one electron. To describe the bonding in H_2Te, use the $1s$ orbitals on the hydrogen atom and two of the $5p$ orbitals on the tellurium atom.

13-49   We use Problem 13-12 to place the 34 electrons of Cl_2 and the 33 electrons of Cl_2^+ into molecular orbitals. Twenty of the electrons occupy the molecular orbitals corresponding to Ne_2, which we can represent as $[Ne_2]$. Using this notation, the ground-state electron configurations of Cl_2 and Cl_2^+ are

$$Cl_2: [Ne_2](4\sigma)^2(4\sigma*)^2(2\pi)^4(5\sigma)^2(2\pi*)^4$$

$$\text{bond order} = \frac{8-6}{2} = 1$$

$$Cl_2^+: [Ne_2](4\sigma)^2(4\sigma*)^2(2\pi)^4(5\sigma)^2(2\pi*)^3$$

$$\text{bond order} = \frac{8-5}{2} = \frac{3}{2}$$

Because the bond order of Cl_2 is less than that of Cl_2^+, we predict that Cl_2 has the longer bond length.

13-51   Because carbon and oxygen are adjacent to one another in the periodic table, we shall use Figure 13-5 to place the electrons in the molecular bond orbitals. There are 14 valence electrons in CO and 13 in CO^+. The ground-state electron configurations are

$$CO: (1\sigma)^2(1\sigma*)^2(2\sigma)^2(2\sigma*)^2(1\pi)^4(3\sigma)^2$$

$$\text{bond order} = \frac{10-4}{2} = 3$$

$$CO^+: (1\sigma)^2(1\sigma*)^2(2\sigma)^2(2\sigma*)^2(1\pi)^4(3\sigma)^1$$

$$\text{bond order} = \frac{9-4}{2} = 2\tfrac{1}{2}$$

Because CO has the greater bond order, CO has a shorter bond length and a greater bond energy than CO^+.

13-53   Table 13-2 shows that the bond order for the ground state of N_2 is 3. The bond order for nitrogen in the given excited state is $(9-5)/2 = 2$. Therefore, the bond length of nitrogen in the excited state is greater than that of nitrogen in the ground state.

13-55   (a)

8 σ bonds and 1 π bond

(b)

6 σ bonds and 1 π bond

(c)

$$H-\overset{\underset{\displaystyle H}{|}}{\overset{\displaystyle H}{|}}{C}-\overset{\underset{\displaystyle H}{|}}{\overset{\displaystyle H}{|}}{C}-C\equiv N\!:$$ 8 σ bonds and 2 π bonds

(d)

$$H-\overset{\underset{\displaystyle H}{|}}{\overset{\displaystyle H}{|}}{C}-\overset{\cdot\cdot}{\underset{\cdot\cdot}{O}}-\overset{\underset{\displaystyle H}{|}}{\overset{\displaystyle H}{|}}{C}-H$$ 8 σ bonds

13-57   There are cis-trans isomers of 1,2-dibromoethene, which have different physical properties:

cis
has a dipole moment

trans
has no net dipole moment

13-59   The Lewis formula for CH_3^+ is

Because there are three pairs of electrons about the carbon atom, it is appropriate to use sp^2 hybrid orbitals on the carbon atom. The three localized bond orbitals are formed by combining one of the sp^2 orbitals with a hydrogen $1s$ orbital to form the three σ bonds of the methyl cation.

13-61   There are $4 + (3 \times 6) + 2 = 24$ valence electrons in CO_3^{2-}. Because there are three sets of electrons about the carbon atom, it is appropriate to use sp^2 hybrid orbitals on the carbon atom. We shall also use sp^2 hybrid orbitals on each of the three oxygen atoms. The σ-bond framework of CO_3^{2-} is

Six of the 24 valence electrons occupy the three σ orbitals to form three σ bonds. Twelve of the valence electrons occupy the remaining six sp^2 orbitals as lone pairs. The remaining six valence electrons occupy three of the π orbitals according to the energy-level diagram given in Problem 13-60. The π electron configuration is $(\pi)^2(\pi^{nb})^4$. The total bond order is four; three from the σ bonds formed from the sp^2 orbitals and one from the bonding π orbital. The fourth bond is shared by the four atoms as indicated in the resonance Lewis formula

$$\left[\begin{array}{c} O \\ \diagdown \\ C \cdots O \\ \diagup \\ O \end{array} \right]^{2-}$$

13-63   The Lewis formula for NO_2^+ is

$$:O=\overset{\oplus}{N}=O:$$

There are $5 + (2 \times 6) - 1 = 16$ valence electrons in NO_2^+. The Lewis formula suggests that NO_2^+ is linear, and because there are three sets of valence electrons about each oxygen atom it is appropriate to use sp hybrid orbitals on the nitrogen atom and sp^2 hybrid orbitals on the oxygen atoms. Four of the valence electrons occupy the two σ-bond orbitals formed by combining the sp orbitals on the nitrogen and one sp^2 orbital on each of the oxygen atoms. Eight of the valence electrons occupy the remaining sp^2 orbitals on the oxygen atoms. We shall place the remaining four valence electrons in the π levels according the energy level diagram in Figure 13-36. The electron configuration is $(\pi)^4$. The total bond order is $2 + 2 = 4$, which is consistent with the Lewis formula for NO_2^+.

E. ANSWERS TO THE SELF-TEST

1. true
2. true
3. molecular
4. false
5. σ
6. is concentrated in the region between the two nuclei
7. is zero in the region between the two nuclei
8. 1 σ^*
9. π
10. bonding orbitals
11. false
12. bond order $= \frac{1}{2}$ (number of electrons in bonding orbitals − number of electrons in antibonding orbitals)
13. false
14. true
15. true
16. unpaired
17. false
18. false
19. true
20. true
21. atomic orbitals on the same atom
22. a $2s$ orbital with one $2p$ orbital on the same atom
23. two
24. 180° from each other
25. an sp hybrid orbital on the Be atom and a $1s$ atomic orbital on the H atom

26. true

27. true

28. a $2s$ orbital with two $2p$ orbitals on the same atom

29. the vertices of an equilateral triangle

30. three

31. true

32. atomic orbitals combined

33. atomic orbitals combined

34. a $2s$ orbital and all three $2p$ orbitals on the same atom

35. the vertices of a tetrahedron

36. four

37. true

38. sp^3

39. an ns orbital, the three np orbitals, and one nd orbital (e.g., $3s$, three $3p$, and one $3d$) on the same atom

40. five

41. the vertices of a trigonal bipyramid

42. an ns orbital, the three np orbitals, and two nd orbitals (e.g., $3s$, three $3p$, and two $3d$ orbitals) on the same atom

43. six

44. the vertices of a regular octahedron

45. σ and π

46. sp^2 and $1s$

47. all the σ bonds in the molecule

48. remaining $2p$ atomic orbital on each carbon atom

49. false

50. true

51. on opposite sides of the double bond

52. false

53. 11-*trans*-retinal

54. σ; π

55. sp; $1s$

56. the two $2p$ atomic orbitals on one carbon atom and the two $2p$ atomic orbital on the other carbon atom

57. true

58. planar . . . hexagonal

59. sp^2

60. combination of $2p$ atomic orbitals on the carbon atoms

61. false

62. delocalized

63. true

64. nonbonding

14 LIQUIDS AND SOLIDS

A. OUTLINE OF CHAPTER 14

14-1   The molecules in solids and liquids are in contact with each other.

The molecules of a solid are close together and restricted to fixed positions in space.

The molecules of a solid vibrate about fixed positions, but they usually do not rotate or translate from site to site.

The molecules in a liquid are close together and are not restricted to fixed positions in space.

The molecules in a liquid vibrate, rotate, and translate throughout the fluid.

Solids and liquids have densities that are about a thousand times greater than the densities of gases.

The molar volume of a liquid is approximately equal to the molar volume of the solid phase of that liquid.

The molar volume of a gas is much larger than the molar volume of a liquid or solid at the same temperature and pressure.

14-2   The processes of melting and boiling appear as horizontal lines on a heating curve.

A heating curve for one mole of water starting as ice at $-10\,^{\circ}\mathrm{C}$ is shown in Figure 14-2.

The molar enthalpy of fusion or the molar heat of fusion ΔH_{fus} is the heat required to melt one mole of a substance.

The molar enthalpy of vaporization or the molar heat of vaporization ΔH_{vap} is the heat required to vaporize one mole of a substance.

The heat absorbed in raising the temperature of a substance from T_1 to T_2 without a change in phase is given by

$$q_P = nC_P(T_2 - T_1) \tag{14-1}$$

where C_P is the molar heat capacity at constant pressure and n is the number of moles.

14-3   Energy is required to melt a solid and to vaporize a liquid.

The energy as heat that is required to melt n moles of a substance is given by

$$q_{fus} = n\Delta H_{fus} \tag{14-2}$$

The enthalpy of fusion is positive because it requires energy to break up the crystal lattice.

The energy as heat that is required to vaporize n moles of a substance is given by

$$q_{vap} = n\Delta H_{vap} \tag{14-3}$$

The enthalpy of vaporization is positive because it requires energy to separate the molecules in a liquid from each other.

The values of the melting point, boiling point, and molar enthalpies of vaporization and fusion of some substances are given in Table 14-3.

Sublimation is the process whereby a solid is converted directly into a gas.

The molar enthalpy of sublimation or the molar heat of sublimation ΔH_{sub} is the energy required to sublime one mole of a substance.

14-4   Van der Waals forces are attractive forces between molecules.

The value of ΔH_{vap} is a measure of how strongly the molecules in a liquid attract each other.

Values of ΔH_{vap} are relatively large for ionic compounds.

Polar molecules attract each other by a dipole-dipole attraction (Figure 14-4).

Water and some other compounds have relatively large values of ΔH_{vap} because of hydrogen bonding (Figures 14-5, 14-6, and 14-7).

Hydrogen bonding is the electrostatic attraction between a hydrogen atom in one molecule and a highly electronegative atom in a neighboring molecule.

All molecules attract each other by London forces.

London forces are instantaneous dipole-dipole attractions caused by the motion of the electrons in the molecules (Figure 14-9).

The strengths of London forces depend on the number of electrons in the two molecules.

The attractive forces between molecules are collectively called van der Waals forces.

The van der Waals forces are compared in Table 14-5.

14-5   Viscosity, surface tension, and dielectric constant are properties of liquids.

The viscosity of a liquid is a measure of its resistance to flow.

Viscosity is denoted by η.

Viscosity decreases with increasing temperature.

The viscosity of a liquid is proportional to its density.

Surface tension is due to the net attractive force toward the interior of a liquid that a molecule at the surface of the liquid experiences.

A liquid whose molecules attract each other strongly has a high surface tension.

The surface tension of a liquid tends to hold a drop of the liquid in a spherical shape.

Surfactants lower the surface tension of a liquid by concentrating at the surface of the liquid.

Capillary action is the rise of a liquid in a thin tube.

A meniscus is the surface of a liquid in a capillary.

For a liquid that wets glass, the meniscus is concave; for a liquid that does not wet glass, the meniscus is convex (Figure 14-16).

Coulomb's law for two charges separated by a distance d in a material medium is given by

$$E = (2.31 \times 10^{-16} \, \text{J} \cdot \text{pm}) \frac{Z_1 Z_2}{Dd} \tag{14-5}$$

where Z_1 and Z_2 are the charges on the two ions and D is the dielectric constant of the medium.

D is greater than one for all materials.

The interaction between two charges in a material medium is always less than that in a vacuum.

The dielectric constant of a liquid depends upon the dipole moment of the molecules in the liquid.

Liquids with relatively large dielectric constants are good solvents for ionic compounds.

The values of the viscosity, surface tension, and dielectric constants for some liquids are given in Table 14-6.

14-6   A liquid has a unique equilibrium vapor pressure at each temperature.

When the rate of evaporation of a liquid is equal to the rate of condensation of its vapor, a liquid-vapor equilibrium is established (Figures 14-18 and 14-19).

The equilibrium between a liquid and its vapor is a dynamic equilibrium.

The equilibrium vapor pressure is the pressure of the vapor in equilibrium with the liquid.

The equilibrium vapor pressure of a liquid increases with temperature (Table 14-7).

The vapor pressure curve is a plot of the equilibrium vapor pressure of a substance versus temperature (Figure 14-21).

The vapor pressure of a liquid depends upon the attractive forces between its constituent molecules.

The normal boiling point of a liquid is the temperature at which its equilibrium vapor pressure is 1 atm.

The boiling point of a liquid decreases with increasing elevation.

14-7   Relative humidity is based on the vapor pressure of water.

The relative humidity is a measure of the amount of water vapor in the atmosphere:

$$\text{relative humidity} = \frac{P_{H_2O}}{P^{\circ}_{H_2O}} \times 100 \tag{14-6}$$

where P_{H_2O} is the partial pressure of the water vapor and $P^{\circ}_{H_2O}$ is the equilibrium vapor pressure at the same temperature (Table 14-7).

The dew point is the air temperature at which the relative humidity reaches 100 percent.

14-8   A phase diagram displays the regions of all the phases of a pure substance simultaneously.

The phase diagrams of water and carbon dioxide are shown in Figures 14-22 and 14-23.

The critical point terminates the vapor pressure curve.

Above the critical point, the gas and liquid phases become indistinguishable.

A gas with a temperature above its critical temperature cannot be liquefied.

The sublimation pressure curve gives the temperature at which the solid and gas phases are in equilibrium with each other at a particular pressure.

The melting point curve gives the temperature at which the solid and liquid phases are in equilibrium with each other at a particular pressure.

Melting points are weakly dependent on pressure.

The vapor pressure curve gives the temperature at which the liquid and gas phases are in equilibrium with each other at a particular pressure.

The triple point is the temperature and pressure at which three phases coexist in equilibrium.

14-9   X-ray diffraction patterns yield information about the structures of crystalline solids.

A crystalline solid has an ordered arrangement of its constituent molecules or ions in the solid state, which is called a crystal lattice.

The crystal lattice of a crystalline solid produces a characteristic pattern when X-rays are passed through the solid onto X-ray film.

The X-ray diffraction pattern produced by a crystal is shown in Figure 14-24.

The unit cell is the smallest subunit of a crystal lattice that can be used to generate the entire lattice.

The three cubic unit cells are simple cubic, body-centered cubic, and face-centered cubic, as shown in Figure 14-28.

The unit cell of many metals is one of the three cubic unit cells.

The crystallographic or effective radius of an atom in a solid can be calculated from the density of the solid and its crystal structure.

The molar volume of the solid is given by

$$V_{mol} = \frac{M}{d} = \frac{\text{molar mass}}{\text{density}} \tag{14-9}$$

The volume of the unit cell of a crystalline solid is given by

$$V_{\text{unit cell}} = \frac{nM}{dN_0} \qquad (14\text{-}10)$$

where n is the number of atoms per unit cell and N_0 is Avogadro's number.

The radius of an atom in the unit cell can be determined from the geometry when the type of unit cell and its volume are known.

Avogadro's number can be determined from the density and dimensions of the unit cell of a solid (Example 14-12).

14-10 Crystals can be classified according to the forces between the constituent particles.

The constituent particles of atomic crystals are atoms.

The constituent particles of ionic crystals are ions of opposite charge.

The lattice structure of ionic crystals depends on the sizes of the cations and anions (Figure 14-29).

The high lattice energy of ionic crystals is due to the total electrostatic interaction energy.

The constituent particles of molecular crystals are molecules.

The forces that hold together molecular crystals are weaker than those in ionic crystals.

Molecular crystals have a variety of crystal structures.

The structure of a single molecule can be determind by X-ray crystallography (Figures 14-31 and 14-32).

Diamond has a covalently bonded tetrahedral network (Figure 14-33).

Graphite has a layered structure (Figure 14-34).

14-11 The electrons in metals are delocalized throughout the crystal.

In one mole of a metal, there are Avogadro's number of molecular orbitals.

The energies of the molecular orbitals in a metal are so close together that they form essentially a continuum of energy levels from the bonding orbital of lowest energy to the antibonding orbital of highest energy.

The molecular orbitals of a metal are delocalized over all the atoms in the metal.

The electrons in the delocalized orbitals in a metal are easily displaced by an electric current, and thus, metals are good conductors of electricity.

Delocalized bonding orbitals are called the valence band.

The delocalized antibonding orbitals are called the conduction band.

In an insulator, all the valence electrons occupy the valence band and the conduction band is empty.

The energy difference between the valence band and the conduction band is called the band gap.

In an insulator, the band gap is so large that the electrons in the valence band are not easily promoted into the conduction band.

In a good conductor, the band gap is zero.

In a semiconductor, the band gap is relatively small.

The bond structures of metals, semiconductors, and insulators are summarized in Figure 14-36.

14-12   Amorphous solids do not have a crystalline structure.

When liquids of high viscosity are cooled rapidly, an amorphous solid, which lacks a high degree of spatial order, is formed.

An amorphous solid lacks a sharp melting point.

The various types of solids are summarized in Table 14-8.

B. SELF-TEST

1. The molecules in a solid move freely throughout the solid. *True/False*

2. The molecules in a liquid have no rotational motion. *True/False*

3. A liquid is easy to compress to smaller volumes. *True/False*

4. When heat is added to a liquid at a constant rate, the temperature will remain constant for a period of time at _____.

5. When heat is added to a solid at a constant rate, the temperature will remain constant for a period of time at _____.

6. The amount of energy as heat required to raise the temperature of a substance is determined by the _____ of the substance.

7. The energy as heat required to melt one mole of a solid is the _____
_____.

8. Energy as heat is required to melt a solid because _____
_____.

9. The energy as heat required to vaporize one mole of a liquid is the _____
_____.

10. Energy is required to vaporize a liquid because _____
_____.

11. The evaporation of water in perspiration (*adds/removes*) heat from the body.

12. A solid cannot be converted to a gas directly. *True/False*

13. The energy required to sublime one mole of a solid is the _____

_____ .

14. There are attractive forces between all molecules or atoms. *True/False*

15. A large value of ΔH_{vap} indicates a strong attraction between the particles in a liquid. *True/False*

16. The values of ΔH_{vap} for ionic compounds are smaller than for molecular compounds. *True/False*

17. The attractive forces between polar molecules are due to _____ attraction.

18. The attractive forces between polar molecules are (*greater/less*) than the attractive forces between nonpolar molecules of similar size.

19. Hydrogen bonding is _____

_____ .

20. The attractive forces between nonpolar molecules are called _____ .

21. The attractive force between nonpolar molecules is due to _____

_____ .

22. When a liquid is placed in a closed container, vapor from the liquid appears in the container. *True/False*

23. When the rate of condensation is equal to the rate of evaporation, the pressure of the vapor above the liquid remains the same. *True/False*

24. The system of a liquid in a closed container is at equilibrium when the rate

_____ .

25. When a system is at equilibrium, no net observable change occurs. *True/False*

26. Define equilibrium vapor pressure. _____

_____

27. The equilibrium vapor pressure of a substance is the same at all temperatures. *True/False*

28. The normal boiling point of a liquid is _____

_____ .

29. Water always boils at 100°C. *True/False*

30. What is meant by relative humidity? _____

_____

31. The dew point is defined as _____

_____ .

32. The viscosity of a liquid is a measure of _____ .

33. A viscous liquid flows (*slowly/fast*).

34. The viscosity of a liquid usually (*increases/decreases*) when the temperature of the liquid increases.

35. The molecules at the surface of a liquid experience a net inward force. *True/False*

36. The surface tension of a liquid tends to (*minimize/maximize*) the surface area of a liquid.

37. A liquid whose molecules attract each other strongly has a (*high/low*) surface tension.

38. A surfactant (*raises/lowers*) the surface tension of a liquid.

39. The meniscus of a liquid that wets glass is (*convex/concave*).

40. The dielectric constant for all materials is (*greater/less*) than one.

41. The interaction between two charges is always (*greater/less*) than that in a vacuum.

42. The dielectric constant of a liquid depends upon the _____ of the molecules in the liquid.

43. Liquids with a relatively large dielectric constant are (*good/poor*) solvents for ionic compounds.

44. A phase diagram is a diagram that shows _____

_____ .

45. A phase diagram can be used to determine the state of a substance at any pressure and temperature. *True/False*

46. The melting point curve separates the _____ phase and the _____ phase in a phase diagram of a substance.

47. The melting point of water changes markedly with pressure. *True/False*

48. The critical temperature is the temperature _____

_____ .

49. A triple point is _____

_____ .

50. A gas can be liquefied at all temperatures by applying sufficient pressure. *True/False*

51. Liquid carbon dioxide does not exist at any temperature and pressure. *True/False*

52. The melting point of all substances decreases with pressure. *True/False*

53. A substance does not have a normal boiling point when the pressure at the solid-liquid-gas triple point is (*greater/lesser*) than 1 atm.

54. A crystalline solid is a solid that has _____

_____ .

55. An X-ray diffraction pattern of a crystal can be used to determine the _____ of the crystal.

56. The unit cells of all crystals are identical. *True/False*

57. The three types of cubic unit cells are _____ , _____ , and _____ .

58. A body-centered unit cell has an atom at the _____ of the unit cell.

59. A face-centered cubic unit cell has an atom at each vertex and at _____ _____ of the unit cell.

60. Atomic or ionic radii can be calculated from crystal structures. *True/False*

61. The molar volume of a solid can be calculated from its _____ and _____ .

62. Avogadro's number can be determined from the density and crystalline structure of a solid. *True/False*

63. The constituent particles of an atomic crystal are _____ .

64. The lattice energy of an ionic crystal is (*large/small*).

65. The different crystal-packing arrangements for NaCl and CsCl are due to _____ .

66. Molecular crystals usually have (*higher/lower*) melting points than ionic crystals.

67. All molecular crystals have the same structure. *True/False*

68. The structure of a single molecule can be determined from the crystal structure of the molecular solid. *True/False*

69. Two crystalline forms of solid carbon are ——————— and ———————.

70. The constituent particles in diamond are held together by ———————
——————————————————————.

71. In one mole of a metal, there are Avogadro's number of molecular orbitals. *True/False*

72. In a metal the energies of the molecular orbitals form a continuous band. *True/False*

73. The molecular orbitals in a crystalline solid are delocalized. *True/False*

74. The valence band is ——————————————————.

75. The conduction band is ——————————————————.

76. An electric current is carried in a solid by the electrons in the (*conduction/valence*) band.

77. The band gap is ——————————————————
——————————————————————.

78. In an insulator, the band gap is (*large/small*).

79. In a semiconductor, the band gap is (*large/small*).

80. When a viscous liquid is cooled rapidly, a (*crystalline/amorphous*) solid is formed.

81. An amorphous solid has a sharp melting point. *True/False*

C. CALCULATIONS YOU SHOULD KNOW HOW TO DO

1. Compute the quantity of heat that is absorbed or evolved in vaporization, fusion, or sublimation. See Example 14-2 and Problems 14-1 through 14-8.

2. Construct a heating curve for a substance that is heated at some given rate. You need the values of the heat capacities of the solid, liquid, and gas phases and the values of ΔH_{fus} and ΔH_{vap}. See Problems 14-9 through 14-12.

3. Calculate the relative humidity and the dew point by using Equation (14-6) and the data in Table 14-7. See Example 14-7 and Problems 14-23 through 14-24.

4. Calculate the crystallographic radius of an atom in a solid using Equation (14-10). See Example 14-11 and Problems 14-35 through 14-38, 14-41, and 14-42.

5. Calculate Avogadro's number, given the density and length of an edge of the unit cell. See Example 14-12 and Problems 14-39 and 14-40.

D. SOLUTIONS TO THE ODD-NUMBERED PROBLEMS

14-1   The number of moles in 5.00 kg of NH_3 is

$$\text{moles of } NH_3 = (5.00 \text{ kg}) \left(\frac{10^3 \text{ g}}{1 \text{ kg}} \right) \left(\frac{1 \text{ mol } NH_3}{17.03 \text{ g } NH_3} \right)$$
$$= 293.6 \text{ mol}$$

The amount of heat absorbed when 293.6 mol of $NH_3(l)$ is vaporized is

$$\text{heat absorbed} = n \, \Delta H_{vap} = (293.6 \text{ mol})(23.4 \text{ kJ} \cdot \text{mol}^{-1}) = 6870 \text{ kJ}$$

14-3   The number of moles in 20.1 g of mercury is

$$n = (20.1 \text{ g}) \left(\frac{1 \text{ mol Hg}}{200.6 \text{ g Hg}} \right) = 0.1002 \text{ mol}$$

The heat *released* when 0.1002 mol of mercury freezes at its melting point is (Table 14-1)

$$q_P = n \, \Delta H_{fus} = (0.1002 \text{ mol})(2.30 \text{ kJ} \cdot \text{mol}^{-1}) = 0.230 \text{ kJ}$$

The heat *released* when 0.1002 mol of mercury is cooled from 298 K to 234 K (the melting point of mercury) is given by

$$q_P = nC_P(T_2 - T_1) = (0.1002 \text{ mol})(28.0 \text{ J} \cdot \text{K}^{-1} \cdot \text{mol}^{-1})(64 \text{ K})$$
$$= 180 \text{ J} = 0.180 \text{ kJ}$$

The total heat released is 0.230 kJ + 0.180 kJ = 0.410 kJ.

14-5   To heat the gallium from 20.0°C to its melting point (29.0°C) requires

$$\text{heat absorbed} = (5.00 \text{ g})(0.37 \text{ J} \cdot \text{K}^{-1} \cdot \text{g}^{-1})(9.0 \text{ K}) = 17 \text{ J}$$

The amount of energy as heat absorbed when it melts is given by

$$\text{heat absorbed} = (5.00 \text{ g}) \left(\frac{1 \text{ mol Ga}}{69.72 \text{ g Ga}} \right) (5.59 \text{ kJ} \cdot \text{mol}^{-1})$$
$$= 0.401 \text{ kJ} = 401 \text{ J}$$

The total quantity of heat absorbed is 17 J + 401 J = 418 J.

14-7   The heat absorbed by the sublimation of 100.0 g of $CO_2(s)$ is

$$q_P = (100.0 \text{ g}) \left(\frac{1 \text{ mol } CO_2}{44.01 \text{ g } CO_2} \right) (25.2 \text{ kJ} \cdot \text{mol}^{-1}) = 57.3 \text{ kJ}$$

The value of ΔH_{sub} for $CO_2(s)$ is given on page 480 of the text.

14-9   Mercury is a solid from 200 K to 234 K. It requires

$$q_P = nC_P(T_2 - T_1) = (7.50 \text{ g}) \left(\frac{1 \text{ mol Hg}}{200.6 \text{ g Hg}} \right) (27.2 \text{ J} \cdot \text{K}^{-1} \cdot \text{mol}^{-1})(34 \text{ K})$$
$$= 34.6 \text{ J}$$

to heat the solid mercury from 200 K to its melting point. If heat is supplied at $100 \text{ J} \cdot \text{min}^{-1}$, then the time required is

$$t = \frac{34.6 \text{ J}}{100 \text{ J} \cdot \text{min}^{-1}} = 0.346 \text{ min} = 20.8 \text{ s}$$

The heat required to melt the mercury is given by

$$q_P = n \, \Delta H_{\text{vap}} = (7.50 \text{ g}) \left(\frac{1 \text{ mol Hg}}{200.6 \text{ g Hg}} \right) (2.30 \text{ kJ} \cdot \text{mol}^{-1})$$
$$= 0.0860 \text{ kJ} = 86.0 \text{ J}$$

At a heating rate of $100 \text{ J} \cdot \text{min}^{-1}$, the time required is

$$t = \frac{86.0 \text{ J}}{100 \text{ J} \cdot \text{min}^{-1}} = 0.860 \text{ min} = 51.6 \text{ s}$$

The heat required to heat the liquid mercury from its melting point (234 K) to its boiling point (630 K) is

$$q_P = nC_P(T_2 - T_1) = (0.0374 \text{ mol})(28.0 \text{ J} \cdot \text{K}^{-1} \cdot \text{mol}^{-1})(630 \text{ K} - 234 \text{ K})$$
$$= 415 \text{ J}$$

and the time required is

$$t = \frac{415 \text{ J}}{100 \text{ J} \cdot \text{min}^{-1}} = 4.15 \text{ min} = 249 \text{ s}$$

The heat required to vaporize the mercury is

$$q_P = n \, \Delta H_{\text{vap}} = (0.0374 \text{ mol})(59.1 \text{ kJ} \cdot \text{mol}^{-1})$$
$$= 2.21 \text{ kJ} = 2210 \text{ J}$$

and the time required is

$$t = \frac{2210 \text{ J}}{100 \text{ J} \cdot \text{min}^{-1}} = 22.1 \text{ min} = 1330 \text{ s}$$

The heat required to heat the mercury vapor from 630 K to 800 K is given by

$$q_P = nC_P(T_2 - T_1) = (0.0374 \text{ mol})(20.8 \text{ J} \cdot \text{K}^{-1} \cdot \text{mol}^{-1})(800 \text{ K} - 630 \text{ K})$$
$$= 132 \text{ J}$$

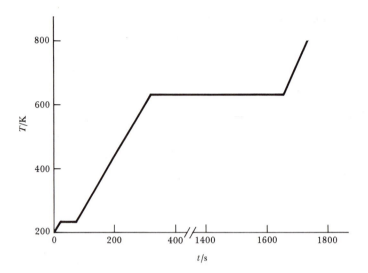

and the time required is

$$t = \frac{132 \text{ J}}{100 \text{ J} \cdot \text{min}^{-1}} = 1.32 \text{ min} = 79.2 \text{ s}$$

The heating curve is shown at the top of the page.

14-11 The quantity of heat absorbed by the NaCl(s) is

$$q_P = (3.00 \text{ kJ} \cdot \text{min}^{-1})(250 \text{ s}) \left(\frac{1 \text{ min}}{60 \text{ s}} \right) = 12.5 \text{ kJ}$$

The number of moles of NaCl(s) is

$$n = (25.0 \text{ g}) \left(\frac{1 \text{ mol NaCl}}{58.44 \text{ g NaCl}} \right) = 0.428 \text{ mol}$$

and the molar enthalpy of fusion is

$$\Delta H_{\text{fus}} = \frac{12.5 \text{ kJ}}{0.428 \text{ mol}} = 29.2 \text{ kJ} \cdot \text{mol}^{-1}$$

14-13 Cl_2 is a homonuclear diatomic molecule and so is nonpolar. The Lewis formula for ClF is

$$: \overset{..}{\underset{..}{Cl}} - \overset{..}{\underset{..}{F}} :$$

Fluorine is more electronegative than chlorine; therefore, the fluorine atom has a small negative charge and the chlorine atom has a small positive charge.

$$\overset{\delta+}{\underset{\cdot\cdot}{:}\overset{\cdot\cdot}{\underset{\cdot\cdot}{Cl}}}\!-\!\overset{\delta-}{\underset{\cdot\cdot}{\overset{\cdot\cdot}{F}}}:$$

The molecule is polar.

The Lewis formula for NF_3 is

$$:\!\overset{\cdot\cdot}{\underset{\cdot\cdot}{F}}\!-\!\overset{\cdot\cdot}{N}\!-\!\overset{\cdot\cdot}{\underset{\cdot\cdot}{F}}\!:$$
$$\underset{\cdot\cdot}{\overset{|}{\underset{\cdot\cdot}{F}}}:$$

Since fluorine is more electronegative than nitrogen, there is a small negative charge on each fluorine atom and a small positive charge on the nitrogen atom.

$$\overset{\delta-}{:\!\overset{\cdot\cdot}{\underset{\cdot\cdot}{F}}}\!-\!\overset{3\delta+}{\overset{\cdot\cdot}{N}}\!-\!\overset{\delta-}{\overset{\cdot\cdot}{\underset{\cdot\cdot}{F}}}\!:$$
$$\underset{\delta-}{\underset{\cdot\cdot}{\overset{|}{\underset{\cdot\cdot}{F}}}}:$$

The molecule is polar because of its trigonal pyramidal shape.

The molecule F_2, being homonuclear, is nonpolar.

14-15 The only ionic compound is KBr. Thus we predict that KBr has the highest boiling point. The Lewis formulas of C_2H_5OH and C_2H_6 are

We see that liquid C_2H_5OH is partially hydrogen-bonded and C_2H_6 is nonpolar. Helium is a smaller molecule than C_2H_6. Therefore, the predicted order is

$$T_b[He] < T_b[C_2H_6] < T_b[C_2H_5OH] < T_b[KBr]$$

14-17 The two nonpolar molecules, CH_4 and C_2H_6, have the lowest molar enthalpies of vaporization. Of the two, the smaller molecule, CH_4, has the lower value of ΔH_{vap}. The two polar molecules, CH_3OH and C_2H_5OH, are hydrogen-bonded and have the highest values of molar enthalpies of vaporization. The larger of the two, C_2H_5OH, has the higher value of ΔH_{vap}. Therefore, the order of the molar enthalpies of vaporization is

$$\Delta H_{vap}[CH_4] < \Delta H_{vap}[C_2H_6] < \Delta H_{vap}[CH_3OH] < \Delta H_{vap}[C_2H_5OH]$$

14-19 Using Figure 14-21, the equilibrium vapor pressure of ethyl alcohol at 60°C is about 0.5 atm. If all the ethyl alcohol (CH_3CH_2OH) were to vaporize, then the pressure would be

$$P = \frac{nRT}{V} = \frac{[(0.75 \text{ g})(1 \text{ mol}/46.07 \text{ g})](0.0821 \text{ L}\cdot\text{atm}\cdot\text{mol}^{-1}\cdot\text{K}^{-1})(333 \text{ K})}{0.400 \text{ L}}$$
$$= 1.1 \text{ atm}$$

This pressure is greater than the equilibrium vapor pressure of ethyl alcohol at 60°C, and so vapor will condense until the vapor pressure is about 0.5 atm. Liquid will be present.

14-21 We use the ideal-gas equation and write

$$P = \frac{nRT}{V} = \left(\frac{nR}{V}\right)T = aT$$

where we have let nR/V be denoted by a constant a. The value of a can be determined by using the fact that $P = 300$ torr at 75.0°C.

$$a = \frac{P}{T} = \frac{300 \text{ torr}}{348 \text{ K}} = 0.862 \text{ torr}\cdot\text{K}^{-1}$$

Thus, we have

$$P = (0.862 \text{ torr}\cdot\text{K}^{-1})T$$

A plot of P versus T from 80.0°C to 40.0°C (assuming no condensation) is shown below.

The pressure calculated from the ideal-gas equation exceeds the equilibrium vapor pressure at around 330 K, or around 60°C (Figure 14-21). Thus, we expect condensation to occur at around 60°C.

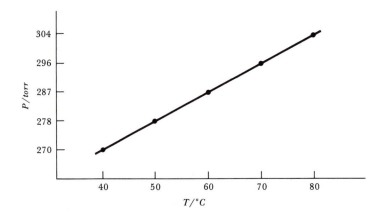

14-23   We can find the partial pressure of the water vapor in the air at the two temperatures from the relative humidity

$$\text{relative humidity} = \frac{P_{H_2O}}{P_{H_2O}^\circ} \times 100$$

$$P_{H_2O} = \frac{(P_{H_2O}^\circ)(\text{relative humidity})}{100}$$

We can find $P_{H_2O}^\circ$ in Table 14-7.
At 20°C

$$P_{H_2O} = \frac{(17.4 \text{ torr})(70)}{100} = 12.2 \text{ torr}$$

The dew point is the temperature at which the partial pressure of the water vapor in the air is equal to the equilibrium vapor pressure. From Table 14-7, we see that the dew point is about 15°C.
At 30°C

$$P_{H_2O} = \frac{(31.6 \text{ torr})(70)}{100} = 22.1 \text{ torr}$$

The dew point is about 24°C.
The 20°C day is much more comfortable.

14-25   The surface area of a sphere of radius r is given by

$$\text{Area} = 4\pi r^2$$

The surface area of a drop 2 mm in diameter is

$$\text{Area} = 4\pi(1 \text{ mm})^2 = 12.6 \text{ mm}^2$$

The energy of the surface is

$$\text{surface tension} = \frac{\text{energy}}{\text{area}}$$

or

$$\text{energy} = \text{surface tension} \times \text{area}$$
$$= (72 \text{ mJ} \cdot \text{m}^{-2})(12.6 \text{ mm}^2)\left(\frac{1 \text{ m}}{10^3 \text{ mm}}\right)^2$$
$$= 9.1 \times 10^{-4} \text{ mJ} = 0.91 \text{ } \mu\text{J}$$

When the drop is split into two drops, the total volume remains the same. We must find the volume of the drop

$$V = \tfrac{4}{3}\pi r^3 = \tfrac{4}{3}\pi(1\ \text{mm})^3 = 4.2\ \text{mm}^3$$

The volume of each smaller drop is $2.1\ \text{mm}^3$. We now can find the radius of each of the small drops.

$$2.1\ \text{mm}^3 = \tfrac{4}{3}\pi r^3$$

$$r^3 = \frac{3}{4\pi}(2.1\ \text{mm}^3) = 0.50\ \text{mm}^3$$

$$r = 0.79\ \text{mm}$$

The surface area of each smaller drop is

$$\text{Area} = 4\pi(0.79\ \text{mm})^2 = 7.8\ \text{mm}^2$$

The total surface area of the smaller drops is

$$\text{Area} = (2)(7.8\ \text{mm}^2) = 15.6\ \text{mm}^2$$

The energy of the surface is

$$\text{energy} = \text{surface tension} \times \text{area}$$
$$= (72\ \text{mJ}\cdot\text{m}^{-2})(15.6\ \text{mm}^2)\left(\frac{1\ \text{m}}{10^3\ \text{mm}}\right)^2$$
$$= 1.1 \times 10^{-3}\ \text{mJ} = 1.1\ \mu\text{J}$$

The energy required to change the drop to two drops is

$$\text{energy} = 1.1\ \mu\text{J} - 0.91\ \mu\text{J} = 0.2\ \mu\text{J}$$

14-27 We shall use Equation (14-4) as developed in Example 14-5 and let 2 represent cyclohexane and 1 represent water.

$$\frac{\eta_2}{\eta_1} = \frac{t_2\, d_2}{t_1\, d_1} = \frac{(2700\ \text{s})(0.942\ \text{g}\cdot\text{mL}^{-1})}{(71\ \text{s})(0.998\ \text{g}\cdot\text{mL}^{-1})} = 36$$

14-29 The Lewis formulas of *trans*-1,2-dichloroethene and *cis*-1,2-dichloroethene are

trans   cis

The dipole moment of the trans isomer is zero and that of the cis isomer is nonzero. Thus, the dielectric constant of *cis*-1,2-dichloroethene is greater than the dielectric constant of *trans*-1,2-dichloroethene.

14-31   (a)   gas   (b)   solid   (c)   liquid   (d)   liquid

14-33   The phase diagram (not to scale) of oxygen is shown below.
The melting point curve slopes to the right, which means that the melting point increases with pressure. Thus solid oxygen does not melt under an applied pressure as water does.

14-35   From Figure 14-28 we see that there is one atom at each corner and one atom at the center of a body-centered cubic lattice. Each of the eight atoms at the corners is shared by seven other unit cells, while the atom at the center belongs entirely to the unit cell. Thus, there is a total of two atoms per unit cell.

14-37   We use Equation (14-10) to calculate the volume of the unit cell.

$$V_{\text{unit cell}} = \frac{nM}{N_0 d}$$

$$= \frac{(4 \text{ atom} \cdot \text{unit cell}^{-1})(107.9 \text{ g} \cdot \text{mol}^{-1})}{(6.022 \times 10^{23} \text{ atom} \cdot \text{mol}^{-1})(10.50 \text{ g} \cdot \text{cm}^{-3})}$$

$$= 6.826 \times 10^{-23} \text{ cm}^3 \cdot \text{unit cell}^{-1}$$

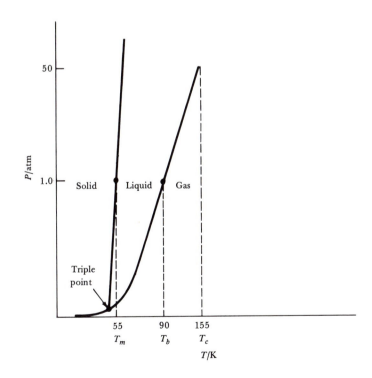

The length of an edge of a unit cell is

$$\text{length} = (6.826 \times 10^{-23} \text{ cm}^3)^{1/3} = 4.087 \times 10^{-8} \text{ cm}$$
$$= 408.7 \text{ pm}$$

14-39   Solving Equation (14-10) for N_0, we have

$$N_0 = \frac{nM}{dV_{\text{unit cell}}}$$

The volume of the unit cell is given by

$$V_{\text{unit cell}} = l^3 = (361.6 \times 10^{-12} \text{ m})^3 = (361.6 \times 10^{-10} \text{ cm})^3$$
$$= 4.728 \times 10^{-23} \text{ cm}^3$$

Therefore,

$$N_0 = \frac{(4 \text{ atom} \cdot \text{unit cell}^{-1})(63.55 \text{ g} \cdot \text{mol}^{-1})}{(8.93 \text{ g} \cdot \text{cm}^{-3})(4.728 \times 10^{-23} \text{ cm}^{-3})}$$
$$= 6.02 \times 10^{23} \text{ atom} \cdot \text{mol}^{-1}$$

14-41   From Figure 14-29, we see that there are four KF units per unit cell. Using Equation (14-10), then, gives

$$V_{\text{unit cell}} = \frac{nM}{dN_0}$$
$$= \frac{(4 \text{ formula unit} \cdot \text{unit cell}^{-1})(58.10 \text{ g} \cdot \text{mol}^{-1})}{(2.481 \text{ g} \cdot \text{cm}^{-3})(6.022 \times 10^{23} \text{ formula unit} \cdot \text{mol}^{-1})}$$
$$= 1.555 \times 10^{-22} \text{ cm}^3 \cdot \text{unit cell}^{-1}$$

The length of an edge of a unit cell is

$$l = (1.555 \times 10^{-22} \text{ cm}^3)^{1/3} = 5.377 \times 10^{-8} \text{ cm}$$

From Figure 14-29a, we see that the nearest-neighbor distance is $l/2$, or 2.689×10^{-8} cm, or 268.9 pm.

14-43   Your body uses energy to melt the snow. It requires

$$q_{\text{P}} = n \, \Delta H_{\text{fus}} = (1.00 \text{ g}) \left(\frac{1 \text{ mol}}{18.02 \text{ g}} \right) (6.01 \text{ kJ} \cdot \text{mol}^{-1}) = 0.334 \text{ kJ}$$

to melt one gram of snow if the temperature of the snow is $0\,^\circ$C.

14-45 $\Delta H_{\text{vap}} = (85 \text{ J} \cdot \text{K}^{-1} \cdot \text{mol}^{-1}) T_b$
$$= (85 \text{ J} \cdot \text{K}^{-1} \cdot \text{mol}^{-1})(373 \text{ K})$$
$$= 31700 \text{ J} \cdot \text{mol}^{-1} = 31.7 \text{ kJ} \cdot \text{mol}^{-1}$$

The actual value of ΔH_{vap} for water is $40.7 \text{ kJ} \cdot \text{mol}^{-1}$. Water is strongly hydrogen-bonded, and so has strong specific intermolecular interactions. Therefore, Trouton's rule does not apply to water.

14-47 Near the triple point, we have the relation

$$\Delta H_{sub} = \Delta H_{fus} + \Delta H_{vap}$$

This relation results from Hess's law.

14-49 Both silicon carbide (SiC) and boron nitride (BN) form a diamondlike covalent crystal network.

14-51 The boiling point of water was used to determine the atmospheric pressure of Lhasa. Plots of atmospheric pressure versus altitude were used to find the altitude corresponding to the atmospheric pressure of Lhasa.

14-53 The vapor pressure of the solid is equal to the vapor pressure of the liquid at the triple point. Thus, we write

$$10.560 - \frac{1640 \text{ K}}{T_t} = 7.769 - \frac{1159 \text{ K}}{T_t}$$

$$2.791 = \frac{481 \text{ K}}{T_t}$$

$$T_t = \frac{481 \text{ K}}{2.791} = 172 \text{ K}$$

The triple point pressure is given by either

$$\log P_t = 7.769 - \frac{1159 \text{ K}}{172 \text{ K}}$$

$$= 7.769 - 6.738 = 1.03$$

or

$$\log P_t = 10.560 - \frac{1640 \text{ K}}{172 \text{ K}} = 1.03$$

and so

$$P_t = 10^{1.03} = 10.7 \text{ torr}$$

14-55 The equilibrium vapor pressure of water at $30°C$ is 31.6 torr (Table 14-7). We can find the partial pressure of water from the definition of relative humidity

$$\text{relative humidity} = \frac{P_{H_2O}}{P^\circ_{H_2O}} \times 100$$

or

$$P_{H_2O} = \frac{(65)(31.6 \text{ torr})}{100} = 21 \text{ torr}$$

The temperature at which the equilibrium vapor pressure is 21 torr is the dew point. From Table 14-7 we see that the dew point is about 23°C.

14-57 We solve Equation (14-10) for n:

$$n = \frac{V_{\text{unit cell}}\, dN_0}{M}$$
$$= \frac{(486 \times 10^{-10} \text{ cm})^3 (3.25 \text{ g}\cdot\text{cm}^{-3})(6.022 \times 10^{23} \text{ formula unit}\cdot\text{mol}^{-1})}{56.08 \text{ g}\cdot\text{mol}^{-1}}$$
$$= 4.00$$

The result tells us that there are four formula units of CaO in a unit cell, and so the unit cell must be the NaCl type.

14-59 The unit cell of cesium chloride is body-centered cubic. The volume of the unit cell of CsCl is

$$V_{\text{unit cell}} = (412.1 \text{ pm})^3 = (4.121 \times 10^{-8} \text{ cm})^3 = 6.999 \times 10^{-23} \text{ cm}^3$$

A unit cell consists of one chloride ion and one cesium ion (Figure 14-29). We solve Equation (14-10) for d:

$$d = \frac{nM}{V_{\text{unit cell}}\, N_0}$$
$$= \frac{(1 \text{ formula unit}\cdot\text{unit cell}^{-1})(168.4 \text{ g}\cdot\text{mol}^{-1})}{(6.999 \times 10^{-23} \text{ cm}^3\cdot\text{unit cell}^{-1})(6.022 \times 10^{23} \text{ formula unit}\cdot\text{mol}^{-1})}$$
$$= 3.995 \text{ g}\cdot\text{cm}^{-3}$$

14-61 A unit cell of a simple cubic lattice is shown in Figure 14-28. Each of the eight atoms is shared by eight unit cells, so there is a total of one atom per unit cell for a simple cubic lattice. The volume of a unit cell of a simple cubic lattice is given by

$$V_{\text{unit cell}} = (334.7 \times 10^{-10} \text{ cm})^3 = 3.749 \times 10^{-23} \text{ cm}^3$$

We solve Equation (14-10) for d:

$$d = \frac{nM}{V_{\text{unit cell}}\, N_0}$$
$$= \frac{(1 \text{ atom}\cdot\text{unit cell}^{-1})(209 \text{ g}\cdot\text{mol}^{-1})}{(3.749 \times 10^{-23} \text{ cm}^3\cdot\text{unit cell}^{-1})(6.022 \times 10^{23} \text{ formula unit}\cdot\text{mol}^{-1})}$$
$$= 9.26 \text{ g}\cdot\text{cm}^{-3}$$

14-63   The equilibrium vapor pressure of water at 25°C is 23.8 torr (Table 14-7). The concentration in moles per liter is given by the ideal-gas equation:

$$\frac{n}{V} = \frac{P}{RT} = \frac{(23.8 \text{ torr})(1 \text{ atm}/760 \text{ torr})}{(0.0821 \text{ L} \cdot \text{atm} \cdot \text{mol}^{-1} \cdot \text{K}^{-1})(298 \text{ K})}$$
$$= 1.28 \times 10^{-3} \text{ mol} \cdot \text{L}^{-1}$$

14-65   We use Equation (14-4) as developed in Example 14-5 and let 2 represent acetone and 1 represent water.

$$\frac{\eta_2}{\eta_1} = \frac{t_2 \, d_2}{t_1 \, d_1} = \frac{(59 \text{ s})(0.791 \text{ g} \cdot \text{mL}^{-1})}{(123 \text{ s})(0.998 \text{ g} \cdot \text{mL}^{-1})} = 0.38$$

14-67   There are three triple points. When heated from 40°C to 200°C at 1 atm, sulfur goes from rhombic to monoclinic to liquid and then to the gaseous state. Sublimation will occur below 10^{-5} atm.

14-69   Because the bond geometry about each carbon atom is tetrahedral, it is appropriate to use sp^3 hybrid orbitals and to describe the carbon-carbon bonds as $C(sp^3)$-$C(sp^3)$.

14-71   The vapor pressure of the solid is equal to the vapor pressure of the liquid at the triple point. Therefore, we write

$$10.646 - \frac{2559.1 \text{ K}}{T_t} = 7.538 - \frac{1511 \text{ K}}{T_t}$$

$$3.108 = \frac{1048 \text{ K}}{T_t}$$

$$T_t = 337.2 \text{ K}$$

The logarithm of the pressure at the triple point is given by

$$\log P_t = 7.538 - \frac{1511 \text{ K}}{337.2 \text{ K}} = 3.057$$

$$P_t = 10^{3.057} = 1140 \text{ torr}$$

E. ANSWERS TO THE SELF-TEST

1.   false

2.   false (They have hindered rotational motion.)

3.   false

4.   the boiling point of the liquid

5.   the melting point of the solid

6.   heat capacity

7.   molar enthalpy of fusion, ΔH_{fus}

8. the forces that hold the crystal into a lattice must be overcome by the thermal motion of the particles

9. molar enthalpy of vaporization, ΔH_{vap}

10. attractive forces between the molecules must be overcome by the thermal motion of the particles

11. removes

12. false

13. molar enthalpy of sublimation, ΔH_{sub}

14. true

15. true

16. false

17. dipole-dipole

18. greater

19. an electrostatic intermolecular attraction between a hydrogen atom in one molecule and a highly electronegative atom (e.g., O or F) in another molecule

20. London forces

21. instantaneous dipole-dipole interactions

22. true

23. true

24. of evaporation is equal to the rate of condensation

25. true

26. The equilibrium vapor pressure is the pressure of the vapor that is in equilibrium with a liquid at a given temperature.

27. false

28. the temperature at which the equilibrium vapor pressure of the liquid is exactly 1 atm

29. false

30. Relative humidity is the water vapor pressure in the air relative to the equilibrium water vapor pressure at that temperature.

31. the temperature at which the relative humidity would be 100% for a given vapor pressure of water

32. the resistance of a liquid to flow

33. slowly

34. decreases

35. true

36. minimize

37. high

38. lowers

39. concave

40. greater

41. less

42. dipole moment

43. good

44. the regions of the various phases of a pure substance simultaneously

45. true

46. solid . . . liquid

47. false (It changes by a small amount.)

48. above which a gas cannot be liquefied regardless of the pressure applied

49. a temperature and pressure at which three phases coexist at equilibrium

50. false

51. false

52. false

53. greater

54. an ordered arrangement of its constituent molecules or ions in the solid state

55. crystalline structure

56. false

57. simple cubic, body-centered cubic, and face-centered cubic

58. the center

59. the center of each face

60. true

61. molar mass and density

62. true

63. atoms

64. large

65. difference in size of the cesium ion and the sodium ion

66. lower

67. false

68. true

69. diamond . . . graphite

70. covalent bonds

71. true

72. true

73. true

74. the bonding molecular orbitals of a crystal

75. the antibonding molecular orbitals of a crystal

76. conduction

77. the energy difference between the valence band and the conduction band

78. large

79. small

80. amorphous

81. false

15

COLLIGATIVE PROPERTIES OF SOLUTIONS

A. OUTLINE OF CHAPTER 15

15-1 Solutes affect the properties of the solvent.

Colligative properties of solutions depend primarily on the ratio of the number of solute particles to the number of solvent molecules in a solution.

Molarity, M, is the number of moles of solute per liter of solution.

The mole fraction X_1 of component 1 in a two-component solution is defined by the relation

$$X_1 = \frac{n_1}{n_1 + n_2} \qquad (15\text{-}2)$$

where n_1 is the number of moles of component 1 and n_2 is the number of moles of component 2 in the solution.

The sum of the mole fraction of all the components in a solution must equal one.

The mass percent of the solvent in a solution is given by

$$\text{mass \%} = \frac{\text{mass of solute}}{\text{mass of solute} + \text{mass of solvent}} \times 100 \qquad (15\text{-}5)$$

Molality, m, is the number of moles of solute per kilogram of solvent or in an equation

$$\text{molality} = \frac{\text{moles of solute}}{\text{kilogram of solvent}} \qquad (15\text{-}6)$$

Solute concentrations expressed as mole fraction, molality, and mass percent are temperature-independent.

Colligative molality m_c is the molality times the number of solute particles per formula unit (Table 15-1).

The major colligative properties are vapor pressure lowering, boiling-point elevation, freezing-point depression, and osmotic pressure.

15-2 The equilibrium partial pressure of a pure liquid always decreases when a substance is dissolved in the liquid.

A solute decreases the rate of evaporation of the solvent and thereby lowers the vapor pressure of the solvent (Figure 15-2).

The equilibrium vapor pressure of the solvent over a solution is proportional to the mole fraction of the solvent in the solution.

In an ideal solution the solvent and solute molecules are randomly distributed.

In an ideal solution the vapor pressure of the solvent P_1 is directly proportional to the mole fraction of the solvent X_1 (Raoult's law).

Raoult's law states that

$$P_1 = X_1 P_1^\circ \tag{15-9}$$

where P_1° is the vapor pressure of the pure solvent at the temperature of the solution.

The vapor pressure lowering is the amount by which the vapor pressure of a solution is less than the vapor pressure of the pure solvent.

The vapor pressure lowering ΔP_1 is given by

$$\Delta P_1 = X_2 P_1^\circ \tag{15-11}$$

where X_2 is the mole fraction of the solute.

15-3 Nonvolatile solutes increase the boiling point of a liquid.

The key to understanding colligative properties is the lowering of the solvent vapor pressure by the solute (Figure 15-3).

The boiling-point elevation is the amount by which the boiling point of a solution exceeds the boiling point of the pure solvent.

The boiling-point elevation, $T_b - T_b^\circ$, is given by the equation

$$T_b - T_b^\circ = K_b m_c \tag{15-14}$$

The boiling-point elevation constant, K_b, depends only on the solvent properties.

The values of some boiling-point elevations are given in Table 15-2.

15-4 Solutes decrease the freezing point of a liquid.

When an aqueous solution begins to freeze, the solid that separates out is usually pure ice.

The freezing-point depression is the amount by which the freezing point of the solution is lower than that of the pure solvent.

The freezing-point depression, $T_f^\circ - T_f$, is given by the equation

$$T_f^\circ - T_f = K_f m_c \qquad (15\text{-}15)$$

The freezing-point depression constant K_f depends only on the solvent properties.

The values of some freezing-point depression constants are given in Table 15-2.

The freezing-point depression effect is the basis of the action of antifreeze.

Ethylene glycol is a commonly used antifreeze.

Equation (15-15) can be written in the form

$$T_f^\circ - T_f = K_f i m \qquad (15\text{-}16)$$

where i is the van't Hoff i-factor.

The van't Hoff i-factor indicates the number of solute particles produced per formula unit that dissolves.

For strong electrolytes, i equals the total number of ions in the formula unit.

For weak electrolytes, i is a measure of the degree of dissociation of the weak electrolyte.

Freezing-point depression measurements can be used to determine the degree of dissociation of weak electrolytes in solution.

15-5 Osmotic pressure requires a semipermeable membrane.

Only water can pass through a semipermeable membrane; the solute cannot.

The escaping tendency of water from pure water is greater than the escaping tendency of water from a solution.

The osmotic pressure is the pressure that must be applied to a solution to raise the escaping tendency of water in the solution to that of pure water (Figure 15-10).

The osmotic pressure π is computed from the expression

$$\pi = RTM_c \qquad (15\text{-}17)$$

where M_c is the colligative molarity, R is the molar gas constant, and T is the Kelvin temperature.

Osmotic pressure is the pressure that must be applied to a solution to increase the vapor pressure of the solvent to a value equal to the vapor pressure of the pure solvent at that temperature.

Reverse osmosis is used to obtain pure water from seawater (Figure 15-11).

Osmotic pressure is a large effect and is used to determine molecular masses of proteins.

Osmotic pressure plays a major role in living systems by keeping cells inflated.

The net flow of water across cell walls is to the side with the larger colligative molarity.

15-6 The components of an ideal solution obey Raoult's law.

A solution of A and B is ideal if the interaction between A and B molecules is the same as that between A molecules and between B molecules.

In an ideal solution of liquids A and B, $P_A = X_A P_A^\circ$ and $P_B = X_B P_B^\circ$ [Equation (15-18)].

The total pressure over an ideal solution composed of A and B is given by

$$P_{total} = P_B^\circ + X_A(P_A^\circ - P_B^\circ) \tag{15-21}$$

A plot of P_{total} versus X_A is a straight line for an ideal solution (Figure 15-17).

A separation of two liquids can be achieved by fractional distillation (Figure 15-18).

There are deviations from Raoult's law if the solution is not ideal.

Negative deviations from ideal behavior occur when the total vapor pressure is lower than that calculated from Equation (15-21) (Figure 15-19a).

Positive deviations from ideal behavior occur when the total vapor pressure is greater than that calculated from Equation (15-21) (Figure 15-19b).

An azeotrope is a solution that distills without a change in composition.

15-7   The solubility of a gas in a liquid is directly proportional to the pressure of the gas over the liquid.

Henry's law is the relationship between the solubility of a gas in a liquid and the partial pressure of the gas.

Henry's law states that

$$P_{gas} = k_h M_{gas} \tag{15-22}$$

where P_{gas} is the partial pressure of the gas, M_{gas} is the molarity of the dissolved gas, and k_h is Henry's law constant.

The values of Henry's law constant for some gases are given in Table 15-3.

The solubility of gases in liquids decreases with increasing temperature.

B. SELF-TEST

1. The colligative properties of a solution depend primarily on _____

_____.

2. The mole fraction of the solute in a two-component solution is given by _____

_____.

3. The sum of the mole fraction of the solute and the mole fraction of the solvent in a two-component solution is one. *True/False*

4. The mass percent concentration of the solute is given by _____

_____ .

5. The molarity of a solution is defined as _____

_____ .

6. The molality of a solution is defined as _____

_____ .

7. The molality of a solution is always the same as the molarity of the solution. *True/False*

8. A 1.0-m aqueous solution of glucose is prepared by dissolving 1.0 mol of glucose in

_____ .

9. Colligative molality is defined as _____

_____ .

10. The colligative molality of an aqueous solution is always equal to the molality of the solution. *True/False*

11. The equilibrium vapor pressure of a pure liquid is always (*increased/decreased*) when a substance is dissolved in the liquid.

12. Explain why the vapor pressure of a solution containing a nonvolatile solute is less than the vapor pressure of the pure solvent at the same temperature. _____

_____

13. The equilibrium vapor pressure of the solvent over a solution is proportional _____

_____ .

14. Raoult's law states that _____

_____ .

15. The equation that expresses Raoult's law is _____ where _____

_____ .

16. A 1.0-m aqueous solution of sucrose has the same equilibrium vapor pressure as a 1.0-m aqueous solution of NaCl at the same temperature. *True/False*

17. Nonvolatile solutes (*increase/decrease*) the boiling point of the solvent.

18. Explain why a nonvolatile solute raises the boiling point of the solvent. _____

_____

_____

19. The boiling-point elevation due to a dissolved solute is proportional to _____ _____.

20. The boiling-point elevation of a solution depends on the number of solute particles dissolved in the solvent. *True/False*

21. The boiling-point elevation constant K_b depends on the nature of the solute. *True/False*

22. Solutes (*increase/decrease*) the freezing point of the solvent.

23. Explain why a solute lowers the freezing point of the solvent. _____ _____ _____

24. The freezing-point depression of a solution due to a solute is proportional to _____.

25. The freezing-point depression of a solution depends on the concentration of solute particles dissolved in the solvent. *True/False*

26. The freezing-point depression constant K_f depends on the nature of the solute. *True/False*

27. An antifreeze is added to water to _____.

28. An antifreeze mixture cannot freeze regardless of the temperature. *True/False*

29. The freezing point of seawater is 0.0°C. *True/False*

30. The van't Hoff *i*-factor for NaCl(*aq*) is 2. *True/False*

31. The van't Hoff *i*-factor for HF(*aq*) is 2. *True/False*

32. Freezing-point depression measurements can be used to measure the _____ _____ of weak electrolytes.

33. If a beaker of water and a beaker of an aqueous solution of NaCl are placed under a Bell jar, then water transfers from the beaker of _____ to the beaker of _____ .

34. If a 0.1-M aqueous solution of NaCl is separated from a 1.0-M aqueous solution of NaCl by a rigid semipermeable membrane, then water passes spontaneously from the _____ solution to the _____ solution.

35. The osmotic pressure of a solution is _____ _____ .

36. The osmotic pressure is given by the equation $\pi =$ _____ .

37. The osmotic pressure of a solution is the equilibrium vapor pressure of the solution. *True/False*

38. The osmotic pressure of a solution depends primarily on the concentration of solute particles dissolved in the solvent. *True/False*

39. Pressure can be used to obtain pure water from seawater. *True/False*

40. Red blood cells will rupture when placed in pure water because _____ _____ .

41. Explain why the osmotic pressure of a protein solution can be used to determine the molecular mass of a protein. _____ _____

42. Boiling-point elevation is useful for determining the formula mass of low molecular mass compounds. *True/False*

43. A solution of two liquids is ideal when _____ _____ .

44. In an ideal solution of two liquids, each component obeys Raoult's law. *True/False*

45. The total vapor pressure over an ideal solution of two volatile liquids is given by $P_{total} =$ _____ .

46. Two volatile liquids cannot be separated by fractional distillation. *True/False*

47. When the total vapor pressure over a solution of two liquids is greater than that calculated using Raoult's law, there is a (*negative/positive*) deviation from ideal behavior.

48. A negative deviation from ideal behavior occurs when the total vapor pressure over a solution of two liquids is less than that calculated using Raoult's law. *True/False*

49. A solution of ethyl alcohol and water is an (*ideal/nonideal*) solution.

50. An azeotrope is a solution that _____ _____ .

51. The solubility of a gas in a liquid depends upon the partial pressure of the gas. *True/False*

52. Henry's law states that _____

_____ .

53. The solubility of a gas in a liquid (*increases/decreases*) with increasing temperature.

54. At a given temperature, the solubility of a gas in a liquid (*doubles/halves*) when the partial pressure of the gas is doubled.

C. CALCULATIONS YOU SHOULD KNOW HOW TO DO

1. Calculate mole fractions. See Example 15-1 and Problems 15-1 through 15-6.

2. Calculate the molality and colligative molality of solutions. See Examples 15-1 and 15-2 and Problems 15-7 through 15-12.

3. Calculate solvent vapor pressure and vapor pressure lowering by using Raoult's law, Equations (15-9) or (15-11). See Example 15-3 and Problems 15-17 through 15-30.

4. Calculate the boiling-point elevation of solutions. See Example 15-4 and Problems 15-31 through 15-36.

5. Calculate the freezing-point depression of solutions and the degree of dissociation of solutes in solution from the freezing-point depression. See Examples 15-5 and 15-6 and Problems 15-37 through 15-48.

6. Calculate the osmotic pressure of solutions and the molecular mass of a solute from the osmotic pressure. See Examples 15-7 and 15-8 and Problems 15-49 through 15-53.

7. Calculate the vapor pressure over an ideal solution of two liquids. See Example 15-9 and Problems 15-55 and 15-56.

8. Calculate the solubility of a gas in a liquid using Henry's law. See Example 15-10 and Problems 15-57 through 15-62.

D. SOLUTIONS TO THE ODD-NUMBERED PROBLEMS

15-1   The mole fraction of water is given by

$$X_{H_2O} = \frac{n_{H_2O}}{n_{H_2O} + n_{C_2H_5OH}}$$

The mole fraction of water is

$$X_{H_2O} = \frac{(80.0 \text{ g})\left(\dfrac{1 \text{ mol } H_2O}{18.02 \text{ g } H_2O}\right)}{(80.0 \text{ g})\left(\dfrac{1 \text{ mol } H_2O}{18.02 \text{ g } H_2O}\right) + (20.0 \text{ g})\left(\dfrac{1 \text{ mol } C_2H_5OH}{46.07 \text{ g } C_2H_5OH}\right)}$$

$$= \frac{4.440 \text{ mol}}{4.440 \text{ mol} + 0.4341 \text{ mol}} = \frac{4.440 \text{ mol}}{4.874 \text{ mol}}$$

$$= 0.911$$

The mole fraction of ethyl alcohol is

$$X_{C_2H_5OH} = \frac{(20.0 \text{ g})\left(\dfrac{1 \text{ mol } C_2H_5OH}{46.07 \text{ g } C_2H_5OH}\right)}{4.440 \text{ mol} + 0.4341 \text{ mol}} = 0.0891$$

15-3 The mole fraction of acetone (A) is given by

$$X_A = \frac{n_A}{n_A + n_{H_2O}} = 0.19$$

In 1.00 kg of solution we have x g of acetone and $(1000 - x)$ g of water. The moles of acetone and water are given by

$$n_A = \frac{x \text{ g}}{58.08 \text{ g} \cdot \text{mol}^{-1}} \qquad n_{H_2O} = \frac{(1000 - x)\text{g}}{18.02 \text{ g} \cdot \text{mol}^{-1}}$$

Thus,

$$\frac{\left(\dfrac{x}{58.08}\right)}{\left(\dfrac{x}{58.08}\right) + \left(\dfrac{1000 - x}{18.02}\right)} = 0.19$$

and

$$\frac{x}{58.08} = \frac{0.19x}{58.08} + \frac{190}{18.02} - \frac{0.19x}{18.02}$$

Multiplying both sides of this equation by 58.08 yields

$$x = 0.19x + 612.4 - 0.612x$$

Solving for x yields

$$x = \frac{612.4}{1.42} = 431$$

The solution is prepared by mixing 431 g of acetone with 569 g of water.

15-5 In a 100.0-mL sample we have 70.0 mL of isopropyl alcohol and 30.0 mL of water. The masses of isopropyl alcohol and water are

$$\text{mass}_{\text{alc}} = (70.0 \text{ mL})(0.785 \text{ g} \cdot \text{mL}^{-1}) = 55.0 \text{ g}$$

$$\text{mass}_{\text{H}_2\text{O}} = (30.0 \text{ mL})(1.000 \text{ g} \cdot \text{mL}^{-1}) = 30.0 \text{ g}$$

The numbers of moles of isopropyl alcohol and water are

$$n_{\text{alc}} = (55.0 \text{ g})\left(\frac{1 \text{ mol C}_3\text{H}_7\text{OH}}{60.09 \text{ g C}_3\text{H}_7\text{OH}}\right) = 0.917 \text{ mol}$$

$$n_{\text{H}_2\text{O}} = (30.0 \text{ g})\left(\frac{1 \text{ mol H}_2\text{O}}{18.02 \text{ g H}_2\text{O}}\right) = 1.665 \text{ mol}$$

The mole fraction of isopropyl alcohol is given by

$$X_{\text{alc}} = \frac{n_{\text{alc}}}{n_{\text{alc}} + n_{\text{H}_2\text{O}}} = \frac{0.915 \text{ mol}}{0.915 \text{ mol} + 1.665 \text{ mol}} = 0.355$$

15-7 A 2.50-m formic-acid-in-acetone solution contains 2.50 mol of formic acid dissolved in 1000 g of acetone. Thus, we dissolve

$$(2.50 \text{ mol})\left(\frac{46.03 \text{ g HCHO}_2}{1 \text{ mol HCHO}_2}\right) = 115 \text{ g}$$

of formic acid in 1000 g of acetone.

15-9 The number of moles of I_2 that correspond to 2.603 g is

$$n = (2.603 \text{ g})\left(\frac{1 \text{ mol I}_2}{253.8 \text{ g I}_2}\right) = 0.01026 \text{ mol}$$

Therefore,

$$\text{molality} = \frac{\text{moles of solute}}{\text{kilograms of solvent}}$$

$$= \frac{0.01026 \text{ mol I}_2}{0.1000 \text{ kg CCl}_4} = 0.1026 \text{ m}$$

15-11 (a) There are two ions per $MgSO_4$ formula unit, because $MgSO_4$ dissociates into $Mg^{2+}(aq)$ and $SO_4^{2-}(aq)$ in water. The colligative molality is thus 2.0 m_c.

  (b) There are three ions per $Cu(NO_3)_2$ formula unit, because $Cu(NO_3)_2$ dissociates into $Cu^{2+}(aq)$ and $2NO_3^-(aq)$ in water. The colligative molality is thus 3.0 m_c.

  (c) There is one solute particle per C_2H_5OH formula unit because C_2H_5OH does not dissociate in water. Thus, the colligative molality is 1.0 m_c.

  (d) There are five ions per $Al_2(SO_4)_3$ formula unit, because $Al_2(SO_4)_3$ dissociates into $2Al^{3+}(aq)$ and $3SO_4^{2-}(aq)$ in water. The colligative molality is thus $5.0 \ m_c$.

15-13 (a) $AgNO_3$: a strong electrolyte; $i = 2$

(b) $MgCl_2$: a strong electrolyte; $i = 3$

(c) K_2SO_4: a strong electrolyte; $i = 3$

15-15 HNO_2, H_2SO_4, and $HClO_4$ are all strong acids and, thus, are strong electrolytes. When HNO_3 or $HClO_4$ dissolves in water, two ions are produced per formula unit, one $H^+(aq)$ ion and one $NO_3^-(aq)$ or $ClO_4^-(aq)$ ion. When H_2SO_4 dissolves in water, three ions are produced per formula unit, two $H^+(aq)$ ions and one $SO_4^{2-}(aq)$ ion. Therefore, H_2SO_4 would be expected to have the largest van't Hoff i-factor.

15-17 The mole fraction of water in the solution is

$$
\begin{aligned}
X_{H_2O} &= \frac{n_{H_2O}}{n_{H_2O} + n_{glucose}} \\[2mm]
&= \frac{(500.0 \ g)\left(\dfrac{1 \ mol \ H_2O}{18.02 \ g \ H_2O}\right)}{(500.0 \ g)\left(\dfrac{1 \ mol \ H_2O}{18.02 \ g \ H_2O}\right) + (20.0 \ g)\left(\dfrac{1 \ mol \ glucose}{180.16 \ g \ glucose}\right)} \\[2mm]
&= 0.996
\end{aligned}
$$

Raoult's law is

$$P_{H_2O} = X_{H_2O} P^\circ_{H_2O}$$

and thus at 37°C we have

$$P_{H_2O} = (0.996)(47.1 \ torr) = 46.9 \ torr$$

The vapor pressure lowering is

$$P^\circ_{H_2O} - P_{H_2O} = 47.1 \ torr - 46.9 \ torr = 0.2 \ torr$$

15-19 The mole fraction of water in the solution is

$$
\begin{aligned}
X_{H_2O} &= \frac{n_{H_2O}}{n_{H_2O} + n_{sucrose}} \\[2mm]
&= \frac{(195 \ g)\left(\dfrac{1 \ mol \ H_2O}{18.02 \ g \ H_2O}\right)}{(195 \ g)\left(\dfrac{1 \ mol \ H_2O}{18.02 \ g \ H_2O}\right) + (20.0 \ g)\left(\dfrac{1 \ mol \ sucrose}{342.30 \ g \ sucrose}\right)} \\[2mm]
&= 0.995
\end{aligned}
$$

From Raoult's law we compute

$$P_{H_2O} = X_{H_2O} P^\circ_{H_2O} = (0.995)(23.76 \ torr) = 23.64 \ torr$$

The vapor pressure lowering is

$$P^\circ_{H_2O} - P_{H_2O} = 23.76 \text{ torr} - 23.64 \text{ torr} = 0.12 \text{ torr}$$

15-21 The mole fraction of ethyl alcohol in the solution is given by

$$X_{alc} = \frac{n_{alc}}{n_{alc} + n_{urea}}$$

Thus,

$$X_{alc} = \frac{\left(\dfrac{100.0 \text{ g}}{46.07 \text{ g}\cdot\text{mol}^{-1}}\right)}{\left(\dfrac{100.0 \text{ g}}{46.07 \text{ g}\cdot\text{mol}^{-1}}\right) + \left(\dfrac{20.0 \text{ g}}{60.06 \text{ g}\cdot\text{mol}^{-1}}\right)} = 0.867$$

The vapor pressure of ethyl alcohol is calculated by using Raoult's law.

$$P_{alc} = X_{alc}P^\circ_{alc} = (0.867)(59.2 \text{ torr}) = 51.3 \text{ torr}$$

The vapor pressure lowering is

$$P^\circ_{alc} - P_{alc} = 59.2 \text{ torr} - 51.3 \text{ torr} = 7.9 \text{ torr}$$

15-23 The number of moles of water in 1000 g (1 kg) is

$$n_{H_2O} = (1000 \text{ g})\left(\frac{1 \text{ mol H}_2\text{O}}{18.02 \text{ g H}_2\text{O}}\right) = 55.49 \text{ mol}$$

(a) The colligative molality of 0.25 m NaCl is 0.50 m_c. Thus, there are 0.50 mol of solute particles in 1000 g of water. The mole fraction of water in the solution is

$$X_{H_2O} = \frac{n_{H_2O}}{n_{H_2O} + n_{solute}} = \frac{55.49 \text{ mol}}{55.49 \text{ mol} + 0.50 \text{ mol}} = 0.9911$$

The partial pressure of water is computed by using Raoult's law:

$$P_{H_2O} = X_{H_2O}P^\circ_{H_2O} = (0.9911)(17.54 \text{ torr}) = 17.38 \text{ torr}$$

The vapor pressure lowering is

$$P^\circ_{H_2O} - P_{H_2O} = 17.54 \text{ torr} - 17.38 \text{ torr} = 0.16 \text{ torr}$$

(b) The colligative molality of 0.25 m CaCl$_2$ is 0.75 m_c. Thus, there are 0.75 mol of solute particles in 1000 g of water. The mole fraction of water in the solution is

$$X_{H_2O} = \frac{55.49 \text{ mol}}{55.49 \text{ mol} + 0.75 \text{ mol}} = 0.9867$$

The vapor pressure of water is

$$P_{H_2O} = X_{H_2O}P°_{H_2O} = (0.9867)(17.54 \text{ torr}) = 17.31 \text{ torr}$$

The vapor pressure lowering is

$$P°_{H_2O} - P_{H_2O} = 17.54 \text{ torr} - 17.31 \text{ torr} = 0.23 \text{ torr}$$

(c)   The colligative molality of 0.25 m sucrose is 0.25 m_c. Thus, there are 0.25 mol of solute particles in 1000 g of water. The mole fraction of water in the solution is

$$X_{H_2O} = \frac{55.49 \text{ mol}}{55.49 \text{ mol} + 0.25 \text{ mol}} = 0.9955$$

The vapor pressure of water is

$$P_{H_2O} = X_{H_2O}P°_{H_2O} = (0.9955)(17.54 \text{ torr}) = 17.46 \text{ torr}$$

The vapor pressure lowering is

$$P°_{H_2O} - P_{H_2O} = 17.54 \text{ torr} - 17.46 \text{ torr} = 0.08 \text{ torr}$$

(d)   The colligative molality of 0.25 m $Al(ClO_4)_3$ is 1.00 m_c. Thus, there is 1.00 mol of solute particles in 1000 g of water. The mole fraction of water in the solution is

$$X_{H_2O} = \frac{55.49 \text{ mol}}{55.49 \text{ mol} + 1.00 \text{ mol}} = 0.9823$$

The vapor pressure of water is computed by using Raoult's law

$$P_{H_2O} = X_{H_2O}P°_{H_2O} = (0.9823)(17.54 \text{ torr}) = 17.23 \text{ torr}$$

The vapor pressure lowering is

$$P°_{H_2O} - P_{H_2O} = 17.54 \text{ torr} - 17.23 \text{ torr} = 0.31 \text{ torr}$$

15-25   We use Equation (15-11)

$$\Delta P_1 = X_2 P°_1$$

with $P°_1 = 0.0313$ atm in each case.

(a) $$m_c = \frac{2.00 \text{ mol } C_{12}H_{22}O_{11}}{1000 \text{ g } H_2O}$$

We must convert the molality to mole fraction. The number of moles of water in 1000 grams is

$$\text{moles of H}_2\text{O} = (1000 \text{ g H}_2\text{O})\left(\frac{1 \text{ mol H}_2\text{O}}{18.02 \text{ g H}_2\text{O}}\right) = 55.49 \text{ mol}$$

$$X_2 = \frac{2.00 \text{ mol C}_{12}\text{H}_{22}\text{O}_{11}}{55.49 \text{ mol H}_2\text{O} + 2.00 \text{ mol C}_{12}\text{H}_{22}\text{O}_{11}} = 0.03479$$

The vapor pressure lowering is

$$\Delta P_1 = (0.03479)(0.0313 \text{ atm}) = 1.09 \times 10^{-3} \text{ atm}$$

(b)
$$m_c = \frac{\left(\dfrac{2 \text{ mol ions}}{1 \text{ mol NaCl}}\right)(2.00 \text{ mol NaCl})}{1000 \text{ g H}_2\text{O}} = \frac{4.00 \text{ mol ions}}{1000 \text{ g H}_2\text{O}}$$

$$X_2 = \frac{4.00 \text{ mol ions}}{55.49 \text{ mol H}_2\text{O} + 4.00 \text{ mol ions}} = 0.06724$$

$$\Delta P_1 = (0.06724)(0.0313 \text{ atm}) = 2.10 \times 10^{-3} \text{ atm}$$

(c)
$$m_c = \frac{\left(\dfrac{3 \text{ mol ions}}{1 \text{ mol CaCl}_2}\right)(2.00 \text{ mol CaCl}_2)}{1000 \text{ g H}_2\text{O}} = \frac{6.00 \text{ mol ions}}{1000 \text{ g H}_2\text{O}}$$

$$X_2 = \frac{6.00 \text{ mol ions}}{55.49 \text{ mol H}_2\text{O} + 6.00 \text{ mol ions}} = 0.09758$$

$$\Delta P_1 = (0.09758)(0.0313 \text{ atm}) = 3.05 \times 10^{-3} \text{ atm}$$

15-27   We use Equation (15-11) to find the mole fraction of sucrose. The value of P_1° at 25°C is 23.76 torr (Table 14-7). Solving Equation (15-11) for the mole fraction gives

$$X_2 = \frac{\Delta P_1}{P_1^\circ} = \frac{3.56 \text{ torr}}{23.76 \text{ torr}} = 0.1498$$

Recall from the definition of mole fraction that

$$X_2 = \frac{\text{moles sucrose}}{\text{moles sucrose} + \text{moles water}}$$

We shall find the concentration of sucrose in the units molality. Thus, we take 1000 g of water; the number of moles in 1000 g of water is given by

$$\text{moles of H}_2\text{O} = (1000 \text{ g H}_2\text{O})\left(\frac{1 \text{ mol H}_2\text{O}}{18.02 \text{ g H}_2\text{O}}\right) = 55.49 \text{ mol}$$

We then have that

$$\frac{\text{moles sucrose}}{\text{moles sucrose} + 55.49 \text{ mol water}} = 0.1498$$

Solving for moles sucrose, we have that

$$\begin{aligned}
\text{moles sucrose} &= (0.1498)(\text{moles sucrose} + 55.49 \text{ mol}) \\
&= 0.1498 \text{ moles sucrose} + 8.312 \text{ mol}
\end{aligned}$$

$$0.8502 \text{ moles sucrose} = 8.312 \text{ mol}$$

$$\text{moles sucrose} = 9.78 \text{ mol}$$

The concentration is

$$m = \frac{9.78 \text{ mol}}{1000 \text{ g water}} = 9.78 \text{ m}$$

15-29 We use Equation (15-11) to find the mole fraction of ethylene glycol. The value of P_1° at $90\,^\circ$C is 526 torr (Table 14-7); thus, the mole fraction of ethylene glycol in the solution is

$$X_2 = \frac{\Delta P_1}{P_1^\circ} = \frac{12.0 \text{ torr}}{526 \text{ torr}} = 0.0228$$

15-31 The colligative molality of the $Sc(ClO_4)_3(aq)$ solution is

$$m_c = (4)(2.0 \text{ m}) = 8.0 \text{ m}_c$$

The boiling-point elevation is

$$\begin{aligned}
T_b - T_b^\circ &= K_b m_c \\
&= (0.52 \text{ K} \cdot \text{m}_c^{-1})(8.0 \text{ m}_c) = 4.2 \text{ K} = 4.2\,^\circ\text{C}
\end{aligned}$$

The boiling point of the solution is

$$T_b = 100.00\,^\circ\text{C} + 4.2\,^\circ\text{C} = 104.2\,^\circ\text{C}$$

15-33 The molality of the solution is

$$m = \frac{(10.0 \text{ g})\left(\dfrac{1 \text{ mol picric acid}}{229.1 \text{ g picric acid}}\right)}{(0.100 \text{ kg cyclohexane})} = 0.436 \text{ m}$$

$$m_c = m = 0.436 \text{ m}_c$$

The boiling-point elevation is (Table 15-2)

$$T_b - T_b^\circ = K_b m_c = (2.79 \text{ K} \cdot \text{m}_c^{-1})(0.436 \text{ m}_c) = 1.22 \text{ K} = 1.22\,^\circ\text{C}$$

The boiling-point (Table 15-2) is

$$T_b = 80.7°C + 1.22°C = 81.9°C$$

15-35 Urea does not dissociate in solution, and thus the colligative molality of the urea is

$$m_c = \frac{(25.0 \text{ g}) \left(\dfrac{1 \text{ mol urea}}{60.06 \text{ g urea}} \right)}{1.500 \text{ kg nitrobenzene}} = 0.278 \ m_c$$

The boiling-point elevation (Table 15-2) is

$$T_b - T_b° = K_b m_c = (5.24 \text{ K} \cdot m_c^{-1})(0.278 \ m_c) = 1.46 \text{ K} = 1.46°C$$

The boiling point of the solution (Table 15-2) is

$$T_b = 210.8°C + 1.46°C = 212.3°C$$

15-37 The molality of the solution is

$$m = \frac{(60.0 \text{ g}) \left(\dfrac{1 \text{ mol glucose}}{180.16 \text{ g glucose}} \right)}{0.200 \text{ kg water}} = 1.665 \ m = 1.665 \ m_c$$

The freezing-point depression (Table 15-2) is

$$T_f° - T_f = K_f m_c = (1.86 \text{ K} \cdot m_c^{-1})(1.665 \ m_c) = 3.10 \text{ K} = 3.10°C$$

The freezing point is

$$T_f = 0.00°C - 3.10°C = -3.10°C$$

15-39 The molality of the solution is

$$m = \frac{(22.0 \text{ g}) \left(\dfrac{1 \text{ mol CCl}_4}{153.81 \text{ g CCl}_4} \right)}{0.800 \text{ kg benzene}} = 0.179 \ m = 0.179 \ m_c$$

The freezing-point depression (Table 15-2) is

$$T_f° - T_f = K_f m_c = (5.10 \text{ K} \cdot m_c^{-1})(0.179 \ m_c) = 0.913 \text{ K} = 0.913°C$$

The freezing point is

$$T_f = 5.50°C - 0.913°C = 4.59°C$$

15-41 Diphenyl and naphthalene do not dissociate in solution, and thus the colligative molality is

$$m_c = \frac{(5.00 \text{ g}) \left(\dfrac{1 \text{ mol diphenyl}}{154.20 \text{ g diphenyl}} \right) + (7.50 \text{ g}) \left(\dfrac{1 \text{ mol naphthalene}}{128.16 \text{ g naphthalene}} \right)}{(0.200 \text{ kg benzene})}$$

$$= 0.455 \ m_c$$

The freezing-point depression of the solution (Table 15-2) is

$$T_f^\circ - T_f = K_b m_c = (5.10 \text{ K} \cdot \text{m}_c^{-1})(0.455 \ \text{m}_c) = 2.32 \text{ K} = 2.32°\text{C}$$

The freezing point of the solution is

$$T_f = 5.50°\text{C} - 2.32°\text{C} = 3.18°\text{C}$$

15-43 We can find the molality of the solution from the freezing-point depression.

$$T_f^\circ - T_f = K_f m_c$$

$$4.43 \text{ K} = (40.0 \text{ K} \cdot \text{m}_c^{-1}) m_c$$

$$m_c = \frac{4.43 \text{ K}}{40.0 \text{ K} \cdot \text{m}_c^{-1}} = 0.111 \ \text{m}_c$$

Because the mass is given, we have the correspondence

$$0.111 \text{ mol} \cdot \text{kg}^{-1} \approx \frac{0.500 \text{ g vitamin K}}{0.0100 \text{ kg camphor}} = 50.0 \text{ g} \cdot \text{kg}^{-1}$$

and therefore

$$0.111 \text{ mol} \approx 50.0 \text{ g}$$

Dividing both sides by 0.111, we have

$$1.00 \text{ mol} \approx 450 \text{ g}$$

The molecular mass of vitamin K is 450.

An alternative solution is to use the definition of molality.

$$m_c = \frac{\text{moles of vitamin K}}{\text{kilograms of camphor}}$$

$$0.111 \text{ m} = \frac{(0.500 \text{ g})/(\text{molar mass})}{(0.0100 \text{ kg})}$$

Solving for the molar mass, we have

$$\text{molar mass} = \frac{0.500 \text{ g}}{(0.111 \text{ mol} \cdot \text{kg}^{-1})(0.0100 \text{ kg})} = 450 \text{ g} \cdot \text{mol}^{-1}$$

The molecular mass of vitamin K is 450.

15-45 We can find the colligative molality of the solution from the freezing-point depression.

$$T_f^\circ - T_f = K_f m_c$$

$$0.00°C - (-57°C) = (1.86 \text{ K} \cdot \text{m}_c^{-1}) m_c$$

$$m_c = \frac{57 \text{ K}}{1.86 \text{ K} \cdot \text{m}_c^{-1}} = 30.6 \text{ m}_c$$

The concentration of $CaCl_2$ in the pond is

$$\text{molality} = \frac{m_c}{3} = \frac{30.6 \text{ m}_c}{3} = 10 \text{ m}$$

15-47 The molality of the solution is

$$m = \frac{(40.7 \text{ g}) \left(\dfrac{1 \text{ mol HgCl}_2}{271.5 \text{ g HgCl}_2} \right)}{0.100 \text{ kg H}_2\text{O}} = 1.50 \text{ m}$$

If $HgCl_2$ were completely dissociated, the colligative molality would be 4.50 m_c. We can find the colligative molality from the freezing-point depression.

$$T_f^\circ - T_f = K_f m_c$$

$$2.83 \text{ K} = (1.86 \text{ K} \cdot \text{m}_c^{-1}) m_c$$

$$m_c = \frac{2.83 \text{ K}}{1.86 \text{ K} \cdot \text{m}_c^{-1}} = 1.52 \text{ m}_c$$

The colligative molality is only slightly greater than the molality indicating that the compound $HgCl_2$ is essentially undissociated in water.

15-49 The osmotic pressure is given by

$$\pi = RTM_c$$

Thus,

$$\pi = (0.0821 \text{ L} \cdot \text{atm} \cdot \text{K}^{-1} \cdot \text{mol}^{-1})(310 \text{ K})(0.25 \text{ mol} \cdot \text{L}^{-1}) = 6.4 \text{ atm}$$

15-51 The concentration of insulin in the aqueous solution is

$$M_c = \frac{\pi}{RT} = \frac{(6.48 \text{ torr}) \left(\dfrac{1 \text{ atm}}{760 \text{ torr}} \right)}{(0.0821 \text{ L} \cdot \text{atm} \cdot \text{K}^{-1} \cdot \text{mol}^{-1})(298 \text{ K})}$$

$$= 3.48 \times 10^{-4} \text{ mol} \cdot \text{L}^{-1}$$

The molecular mass can be calculated from the concentration. We have the correspondence

$$3.48 \times 10^{-4} \text{ mol} \cdot \text{L}^{-1} \approx \frac{2.00 \times 10^{-2} \text{ g}}{0.0100 \text{ L}} = 2.00 \text{ g} \cdot \text{L}^{-1}$$

and, therefore,

$$3.48 \times 10^{-4} \text{ mol} \approx 2.00 \text{ g}$$

Dividing both sides by 3.48×10^{-4}, we have

$$1.00 \text{ mol} \approx 5.75 \times 10^{3} \text{ g}$$

The molecular mass of insulin is about 5750.

15-53   The concentration of a solution for which an applied pressure of 100 atm is just sufficient to cause reverse osmosis is

$$M_c = \frac{\pi}{RT} = \frac{100 \text{ atm}}{(0.0821 \text{ L} \cdot \text{atm} \cdot \text{K}^{-1} \cdot \text{mol}^{-1})(293 \text{ K})}$$
$$= 4.16 \text{ mol} \cdot \text{L}^{-1} = 4.16 \text{ M}_c$$

The concentration of seawater is 1.1 M_c. The number of moles of ions in the seawater will remain the same after reverse osmosis:

moles of ions before reverse osmosis = moles of ions after reverse osmosis

$$M_B V_B = M_A V_A$$
$$(1.1 \text{ M}_c)V_B = (4.16 \text{ M}_c)V_A$$

We want 10 L of fresh water, that is, $\Delta V = 10$ L:

$$V_B = 10 \text{ L} + V_A \qquad \text{or} \qquad V_A = V_B - 10 \text{ L}$$
$$(1.1 \text{ M}_c)V_B = (4.16 \text{ M}_c)(V_B - 10 \text{ L})$$

Solving for V_B, we have

$$41.6 \text{ L} = 3.1 V_B$$
$$V_B = \frac{41.6 \text{ L}}{3.1} = 13 \text{ L}$$

15-55   (a)   The total pressure over the solution is given by

$$P_{\text{total}} = P_B + P_T$$

where P_B and P_T denote the partial pressures of benzene and toluene, respectively. Assuming an ideal solution, we use Raoult's law for both components.

$$P_{total} = X_B P_B^\circ + X_T P_T^\circ$$
$$= (0.250)(768 \text{ torr}) + (0.750)(293 \text{ torr})$$
$$= 412 \text{ torr}$$

(b) The mole fraction of benzene in the vapor phase is given by

$$Y_B = \frac{n_B}{n_{total}} = \frac{(P_B V/RT)}{(P_{total} V/RT)} = \frac{P_B}{P_{total}} = \frac{X_B P_B^\circ}{P_{total}}$$
$$= \frac{(0.250)(768 \text{ torr})}{412 \text{ torr}}$$
$$= 0.466$$

15-57 From Henry's law we have

$$P_{gas} = k_h M_{gas}$$

Thus,

$$M_{gas} = \frac{P_{gas}}{k_h}$$

The Henry's law constant for N_2 is $1.6 \times 10^3 \text{ atm} \cdot M^{-1}$ (see Table 15-3):

$$M_{N_2} = \frac{0.79 \text{ atm}}{1.6 \times 10^3 \text{ atm} \cdot M^{-1}} = 4.9 \times 10^{-4} \text{ M}$$

15-59 Solving Henry's law for the concentration of CO_2, we have that

$$M_{CO_2} = \frac{P_{CO_2}}{k_h}$$
$$= \frac{2.0 \text{ atm}}{29 \text{ atm} \cdot M^{-1}} = 0.069 \text{ M}$$

15-61 The partial pressure of O_2 (Table 15-3) is

$$P_{O_2} = k_h M_{O_2} = (7.8 \times 10^2 \text{ atm} \cdot M^{-1})(1.28 \times 10^{-3} \text{ M}) = 1.0 \text{ atm}$$

Air is 20 percent O_2; the air pressure when the partial pressure of O_2 is 1.0 atm is

$$(0.20)P_{air} = P_{O_2}$$
$$P_{air} = \frac{1.0 \text{ atm}}{0.20} = 5.0 \text{ atm}$$

The increase in pressure is 4.0 atm. The depth of the dive is

$$\left(\frac{33 \text{ ft}}{1 \text{ atm}}\right)(4.0 \text{ atm}) = 130 \text{ ft}$$

15-63   The concentration of immunoglobulin G in the aqueous solution is

$$M_c = \frac{\pi}{RT} = \frac{(0.619 \text{ torr})\left(\dfrac{1 \text{ atm}}{760 \text{ torr}}\right)}{(0.0821 \text{ L} \cdot \text{atm} \cdot \text{K}^{-1} \cdot \text{mol}^{-1})(298 \text{ K})}$$
$$= 3.33 \times 10^{-5} \text{ mol} \cdot \text{L}^{-1}$$

Thus, we have the correspondence

$$3.33 \times 10^{-5} \text{ mol} \cdot \text{L}^{-1} \approx \frac{0.500 \text{ g IgG}}{0.100 \text{ L solution}} = 5.00 \text{ g} \cdot \text{L}^{-1}$$

or

$$3.33 \times 10^{-5} \text{ mol} \approx 5.00 \text{ g}$$

Dividing both sides by 3.33×10^{-5}, we have

$$1.00 \text{ mol} \approx 1.50 \times 10^5 \text{ g}$$

The molecular mass of immunoglobulin G is about 150,000.

15-65   Ethyl alcohol (boiling point, 78°C) is a temporary antifreeze because its equilibrium vapor pressure is much greater than one atmosphere at 100°C. Thus, ethyl alcohol is much more readily lost by evaporation from the coolant system than a relatively high boiling liquid like ethylene glycol (boiling point, 197°C).

15-67   The number of moles of sucrose in two teaspoons is

$$n = (2.00 \text{ teaspoons})\left(\frac{14 \text{ g}}{3 \text{ teaspoons}}\right)\left(\frac{1 \text{ mol sucrose}}{342.3 \text{ g sucrose}}\right)$$
$$= 0.0273 \text{ mol}$$

The mass of water in one cup is

$$\text{mass} = dV = (1.00 \text{ g} \cdot \text{mL}^{-1})(1.00 \text{ cup})\left(\frac{0.946 \text{ L}}{4 \text{ cups}}\right)\left(\frac{1000 \text{ mL}}{1 \text{ L}}\right)$$
$$= 237 \text{ g} = 0.237 \text{ kg}$$

The molality is

$$\text{molality} = \frac{\text{moles of solute}}{\text{kilogram of solvent}}$$
$$= \frac{0.0273 \text{ mol sucrose}}{0.237 \text{ kg H}_2\text{O}} = 0.115 \text{ m}$$

15-69   The freezing point is $-36\,°C$, so the freezing-point depression is

$$T_f^° - T_f = 0.00\,°C - (-36\,°C) = 36\,°C = 36\ K$$

The colligative molality of the solution is

$$T_f^° - T_f = K_b m_c$$
$$36\ K = (1.86\ K \cdot m_c^{-1}) m_c$$
$$m_c = 19\ m_c = 19\ m$$

The boiling point of the 19-m_c solution is given by

$$T_b = T_b^° + K_b m_c$$
$$= 100.0\,°C + (0.52\ K \cdot m_c^{-1})(19\ m_c)$$
$$= 110\,°C$$

15-71   The mole fractions are calculated as follows. For methyl alcohol we have

$$X_{met} = \frac{n_{met}}{n_{met} + n_{eth} + n_{H_2O}}$$

$$= \frac{\left(\dfrac{0.305\ g}{32.04\ g \cdot mol^{-1}}\right)}{\left(\dfrac{0.305\ g}{32.04\ g \cdot mol^{-1}}\right) + \left(\dfrac{0.275\ g}{46.07\ g \cdot mol^{-1}}\right) + \left(\dfrac{10.0\ g}{18.02\ g \cdot mol^{-1}}\right)}$$

$$= \frac{0.00952\ mol}{0.00952\ mol + 0.00597\ mol + 0.5549\ mol} = 0.0167$$

For ethyl alcohol we have

$$X_{eth} = \frac{0.00597\ mol}{0.00952\ mol + 0.00597\ mol + 0.5549\ mol} = 0.0105$$

The molalities are calculated as follows

$$m_{met} = \frac{0.00952\ mol\ CH_3OH}{0.0100\ kg\ H_2O} = 0.952\ m$$

$$m_{eth} = \frac{0.00597\ mol\ CH_3CH_2OH}{0.0100\ kg\ H_2O} = 0.597\ m$$

15-73   There will be a net flow of water from the solution with the lower osmotic pressure to the solution with the higher osmotic pressure. Because $\pi = RTM_c$, the net flow of water will be from the solution with the lower value of M_c to the solution with the higher value of M_c.

(a)   Because M_c $(=0.20\ M_c)$ is the same in both cases, there is no net flow.

(b)   The values of M_c are

$$M_c \text{ [for 0.10 M Al(NO}_3)_3] = 4 \times 0.10 \text{ M} = 0.40 \text{ M}_c$$
$$M_c \text{ [for 0.20 M NaNO}_3] = 2 \times 0.20 \text{ M} = 0.40 \text{ M}_c$$

Because M_c is the same in both cases, there is no net flow.

(c)   Because $M_c = 0.30 \text{ M}_c$ for 0.10 M $CaCl_2$ and $M_c = 1.50 \text{ M}_c$ for 0.50 M $CaCl_2$, the net flow of water will be from the 0.30 M_c solution to the 1.50 M_c solution.

15-75   Consider a 100.0-mL sample of the solution. The total mass of the solution is

$$(100.0 \text{ mL})(1.101 \text{ g} \cdot \text{mL}^{-1}) = 110.1 \text{ g}$$

The mass of glycerol is

$$(110.1 \text{ g})(0.400) = 44.0 \text{ g}$$

The mass of water is

$$(110.1 \text{ g})(0.600) = 66.1 \text{ g}$$

The molarity of glycerol in the solution is

$$M = \frac{(44.0 \text{ g})\left(\dfrac{1 \text{ mol glycerol}}{92.09 \text{ g glycerol}}\right)}{(0.100 \text{ L})} = 4.78 \text{ M}$$

The molality of glycerol in the solution is

$$m = \frac{(44.0 \text{ g})\left(\dfrac{1 \text{ mol glycerol}}{92.09 \text{ g glycerol}}\right)}{(0.0661 \text{ kg H}_2\text{O})} = 7.23 \text{ m}$$

The molality at $0°C$ is the same as at $20°C$, because molality is independent of temperature.

15-77   (a)   The molality of the $K_2SO_4(aq)$ solution is

$$m = \frac{(5.00 \text{ g})\left(\dfrac{1 \text{ mol K}_2\text{SO}_4}{174.26 \text{ g K}_2\text{SO}_4}\right)}{(0.250 \text{ kg H}_2\text{O})} = 0.115 \text{ m}$$

In water K_2SO_4 dissociates into $2K^+(aq) + SO_4^{2-}(aq)$; thus the colligative molality is

$$m_c = 3 \ m = 0.345 \ m_c$$

The freezing point of the solution is

$$T_f = T_f^\circ - K_f m_c$$
$$= 0.00°C - (1.86 \text{ K}\cdot\text{m}_c^{-1})(0.345 \text{ m}_c) = -0.64°C$$

The boiling point of the solution is

$$T_b = T_b^\circ + K_b m_c$$
$$= 100.00°C + (0.52 \text{ K}\cdot\text{m}_c^{-1})(0.345 \text{ m}_c) = 100.18°C$$

(b) The compound C_2H_5OH does not dissociate in aqueous solution, so $m = m_c$ and

$$m_c = \frac{(5.00 \text{ g})\left(\dfrac{1 \text{ mol } C_2H_5OH}{46.07 \text{ g } C_2H_5OH}\right)}{(0.250 \text{ kg } H_2O)} = 0.434 \text{ m}_c$$

The freezing point of the solution is

$$T_f = T_f^\circ - K_f m_c$$
$$= 0.00 - (1.86 \text{ K}\cdot\text{m}_c^{-1})(0.434 \text{ m}_c) = -0.81°C$$

The boiling point of the solution is

$$T_b = T_b^\circ + K_b m_c$$
$$= 100.00 + (0.52 \text{ K}\cdot\text{m}_c^{-1})(0.434 \text{ m}_c) = 100.23°C$$

(c) Aniline does not dissociate in camphor, and thus $m = m_c$.

$$m_c = \frac{(1.00 \text{ g})\left(\dfrac{1 \text{ mol aniline}}{93.13 \text{ g aniline}}\right)}{(0.0500 \text{ kg camphor})} = 0.215 \text{ m}_c$$

The freezing point of the solution is

$$T_f = T_f^\circ - K_f m_c$$
$$= 179.8°C - (40.0 \text{ K}\cdot\text{m}_c^{-1})(0.215 \text{ m}_c) = 171.2°C$$

The boiling point of the solution is

$$T_b = T_b^\circ - K_b m_c$$
$$= 208.0°C + (5.95 \text{ K}\cdot\text{m}_c^{-1})(0.215 \text{ m}_c) = 209.3°C$$

15-79 The mass of carbon disulfide is

$$\text{mass} = dV = (1.261 \text{ g}\cdot\text{mL}^{-1})(100 \text{ mL}) = 126.1 \text{ g} = 0.1261 \text{ kg}$$

Let M be the formula mass of phosphorus and P_x be the chemical formula of phosphorus. The molality of the solution is

$$m = \frac{(2.74 \text{ g}) \left(\dfrac{1 \text{ mol } P_x}{M \text{ g } P_x} \right)}{0.1261 \text{ kg}} = \frac{21.74}{M} \text{ m}$$

We can find the colligative molality from the boiling point elevation of carbon disulfide.

$$m_c = \frac{T_b - T_b^\circ}{K_b} = \frac{46.71\,^\circ\text{C} - 46.30\,^\circ\text{C}}{2.34 \text{ K} \cdot m_c^{-1}} = 0.175 \text{ m}_c$$

Assume that $m = m_c$. We have the equality

$$\frac{21.74}{M} \text{ m}_c = 0.175 \text{ m}_c$$

$$M = \frac{21.74}{0.175} = 124$$

The atomic mass of phosphorus is 30.97 and

$$\frac{124}{30.97} = 4.00$$

Thus, there are four phosphorus atoms per formula unit and the molecular formula of phosphorus is P_4.

15-81   The mass of the solution prepared by dissolving 2.00 mol of NaOH in 1000 g of H_2O is

$$\text{mass}_{sol} = \text{mass}_{NaOH} + \text{mass}_{H_2O}$$

Thus,

$$\text{mass}_{sol} = (2.00 \text{ mol}) \left(\frac{40.00 \text{ g NaOH}}{1 \text{ mol NaOH}} \right) + 1000 \text{ g} = 1080 \text{ g}$$

The volume of the solution can be found from the density

$$V = \frac{1080 \text{ g}}{1.22 \text{ g} \cdot \text{mL}^{-1}} = 885 \text{ mL} = 0.885 \text{ L}$$

The molarity is

$$\text{molarity} = \frac{2.00 \text{ mol NaOH}}{0.885 \text{ L solution}} = 2.26 \text{ M}$$

15-83 The molality of the solution is

$$\text{molality} = \frac{\text{moles of solute}}{\text{kilograms of solvent}}$$

$$= \frac{(50.0 \text{ g}) \left(\dfrac{1 \text{ mol ethylene glycol}}{62.07 \text{ g ethylene glycol}} \right)}{0.0500 \text{ kg water}} = 16.1 \text{ m}$$

$$m_c = 16.1 \text{ m}_c$$

The boiling point is

$$T_b = T_b^\circ + K_b m_c = 100.00°C + (0.52 \text{ K} \cdot \text{m}_c^{-1})(16.1 \text{ m}_c) = 108.4°C$$

15-85 In a 100.0-g sample of the sulfuric acid we have 98 g of H_2SO_4 and 2.0 g of water. The volume of the 100.0 g sample is

$$V = \frac{m}{d} = \frac{100.0 \text{ g}}{1.84 \text{ g} \cdot \text{mL}^{-1}} = 54.3 \text{ mL}$$

The molarity of the solution is

$$M = \frac{n}{V} = \frac{(98 \text{ g}) \left(\dfrac{1 \text{ mol } H_2SO_4}{98.08 \text{ g } H_2SO_4} \right)}{(0.0543 \text{ L solution})} = 18 \text{ M}$$

15-87 We have from the definition of mole fraction that

$$X_2 = \frac{\text{moles of solute}}{\text{moles of solute} + \text{moles of water}}$$

From the definition of colligative molality, we have

$$m_c = \frac{\text{moles of solute}}{1000 \text{ grams of water}}$$

If we take 1000 g of water, then

$$\text{moles of solute} = m_c$$

$$\text{moles of } H_2O = (1000 \text{ g } H_2O)\left(\frac{1 \text{ mol } H_2O}{18.02 \text{ g } H_2O} \right) = 55.5 \text{ mol}$$

Thus we have that

$$X_2 = \frac{m_c}{m_c + 55.5 \text{ mol} \cdot \text{kg}^{-1}}$$

If the solution is dilute, then m_c is much smaller than $55.5 \ \text{mol} \cdot \text{kg}^{-1}$ and so we can neglect m_c in the denominator.

$$X_2 \approx \frac{m_c}{55.5 \ \text{mol} \cdot \text{kg}^{-1}}$$

If $m_c = 1.00 \ m_c$, then the mole fraction is

$$X_2 = \frac{1.00 \ m_c}{1.00 \ m_c + 55.5 \ \text{mol} \cdot \text{kg}^{-1}} = 0.0177$$

and the estimated mole fraction is

$$X_2 = \frac{1.00 \ m_c}{55.5 \ \text{mol} \cdot \text{kg}^{-1}} = 0.0180$$

The error is

$$\text{error} = \frac{0.0180 - 0.0177}{0.0177} \times 100 = 1.7\%$$

15-89   We use the van't Hoff equation, Equation (15-16), and solve for im.

$$im = \frac{T_f^\circ - T_f}{K_f} = \frac{3.74 \ \text{K}}{1.86 \ \text{K} \cdot \text{m}_c^{-1}} = 2.01 \ m_c$$

where we obtained K_f from Table 15-2. Solving for i, the number of solute particles per formula unit, we have

$$i = \frac{m_c}{m} = \frac{2.01}{1.00} = 2.01$$

E. ANSWERS TO THE SELF-TEST

1. the ratio of the number of solute particles to the number of solvent molecules, or to the colligative molality of the solution

2. moles of solute divided by the total number of moles present

3. true

4. mass of the solute divided by the mass of the solution (mass of the solute plus the mass of the solvent) times 100

5. the number of moles of solute per liter of solution

6. the number of moles of solute per kilogram of solvent

7. false

8. 1000 g (1.000 kg) of water

9. the molality times the number of solute particles per formula unit

10. false

11. decreased

12. The solute molecules decrease the rate of evaporation of the solvent relative to that of the pure solvent.

13. the mole fraction of the solvent

14. the vapor pressure of the solvent is proportional to the mole fraction of the solvent

15. $P_1 = X_1 P_1^\circ$ . . . P_1 is the vapor pressure of the solvent, X_1 is the mole fraction of the solvent, and P_1° is the equilibrium vapor pressure of the pure solvent

16. false

17. increase

18. The solute lowers the vapor pressure of the solvent, and thus the temperature must be raised to increase the vapor pressure back to 1.0 atm, where the solutions boils.

19. the colligative molality

20. true

21. false

22. decrease

23. The solute decreases the rate of crystallization of the solvent (the solute lowers the escaping tendency of the solvent).

24. the colligative molality of the solute

25. true

26. false

27. decreases the freezing point of water

28. false

29. false

30. true

31. false

32. degree of dissociation

33. pure water; NaCl solution

34. 0.1 M . . . 1.0 M

35. the pressure required to increase the escaping tendency of the solvent in a solution to a value equal to that of the pure solvent

36. RTM_c

37. false

38. true

39. true

40. of the osmotic pressure that develops in the cells (Water enters the cells, thereby expanding the cell.)

41. Osmotic pressure is a large effect, and only a relatively small protein concentration is necessary for a molecular mass determination.

42. false (Boiling-point elevation is a relatively small effect except for high solute concentrations.)

43. the interactions between unlike molecules are the same as those between like molecules

44. true

45. $P_B^\circ + X_A(P_A^\circ - P_B^\circ)$

46. false

47. positive

48. true

49. nonideal

50. distills without a change in composition

51. true

52. the solubility of a gas in a liquid is proportional to the partial pressure of the gas

53. decreases

54. doubles

16 RATES AND MECHANISMS OF CHEMICAL REACTIONS

A. OUTLINE OF CHAPTER 16

16-1 A reaction rate tells us how fast a quantity of reactant or product is changing with time.

A reaction rate can be defined as the average rate at which a product is consumed.

The reaction rate is given by

$$\frac{\text{change in concentration of the product}}{\text{elapsed time}} = \frac{\Delta\,[\text{product}]}{\Delta t}$$

The reaction rate can be defined in terms of the rate at which a reactant is consumed.

The reaction rate is given by

$$-\frac{\text{change in concentration of reactant}}{\text{elapsed time}} = -\frac{\Delta\,[\text{reactant}]}{\Delta t}$$

The rate of consumption of a reactant is related to the rate of production of a product by the balancing coefficients of the equation that describes the reaction.

The units of reaction rate are moles per liter per unit time; for example, $\text{mol} \cdot \text{L}^{-1} \cdot \text{s}^{-1}$ or $\text{M} \cdot \text{s}^{-1}$.

16-2 The rate law of a reaction can be determined by the method of initial rates.

The rate law of a reaction is the equation that gives the dependence of the reaction rate on the concentrations of one or more of the reactants.

The reaction rate constant k is the proportionality constant between the reaction rate and the concentration terms on which the rate depends.

Rate laws must be determined experimentally.

There is no necessary relationship between the balancing coefficients in a chemical equation and the order of the various species concentrations in the reaction rate law.

The initial reaction rate is the reaction rate measured over a time interval that is short enough that the reactant concentrations do not vary appreciably from their initial values.

A reaction rate is said to be first order in the concentration of a species when the rate is proportional to the first power of the concentration of that species.

If the initial concentration of a reactant is doubled and the initial rate doubles, then the rate law is first order in the concentration of that reactant.

A first-order rate law is of the following form:

$$\text{rate} = k[A]$$

The units of the rate constant for a first-order reaction are time^{-1}, for example, s^{-1}.

A second-order rate law is of the following form:

$$\text{rate} = k[A]^2 \qquad \text{or} \qquad \text{rate} = k[A][B]$$

The units of a second-order rate constant are $M^{-1} \cdot \text{time}^{-1}$, for example, $M^{-1} \cdot s^{-1}$.

If the initial concentration of a reactant is doubled and the initial rate increases by a factor of 4, then the rate law is second order in the concentration of that reactant.

If a reaction rate law is of the form

$$\text{rate} = k[A]^x[B]^y[C]^z$$

then the rate law is x order in [A], y order in [B], and z order in [C]. The overall order is $x + y + z$.

The units of the rate constant depend on the order of the reaction.

A reaction rate is said to be zero order in the concentration of a species when the rate does not depend on the concentration of that species.

A large rate constant implies a rapid reaction; a small rate constant implies a slow reaction.

16-3   A plot of ln [A] versus time gives a straight line for a first-order reaction.

If the rate law is rate $= k[A]$, then the dependence of [A] on time is given by

$$\ln [A] = \ln [A]_0 - kt \qquad\qquad (16\text{-}10)$$

where $[A]_0$ is the concentration of A at time $t = 0$.

The natural logarithms, ln, are logarithms to the base e.

Another form of Equation (16-10) is

$$\ln \frac{[A]}{[A]_0} = -kt \qquad (16\text{-}13)$$

For a first-order reaction, a plot of $\ln [A]$ versus t is linear (Figure 16-2).

If the rate law is not first order in $[A]$, then a plot of $[A]$ versus t will not be linear.

16-4   The half-life for a first-order reaction is independent of the initial concentration.

The half-life $t_{1/2}$ is the time required for the concentration of a reactant to decrease by a factor of 2 (Figure 16-3).

The half-life for a first-order reaction is related to the rate constant by

$$t_{1/2} = \frac{0.693}{k} \qquad (16\text{-}14)$$

If the half-life for a reactant is independent of the initial concentration of the reactant, then the reaction rate law is first order.

The larger the rate constant, the smaller is the half-life, which implies a rapid reaction.

The smaller the rate constant, the longer is the half-life, which implies a slow reaction.

16-5   A plot of $1/[A]$ versus time is linear for a second-order reaction.

If the rate law is rate $= k[A]^2$, then the dependence of $[A]$ on time is given by

$$\frac{1}{[A]} = \frac{1}{[A]_0} + kt \qquad (16\text{-}17)$$

where $[A]_0$ is the concentration of A at time $t = 0$.

For a second-order reaction, a plot of $1/[A]$ versus t results in a straight line (Figure 16-4).

16-6   The half-life of a second-order reaction depends upon the initial concentration.

For a second-order reaction, the time it takes for the concentration of A to decrease by one-half depends on the initial value of the concentration of A.

The half-life $t_{1/2}$ of a second-order reaction depends on the rate constant k and the initial value of the concentration of A, $[A]_0$ according to

$$t_{1/2} = \frac{1}{k[A]_0} \qquad (16\text{-}20)$$

16-7   Many reactions involve more than one step.

The rate law for an elementary process can be deduced from the stoichiometry of the reaction.

Molecules must collide before they can react.

In general a rate law cannot be deduced from the reaction stoichiometry.

An elementary process is a chemical reaction that occurs in a single step.

A series of elementary processes that add up to give the overall reaction is called a reaction mechanism.

An intermediate is a species that appears in the reaction mechanism but does not appear as a reactant or a product in the overall reaction.

The presence of an intermediate often can be detected experimentally.

16-8 Some reaction mechanisms have a rate-determining step.

The rate of an overall reaction can be no faster than the rate of the slowest step.

If one step in a reaction mechanism is much slower than any of the other steps, then that step controls the overall reaction rate.

A slow step that controls the overall rate of a reaction is called a rate-determining step.

The overall reaction rate law is given by the rate law of the rate-determining step.

16-9 Reactants must surmount an energy barrier in order to react.

The collision theory of reaction rates postulates that only the more energetic collisions that occur with the correct relative orientations of the molecules lead to a reaction.

Most collisions do not lead to reaction.

The reaction rate for the elementary process

$$A + B \longrightarrow C + D$$

is given by

$$\text{rate} = \begin{pmatrix} \text{collision} \\ \text{frequency} \end{pmatrix} \times \begin{pmatrix} \text{fraction of} \\ \text{collisions with} \\ \text{the required} \\ \text{energy} \end{pmatrix} \times \begin{pmatrix} \text{fraction of collisions} \\ \text{in which molecules} \\ \text{have the required} \\ \text{relative orientation} \end{pmatrix}$$

The greater the concentration of reactants A and B, the greater will be the number of collisions; thus, the collision frequency is proportional to [A][B].

The faster the molecules are moving, the more frequently will they collide; thus the collision frequency increases with temperature.

A plot of the number of collisions per second of the A and B molecules versus the kinetic energy of the colliding molecules is shown in Figure 16-9.

The activation energy E_a is the minimum energy necessary to achieve a reaction between the colliding molecules (Figure 16-10).

Two molecules can react only if they have the correct relative orientation (Figure 16-8).

An activation energy diagram shows the energies involved in the pathway of the reaction of A and B to give C and D (Figure 16-11).

A transition state represents the state of highest energy in passing from reactants to products (Figure 16-12).

A transition state has a very short lifetime and cannot be isolated or detected by ordinary means.

16-10 The Arrhenius equation describes the temperature dependence of a reaction rate constant.

From the collision theory of reaction rates and the kinetic theory of gases, the temperature dependence of the rate constant is given by the Arrhenius equation:

$$k = Ae^{-E_a/RT} \qquad (16\text{-}25)$$

A second form of the Arrhenius equation is

$$\ln\left(\frac{k_2}{k_1}\right) = \frac{E_a}{R}\left(\frac{T_2 - T_1}{T_1 T_2}\right) \qquad (16\text{-}27)$$

where k_2 and k_1 are the rate constants at the Kelvin temperatures T_2 and T_1, respectively; E_a is the activation energy; and R is the gas constant.

Reaction rate constants increase with increasing temperature because the activation energy is positive.

16-11 A catalyst is a substance that increases the reaction rate but is not consumed in the reaction.

A catalyst is a reaction rate facilitator that acts by providing a different and faster pathway to the products.

A catalyzed reaction pathway has a lower activation energy than the uncatalyzed reaction (Figure 16-13).

A catalyst increases the rates of both the forward and reverse reactions.

A catalyst does not affect the final amounts of reactants and products.

The rate law for a catalyzed reaction may be different from that of the uncatalyzed reaction.

A homogeneous catalyst is in the same phase as the reaction mixture.

A heterogeneous catalyst is in a different phase than the reaction mixture.

Platinum metal is a widely used heterogeneous catalyst (Figure 16-15).

Enzymes are proteins that catalyze reactions in living systems.

The lock-and-key theory of enzyme activity involves the postulate that the shape of the enzyme at the binding site allows only the substrate to bind to the enzyme (Figure 16-17).

The substrate is the substance that is reacting.

B. SELF-TEST

Questions 1 through 6 refer to the reaction

$$2H_2O_2(aq) \rightleftharpoons 2H_2O(l) + O_2(g)$$

1. The rate at which oxygen is produced is defind as _____
_____ .

2. The reaction rate for the decomposition of hydrogen peroxide is always positive. *True/False*

3. The rate of the decomposition of hydrogen peroxide can be determined by measuring the increase in pressure due to oxygen. *True/False*

4. The rate of the decomposition of hydrogen peroxide can be determined by measuring the concentration of H_2O_2 at various times during the reaction. *True/False*

5. The rate for the decomposition of hydrogen peroxide can be expressed in terms of the concentration of _____ or of _____ .

6. From the stoichiometry of the decomposition of hydrogen peroxide, the reaction rate law must be rate = $k[H_2O_2]^2$. *True/False*

7. Reaction rates can be expressed in the units $M \cdot min^{-1}$. *True/False*

8. The rate law of a reaction is the equation that gives the dependence of the rate on _____ .

9. The method of initial rates is used to determine _____ .

10. If a reaction rate law is first order in reactant A, then the initial reaction rate (*remains the same/doubles/quadruples*) when the concentration of A is doubled and the concentrations of all other reactants remain the same.

11. The rate law of a first-order reaction is rate = _____ .

12. The units of the rate constant for a first-order reaction are _____ .

13. Given that the decomposition of hydrogen peroxide is a first-order reaction in H_2O_2, the rate law is rate = _____ .

14. A large rate constant implies a (*fast/slow*) reaction.

15. If a reaction rate law is second order in reactant B, then the initial reaction rate (*remains the same/doubles/quadruples*) when the concentration of B is doubled and the concentrations of all other reactants remain the same.

16. The rate law of a second-order reaction is rate = _____ or rate = _____.

17. The units of the rate constant for a second-order reaction are _____.

18. The reaction $H^+(aq) + OH^-(aq) \rightleftharpoons H_2O(l)$ is first order in both $[H^+]$ and $[OH^-]$. The rate law is rate = _____.

19. If a reaction is first order, then the dependence of the concentration of A on time is given by $\ln[A]$ = _____.

20. The decomposition of H_2O_2 is a first-order reaction in H_2O_2. A plot of _____ _____ versus time is a straight line.

21. If the plot of the natural logarithm of the concentration of a reactant A versus time is not a straight line, then the reaction is not a first-order reaction. *True/False*

22. The half-life of a reactant is defined as _____ _____.

23. The half-life for a first-order reaction is (*independent of, dependent on*) the initial concentration of the reactant.

24. If the rate law for the decomposition of hydrogen peroxide is rate = $k[H_2O_2]$, then the time it takes for hydrogen peroxide to decrease from 0.500 M to 0.250 M is the same as the time to decrease from 0.250 M to 0.125 M. *True/False*

25. The half-life of a first-order reaction is related to the rate constant for the reaction. *True/False*

26. The larger the rate constant, the (*smaller/larger*) will be the half-life of a reaction.

27. The larger the half-life, the (*faster/slower*) will be the reaction.

28. The rate law for the reaction

$$NO_2(g) + CO(g) \rightleftharpoons NO(g) + CO_2(g)$$

is rate = $k[NO_2]^2$. The time it takes for $[NO_2]$ to decrease from 0.500 M to 0.250 M is the same as the time to decrease from 0.250 M to 0.125 M. *True/False*

29. For a second-order reaction, a plot of _____ versus time is a straight line.

30. An elementary process is a process that occurs in one or more steps. *True/False*

31. The rate law for an elementary process can be written from the stoichiometry of the reaction. *True/False*

32. Most reactions are elementary processes. *True/False*

33. A series of elementary processes that add up to give the overall reaction stoichiometry is called the reaction _____.

34. A reaction mechanism may involve intermediate species that do not appear in the overall reaction. *True/False*

35. The overall reaction rate may be controlled by a rate-determining step, which is the fastest step in the reaction mechanism. *True/False*

36. The sum of the elementary processes for a reaction need not add up to the overall chemical equation. *True/False*

37. All collisions between reactant molecules lead to reaction. *True/False*

38. The collision frequency depends upon the concentrations of the reactants. *True/False*

39. The collision frequency depends upon the temperature of the reactants. *True/False*

40. The activation energy is the minimum amount of kinetic energy that reactant molecules must have in order to react. *True/False*

41. An activation energy diagram shows the energy barrier that reactants must overcome in order to form the products in a reaction. *True/False*

42. A transition state represents the state of (*lowest/highest*) energy in passing from reactants to products.

43. A transition state is (*stable/unstable*).

44. A transition state can be isolated. *True/False*

45. Rate constants increase as the temperature increases. *True/False*

46. The temperature dependence of a rate constant is given by the _____ equation.

47. The equation that gives the relationship of the rate constant at two temperatures is

_____ = _____.

48. A catalyst acts by providing a new reaction pathway with a (*lower/higher*) activation energy.

49. A catalyst increases the rate of production of products without affecting the rate of the reverse reaction. *True/False*

50. Platinum metal is used as a catalyst for the reaction

$$C_2H_4(g) + H_2(g) \rightleftharpoons C_2H_6(g)$$

In this reaction, platinum is a (*homogeneous/heterogeneous*) catalyst.

51. The rate law for the catalyzed reaction must be the same as that of the uncatalyzed reaction. *True/False*

52. Enzymes are proteins that enable reactions to take place in living systems. *True/False*

53. The shape of an enzyme is important to its activity as a catalyst. *True/False*

C. *CALCULATIONS YOU SHOULD KNOW HOW TO DO*

1. Express the reaction rate in terms of changes in the reactant or product concentrations. See Examples 16-1 and 16-2 and Problems 16-1 and 16-2.

2. Calculate the reaction rate from the changes in the concentrations of reactants or products. See Example 16-3 and Problems 16-5 through 16-10.

3. Determine the reaction rate law by using the method of initial rates. See Examples 16-5 and 16-6 and Problems 16-11 through 16-20.

4. Determine the amount of reactant for a first-order reaction remaining after a given amount of time using Equation (16-13). See Example 16-7 and Problems 16-21 through 16-28.

5. Calculate the rate constant for a second-order reaction using the slope of the line obtained by plotting $1/[A]$ versus time or by using Equation (16-17). Calculate the amount of reactant remaining after the reaction has taken place for a given length of time using Equation (16-17). See Examples 16-9 and 16-10 and Problems 16-29 through 16-34.

6. Determine the rate law from the reaction mechanism. See Example 16-13 and Problems 16-37 and 16-38.

7. Use the Arrhenius equation, Equation (16-27), to find the rate constant at some other temperature or to find the activation energy given the rate constants at two different temperatures. See Examples 16-15, and Problems 16-39 through 16-44.

Natural Logarithms

Common logarithms are logarithms to the base 10 and are designated by log. Natural logarithms are logarithms to the base e and are designated by ln. Natural logarithms arise naturally in calculus. The number e is an irrational number whose value is

$$e = 2.718281 \cdots \tag{1}$$

By definition, if

$$x = \ln y \tag{2}$$

then

$$y = e^x \tag{3}$$

Equation (3) is called the inverse of Equation (2).

Even if you have not taken a course in calculus, you should have little trouble with calculations involving natural logarithms. The functions ln y and e^x occur on all hand calculators nowadays. For example, by entering "2" and pushing the "e^x" key, you obtain

$$e^2 = 7.389056 \cdots$$

By entering "2" followed by the "change sign" key and then the "e^x" key, you obtain

$$e^{-2} = 0.135335 \cdots$$

Note that $e^{-2} = 1/e^2$, as you might expect. In fact, the mathematical properties of e^x and natural logarithms are similar to those of 10^x and log y. For example,

$$\ln (ab) = \ln a + \ln b \tag{4}$$

$$\ln \left(\frac{a}{b}\right) = \ln a - \ln b \tag{5}$$

$$\ln (a^n) = n \ln a \tag{6}$$

$$\ln (\sqrt[n]{a}) = \ln (a^{1/n}) = \left(\frac{1}{n}\right) \ln a \tag{7}$$

and

$$e^{-x} = \frac{1}{e^x} \tag{8}$$

$$e^a e^b = e^{a+b} \tag{9}$$

$$\frac{e^a}{e^b} = e^{a-b} \tag{10}$$

$$(e^x)^n = e^{nx} \tag{11}$$

Furthermore, from Equation (10) with $a = b$, we get

$$e^0 = \frac{e^a}{e^a} = 1$$

and so

$$\ln(1) = \ln e^0 = (0) \ln e = 0$$

Furthermore, from Equation (5) with $a = 1$, we find that

$$\ln y > 0 \qquad \text{if } y > 1$$
$$\ln y < 0 \qquad \text{if } y < 1$$

just as in the case of common logarithms.

Exercise:

Using your hand calculator, determine the following quantities:

(a) $e^{0.37}$ (c) $\ln(4.07)$

(b) $e^{-6.02}$ (d) $\ln(0.00965)$

Answers:

(a)   1.45   (c)   1.40

(b) 2.43×10^{-3} (d) -4.64

Exercise:

Given that (a) $\ln y = 3.065$ and (b) $\ln y = -0.605$, determine y.

Answer:

(a) $y = e^{3.065} = 21.43$ and (b) $y = e^{-0.605} = 0.546$

D. SOLUTIONS TO THE ODD-NUMBERED PROBLEMS

16-1   There are two moles of NOCl in the balanced equation; therefore, the rate in terms of NOCl is given as

$$\text{rate} = -\frac{1}{2} \frac{\Delta[\text{NOCl}]}{\Delta t}$$

Note that a minus sign is placed in front of the expression to ensure that the rate is a positive quantity. There are two moles of NO produced in the balanced equation; thus, we have that

$$\text{rate} = \frac{1}{2} \frac{\Delta[\text{NO}]}{\Delta t}$$

Note that the rate will be positive because [NO] is increasing. There is one mole of Cl_2 produced in the balanced equation; thus,

$$\text{rate} = \frac{\Delta[Cl_2]}{\Delta t}$$

16-3 (a) Use a spectrophotometer to measure the decrease in the yellow color due to $I_2(aq)$ as a function of time.

(b) Measure the total pressure in the reaction vessel as a function of time. Because there are more moles of gaseous products than reactants, the total pressure increases as the reaction proceeds.

16-5 We have

$$\text{rate} = (3.0 \times 10^6 \text{ M}^{-1} \cdot \text{s}^{-1})[O_3][\text{NO}]$$

Thus, the rate is

$$\begin{aligned} \text{rate} &= (3.0 \times 10^6 \text{ M}^{-1} \cdot \text{s}^{-1})(6.0 \times 10^{-4} \text{ M})(4.0 \times 10^{-5} \text{ M}) \\ &= 7.2 \times 10^{-2} \text{ M} \cdot \text{s}^{-1} \end{aligned}$$

16-7 The rate of production of O_2 is three-halves as great as the rate of loss of O_3 because three O_2 molecules are produced by the consumption of two O_3 molecules. Thus,

$$\begin{aligned} \frac{\Delta P_{O_2}}{\Delta t} &= -\left(\frac{3}{2}\right)\left(\frac{\Delta P_{O_3}}{\Delta t}\right) = \left(\frac{3}{2}\right)(5.0 \times 10^{-4} \text{ atm} \cdot \text{s}^{-1}) \\ &= 7.5 \times 10^{-4} \text{ atm} \cdot \text{s}^{-1} \end{aligned}$$

16-9 The average reaction rate is given by

$$\text{rate} = -\left(\frac{1}{2}\right)\left(\frac{\Delta[N_2O_5]}{\Delta t}\right) = -\left(\frac{1}{2}\right)\left(\frac{[N_2O_5]_2 - [N_2O_5]_1}{t_2 - t_1}\right)$$

where the factor of $\frac{1}{2}$ arises because of the balancing coefficient of 2 in front of N_2O_5. Application of this equation to the data yields the following results. For the time interval from 0 to 175 s we have

$$\text{rate} = -\left(\frac{1}{2}\right)\left(\frac{1.32 \text{ M} - 1.48 \text{ M}}{175 \text{ s} - 0}\right) = 4.6 \times 10^{-4} \text{ M} \cdot \text{s}^{-1}$$

For the time interval 175 s to 506 s we have

$$\text{rate} = -\left(\frac{1}{2}\right)\left(\frac{1.07 \text{ M} - 1.32 \text{ M}}{506 \text{ s} - 175 \text{ s}}\right) = 3.8 \times 10^{-4} \text{ M} \cdot \text{s}^{-1}$$

For the time interval 506 s to 845 s we have

$$\text{rate} = -\left(\frac{1}{2}\right)\left(\frac{0.87 \text{ M} - 1.07 \text{ M}}{845 \text{ s} - 506 \text{ s}}\right) = 2.9 \times 10^{-4} \text{ M} \cdot \text{s}^{-1}$$

Note that the reaction rate decreases as the reaction proceeds.

16-11   The rate of the reaction doubles as $[SO_2Cl_2]_0$ is doubled. Assuming that the rate law does not vary with time, we deduce that the order of the reaction is first order. The first-order reaction rate law is

$$\text{rate} = k[SO_2Cl_2]$$

16-13   Using Runs 1 and 2, we write

$$\frac{\text{rate}_2}{\text{rate}_1} = \frac{4.00 \times 10^{-30} \text{ M} \cdot \text{s}^{-1}}{2.40 \times 10^{-30} \text{ M} \cdot \text{s}^{-1}} = \frac{(0.55 \text{ M})^x}{(0.33 \text{ M})^x} = \left(\frac{0.55}{0.33}\right)^x$$

or

$$1.67 = (1.67)^x$$

Therefore, $x = 1$ and the reaction is first order.
   We evaluate the rate constant using Run 1 (arbitrarily):

$$k = \frac{\text{rate}_0}{[C_2H_5Cl]_0} = \frac{2.40 \times 10^{-30} \text{ M} \cdot \text{s}^{-1}}{0.33 \text{ M}} = 7.3 \times 10^{-30} \text{ s}^{-1}$$

The rate law is

$$\text{rate} = (7.3 \times 10^{-30} \text{ s}^{-1})[C_2H_5Cl]$$

We could use any pair of data to determine the rate law.

16-15   Using Runs 1 and 3 (arbitrarily), we write

$$\frac{\text{rate}_3}{\text{rate}_1} = \frac{1.18 \times 10^{-5} \text{ M} \cdot \text{s}^{-1}}{1.75 \times 10^{-6} \text{ M} \cdot \text{s}^{-1}} = \frac{(0.65 \text{ M})^x}{(0.25 \text{ M})^x} = \left(\frac{0.65}{0.25}\right)^x$$

or

$$6.74 = (2.6)^x$$

Therefore, $x = 2$ and the reaction is second order.

Let's use Run 2 to evaluate the rate constant:

$$k = \frac{\text{rate}_0}{[\text{NOCl}]_0^2} = \frac{4.94 \times 10^{-6} \text{ M} \cdot \text{s}^{-1}}{(0.42 \text{ M})^2} = 2.8 \times 10^{-5} \text{ M}^{-1} \cdot \text{s}^{-1}$$

The rate law is

$$\text{rate} = (2.8 \times 10^{-5} \text{ M}^{-1} \cdot \text{s}^{-1})[\text{NOCl}]^2$$

16-17 When $[\text{Cr}(\text{H}_2\text{O})_6^{3+}]_0$ is increased by a factor of 10 and $[\text{SCN}^-]_0$ remains constant, the initial rate increases by a factor of 10. Thus, the rate is first order in $[\text{Cr}(\text{H}_2\text{O})_6^{3+}]$. When $[\text{SCN}^-]$ is increased by a factor of $0.5/0.2 = 2.5$ and $[\text{Cr}(\text{H}_2\text{O})_6^{3+}]_0$ remains the same, the initial rate increases by a factor of 2.5 ($1.5 \times 10^{-9}/6.0 \times 10^{-10}$). Assuming that the rate law does not vary with time, we deduce that the rate is first order in $[\text{SCN}^-]$. The rate law is

$$\text{rate} = k[\text{Cr}(\text{H}_2\text{O})_6^{3+}][\text{SCN}^-]$$

We can calculate the rate constant by using the data from any of the runs. The value of the rate constant is given by

$$k = \frac{\text{rate}}{[\text{Cr}(\text{H}_2\text{O})_6^{3+}][\text{SCN}^-]}$$

$$k = \frac{2.0 \times 10^{-11} \text{ M} \cdot \text{s}^{-1}}{(1.0 \times 10^{-4} \text{ M})(0.10 \text{ M})} = 2.0 \times 10^{-6} \text{ M}^{-1} \cdot \text{s}^{-1}$$

$$k = \frac{2.0 \times 10^{-10} \text{ M} \cdot \text{s}^{-1}}{(1.0 \times 10^{-3} \text{ M})(0.10 \text{ M})} = 2.0 \times 10^{-6} \text{ M}^{-1} \cdot \text{s}^{-1}$$

$$k = \frac{6.0 \times 10^{-10} \text{ M} \cdot \text{s}^{-1}}{(1.5 \times 10^{-3} \text{ M})(0.20 \text{ M})} = 2.0 \times 10^{-6} \text{ M}^{-1} \cdot \text{s}^{-1}$$

$$k = \frac{1.5 \times 10^{-9} \text{ M} \cdot \text{s}^{-1}}{(1.5 \times 10^{-3} \text{ M})(0.50 \text{ M})} = 2.0 \times 10^{-6} \text{ M}^{-1} \cdot \text{s}^{-1}$$

16-19 Using Runs 1 and 2, we have

$$\frac{\text{rate}_2}{\text{rate}_1} = \frac{4.40 \times 10^4 \text{ M} \cdot \text{s}^{-1}}{2.61 \times 10^4 \text{ M} \cdot \text{s}^{-1}} = \frac{(1.10 \text{ M})^x}{(0.65 \text{ M})^x} = \left(\frac{1.10}{0.65}\right)^x$$

$$1.699 = (1.69)^x$$

Therefore, $x = 1$ and the reaction is first order in $[\text{NO}_2]$.
  Using Runs 2 and 3, we have

$$\frac{1.32 \times 10^5 \text{ M} \cdot \text{s}^{-1}}{4.40 \times 10^4 \text{ M} \cdot \text{s}^{-1}} = \frac{(1.70 \text{ M})(1.55 \text{ M})^x}{(1.10 \text{ M})(0.80 \text{ M})^x} = 1.55 \left(\frac{1.55}{0.80}\right)^x$$

or

$$1.94 = (1.94)^x$$

Therefore, $x = 1$ and the reaction is first order in $[O_3]$.
   We use Run 2 (arbitrarily) to evaluate k:

$$k = \frac{4.40 \times 10^4 \text{ M} \cdot \text{s}^{-1}}{(1.10 \text{ M})(0.80 \text{ M})} = 5.0 \times 10^4 \text{ M}^{-1} \cdot \text{s}^{-1}$$

The rate law is

$$\text{rate} = (5.0 \times 10^4 \text{ M}^{-1} \cdot \text{s}^{-1})[NO_2][O_3]$$

16-21   The half-life of the reaction is given by

$$t_{1/2} = \frac{0.693}{k}$$

$$= \frac{0.693}{2.2 \times 10^{-5} \text{ s}^{-1}} = 3.15 \times 10^4 \text{ s}$$

The number of half-lives in 5.0 hours is

$$\frac{(5.0 \text{ h})\left(\dfrac{60 \text{ min}}{1 \text{ h}}\right)\left(\dfrac{60 \text{ s}}{1 \text{ min}}\right)}{3.15 \times 10^4 \text{ s/half-life}} = 0.571 \text{ half-life}$$

The fraction of SO_2Cl_2 remaining after 0.571 half-lives is given by

$$\frac{[A]}{[A]_0} = \left(\frac{1}{2}\right)^n$$

Thus,

$$\text{fraction remaining} = \frac{[SO_2Cl_2]}{[SO_2Cl_2]_0} = \left(\frac{1}{2}\right)^{0.571}$$
$$= 0.67$$

An alternative solution is to use the equation

$$\ln \frac{[A]}{[A]_0} = -kt$$

where $[A]/[A]_0$ is the fraction of SO_2Cl_2 remaining. Thus,

$$\ln \frac{[A]}{[A]_0} = -(2.2 \times 10^{-5} \text{ s}^{-1})\left(\frac{60 \text{ s}}{1 \text{ min}}\right)\left(\frac{60 \text{ min}}{1 \text{ h}}\right)(5.0 \text{ h})$$
$$= -0.396$$

Taking antilogarithms, we have

$$\text{fraction remaining} = \frac{[A]}{[A]_0} = e^{-0.396} = 0.67$$

16-23 The fraction of a reactant remaining after time t is given by

$$\ln \frac{[A]}{[A]_0} = -kt$$

Substituting the values of k and t, we have

$$\ln \frac{[A]}{[A]_0} = -(4.0 \times 10^{-4}\ \text{s}^{-1}) \left(\frac{3600\ \text{s}}{1\ \text{h}}\right)(1\ \text{h})$$
$$= -1.44$$

Taking antilogarithms, we have

$$\text{fraction remaining} = \frac{[A]}{[A]_0} = e^{-1.44} = 0.24$$

16-25 If the reaction rate is first order, then a plot of $\ln [S_2O_8^{2-}]$ versus time is a straight line.

t/min	$\ln [S_2O_8^{2-}]/\text{M}$
0	-2.30
17	-3.00
34	-3.69
51	-4.42

A plot of $\ln [S_2O_8^{2-}]$ versus time is a straight line, as shown in the figure on page 346. Therefore, the reaction is first order:

$$\text{rate} = k[S_2O_8^{2-}]$$

The equation for the straight line is

$$\ln [S_2O_8^{2-}] = \ln [S_2O_8^{2-}]_0 - kt$$

We can use the data at any time point to calculate k. At time $t = 17$ min, we have

$$-3.00 = -2.30 - k(17\ \text{min})$$

$$k = \frac{0.70}{17\ \text{min}} = 0.041\ \text{min}^{-1}$$

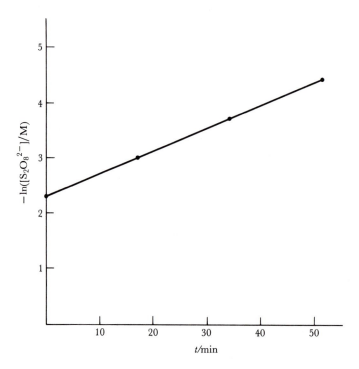

An alternative method to calculate k is as follows:

$[S_2O_8^{2-}]$/M	t/min	$t_{1/2}$/min
0.100	0	
0.050	17	17
0.025	34	17
0.012	51	17

The $[S_2O_8^{2-}]$ is reduced by a factor of 2 every 17 min, so the half-life is 17 min. We can calculate k from the expression

$$t_{1/2} = \frac{0.693}{k}$$

$$k = \frac{0.693}{t_{1/2}} = \frac{0.693}{17 \text{ min}} = 0.041 \text{ min}^{-1}$$

16-27 We first make a table of the natural logarithm of the concentration of C_2H_4O as a function of time.

t/min	$\ln\,([C_2H_4O]/M)$
0	-2.453
50	-3.068
72	-3.338
93	-3.597
130	-4.051

A plot of the logarithm of $[C_2H_4O]$ versus time is a straight line as shown in the figure below; therefore, the reaction is first order. We can determine the value of the rate constant from the slope of the line, or we can use Equation (16-13) and the value of the concentration of C_2H_4O at time equals zero and that at any other time.

$$\ln\frac{[C_2H_4O]}{[C_2H_4O]_0} = -kt$$

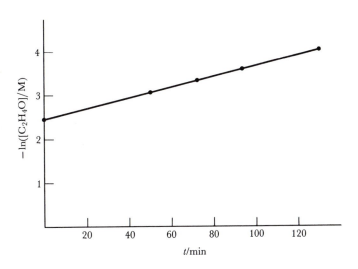

For example, substituting in the values for the first ($t = 0$ min) and third ($t = 42$ min) data points, we have that

$$\ln \frac{0.0355 \text{ M}}{0.0860 \text{ M}} = -(72 \text{ min} - 0 \text{ min})k$$

$$-0.885 = -(72 \text{ min})k$$

$$k = \frac{0.885}{72 \text{ min}} = 0.012 \text{ min}^{-1}$$

16-29 To find the concentration of CH_3CHO 5.0 min later, we use the second-order rate equation, Equation (16-17).

$$\frac{1}{[CH_3 \text{ CHO}]} - \frac{1}{[CH_3 \text{ CHO}]_0} = kt$$

Substituting in the values of $[CH_3 \text{ CHO}]_0$, k, and t yields

$$\frac{1}{[CH_3 \text{ CHO}]} - \frac{1}{0.012 \text{ M}} = (0.105 \text{ M}^{-1} \cdot \text{s}^{-1})(5.0 \text{ min}) \left(\frac{60 \text{ s}}{1 \text{ min}} \right)$$

$$\frac{1}{[CH_3 \text{ CHO}]} - 83.3 \text{ M}^{-1} = 31.5 \text{ M}^{-1}$$

$$\frac{1}{[CH_3 \text{ CHO}]} = (31.5 + 83.3)\text{M}^{-1} = 114.8 \text{ M}^{-1}$$

Solving for $[CH_3 \text{ CHO}]$ yields

$$[CH_3 \text{ CHO}] = \frac{1}{114.8 \text{ M}^{-1}} = 8.7 \times 10^{-3} \text{ M}$$

16-31 We first set up a table of $1/[NO_2]$ as a function of time.

t/s	$1/[NO_2]/\text{M}^{-1}$
0	12.0
4.2	15.0
7.9	17.6
11.4	20.1
15.0	22.7

A plot of $1/[NO_2]$ versus time is a straight line, as shown in the figure below. We can calculate the value of the rate constant from the slope of the line or by using Equation (16-17) and the value of the concentration of C_2H_4O at time equals zero and that at any other time.

$$\frac{1}{[NO_2]} - \frac{1}{[NO_2]_0} = kt$$

For example, substituting in the values for the first ($t = 0$ s) and third ($t = 7.9$ s) data points yields

$$17.6 \text{ M}^{-1} - 12.0 \text{ M}^{-1} = k(7.9 \text{ s} - 0 \text{ s})$$

Solving for k, we have that

$$k = \frac{5.6 \text{ M}^{-1}}{7.9 \text{ s}} = 0.71 \text{ M}^{-1} \cdot \text{s}^{-1}$$

16-33 The value of the half-life of a second-order reaction is given by

$$t_{1/2} = \frac{1}{k[A]_0}$$

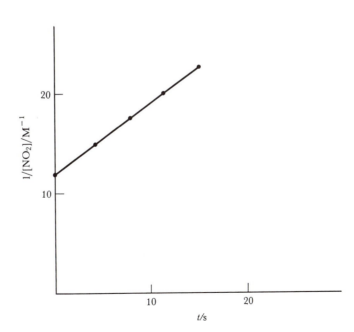

Thus, the rate constant is equal to

$$k = \frac{1}{t_{1/2}[N_2O]_0}$$

The half-life of the reaction is 4500 s when $[N_2O]_0$ is 2.0×10^{-2} M; therefore,

$$k = \frac{1}{(4500 \text{ s})(2.0 \times 10^{-2} \text{ M})} = 1.1 \times 10^{-2} \text{ M}^{-1} \cdot \text{s}^{-1}$$

16-35 (a) rate $= k[N_2O][O]$

(b) rate $= k[O][O_3]$

(c) rate $= k[ClCO][Cl_2]$

16-37 The rate law is determined by the slow elementary step. The rate law for the slow step is

$$\text{rate} = k[CO_2][OH^-]$$

The experimental rate law is

$$\text{rate} = k[CO_2][OH^-]$$

Thus, the mechanism is consistent with the rate equation, because the mechanism leads to the same rate law as is found experimentally.

16-39 The Arrhenius equation is

$$\ln\left(\frac{k_2}{k_1}\right) = \frac{E_a}{R}\left(\frac{T_2 - T_1}{T_1 T_2}\right)$$

Thus, substituting in the values of k_1, k_2, T_1, T_2, and R, we have

$$\ln\left(\frac{9.15 \times 10^{-4} \text{ s}^{-1}}{2.35 \times 10^{-4} \text{ s}^{-1}}\right) = \frac{E_a}{(8.314 \text{ J} \cdot \text{mol}^{-1} \cdot \text{K}^{-1})}\left(\frac{(303 \text{ K} - 293 \text{ K})}{(303 \text{ K})(293 \text{ K})}\right)$$

$$1.36 = E_a(1.35 \times 10^{-5} \text{ J}^{-1} \cdot \text{mol})$$

Solving for E_a, we have

$$E_a = \frac{1.36}{1.35 \times 10^{-5} \text{ J}^{-1} \cdot \text{mol}} = 1.0 \times 10^5 \text{ J} \cdot \text{mol}^{-1}$$

$$= 100 \text{ kJ} \cdot \text{mol}^{-1}$$

16-41 The Arrhenius equation is

$$\ln\left(\frac{k_2}{k_1}\right) = \frac{E_a}{R}\left(\frac{T_2 - T_1}{T_1 T_2}\right)$$

Thus, substituting in the values of k_1, T_1, T_2, E_a, and R, we have

$$\ln \left(\frac{k_2}{6.07 \times 10^{-8} \text{ s}^{-1}} \right) = \frac{(262 \times 10^3 \text{ J} \cdot \text{mol}^{-1})}{(8.314 \text{ J} \cdot \text{mol}^{-1} \text{K}^{-1})} \left(\frac{(800 \text{ K} - 600 \text{ K})}{(600 \text{ K})(800 \text{ K})} \right)$$
$$= 13.13$$

Taking the antilogarithm of both sides, we have

$$\frac{k_2}{6.07 \times 10^{-8} \text{ s}^{-1}} = e^{13.13} = 5.04 \times 10^5$$

or solving for k_2,

$$k_2 = (5.04 \times 10^5)(6.07 \times 10^{-8} \text{ s}^{-1})$$
$$= 3.06 \times 10^{-2} \text{ s}^{-1}$$

16-43 The Arrhenius equation is

$$\ln \left(\frac{k_2}{k_1} \right) = \frac{E_a}{R} \left(\frac{T_2 - T_1}{T_1 T_2} \right)$$

The half-life, $t_{1/2}$, is related to the rate constant of a first-order process by

$$t_{1/2} = \frac{0.693}{k}$$

Thus, the Arrhenius equation in terms of $t_{1/2}$ is

$$\ln \left(\frac{t_{1/2,1}}{t_{1/2,2}} \right) = \frac{E_a}{R} \left(\frac{T_2 - T_1}{T_1 T_2} \right)$$

Thus,

$$\ln \left(\frac{4.5 \text{ h}}{t_{1/2,2}} \right) = \frac{(586 \times 10^3 \text{ J} \cdot \text{mol}^{-1})(310.2 \text{ K} - 302.8 \text{ K})}{(8.314 \text{ J} \cdot \text{mol}^{-1} \cdot \text{K}^{-1})(310.2 \text{ K})(302.8 \text{ K})}$$
$$= 5.55$$

Taking the antilogarithm of both sides, we have

$$\frac{4.5 \text{ h}}{t_{1/2,2}} = e^{5.55} = 260$$

or

$$t_{1/2} = \frac{4.5 \text{ h}}{260} = 0.017 \text{ h at } 37°\text{C}$$

16-45 A catalyst cannot affect the final concentrations in a chemical reaction. A catalyst provides an alternate pathway from the reactants to the products and from the products to the reactants. Thus, the rates of the forward and reverse reactions are equally affected, so that the final concentrations are unaffected.

16-47 (a) The catalysts are $H^+(aq)$ and $Br^-(aq)$. Although neither appears in the overall reaction, both appear in the rate law.
　　　(b) The overall order of the reaction is third order.
　　　(c) The reaction is first order in $[H_2O_2]$:

$$\text{rate} = k[H^+]_0[Br^-]_0[H_2O_2]$$

where the zero subscripts indicate that the concentrations of $H^+(aq)$ and $Br^-(aq)$ are constant. The effective first-order rate constant is

$$k_{eff} = k[H^+]_0[Br^-]_0$$
$$= (1.0 \times 10^3 \text{ M}^{-2} \cdot \text{s}^{-1})(1.00 \times 10^{-3} \text{ M})(1.00 \times 10^{-3} \text{ M})$$
$$= (1.0 \times 10^{-3} \text{ s}^{-1})$$

and the half-life of the $H_2O_2(aq)$ is

$$t_{1/2} = \frac{0.693}{1.0 \times 10^{-3} \text{ M}} = 693 \text{ s}$$

A plot of $[H_2O_2]$ versus time looks like Figure 16-3 for $[N_2O_5]$, but with a half-life of 693 s.

16-49 Run the reaction with all initial concentrations the same in vessels with different amounts of reaction vessel wall area in contact with the reaction mixture. If the reaction rate increases as the wall surface in contact with the reaction mixture increases, then the reaction is catalyzed by the wall.

16-51 The mechanism outlined in Figure 16-15 can be written as

$$O_2(g) \longrightarrow O_2(surface) \qquad \text{fast}$$
$$O_2(surface) \longrightarrow 2O(surface) \qquad \text{slow}$$
$$SO_2(g) + O(surface) \longrightarrow SO_3(g) \qquad \text{fast}$$

Except at very low pressure, the platinum surface will be completely covered with oxygen molecules. Thus, the value of $[O_2(surface)]$ will be essentially constant and the rate-determining step will be independent of the pressure of $O_2(g)$. Furthermore, the oxygen atoms react rapidly with SO_2 molecules except when the number of SO_2 molecules present is very small (when the pressure of SO_2 is very low).

16-53 The total pressure in the reaction vessel is

$$P_{total} = P_{CO} + P_{CO_2}$$

If P°_{CO} is the initial pressure of CO, then the partial pressure of CO is equal to

$$P_{CO} = P^\circ_{CO} - 2P_{CO_2}$$

because each mole of CO_2 produced consumes 2 mol of CO. Thus, substituting $P_{CO_2} = P_{total} - P_{CO}$ into this equation yields

$$P_{CO} = P_{CO}^\circ - 2(P_{total} - P_{CO})$$

Solving for P_{CO} yields

$$P_{CO} = 2P_{total} - P_{CO}^\circ = 2P_{total} - 250 \text{ torr}$$

Using the data given in the problem and the preceding equation, we obtain

t/s	$P_{CO}/torr$
0	250
398	226
1002	198
1801	170

The average reaction rates $(\Delta P_{CO}/\Delta t)$ for each of these three time intervals are as follows. For 0 to 398 s,

$$\text{rate} = -\frac{\Delta P_{CO}}{\Delta t} = -\frac{226 \text{ torr} - 250 \text{ torr}}{398 \text{ s} - 0 \text{ s}} = 0.0603 \text{ torr} \cdot \text{s}^{-1}$$

For 398 to 1002 s,

$$\text{rate} = -\frac{\Delta P_{CO}}{\Delta t} = -\frac{198 \text{ torr} - 226 \text{ torr}}{1002 \text{ s} - 398 \text{ s}} = 0.0464 \text{ torr} \cdot \text{s}^{-1}$$

For 1002 to 1801 s,

$$\text{rate} = -\frac{\Delta P_{CO}}{\Delta t} = -\frac{170 \text{ torr} - 198 \text{ torr}}{1801 \text{ s} - 1002 \text{ s}} = 0.0350 \text{ torr} \cdot \text{s}^{-1}$$

The reaction rate law is

$$\text{rate} = kP_{CO}^x$$

We can find x by taking the ratio of the rates over any two time intervals.

$$\frac{\text{rate}_2}{\text{rate}_1} = \left(\frac{P_{CO_2}}{P_{CO_1}}\right)^x$$

We now substitute in the values for the average rates and the average values of P_{CO} over each time interval and try $x = 1$, $x = 2$, and $x = 3$ to find the value of x that satisfies the above equation. We find that x is equal to 2. Thus, the reaction is second order.

$$\text{rate} = kP_{CO}^2$$

We can calculate the value of k by taking the value of P_{CO} as the average value over the time interval for which we know the average rate. Thus, we have that

$$k = \frac{0.0603 \text{ torr} \cdot \text{s}^{-1}}{\left(\dfrac{250 \text{ torr} + 226 \text{ torr}}{2}\right)^2} = 1.1 \times 10^{-6} \text{ torr}^{-1} \cdot \text{s}^{-1}$$

$$= \frac{0.0464 \text{ torr} \cdot \text{s}^{-1}}{\left(\dfrac{226 \text{ torr} + 198 \text{ torr}}{2}\right)^2} = 1.0 \times 10^{-6} \text{ torr}^{-1} \cdot \text{s}^{-1}$$

$$= \frac{0.0350 \text{ torr} \cdot \text{s}^{-1}}{\left(\dfrac{198 \text{ torr} + 170 \text{ torr}}{2}\right)^2} = 1.0 \times 10^{-6} \text{ torr}^{-1} \cdot \text{s}^{-1}$$

We could also have used the second-order rate equation, Equation (16-17), to calculate k. The rate law for the reaction is

$$\text{rate} = (1.0 \times 10^{-6} \text{ torr}^{-1} \cdot \text{s}^{-1})P_{CO}^2$$

16-55 (a) The initial rate is

$$\begin{aligned}\text{rate} &= (2.99 \times 10^6 \text{ M}^{-1} \cdot \text{s}^{-1})(2.0 \times 10^{-6} \text{ M})(6.0 \times 10^{-5} \text{ M}) \\ &= 3.6 \times 10^{-4} \text{ M} \cdot \text{s}^{-1}\end{aligned}$$

(b) The rate of production of NO_2 is

$$\text{rate} = \frac{\Delta[NO_2]}{\Delta t} = 3.6 \times 10^{-4} \text{ M} \cdot \text{s}^{-1}$$

The amount of NO_2 produced in 1 h is

$$[NO_2] = \text{rate} \times \text{time}$$
$$= (3.6 \times 10^{-4} \text{ M} \cdot \text{s}^{-1})\left(\frac{60 \text{ min}}{1 \text{ h}}\right)\left(\frac{60 \text{ s}}{1 \text{ min}}\right)(1 \text{ h})$$
$$= 1.3 \text{ mol} \cdot \text{L}^{-1}$$

16-57 We see from the data that the number of bacteria doubles every 15 min. The doubling time is independent of the number of bacteria; thus, the rate law is first order [see also the plot of ln (number of bacteria) versus t on the next page].

$$\text{rate of production} = k(\text{number of bacteria})$$

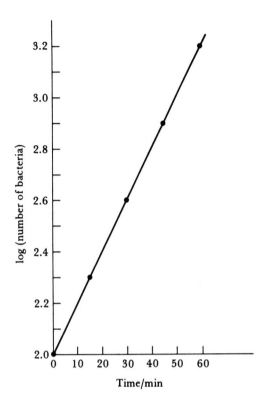

The half-life is equal to 15 min. There are

$$\frac{120 \text{ min}}{15 \text{ min/half-life}} = 8.0 \text{ half-lives}$$

in 2 h. The number of bacteria after n half-lives is given by

number of bacteria = (initial number of bacteria)$(2)^n$

or after 8 half-lives

number of bacteria = $(100)(2)^8 = 2.56 \times 10^4$

The rate constant is

$$k = \frac{0.693}{t_{1/2}} = \frac{0.693}{15 \text{ min}} = 4.6 \times 10^{-2} \text{ min}^{-1}$$

16-59 The fraction of a reactant remaining after time t is given by

$$\ln \frac{[A]}{[A]_0} = -kt$$

The fraction of material retained is $0.10 = 100 - 0.90$. The value of the rate constant is given by

$$k = \frac{0.693}{t_{1/2}} = \frac{0.693}{70 \text{ day}} = 0.0099 \text{ day}^{-1}$$

Thus,

$$\ln (0.10) = -(0.0099 \text{ day}^{-1})t$$

$$t = -\frac{2.30}{0.0099 \text{ day}^{-1}} = 230 \text{ day}$$

16-61 The formation of a covalent bond from two radicals does not involve any bond-breaking process. For example, for two $CH_3 \cdot$ radicals,

$$2H_3C \cdot \longrightarrow H_3C-CH_3$$

Thus $E_a \simeq 0$. The only limitation to the reaction would be the orientations of the two radicals as they collide.

16-63 The reaction rate is given by

$$\text{rate} = -\frac{\Delta P_{SO_2Cl_2}}{\Delta t}$$

Thus, the rate over the interval 0 to 5000 s is

$$\text{rate} = -\frac{680 \text{ torr} - 760 \text{ torr}}{5000 \text{ s} - 0 \text{ s}} = 1.6 \times 10^{-2} \text{ torr} \cdot \text{s}^{-1}$$

The rate over the interval 5000 s to 10,000 s is

$$\text{rate} = -\frac{610 \text{ torr} - 680 \text{ torr}}{10,000 \text{ s} - 5000 \text{ s}} = 1.4 \times 10^{-2} \text{ torr} \cdot \text{s}^{-1}$$

We assume that the rate law is of the form

$$\text{rate} = kP_{SO_2Cl_2}^x$$

We can find x and k by dividing the rate by $P_{SO_2Cl_2}^x$ for various values of x. The value of x for which k is the same for all time intervals is the correct value of x. We take the value of $P_{SO_2Cl_2}$ as the average value over the time interval for which we know the average rate. Thus with $x = 1$, we have for the two time intervals

$$k = \frac{\text{rate}}{P_{SO_2Cl_2}} = \frac{1.6 \times 10^{-2} \text{ torr} \cdot \text{s}^{-1}}{720 \text{ torr}} = 2.2 \times 10^{-5} \text{ s}^{-1}$$

$$k = \frac{\text{rate}}{P_{SO_2Cl_2}} = \frac{1.4 \times 10^{-2} \text{ torr} \cdot \text{s}^{-1}}{645 \text{ torr}} = 2.2 \times 10^{-5} \text{ s}^{-1}$$

The agreement between the k values tells us that the rate law is first order in the pressure of SO_2Cl_2; that is,

$$\text{rate} = kP_{SO_2Cl_2} = (2.2 \times 10^{-5} \ s^{-1})P_{SO_2Cl_2}$$

16-65 From the rate law we compute the initial rate for Run 1

$$\text{rate} = (5.0 \times 10^{-3} \ M^{-1} \cdot s^{-1})(0.20 \ M)(0.20 \ M) = 2.0 \times 10^{-4} \ M \cdot s^{-1}$$

For Run 2 we have

$$4.0 \times 10^{-4} \ M \cdot s^{-1} = (5.0 \times 10^{-3} \ M^{-1} \cdot s^{-1})(0.20 \ M)[I^-]_0$$

Thus,

$$[I^-]_0 = \frac{4.0 \times 10^{-4} \ M \cdot s^{-1}}{(5.0 \times 10^{-3} \ M^{-1} \cdot s^{-1})(0.20 \ M)} = 0.40 \ M$$

For Run 3 we have

$$8.0 \times 10^{-4} \ M \cdot s^{-1} = (5.0 \times 10^{-3} \ M^{-1} \cdot s^{-1})[C_2H_4Br_2]_0(0.20 \ M)$$

Thus,

$$[C_2H_4Br_2]_0 = \frac{8.0 \times 10^{-4} \ M \cdot s^{-1}}{(5.0 \times 10^{-3} \ M^{-1} \cdot s^{-1})(0.20 \ M)} = 0.80 \ M$$

16-67 The production of bacteria is a first-order reaction with a certain doubling time. Thus, the number of bacteria present after n doubling times is given by

$$\text{number of bacteria} = (\text{number of bacteria})_0(2)^n$$

The number of doubling times in 10 days is

$$n = \frac{(10 \ \text{day})\left(\dfrac{24 \ h}{1 \ \text{day}}\right)}{39 \ h} = 6.15$$

Thus, if there are 20,000 bacteria per milliliter present initially, after 10 days at $40°F$

$$\text{number of bacteria} = (20{,}000)(2)^{6.15} = 1.4 \times 10^6 \text{ per milliliter}$$

16-69 The rate of loss of neurons is given as constant; thus, the rate law is

$$\text{rate} = k = 2 \times 10^5 \text{ neurons} \cdot \text{day}^{-1}$$

Notice that the rate law is independent of the number of neurons. That is, the rate law is zero order in the concentration of neurons. The number of days required for $2 \times 10^{10} \times 0.20 = 4 \times 10^9$ neurons to be lost is

$$(2 \times 10^5 \text{ neurons} \cdot \text{day}^{-1}) (\text{number of days}) = 4 \times 10^9 \text{ neurons}$$

Thus, the number of days is

$$\frac{4 \times 10^9 \text{ neurons}}{2 \times 10^5 \text{ neurons} \cdot \text{day}^{-1}} = 2 \times 10^4 \text{ days}$$

Thus, the age in years at which the number of neurons is 80 percent of the original value is

$$(2 \times 10^4 \text{ days}) \left(\frac{1 \text{ y}}{365 \text{ days}} \right) = 55 \text{ y}$$

$$\text{age} = 55 \text{ y} + 30 \text{ y} = 85 \text{ y}$$

16-71   We first must calculate $\ln k$ and $1/T$.

$\ln (k/\text{s}^{-1})$	$(1/T)/10^{-3} \text{ K}^{-1}$
-14.06	3.66
-10.27	3.36
-7.60	3.14
-5.32	2.96

A plot of $\ln (k/\text{s}^{-1})$ versus $1/T$ is a straight line, as shown on the facing page. To estimate k at 50°C (323 K), we read the value of $\ln (k/\text{s}^{-1})$ at $1/(323 \text{ K}) = 3.10 \times 10^{-3} \text{ K}^{-1}$ from the plot. We see that

$$\ln (k/\text{s}^{-1}) = -7.4$$

Solving for k yields

$$k = 6.1 \times 10^{-4} \text{ s}^{-1}$$

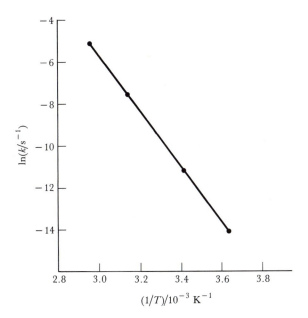

We can calculate E_a by using the Arrhenius equation and any pair of data.

$$\ln\left(\frac{k_2}{k_1}\right) = \frac{E_a}{R}\left(\frac{T_2 - T_1}{T_1 T_2}\right)$$

$$\ln\left(\frac{49.8 \times 10^{-5}\text{ s}^{-1}}{3.46 \times 10^{-5}\text{ s}^{-1}}\right) = \frac{E_a(318\text{ K} - 298\text{ K})}{(8.314\text{ J}\cdot\text{mol}^{-1}\cdot\text{K}^{-1})(318\text{ K})(298\text{ K})}$$

$$2.67 = (2.54 \times 10^{-5}\text{ J}^{-1}\cdot\text{mol})E_a$$

$$E_a = \frac{2.67}{2.54 \times 10^{-5}\text{ J}^{-1}\cdot\text{mol}} = 1.1 \times 10^5\text{ J}\cdot\text{mol}^{-1} = 110\text{ kJ}\cdot\text{mol}^{-1}$$

16-73   We first calculate the value of the rate constant at 700 K.

$$\ln k = 33.53 - \frac{2.30 \times 10^5\text{ J}\cdot\text{mol}^{-1}}{RT}$$

$$= 33.53 - \frac{2.30 \times 10^5\text{ J}\cdot\text{mol}^{-1}}{(8.314\text{ J}\cdot\text{mol}^{-1}\cdot\text{K}^{-1})(700\text{ K})} = -5.99$$

Taking the antilogarithm of both sides yields

$$k = 2.50 \times 10^{-3} \text{ min}^{-1}$$

We calculate the fraction of CH_3CH_2Br remaining using Equation (16-13).

$$\ln \left(\frac{[CH_3CH_2Br]}{[CH_3CH_2Br]_0} \right) = \ln (\text{fraction remaining}) = -kt$$
$$= -(2.50 \times 10^{-3} \text{ min}^{-1})(30.0 \text{ min})$$
$$= -0.0751$$

Taking the antilogarithm of both sides yields

$$\text{fraction remaining} = 0.928$$

16-75   According to Tables 13-1 and 13-2, the bond dissociation energies of H_2 and F_2 are 436 $kJ \cdot mol^{-1}$ and 154 $kJ \cdot mol^{-1}$, respectively. Consequently, the dissociation of H_2 is expected to have the higher activation energy.

16-77   We use the Arrhenius equation

$$\ln \left(\frac{k_2}{k_1} \right) = \frac{E_a}{R} \left(\frac{T_2 - T_1}{T_1 T_2} \right)$$

We are given that $k_2 = 2k_1$. Thus,

$$\ln \left(\frac{2k_1}{k_1} \right) = \frac{E_a(303 \text{ K} - 293 \text{ K})}{(8.314 \text{ J} \cdot mol^{-1} \cdot K^{-1})(303 \text{ K})(293 \text{ K})}$$
$$0.693 = E_a(1.35 \times 10^{-5} \text{ J}^{-1} \cdot mol)$$
$$E_a = 51,000 \text{ J} \cdot mol^{-1} = 51 \text{ kJ} \cdot mol^{-1}$$

16-79   The rate law is given as

$$\text{rate} = \frac{d[A]}{dt} = -k[A]^2$$

which we can rearrange to

$$\frac{d[A]}{[A]^2} = -k\,dt$$

If we integrate both sides of this equation, then we obtain

$$-\frac{1}{[A]} = -kt + \text{constant}$$

We are given that $[A] = [A]_0$ at time $t = 0$. Therefore,

$$-\frac{1}{[A]_0} = -k(0) + \text{constant}$$

$$\text{constant} = -\frac{1}{[A]_0}$$

Thus, the final equation is

$$\frac{1}{[A]} = \frac{1}{[A]_0} + kt$$

16-81   We use the Arrhenius equation

$$\ln\left(\frac{k_2}{k_1}\right) = \frac{E_a}{R}\left(\frac{T_2 - T_1}{T_1 T_2}\right)$$

where $k_2 = 2k_1$ and $T_1 = 293$ K.

$$\ln\left(\frac{2k_1}{k_1}\right) = \frac{(50.0 \times 10^3\,\text{J}\cdot\text{mol}^{-1})(T_2 - 293\,\text{K})}{(8.314\,\text{J}\cdot\text{mol}^{-1}\cdot\text{K}^{-1})(293\,\text{K})(T_2)}$$

$$0.693 = \frac{(20.53)(T_2 - 293\,\text{K})}{T_2}$$

$$0.693\,T_2 = 20.53T_2 - 6014\,\text{K}$$

$$T_2 = 303\,\text{K} = 30\,^\circ\text{C}$$

16-83
$$\text{rate} = k[A]^3$$
$$M \cdot s^{-1} = kM^3$$
$$k = \frac{M \cdot s^{-1}}{M^3} = M^{-2} \cdot s^{-1}$$

16-85   We first make a table of ln k versus $1/T$ for the data given and then plot the new data.

ln (k/s^{-1})	$1/T/K^{-1}$
-10.27	3.36×10^{-3}
-8.91	3.25×10^{-3}
-7.78	3.14×10^{-3}
-6.50	3.05×10^{-3}
-5.32	2.96×10^{-3}

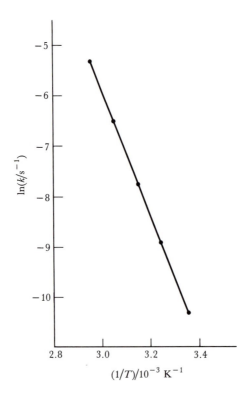

The figure on the opposite page shows that the plot of $\ln k$ versus $1/T$ is a straight line. We can determine the slope of the line using any two data points.

$$\text{slope} = \frac{\ln k_2 - \ln k_1}{\dfrac{1}{T_2} - \dfrac{1}{T_1}}$$

$$= \frac{-6.50 - (-8.91)}{3.05 \times 10^{-3} \text{ K}^{-1} - 3.25 \times 10^{-3} \text{ K}^{-1}} = -12{,}050 \text{ K}$$

The slope of the line is related to the activation energy by [Equation (16-26)]

$$\text{slope} = -\frac{E_a}{R}$$

Thus, we have that

$$E_a = -(\text{slope})(R) = -(-12{,}050 \text{ K})(8.314 \text{ J} \cdot \text{mol}^{-1} \cdot \text{K}^{-1})$$
$$= 100{,}000 \text{ J} \cdot \text{mol}^{-1} = 100 \text{ kJ} \cdot \text{mol}^{-1}$$

E. ANSWERS TO THE SELF-TEST

1. $\Delta[O_2]/\Delta t$
2. true
3. true
4. true
5. $H_2O_2 \ldots O_2$
6. false
7. true
8. the concentrations of the reactants
9. the rate law of a reaction
10. doubles
11. $k[A]$
12. time^{-1}
13. $k[H_2O_2]$
14. fast
15. quadruples

16. $k[A]^2 \ldots k[A][B]$
17. $M^{-1} \cdot \text{time}^{-1}$
18. $k[OH^-][H^+]$
19. $\ln [A]_0 - kt$
20. $\ln [H_2O_2]$
21. true
22. time it takes for the concentration of the reactant to decrease by one-half
23. independent of
24. true
25. true
26. smaller
27. slower
28. false

29. 1/[A]

30. false (It occurs in one step.)

31. true

32. false

33. mechanism

34. true

35. false (It is controlled by the slowest step.)

36. false

37. false

38. true

39. true

40. true

41. true

42. highest

43. unstable

44. false

45. true

46. Arrhenius

47. $\ln\left(\dfrac{k_2}{k_1}\right) = \dfrac{E_a}{R}\left(\dfrac{T_2 - T_1}{T_1 T_2}\right)$

48. lower

49. false

50. heterogeneous

51. false

52. true

53. true

17 CHEMICAL EQUILIBRIUM

A. *OUTLINE OF CHAPTER 17*

17-1   A chemical equilibrium is a dynamic equilibrium.

Chemical equilibrium is attained when the forward reaction rate equals the reverse reaction rate.

A state of chemical equilibrium exists when there is no net change in the concentrations of the reactants and products of the reaction.

At equilibrium, the forward reaction and the reverse reaction are still taking place.

A dynamic equilibrium is an equilibrium in which no net change occurs although the forward and reverse processes are still taking place.

Double arrows, $\rightleftharpoons$, denote a reaction equilibrium.

17-2   A chemical equilibrium can be attained from either direction.

Initial concentrations are denoted by the subscript zero, as in $[N_2O_4]_0$.

At equilibrium the reactant and the product concentrations remain constant with time (Figure 17-4).

When stoichiometrically equivalent amounts of reactants or products react, the same equilibrium is attained starting from either the reactant side (left) or the product side (right) of the equation for the reaction (Table 17-1).

17-3   The equilibrium constant for a chemical equation is equal to the ratio of product concentration terms to reactant concentration terms.

The equilibrium-constant expression for the general equation

$$aA(g) + bB(soln) + cC(s) \rightleftharpoons xX(g) + yY(soln) + zZ(l)$$

is given by the law of concentration action (Guldberg and Waage) as

$$K_c = \frac{[X]^x[Y]^y}{[A]^a[B]^b} \qquad (17\text{-}6)$$

where K_c is the equilibrium constant.

The value of K_c is equal to a constant at a given temperature.

The law of concentration action states that the equilibrium-constant expression for a chemical reaction is formulated as the ratio of product equilibrium concentrations to reactant equilibrium concentrations, with each concentration factor raised to a power equal to the stoichiometric coefficient of that species in the balanced equation for the reaction.

Pure liquid and solids, whose concentrations cannot be varied, do not appear in the equilibrium-constant expression.

The subscript c in K_c denotes that the equilibrium constant is expressed in terms of concentrations.

The law of concentration action tells us how to write the equilibrium-constant expression for any balanced chemical equation.

When the value of K_c is large, the equilibrium lies far to the right.

Reactions with large values of K_c give large amounts of product compared to reactants at equilibrium.

When the value of K_c is small, the equilibrium lies far to the left.

Reaction with small values of K_c give little product compared to reactants at equilibrium.

17-4   Equilibrium constants can be expressed in terms of partial pressures.

For reactions involving only gases, it is often more convenient to express the equilibrium constant in terms of gas partial pressures.

An equilibrium-constant expression written in terms of gas partial pressures is denoted by K_p.

The relation between K_c and K_p for a gas reaction is obtained by using the K_c expression and the relation $P = [gas]RT$ (Example 17-4).

17-5   Equilibrium constants are used in a variety of calculations.

The reaction stoichiometry is used to find the relationship between the initial concentration and the equilibrium concentration of a reactant or a product (Example 17-5).

The solutions to the quadratic equation $ax^2 + bx + c = 0$ are

$$x = \frac{-b \pm \sqrt{b^2 - 4ac}}{2a}$$

The equilibrium concentrations of reactants and products can be found from the initial concentrations, the equilibrium-constant expression, and the value of the equilibrium constant.

17-6 Equilibrium constants for chemical equations can be combined to obtain equilibrium constants for other equations.

The equilibrium constant for the reverse equation is equal to the reciprocal of the equilibrium constant for the forward equation, $K_r = 1/K_f$ [Equation (17-9)].

If we add two chemical equations, then the equilibrium constant for the resulting equation is equal to the product of the equilibrium constants for the two equations that are added together:

$$K_3 = K_1 K_2 \qquad\qquad (17\text{-}13)$$

If we multiply a chemical equation by some number n, then the equilibrium constant of the resulting equation will be K_1^n.

17-7 Le Châtelier's principle is used to predict the direction of shift in a chemical reaction displaced from equilibrium.

Le Châtelier's principle states that if a chemical reaction at equilibrium is subjected to a change in conditions that displaces it from equilibrium, then the reaction adjusts toward a new equilibrium state. The reaction proceeds in the direction that, at least partially, offsets the change in conditions.

The conditions whose change can affect a reaction equilibrium are
(a) The concentration of a reactant or product
(b) The reaction volume or the applied pressure
(c) The temperature

If the equilibrium of a reaction is disturbed by injecting more of a product, then the reaction equilibrium shifts from right to left toward the reactant side of the equation.

If the equilibrium of a reaction is disturbed by injecting more of a reactant, then the reaction equilibrium shifts from left to right toward the product side of the equation.

A decrease in volume (or increase in applied pressure) shifts a reaction equilibrium toward the side with the smaller number of moles of gas.

The allotropic form of a substance with the smaller molar volume is the stable form at high pressure.

17-8 Le Châtelier's principle has a quantitative basis.

When an equilibrium is disturbed, the new equilibrium conditions can be found from the original equilibrium conditions, the equilibrium-constant expression, and the application of Le Châtelier's principle.

When an amount of reactant or product is injected into the equilibrium mixture, the new equilibrium concentrations are found using K_c.

When a change in pressure disturbs the equilibrium mixture, the new equilibrium pressures are found using K_p.

17-9 An increase in temperature shifts a reaction equilibrium in the direction in which heat is absorbed.

Endothermic reactions ($\Delta H_{rxn}^{\circ} > 0$) shift to the right with increasing temperature and shift to the left with decreasing temperature.

Exothermic reactions ($\Delta H_{rxn}^{\circ} < 0$) shift to the left with increasing temperature and shift to the right with decreasing temperature.

A change in temperature changes the value of the equilibrium constant.

For an endothermic reaction, the value of K increases with increasing temperature.

For an exothermic reaction, the value of K decreases with increasing temperature.

The Haber process is the commercial production of ammonia from hydrogen gas and nitrogen gas at 500°C and 300 atm.

17-10   Chemical reactions always proceed toward completion.

The reaction quotient Q_c has the same algebraic form as the equilibrium constant expression K_c, but the concentration values inserted in Q_c need not be equilibrium values.

The value of Q_c depends on how the system is prepared.

The value of the ratio Q_c/K_c is used to determine the direction in which a reaction system proceeds toward equilibrium.

If $Q_c/K_c < 1$, then the reaction proceeds left to right to equilibrium.

If $Q_c/K_c > 1$, then the reaction proceeds right to left to equilibrium.

A reaction that is not at equilibrium proceeds to equilibrium in the direction such that Q_c approaches K_c in magnitude.

At equilibrium, $Q_c = K_c$.

The direction of reaction spontaneity is always toward equilibrium.

B. SELF-TEST

1. Chemical equilibrium is attained when the rates of the forward reaction and the reverse reaction are zero. *True/False*

2. At equilibrium, the concentrations of reactants increase and the concentrations of the products decrease. *True/False*

3. A chemical equilibrium is a (*dynamic/static*) equilibrium.

4. A chemical equilibrium can be approached from the reactant side or from the product side of a reaction. *True/False*

5. At equilibrium the rate of the forward reaction equals _____

_____.

6. If we start a reaction with only reactants, then in time the concentrations of the reactants will (*decrease/increase/stay the same*).

7. If we start a reaction with only reactants, then in time the concentrations of products will (*decrease/increase/stay the same*).

8. At equilibrium the concentrations of the reactants and products are (*constant/increasing/decreasing*).

9. The value of the equilibrium constant of a reaction depends on the initial values of the concentrations of the reactants. *True/False*

10. The value of the equilibrium constant of a reaction depends on the direction in which the equilibrium is approached. *True/False*

11. The equilibrium product concentrations appear in the (*numerator/denominator*) of the equilibrium-constant expression.

12. The equilibrium reactant concentrations appear in the (*numerator/denominator*) of the equilibrium-constant expression.

13. Each equilibrium concentration in the equilibrium-constant expression is raised to a power equal to its _____ in the balanced equation.

14. Pure solid reactants and products do not appear in the equilibrium-constant expression. *True/False*

15. For the balanced chemical equation

$$a A(g) + b B(soln) + c C(s) \rightleftharpoons x X(g) + y Y(soln) + z Z(l)$$

the equilibrium-constant expression is given by $K_c =$ _____.

16. A small value of K_c means that the equilibrium concentrations of the _____ are small compared to the equilibrium concentrations of the _____.

17. When the value of K_c is small, the equilibrium lies on the (*reactant/product*) side of the chemical equilibrium.

18. A large value of K_c means that the equilibrium concentrations of the _____ are small compared to the equilibrium concentrations of the _____.

19. When the value of K_c is large, the equilibrium lies on the (*reactant/product*) side of the chemical equilibrium.

20. The equilibrium constant of a gas-phase reaction can be expressed in terms of _____ or _____.

21. For reactions involving gases, the equilibrium constant cannot be expressed in terms of pressure. *True/False*

22. The equilibrium constant K_c has units in terms of _____.

23. The equilibrium constant K_p has units in terms of _____.

24. If the equilibrium constant for a forward reaction is K_f, then the equilibrium constant for the reverse reaction is given by $K_r = $ _____.

25. If a chemical equation can be written as the sum of two equations whose equilibrium constants are K_1 and K_2, then the equilibrium constant of the new equation is given by $K_3 = $ _____.

26. If a chemical equation is multiplied by 2, then the equilibrium constant of the resulting equation will be _____.

27. A change in the conditions of a reaction at equilibrium can cause the chemical system to shift to a new equilibrium state. *True/False*

28. Le Châtelier's principle states that _____
_____
_____.

29. The conditions that can affect a reaction equilibrium are _____
_____ , _____ , and _____
_____.

30. If additional gaseous product is injected into an equilibrium mixture, then the reaction equilibrium will shift from _____ to _____.

31. If additional solid product is injected into an equilibrium mixture, then the reaction equilibrium is unaffected. *True/False*

32. A decrease in the volume of a reaction vessel at equilibrium shifts the equilibrium toward the side of the reaction with _____.

33. The reaction equilibrium $H_2(g) + I_2(g) \rightleftharpoons 2HI(g)$ is unaffected by a change in the volume in which the reaction takes place. *True/False*

34. If additional product is injected into an equilibrium mixture, then the new equilibrium conditions cannot be calculated. *True/False*

35. If additional product is injected into an equilibrium mixture, Le Châtelier's principle can be used to find the new equilibrium conditions. *True/False*

36. An increase in temperature always shifts the reaction equilibrium toward the product side of a reaction. *True/False*

37. The equilibrium for an exothermic reaction shifts to the *(left/right)* with increasing temperature.

38. The equilibrium for an endothermic reaction shifts to the *(left/right)* with increasing temperature.

39. The equilibrium for an exothermic reaction shifts to the *(left/right)* with decreasing temperature.

40. The equilibrium for an endothermic reaction shifts to the *(left/right)* with decreasing temperature.

41. A change in temperature has no effect on the value of the equilibrium constant. *True/False*

42. The value of K *(increases, decreases)* with increasing temperature for an endothermic reaction.

43. Ammonia is produced commercially by the ——————— process.

44. The ammonia production reaction is ———————————————.

45. Chemical reactions always proceed toward ———————.

46. The reaction quotient has the same algebraic form as the equilibrium-constant expression for the reaction. *True/False*

47. For the balanced chemical equation

$$aA(g) + bB(soln) + cC(s) \rightleftharpoons xX(g) + yY(soln) + zZ(l)$$

the value of the reaction quotient is given by

$$Q_c = \text{———————}$$

where the subscript zeros denote ———————.

48. Any values of the concentrations of the reactants and products may be used in the reaction quotient. *True/False*

49. The value of Q_c for a given reaction is a constant. *True/False*

50. The value of the ratio Q_c/K_c can be used to predict the direction in which a system will proceed spontaneously toward equilibrium. *True/False*

51. When the value of Q_c/K_c is less than 1, the reaction proceeds from _____ to _____ toward equilibrium.

52. When the value of Q_c/K_c is greater than 1, the reaction proceeds from _____ to _____ toward equilibrium.

53. A system that is not in equilibrium proceeds toward equilibrium in the direction in which Q_c approaches _____ .

54. At equilibrium $Q_c = K_c$. *True/False*

C. *CALCULATIONS YOU SHOULD KNOW HOW TO DO*

1. Use the law of concentration action to write expressions for K_c and K_p. See Examples 17-2 and 17-4 and Problems 17-3 through 17-8.

2. Calculate the values of equilibrium constants from initial or equilibrium concentrations. See Example 17-3 and Problems 17-9 through 17-16.

3. Calculate equilibrium concentrations (or pressures) using K_c (or K_p) expressions together with initial concentrations (or pressures). See Examples 17-5, 17-7, and 17-8 and Problems 17-17 through 17-30.

4. Calculate an equilibrium constant for an equation from other equilibrium constants. See Example 17-9 and Problems 17-31 through 17-34.

5. Use Le Châtelier's principle to predict the effects of change of conditions on the reaction equilibrium. See Examples 17-10, 17-12, and 17-13 and Problems 17-35 through 17-42.

6. Use Le Châtelier's principle to calculate the new equilibrium conditions after a change in conditions of the reaction equilibrium. See Example 17-11 and Problems 17-43 through 17-48.

7. Use Q_c/K_c to determine the direction in which a reaction proceeds toward equilibrium. See Example 17-14 and Problems 17-49 through 17-54.

The Quadratic Formula

The standard form for a quadratic equation in x is

$$ax^2 + bx + c = 0 \tag{1}$$

where a, b, and c are constants. The two solutions to the quadratic equation are

$$x = \frac{-b \pm \sqrt{b^2 - 4ac}}{2a} \tag{2}$$

Equation (2) is called the quadratic formula. The quadratic formula is used to obtain the solutions to a quadratic equation in the standard form, that is, $ax^2 + bx + c = 0$. For example, let's find the solutions to the quadratic equation

$$2x^2 - 2x - 3 = 0$$

In this case $a = 2$, $b = -2$, and $c = -3$, and Equation (2) gives

$$x = \frac{-(-2) \pm \sqrt{(-2)^2 - 4(2)(-3)}}{2(2)}$$

$$= \frac{2 \pm \sqrt{4 + 24}}{4}$$

$$= \frac{2 \pm 5.292}{4}$$

$$= 1.823 \text{ and } -0.823$$

To use the quadratic formula to solve a quadratic equation, it is first necessary to put the quadratic equation in the standard form so that we know the values of the constants a, b, and c to use in Equation (2). For example, consider the problem of solving for x in the quadratic equation

$$\frac{x^2}{0.35 - x} = 0.100$$

To identify the constants a, b, and c, we must write this equation in the standard form of the quadratic equation. Multiplying both sides by $0.35 - x$ yields

$$x^2 = (0.35 - x)(0.100)$$

or

$$x^2 = 0.035 - 0.100x$$

Rearrangement to the standard quadratic form yields

$$x^2 + 0.100x - 0.035 = 0$$

Thus $a = 1$, $b = 0.100$, and $c = -0.035$. Using Equation (2) we have

$$x = \frac{-0.100 \pm \sqrt{(0.10)^2 - 4(1)(-0.035)}}{2(1)}$$

from which we compute

$$x = \frac{-0.100 \pm \sqrt{0.15}}{2}$$

$$= \frac{-0.100 \pm 0.387}{2}$$

Thus the solutions for x are

$$x = \frac{-0.100 + 0.387}{2} = 0.144$$

and

$$x = \frac{-0.100 - 0.387}{2} = -0.244$$

If x represents a concentration or gas pressure, then the only physically possible value of x is $+0.144$ because concentrations and pressures cannot have negative values.

Exercises

Solve the following equations for x.

1 $x^2 - 2x - 1 = 0$

2 $0.600x^2 - x - 0.450 = 0$

3 $x^2 + 2x - 0.285 = 0$

4 $\dfrac{x^2}{0.020 - x} = 0.0100$

5 $\dfrac{x^2}{0.150 - x} = 0.0250$

Answers

1 2.414; -0.414

2 2.04; -0.369

3 0.134; -2.13

4 0.010; -0.020

5 0.0500; -0.0750

D. SOLUTIONS TO THE ODD-NUMBERED PROBLEMS

17-1   The number of moles per liter of $SbCl_5$ that were consumed in the reaction is

$$\left(\begin{array}{c} \text{moles per liter} \\ \text{of } SbCl_5 \text{ consumed} \end{array} \right) = 0.165 \text{ M} - 0.135 \text{ M} = 0.030 \text{ M}$$

From the reaction stoichiometry, we have that the number of moles per liter of $SbCl_3$ produced and of Cl_2 produced is equal to the number of moles per liter of $SbCl_5$ consumed. Thus, at equilibrium

$$[SbCl_3] = 0.0955 \text{ M} + 0.030 \text{ M} = 0.126 \text{ M}$$

$$[Cl_2] = 0.210 \text{ M} + 0.030 \text{ M} = 0.240 \text{ M}$$

17-3   Each product concentration factor, raised to a power equal to its balancing coefficient, appears in the numerator of the K_c expression, and each reactant concentration factor, raised to a power equal to its balancing coefficient, appears in the denominator of the K_c expression. Pure solids and liquids do not appear in the K_c expression.

(a) $K_c = \dfrac{[CO_2]}{[CO]}$ units $= \dfrac{M}{M} =$ unitless

(b) $K_c = \dfrac{[C_{10}H_{12}]}{[C_5H_6]^2}$ units $= \dfrac{M}{M^2} = M^{-1}$

(c) $K_c = \dfrac{[NO_2]^4[O_2]}{[N_2O_5]^2}$ units $= \dfrac{M^4M}{M^2} = M^3$

17-5   (a) $K_c = \dfrac{[SO_2][Cl_2]}{[SO_2Cl_2]}$ units $= \dfrac{(M)(M)}{M} = M$

(b) $K_c = \dfrac{[O_2]}{[H_2O_2]^2}$ units $= \dfrac{M}{M^2} = M^{-1}$

(c) $K_c = \dfrac{1}{[H_2O]^3}$ units $= \dfrac{1}{M^3} = M^{-3}$

17-7   (a) $K_p = \dfrac{P_{SO_2}P_{Cl_2}}{P_{SO_2Cl_2}}$ units $=$ torr or atm

(b) $K_p = \dfrac{P_{O_2}}{P_{H_2O_2}^2}$ units $=$ torr^{-1} or atm^{-1}

(c) $K_p = \dfrac{1}{P_{H_2O}^3}$ units $=$ torr^{-3} or atm^{-3}

17-9   The K_c expression for the equation is

$$K_c = \frac{[CO][Cl_2]}{[COCl_2]}$$

We now set up a table of initial and equilibrium concentrations.

	$COCl_2(g)$	$\rightleftharpoons CO(g)$	$+ Cl_2(g)$
Initial concentrations	0.500 M	0	0
Equilibrium concentrations	0.500 M – 0.046 M	0.046 M	0.046 M

Note that at equilibrium $[CO] = [Cl_2]$ because CO and Cl_2 are formed in a $1:1$ ratio on decomposition of $COCl_2$. Substitution of the equilibrium concentrations into the K_c expression yields

$$K_c = \frac{(0.046 \text{ M})(0.046 \text{ M})}{(0.454 \text{ M})} = 4.7 \times 10^{-3} \text{ M at } 527°C$$

17-11   The equilibrium-constant expression for the reaction is

$$K_p = P_{NH_3}^2$$

The equilibrium pressure of $NH_3(g)$ in atmospheres is

$$P_{NH_3} = (62 \text{ torr})\left(\frac{1 \text{ atm}}{760 \text{ torr}}\right) = 8.2 \times 10^{-2} \text{ atm}$$

Substituting this value of P_{NH_3} into K_p yields

$$K_p = (8.2 \times 10^{-2} \text{ atm})^2 = 6.7 \times 10^{-3} \text{ atm}^2$$

17-13   The equilibrium-constant expression for the reaction is

$$K_c = \frac{[HI]^2}{[H_2][I_2]}$$

We now set up a table of the initial concentrations and the equilibrium concentrations. For every H_2 (or I_2) molecule that reacts, two molecules of HI are produced. The decrease in the concentration of H_2 (or I_2) is therefore one half the increase in the concentration of HI.

	$H_2(g)$	$+ I_2(g)$	$\rightleftharpoons 2HI(g)$
Initial concentration	$\dfrac{1.00 \text{ mol}}{2.00 \text{ L}} = 0.500 \text{ M}$	0.500 M	0
Equilibrium concentration	$0.500 \text{ M} - \left(\dfrac{0.780}{2}\right) \text{ M}$ $= 0.110 \text{ M}$	$0.500 \text{ M} - \left(\dfrac{0.780}{2}\right) \text{ M}$ $= 0.110 \text{ M}$	$\dfrac{1.56 \text{ mol}}{2.00 \text{ L}} = 0.780 \text{ M}$

Substituting the values of the equilibrium concentrations into the K_c expression, we find that

$$K_c = \frac{[HI]^2}{[H_2][I_2]} = \frac{(0.780 \text{ M})^2}{(0.110 \text{ M})^2} = 50.3$$

17-15   We set up a pressure table.

	$2NOCl(g)$	$\rightleftharpoons 2NO(g)$ $+$ $Cl_2(g)$	
Initial pressure	2.75 atm	0	0
Equilibrium pressure	2.75 atm $- P_{NO}$	P_{NO}	$P_{Cl_2} = \dfrac{P_{NO}}{2}$

From Dalton's law of partial pressures, we have that

$$P_{tot} = P_{NOCl} + P_{NO} + P_{Cl_2} = 3.58 \text{ atm}$$

Substituting in the expressions from the pressure table, we obtain

$$2.75 - P_{NO} + P_{NO} + \frac{P_{NO}}{2} = 3.58 \text{ atm}$$

Solving for P_{NO}, we have that

$$\frac{P_{NO}}{2} = 0.83 \text{ atm}$$

$$P_{NO} = 1.66 \text{ atm}$$

At equilibrium,

$$P_{Cl_2} = \frac{1.66 \text{ atm}}{2} = 0.83 \text{ atm}$$

$$P_{NOCl} = 2.75 \text{ atm} - 1.66 \text{ atm} = 1.09 \text{ atm}$$

The equilibrium-constant expression for this reaction is

$$K_p = \frac{P_{NO}^2 P_{Cl_2}}{P_{NOCl}^2} = \frac{(1.66 \text{ atm})^2 (0.83 \text{ atm})}{(1.09 \text{ atm})^2} = 1.9 \text{ atm}$$

17-17   The K_c expression for the equation is

$$K_c = \frac{[Ni(CO)_4]}{[CO]^4} = 5.0 \times 10^4 \text{ M}^{-3}$$

At equilibrium we have

$$\frac{(0.85 \text{ M})}{[CO]^4} = 5.0 \times 10^4 \text{ M}^{-3}$$

Solving for [CO] yields

$$[CO] = \left(\frac{0.85 \text{ M}}{5.0 \times 10^4 \text{ M}^{-3}}\right)^{1/4} = 0.064 \text{ M}$$

17-19   From the law of concentration action we have

$$K_c = \frac{[PCl_3][Cl_2]}{[PCl_5]}$$

We set up a table of initial concentrations and equilibrium concentrations. Let x be the number of moles per liter of PCl_3 that is produced by the decomposition of PCl_5. From the reaction stoichiometry, $[PCl_3] = [Cl_2] = x$ and $[PCl_5] = 0.25 \text{ M} - x$ at equilibrium.

	$PCl_5(g)$	$\rightleftharpoons PCl_3(g) +$	$Cl_2(g)$
Initial concentrations	$\dfrac{0.50 \text{ mol}}{2.0 \text{ L}} = 0.25 \text{ M}$	0	0
Equilibrium concentrations	$0.25 - x$	x	x

Substituting the equilibrium concentration expressions into the K_c expression, we have

$$K_c = \frac{(x)(x)}{0.25 \text{ M} - x} = \frac{x^2}{0.25 \text{ M} - x} = 1.8 \text{ M}$$

or

$$x^2 = (1.8 \text{ M})(0.25 \text{ M} - x) = 0.45 \text{ M}^2 - (1.8 \text{ M})x$$

We rearrange this equation to the standard quadratic form.

$$x^2 + (1.8 \text{ M})x - 0.45 \text{ M}^2 = 0$$

Using the quadratic formula, we have

$$x = \frac{-1.8 \text{ M} \pm \sqrt{(1.8 \text{ M})^2 - (4)(1)(-0.45 \text{ M}^2)}}{(2)(1)}$$

Taking the positive root, we obtain

$$x = 0.22 \text{ M}$$

Therefore, at equilibrium, $[PCl_3] = 0.22$ M, $[Cl_2] = 0.22$ M, and $[PCl_5] = 0.25$ M $- 0.22$ M $= 0.03$ M.

17-21 From the law of concentration action we have

$$K_c = 2.5 \times 10^4 = \frac{[HCl]^2}{[H_2][Cl_2]}$$

Let x be the number of moles of H_2 or of Cl_2 that react. From the reaction stoichiometry, the number of moles of HCl produced is $2x$. Let V be the reaction volume. Thus the initial and equilibrium concentrations are

	$H_2(g)$	+	$Cl_2(g)$	$\rightleftharpoons$ 2HCl(g)
Initial concentration	$\dfrac{0.50 \text{ mol}}{V}$		$\dfrac{0.50 \text{ mol}}{V}$	0
Equilibrium concentration	$\dfrac{0.50 \text{ mol} - x}{V}$		$\dfrac{0.50 \text{ mol} - x}{V}$	$\dfrac{2x}{V}$

Substituting the equilibrium concentrations into the K_c expression yields

$$\frac{\left(\dfrac{2x}{V}\right)^2}{\left(\dfrac{0.50 \text{ mol} - x}{V}\right)^2} = 2.5 \times 10^4$$

Taking the square root of both sides yields

$$\frac{\left(\dfrac{2x}{V}\right)}{\left(\dfrac{0.50 \text{ mol} - x}{V}\right)} = 1.58 \times 10^2$$

After canceling V, the volume, in the denominator and the numerator, we have

$$2x = 158(0.50 \text{ mol} - x)$$

Collecting like terms yields

$$160x = 79 \text{ mol}$$

and solving for x gives

$$x = 0.49 \text{ mol}$$

The number of moles of HCl at equilibrium is $2x$, or 0.98 mol.

17-23 From the law of concentration action, we have

$$K_c = [\text{NH}_3][\text{H}_2\text{S}] = 1.81 \times 10^{-4} \text{ M}^2$$

We need K_p in order to calculate equilibrium pressures. The relation between K_c and K_p is

$$K_c = \left(\frac{P_{\text{NH}_3}}{RT}\right)\left(\frac{P_{\text{H}_2\text{S}}}{RT}\right) = \left(\frac{1}{RT}\right)^2 K_p$$

or

$$K_p = (RT)^2 K_c$$

Thus,

$$K_p = [(0.0821 \text{ L} \cdot \text{atm} \cdot \text{K}^{-1} \cdot \text{mol}^{-1})(298 \text{ K})]^2(1.81 \times 10^{-4} \text{ mol}^2 \cdot \text{L}^{-2})$$
$$= 0.108 \text{ atm}^2$$

From the reaction stoichiometry we have $P_{\text{NH}_3} = P_{\text{H}_2\text{S}}$ at equilibrium. Let x be the equilibrium pressure of NH_3. Thus,

$$K_p = (x)(x) = x^2 = 0.108 \text{ atm}^2$$

Taking the square root yields

$$x = 0.329 \text{ atm}$$

At equilibrium $P_{\text{H}_2\text{S}} = P_{\text{NH}_3} = 0.329$ atm. The total pressure is the sum of the partial pressures of H_2S and NH_3

$$P_{\text{tot}} = P_{\text{H}_2\text{S}} + P_{\text{NH}_3} = 0.329 \text{ atm} + 0.329 \text{ atm} = 0.658 \text{ atm}$$

17-25 From the law of concentration action we have

$$K = \frac{[\text{Cl}_2][\text{I}_2]}{[\text{ICl}]^2} = 0.11$$

We set up a table of initial concentrations and equilibrium concentrations. Let x be the number of moles per liter of I_2 or of Cl_2 that react. From the reaction stoichiometry the number of moles per liter of ICl produced is $2x$. (Each mole of I_2 or Cl_2 that reacts produces two moles of ICl.)

	$2ICl(g) \rightleftharpoons I_2(g)$	$+$	$Cl_2(g)$
Initial concentration	0	$\dfrac{0.65 \text{ mol}}{1.5 \text{ L}} = 0.43 \text{ M}$	$\dfrac{0.33 \text{ mol}}{1.5 \text{ L}} = 0.22 \text{ M}$
Equilibrium concentration	$2x$	$0.43 \text{ M} - x$	$0.22 \text{ M} - x$

Substituting the equilibrium concentration expressions in the K expression yields

$$\frac{(0.43 \text{ M} - x)(0.22 \text{ M} - x)}{(2x)^2} = 0.11$$

Multiplying out and collecting terms yields

$$0.56x^2 - (0.65 \text{ M})x + (0.0946 \text{ M}^2) = 0$$

The solutions of this quadratic equation are

$$x = \frac{+0.65 \pm \sqrt{(0.65 \text{ M})^2 - (4)(0.56)(0.0946 \text{ M}^2)}}{2(0.56)}$$

The two roots are $x = 0.17$ M and $x = 0.99$ M. We reject the 0.99-M root as physically impossible because it gives negative values for $[I_2]$ and $[Cl_2]$. Thus, we have at equilibrium

$$[Cl_2] = 0.22 \text{ M} - 0.17 \text{ M} = 0.05 \text{ M}$$

$$[I_2] = 0.43 \text{ M} - 0.17 \text{ M} = 0.26 \text{ M}$$

$$[ICl] = 2(0.17 \text{ M}) = 0.34 \text{ M}$$

17-27 From the law of concentration action we have

$$K_p = \frac{P_{NO_2}^2}{P_{N_2O_4}} = 4.90 \text{ atm}$$

We know from Dalton's law of partial pressures that

$$P_{tot} = P_{NO_2} + P_{N_2O_4} = 1.45 \text{ atm}$$

Solving for P_{NO_2}, we have

$$P_{NO_2} = 1.45 \text{ atm} - P_{N_2O_4}$$

Substituting this expression for P_{NO_2} into the K_p expression, we have

$$\frac{(1.45 \text{ atm} - P_{N_2O_4})^2}{P_{N_2O_4}} = \frac{2.10 \text{ atm}^2 - (2.90 \text{ atm})P_{N_2O_4} + P_{N_2O_4}^2}{P_{N_2O_4}} = 4.90 \text{ atm}$$

or

$$2.10 \text{ atm}^2 - (2.90 \text{ atm})P_{N_2O_4} + P_{N_2O_4}^2 = (4.90 \text{ atm})P_{N_2O_4}$$

Rearranging to the standard form of a quadratic equation, we have

$$P_{N_2O_4}^2 - (7.80 \text{ atm})P_{N_2O_4} + 2.10 \text{ atm}^2 = 0$$

The solutions to the above equation from the quadratic formula are

$$P_{N_2O_4} = \frac{7.80 \text{ atm} \pm \sqrt{60.84 \text{ atm}^2 - (4)(1)(2.10 \text{ atm}^2)}}{2}$$

$$= \frac{7.80 \text{ atm} \pm 7.24 \text{ atm}}{2} = 7.52 \text{ atm and } 0.28 \text{ atm}$$

We can rule out the value 7.52 atm, because it is larger than the total pressure. At equilibrium $P_{N_2O_4} = 0.28$ atm and $P_{NO_2} = 1.45$ atm $- 0.28$ atm $= 1.17$ atm.

17-29   From the law of concentration action we have

$$K_p = \frac{P_{CO_2}}{P_{CO}} = 600$$

The total pressure is equal to the sum of the partial pressures of $CO(g)$ and $CO_2(g)$:

$$P_{tot} = P_{CO} + P_{CO_2} = 1.80 \text{ atm}$$

Solving for P_{CO_2}, we have

$$P_{CO_2} = 1.80 \text{ atm} - P_{CO}$$

Substituting the expression for P_{CO_2} into the K_p expression, we have that

$$\frac{1.80 \text{ atm} - P_{CO}}{P_{CO}} = 600$$

Thus,

$$P_{CO} = \frac{1.80 \text{ atm}}{601} = 3.00 \times 10^{-3} \text{ atm}$$

The partial pressure of CO_2 is

$$P_{CO_2} = 1.80 \text{ atm} - P_{CO} = 1.80 \text{ atm} - 0.0030 \text{ atm} = 1.80 \text{ atm}$$

17-31 The equation

$$CH_4(g) + 2H_2O(g) \rightleftharpoons CO_2(g) + 4H_2(g)$$

is the sum of the two given equations. Thus, the equilibrium constant is given by

$$K = K_1 K_2 = (1.44)(25.6 \text{ atm}^2) = 36.9 \text{ atm}^2$$

17-33 For the reverse of Equation (1) multiplied through by 2,

$$2Cl_2(g) + 2MgO(s) \rightleftharpoons 2MgCl_2(s) + O_2(g) \qquad (3)$$

we have

$$K_3 = \frac{1}{K_1^2} = \left(\frac{1}{2.95 \text{ atm}^{1/2}}\right)^2 = 0.115 \text{ atm}^{-1}$$

For Equation (2) multiplied by 2,

$$2MgCl_2(s) + 2H_2O(g) \rightleftharpoons 2MgO(s) + 4HCl(g) \qquad (4)$$

we have

$$K_4 = (8.40 \text{ atm})^2 = 70.56 \text{ atm}^2$$

Adding Equations (3) and (4) yields

$$2Cl_2(g) + 2H_2O(g) \rightleftharpoons 4HCl(g) + O_2(g) \qquad (5)$$

Thus,

$$K_5 = K_3 K_4 = (0.115 \text{ atm}^{-1})(70.56 \text{ atm}^2) = 8.11 \text{ atm}$$

17-35 (a) An increase in P_{H_2O} will cause a shift in the reaction equilibrium from right to left, because this is the direction in which P_{H_2O} will decrease. Thus, P_{CO} will decrease and P_{CO_2} will increase.

(b) The number of moles of gas is the same on both sides of the equation, and thus an increase in the reaction volume will not change the equilibrium pressures of the reactants or products.

17-37 (a) $\rightarrow$ The reaction is exothermic. A decrease in temperature shifts the equilibrium to the right.

(b) $\rightarrow$ A shift to the right causes a decrease in the number of moles per unit volume.

(c) $\leftarrow$ A shift in equilibrium to the left will increase the pressure of H_2.

(d) $\leftarrow$ A shift in equilibrium to the left will decrease the pressure of CH_4.

(e)   no change   The amount of $C(s)$ does not affect the equilibrium constant for the reaction and thus the equilibrium concentrations are independent of the amount of $C(s)$.

17-39   (a)   ←   The reaction is exothermic, and thus an increase in temperature makes the evolution of heat less favorable and shifts the equilibrium to the left.

(b)   ←   There are more moles of gas on the left than on the right, and thus a shift to the left will partially offset the decrease in the number of moles per unit volume.

(c)   ←   The decrease of $[O_2]$ will be offset partially by a shift to the left.

(d)   →   The increase of $[SO_2]$ will be partially offset by a shift to the right.

17-41   (1)

$$K_c = \frac{[CO][H_2]}{[H_2O]}$$

An increase in temperature shifts the reaction equilibrium to the right (endothermic reaction). A decrease in reaction volume shifts the reaction equilibrium to the left ($\Delta n_{gas} = +1$).

(2)

$$K_c = \frac{[CO_2][H_2]}{[CO][H_2O]}$$

An increase in temperature shifts the reaction equilibrium to the left (exothermic reaction). A decrease in reaction volume has no effect ($\Delta n_{gas} = 0$).

(3)

$$K_c = \frac{[H_2O][CH_4]}{[CO][H_2]^3}$$

An increase in temperature shifts the equilibrium to the left (exothermic reaction). A decrease in the reaction volume shifts the equilibrium to the right ($\Delta n_{gas} = -2$).

17-43   We first must calculate the value of the equilibrium constant K_p.

$$K_p = \frac{P_{NO_2}^2}{P_{N_2O_4}} = \frac{(393 \text{ torr})^2}{292 \text{ torr}} = 529 \text{ torr}$$

The new initial pressure of P_{NO_2} is given by

$$P_{tot} = P_{NO_2} + P_{N_2O_4}$$

Substituting the given values into this relation gives

$$812 \text{ torr} = P_{NO_2} + 292 \text{ torr}$$

$$P_{NO_2} = 520 \text{ torr}$$

To determine the new equilibrium conditions, we set up a pressure table.

	$N_2O_4(g)$	$\rightleftharpoons$	$2NO_2(g)$
Initial pressure	292 torr		520 torr
Equilibrium pressure	292 torr $+ x$		520 torr $- 2x$

We know from Le Châtelier's principle that x will be a positive quantity. We substitute these equilibrium pressures into the equilibrium-constant expression to get

$$\frac{(520 \text{ torr} - 2x)^2}{292 \text{ torr} + x} = 529 \text{ torr}$$

This expression can be written out as

$$270{,}400 \text{ torr}^2 - (2080 \text{ torr})x + 4x^2 = 154{,}500 \text{ torr}^2 + (529 \text{ torr})x$$

or

$$4x^2 - (2609 \text{ torr})x + 115{,}900 \text{ torr}^2 = 0$$

We use the quadratic formula for the roots to this equation.

$$x = \frac{-(-2609 \text{ torr}) \pm \sqrt{6{,}806{,}881 \text{ torr}^2 - 1{,}854{,}400 \text{ torr}^2}}{8}$$

The two roots are

$$x = 604 \text{ torr} \quad \text{and} \quad 48 \text{ torr}$$

We reject the first root because the value of the pressure change is greater than the total pressure. Thus, the equilibrium partial pressures are

$$P_{N_2O_4} = 292 \text{ torr} + 48 \text{ torr} = 340 \text{ torr}$$
$$P_{NO_2} = 520 \text{ torr} - (2)(48 \text{ torr}) = 424 \text{ torr}$$

As a final check, note that

$$\frac{[NO_2]^2}{[N_2O_4]} = \frac{(424 \text{ torr})^2}{340 \text{ torr}} = 529 \text{ torr}$$

17-45   The initial total pressure is

$$P_{tot} = P_{NH_3} + P_{H_2S}$$

However, $P_{NH_3} = P_{H_2S}$. Therefore,

$$P_{NH_3} = \frac{P_{tot}}{2} = \frac{0.664 \text{ atm}}{2} = 0.332 \text{ atm}$$

The value of the equilibrium constant is

$$K_p = P_{NH_3}P_{H_2S} = (0.332 \text{ atm})^2 = 0.110 \text{ atm}^2$$

The new total pressure is given by

$$0.906 \text{ atm} = P_{NH_3} + 0.332 \text{ atm}$$

Thus the new initial pressure of NH_3 is

$$P_{NH_3} = 0.574 \text{ atm}$$

To determine the new equilibrium conditions, we set up a pressure table:

	$NH_4HS(s) \rightleftharpoons$	$NH_3(g)$	+	$H_2S(g)$
Initial pressure	—	0.574 atm		0.332 atm
Equilibrium pressure	—	0.574 atm − x		0.332 atm − x

We know from Le Châtelier's principle that x will be a positive quantity. We substitute these equilibrium values into the equilibrium-constant expression to get

$$(0.574 \text{ atm} - x)(0.332 \text{ atm} - x) = 0.110 \text{ atm}^2$$

This expression can be written out as

$$x^2 - 0.902 \text{ atm } x + 0.081 \text{ atm}^2 = 0$$

The two roots of this equation are

$$x = 0.80 \text{ atm} \quad \text{and} \quad 0.099 \text{ atm}$$

We reject the first root because it leads to negative values of P_{NH_3} and P_{H_2S}. Using $x = 0.099$ atm, we obtain

$$P_{NH_3} = 0.574 \text{ atm} - 0.099 \text{ atm} = 0.475 \text{ atm}$$
$$P_{H_2S} = 0.332 \text{ atm} - 0.099 \text{ atm} = 0.233 \text{ atm}$$

As a final check, note that $P_{NH_3}P_{H_2S} = (0.475\ atm)(0.233\ atm) = 0.111\ atm^2 = K_p$.

17-47   The value of the equilibrium constant is

$$K_p = \frac{P_{CO}P_{H_2O}}{P_{CO_2}P_{H_2}} = \frac{(512\ torr)(77\ tor)}{(384\ torr)(192\ torr)} = 0.535$$

Because the volume has doubled, the partial pressure of each gas is halved in value. The new partial pressures are

$$P_{CO} = 256\ torr$$
$$P_{H_2O} = 38.5\ torr$$
$$P_{CO_2} = 192\ torr$$
$$P_{H_2} = 96.0\ torr$$

Because the total number of moles of gas is the same on both sides of the chemical equation for this reaction, these values are also the equilibrium values. Note that they satisfy the equilibrium-constant expression.

$$K_p = \frac{(256\ torr)(38.5\ torr)}{(192\ torr)(96.0\ torr)} = 0.535$$

17-49   The equilibrium-constant expression is

$$K_c = \frac{[SO_3]^2}{[SO_2]^2[O_2]} = 13\ M^{-1}$$

The Q_c expression is

$$Q_c = \frac{[SO_3]_0^2}{[SO_2]_0^2[O_2]_0}$$

The value of Q_c is calculated from the given initial concentrations:

Q_c		Q_c/K_c	Direction that the reaction proceeds toward equilibrium
(a) $\dfrac{(0.10\ M)^2}{(0.40\ M)^2(0.20\ M)}$	$= 0.31\ M^{-1}$	$\dfrac{0.31\ M^{-1}}{13\ M^{-1}} = 0.024 < 1$	$\longrightarrow$
(b) $\dfrac{(0.30\ M)^2}{(0.05\ M)^2(0.10\ M)}$	$= 360\ M^{-1}$	$\dfrac{360\ M^{-1}}{13\ M^{-1}} = 28 > 1$	$\longleftarrow$

17-51   The Q_p expression is

$$Q_p = \frac{(P_{CO})_0(P_{H_2O})_0}{(P_{H_2})_0(P_{CO_2})_0}$$

Thus,

$$Q_p = \frac{(1.25 \text{ atm})(0.10 \text{ atm})}{(0.55 \text{ atm})(0.20 \text{ atm})} = 1.1$$

Because Q_p does not equal K_p, the reaction is not at equilibrium

$$\frac{Q_p}{K_p} = \frac{1.1}{1.59} = 0.69 < 1$$

Because the value of Q_p/K_p is less than 1, the reaction proceeds from left to right toward equilibrium.

17-53   The Q_p expression for the reaction is

$$Q_p = \frac{(P_{SO_3})_0^2}{(P_{SO_2})_0^2(P_{O_2})_0}$$

Thus,

$$Q_p = \frac{(0.20 \text{ atm})^2}{(0.30 \text{ atm})^2(0.50 \text{ atm})} = 0.89 \text{ atm}^{-1}$$

$$\frac{Q_p}{K_p} = \frac{0.89 \text{ atm}^{-1}}{0.14 \text{ atm}^{-1}} = 6.4$$

Because $Q_p/K_p > 1$, the reaction proceeds from right to left toward equilibrium.

17-55   The K_p expression for the equation is

$$K_p = P_{NH_3}^2 = 6.66 \times 10^{-3} \text{ atm}^2 \qquad \text{(at 20°C)}$$

Thus, the equilibrium partial pressure of $NH_3(g)$ at 20°C is

$$P_{NH_3} = (6.66 \times 10^{-3} \text{ atm}^2)^{1/2} = 8.16 \times 10^{-2} \text{ atm}$$

17-57   From the law of concentration action we have

$$K_c = \frac{[I_2][H_2]}{[HI]^2}$$

Let x be the number of moles per liter of H_2 that are produced from the decomposition of HI. Each mole of H_2 produced requires that two moles of HI decompose. We will set up a table of initial concentrations and equilibrium concentrations.

	2HI(g)	⇌	H₂(g)	+ I₂(g)
Initial concentrations	3.52 M + 1.00 M = 4.52 M		0.42 M	0.42 M
Equilibrium concentrations	4.52 M − 2x		0.42 M + x	0.42 M + x

Notice that the initial concentration of HI is the equilibrium concentration *plus* the concentration of the added HI. Substituting the equilibrium concentration expressions in the K_c expression, we have

$$K_c = \frac{(0.42 \text{ M} + x)(0.42 \text{ M} + x)}{(4.52 \text{ M} - 2x)^2}$$

We can calculate the value of K_c from the initial equilibrium concentrations.

$$K_c = \frac{(0.42 \text{ M})(0.42 \text{ M})}{(3.52 \text{ M})^2} = 0.014$$

We now have

$$\frac{(0.42 \text{ M} + x)^2}{(4.52 \text{ M} - 2x)^2} = 0.014$$

Taking the square root of both sides, we obtain

$$\frac{0.42 \text{ M} + x}{4.52 \text{ M} - 2x} = 0.12$$

or

$$0.42 \text{ M} + x = (0.12)(4.52 \text{ M} - 2x) = 0.54 \text{ M} - 0.24x$$

$$1.24x = 0.12 \text{ M}$$

$$x = \frac{0.12 \text{ M}}{1.24} = 0.097 \text{ M}$$

Thus, in the new equilibrium state we have

$$[\text{HI}] = 4.52 \text{ M} - 2x = 4.33 \text{ M}$$
$$[\text{H}_2] = 0.42 \text{ M} + x = 0.52 \text{ M}$$
$$[\text{I}_2] = 0.42 \text{ M} + x = 0.52 \text{ M}$$

17-59 (a)   No change: There are the same number of moles of gaseous species on each side of the reaction.
  (b)   ←   The reaction is exothermic.
  (c)   →   A shift to the right will decrease P_{NO_2}.
  (d)   →   A shift to the right will decrease P_{SO_3} and P_{NO}.

17-61   From the law of concentration action we have

$$K_c = \frac{[H_2O]^2}{[H_2]^2}$$

We can find the value of K_c by substituting the values of the equilibrium concentrations.

$$K_c = \frac{(0.25 \text{ M})^2}{(0.25 \text{ M})^2} = 1.0$$

We set up a table of initial concentrations and equilibrium concentrations. Let x be the number of moles per liter of H_2O produced by the reaction of SnO_2 and H_2. One mole of H_2 reacts to produce one mole of H_2O.

	$SnO_2(s)$ +	$2H_2(g)$	$\rightleftharpoons$ Sn(s) +	$2H_2O(g)$
Initial concentration	—	0.50 M	—	0.25 M
Equilibrium concentration	—	0.50 M − x	—	0.25 M + x

Substituting the equilibrium concentration expressions in the K_c expression, we have

$$K_c = \frac{(0.25 \text{ M} + x)^2}{(0.50 \text{ M} - x)^2} = 1.0$$

Taking the square root of both sides, we have that

$$\frac{0.25 \text{ M} + x}{0.50 \text{ M} - x} = 1.0$$

or

$$0.25 \text{ M} + x = (1.0)(0.50 \text{ M} - x) = 0.50 \text{ M} - x$$

$$2x = 0.25 \text{ M}$$

$$x = \frac{0.25 \text{ M}}{2} = 0.125 \text{ M}$$

Thus, at the new equilibrium state we have

$$[H_2O] = 0.25 \text{ M} + 0.125 \text{ M} = 0.38 \text{ M}$$
$$[H_2] = 0.50 \text{ M} - 0.125 \text{ M} = 0.38 \text{ M}$$

17-63 The K_p expression for the equation is

$$K_p = P_{NH_3}^2 P_{CO_2}$$

The total pressure is given by

$$P_{tot} = P_{NH_3} + P_{CO_2}$$

but from the reaction stoichiometry

$$P_{NH_3} = 2P_{CO_2}$$

Thus,

$$P_{tot} = 3P_{CO_2}$$

or

$$P_{CO_2} = \frac{1}{3} P_{tot}$$

Therefore

$$K_p = (2P_{CO_2})^2 P_{CO_2} = 4P_{CO_2}^3$$
$$= 4\left(\frac{1}{3} P_{tot}\right)^3$$
$$= \frac{4}{27} P_{tot}^3$$

17-65 For the general equation

$$aA(g) + bB(g) \rightleftharpoons cC(g) + dD(g)$$

we have for K_c

$$K_c = \frac{[C]^c[D]^d}{[A]^a[B]^b} \qquad (1)$$

From the ideal-gas equation,

$$P_iV = n_iRT$$

we have

$$P_i = \left(\frac{n_i}{V}\right)RT = [i]RT \qquad (2)$$

where $[i]$ is the concentration of the gas i in $mol \cdot L^{-1}$. Substitution of Equation (2) into Equation (1) yields

$$K_c = \frac{\left(\dfrac{P_C}{RT}\right)^c \left(\dfrac{P_D}{RT}\right)^d}{\left(\dfrac{P_A}{RT}\right)^a \left(\dfrac{P_B}{RT}\right)^b}$$

$$= \frac{P_C^c P_D^d}{P_A^a P_B^b} (RT)^{a+b-c-d} = K_p (RT)^{-\Delta n}$$

where $\Delta n = c + d - a - b$. Solving for K_p, we obtain

$$K_p = K_c (RT)^{\Delta n}$$

17-67   (a)   The K expression for the equation is

$$K_c = \frac{[HI]^2}{[H_2][I_2]} = \frac{P_{HI}^2}{P_{H_2} P_{I_2}} = 85$$

With $P_{HI} = P_{I_2} = P_{H_2}$ we have

$$Q_c = \frac{(P_{HI})_0^2}{(P_{H_2})_0 (P_{I_2})_0} = \frac{(P_{HI})_0^2}{(P_{HI})_0 (P_{HI})_0} = 1 \neq K_c$$

and thus equilibrium is not possible with these concentrations.

(b)   The initial value of $[HI]$ is

$$[HI] = \frac{(5.0 \text{ g}) \left(\dfrac{1 \text{ mol}}{127.9 \text{ g}}\right)}{(2.00 \text{ L})} = 0.0195 \text{ M}$$

The initial and equilibrium concentrations are as follows:

	$H_2(g)$ + $I_2(g)$ $\rightleftharpoons$ 2HI(g)		
Initial concentrations	0	0	0.0195 M
Equilibrium concentrations	x	x	0.0195 M $- 2x$

Thus,

$$\frac{(0.0195 \text{M} - 2x)^2}{x^2} = 85$$

Take the square root of both sides of this equation to get

$$\frac{0.0195 \text{ M} - 2x}{x} = 9.22$$

and solve for x

$$x = \frac{0.0195 \text{ M}}{11.22} = 0.00174 \text{ M}$$

Thus, at equilibrium we have

$$[\text{H}_2] = [\text{I}_2] = 1.7 \times 10^{-3} \text{ M}$$
$$[\text{HI}] = 0.0195 \text{ M} - 2(0.00174 \text{ M}) = 0.016 \text{ M}$$

17-69   The initial number of moles of O_2 is computed by using the ideal-gas equation.

$$n = \frac{PV}{RT} = \frac{(1.00 \text{ atm})(2.00 \text{ L})}{(0.0821 \text{ L} \cdot \text{atm} \cdot \text{K}^{-1} \cdot \text{mol}^{-1})(298 \text{ K})}$$
$$= 0.0817 \text{ mol}$$

The number of moles of MgCl_2 is

$$n = \frac{50 \text{ g}}{95.21 \text{ g} \cdot \text{mol}^{-1}} = 0.525 \text{ mol}$$

Thus, O_2 is the limiting reactant. The K_p expression is

$$K_p = \frac{P_{\text{Cl}_2}}{P_{\text{O}_2}^{1/2}} = 1.75 \text{ atm}^{1/2}$$

The initial pressure of O_2 at 823 K is

$$P_{\text{O}_2} = (1.00 \text{ atm})\left(\frac{823 \text{ K}}{298 \text{ K}}\right) = 2.76 \text{ atm}$$

If we let $x = P_{\text{Cl}_2}$ at equilibrium, then we have

$$\frac{x}{\left(2.76 \text{ atm} - \dfrac{x}{2}\right)^{1/2}} = 1.75 \text{ atm}^{1/2}$$

Squaring both sides of the equation yields

$$\frac{x^2}{2.76 \text{ atm} - \dfrac{x}{2}} = 3.06 \text{ atm}$$

or

$$x^2 + (1.53 \text{ atm})x - 8.45 \text{ atm}^2 = 0$$

The solutions to the quadratic equation are

$$x = \frac{-1.53 \text{ atm} \pm \sqrt{(1.53 \text{ atm})^2 + 4(8.45 \text{ atm}^2)}}{2}$$

$$x = 2.24 \text{ atm}$$

Thus, at equilibrium

$$P_{Cl_2} = 2.24 \text{ atm}$$

$$P_{O_2} = 2.76 \text{ atm} - \frac{2.24 \text{ atm}}{2} = 1.64 \text{ atm}$$

17-71   High pressure favors the more dense form. Thus, the brown form is more stable at high pressure.

17-73   The value of the equilibrium constant for the reaction whose equation is

$$2SO_2(g) + O_2(g) \rightleftharpoons 2SO_3(g)$$

is given by

$$K_p = (1.8 \times 10^{12} \text{ atm}^{-1/2})^2 = 3.24 \times 10^{24} \text{ atm}^{-1}$$

The equilibrium-constant expression for the above reaction is

$$K_p = \frac{P_{SO_3}^2}{P_{SO_2}^2 P_{O_2}}$$

The pressure of a gas is related to the concentration of the gas by $P = [gas]\,RT$. Substituting in the expressions for the partial pressure of each gas yields

$$3.24 \times 10^{24} \text{ atm}^{-1} = \frac{[SO_3]^2(RT)^2}{[SO_2]^2(RT)^2[O_2]\,RT} = \frac{[SO_3]^2}{[SO_2]^2[O_2]\,RT} = \frac{K_c}{RT}$$

$$K_c = (RT)(3.24 \times 10^{24} \text{ atm}^{-1})$$
$$= (0.0821 \text{ L·atm·mol}^{-1} \cdot \text{K}^{-1})(583 \text{ K})(3.24 \times 10^{24} \text{ atm}^{-1})$$
$$= 1.6 \times 10^{26} \text{ M}^{-1}$$

17-75   The value of the equilibrium constant is

$$K_p = \frac{P_{COCl_2}}{P_{CO}P_{Cl_2}} = \frac{0.142 \text{ atm}}{(1.08 \text{ atm})^2} = 0.122 \text{ atm}^{-1}$$

The total pressure is given by

$$P_{tot} = P_{COCl_2} + P_{CO} + P_{Cl_2}$$

or, initially after the addition of $CO(g)$,

$$3.27 \text{ atm} = 0.142 \text{ atm} + P_{CO} + 1.08 \text{ atm}$$
$$P_{CO} = 2.05 \text{ atm}$$

We set up a pressure table to calculate the new equilibrium pressures.

	$CO(g)$ +	$Cl_2(g)$	$\rightleftharpoons COCl_2(g)$
Initial pressure	2.05 atm	1.08 atm	0.142 atm
Equilibrium pressure	2.05 atm $- x$	1.08 atm $- x$	0.142 atm $+ x$

where we know from Le Châtelier's principle that x is a positive number. Substituting in the above expressions in the equilibrium-constant expression yields

$$K_p = \frac{0.142 \text{ atm} + x}{(2.05 \text{ atm} - x)(1.08 \text{ atm} - x)} = 0.122 \text{ atm}^{-1}$$
$$0.122 \text{ atm}^{-1} \, x^2 - 1.382x + 0.128 \text{ atm} = 0$$

The two roots are

$$x = 0.094 \text{ atm} \quad \text{and} \quad 11.2 \text{ atm}$$

We reject 11.2 atm because it exceeds the total pressure, and so the partial pressures are

$$P_{COCl_2} = 0.142 \text{ atm} + 0.094 \text{ atm} = 0.236 \text{ atm}$$
$$P_{CO} = 2.05 \text{ atm} - 0.094 \text{ atm} = 1.96 \text{ atm}$$
$$P_{Cl_2} = 1.08 \text{ atm} - 0.094 \text{ atm} = 0.99 \text{ atm}$$
$$P_{tot} = 3.19 \text{ atm}$$

Note that

$$\frac{P_{COCl_2}}{P_{CO}P_{Cl_2}} = \frac{0.236 \text{ atm}}{(1.96 \text{ atm})(0.99 \text{ atm})} = 0.122 \text{ atm}^{-1}$$

17-77 From Problem 17-76 we have that $K = 54.4$. The concentration table is

	$H_2(g)$	$+ I_2(g)$	$\rightleftharpoons 2HI(g)$
Initial concentration	$\dfrac{2.00 \text{ mol}}{V}$	$\dfrac{2.00 \text{ mol}}{V}$	0
Equilibrium concentration	$\dfrac{2.00 \text{ mol} - x}{V}$	$\dfrac{2.00 \text{ mol} - x}{V}$	$\dfrac{2x}{V}$

The equilibrium-constant expression is

$$K = \frac{[\text{HI}]^2}{[\text{H}_2][\text{I}_2]} = \frac{\left(\dfrac{2x}{V}\right)^2}{\left(\dfrac{2.00 \text{ mol} - x}{V}\right)\left(\dfrac{2.00 \text{ mol} - x}{V}\right)} = 54.4$$

$$\frac{(2x)^2}{(2.00 \text{ mol} - x)^2} = 54.4$$

Taking the square root of both sides yields

$$\frac{2x}{2.00 \text{ mol} - x} = 7.376$$

or

$$x = 1.57 \text{ mol}$$

The percentage of $I_2(g)$ that reacts is given by

$$\% \text{ reacted} = \frac{1.57 \text{ mol}}{2.00 \text{ mol}} \times 100 = 78.7\%$$

17-79 We set up the pressure table.

	$CO_2(g)$	$+ H_2(g)$	$\rightleftharpoons CO(g)$	$+ H_2O(g)$
Initial pressure	26.1 torr	26.1 torr	0	0
Equilibrium pressure	26.1 torr $- x$	26.1 torr $- x$	x	x

where we know from Le Châtelier's principle that x is a positive number. The equilibrium-constant expression is

$$K = \frac{P_{CO}P_{H_2O}}{P_{CO_2}P_{H_2}} = \frac{(x)(x)}{(26.1 \text{ torr} - x)(26.1 \text{ torr} - x)}$$

$$= \frac{x^2}{(26.1 \text{ torr} - x)^2} = 0.719$$

Taking the square root of both sides yields

$$\frac{x}{26.1 \text{ torr} - x} = 0.848$$

$$x = 12.0 \text{ torr}$$

The partial pressures are

$$P_{CO} = P_{H_2O} = 12.0 \text{ torr}$$

$$P_{CO_2} = P_{H_2} = 26.1 \text{ torr} - 12.0 \text{ torr} = 14.1 \text{ torr}$$

Note that

$$\frac{P_{CO}P_{H_2O}}{P_{CO_2}P_{H_2}} = \frac{(12.0 \text{ torr})(12.0 \text{ torr})}{(14.1 \text{ torr})(14.1 \text{ torr})} = 0.72$$

17-81 Because equal molar quantities of both $NH_3(g)$ and $HCl(g)$ are involved, we have initially

$$P_{NH_3} = P_{HCl} = \frac{8.76 \text{ atm}}{2} = 4.38 \text{ atm}$$

We set up a pressure table.

	$NH_4Cl(s) \rightleftharpoons NH_3(g)$	$+$	$HCl(g)$
Initial pressure	—	4.38 atm	4.38 atm
Equilibrium pressure	—	4.38 atm $- x$	4.38 atm $- x$

The equilibrium-constant expression is

$$K_p = P_{NH_3}P_{HCl} = (4.38 \text{ atm} - x)^2 = 5.67 \times 10^{-2} \text{ atm}^2$$

Taking the square root of both sides yields

$$4.38 \text{ atm} - x = 0.238 \text{ atm}$$

$$x = 4.14 \text{ atm}$$

The partial pressures of the two gases at equilibrium are

$$P_{NH_3} = P_{HCl} = 4.38 \text{ atm} - 4.14 \text{ atm} = 0.24 \text{ atm}$$

Now according to the chemical equation, the number of moles of $NH_4Cl(s)$ produced is equal to the number of moles of $NH_3(g)$ or $HCl(g)$ that react. We use the ideal-gas law to find the number of moles of $NH_4Cl(s)$ produced.

$$n = \frac{PV}{RT} = \frac{(4.14 \text{ atm})(2.00 \text{ L})}{(0.0821 \text{ L} \cdot \text{atm} \cdot \text{mol}^{-1} \cdot \text{K}^{-1})(573 \text{ K})} = 0.176 \text{ mol}$$

The mass of $NH_4Cl(s)$ is

$$\text{mass} = (0.176 \text{ mol } NH_4Cl) \left(\frac{53.49 \text{ g } NH_4Cl}{1 \text{ mol } NH_4Cl} \right) = 9.41 \text{ g}$$

17-83 We first calculate the value of the equilibrium constant from the initial equilibrium concentrations.

$$K_c = \frac{[CO_2][H_2]}{[CO][H_2O]} = \frac{\left(\dfrac{0.80 \text{ mol}}{V}\right)\left(\dfrac{0.20 \text{ mol}}{V}\right)}{\left(\dfrac{0.10 \text{ mol}}{V}\right)\left(\dfrac{0.40 \text{ mol}}{V}\right)} = 4.0$$

where we have let V be the volume of the container. We see that the volume cancels in the equilibrium-constant expression, and so we can use just the number of moles of each constituent. Let x be the number of moles of $CO_2(g)$ to be added. Note that the increase in the number of moles of $CO(g)$ is 0.20 mol $-$ 0.10 mol = 0.10 mol. The concentration table is

	$CO(g)$	$+ \; H_2O(g)$	$\rightleftharpoons CO_2(g)$	$+ \; H_2(g)$
Initial concentration	0.10 mol	0.40 mol	0.80 mol $+ x$	0.20 mol
Equilibrium concentration	0.20 mol	0.40 mol $+$ 0.10 mol	0.80 mol $+ x -$ 0.10 mol	0.20 mol $-$ 0.10 mol
or	0.20 mol	0.50 mol	0.70 mol $+ x$	0.10 mol

Substituting the above expressions into the equilibrium-constant expression yields

$$K_c = \frac{(0.70 \text{ mol} + x)(0.10 \text{ mol})}{(0.20 \text{ mol})(0.50 \text{ mol})} = 4.0$$

Solving for x,

$$0.070 \text{ mol}^2 + (0.10 \text{ mol})x = 0.40 \text{ mol}^2$$

yields

$$x = 3.3 \text{ mol}$$

We must add 3.3 mol of $CO_2(g)$.

17-85   From Dalton's law of partial pressures,

$$P_{tot} = P_{NH_3} + P_{H_2S}$$

Because we started with pure $NH_4HS(s)$, at equilibrium

$$P_{NH_3} = P_{H_2S}$$

Thus,

$$P_{NH_3} = \frac{P_{tot}}{2} = \frac{0.659 \text{ atm}}{2} = 0.3295 \text{ atm}$$

$$K_p = P_{NH_3} P_{H_2S} = (0.3295 \text{ atm})^2 = 0.109 \text{ atm}^2$$

The number of moles of $NH_3(g)$ produced can be found using the ideal-gas equation.

$$n = \frac{PV}{RT} = \frac{(0.3295 \text{ atm})(3.0 \text{ L})}{(0.0821 \text{ L} \cdot \text{atm} \cdot \text{mol}^{-1} \cdot \text{K}^{-1})(298 \text{ K})} = 0.0404 \text{ mol}$$

The number of moles of $NH_4HS(s)$ that reacted is

$$\text{moles of NH}_4\text{HS reacted} = (0.0404 \text{ mol NH}_3) \left(\frac{1 \text{ mol NH}_4\text{HS}}{1 \text{ mol NH}_3} \right) = 0.0404 \text{ mol}$$

The number of moles of $NH_4HS(s)$ initially present is

$$\text{moles of NH}_4\text{HS initially} = (5.2589 \text{ g}) \left(\frac{1 \text{ mol NH}_4\text{HS}}{51.11 \text{ g NH}_4\text{HS}} \right) = 0.1029 \text{ mol}$$

The percentage of $NH_4HS(s)$ that reacted is

$$\% \text{ reacted} = \frac{0.0404 \text{ mol}}{0.1029 \text{ mol}} \times 100 = 39\%$$

17-87 From Dalton's law of partial pressures

$$P_{\text{tot}} = P_{HC_2H_5O_2} + P_{(HC_2H_5O_2)_2} = 1.50 \text{ atm}$$

Thus,

$$P_{HC_2H_5O_2} = 1.50 \text{ atm} - P_{(HC_2H_5O_2)_2}$$

The equilibrium-constant expression for the reaction is

$$K_p = \frac{P_{(HC_2H_5O_2)_2}}{P^2_{HC_2H_5O_2}} = 3.72 \text{ atm}^{-1}$$

Substituting in the expression for $P_{HC_2H_5O_2}$, we have

$$\frac{P_{(HC_2H_5O_2)_2}}{(1.50 \text{ atm} - P_{(HC_2H_5O_2)_2})^2} = 3.72 \text{ atm}^{-1}$$

Rearranging, we obtain

$$(3.72 \text{ atm}^{-1})(P_{(HC_2H_5O_2)_2})^2 - (12.16)P_{(HC_2H_5O_2)_2} + 8.37 \text{ atm} = 0$$

The two roots are

$$2.28 \text{ atm} \quad \text{and} \quad 0.985 \text{ atm}$$

We reject the root 2.28 atm because it is greater than the original total pressure. Thus,

$$P_{(HC_2H_5O_2)_2} = 0.985 \text{ atm}$$

E. ANSWERS TO THE SELF-TEST

1. false
2. false
3. dynamic
4. true
5. the rate of the reverse reaction

6. decrease
7. increase
8. constant
9. false
10. false

11. numerator

12. denominator

13. stoichiometric or balancing coefficient

14. true

15. $\dfrac{[X]^x[Y]^y}{[A]^a[B]^b}$

16. products . . . reactants

17. reactant

18. reactants . . . products

19. product

20. concentrations or partial pressures

21. false

22. concentration or molarity

23. pressure, such as torr or atm

24. $1/K_f$

25. $K_1 K_2$

26. K_1^2

27. true

28. if a chemical reaction at equilibrium is subjected to a change in conditions that displaces it from equilibrium, then the reaction adjusts toward a new equilibrium state. The reaction proceeds in the direction that at least partially offsets the change in conditions

29. volume (or pressure), addition of product or reactant, temperature

30. right to left

31. true

32. the fewer number of moles of gases

33. true

34. false

35. true

36. false

37. left

38. right

39. right

40. left

41. false

42. increases

43. Haber

44. $N_2(g) + H_2(g) \underset{300\ atm}{\overset{500°C}{\rightleftharpoons}} 2NH_3(g)$

45. equilibrium

46. true

47. $\dfrac{[X]_0^x[Y]_0^y}{[A]_0^a[B]_0^b}$ . . . arbitrary concentrations

48. true

49. false

50. true

51. left to right

52. right to left

53. K_c in magnitude

54. true

18   ACIDS AND BASES I

A. OUTLINE OF CHAPTER 18

18-1   An acid is a proton donor and a base is a proton acceptor.

In the Arrhenius acid-base classification, an acid produces $H^+(aq)$ and a base produces $OH^-(aq)$ in aqueous solution.

In the Brønsted-Lowry acid-base classification, an acid is a proton donor and a base is a proton acceptor.

In aqueous solution an acid donates a proton to water, which acts as a base, to produce the hydronium ion $H_3O^+(aq)$.

The hydronium ion is a hydrated photon (Figure 18-1).

A proton-transfer reaction is a reaction involving the transfer of a proton from one molecule to another.

A proton-transfer reaction is also called a protonation reaction.

18-2   In an aqueous solution the product of the ion concentrations $[H_3O^+][OH^-]$ is a constant.

The equilibrium-constant expression for the equation of the autoprotonation reaction

$$H_2O(l) + H_2O(l) \rightleftharpoons H_3O^+(aq) + OH^-(aq)$$

is given by

$$K_w = [H_3O^+][OH^-] = 1.00 \times 10^{-14} \text{ M}^2 \qquad \text{at } 25°C \qquad (18\text{-}3)$$

K_w is called the ion product constant of water.

In a neutral solution, $[H_3O^+] = [OH^-]$.

In an acidic solution, $[H_3O^+] > [OH^-]$.

In a basic solution, $[OH^-] > [H_3O^+]$.

18-3   Strong acids and strong bases are completely dissociated in aqueous solutions.

Strong acids transfer all their dissociable protons to water molecules.

Acids and bases that are incompletely dissociated in water are called weak acids and weak bases.

Most acids and bases are weak.

Some strong acids and bases are given in Table 18-1.

18-4   Almost all organic acids are weak acids.

Carboxylic acids have the general formula RCOOH, where R is a hydrogen atom or an alkyl group.

The —COOH group is the carboxyl group.

The anion formed by a carboxylic acid is called a carboxylate ion and has the general formula $RCOO^-$.

The carboxylate ion, $RCOO^-$, is stabilized by charge delocalization.

18-5   pH is a measure of the acidity of an aqueous solution.

The pH of a solution is defined as

$$pH = -\log [H_3O^+] \tag{18-4}$$

By definition,

$$pOH = -\log [OH^-] \tag{18-5}$$

At 25°C the pH and pOH are related by the expression

$$pH + pOH = 14.00 \tag{18-6}$$

The pH of most aqueous solutions lies in the range 0 to 14, although pH values outside this range are occasionally encountered.

A change in pH of one unit corresponds to a 10-fold change in $[H_3O^+]$.

In a neutral aqueous solution at 25°C, pH = 7.0.

In an acidic aqueous solution at 25°C, pH < 7.0.

In a basic aqueous solution at 25°C, pH > 7.0.

18-6   Weak acids and weak bases are dissociated only partially in water.

The value of $[H_3O^+]$ can be found from a pH measurement.

$$[H_3O^+] = 10^{-pH} \tag{18-7}$$

The percent dissociation of an acid is given by

$$\% \text{ dissociation} = \frac{[H_3O^+]}{[\text{acid}]_0} \times 100$$

where $[\text{acid}]_0$ is the stoichiometric concentration of the acid.

Stoichiometric concentration is the concentration at which a solution is prepared.

Solutions of weak acids and bases are poor conductors of electricity.

18-7 Acids with large values of K_a are stronger than acids with smaller values of K_a.

For the acid-dissociation equilibrium

$$HB(aq) + H_2O(l) \rightleftharpoons H_3O^+(aq) + B^-(aq)$$

the acid-dissociation constant K_a is given by

$$K_a = \frac{[H_3O^+][B^-]}{[HB]}$$

The pH of an aqueous solution of a weak acid can be found from the value of K_a and the K_a expression.

The material balance condition means that the anion must be either in the form of the undissociated acid or the aqueous anion.

A concentration table contains the initial concentrations and equilibrium concentrations of all the species in an aqueous solution.

The percent dissociation of an acid increases as the concentration of the acid decreases (Figure 18-5).

By definition,

$$pK_a = -\log K_a \tag{18-15}$$

The smaller the pK_a value for an acid, the stronger is the acid.

Some K_a and pK_a values are given in Table 18-3.

18-8 The method of successive approximations is often used in solving acid-base equilibrium problems.

In the first approximation, it is assumed that the value of $[H_3O^+]$ is small relative to the initial concentration of a weak acid; that is,

$$[acid]_0 - [H_3O^+] \approx [acid]_0$$

In the second approximation, the value of $[H_3O^+]$ found in the first approximation is used in acid-dissociation-constant expression.

18-9 Bases with large values of K_b are stronger than bases with smaller values of K_b.

Ammonia is a base in aqueous solution because it reacts with water to accept a proton.

Amines are compounds in which the hydrogen atoms in ammonia are substituted by hydrocarbon groups.

For the base-protonation equilibrium

$$B(aq) + H_2O(l) \rightleftharpoons HB^+(aq) + OH^-(aq)$$

the base-protonation constant K_b is given by

$$K_b = \frac{[HB^+][OH^-]}{[B]}$$

The pH of an aqueous solution of a weak base can be found from the value of K_b and the K_b expression.

By definition,

$$pK_b = -\log K_b \qquad (18\text{-}16)$$

The smaller the value of pK_b, the stronger is the base.

Some K_b and pK_b values are given in Table 18-4.

18-10   The pair of species HB, B^- is called a conjugate acid-base pair.

For the dissociation of an acid in water

$$HB(aq) + H_2O(l) \rightleftharpoons H_3O^+(aq) + B^-(aq)$$

B^- is the conjugate base of HB.

For the dissociation of a base in water

$$B^-(aq) + H_2O(l) \longrightarrow HB(aq) + OH^-(aq)$$

HB is the conjugate acid of B^-.

For a conjugate acid-base pair,

$$K_a K_b = K_w \qquad (18\text{-}23)$$

or at 25°C

$$pK_a + pK_b = 14.00 \qquad (18\text{-}24)$$

The anion of a weak acid is itself a weak base.

The values of K_a, K_b, pK_a, and pK_b for a number of conjugate acid-base pairs are listed in Table 18-5.

18-11   Aqueous solutions of many salts are either acidic or basic.

Various ions react with water to produce hydronium ions or hydroxide ions.

The neutral ions, basic anions, acidic anions, and acidic cations are described on pages 660 and 661 of the text.

The acidic, neutral, or basic properties of a number of ions are given in Table 18-6.

Many metal ions are acidic in aqueous solution because one of the water molecules attached to the aqueous ion transfers a proton to an unattached water molecule.

The values of K_a and pK_a for various cations in aqueous solution at 25°C are listed in Table 18-7.

The pH of aqueous salt solutions can be found from K_a or K_b expressions.

18-12 A Lewis acid is an electron-pair acceptor.

In the Lewis acid-base classification, an acid is an electron-pair acceptor and a base is an electron-pair donor.

In general, an electron-deficient species can act as a Lewis acid and a species with a lone pair of electrons can act as a Lewis base.

Boric acid, $B(OH)_3$, is a Lewis acid, but not a Brønsted-Lowry acid, because it accepts a hydroxide ion rather than donates a proton.

B. SELF-TEST

1. In the Brønsted-Lowry acid-base classification, an acid is _____.

2. In the Brønsted-Lowry acid-base classification, a base is _____.

3. Ammonia, NH_3, is an example of an acid. *True/False*

4. The hydronium ion is _____.

5. In aqueous solution the proton exists as a bare proton. *True/False*

6. In aqueous solution the concentration of $OH^-(aq)$ does not depend on the concentration of $H_3O^+(aq)$. *True/False*

7. The concentration of $OH^-(aq)$ in a neutral aqueous solution at 25°C is _____ _____.

8. The ion-product constant of water is given by the expression $K_w =$ _____.

9. The value of K_w at 25°C is _____.

10. The concentration of $OH^-(aq)$ in an aqueous solution is 2.5×10^{-3} M; the solution is (*acidic/basic/neutral*).

11. The concentration of $H_3O^+(aq)$ in an aqueous solution is 2.5×10^{-11} M; the solution is (*acidic/basic/neutral*).

12. A strong acid is (*completely/partially*) dissociated in aqueous solution. *True/False*

13. A strong base is (*completely/partially*) dissociated in aqueous solution. *True/False*

14. An aqueous solution of HBr contains significant quantities of the species HBr(*aq*), $Br^-(aq)$, and $H_3O^+(aq)$. *True/False*

15. An aqueous solution of KOH contains the ionic species ———————————— ———————————— .

16. A carboxylic acid has the general formula ———————— .

17. Most carboxylic acids are (*weak/strong*) acids.

18. A carboxylic acid reacts with water to form a hydronium ion and a ———————— ion.

19. The negative charge on a carboxylate ion is distributed over both oxygen atoms. *True/False*

20. The pH of a solution is defined as ———————————————————— .

21. The pH scale compresses the wide range of the concentrations of $H_3O^+(aq)$ that occur. *True/False*

22. Measurement of the pH of a solution can be used to determine the concentration of $H_3O^+(aq)$ in the solution. *True/False*

23. The concentration of $OH^-(aq)$ in an aqueous solution cannot be determined from pH measurements. *True/False*

24. The pH of pure water or of a neutral aqueous solution at 25°C is ———————— .

25. Acidic solutions have pH values (*less/greater*) than 7.00.

26. Basic solutions have pH values (*less/greater*) than 7.00.

27. A weak acid is (*completely/partially*) dissociated in water.

28. A 0.10-M aqueous solution of the weak acid, HF, has the same pH as a 0.10-M aqueous solution of HCl. *True/False*

29. An aqueous solution of acetic acid, CH_3COOH, contains no undissociated acid. *True/False*

30. An acid-dissociation reaction is an example of a ———————————— reaction.

31. The stoichiometric concentration of an acid is the equilibrium concentration of the acid in aqueous solution. *True/False*

32. An aqueous solution of nitrous acid, $HNO_2(aq)$, contains the species ———————— ———————————————————— .

33. The equation for the acid dissociation reaction of hydrofluoric acid, HF(aq), is

_____ .

34. The acid-dissociation-constant expression for hydrofluoric acid, HF(aq), is $K_a =$

_____ .

35. The value of K_a indicates the extent of an acid's dissociation in water. *True/False*

36. A weak base is (*completely/partially*) protonated in aqueous solution.

37. A 0.10-M aqueous solution of ammonia, NH_3, has the same pH as a 0.10-M aqueous solution of sodium hydroxide, NaOH. *True/False*

38. The protonation equation for ammonia, NH_3, in water is _____

_____ .

39. The base-protonation equilibrium constant expression for ammonia, NH_3, is given by $K_b =$ _____ .

40. The value of the pK_a of an acid is given by $pK_a =$ _____ .

41. The larger the pK_a value of an acid, the (*stronger/weaker*) is the acid.

42. The value of the pK_b of a base is given by $pK_b =$ _____ .

43. The larger the value of pK_b of a base, the (*stronger/weaker*) is the base.

44. The conjugate (*acid/base*) of $NH_3(aq)$ is $NH_4^+(aq)$.

45. The conjugate (*acid/base*) of HF(aq) is $F^-(aq)$.

46. The base-protonation constant K_b for the conjugate base of an acid is related to the acid-dissociation constant K_a of the acid by the relation $K_b =$ _____ .

47. The conjugate base of a strong monoprotic strong acid is a(n)(*acidic/basic/neutral*) ion.

48. The aqueous solution of a salt is always neutral. *True/False*

49. The anion of a weak acid is a (*neutral/acidic/basic*) anion.

50. The cation of a strong base is a (*neutral/acidic/basic*) cation.

51. The conjugate acid of a weak base is an acidic cation. *True/False*

52. An anion may be acidic because the anion contains an acidic proton. *True/False*

53. There are no (*basic/acidic*) cations.

54. A metal ion may be acidic because one of the water molecules attached to it may transfer a proton to an unattached water molecule. *True/False*

55. In the Lewis acid-base classification an acid is ————————————————.

56. In the Lewis acid-base classification a base is ————————————————.

57. An electron-deficient species can act as a Lewis (*acid/base*).

58. A species with a lone pair of electrons can act as a Lewis (*acid/base*).

C. CALCULATIONS YOU SHOULD KNOW HOW TO DO

1. Calculate the concentrations of various ionic species in solutions of strong acids or strong bases. See Examples 18-1 and 18-2 and Problems 18-1 through 18-4.

2. Calculate the pH of an aqueous solution of a strong base or a strong acid. See Examples 18-4 and 18-5 and Problems 18-5 through 18-10. (The use of logarithms is explained in the following section.)

3. Calculate $[H_3O^+]$ or $[OH^-]$ from the pH. See Example 18-7 and Problems 18-11 through 18-16.

4. Calculate the value of K_a or K_b from the pH of the solution. See Problems 18-17 through 18-20 and 18-29 and 18-30.

5. Given the pH, calculate the percentage of weak-acid molecules that are dissociated or the percentage of weak-base molecules that are protonated. See Example 18-8 and Problems 18-21 and 18-22.

6. Calculate the pH of an acidic solution given the value of K_a. Use the method of successive approximations in solving this type of problem. See Examples 18-9 and 18-11 and Problems 18-23 through 18-28.

7. Calculate the pH of a basic solution given the value of K_b. See Example 18-12 and Problems 18-31 through 18-34.

8. Calculate K_a given K_b or calculate K_b given K_a for conjugate acid-base pairs. See Example 18-14 and Problems 18-45 through 18-48.

10. Calculate the pH of aqueous salt solutions. See Example 18-16 and Problems 18-55 through 18-64.

A Review of Common Logarithms

Recall from Appendix A of the text that $100 = 10^2$, $1000 = 10^3$, and so on. Also recall that

$$\sqrt{10} = 10^{1/2} = 10^{0.50} = 3.16$$

By taking the square root of both sides of

$$10^{0.50} = 3.16$$

we find that

$$\sqrt{10^{0.50}} = 10^{(1/2)(0.50)} = 10^{0.25} = \sqrt{3.16} = 1.78$$

Furthermore, because

$$(10^x)(10^y) = 10^{x+y}$$

we can write

$$10^{0.25} \times 10^{0.50} = 10^{0.75} = (3.16)(1.78) = 5.62$$

By continuing this process, we would be able to express any number y as

$$y = 10^x \qquad (1)$$

The number x to which 10 must be raised to get y is called the *common logarithm* of y and is written as

$$x = \log y \qquad (2)$$

Equations (1) and (2) are equivalent.

For example, we have shown above that

$$\log 1.78 = 0.25$$
$$\log 3.16 = 0.50$$
$$\log 5.62 = 0.75$$
$$\log 10.00 = 1.00$$

You can find the common logarithm of any number using a hand calculator. You simply enter the number and push the LOG key. For example, to find the logarithm of 42,500, simply enter 42,500 and push the LOG key to obtain 4.6284. To find the common logarithm of 0.000465, enter 0.000465 and push the LOG key to obtain -3.333.

Because logarithms are exponents ($y = 10^x$), logarithms have certain special properties, such as

$$\log ab = \log a + \log b \qquad (3)$$

$$\log \frac{a}{b} = \log a - \log b \qquad (4)$$

$$\log a^n = n \log a \qquad (5)$$

$$\log \sqrt[n]{a} = \log a^{1/n} = \frac{1}{n} \log a \qquad (6)$$

If we let $a = 1$ in Equation (4), then we have

$$\log \frac{1}{b} = \log 1 - \log b$$

or, because $\log 1 = 0$,

$$\log \frac{1}{b} = -\log b \qquad (7)$$

Thus, we change the sign of the logarithm by taking the reciprocal of its argument.
   Up to this point we have found the value of x in

$$y = 10^x$$

when y is given. It is often necessary to find the value of y when x is given. Because x is called the logarithm of y, y is called the *antilogarithm* of x. For example, suppose that $x = 6.1303$ and we wish to find y. We write

$$y = 10^{6.1303}$$

You can find the antilogarithm of any number using a hand calculator. On some calculators, you enter the number and press the INV key (for inverse) and then the LOG key. On other calculators, you enter the number and press the 10^x key. Using a hand calculator, the antilogarithm of 6.1303 is

$$10^{6.1303} = 1.350 \times 10^6$$

Exercises

1 Find the logarithms of the following numbers:

(a) 3.12×10^{-10} **(b)** 8.06×10^5

(c) 12.3   **(d)** 6.63×10^{-12}

(e) 556,000   **(f)** 0.0000291

2 Find the antilogarithms of the following numbers:

(a) 4.316   **(b)** 0.711

(c) -5.2573 **(d)** -1.6289

Answers

1 **(a)** -9.506 **(b)** 5.906

       (c) 1.090   **(d)** -11.179

       (e) 5.745   **(f)** -4.536

2 **(a)** 2.07×10^4 **(b)** 5.14

       (c) 5.530×10^{-6} **(d)** 2.350×10^{-2}

D. SOLUTIONS TO THE ODD-NUMBERED PROBLEMS

18-1 Because $HClO_4$ is a strong acid in water (see Table 18-1), it is completely dissociated, and thus,

$$[H_3O^+] = 0.150 \text{ M} \quad \text{and} \quad [ClO_4^-] = 0.150 \text{ M}$$

We can calculate $[OH^-]$ from the K_w expression.

$$K_w = [H_3O^+][OH^-] = 1.00 \times 10^{-14} \text{ M}^2$$

Solving for $[OH^-]$, we get

$$[OH^-] = \frac{1.00 \times 10^{-14} \text{ M}^2}{[H_3O^+]} = \frac{1.00 \times 10^{-14} \text{ M}^2}{0.150 \text{ M}} = 6.67 \times 10^{-14} \text{ M}$$

Because $[H_3O^+] > [OH^-]$, the solution is acidic.

18-3 We first must find the number of moles in 2.00 g of TlOH.

$$n = (2.00 \text{ g}) \left(\frac{1 \text{ mol TlOH}}{221.4 \text{ g TlOH}} \right) = 0.00903 \text{ mol}$$

The molarity of the solution is

$$\text{molarity} = \frac{\text{moles of solute}}{\text{volume of solution}} = \frac{0.00903 \text{ mol}}{0.500 \text{ L}} = 0.0181 \text{ M}$$

Because TlOH is a strong base (Table 18-1), it is completely dissociated in
aqueous solution, and thus,

$$[OH^-] = 1.81 \times 10^{-2} \text{ M} \quad \text{and} \quad [Tl^+] = 1.81 \times 10^{-2} \text{ M}$$

We can calculate $[H_3O^+]$ from the K_w expression

$$K_w = [H_3O^+][OH^-] = 1.00 \times 10^{-14} \text{ M}^2$$

to get

$$[H_3O^+] = \frac{1.00 \times 10^{-14} \text{ M}^2}{[OH^-]} = \frac{1.00 \times 10^{-14} \text{ M}^2}{1.81 \times 10^{-2} \text{ M}} = 5.52 \times 10^{-13} \text{ M}$$

18-5 Because HNO_3 is a strong acid in solution, it is completely dissociated, and thus,

$$[H_3O^+] = 0.020 \text{ M}$$

The pH is given by

$$pH = -\log [H_3O^+] = -\log (0.020) = 1.70$$

Because pH < 7, the solution is acidic.

18-7   Because both HCl and HBr are strong acids, they are completely dissociated in water. Thus, from HCl,

$$[H_3O^+] = 0.035 \text{ M}$$

and from HBr

$$[H_3O^+] = 0.045 \text{ M}$$

The total concentration of $H_3O^+(aq)$ is

$$[H_3O^+] = 0.035 \text{ M} + 0.045 \text{ M} = 0.080 \text{ M}$$

The pH of the solution is

$$pH = -\log [H_3O^+] = -\log (0.080) = 1.10$$

The solution is acidic because the pH $<$ 7.
The pOH of the solution is

$$pOH = 14.00 - pH = 14.00 - 1.10 = 12.90$$

18-9   We first calculate the number of moles of KOH in 2.00 g:

$$n = (2.00 \text{ g}) \left(\frac{1 \text{ mol KOH}}{56.11 \text{ g KOH}} \right) = 3.564 \times 10^{-2} \text{ mol}$$

The molarity of the solution is

$$\text{molarity} = \frac{\text{moles of solute}}{\text{volume of solution}} = \frac{3.564 \times 10^{-2} \text{ mol}}{0.500 \text{ L}} = 7.13 \times 10^{-2} \text{ M}$$

Because KOH is a strong base, it is completely dissociated, and thus,

$$[OH^-] = 7.13 \times 10^{-2} \text{ M}$$

The pOH of the solution is

$$pOH = -\log [OH^-] = -\log (7.13 \times 10^{-2}) = 1.15$$

The pH of the solution is

$$pH = 14.00 - pOH = 14.00 - 1.15 = 12.85$$

18-11   We calculate the value of $[H_3O^+]$ using Equation (18-7),

$$[H_3O^+] = 10^{-pH}$$

The pH of the muscle fluids is given as 6.8, and so

$$[H_3O^+] = 10^{-6.8} = 1.6 \times 10^{-7} \text{ M}$$

If you do not use a calculator to find $10^{-6.8}$, then write

$$[H_3O^+] = 10^{-6.8} = 10^{0.20} \times 10^{-7} = 1.6 \times 10^{-7} \text{ M}$$

18-13   We use Equation (18-7). The pH is 1.0, and so

$$[H_3O^+] = 10^{-pH} = 10^{-1.0} = 0.1 \text{ M}$$

Because HCl is a strong acid, the concentration of HCl in the stomach when the pH = 1.0 is

$$[HCl] = [H_3O^+] = 0.1 \text{ M}$$

18-15   We use Equation (18-7). The pH of human blood is 7.4, and so

$$[H_3O^+] = 10^{-pH} = 10^{-7.4} = 4 \times 10^{-8} \text{ M}$$

We can calculate $[OH^-]$ from the K_w expression.

$$[OH^-] = \frac{1.00 \times 10^{-14} \text{ M}^2}{[H_3O^+]} = \frac{1.00 \times 10^{-14} \text{ M}^2}{4 \times 10^{-8} \text{ M}} = 3 \times 10^{-7} \text{ M}$$

18-17   The equation for the reaction is

$$CH_3CH_2COOH(aq) + H_2O(l) \rightleftharpoons H_3O^+(aq) + CH_3CH_2COO^-(aq)$$

The acid-dissociation-constant expression is

$$K_a = \frac{[H_3O^+][CH_3CH_2COO^-]}{[CH_3CH_2COOH]}$$

We can find the value of $[H_3O^+]$ from the pH of the solution. We have

$$[H_3O^+] = 10^{-pH} = 10^{-3.09} = 8.13 \times 10^{-4} \text{ M}$$

From the reaction stoichiometry at equilibrium, $[H_3O^+] = [CH_3CH_2COO^-]$, because we started with only CH_3CH_2COOH. At equilibrium

$$CH_3CH_2COOH = 0.050 \text{ M} - [H_3O^+] = 0.050 \text{ M} - 0.000813 \text{ M} = 0.049 \text{ M}$$

Substituting the values of the concentrations of H_3O^+ (aq), $CH_3CH_2COO^-$ (aq), and CH_3CH_2COOH (aq) into the K_a expression, we have

$$K_a = \frac{(8.13 \times 10^{-4} \text{ M})(8.13 \times 10^{-4} \text{ M})}{(0.049 \text{ M})}$$
$$= 1.3 \times 10^{-5} \text{ M}$$

18-19   The equation for the reaction is

$$CH_3COOH(aq) + H_2O(l) \rightleftharpoons H_3O^+(aq) + CH_3COO^-(aq)$$

The acid-dissociation-constant expression is

$$K_a = \frac{[H_3O^+][CH_3COO^-]}{[CH_3COOH]}$$

We can find the value of $[H_3O^+]$ from the pH:

$$[H_3O^+] = 10^{-pH} = 10^{-3.39} = 4.07 \times 10^{-4} \text{ M}$$

We can set up a table of initial concentrations and equilibrium concentrations.

	CH$_3$COOH(*aq*)	+	H$_2$O(*l*)	⇌	H$_3$O$^+$(*aq*)	+	CH$_3$COO$^-$(*aq*)
Initial concentration	1.00×10^{-2} M		—		≈ 0		0
Equilibrium concentration	1.00×10^{-2} M $- [H_3O^+]$ $= 0.96 \times 10^{-2}$ M		—		4.07×10^{-4} M		$[CH_3COO^-] = [H_3O^+]$ $= 4.07 \times 10^{-4}$ M

Substituting in the values of the equilibrium concentrations in the K_a expression, we have

$$K_a = \frac{(4.07 \times 10^{-4} \text{ M})(4.07 \times 10^{-4} \text{ M})}{0.96 \times 10^{-2} \text{ M}}$$
$$= 1.7 \times 10^{-5} \text{ M}$$

18-21 The reaction is

$$\text{acetylsalicylic acid}(aq) + H_2O(l) \rightleftharpoons H_3O^+(aq) + \text{acetylsalicylate } (aq)$$

or

$$\text{acid}(aq) + H_2O(l) \rightleftharpoons H_3O^+(aq) + \text{anion } (aq)$$

The acid-dissociation-constant expression is

$$K_a = \frac{[\text{anion}][H_3O^+]}{[\text{acid}]} = 2.75 \times 10^{-5} \text{ M}$$

The ratio of the dissociated acid to the undissociated acid is given by

$$\frac{[\text{anion}]}{[\text{acid}]} = \frac{K_a}{[H_3O^+]} = \frac{2.75 \times 10^{-5} \text{ M}}{[H_3O^+]}$$

The concentration of $H_3O^+(aq)$ is

$$[H_3O^+] = 10^{-pH} = 10^{-1.5} = 3.2 \times 10^{-2} \text{ M}$$

and so

$$\frac{[\text{anion}]}{[\text{acid}]} = \frac{2.75 \times 10^{-5} \text{ M}}{3.2 \times 10^{-2} \text{ M}} = 9 \times 10^{-4}$$

18-23   The equation is

$$C_6H_5COOH(aq) + H_2O(l) \rightleftharpoons H_3O^+(aq) + C_6H_5COO^-(aq)$$

We can set up a table of initial and equilibrium concentrations.

	$C_6H_5COOH(aq)$	$+$ $H_2O(l)$ $\rightleftharpoons$ $H_3O^+(aq)$	$+$ $C_6H_5COO^-(aq)$
Initial concentration	0.020 M	— ≈ 0	0
Equilibrium concentration	$0.020 \text{ M} - [H_3O^+]$	— $[H_3O^+]$	$[C_6H_5COO^-] = [H_3O^+]$

Substituting the equilibrium concentration expressions in the K_a expression, we have

$$K_a = \frac{[H_3O^+][C_6H_5COO^-]}{[C_6H_5COOH]} = \frac{[H_3O^+]^2}{0.020 \text{ M} - [H_3O^+]} = 6.46 \times 10^{-5} \text{ M}$$

We shall use the method of successive approximations to find $[H_3O^+]$. We ignore $[H_3O^+]$ in the denominator to obtain

$$\frac{[H_3O^+]^2}{0.020 \text{ M}} = 6.46 \times 10^{-5} \text{ M}$$

Solving for $[H_3O^+]$, we have

$$[H_3O^+] = 1.14 \times 10^{-3} \text{ M}$$

The second approximation is

$$\frac{[H_3O^+]^2}{0.020 \text{ M} - 1.14 \times 10^{-3} \text{ M}} = 6.46 \times 10^{-5} \text{ M}$$

which gives

$$[H_3O^+] = 1.10 \times 10^{-3} \text{ M}$$

The third approximation gives $[H_3O^+] = 1.10 \times 10^{-3}$ M. Thus,

$$[C_6H_5COO^-] = [H_3O^+] = 1.10 \times 10^{-3} \text{ M}$$
$$[C_6H_5COOH] = 0.020 \text{ M} - 1.10 \times 10^{-3} \text{ M} = 0.019 \text{ M}$$
$$\text{pH} = -\log (1.10 \times 10^{-3}) = 2.96$$

18-25   The equation is

$$Cl_3CCOOH(aq) + H_2O(l) \rightleftharpoons H_3O^+(aq) + Cl_3CCOO^-(aq)$$

We can set up a table of initial and equilibrium concentrations.

	$Cl_3CCOOH(aq)$	+	$H_2O(l) \rightleftharpoons H_3O^+(aq)$	+	$Cl_3CCOO_2^-(aq)$
Initial concentration	0.030 M	—	≈ 0		0
Equilibrium concentration	0.030 M $- [H_3O^+]$	—	$[H_3O^+]$		$[Cl_3CCOO^-] = [H_3O^+]$

Substituting the equilibrium concentration expressions in the K_a expression, we have

$$K_a = \frac{[H_3O^+][Cl_3CCOO^-]}{[Cl_3CCOOH]} = \frac{[H_3O^+]^2}{0.030 \text{ M} - [H_3O^+]} = 2.3 \times 10^{-1} \text{ M}$$

We cannot use the method of successive approximations because the value of K_a is too large. Rewriting the equation in the form of the standard quadratic equation gives

$$[H_3O^+]^2 + 2.3 \times 10^{-1} \text{ M } [H_3O^+] - 6.9 \times 10^{-3} \text{ M}^2 = 0$$

The quadratic formula gives

$$[H_3O^+] = \frac{-2.3 \times 10^{-1} \text{ M} \pm \sqrt{5.29 \times 10^{-2} \text{ M}^2 - (4)(1)(-6.9 \times 10^{-3} \text{ M}^2)}}{(2)(1)}$$
$$= \frac{-2.3 \times 10^{-1} \text{ M} \pm 2.8 \times 10^{-1} \text{ M}}{2}$$
$$= 2.5 \times 10^{-2} \text{ M} \quad \text{and} \quad -0.26 \text{ M}$$

We reject the negative root and write

$$\text{pH} = -\log\,[\text{H}_3\text{O}^+] = -\log\,(2.5 \times 10^{-2}) = 1.6$$
$$[\text{Cl}_3\text{COO}^-] = [\text{H}_3\text{O}^+] = 2.5 \times 10^{-2}\ \text{M}$$
$$[\text{Cl}_3\text{COOH}] = 0.030\ \text{M} - 0.025\ \text{M} = 0.005\ \text{M}$$

18-27   We can set up a table of initial and equilibrium concentrations.

	$\text{HO}_3\text{SNH}_2(aq)$	$+$	$\text{H}_2\text{O}(l)$ $\rightleftharpoons$	$\text{H}_3\text{O}^+(aq)$	$+$	$\text{O}_3\text{SNH}_2^-(aq)$
Initial concentration	0.040 M		—	≈ 0		0
Equilibrium concentration	$0.040\ \text{M} - [\text{H}_3\text{O}^+]$		—	$[\text{H}_3\text{O}^+]$		$[\text{O}_3\text{SNH}_2^-] = [\text{H}_3\text{O}^+]$

Substituting the equilibrium concentration expressions in the K_a expression, we have

$$K_a = \frac{[\text{H}_3\text{O}^+][\text{O}_3\text{SNH}_2^-]}{[\text{HO}_3\text{SNH}_2]} = \frac{[\text{H}_3\text{O}^+]^2}{0.040\ \text{M} - [\text{H}_3\text{O}^+]} = 0.10\ \text{M}$$

or

$$[\text{H}_3\text{O}^+]^2 + 0.10\ \text{M}\,[\text{H}_3\text{O}^+] - 0.0040\ \text{M}^2 = 0$$

The quadratic formula gives

$$[\text{H}_3\text{O}^+] = \frac{-0.10\ \text{M} \pm \sqrt{0.010\ \text{M}^2 - (4)(1)(-0.0040\ \text{M}^2)}}{(2)(1)}$$
$$= \frac{-0.10\ \text{M} \pm 0.16\ \text{M}}{2}$$
$$= 0.03\ \text{M} \quad \text{and} \quad -0.13\ \text{M}$$

We reject the negative root and write

$$\text{pH} = -\log\,[\text{H}_3\text{O}^+] = -\log\,(0.03) = 1.5$$
$$[\text{O}_3\text{SNH}_2^-] = [\text{H}_3\text{O}^+] = 0.03\ \text{M}$$
$$[\text{HO}_3\text{SNH}_2] = 0.040\ \text{M} - 0.03\ \text{M} = 0.01\ \text{M}$$

18-29   The equation for the reaction is

$$NH_3(aq) + H_2O(l) \rightleftharpoons NH_4^+(aq) + OH^-(aq)$$

The pOH of the solution is given by

$$pOH = 14.00 - pH = 14.00 - 11.12 = 2.88$$

and the concentration of $OH^-(aq)$ is

$$[OH^-] = 10^{-pOH} = 10^{-2.88} = 1.32 \times 10^{-3}\ M$$

We can set up a table of initial and equilibrium concentrations.

	$NH_3(aq)$	+	$H_2O(l) \rightleftharpoons NH_4^+(aq)$	+	$OH^-(aq)$
Initial concentration	0.100 M	—	0		≈ 0
Equilibrium concentration	0.100 M − [OH⁻] = 9.9×10^{-2} M	—	$[NH_4^+] = [OH^-] = 1.32 \times 10^{-3}$ M		$[OH^-] = 1.32 \times 10^{-3}$ M

The base protonation constant expression is

$$K_b = \frac{[NH_4^+][OH^-]}{[NH_3]} = \frac{(1.32 \times 10^{-3}\ M)(1.32 \times 10^{-3}\ M)}{9.9 \times 10^{-2}\ M}$$
$$= 1.8 \times 10^{-5}\ M$$

18-31   The equation is

$$C_5H_5N(aq) + H_2O(l) \rightleftharpoons C_5H_5NH^+(aq) + OH^-(aq)$$

We can set up a table of initial and equilibrium concentrations.

	$C_5H_5N(aq)$	+	$H_2O(l) \rightleftharpoons C_5H_5NH^+(aq)$	+	$OH^-(aq)$
Initial concentration	0.300 M	—	0		≈ 0
Equilibrium concentration	0.300 M − [OH⁻]	—	$[C_5H_5NH^+] = [OH^-]$		[OH⁻]

The expression for K_b is

$$K_b = \frac{[C_5H_5NH^+][OH^-]}{[C_5H_5N]} = \frac{[OH^-]^2}{0.300 \text{ M} - [OH^-]} = 1.46 \times 10^{-9} \text{ M}$$

Because K_b is so small, we expect that $[OH^-]$ will be small and thus negligible compared with 0.300 M. The expression or K_b becomes

$$\frac{[OH^-]^2}{0.300 \text{ M}} \approx 1.46 \times 10^{-9} \text{ M}$$

$$[OH^-]^2 \approx 4.38 \times 10^{-10} \text{ M}^2$$

$$[OH^-] \approx 2.09 \times 10^{-5} \text{ M}$$

We can see that $[OH^-]$ is much smaller than 0.300 M. The method of successive approximations also confirms this result. The pOH of the solution is given by

$$\text{pOH} = -\log[OH^-] = -\log(2.09 \times 10^{-5}) = 4.68$$

and the pH is

$$\text{pH} = 14.00 - \text{pOH} = 14.00 - 4.68 = 9.32$$

18-33   We can set up a table of initial and equilibrium concentrations.

	$(CH_3)_2NH(aq)$	+	$H_2O(l)$	$\rightleftharpoons$ $(CH_3)_2NH_2^+(aq)$	+	$OH^-(aq)$
Initial concentration	0.060 M		—	0		≈ 0
Equilibrium concentration	0.060 M $- [OH^-]$		—	$[(CH_3)_2NH_2^+] = [OH^-]$		$[OH^-]$

The expression for K_b is

$$K_b = \frac{[(CH_3)_2NH_2^+][OH^-]}{[(CH_3)_2NH]} = \frac{[OH^-]^2}{0.060 \text{ M} - [OH^-]} = 5.81 \times 10^{-4} \text{ M}$$

The value of K_b is not small enough to ignore $[OH^-]$ in the denominator, and so we use the quadratic formula.

$$[OH^-]^2 + 5.81 \times 10^{-4} \text{ M}[OH^-] - 3.49 \times 10^{-5} \text{ M}^2 = 0$$

The quadratic formula gives

$$[OH^-] = \frac{-5.81 \times 10^{-4}\ M \pm \sqrt{3.38 \times 10^{-7}\ M^2 - (4)(1)(-3.49 \times 10^{-5}\ M^2)}}{(2)(1)}$$

$$= \frac{-5.81 \times 10^{-4}\ M \pm 1.18 \times 10^{-2}\ M}{2}$$

$$= 5.61 \times 10^{-3}\ M \qquad (\text{and} -0.00619\ M)$$

We could also have used the method of successive approximations. The pOH of the solution is given by

$$pOH = -\log[OH^-] = -\log(5.61 \times 10^{-3}) = 2.25$$

and so the pH is

$$pH = 14.00 - pOH = 14.00 - 2.25 = 11.75$$

18-35 (a) The equilibrium is shifted from right to left.
      (b) The equilibrium is shifted from left to right.
      (c) The equilibrium is shifted from left to right.
      (d) The equilibrium is shifted from left to right, because there are more solute particles on the right than on the left.

18-37 (a) The equilibrium shifts from right to left.
      (b) The equilibrium is not affected. Because $\Delta H^\circ_{rxn} \approx 0$, the equilibrium constant does not change with temperature.
      (c) The equilibrium shifts from right to left.
      (d) The equilibrium shifts from left to right because the added $NH_3(aq)$ reacts with $H_3O^+(aq)$ according to

$$NH_3(aq) + H_3O^+(aq) \rightleftharpoons NH_4^+(aq) + H_2O(l)$$

      (e) The equilibrium shifted from right to left because $[H_3O^+]$ has increased.

18-39 (a) $C_6H_5COOH(aq) + H_2O(l) \rightleftharpoons H_3O^+(aq) + C_6H_5COO^-(aq)$

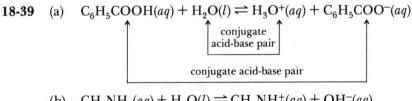

      (b) $CH_3NH_2(aq) + H_2O(l) \rightleftharpoons CH_3NH_3^+(aq) + OH^-(aq)$

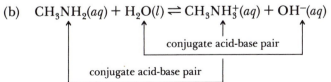

(c) $HCOOH(aq) + H_2O(l) \rightleftharpoons H_3O^+(aq) + HCOO^-(aq)$

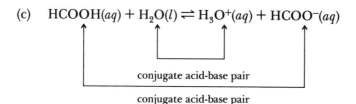

conjugate acid-base pair

conjugate acid-base pair

18-41 (a) $ClO^-(aq)$ (b) $NH_3(aq)$
       (c) $N_3^-(aq)$ (d) $S^{2-}(aq)$

18-43 (a) Acid; conjugate base is $CNO^-(aq)$.
       (b) Base; conjugate acid is $HOBr(aq)$.
       (c) Acid; conjugate base is $ClO_3^-(aq)$.
       (d) Acid; conjugate base is $CH_3NH_2(aq)$.
       (e) Base; conjugate acid is $ClNH_3^+(aq)$.
       (f) Base; conjugate acid is $HONH_3^+(aq)$.

18-45 We have that $K_b = \dfrac{K_w}{K_a}$.

(a) $K_b = \dfrac{1.00 \times 10^{-14} \text{ M}^2}{1.34 \times 10^{-5} \text{ M}} = 7.46 \times 10^{-10}$ M for $CH_3CH_2COO^-$

(b) $K_b = \dfrac{1.00 \times 10^{-14} \text{ M}^2}{6.76 \times 10^{-4} \text{ M}} = 1.48 \times 10^{-11}$ M for F^-

(c) $K_b = \dfrac{1.00 \times 10^{-14} \text{ M}^2}{5.71 \times 10^{-10} \text{ M}} = 1.75 \times 10^{-5}$ M for NH_3

(d) $K_b = \dfrac{1.00 \times 10^{-14} \text{ M}^2}{6.32 \times 10^{-8} \text{ M}} = 1.58 \times 10^{-7}$ M for HPO_4^{2-}

18-47 (a) The equation is the sum of two equations:

(1) $HCNO(aq) + H_2O(l) \rightleftharpoons H_3O^+(aq) + CNO^-(aq)$

$$K_1 = K_a = 2.19 \times 10^{-4} \text{ M}$$

(2) $NO_2^-(aq) + H_3O^+(aq) \rightleftharpoons HNO_2(aq) + H_2O(l)$

$$K_2 = \frac{1}{K_a} = \frac{1}{4.47 \times 10^{-4} \text{ M}}$$

The equilibrium constant for an equation that is equal to the sum of two other equations is equal to the product of the equilibrium constants for the two equations. Thus,

$$K = K_1 K_2 = \frac{2.19 \times 10^{-4} \text{ M}}{4.47 \times 10^{-4} \text{ M}} = 0.490$$

(b)   Proceeding as in part (a), we have

    (1) $NH_4^+(aq) + H_2O(l) \rightleftharpoons NH_3(aq) + H_3O^+(aq)$

$$K_1 = K_a = 5.71 \times 10^{-10} \text{ M}$$

    (2) $HCOO^-(aq) + H_3O^+(aq) \rightleftharpoons H_2O(l) + HCOOH(aq)$

$$K_2 = \frac{1}{K_a} = \frac{1}{1.78 \times 10^{-4} \text{ M}}$$

and thus,

$$K = K_1 K_2 = \frac{5.71 \times 10^{-10} \text{ M}}{1.78 \times 10^{-4} \text{ M}} = 3.21 \times 10^{-6}$$

18-49   (a)   acidic (acidic cation, neutral anion)

$$Al(H_2O)_6^{3+}(aq) + H_2O(l) \rightleftharpoons Al(OH)(H_2O)_5^{2+}(aq) + H_3O^+(aq)$$

    (b)   acidic (acidic cation, neutral anion)

$$NH_4^+(aq) + H_2O(l) \rightleftharpoons NH_3(aq) + H_3O^+(aq)$$

    (c)   basic (neutral cation, basic anion)

$$HCO_3^-(aq) + H_2O(l) \rightleftharpoons H_2CO_3(aq) + OH^-(aq)$$

    (d)   basic (neutral cation, basic anion)

$$CNO^-(aq) + H_2O(l) \rightleftharpoons HCNO(aq) + OH^-(aq)$$

18-51   (a)   neutral cation, basic anion; basic solution
    (b)   neutral cation, neutral anion; neutral solution
    (c)   neutral cation, basic anion; basic solution
    (d)   acidic cation, neutral anion; acidic solution

18-53   The reaction that takes place in producing certain soaps is

$$NaOH(aq) + CH_3(CH_2)_{16}COOH(s) \rightleftharpoons \underset{\text{soap}}{NaCH_3(CH_2)_{16}COO(s)} + H_2O(l)$$

Soap is made up of a neutral cation and a basic anion. The anion is the conjugate base of a weak acid and is a weak base. The soap solution is basic.

18-55   The salt NaClO dissociates completely in water to yield $Na^+(aq)$ and $ClO^-(aq)$. The reaction of the anion with water is

$$ClO^-(aq) + H_2O(l) \rightleftharpoons HClO(aq) + OH^-(aq)$$

The value of the equilibrium constant is $K_b = 3.33 \times 10^{-7}$ M (Table 18-5). We can set up a table of initial and equilibrium concentrations.

	ClO$^-$(aq)	+	H$_2$O(l) $\rightleftharpoons$ HClO(aq)	+	OH$^-$(aq)
Initial concentration	0.030 M	—	0		≈ 0
Equilibrium concentration	0.030 M − [OH$^-$]	—	[HClO] = [OH$^-$]		[OH$^-$]

The expression for K_b is

$$K_b = \frac{[HClO][OH^-]}{[ClO^-]} = \frac{[OH^-]^2}{0.030 \text{ M} - [OH^-]} = 3.33 \times 10^{-7} \text{ M}$$

We can neglect [OH$^-$] relative to 0.050 M because K_b is very small. We have

$$\frac{[OH^-]^2}{0.030 \text{ M}} \approx 3.33 \times 10^{-7} \text{ M}$$

Solving for [OH$^-$], we have

$$[OH^-] = 9.99 \times 10^{-5} \text{ M}$$

The method of successive approximations yields [OH$^-$] = 9.98×10^{-5} M, and so

$$[HClO] = 9.98 \times 10^{-5} \text{ M}$$

Using the ion product constant of water, we obtain

$$[H_3O^+] = \frac{1.00 \times 10^{-14} \text{ M}^2}{[OH^-]} = \frac{1.00 \times 10^{-14} \text{ M}^2}{9.98 \times 10^{-5} \text{ M}} = 1.00 \times 10^{-10} \text{ M}$$

The pH of the solution is

$$pH = -\log [H_3O^+] = -\log (1.00 \times 10^{-10}) = 10.00$$

18-57 The equation is

$$CNO^-(aq) + H_2O(l) \rightleftharpoons HCNO(aq) + OH^-(aq)$$

The value of the equilibrium constant is $K_b = 4.57 \times 10^{-11}$ M (Table 18-5). We can set up a table of the initial and equilibrium concentrations.

	CNO⁻(aq)	+	H₂O(l) ⇌	HCNO(aq)	+	OH⁻(aq)
Initial concentration	0.20 M		—	0		≈0
Equilibrium concentration	0.20 M − [OH⁻]		—	[HCNO] = [OH⁻]		[OH⁻]

The K_b expression is

$$K_b = \frac{[HCNO][OH^-]}{[CNO^-]} = \frac{[OH^-]^2}{0.20\ M - [OH^-]} = 4.57 \times 10^{-11}\ M$$

Neglecting [OH⁻] with respect to 0.20 M, we have

$$\frac{[OH^-]^2}{0.20\ M} = 4.57 \times 10^{-11}\ M$$

$$[OH^-] = 3.02 \times 10^{-6}\ M$$

and

$$[HCNO] = 3.02 \times 10^{-6}\ M$$

and

$$[CNO^-] = 0.20\ M - [OH^-] = 0.20\ M - 3.02 \times 10^{-6}\ M = 0.20\ M$$

Using the ion product constant of water, we obtain

$$[H_3O^+] = \frac{1.00 \times 10^{-14}\ M^2}{[OH^-]} = \frac{1.00 \times 10^{-14}\ M^2}{3.02 \times 10^{-6}\ M} = 3.31 \times 10^{-9}\ M$$

$$pH = -\log [H_3O^+] = -\log (3.31 \times 10^{-9}) = 8.48$$

18-59 We can set up a table of the initial and equilibrium concentrations.

	C₅H₅NH⁺(aq)	+	H₂O(l) ⇌	H₃O⁺(aq)	+	C₅H₅N(aq)
Initial concentration	0.30 M		—	≈0		0
Equilibrium concentration	0.30 M − [H₃O⁺]		—	[H₃O⁺]		[C₅H₅N] = [H₃O⁺]

The K_a expression is (Table 18-5)

$$K_a = \frac{[H_3O^+][C_5H_5N]}{[C_5H_5NH^+]} = \frac{[H_3O^+]^2}{0.30\ M - [H_3O^+]} = 6.92 \times 10^{-6}\ M$$

Neglecting $[H_3O^+]$ with respect to 0.30 M, we have

$$\frac{[H_3O^+]^2}{0.30\ M} = 6.92 \times 10^{-6}\ M$$

$$[H_3O^+] = 1.44 \times 10^{-3}\ M$$

This value of $[H_3O^+]$ is confirmed by successive approximations. The pH of the solution is

$$pH = -\log[H_3O^+] = -\log(1.44 \times 10^{-3}) = 2.84$$

18-61   The number of moles in 23.7 g of NH_4ClO_4 is

$$n = (23.7\ \text{g})\left(\frac{1\ \text{mol}\ NH_4ClO_4}{117.49\ \text{g}\ NH_4ClO_4}\right) = 0.202\ \text{mol}$$

The molarity of a saturated $NH_4ClO_4(aq)$ solution is

$$\text{molarity} = \frac{0.202\ \text{mol}}{0.100\ \text{L}} = 2.02\ M$$

We can set up a table of initial and equilibrium concentrations.

	$NH_4^+(aq)$	+	$H_2O(l) \rightleftharpoons H_3O^+(aq)$	+ $NH_3(aq)$
Initial concentration	2.02 M	—	≈ 0	0
Equilibrium concentration	$2.02\ M - [H_3O^+]$	—	$[H_3O^+]$	$[NH_3] = [H_3O^+]$

The K_a expression is (Table 18-5)

$$K_a = \frac{[H_3O^+][NH_3]}{[NH_4^+]} = \frac{[H_3O^+]^2}{2.02\ M - [H_3O^+]} = 5.71 \times 10^{-10}\ M$$

Neglecting $[H_3O^+]$ relative to 2.02 M, we have

$$\frac{[H_3O^+]^2}{2.02 \text{ M}} \approx 5.71 \times 10^{-10} \text{ M}$$

$$[H_3O^+] \approx 3.40 \times 10^{-5} \text{ M}$$

The value of $[H_3O^+]$ is confirmed by successive approximations. The pH of the solution is

$$pH = -\log [H_3O^+] = -\log (3.40 \times 10^{-5}) = 4.47$$

18-63   In aqueous solution Fe^{3+} exists as $[Fe(H_2O)_6]^{3+}(aq)$.

	$[Fe(H_2O)_6]^{3+}(aq)$ + $H_2O(l)$ $\rightleftharpoons$ $H_3O^+(aq)$ + $[Fe(OH)(H_2O)_5]^{2+}(aq)$			
Initial concentration	0.20 M	—	≈ 0	0
Equilibrium concentration	0.20 M $-$ $[H_3O^+]$	—	$[H_3O^+]$	$[[Fe(OH)(H_2O)_5]^{2+}]$ $= [H_3O^+]$

The K_a expression is

$$K_a = \frac{[H_3O^+][[Fe(OH)(H_2O)_5]^{2+}]}{[Fe(H_2O)_6]^{3+}} = \frac{[H_3O^+]^2}{0.20 \text{ M} - [H_3O^+]} = 1.0 \times 10^{-3} \text{ M}$$

Successive approximations are 0.014 M, 0.0136 M, 0.0137 M, and 0.0137 M. Thus,

$$[H_3O^+] = 0.0137 \text{ M}$$

and the pH of the solution is

$$pH = -\log [H_3O^+] = -\log (0.0137) = 1.86$$

18-65   (a)   HCl yields $H^+(aq)$ in aqueous solution and thus is an Arrhenius acid. HCl is a proton donor and thus is a Brønsted-Lowry acid.

(b) $AlCl_3$ yields $H^+(aq)$ in aqueous solution and thus is an Arrhenius acid. $Al(H_2O)_6^{3+}(aq)$ is a proton donor and thus is a Brønsted-Lowry acid. $AlCl_3$ is an electron-deficient species and is a Lewis acid.

(c) BCl_3 is neither an Arrhenius acid nor a Brønsted-Lowry acid. BCl_3 can act as an electron-pair acceptor and hence is a Lewis acid.

18-67 (a) CH_3OCH_3 has two lone pairs of electrons and so can act as an electron-pair donor.

$$H-\overset{\overset{\displaystyle H}{|}}{\underset{\underset{\displaystyle H}{|}}{C}}-\overset{..}{\underset{..}{O}}-\overset{\overset{\displaystyle H}{|}}{\underset{\underset{\displaystyle H}{|}}{C}}-H$$

CH_3OCH_3 is a Lewis base.

(b) $GaCl_3$ is an electron-deficient species and thus is a Lewis acid.

(c) H_2O has two lone pairs of electrons and so can act as an electron-pair donor. H_2O is a Lewis base.

18-69 The acid dissociation constant expression is

The ratio of the concentrations of the anion to the undissociated acid is

We can find the value of $[H_3O^+]$ from the pH of the solution.

$$[H_3O^+] = 10^{-pH} = 10^{-7.4} = 4.0 \times 10^{-8} \text{ M}$$

The value of the ratio is

$$\text{ratio} = \frac{1.1 \times 10^{-4} \text{ M}}{4.0 \times 10^{-8} \text{ M}} = 2.8 \times 10^3$$

Most of the acid is dissociated at this pH.

18-71 The number of moles of benzoic acid in 6.15 g is

$$n = (6.15 \text{ g}) \left(\frac{1 \text{ mol C}_6\text{H}_5\text{COOH}}{122.12 \text{ g C}_6\text{H}_5\text{COOH}} \right) = 0.05036 \text{ mol}$$

The molarity of the solution is

$$\text{molarity} = \frac{\text{moles of solute}}{\text{volume of solution}} = \frac{0.05036 \text{ mol}}{0.600 \text{ L}} = 0.0839 \text{ M}$$

We can set up a table of initial and equilibrium concentrations.

	$C_6H_5COOH(aq)$ +	$H_2O(l) \rightleftharpoons$	$H_3O^+(aq)$ +	$C_6H_5COO^-(aq)$
Initial concentration	0.0839 M	—	≈ 0	0
Equilibrium concentration	$0.0839 \text{ M} - [H_3O^+]$	—	$[H_3O^+]$	$[C_6H_5COO^-] = [H_3O^+]$

Substituting the equilibrium concentration expressions in the K_a expression, we have

$$K_a = \frac{[H_3O^+][C_6H_5COO^-]}{[C_6H_5COOH]} = \frac{[H_3O^+]^2}{0.0839 \text{ M} - [H_3O^+]} = 6.46 \times 10^{-5} \text{ M}$$

The method of successive approximations yields the following: 2.33×10^{-3} M, 2.30×10^{-3} M, and 2.30×10^{-3} M. Thus,

$$[H_3O^+] = 2.30 \times 10^{-3} \text{ M}$$

and the pH of the solution is

$$pH = -\log [H_3O^+] = -\log (2.30 \times 10^{-3}) = 2.64$$

18-73   The pOH of the solution is given by

$$pOH = 14.00 - pH = 14.00 - 10.52 = 3.48$$

and the value of $[OH^-]$ is given by

$$[OH^-] = 10^{-pOH} = 10^{-3.48} = 3.31 \times 10^{-4} \text{ M}$$

Because one mole of $Mg(OH)_2$ yields two moles of OH^- in water, the concentration of $Mg(OH)_2$ is

$$[Mg(OH)_2] = \tfrac{1}{2}[OH^-] = 1.655 \times 10^{-4} \text{ M}$$

Solubility often is expressed as the number of grams per 100 mL of solution. The number of moles in 100 mL of solution is

$$n = \text{molarity} \times \text{volume} = (1.655 \times 10^{-4} \text{ mol} \cdot \text{L}^{-1})(0.100 \text{ L})$$
$$= 1.655 \times 10^{-5} \text{ mol}$$

The mass in 1.655×10^{-5} mol of $Mg(OH)_2$ is

$$\text{mass} = (1.655 \times 10^{-5} \text{ mol}) \left(\frac{58.33 \text{ g Mg(OH)}_2}{1 \text{ mol Mg(OH)}_2} \right) = 9.66 \times 10^{-4} \text{ g}$$

The solubility of $Mg(OH)_2$ is 9.66×10^{-4} g per 100 mL of solution.

18-75   The number of moles in two 5-grain aspirin tablets is

$$n = (2)(324 \text{ mg}) \left(\frac{1 \text{ g}}{1000 \text{ mg}} \right) \left(\frac{1 \text{ mol aspirin}}{180.15 \text{ g aspirin}} \right) = 0.00360 \text{ mol}$$

The molarity of the solution is

$$\text{molarity} = \frac{\text{moles of solute}}{\text{volume of solution}} = \frac{0.00360 \text{ mol}}{0.500 \text{ L}} = 0.00720 \text{ M}$$

We can set up a table of initial and equilibrium concentrations.

	acetylsalicylic acid (*aq*) + $H_2O(l)$ $\rightleftharpoons$ H_3O^+(*aq*) + acetylsalicylate (*aq*)			
Initial concentration	0.00720 M	—	≈ 0	0
Equilibrium concentration	0.00720 M $-$ $[H_3O^+]$	—	$[H_3O^+]$	$[H_3O^+]$

Substituting the equilibrium concentration expression in the K_a expression, we have

$$K_a = \frac{[H_3O^+]^2}{0.00720 \text{ M} - [H_3O^+]} = 2.75 \times 10^{-5} \text{ M}$$

The method of successive approximations yields the following: 4.45×10^{-4} M, 4.31×10^{-4} M, and 4.31×10^{-4} M. Thus,

$$[H_3O^+] = 4.31 \times 10^{-4} \text{ M}$$

The pH of the solution is

$$pH = -\log [H_3O^+] = -\log (4.31 \times 10^{-4}) = 3.37$$

18-77 If we take $[H_3O^+] = 2.60 \times 10^{-8}$ M, then we calculate for the pH:

$$pH = -\log (2.60 \times 10^{-8}) = 7.58$$

But this answer is wrong, because an acidic solution at $25\,°C$ has pH < 7.00. The given concentration of HCl(aq) is so low that we cannot neglect the dissociation of water as an important source of $H_3O^+(aq)$. Thus, we have

$$2H_2O(l) \rightleftharpoons H_3O^+(aq) + OH^-(aq)$$

At equilibrium, $[H_3O^+] = 2.60 \times 10^{-8}$ M $+ [OH^-]$. Using the K_w expression, we have

$$K_w = 1.00 \times 10^{-14} \text{ M}^2 = [H_3O^+][OH^-]$$
$$= (2.60 \times 10^{-8} \text{ M} + [OH^-])[OH^-]$$

We have a quadratic equation in $[OH^-]$:

$$[OH^-]^2 + (2.60 \times 10^{-8} \text{ M})[OH^-] - 1.00 \times 10^{-14} \text{ M}^2 = 0$$

and

$$[OH^-] = \frac{-2.60 \times 10^{-8} \text{ M} \pm \sqrt{6.76 \times 10^{-16} \text{ M}^2 + 4.00 \times 10^{-14} \text{ M}^2}}{(2)}$$

$$= 8.78 \times 10^{-8} \text{ M} \qquad (\text{and} -1.13 \times 10^{-7} \text{ M})$$

Thus, we have

$$[H_3O^+] = 2.60 \times 10^{-8} \text{ M} + 8.78 \times 10^{-8} \text{ M}$$
$$= 1.138 \times 10^{-7} \text{ M}$$

and

$$pH = -\log (1.138 \times 10^{-7} \text{ M}) = 6.94$$

Note that the pH < 7.00; the solution is acidic.

18-79   Consider this problem by analogy with water, for which we have

$$2H_2O(l) \rightleftharpoons H_3O^+(aq) + OH^-(aq)$$

The strongest acid that can exist in appreciable concentrations in water is $H_3O^+(aq)$, because any stronger acid, for example, HNO_3, will be deprotonated essentially completely by $H_2O(l)$. The strongest base that can exist in appreciable concentrations in water is $OH^-(aq)$, because any stronger base, for example, $O^{2-}(aq)$, will be protonated essentially completely by $H_2O(l)$. Thus, in liquid ammonia, the strongest acid that can exist in appreciable concentrations is $NH_4^+(amm)$, and the strongest base that can exist in appreciable concentrations is $NH_2^-(amm)$. It is of interest to note, for example, that acetic acid is a strong acid in $NH_3(l)$.

18-81   The acid-dissociation-constant expression for uric acid is

$$K_a = \frac{[H_3O^+][\text{urate}]}{[\text{uric acid}]} = 4 \times 10^{-6} \text{ M}$$

from which we obtain

$$\frac{[\text{urate}]}{[\text{uric acid}]} = \frac{4 \times 10^{-6} \text{ M}}{[H_3O^+]}$$

Thus, [urate] > [uric acid] if 4×10^{-6} M > $[H_3O^+]$

$$pH > -\log(4 \times 10^{-6}) = 5.4$$

18-83   The pOH of the solution is given by

$$pOH = 14.00 - pH = 14.00 - 13.50 = 0.50$$

and so

$$[OH^-] = 10^{-pOH} = 10^{-0.50} = 0.32 \text{ M}$$

Because 1 mol of $Sr(OH)_2$ (aq) yields 2 mol of OH^- (aq), the concentration of $Sr(OH)_2$ (aq) is

$$[Sr(OH)_2] = \tfrac{1}{2}[OH^-] = 0.16 \text{ M}$$

Solubility is often expressed as the number of grams per 100 mL of solution. The number of moles in 100 mL is

$$n = \text{molarity} \times \text{volume} = (0.16 \text{ mol} \cdot \text{L}^{-1})(0.100 \text{ L})$$
$$= 1.6 \times 10^{-2} \text{ mol}$$

The mass corresponding to 1.6×10^{-2} mol of $Sr(OH)_2$ is

$$\text{mass of } Sr(OH)_2 = (1.6 \times 10^{-2} \text{ mol}) \left(\frac{121.64 \text{ g } Sr(OH)_2}{1 \text{ mol } Sr(OH)_2} \right) = 1.9 \text{ g}$$

The solubility of $Sr(OH)_2$ is 1.9 g per 100 mL of solution.

18-85 We set up a concentration table.

	$HCOOH(aq)$	$+$	$H_2O(l) \rightleftharpoons$	$HCOO^-(aq)$	$+$	$H_3O^+(aq)$
Initial concentration	0.150 M		—	0		≈ 0
Equilibrium concentration	$0.150 \text{ M} - [H_3O^+]$		—	$[H_3O^+]$		$[H_3O^+]$

The K_a expression is (Table 18-3)

$$K_a = \frac{[HCOO^-][H_3O^+]}{[HCOOH]} = \frac{[H_3O^+]^2}{0.150 \text{ M} - [H_3O^+]} = 1.78 \times 10^{-4} \text{ M}$$

The method of successive approximations yields

$$[H_3O^+] = 5.08 \times 10^{-3} \text{ M}$$
$$pH = -\log (5.08 \times 10^{-3}) = 2.29$$

The concentrations of the species in the solution are

$$[HCOO^-] = 5.08 \times 10^{-3} \text{ M}$$
$$[HCOOH] = 0.150 \text{ M} - 5.08 \times 10^{-3} \text{ M} = 0.145 \text{ M}$$
$$[OH^-] = \frac{1.00 \times 10^{-14} \text{ M}^2}{5.08 \times 10^{-3} \text{ M}} = 1.97 \times 10^{-12} \text{ M}$$

18-87 We set up a concentration table.

	$(CH_3)_2NH(aq)$	$+$	$H_2O(l) \rightleftharpoons$	$(CH_3)_2NH_2^+(aq)$	$+$	$OH^-(aq)$
Initial concentration	0.150 M		—	0		≈ 0
Equilibrium concentration	$0.150 \text{ M} - [OH^-]$		—	$[OH^-]$		$[OH^-]$

The K_b expression is (Table 18-4)

$$K_a = \frac{[(CH_3)_2NH_2^+][OH^-]}{[(CH_3)_2NH]} = \frac{[OH^-]^2}{0.150\ M - [OH^-]} = 5.81 \times 10^{-4}\ M$$

The method of successive approximations yields after four iterations

$$[OH^-] = 9.05 \times 10^{-3}\ M$$

The concentrations of the species in the solution are

$$[(CH_3)_2NH_2^+] = 9.05 \times 10^{-3}\ M$$

$$[(CH_3)_2NH] = 0.150\ M - 9.05 \times 10^{-3}\ M = 0.141\ M$$

$$[H_3O^+] = \frac{1.00 \times 10^{-14}\ M^2}{9.05 \times 10^{-3}\ M} = 1.10 \times 10^{-12}\ M$$

$$pH = -\log(1.10 \times 10^{-12}) = 11.96$$

18-89   We set up a concentration table.

	HCNO(aq)	+	H₂O(l)	⇌ CNO⁻(aq) +	H₃O⁺(aq)
Initial concentration	0.100 M		—	0	≈0
Equilibrium concentration	0.100 M − [H₃O⁺]		—	[H₃O⁺]	[H₃O⁺]

The K_a expression is (Table 18-3)

$$K_a = \frac{[CNO^-][H_3O^+]}{[HCNO]} = \frac{[H_3O^+]^2}{0.100\ M - [H_3O^+]} = 2.19 \times 10^{-4}\ M$$

The method of successive approximations yields after three iterations

$$[H_3O^+] = 4.57 \times 10^{-3}\ M$$

$$[CNO^-] = 4.57 \times 10^{-3}\ M$$

The percent of the acid that is dissociated is

$$\% \text{ dissociated} = \frac{[CNO^-]}{[HCNO]_0} \times 100 = \frac{4.57 \times 10^{-3}\ M}{0.100\ M} \times 100$$

$$= 4.57\%$$

E. ANSWERS TO THE SELF-TEST

1. a proton donor

2. a proton acceptor

3. false

4. $H_3O^+(aq)$

5. false

6. false

7. 1.00×10^{-7} M

8. $[H_3O^+][OH^-]$

9. 1.00×10^{-14} M^2

10. basic

11. basic

12. completely

13. completely

14. false

15. $K^+(aq)$, $OH^-(aq)$, and $H_3O^+(aq)$

16. RCOOH

17. weak

18. carboxylate

19. true

20. $-\log[H_3O^+]$ or the negative of the common logarithm of the concentration of the hydronium ion

21. true

22. true

23. false

24. 7.00

25. less

26. greater

27. partially

28. false

29. false

30. proton transfer or protonation

31. false

32. $HNO_2(aq)$, $H_3O^+(aq)$, $NO_2^-(aq)$, and $OH^-(aq)$

33. $HF(aq) + H_2O(l) \rightleftharpoons$
     $\qquad\qquad H_3O^+(aq) + F^-(aq)$

34. $[H_3O^+][F^-]/[HF]$

35. true

36. partially

37. false

38. $NH_3(aq) + H_2O(l) \rightleftharpoons$
     $\qquad\qquad NH_4^+(aq) + OH^-(aq)$

39. $[NH_4^+][OH^-]/[NH_3]$

40. $-\log K_a$

41. weaker

42. $-\log K_b$

43. weaker

44. acid

45. base

46. K_w/K_a

47. neutral

48. false

49. basic

50. neutral

51. true

52. true

53. basic

54. true

55. an electron-pair acceptor

56. an electron-pair donor

57. acid

58. base

19 ACIDS AND BASES II

A. OUTLINE OF CHAPTER 19

19-1 The Henderson-Hasselbalch equation can often be used to calculate the pH of a buffer solution.

A solution that is resistant to changes in pH upon the addition of an acid or a base is called a buffer.

A solution of an acid and its conjugate base can act as a buffer.

The pH of a buffer can be calculated using the Henderson-Hasselbalch equation:

$$pH = pK_a + \log \left(\frac{[B^-]_0}{[HB]_0} \right) \qquad (19\text{-}5)$$

where the subscript zero indicates stoichiometric concentrations.

For the Henderson-Hasselbalch equation to be valid, both K_a and K_b must be less than 10^{-3} M, and in addition, the ratio of $[B^-]_0$ to $[HB]_0$ must be between 0.1 and 10 M.

19-2 A buffer solution suppresses pH changes when acid or base is added.

Calculations showing the small change in pH of a buffer solution when either acid or base is added are presented.

The capacity of a buffer to resist changes is not unlimited.

A buffer solution resists changes in pH upon dilution with solvent.

19-3 An indicator is used to signal the equivalence point of a titration.

The titration of an acid with a base involves the addition of the base to a given volume of the acid until all the acid has reacted.

A titration curve of a strong acid with a strong base is shown in Figure 19-4.

At the equivalence point in a titration, stoichiometrically equivalent amounts of acid and base have reacted.

At the equivalence point in the titration of a strong monoprotic acid with a strong base or a strong base with a strong monoprotic acid, the pH of the solution is 7.

The reaction between an acid and base goes essentially to completion; the value of the equilibrium constant is $K = 1/K_w = 1.00 \times 10^{14}$ M^{-1} at 25°C.

An indicator is a weak organic acid whose color varies with pH.

An indicator signals the equivalence point in a titration experiment by changing color.

The acid-dissociation equation of an indicator is represented as

$$\text{HIn}(aq) + \text{H}_2\text{O}(l) \rightleftharpoons \text{H}_3\text{O}^+(aq) + \text{In}^-(aq)$$
$$\text{(one color)} \qquad\qquad\qquad\qquad\qquad \text{(another color)}$$

The acid-dissociation constant K_{ai} for the above equation is

$$K_{ai} = \frac{[\text{H}_3\text{O}^+][\text{In}^-]}{[\text{HIn}]} \tag{19-8}$$

At the color transition point, pH $\approx$ pK_{ai}.

The pK_{ai} values and color changes of some indicators are listed in Table 19-2 and displayed in Figure 19-6.

The end point of a titration is the point at which the indicator changes color.

The end point is the experimental estimate of the equivalence point.

19-4 The pH changes abruptly at the equivalence point of the titration of a strong acid with a strong base.

Calculations of pH for the titration curve of the strong acid, HCl(aq) with the strong base, NaOH(aq) (shown in Figure 19-4) are presented.

19-5 Weak acids can be titrated with strong bases.

A titration curve of a weak acid with a strong base is shown in Figure 19-7.

The equivalence point in the titration of a weak acid with a strong base occurs at a pH greater than 7.

The pH at the equivalence point in the titration of a weak acid with a strong base can be calculated using K_b for the anion of the weak acid.

19-6 pH = pK_a at the midpoint in the titration of a weak acid with a strong base.

Calculation of the midpoint on a titration curve of a weak acid with a strong base is discussed.

Figure 19-8 shows the concentrations of undissociated acetic acid and acetate ion plotted against the volume of added base in the titration of CH$_3$COOH(aq) with NaOH(aq).

19-7 Weak bases can be titrated with strong acids.

A titration curve of a weak base with a strong acid is shown in Figure 19-9.

The equivalence point in the titration of a weak base with a strong acid occurs at a pH less than 7.

The titration curve of a weak base with a strong acid can be sketched by calculating the initial pH, the pH at the midpoint, the pH at the equivalence point, and a point after the equivalence point.

19-8 Polyprotic acids can donate more than one proton.

A polyprotic acid has more than one acid-dissociation constant.

Each successive acid-dissociation constant of an inorganic polyprotic oxyacid is about 10^{-5} times the value of the preceding one.

The dissociation constants of some polyprotic acids are given in Table 19-4.

When calculating the pH of an aqueous solution of a polyprotic acid, usually only the first dissociation constant need be used and the other dissociation constants may be neglected.

The titration curve of the polyprotic acid, H_3PO_4, with $KOH(aq)$ is shown in Figure 19-11.

B. SELF-TEST

1. A buffer is a solution that can resist changes in _____.

2. A solution that contains a mixture of a weak acid and its conjugate base can be used as a buffer. *True/False*

3. A buffer is resistant to changes in pH upon the addition of an acid but not upon the addition of a base. *True/False*

4. The pH of a buffer can be estimated by using the _____ _____ equation.

5. A buffer resists changes in pH regardless of the amount of acid or base added to the solution. *True/False*

6. A buffer resists changes in pH upon dilution with solvent. *True/False*

7. In a titration of a solution of $HCl(aq)$ with a solution of $NaOH(aq)$, the solution of _____ is added slowly to the solution of _____.

8. In a titration of a solution of $HCl(aq)$ with a solution of $KOH(aq)$, the titrant is _____.

9. The titration curve of a titration of a solution of $HCl(aq)$ with a solution of $NaOH(aq)$ is a plot of _____ versus _____.

10. The equivalence point of a titration of a solution of HCl(aq) with a solution of NaOH(aq) is the point _____

_____ .

11. An indicator is a weak organic (*acid/base*) that changes color with pH.

12. Indicators come in a variety of colors. *True/False*

13. Indicators can be used to estimate the _____ of a solution from the colors of the indicators.

14. The pH at which an indicator changes color is approximately equal to _____ of the indicator.

15. Methyl orange is red in the acid form and yellow in the base form. Methyl orange is _____ in the pH transition region.

16. In the titration of a solution of HCl(aq) with a solution of NaOH(aq), phenolphthalein is added to the solution of HCl(aq) to signal _____ .

17. The end point of a titration is indicated by _____ .

18. In the titration of a solution of HCl(aq) with a solution of NaOH(aq), the pH changes very little around the equivalence point. *True/False*

19. The equivalence point of an acid-base titration always occurs at pH = 7.0. *True/False*

20. In the titration of a solution of acetic acid, $CH_3COOH(aq)$, with a solution of NaOH(aq), the equivalence point occurs at a pH greater than 7.0. *True/False*

21. Explain why the equivalence point of a titration of a solution of $CH_3COOH(aq)$ with a solution of NaOH(aq) does not occur at pH = 7.0. _____

_____

22. The equilibrium constant for the reaction of a weak acid such as acetic acid with a strong base is very large. *True/False*

23. At the midpoint in the titration of a weak acid with a strong base, the pH is equal to _____ .

24. In the titration of acetic acid with a strong base, the concentration of acetic acid before the equivalence point (*increases/decreases*) with increasing amounts of base.

25. In the titration of acetic acid with a strong base, the concentration of the acetate ion before the equivalence point (*increases/decreases*) with increasing amounts of base.

26. In the titration of acetic acid with a strong base, the concentration of acetate ion after the equivalence point (*increases/decreases*) with increasing amounts of base.

27. In the titration of acetic acid with a strong base, the concentration of acetic acid after the equivalence point is essentially zero with increasing amounts of base. *True/False*

28. In the titration of a solution of ammonia, $NH_3(aq)$, with the solution of $HCl(aq)$, the equivalence point occurs at a pH greater than 7.0. *True/False*

29. Explain why the equivalence point of a titration of a solution of $NH_3(aq)$ with a solution of $HCl(aq)$ does not occur at pH = 7.0. _____

_____

30. The titration curve of a weak base with a strong acid can be sketched by calculating

_____ , _____ , _____

_____ , and _____ .

31. A triprotic acid has (*one/two/three*) acid-dissociation constants.

32. The successive acid-dissociation constants of a polyprotic acid have similar values. *True/False*

33. When calculating the pH of an aqueous solution of a polyprotic acid, usually only the first acid-dissociation constant need be used. *True/False*

34. In the titration of $H_3PO_4(aq)$ with $KOH(aq)$, there are (*one/two/three*) equivalence points.

C. CALCULATIONS YOU SHOULD KNOW HOW TO DO

1. Calculate the pH of a buffer using the Henderson-Hasselbalch equation. See Example 19-1 and Problems 19-1 through 19-6.

2. Use the Henderson-Hasselbalch equation to calculate the pH of a buffer before and after the addition of small amounts of strong acid or base. See Example 19-2 and Problems 19-7 through 19-12.

3. Calculate the points on a titration curve of a strong acid with a strong base. See Section 19-4 of the text and Problems 19-21 through 19-28.

4. Calculate certain points on a titration curve of a weak acid with a strong base. You should be able to calculate the initial pH, the pH at the equivalence point, the midpoint, and a point after the equivalence point. See Sections 19-5 and 19-6 of the text and Problems 19-29 through 19-32.

5. Calculate certain points on a titration curve of a weak base with a strong acid. You should be able to calculate the initial pH, the pH at the equivalence point, the midpoint, and a point after the equivalence point. See Section 19-7 of the text and Problems 19-33 and 19-34.

6. Determine the molecular mass of an unknown acid or base by titrating to the equivalence point. See Problems 19-35, 19-36, 19-39, and 19-40.

7. Calculate the pH of an aqueous solution of a polyprotic acid. See Example 19-11 and Problems 19-41 and 19-42.

D. SOLUTIONS TO THE ODD-NUMBERED PROBLEMS

19-1   We first identify the acid and the base:

acid: $CH_3COOH(aq)$ base: $CH_3COO^-(aq)$ (from $NaCH_3COO(aq)$)

The stoichiometric concentrations of the acid and base forms are

$$[acid]_0 = 0.050 \text{ M} \qquad [base]_0 = 0.050 \text{ M}$$

The pK_a of acetic acid (Table 18-3) is

$$pK_a = 4.76$$

From the Henderson-Hasselbalch equation we have

$$pH = pK_a + \log \frac{[B^-]_0}{[HB]_0} = 4.76 + \log \left(\frac{0.050 \text{ M}}{0.050 \text{ M}} \right)$$
$$= 4.76 + 0.00 = 4.76$$

19-3   The stoichiometric concentrations of the acid and base forms are

$$[HB]_0 = [HCOOH]_0 = 0.15 \text{ M}$$
$$[B^-]_0 = [HCOO^-]_0 = 0.25 \text{ M (from NaHCOO}(aq))$$

The pK_a of formic acid is (Table 18-3)

$$pK_a = 3.75$$

From the Henderson-Hasselbalch equation we have

$$pH = pK_a + \log \frac{[B^-]_0}{[HB]_0} = 3.75 + \log \left(\frac{0.25 \text{ M}}{0.15 \text{ M}} \right)$$
$$= 3.75 + 0.22 = 3.97$$

19-5   The stoichiometric concentrations of the acid and base forms are

$$[\text{acid}]_0 = [C_5H_5NH^+]_0 = 0.250 \text{ M} \qquad (\text{from } C_5H_5HCl \ (aq))$$
$$[\text{base}]_0 = [C_5H_5N]_0 = 0.200 \text{ M}$$

The pK_a of pyridinium chloride is (Table 18-5)

$$pK_a = 5.16$$

From the Henderson-Hasselbalch equation we have

$$pH = pK_a + \log \frac{[\text{base}]_0}{[\text{acid}]_0} = 5.16 + \log \left(\frac{0.200 \text{ M}}{0.250 \text{ M}}\right)$$
$$= 5.16 - 0.097 = 5.06$$

19-7   The number of moles in 1.00 g of KOH is

$$n = (1.00 \text{ g}) \left(\frac{1 \text{ mol KOH}}{56.11 \text{ g KOH}}\right) = 0.0178 \text{ mol}$$

The $OH^-(aq)$ reacts with $NH_4^+(aq)$ in the buffer via the reaction

$$NH_4^+(aq) + OH^-(aq) \longrightarrow NH_3(aq) + H_2O(l)$$

(a)   The number of moles of ammonium ion in the buffer solution before adding KOH is

$$\text{moles of } NH_4^+ = MV = (0.10 \text{ mol} \cdot L^{-1})(0.500 \text{ L}) = 0.050 \text{ mol}$$

The number of moles of NH_4^+ after the addition of 0.0178 mol of KOH is

$$\text{moles of } NH_4^+ = \text{moles of } NH_4^+ \text{ before} - \text{moles of } OH^- \text{ added}$$
$$= 0.050 \text{ mol} - 0.0178 \text{ mol} = 0.032 \text{ mol}$$

The number of moles of ammonia in the buffer solution before adding KOH is

$$\text{moles of } NH_3 = MV = (0.10 \text{ M})(0.500 \text{ L}) = 0.050 \text{ mol}$$

The number of moles of NH_3 after the addition of 0.0178 mol of KOH is

$$\text{moles of } NH_3 = \text{moles of } NH_3 \text{ before} + \text{moles of } OH^- \text{ added}$$
$$= 0.050 \text{ mol} + 0.0178 \text{ mol} = 0.068 \text{ mol}$$

The initial pH of the buffer is

$$pH = pK_a + \log \frac{[\text{base}]_0}{[\text{acid}]_0} = 9.24 + \log \left(\frac{0.10 \text{ M}}{0.10 \text{ M}}\right) = 9.24$$

The final pH of the buffer is

$$pH = pK_a + \log \frac{[base]_0}{[acid]_0} = 9.24 + \log \left(\frac{0.068 \text{ mol}}{0.032 \text{ mol}} \right) = 9.57$$

The change in pH is $9.57 - 9.24 = 0.33$.

(b)   The number of moles of ammonium ion in the buffer solution before adding KOH is

$$\text{moles of NH}_4^+ = MV = (1.00 \text{ mol} \cdot \text{L}^{-1})(0.500 \text{ L}) = 0.0500 \text{ mol}$$

The number of moles of $NH_4^+(aq)$ after the addition of 0.0178 mol of KOH is

$$\text{moles of NH}_4^+ = \text{moles of NH}_4^+ \text{ before} - \text{moles of OH}^- \text{ added}$$
$$= 0.500 \text{ mol} - 0.0178 \text{ mol} = 0.482 \text{ mol}$$

The number of moles of ammonia in the buffer solution before adding KOH is

$$\text{moles of NH}_3 = MV = (1.00 \text{ mol} \cdot \text{L}^{-1})(0.500 \text{ L}) = 0.500 \text{ mol}$$

The number of moles of $NH_3(aq)$ after the addition of 0.0178 mol of KOH is

$$\text{moles of NH}_3 = \text{moles of NH}_3 \text{ before} + \text{moles of KOH added}$$
$$= 0.500 \text{ mol} + 0.0178 \text{ mol} = 0.518 \text{ mol}$$

The final pH of the buffer is

$$pH = pK_a + \log \frac{[base]_0}{[acid]_0} = 9.24 + \log \left(\frac{0.518 \text{ mol}}{0.482 \text{ mol}} \right) = 9.27$$

The change in pH is $9.27 - 9.24 = 0.03$. The pH change in the more concentrated buffer solution is much less than the pH change in the more dilute buffer solution.

19-9   The pH before addition of the HCl is calculated by using the Henderson-Hasselbalch equation.

$$pH = pK_a + \log \frac{[base]_0}{[acid]_0}$$
$$= -\log (1.30 \times 10^{-5}) + \log \left(\frac{0.0150 \text{ M}}{0.0200 \text{ M}} \right)$$
$$= 4.89 - 0.12 = 4.77$$

The equation for the reaction upon addition of acid is

$$CH_3CH_2COO^-(aq) + H_3O^+(aq) \rightleftharpoons CH_3CH_2COOH(aq) + H_2O(l)$$

The pH after the addition of 2.0 mmol of HCl is computed as follows.

$$\text{millimoles of acid} = (1000 \text{ mL})(0.0200 \text{ M}) + 2.0 \text{ mmol} = 22.0 \text{ mmol}$$
$$\text{millimoles of base} = (1000 \text{ mL})(0.0150 \text{ M}) - 2.0 \text{ mmol} = 13.0 \text{ mmol}$$

Thus, the pH is

$$pH = 4.87 + \log\left[\frac{(13.0 \text{ mmol}/V)}{(22.0 \text{ mmol}/V)}\right]$$
$$= 4.89 - 0.23 = 4.66$$

and the change in pH is

$$\Delta(pH) = 4.66 - 4.77 = -0.11$$

19-11   The total initial number of moles of acid is

$$\begin{pmatrix} \text{moles of acid} \\ \text{available} \end{pmatrix} = (2.16 \text{ g})\left(\frac{1 \text{ mol } CH_3CH_2COOH}{74.08 \text{ g } CH_3CH_2COOH}\right) = 0.0292 \text{ mol}$$

The number of moles of base added is

$$\begin{pmatrix} \text{moles of base} \\ \text{added} \end{pmatrix} = (0.56 \text{ g})\left(\frac{1 \text{ mol NaOH}}{40.0 \text{ g NaOH}}\right) = 0.014 \text{ mol}$$

The added sodium hydroxide reacts with the propionic acid to produce sodium propionate; thus, the stoichiometric concentrations of propionic acid and propionate ion are as follows.

$$[CH_3CH_2COOH]_0 = \frac{0.0292 \text{ mol} - 0.014 \text{ mol}}{0.100 \text{ L}} = 0.152 \text{ M}$$

$$[CH_3CH_2COO^-]_0 = \frac{0.014 \text{ mol}}{0.100 \text{ L}} = 0.14 \text{ M}$$

The pH of the resulting buffer is calculated by using the Henderson-Hasselbalch equation.

$$pH = pK_a + \log\frac{[\text{base}]_0}{[\text{acid}]_0}$$
$$= 4.89 + \log\left(\frac{0.14 \text{ M}}{0.152 \text{ M}}\right) = 4.85$$

19-13   Inspection of Figure 19-6 shows that the pH at which both Nile blue and thymol blue are blue is about 9.5 ± 0.5.

19-15   From Figure 19-6 we see that the pH at which bromcresol purple is yellow and bromcresol green is green is about 4.5 ± 0.5.

19-17   We see from Figure 19-6 that the middle of the transition color range of bromcresol green is pH = 5. When bromcresol green is added to the medium, a green color indicates that the pH is around 5. When the color changes, the pH is either too high or too low. Methyl red would also be suitable.

19-19   At the color transition point,

$$pH \approx pK_{ai}$$

In the case of bromcresol green, the middle of the color change occurs at pH ≈ 4.5. Thus

$$pK_{ai} = 4.5$$
$$K_{ai} = 10^{-4.5} = 3 \times 10^{-5} M$$

19-21   (a)   The net ionic equation for the reaction is

$$H_3O^+(aq) + OH^-(aq) \longrightarrow 2H_2O(l)$$

The millimoles of base available is

$$(25.0 \text{ mL})(0.200 \text{ M}) = 5.00 \text{ mmol}$$

The millimoles of acid added is

$$(20.0 \text{ mL})(0.200 \text{ M}) = 4.00 \text{ mmol}$$

Thus, the base is in excess; the excess millimoles of base is

$$(5.00 \text{ mmol}) - (4.00 \text{ mmol}) = 1.00 \text{ mmol}$$

The concentration of $OH^-(aq)$ in the final solution is

$$[OH^-] = \frac{1.00 \text{ mmol}}{(20.0 \text{ mL} + 25.0 \text{ mL})} = 0.0222 \text{ M}$$

The pOH of the solution is

$$pOH = -\log [OH^-] = 1.65$$

The pH of the solution is

$$pH = 14.00 - pOH = 14.00 - 1.65 = 12.35$$

(b)   Proceeding as in part (a), we have

$$\left(\begin{array}{c} \text{millimoles of } OH^-(aq) \\ \text{available} \end{array} \right) = (30.0 \text{ mL})(0.350 \text{ M}) = 10.50 \text{ mmol}$$

$$\left(\begin{array}{c} \text{millimoles of } H_3O^+(aq) \\ \text{added} \end{array} \right) = (20.0 \text{ mL})(0.200 \text{ M}) = 4.00 \text{ mmol}$$

$$[OH^-] = \frac{(10.50 \text{ mmol} - 4.00 \text{ mmol})}{(20.0 \text{ mL} + 30.0 \text{ mL})} = 0.13 \text{ M}$$

The pOH of the solution is

$$pOH = -\log{(0.13)} = 0.89$$

The pH of the solution is

$$pH = 14.00 - pOH = 14.00 - 0.89 = 13.11$$

19-23 (a) The equivalence point is the point at which the number of moles of base added is equal to the number of moles of acid initially present.

$$\left(\begin{array}{c}\text{millimoles of } H_3O^+(aq)\\ \text{available}\end{array}\right) = (50.0 \text{ mL})(0.200 \text{ M}) = 10.0 \text{ mmol}$$

The volume in milliliters of 0.100 M NaOH(aq) that contains 10.0 mmol of $OH^-(aq)$ is

$$(0.100 \text{ } M)V = 10.0 \text{ mmol}$$
$$V = 100 \text{ mL}$$

(b) Proceeding as in part (a), we have

$$\left(\begin{array}{c}\text{millimoles of } H_3O^+(aq)\\ \text{available}\end{array}\right) = (30.0 \text{ mL})(0.150 \text{ M}) = 4.50 \text{ mmol}$$

The volume of 0.100 M NaOH(aq) that contains 4.50 mmol is

$$V(0.100 \text{ M}) = 4.50 \text{ mmol}$$
$$V = 45.0 \text{ mL}$$

19-25 The $OH^-(aq)$ concentration in a 0.100 M NaOH(aq) solution is

$$[OH^-] = 0.100 \text{ M}$$

The pOH of the solution is

$$pOH = -\log{[OH^-]} = -\log{(0.100)} = 1.00$$

and the pH is

$$pH = 14.00 - pOH = 13.00$$

Because NaOH is a strong base and HCl is a strong acid, the pH at the equivalence point [50.0 mL of HCl(aq) added] is 7.0. The pH when 100.0 mL of HCl(aq) is added can be obtained by realizing that once the equivalence point is reached, any additional HCl(aq) remains as $H_3O^+(aq)$. Thus, when 100.0 mL of HCl(aq) is added there are

$$\begin{aligned}\text{millimoles of } H_3O^+(aq) &= MV_{\text{excess}}\\ &= (0.100 \text{ M})(50.0 \text{ mL})\\ &= 5.00 \text{ mmol}\end{aligned}$$

The total volume of the solution is $50.0 \text{ mL} + 100.0 \text{ mL} = 150.0 \text{ mL}$, and so $[H_3O^+]$ is given by

$$[H_3O^+] = \frac{5.00 \text{ mmol}}{150.0 \text{ mL}} = 0.0333 \text{ M}$$

The pH is

$$pH = -\log (0.0333) = 1.48$$

The titration curve is shown below.

19-27 At the equivalence point we have the condition

$$\text{moles of acid} = \text{moles of base}$$

or

$$M_a V_a = M_b V_b$$

Therefore

$$M_a = \frac{M_b V_b}{V_a} = \frac{(0.165 \text{ M})(35.6 \text{ mL})}{25.0 \text{ mL}} = 0.235 \text{ M}$$

Note that we can use the units mL because the volume units cancel.

19-29 The reaction is

$$\underset{\text{cacodylic acid}}{HC_2H_6AsO_2(aq)} + OH^-(aq) \rightleftharpoons \underset{\text{cacodylate}}{C_2H_6AsO_2^-(aq)} + H_2O(l)$$

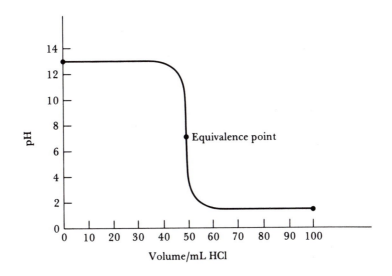

The volume of NaOH(aq) added to reach the equivalence point is given by

$$V_b = \frac{M_a V_a}{M_b} = \frac{(0.100 \text{ M})(25.0 \text{ mL})}{(0.095 \text{ M})} = 26.3 \text{ mL}$$

At the equivalence point

$$\text{millimoles of } C_2H_6AsO_2^- = \text{initial millimoles of } HC_2H_6AsO_2$$
$$= MV = (0.100 \text{ M})(25.0 \text{ mL}) = 2.50 \text{ mmol}$$

The volume of the solution at the equivalence point is 25.0 mL + 26.3 mL = 51.3 mL. The concentration of cacodylate is

$$[C_2H_6AsO_2^-] = \frac{2.50 \text{ mmol}}{51.3 \text{ mL}} = 4.87 \times 10^{-2} \text{ M}$$

The cacodylate ion is a weak base because it is the conjugate base of a weak acid. The reaction of the cacodylate ion with water is described by the equation

$$C_2H_6AsO_2^-(aq) + H_2O(l) \rightleftharpoons HC_2H_6ASO_2(aq) + OH^-(aq)$$

The value of K_b is

$$K_b = \frac{K_w}{K_a} = \frac{1.00 \times 10^{-14} \text{ M}^2}{5.4 \times 10^{-7} \text{ M}} = 1.85 \times 10^{-8} \text{ M}$$

At equilibrium we have

$$[HC_2H_6AsO_2] = [OH^-]$$
$$[C_2H_6AsO_2^-] = 4.87 \times 10^{-2} \text{ M} - [OH^-]$$

We shall calculate the value of $[OH^-]$ using the expression for K_b. The expression for K_b is

$$K_b = \frac{[HC_2H_6AsO_2][OH^-]}{[C_2H_6AsO_2^-]} = \frac{[OH^-]^2}{4.87 \times 10^{-2} \text{ M} - [OH^-]} = 1.85 \times 10^{-8} \text{ M}$$

Neglecting $[OH^-]$ compared to 4.87×10^{-2} M, we have

$$\frac{[OH^-]^2}{4.87 \times 10^{-2} \text{ M}} = 1.85 \times 10^{-8} \text{ M}$$
$$[OH^-]^2 = 9.01 \times 10^{-10} \text{ M}^2$$
$$[OH^-] = 3.00 \times 10^{-5} \text{ M}$$

The pOH of the solution is

$$\text{pOH} = -\log [OH^-] = -\log (3.00 \times 10^{-5}) = 4.52$$

and the pH is

$$pH = 14.00 - pOH = 14.00 - 4.52 = 9.48$$

Referring to Figure 19-6, we see that thymolphthalein or phenolphthalein is a suitable indicator.

19-31 (a)   The reaction equilibrium is

$$CH_3COOH(aq) + H_2O(l) \rightleftharpoons H_3O^+(aq) + CH_3COO^-(aq)$$

The K_a expression is (Table 18-3)

$$K_a = \frac{[H_3O^+][CH_3COO^-]}{[CH_3COOH]} = 1.74 \times 10^{-5} \text{ M}$$

Before any base is added we have

$$[H_3O^+] = [CH_3COO^-]$$

and

$$[CH_3COOH] = 0.200 \text{ M} - [H_3O^+]$$

Substituting these expressions into the K_a expression yields

$$\frac{[H_3O^+]^2}{0.200 \text{ M} - [H_3O^+]} = 1.74 \times 10^{-5} \text{ M}$$

Assuming $[H_3O^+] \ll 0.200$ M yields

$$[H_3O^+] = 1.86 \times 10^{-3} \text{ M}$$

The value is confirmed by the method of successive approximations. The pH of the solution is

$$pH = -\log(1.86 \times 10^{-3}) = 2.73$$

(b)   The equation for the reaction is

$$CH_3COOH(aq) + OH^-(aq) \longrightarrow H_2O(l) + CH_3COO^-(aq)$$

The millimoles of base added is

$$\left(\begin{array}{c}\text{millimoles of OH}^-(aq) \\ \text{added}\end{array}\right) = (5.00 \text{ mL})(0.200 \text{ M}) = 1.00 \text{ mmol}$$

The millimoles of acid available is

$$\left(\begin{array}{c}\text{millimoles of CH}_3COOH(aq) \\ \text{available}\end{array}\right) = (25.00 \text{ mL})(0.200 \text{ M}) = 5.00 \text{ mmol}$$

The millimoles of $CH_3COO^-(aq)$ is equal to the millimoles of $OH^-(aq)$ added because the acid is in excess. Thus, we have for the stoichiometric concentrations

$$[CH_3COO^-]_0 = \frac{1.00 \text{ mmol}}{(25.00 \text{ mL} + 5.00 \text{ mL})} = 0.0333 \text{ M}$$

$$[CH_3COOH]_0 = \frac{(5.00 \text{ mmol} - 1.00 \text{ mmol})}{(25.00 \text{ mL} + 5.00 \text{ mL})} = 0.133 \text{ M}$$

The acid-dissociation equilibrium is

$$CH_3COOH(aq) + H_2O(l) \rightleftharpoons H_3O^+(aq) + CH_3COO^-(aq)$$
$$0.133 \text{ M} - [H_3O^+] \qquad\qquad\qquad [H_3O^+] \qquad 0.0333 \text{ M} + [H_3O^+]$$

Substituting these values into the K_a expression yields

$$\frac{[H_3O^+](0.0333 \text{ M} + [H_3O^+])}{0.133 \text{ M} - [H_3O^+]} = 1.74 \times 10^{-5} \text{ M}$$

We shall use the method of successive approximations to solve for the value of $[H_3O^+]$.

$$[H_3O^+] = 6.93 \times 10^{-5} \text{ M}$$

The pH of the solution is

$$pH = -\log [H_3O^+] = -\log (6.93 \times 10^{-5}) = 4.16$$

(c)   Proceeding as in part (b), we have for the stoichiometric concentrations

$$[CH_3COO^-]_0 = \frac{(12.50 \text{ mL})(0.200 \text{ M})}{(25.00 \text{ mL} + 12.50 \text{ mL})} = 0.0667 \text{ M}$$

$$[CH_3COOH]_0 = \frac{(5.00 \text{ mmol}) - (12.50 \text{ mL})(0.200 \text{ M})}{(25.00 \text{ mL} + 12.50 \text{ mL})}$$
$$= 0.0667 \text{ M}$$

Using the K_a expression we have

$$\frac{[H_3O^+](0.0667 \text{ M} + [H_3O^+])}{0.0667 \text{ M} - [H_3O^+]} = 1.74 \times 10^{-5} \text{ M}$$

The method of successive approximations yields

$$[H_3O^+] = 1.74 \times 10^{-5} \text{ M}$$

The pH of the solution is

$$pH = -\log [H_3O^+] = -\log (1.74 \times 10^{-5}) = 4.76$$

(d)   The millimoles of base added is

$$(25.00 \text{ mL})(0.200 \text{ M}) = 5.00 \text{ mmol}$$

which is equal to the millimoles of acid originally present; thus, the solution is at the equivalence point.

$$[CH_3COO^-]_0 = \frac{5.00 \text{ mmol}}{50.00 \text{ mL}} = 0.100 \text{ M}$$

At the equivalence point we have a 0.100 M solution of $NaCH_3COO(aq)$. The relevant equilibrium is the $CH_3COO^-(aq)$ protonation reaction

$$CH_3COO^-(aq) + H_2O(l) \rightleftharpoons CH_3COOH(aq) + OH^-(aq)$$

The value of K_b is (Table 18-5)

$$K_b = \frac{[CH_3COOH][OH^-]}{[CH_3COO^-]} = 5.75 \times 10^{-10} \text{ M}$$

At equilibrium we have from the equation stoichiometry

$$[CH_3COOH] = [OH^-]$$

and

$$[CH_3COO^-] = 0.100 \text{ M} - [OH^-]$$

Substituting these expressions into the K_b expression yields

$$\frac{[OH^-]^2}{0.100 \text{ M} - [OH^-]} = 5.75 \times 10^{-10} \text{ M}$$

The method of successive approximations yields

$$[OH^-] = 7.58 \times 10^{-6} \text{ M}$$

The pOH is

$$pOH = -\log(7.58 \times 10^{-6}) = 5.12$$

and the pH is

$$pH = 14.00 - pOH = 14.00 - 5.12 = 8.88$$

(e)   In this case, the solution is beyond the equivalence point [see part (d)]. The stoichiometric concentration of $CH_3COO^-(aq)$ is

$$[CH_3COO^-]_0 = \frac{5.00 \text{ mmol}}{25.00 \text{ mL} + 26.00 \text{ mL}}$$
$$= 0.0980 \text{ M}$$

The stoichiometric concentration of $OH^-(aq)$ is equal to

$$[OH^-]_0 = \frac{(\text{total mmol base added}) - (\text{total mmol acid initially})}{\text{total volume}}$$

$$= \frac{(26.00 \text{ mL})(0.200 \text{ M}) - (5.00 \text{ mmol})}{(51.00 \text{ mL})}$$

$$= 0.00392 \text{ M}$$

The equilibrium is

$$CH_3COO^-(aq) + H_2O(l) \rightleftharpoons CH_3COOH(aq) + OH^-(aq)$$

| 0.0980 M | [CH_3COOH] | 0.00392 M |
| $- [CH_3COOH]$ | | $+ [CH_3COOH]$ |

Using the K_b expression, we have

$$\frac{[CH_3COOH](0.00392 \text{ M} + [CH_3COOH])}{(0.0980 \text{ M} - [CH_3COOH])} = 5.75 \times 10^{-10} \text{ M}$$

Neglecting $[CH_3COOH]$ relative to 0.00392 and 0.0980 yields

$$[CH_3COOH] \approx \frac{5.75 \times 10^{-10} \text{ M}(0.0980 \text{ M})}{(0.00392 \text{ M})}$$

$$= 1.44 \times 10^{-8} \text{ M}$$

Thus, we have

$$[OH^-] = 0.00392 \text{ M} + 1.44 \times 10^{-8} \text{ M} = 0.00392 \text{ M}$$

The pOH of the solution is

$$pOH = -\log (0.00392) = 2.41$$

The pH of the solution is

$$pH = 14.00 - 2.41 = 11.59$$

19-33 (a) Before any $HBr(aq)$ is added we have a 0.150-M solution of pyridine.

$$C_5H_5N(aq) + H_2O(l) \rightleftharpoons C_5H_5NH^+(aq) + OH^-(aq)$$

The K_b expression is (Table 18-4)

$$K_b = \frac{[C_5H_5NH^+][OH^-]}{[C_5H_5N]} = 1.46 \times 10^{-9} \text{ M}$$

At equilibrium we have, from the reaction stoichiometry,

$$[C_5H_5NH^+] = [OH^-]$$
$$[C_5H_5N] = 0.150 \text{ M} - [OH^-]$$

Substituting these expressions into the K_b expression yields

$$\frac{[OH^-]^2}{0.150\ M - [OH^-]} = 1.46 \times 10^{-9}\ M$$

The method of successive approximations yields

$$[OH^-] = 1.48 \times 10^{-5}\ M$$

The pOH is

$$pOH = -\log\,[OH^-] = -\log\,(1.48 \times 10^{-5}) = 4.83$$

and the pH is

$$pH = 14.00 - pOH = 14.00 - 4.83 = 9.17$$

(b)   The equation for the reaction that occurs is

$$C_5H_5N(aq) + H_3O^+(aq) \longrightarrow C_5H_5NH^+(aq) + H_2O(l)$$

After the addition of 10.0 mL of 0.150 M HBr, we have the following stoichiometric concentrations.

$$[C_5H_5NH^+]_0 = \frac{(10.0\ mL)(0.150\ M)}{25.0\ mL + 10.0\ mL} = 0.0429\ M$$

$$[C_5H_5N]_0 = \frac{(25.0\ mL)(0.150\ M) - (10.0\ mL)(0.150\ M)}{(25.0\ mL + 10.0\ mL)} = 0.0643\ M$$

Thus, we have at equilibrium

$$\begin{array}{cccc}
C_5H_5N(aq) & +H_2O(l) \rightleftharpoons & C_5H_5NH^+(aq) & + OH^-(aq) \\
0.0643\ M - [OH^-] & & 0.0429\ M + [OH^-] & [OH^-]
\end{array}$$

and

$$K_b = \frac{[OH^-](0.0429\ M + [OH^-])}{0.0643\ M - [OH^-]} = 1.46 \times 10^{-9}\ M$$

Assuming $[OH^-] \ll 0.0429$ or 0.0643 yields

$$[OH^-] \approx \frac{(1.46 \times 10^{-9}\ M)(0.0643\ M)}{(0.0429\ M)} = 2.19 \times 10^{-9}\ M$$

This result for $[OH^-]$ confirms the above assumption. The pOH of the solution is

$$pOH = -\log\,[OH^-] = -\log\,(2.19 \times 10^{-9}) = 8.66$$

and the pH is

$$pH = 14.00 - pOH = 14.00 - 8.66 = 5.34$$

(c)   The problem is similar to part (b), thus

$$[C_5H_5NH^+]_0 = \frac{(24.0 \text{ mL})(0.150 \text{ M})}{(25.0 \text{ mL} + 24.0 \text{ mL})} = 0.0735 \text{ M}$$

and

$$[C_5H_5N]_0 = \frac{(25.0 \text{ mL})(0.150 \text{ M}) - (24.0 \text{ mL})(0.150 \text{ M})}{(25.0 \text{ mL} + 24.0 \text{ mL})} = 0.00306 \text{ M}$$

The K_b expression is

$$K_b = \frac{[OH^-](0.0735 \text{ M} + [OH^-])}{0.00306 \text{ M} - [OH^-]} = 1.46 \times 10^{-9} \text{ M}$$

Assuming $[OH^-] \ll 0.00306$ or 0.0735 M, we have

$$[OH^-] \approx \frac{(1.46 \times 10^{-9} \text{ M})(3.06 \times 10^{-3} \text{ M})}{(7.35 \times 10^{-2} \text{ M})}$$
$$= 6.08 \times 10^{-11} \text{ M}$$

The pH of the solution is

$$pH = 14.00 - pOH = 14.00 + \log (6.08 \times 10^{-11}) = 3.78$$

(d)   The millimoles of HBr(aq) added is

$$(25.0 \text{ mL})(0.150 \text{ M}) = 3.75 \text{ mmol}$$

The total millimoles of C_5H_5N available initially is

$$(25.0 \text{ mL})(0.150 \text{ M}) = 3.75 \text{ mmol}$$

Thus, the resulting solution is at the equivalence point. The solution consists of $C_5H_5NH^+Br^-(aq)$. The stoichiometric concentration of $C_5H_5NH^+(aq)$ is

$$[C_5H_5NH^+]_0 = \frac{3.75 \text{ mmol}}{50.0 \text{ mL}} = 0.0750 \text{ M}$$

At equilibrium we have

$$C_5H_5NH^+(aq) \; + H_2O(l) \Longrightarrow C_5H_5N(aq) + H_3O^+(aq)$$
$$0.0750 \text{ M} - [H_3O^+] \qquad\qquad\qquad [H_3O^+] \qquad\quad [H_3O^+]$$

Substitution of these values into the K_a expression (Table 18-5) yields

$$K_a = \frac{[C_5H_5N][H_3O^+]}{[C_5H_5NH^+]} = \frac{[H_3O^+]^2}{0.0750 \text{ M} - [H_3O^+]} = 6.85 \times 10^{-6} \text{ M}$$

The method of successive approximations yields

$$[H_3O^+] = 7.13 \times 10^{-4} \text{ M}$$

The pH of the solution is

$$pH = -\log[H_3O^+] = -\log(7.13 \times 10^{-4}) = 3.15$$

(e) The resulting solution is beyond the equivalence point [see part (d)]. The stoichiometric concentrations are

$$[C_5H_5NH^+]_0 = \frac{3.75 \text{ mmol}}{25.0 \text{ mL} + 26.0 \text{ mL}} = 0.0735 \text{ M}$$

$$[H_3O^+]_0 = \frac{(26.0 \text{ mL})(0.150 \text{ M}) - (3.75 \text{ mmol})}{(25.0 \text{ mL} + 26.0 \text{ mL})} = 0.00294 \text{ M}$$

Thus, we have

$$\underset{0.0735 \text{ M} - [C_5H_5N]}{C_5H_5NH^+(aq)} + H_2O(l) \rightleftharpoons \underset{[C_5H_5N]}{C_5H_5N(aq)} + \underset{0.00294 \text{ M} + [C_5H_5N]}{H_3O^+(aq)}$$

Substitution of these values into the K_a expression yields

$$\frac{(0.00294 \text{ M} + [C_5H_5N])[C_5H_5N]}{0.0735 \text{ M} - [C_5H_5N]} = 6.85 \times 10^{-6} \text{ M}$$

Assuming $[C_5H_5N] \ll 0.00294$ M or 0.0735 M, we obtain

$$[C_5H_5N] \approx \frac{(6.85 \times 10^{-6} \text{ M})(0.0735 \text{ M})}{(0.00294 \text{ M})}$$
$$\approx 1.71 \times 10^{-4} \text{ M}$$

Because this value of $[C_5H_5N]$ is not small compared to 0.00294 M, we use the method of successive approximations. Thus,

$$[C_5H_5N] \approx \frac{(6.85 \times 10^{-6} \text{ M})(0.0735 \text{ M} - 1.71 \times 10^{-4} \text{ M})}{(0.00294 \text{ M} + 1.71 \times 10^{-4} \text{ M})}$$
$$\approx 1.61 \times 10^{-4} \text{ M}$$

The next approximation is

$$[C_5H_5N] \approx \frac{(6.85 \times 10^{-6} \text{ M})(0.0735 \text{ M} - 1.61 \times 10^{-4} \text{ M})}{(0.00294 \text{ M} + 1.61 \times 10^{-4} \text{ M})}$$
$$\approx 1.62 \times 10^{-4} \text{ M}$$

Further approximations confirm this value of $[C_5H_5N]$; thus, we have

$$[H_3O^+] = 0.00294 \text{ M} + 1.62 \times 10^{-4} \text{ M} = 0.00310 \text{ M}$$

The pH of the solution is

$$pH = \log [H_3O^+] = -\log (0.00310) = 2.51$$

19-35   At the equivalence point

$$\text{moles of acid} = \text{moles of base added}$$

The number of moles of acid is given by

$$\text{moles of acid} = M_b V_b = (0.250 \text{ mol} \cdot \text{L}^{-1})(0.0341 \text{ L})$$
$$= 0.00853 \text{ mol}$$

We have the correspondence

$$1.50 \text{ g ascorbic acid} \backsimeq 0.00853 \text{ mol ascorbic acid}$$

Dividing by 0.00853, we have

$$176 \text{ g ascorbic acid} \backsimeq 1 \text{ mol ascorbic acid}$$

Thus, the molecular mass of ascorbic acid, vitamin C, is 176.

19-37   Because oxalic acid is a diprotic acid (Table 19-4), it takes two moles of NaOH to neutralize one mole of oxalic acid. We have

$$\text{moles of NaOH} = \left(\frac{2 \text{ mol NaOH}}{1 \text{ mol oxalic acid}} \right) (\text{moles of oxalic acid})$$

$$M_b V_b = 2 M_a V_a$$

$$V_b = \frac{2 M_a V_a}{M_b} = \frac{(2)(0.10 \text{ M})(25.0 \text{ mL})}{(0.10 \text{ M})} = 50.0 \text{ mL}$$

19-39   The number of moles of NaOH(aq) used to neutralize the oxalic acid is

$$\text{moles of NaOH} = MV = (0.250 \text{ mol} \cdot \text{L}^{-1})(0.0444 \text{ L}) = 0.0111 \text{ mol}$$

Because oxalic acid is a diprotic acid, it takes two moles of NaOH to neutralize one mole of oxalic acid. Thus, the number of moles of oxalic acid is

$$\text{moles of oxalic acid} = \left(\frac{1 \text{ mol oxalic acid}}{2 \text{ mol NaOH}} \right) (0.0111 \text{ mol NaOH})$$

$$= 5.55 \times 10^{-3} \text{ mol}$$

Thus, we have the correspondence

$$0.500 \text{ g oxalic acid} \approx 0.00555 \text{ mol oxalic acid}$$

Dividing both sides by 0.00555, we have

$$90.1 \text{ g oxalic acid} \approx \text{ one mole of oxalic acid}$$

The molecular mass of oxalic acid is 90.1.

19-41 The reaction is

$$H_3AsO_4(aq) + H_2O(l) \rightleftharpoons H_3O^+(aq) + H_2AsO_4^-(aq)$$

The value of K_{a1} is

$$K_{a1} = 10^{-pK_{a1}} = 10^{-2.22} = 6.03 \times 10^{-3} \text{ M}$$

The K_{a1} expression is

$$K_{a1} = \frac{[H_3O^+][H_2AsO_4^-]}{[H_3AsO_4]} = \frac{[H_3O^+]^2}{0.100 \text{ M} - [H_3O^+]} = 6.03 \times 10^{-3} \text{ M}$$

or

$$[H_3O^+]^2 + 6.03 \times 10^{-3} \text{ M } [H_3O^+] - 6.03 \times 10^{-4} \text{ M}^2 = 0$$

The quadratic formula gives

$$[H_3O^+] = \frac{-6.03 \times 10^{-3} \text{ M} \pm \sqrt{3.64 \times 10^{-5} \text{ M}^2 - (4)(1)(-6.03 \times 10^{-4} \text{ M}^2)}}{(2)(1)}$$

$$= \frac{-6.03 \times 10^{-3} \text{ M} \pm 4.95 \times 10^{-2} \text{ M}}{2}$$

$$= 2.17 \times 10^{-2} \text{ M} \qquad (\text{and} -0.0278 \text{ M})$$

The pH of the solution is

$$pH = -\log [H_3O^+] = -\log (2.17 \times 10^{-2}) = 1.66$$

19-43 The number of moles in 1.00 g of $Mg(OH)_2$ is

$$\text{moles of } Mg(OH)_2 = (1.00 \text{ g})\left(\frac{1 \text{ mol } Mg(OH)_2}{58.33 \text{ g } Mg(OH)_2}\right) = 0.0171 \text{ mol}$$

At neutralization (the equivalence point)

$$\text{moles of } OH^-(aq) = \text{moles of } H_3O^+(aq)$$

There are two moles of $OH^-(aq)$ per mole of $Mg(OH)_2(aq)$, and so we have

$$\text{moles of } OH^-(aq) = \left(\frac{2 \text{ mol } OH^-}{1 \text{ mol } Mg(OH)_2}\right)(0.0171 \text{ mol } Mg(OH)_2)$$
$$= 0.0343 \text{ mol}$$

The volume of HCl (aq) that is neutralized by 0.0343 mol of $OH^-(aq)$ is

$$V_a = \frac{0.0343 \text{ mol}}{M_a} = \frac{0.0343 \text{ mol}}{0.10 \text{ mol} \cdot L^{-1}} = 0.34 \text{ L} = 340 \text{ mL}$$

19-45 Both acids have only one dissociable proton per molecule. Thus, at the equivalence point

$$\text{moles of acid} = \text{moles of base} = M_b V_b$$
$$= (1.00 \text{ mol} \cdot L^{-1})(0.01549 \text{ L}) = 0.01549 \text{ mol}$$

We have the correspondence

$$1.89 \text{ g acid} \approx 0.01549 \text{ mol acid}$$

Dividing both sides by 0.01549, we have

$$122 \text{ g} \approx 1 \text{ mol}$$

Thus, the molecular mass of the acid is 122. The molecular mass of benzoic acid, C_6H_5COOH, is 122.12; the molecular mass of chlorobenzoic acid, $C_6H_4ClCOOH$, is 156.56. The acid must be benzoic acid.

19-47 One equilibrium that is established in the solution is

$$CH_3COOH(aq) + H_2O(l) \rightleftharpoons H_3O^+(aq) + CH_3COO^-(aq)$$
$$K_a = \frac{[CH_3COO^-][H_3O^+]}{[CH_3COOH]} = 1.74 \times 10^{-5} \text{ M}$$

We can set up a table of initial and equilibrium concentrations.

	$CH_3COOH(aq)$	$+ H_2O(l) \rightleftharpoons$	$CH_3COO^-(aq)$	$+ H_3O^+(aq)$
Initial concentration	0.100 M	—	0.100 M	≈ 0
Equilibrium concentration	$0.100 \text{ M} - [H_3O^+]$	—	$0.100 \text{ M} + [H_3O^+]$	$[H_3O^+]$

Note that $[CH_3COO^-]$ is not equal to $[H_3O^+]$ in this case. The K_a expression is

$$K_a = \frac{(0.100 \text{ M} + [H_3O^+])[H_3O^+]}{0.100 \text{ M} - [H_3O^+]} = 1.74 \times 10^{-5} \text{ M}$$

Using the method of successive approximations, we find that

$$[H_3O^+] = 1.74 \times 10^{-5} \text{ M}$$

The equilibrium concentrations are

$$[CH_3COOH] = 0.100 \text{ M} - 1.74 \times 10^{-5} \text{ M} = 0.100 \text{ M}$$
$$[CH_3COO^-] = 0.100 \text{ M} + 1.74 \times 10^{-5} \text{ M} = 0.100 \text{ M}$$

and so we see that

$$[CH_3COOH] \approx [CH_3COOH]_0$$
$$[CH_3COO^-] \approx [CH_3COO^-]_0$$

These relations arise because $[H_3O^+]$ is negligible compared to either $[CH_3COOH]$ or $[CH_3COO^-]$.

19-49 (a)   The two equations corresponding to K_{a1} and K_{a2} are

$$H_2CO_3(aq) + H_2O(l) \rightleftharpoons H_3O^+(aq) + HCO_3^-(aq) \qquad K_{a1}$$
$$HCO_3^-(aq) + H_2O(l) \rightleftharpoons H_3O^+(aq) + CO_3^{2-}(aq) \qquad K_{a2}$$

Add these two equations to get

$$H_2CO_3(aq) + 2H_2O(l) \rightleftharpoons 2H_3O^+(aq) + CO_3^{2-}(aq) \qquad K = K_{a1}K_{a2}$$

and

$$K = \frac{[H_3O^+]^2[CO_3^{2-}]}{[H_2CO_3]} = K_{a1}K_{a2}$$

But according to the stoichiometry of the equation

$$2HCO_3^-(aq) \rightleftharpoons CO_3^{2-}(aq) + H_2CO_3(aq)$$

We have $[CO_3^{2-}] = [H_2CO_3]$, and so

$$\frac{[H_3O^+]^2[\cancel{CO_3^{2-}}]}{[\cancel{H_2CO_3}]} = K_{a1}K_{a2}$$
$$[H_3O^+]^2 = K_{a1}K_{a2}$$
$$[H_3O^+] = (K_{a1}K_{a2})^{1/2}$$

Taking the negative logarithm of both sides of this equation yields

$$-\log\,[H_3O^+] = \frac{1}{2}\,(-\log K_{a1} - \log K_{a2})$$

or

$$pH = \frac{1}{2}\,(pK_{a1} + pK_{a2})$$

From Table 19-4 we have

$$pH = \frac{1}{2}\,(6.35 + 10.33) = 8.34$$

(b) A $NaHCO_3(aq)$ solution acts as a buffer through the reactions described by the following equations.

$NaHCO_3$ acting as a base:

$$HCO_3^-(aq) + H_3O^+(aq) \rightleftharpoons H_2CO_3(aq) + H_2O(l)$$

$NaHCO_3$ acting as an acid:

$$HCO_3^-(aq) + OH^-(aq) \rightleftharpoons CO_3^{2-}(aq) + H_2O(l)$$

19-51
$$3HCO_3^-(aq) + H_3C_6H_5O_7(aq) \rightleftharpoons 3H_2CO_3(aq) + C_6H_5O_7^{3-}(aq)$$
$$\updownarrow$$
$$3CO_2(aq) + 3H_2O(l)$$
$$\updownarrow$$
$$3CO_2(g)$$

or

$$3HCO_3^-(aq) + H_3C_6H_5O_7(aq) \rightleftharpoons 3CO_2(g) + C_6H_5O_7^{3-}(aq) + 3H_2O(l)$$

19-53 The number of moles in 500 mg of $Al(OH)_3$ is

$$\text{moles of } Al(OH)_3 = (0.500\text{ g})\left(\frac{1\text{ mol } Al(OH)_3}{78.00\text{ g } Al(OH)_3}\right) = 0.00641\text{ mol}$$

At neutralization

$$\text{moles of } OH^-(aq) = \text{moles of } H_3O^+(aq)$$

There are three moles of $OH^-(aq)$ per mole of $Al(OH)_3(s)$, and so

$$\left(\frac{3\text{ mol } OH^-}{1\text{ mol } Al(OH)_3}\right)(0.00641\text{ mol } Al(OH)_3) = M_aV_a = (0.10\text{ mol}\cdot L^{-1})V_a$$

$$V_a = \frac{0.0192\text{ mol}}{0.10\text{ mol}\cdot L^{-1}} = 0.19\text{ L} = 190\text{ mL}$$

19-55   At the equivalence point

$$\text{moles of acid} = \text{moles of base} = M_bV_b$$
$$= (0.100 \text{ mol} \cdot \text{L}^{-1})(0.0624 \text{ L}) = 0.00624 \text{ mol}$$

We have the corresponence

$$0.550 \text{ g butyric acid} \backsimeq 0.00624 \text{ mol butyric acid}$$

Dividing both sides by 0.00624, we have

$$88.1 \text{ g butyric acid} \backsimeq 1 \text{ mol butyric acid}$$

The molecular mass of butyric acid is 88.1.

19-57   Measure the pH of the original solution. Dilute the solution and measure the pH of the diluted solution. The pH of a buffer would not be affected by dilution.

19-59   At the equivalence point, the number of moles of base is equal to the number of moles of acid.

$$\text{moles of acid} = MV = (0.150 \text{ M})(0.0690 \text{ L}) = 0.01035 \text{ mol}$$

Thus,

$$0.01035 \text{ mol acid} \backsimeq 1.20 \text{ g acid}$$

and

$$1 \text{ mol acid} \backsimeq \frac{1.20 \text{ g acid}}{0.01035 \text{ acid}} = 116 \text{ g acid}$$

The molecular mass of the acid is 116.

19-61   The number of moles of base used is

$$\text{moles of base} = MV = (0.135 \text{ M})(0.1472 \text{ L}) = 0.0199 \text{ mol}$$

The number of moles of citric acid available is

$$\text{moles of } C_6H_8O_7 = (1.270 \text{ g}) \left(\frac{1 \text{ mol } C_6H_8O_7}{192.12 \text{ g } C_6H_8O_7} \right) = 0.006610 \text{ mol}$$

Thus,

$$\left(\frac{0.0199 \text{ mol OH}^-}{0.006610 \text{ mol } C_6H_8O_7} \right) \left(\frac{1 \text{ mol } H_3O^+}{1 \text{ mol OH}^-} \right) = \frac{3.01 \text{ mol } H_3O^+}{1 \text{ mol } C_6H_8O_7}$$

There are three dissociable protons per molecule of citric acid. We normally write the formula of citric acid as $H_3C_6H_5O_7$.

19-63 Because the pH is greater than the pK_{a2}, the relevant equilibrium is

$$\underset{0.125\ \text{M} - [\text{C}_2\text{O}_4^{2-}]}{\text{HC}_2\text{O}_4^-(aq)} + \text{H}_2\text{O}(l) \rightleftharpoons \underset{[\text{H}_3\text{O}^+]}{\text{H}_3\text{O}^+(aq)} + \underset{[\text{C}_2\text{O}_4^{2-}]}{\text{C}_2\text{O}_4^{2-}(aq)}$$

Thus,

$$K_{a2} = 10^{-pK_{a2}} = 10^{-4.27} = 5.37 \times 10^{-5}\ \text{M} = \frac{[\text{H}_3\text{O}^+][\text{C}_2\text{O}_4^{2-}]}{[\text{HC}_2\text{O}_4^-]}$$

At pH = 5.00 we have

$$[\text{H}_3\text{O}^+] = 10^{-pH} = 10^{-5.00} = 1.00 \times 10^{-5}\ \text{M}$$

Combining this result with the K_{a2} expression yields

$$\frac{(1.00 \times 10^{-5}\ \text{M})[\text{C}_2\text{O}_4^{2-}]}{0.125\ \text{M} - [\text{C}_2\text{O}_4^{2-}]} = 5.37 \times 10^{-5}\ \text{M}$$

Thus,

$$(1.00 \times 10^{-5}\ \text{M})[\text{C}_2\text{O}_4^{2-}] = 6.71 \times 10^{-6}\ \text{M}^2 - (5.37 \times 10^{-5}\ \text{M})[\text{C}_2\text{O}_4^{2-}]$$

and

$$[\text{C}_2\text{O}_4^{2-}] = \frac{6.71 \times 10^{-6}\ \text{M}^2}{6.37\ 10^{-5}\ \text{M}} = 0.105\ \text{M}$$

We have a material balance condition

$$0.125\ \text{M} = [\text{H}_2\text{C}_2\text{O}_4] + [\text{HC}_2\text{O}_4^-] + [\text{C}_2\text{O}_4^{2-}]$$

Because the pH is greater than pK_{a2}, we can ignore K_{a1}. We assume that $[\text{H}_2\text{C}_2\text{O}_4]$ will be much smaller than $[\text{HC}_2\text{O}_4^-]$. Thus neglecting $[\text{H}_2\text{C}_2\text{O}_4]$, we have

$$[\text{HC}_2\text{O}_4^-] = 0.125\ \text{M} - [\text{C}_2\text{O}_4^{2-}] = 0.125\ \text{M} - 0.105\ \text{M} = 0.020\ \text{M}$$

We now calculate the value of $[\text{H}_2\text{C}_2\text{O}_4]$ from the K_{a1} expression

$$\text{H}_2\text{C}_2\text{O}_4(aq) + \text{H}_2\text{O}(l) \rightleftharpoons \text{HC}_2\text{O}_4^-(aq) + \text{H}_3\text{O}^+(aq)$$

$$K_{a1} = 10^{-pK_{a1}} = 10^{-1.27} = 5.37 \times 10^{-2}\ \text{M} = \frac{[\text{HC}_2\text{O}_4^-][\text{H}_3\text{O}^+]}{[\text{H}_2\text{C}_2\text{O}_4]}$$

We know that $[\text{H}_3\text{O}^+] = 1.00 \times 10^{-5}\ \text{M}$ and $[\text{HC}_2\text{O}_4^-] = 0.020\ \text{M}$; thus,

$$5.37 \times 10^{-2}\ \text{M} = \frac{(0.020\ \text{M})(1.00 \times 10^{-5}\ \text{M})}{[\text{H}_2\text{C}_2\text{O}_4]}$$

from which we compute

$$[H_2C_2O_4] = 3.7 \times 10^{-6} \text{ M}$$

which is negligible compared to $[HC_2O_4^-]$ and $[C_2O_4^{2-}]$ and thus confirms our earlier assumption that $[H_2C_2O_4]$ is negligible compared to $[HC_2O_4^-]$ and $[C_2O_4^{2-}]$.

19-65 We shall estimate the pH of the solution by using the Henderson-Hasselbalch equation.

$$pH = pK_a + \log \frac{[\text{base}]_0}{[\text{acid}]_0}$$

We see that the conjugate acid is $H_2PO_4^-(aq)$ and the conjugate base is $HPO_4^{2-}(aq)$. The value of pK_a for $H_2PO_4^-$ is

$$pK_a = -\log K_a = -\log (6.2 \times 10^{-8}) = 7.21$$

We can now do each part in turn:

(a) $[\text{acid}]_0 = 0.050 \text{ M}$ $[\text{base}]_0 = 0.050 \text{ M}$

$$pH = 7.21 + \log \left(\frac{0.050 \text{ M}}{0.050 \text{ M}} \right) = 7.21 + 0.00 = 7.21$$

(b) $[\text{acid}]_0 = 0.050 \text{ M}$ $[\text{base}]_0 = 0.10 \text{ M}$

$$pH = 7.21 + \log \left(\frac{0.10 \text{ M}}{0.050 \text{ M}} \right) = 7.21 + 0.30 = 7.51$$

(c) $[\text{acid}]_0 = 0.10 \text{ M}$ $[\text{base}]_0 = 0.050 \text{ M}$

$$pH = 7.21 + \log \left(\frac{0.050 \text{ M}}{0.10 \text{ M}} \right) = 7.21 - 0.30 = 6.91$$

19-67 If we use equal concentrations of conjugate acid and base, then

$$pH = pK_a + \log \frac{[\text{base}]_0}{[\text{acid}]_0}$$
$$= pK_a$$

To obtain a pH buffered at 3.70, we want $pK_a \approx 3.70$. From Table 18-5 we find that $pK_a = 3.75$ for formic acid, and so a solution of equal concentrations of $HCOOH(aq)$ and $NaHCOO(aq)$ would act as a buffer at $pH = 3.70$.

19-69 At the equivalence point

$$\text{moles of acid} = \text{moles of base}$$

The concentration of acetic acid in the vinegar solution is given by

$$M_a = \frac{M_b V_b}{V_a} = \frac{(0.400 \text{ M})(38.5 \text{ mL})}{(21.0 \text{ mL})} = 0.733 \text{ M}$$

We now calculate the mass percentage of the acetic acid. For convenience, consider a 100-mL sample of vinegar. The number of moles of acetic acid in 100 mL of vinegar is

$$n = MV = (0.733 \text{ mol} \cdot \text{L}^{-1})(0.100 \text{ L}) = 0.0733 \text{ mol}$$

The mass of acetic acid in 100 mL of vinegar is

$$\text{mass of CH}_3\text{COOH} = (0.0733 \text{ mol}) \left(\frac{60.05 \text{ g CH}_3\text{COOH}}{1 \text{ mol CH}_3\text{COOH}} \right) = 4.40 \text{ g}$$

The mass of 100 mL of vinegar solution is

$$\text{mass of vinegar} = dV = (1.060 \text{ g} \cdot \text{mL}^{-1})(100 \text{ mL}) = 106.0 \text{ g}$$

The mass percentage of acetic acid is

$$\text{mass \%} = \frac{\text{mass of acetic acid}}{\text{mass of vinegar solution}} \times 100$$

$$= \frac{4.40 \text{ g}}{106.0 \text{ g}} \times 100 = 4.15\%$$

19-71   The titration reaction is

$$\text{NH}_3(aq) + \text{H}_3\text{O}^+(aq) \longrightarrow \text{NH}_4^+(aq) + \text{H}_2\text{O}(l)$$

The volume of acid required to reach the equivalence point is

$$M_a V_a = M_b V_b$$

$$V_a = \frac{(0.125 \text{ M})(50.0 \text{ mL})}{(0.175 \text{ M})} = 35.7 \text{ mL}$$

The total volume at the equivalence point is

$$50.0 \text{ mL} + 35.7 \text{ mL} = 85.7 \text{ mL}$$

The stoichiometric concentration of $\text{NH}_4^+(aq)$ at the equivalence point is

$$[\text{NH}_4^+]_0 = \frac{(50.0 \text{ mL})(0.125 \text{ M})}{(85.7 \text{ mL})} = 0.0729 \text{ M}$$

The reaction equilibrium at the equivalence point is

$$\underset{0.0729 \text{ M} - [\text{H}_3\text{O}^+]}{\text{NH}_4^+(aq)} + \text{H}_2\text{O}(l) \rightleftharpoons \underset{[\text{H}_3\text{O}^+]}{\text{NH}_3(aq)} + \underset{[\text{H}_3\text{O}^+]}{\text{H}_3\text{O}^+(aq)}$$

Substitution of these values into the K_a expression (Table 18-5) yields

$$K_a = \frac{[H_3O^+][NH_3]}{[NH_4^+]} = \frac{[H_3O^+]^2}{0.0729\ M - [H_3O^+]} = 5.71 \times 10^{-10}\ M$$

The method of successive approximations yields

$$[H_3O^+] = 6.45 \times 10^{-6}\ M$$

The pH of the solution is

$$pH = -\log [H_3O^+] = -\log (6.45 \times 10^{-6}) = 5.19$$

19-73 We set up a concentration table.

	H$_2$Se(aq)	+	H$_2$O(l) $\rightleftharpoons$ HSe$^-$(aq)	+ H$_3$O$^+$(aq)
Initial concentration	0.250 M	—	0	≈ 0
Equilibrium concentration	0.250 M $-$ [H$_3$O$^+$]	—	[H$_3$O$^+$]	[H$_3$O$^+$]

The values of the equilibrium constants are (Table 19-4)

$$pK_{a1} = 3.89 \quad K_{a1} = 1.29 \times 10^{-4}\ M$$
$$pK_{a2} = 11.0 \quad K_{a2} = 1.0 \times 10^{-11}\ M$$

We can ignore the second acid dissociation because the value of K_{a2} is much less than K_{a1}. Thus, we have that

$$K_{a1} = \frac{[HSe^-][H_3O^+]}{[H_2Se]} = \frac{[H_3O^+]^2}{0.250\ M - [H_3O^+]} = 1.29 \times 10^{-4}\ M$$

The method of successive approximations yields

$$[H_3O^+] = 5.61 \times 10^{-3}\ M$$
$$pH = -\log (5.61 \times 10^{-3}) = 2.25$$
$$[HSe^-] = 5.61 \times 10^{-3}\ M$$
$$[H_2Se] = 0.250\ M - 5.61 \times 10^{-3}\ M = 0.244\ M$$

From the second dissociation equilibrium, we have that

$$[Se^{2-}] = \frac{K_{a2}[HSe^-]}{[H_3O^+]} = \frac{(1.0 \times 10^{-11} \text{ M})(5.61 \times 10^{-3} \text{ M})}{5.61 \times 10^{-3} \text{ M}}$$

$$= 1.0 \times 10^{-11} \text{ M}$$

$$[OH^-] = \frac{1.00 \times 10^{-14} \text{ M}^2}{5.61 \times 10^{-3} \text{ M}} = 1.78 \times 10^{-12} \text{ M}$$

19-75   The equations for the acid dissociations are

$$H_2C_2O_4(aq) + H_2O(l) \rightleftharpoons HC_2O_4^-(aq) + H_3O^+(aq)$$
$$HC_2O_4^-(aq) + H_2O(l) \rightleftharpoons C_2O_4^{2-}(aq) + H_3O^+(aq)$$

The "C_2O_4" balance is

$$0.200 \text{ M} = [H_2C_2O_4] + [HC_2O_4^-] + [C_2O_4^{2-}]$$

The electroneutrality condition is

$$[H_3O^+] = [HC_2O_4^-] + 2[C_2O_4^{2-}] + [OH^-]$$

19-77   The NaOH(aq) added reacts with $CH_3COOH(aq)$ according to the equation

$$CH_3COOH(aq) + OH^-(aq) \rightarrow CH_3COO^-(aq) + H_2O(l)$$

Thus, the concentration of acid decreases and the concentration of base increases. We take the buffer to fail when

$$\frac{[\text{base}]_o}{[\text{acid}]_o} = 10$$

After the addition of NaOH(aq),

$$[\text{base}]_o = \frac{\text{initial millimoles of } CH_3COO^- + \text{millimoles of NaOH added}}{\text{total volume}}$$

$$[\text{acid}]_o = \frac{\text{initial millimoles of } CH_3COOH - \text{millimoles of NaOH added}}{\text{total volume}}$$

Thus, we have

$$\frac{\dfrac{\text{initial millimoles of } CH_3COO^- + \text{millimoles of NaOH added}}{\text{total volume}}}{\dfrac{\text{initial millimoles of } CH_3COOH - \text{millimoles of NaOH added}}{\text{total volume}}} = 10$$

The original number of millimoles of acid and base are

initial millimoles of $CH_3COO^- = MV = (0.050 \text{ M})(100 \text{ mL}) = 5.0 \text{ mmol}$
initial millimoles of $CH_3COOH = MV = (0.050 \text{ M})(100 \text{ mL}) = 5.0 \text{ mmol}$

Thus, we have

$$\frac{5.0 \text{ mmol} + \text{millimoles NaOH added}}{5.0 \text{ mmol} - \text{millimoles NaOH added}} = 10$$

$$\text{millimoles of NaOH added} = 4.1 \text{ mmol}$$

The volume of $NaOH(aq)$ added is

$$V = \frac{4.1 \text{ mmol}}{0.10 \text{ M}} = 41 \text{mL}$$

19-79 For $NH_4^+(aq)$, $pK_a = 9.24$ (Table 18-5). Using the Henderson-Hasselbalch equation,

$$pH = pK_a + \log \frac{[\text{base}]_0}{[\text{acid}]_0}$$

we have

$$\log \frac{[\text{base}]_0}{[\text{acid}]_0} = 9.50 - 9.24 = 0.26$$

But

$$[\text{base}]_0 = [NH_3]_0 \quad \text{and} \quad [\text{acid}]_0 = [NH_4^+]_0$$

thus,

$$\frac{[NH_3]_0}{[NH_4^+]_0} = 10^{0.26} = 1.82$$

and

$$[NH_4^+]_0 = \frac{[NH_3]_0}{1.82} = \frac{0.200 \text{ M}}{1.82} = 0.110 \text{ M}$$

The number of moles of NH_4Cl required is

$$n = MV = (0.110 \text{ M})(1.00 \text{ L}) = 0.110 \text{ mol}$$

and the mass of NH_4Cl is

$$\text{mass of } NH_4Cl = (0.110 \text{ mol}) \left(\frac{53.49 \text{ g } NH_4Cl}{1 \text{ mol } NH_4Cl} \right) = 5.88 \text{ g}$$

19-81   The Henderson-Hasselbalch equation is

$$pH = pK_a + \log \frac{[B^-]_0}{[HB]_0}$$

The value of pK_a for the acid-base conjugate pair NH_4^+/NH_3 is 9.24 (Table 18-5). Thus,

$$9.2 = 9.24 + \log \frac{[NH_3]_0}{[NH_4^+]_0}$$

$$\log \frac{[NH_3]_0}{[NH_4^+]_0} \approx 0$$

Taking the logarithm of both sides yields

$$\frac{[NH_3]_0}{[NH_4^+]_0} \approx 1$$

If we neutralize one half of the $NH_3(aq)$, then the concentration of $NH_3(aq)$ will equal the concentration of $NH_4^+(aq)$. Therefore we should add a volume of 0.10 M $HCl(aq)$ that is equal to one half the volume of the original $NH_3(aq)$ solution.

19-83   The Henderson-Hasselbalch equation is

$$pH = pK_a + \log \frac{[B^-]_0}{[HB]_0}$$

The number of moles of NH_4Cl added is given by

$$\text{moles of } NH_4Cl = (25.0 \text{ g}) \left(\frac{1 \text{ mol } NH_4Cl}{53.49 \text{ g } NH_4Cl} \right) = 0.467 \text{ mol}$$

Thus, the stoichiometric concentrations of acid and base are

$$[NH_4^+] = \frac{0.467 \text{ mol}}{0.300 \text{ L}} = 1.56 \text{ M} = [HB]_0$$

$$[NH_3] = 0.500 \text{ M} = [B^-]_0$$

The value of pK_a for the acid-base conjugate pair is 9.24 (Table 18-5).

$$pH = 9.24 + \log \left(\frac{0.500 \text{ M}}{1.56 \text{ M}} \right) = 8.75$$

19-85   The equation for the reaction is

$$HCl(aq) + NO_2^-(aq) \longrightarrow Cl^-(aq) + HNO_2(aq)$$

The number of moles of HCl(aq) added is

$$\text{moles of HCl} = (0.650 \text{ M})(0.0500 \text{ L}) = 0.0325 \text{ mol}$$

In the original solution, we have that

$$\text{moles of HNO}_2 = (1.00 \text{ M})(1.00 \text{ L}) = 1.00 \text{ mol}$$
$$\text{moles of NO}_2^- = (1.00 \text{ M})(1.00 \text{ L}) = 1.00 \text{ mol}$$

After the addition of 50 mL of HCl(aq), the number of moles of $HNO_2(aq)$ and $NO_2^-(aq)$ is

$$\text{moles of NO}_2^- = 1.00 \text{ mol} - 0.0325 \text{ mol} = 0.97 \text{ mol}$$
$$\text{moles of HNO}_2 = 1.00 \text{ mol} + 0.0325 \text{ mol} = 1.03 \text{ mol}$$

The concentrations of $HNO_2(aq)$ and $NO_2^-(aq)$ in the final solution are

$$[NO_2^-] = \frac{0.97 \text{ mol}}{1.05 \text{ L}} = 0.92 \text{ M}$$

$$[HNO_2] = \frac{1.03 \text{ mol}}{1.05 \text{ L}} = 0.98 \text{ M}$$

The value of pK_a for the acid-base conjugate pair HNO_2/NO_2^- is 3.35 (Table 18-5). The Henderson-Hasselbalch equation is

$$pH = pK_a + \log \frac{[NO_2^-]_0}{[HNO_2]_0}$$

The pH of the original solution was

$$pH = 3.35 + \log \left(\frac{1.00 \text{ M}}{1.00 \text{ M}} \right) = 3.35$$

The pH of the final solution is

$$pH = 3.35 + \log \left(\frac{0.92 \text{ M}}{0.98 \text{ M}} \right) = 3.32$$

The change in pH is $3.35 - 3.32 = 0.03$ pH units.

E. ANSWERS TO THE SELF-TEST

1. pH
2. true
3. false
4. Henderson-Hasselbalch
5. false
6. true

7. NaOH(aq); HCl(aq)

8. KOH(aq)

9. pH . . . the volume of NaOH(aq) added

10. at which the number of moles of NaOH(aq) that has been added is exactly equal to the number of moles of HCl(aq) initially present

11. acid

12. true

13. pH

14. the pK_a

15. orange

16. the equivalence point of the titration

17. a change in color of the indicator

18. false

19. false

20. true

21. In this case, the salt, sodium acetate, NaCH$_3$COO(aq), present at the equivalence point is a basic salt because Na$^+$(aq) is a neutral cation and CH$_3$COO$^-$(aq) is basic anion.

22. true

23. pK_a

24. decreases

25. increases

26. decreases

27. true

28. false

29. The salt, NH$_4$Cl(aq), present at the equivalence point is an acidic salt because NH$_4^+$(aq) is an acidic cation and Cl$^-$(aq) is a neutral anion.

30. the initial pH, the midpoint, the equivalence point, and a point after the equivalence point

31. three

32. false

33. true

34. three

20

SOLUBILITY AND PRECIPITATION REACTIONS

A. OUTLINE OF CHAPTER 20

20-1 The law of concentration action governs the equilibrium between an ionic solid and its constituent ions in solution.

Application of the law of concentration action yields the equilibrium-constant expression for the solubility-product constant, K_{sp}.

The solubility of an ionic solid can be computed by using the K_{sp} expression.

The solubility of a salt is the maximum quantity of the salt that can be dissolved in a given volume of solution at equilibrium.

Generally for a salt with the formula $A_x B_y$, the solubility product constant is given by

$$K_{sp} = [A]^x[B]^y \qquad (20\text{-}4)$$

and the solubility is given by

$$s = \frac{[A]}{x} = \frac{[B]}{y} \qquad (20\text{-}5)$$

The values of K_{sp} for various salts are given in Table 20-1.

20-2 The solubility of an ionic solid is decreased when a common ion is present in the solution.

At equilibrium, the K_{sp} expression for an ionic solid must always hold if the solid phase is in contact with the solution.

The decrease in salt solubility arising from the presence of a common ion in the solution is understood readily in terms of Le Châtelier's principle.

The lowering of the solubility of a salt by the presence of a common ion in the solution is called the common ion effect.

20-3   The solubility of an ionic solid is increased by the formation of a soluble complex ion.

A complex ion is a metal ion with small molecules or ions attached to it.

The equilibrium constant for a complexation reaction is designated by K_{comp}.

The equilibrium constant for the solubility of a solid that forms a complex ion is given by $K = K_{sp}K_{comp}$.

When K is much greater than K_{sp}, all the ions will be in the form of the complex ion in solution.

20-4   Salts of weak acids are more soluble in acidic solutions than in neutral or basic solutions.

The increase in solubility of a salt of a weak acid with increased $[H_3O^+]$ is a consequence of the formation of the conjugate acid of the anion of the salt, as shown in Equation (20-18).

The effect of pH on the solubility of a salt of a weak acid is shown in Figure 20-3.

20-5   The magnitude of the ratio Q_{sp}/K_{sp} is used to predict whether an ionic solid can precipitate.

The Q_{sp} expression has the same algebraic form as the K_{sp} expression, but the concentration values used in the Q_{sp} expression need not be equilibrium values.

If $Q_{sp} > K_{sp}$, then precipitation will occur.

If $Q_{sp} < K_{sp}$, then no precipitation will occur.

If $Q_{sp} = K_{sp}$, then the ionic solid is in equilibrium with the solution.

Criteria for the formation of a precipitate from a solution prepared from the constituent ions are summarized in Table 20-2.

20-6   It is often possible to separate one compound from another by selective precipitation.

Selective precipitation can be used to separate one ion from another.

Two ions may be separated by the formation of salts with different solubilities; the less soluble salt will precipitate first.

20-7   Sulfides that are insoluble in water can be separated by adjustment of solution pH.

Qualitative analysis is the identification of the cations and anions in an unknown solution.

Because $S^{2-}(aq)$ is the conjugate base of a weak acid, $HS^-(aq)$, the concentration of $S^{2-}(aq)$ can be controlled by controlling the concentration of $H_3O^+(aq)$.

The solubility of $H_2S(aq)$ in a saturated solution is 0.10 M at 25°C.

The value of $[S^{2-}]$ when $[H_2S] = 0.10$ M is given by

$$[S^{2-}] = \frac{1.1 \times 10^{-21} \text{ M}^3}{[H_3O^+]^2} \qquad \text{at } 25°C \qquad (20\text{-}21)$$

The value of $[S^{2-}]$ in a solution is controlled by controlling the pH of the solution with suitable buffers.

The solubility of metal sulfides can be controlled by controlling $[S^{2-}]$ in a solution.

20-8   Some metal cations can be separated from a mixture by the formation of an insoluble hydroxide of one of them.

The solubility of a metal hydroxide depends on both the value of K_{sp} and the pH of the solution (Figure 20-5).

20-9   Amphoteric hydroxides dissolve in both highly acidic and highly basic solutions.

Amphoteric hydroxides are insoluble in neutral solution.

Most metal oxides and hydroxides react with strong acids to form the metal salt of the acid in water.

Amphoteric hydroxides dissolve in highly basic solutions by the formation of soluble hydroxy complexes; for example,

$$Al(OH)_3(s) + OH^-(aq) \rightleftharpoons Al(OH)_4^-(aq)$$

The values of the equilibrium constants for the formation of hydroxy complexes of some amphoteric hydroxides are given in Table 20-4.

The total solubility of an amphoteric hydroxide at any given pH is given by the sum of the concentration of the metal ion and the concentration of the hydroxy complex ion.

The solubility of an amphoteric metal hydroxide as a function of pH is shown in Figure 20-7.

Example 20-12 illustrates the calculation of the solubility of an amphoteric hydroxide in basic, neutral, or acidic solution.

20-10   Qualitative analysis is the identification of the species present in a sample.

An essential feature of a qualitative analysis scheme is the successive removal of subgroups of the ions by precipitation reactions.

The sample to be analyzed is called the unknown.

A reagent is added that precipitates certain cations but not others.

The supernatant solution is the solution that remains after the removal of the precipitates that result from the addition of a precipitating reagent to the unknown solution.

B. SELF-TEST

1. The law of concentration action can be applied to the equilibrium between a solid and its constituent ions in aqueous solution. *True/False*

2. The equilibrium constant for the equilibrium between a solid and its constituent ions in water is called its _____.

3. The solubility product constant expression for calcium carbonate, $CaCO_3$, is $K_{sp} =$

_____ .

4. The value of K_{sp} determines the solubility of an ionic solid in pure water. *True/False*

5. The solubility of $PbCl_2(s)$ in water is equal to the concentration of $Pb^{2+}(aq)$ if $PbCl_2(s)$ is the only source of $Pb^{2+}(aq)$. *True/False*

6. The solubility of $Fe(OH)_2(s)$ in water is equal to the concentration of $OH^-(aq)$. *True/False*

7. The solubility-product constant expression for $AgCl(s)$ is not $K_{sp} = [Ag^+][Cl^-]$ when $AgCl(s)$ is in equilibrium with a 0.25-M $NaCl(aq)$ solution. *True/False*

8. The common-ion effect is _____

_____ .

9. The solubility of $AgCl(s)$ is (*increased/decreased*) by the addition of $AgNO_3$ to the solution.

10. The solubility of $AgCl(s)$ is (*increased/decreased*) by the addition of $NH_3(aq)$ to the solution.

11. The increased solubility of $AgCl(s)$ in $NH_3(aq)$ relative to the solubility of $AgCl(s)$ in

pure water is a result of the reaction _____

_____ .

12. The equilibrium constant for the dissolution of $AgCl(s)$ in $NH_3(aq)$ is related to the solubility product of $AgCl(s)$ and the equilibrium constant for the complexation reaction

by $K =$ _____ .

13. The equilibrium constant for the dissolution of $AgCl(s)$ in $NH_3(aq)$, K, is much (*larger/smaller*) than the solubility product constant of $AgCl(s)$.

14. The solubility of a salt of a weak acid depends on the pH of the solution. *True/False*

15. Salts of weak acids are more soluble the (*lower/higher*) the pH of the solution.

16. Using a chemical equation, explain why the solubility of silver benzoate,

AgC_6H_5COO, increases as the pH of the solution decreases. _____

_____

17. Any values of the concentrations of the species involved may be used in the concentration quotient. *True/False*

18. If the value of the ratio Q_{sp}/K_{sp} for the equation $Cu(OH)_2(s) \rightleftharpoons Cu^{2+}(aq) + 2OH^-(aq)$ is 10 when a solution of $NaOH(aq)$ is mixed with a solution of $Cu(NO_3)_2(aq)$, then a precipitate of _____ (will/will not) form.

19. If the value of the ratio Q_{sp}/K_{sp} for the equation $Ag_2(SO_4)(s) \rightleftharpoons 2Ag^+(aq) + SO_4^{2-}(aq)$ is 0.032 when a solution of $AgNO_3(aq)$ is added to a solution of $K_2SO_4(aq)$, then a precipitate of _____ (will/will not) form.

20. The solubility of a metal sulfide depends on the pH of the solution. *True/False*

21. The concentration of $S^{2-}(aq)$ in an aqueous solution saturated with $H_2S(aq)$ depends on the pH of the solution. *True/False*

22. The concentration of $S^{2-}(aq)$ in an aqueous solution saturated with $H_2S(aq)$ is related to the concentration of $H_3O^+(aq)$ by $[S^{2-}] = $ _____ .

23. Two metal sulfides may be separated by adjusting the pH of the solution. *True/False*

24. All metal sulfides are insoluble in acidic solution. *True/False*

25. The solubility of an amphoteric hydroxide depends upon the pH of the solution. *True/False*

26. Amphoteric hydroxides are more soluble in both strongly acidic and basic solutions than in pure water. *True/False*

27. Most metal oxides and hydroxides react with strong acids to form water and

_____ .

28. Aluminum hydroxide, $Al(OH)_3,(s)$, is insoluble at any pH. *True/False*

29. The equation for the solubility reaction for $Al(OH)_3(s)$ in water is _____

_____ .

30. The equation for the reaction for $Al(OH)_3(s)$ dissolving in a highly acidic aqueous solution is _____ .

31. The equilibrium constant for the reaction given in Question 30 is related to the solubility product constant for $Al(OH)_3(s)$ and the ion product of water. *True/False*

32. The equation for the reaction for $Al(OH)_3(s)$ dissolving in a highly basic solution is

_____ .

33. Qualitative analysis is the determination of the _____

_____ in a sample.

34. Metal ions may be separated by selective precipitation. *True/False*

35. In a solution of $AgNO_3(aq)$ and $KNO_3(aq)$, $Ag^+(aq)$ can be separated from $K^+(aq)$ by adding $HCl(aq)$ to precipitate _____ as _____.

36. Many metal cations form insoluble metal sulfides. *True/False*

37. The steps in a qualitative analysis scheme may be carried out in any order. *True/False*

38. A qualitative analysis scheme often utilizes the differences in solubilities at various pH values. *True/False*

C. CALCULATIONS YOU SHOULD KNOW HOW TO DO

1. Calculate the solubility of a salt given the value of K_{sp} of the salt. See Examples 20-1 and 20-2 and Problems 20-7, 20-8, 20-11, and 20-12.

2. Calculate the value of K_{sp} for a salt given the solubility of the salt. See Example 20-3 and Problems 20-9, 20-10, 20-13, and 20-14.

3. Calculate the solubility of a salt in an aqueous solution containing a common ion. See Example 20-4 and Problems 20-15 through 20-22.

4. Calculate the solubility of a salt in an aqueous solution in which it forms a complex ion. See Example 20-5 and Problems 20-23 through 20-28.

5. Calculate the solubility of a salt of a weak acid at a given pH. See Section 20-4 and Problems 20-34.

6. Calculate the value of the concentration quotient Q_{sp} and use Q_{sp}/K_{sp} to determine whether precipitation will occur. See Examples 20-7 and 20-8 and Problems 20-35 through 20-40.

7. Calculate the solubilities of metal sulfides at a given pH using Equation (20-21). See Example 20-10 and Problems 20-47 through 20-50.

8. Calculate the solubility of metal hydroxides at a given pH. See Example 20-11 and Problems 20-33, 20-45, and 20-46.

9. Calculate the solubility of amphoteric hydroxides at a given pH. See Example 20-12 and Problems 20-51 and 20-52.

D. SOLUTIONS TO THE ODD-NUMBERED PROBLEMS

20-1 (a) insoluble, rule 3   (b) soluble, rule 2
         (c) soluble, rule 1   (d) soluble, rule 1
         (e) insoluble, rule 5

20-3 (a) insoluble, rule 5   (b) soluble, rule 2
(c) soluble, rule 4   (d) insoluble, rule 6

20-5 (a) CuS is insoluble by rule 5.

$$CuCl_2(aq) + Na_2S(aq) \longrightarrow CuS(s) + 2NaCl(aq)$$
$$Cu^{2+}(aq) + S^{2-}(aq) \longrightarrow CuS(s)$$

(b) MgCO_3 is insoluble by rule 5.

$$MgBr_2(aq) + K_2CO_3(aq) \longrightarrow MgCO_3(s) + 2KBr(aq)$$
$$Mg^{2+}(aq) + CO_3^{2-}(aq) \longrightarrow MgCO_3(s)$$

(c) BaSO_4 is insoluble by rule 6.

$$BaCl_2(aq) + K_2SO_4(aq) \longrightarrow BaSO_4(s) + 2KCl(aq)$$
$$Ba^{2+}(aq) + SO_4^{2-}(aq) \longrightarrow BaSO_4(s)$$

(d) Hg_2Cl_2 is insoluble by rule 3.

$$Hg_2(NO_3)_2(aq) + 2KCl(aq) \longrightarrow Hg_2Cl_2(s) + 2KNO_3(aq)$$
$$Hg_2^{2+}(aq) + 2Cl^-(aq) \longrightarrow Hg_2Cl_2(s)$$

20-7 The solubility equilibrium is

$$PbCrO_4(s) \rightleftharpoons Pb^{2+}(aq) + CrO_4^{2-}(aq)$$

The K_{sp} expression is

$$K_{sp} = [Pb^{2+}][CrO_4^{2-}] = 2.8 \times 10^{-13} \text{ M}^2$$

If PbCrO_4 is equilibrated with pure water, then at equilibrium we have

$$[Pb^{2+}] = [CrO_4^{2-}] = s$$

where s is the solubility of PbCrO_4 in pure water. Thus,

$$K_{sp} = (s)(s) = s^2 = 2.8 \times 10^{-13} \text{ M}^2$$
$$s = 5.3 \times 10^{-7} \text{ M}$$

The solubility in grams per liter is

$$s = (5.3 \times 10^{-7} \text{ mol} \cdot \text{L}^{-1}) \left(\frac{323.2 \text{ g PbCrO}_4}{1 \text{ mol PbCrO}_4} \right) = 1.7 \times 10^{-4} \text{ g} \cdot \text{L}^{-1}$$

20-9 The solubility equilibrium is

$$AgBr(s) \rightleftharpoons Ag^+(aq) + Br^-(aq)$$

The solubility of AgBr is

$$(1.33 \times 10^{-4} \text{ g} \cdot \text{L}^{-1}) \left(\frac{1 \text{ mol AgBr}}{187.8 \text{ g AgBr}} \right) = 7.082 \times 10^{-7} \text{ M}$$

From the reaction stoichiometry, we have

$$\text{solubility of AgBr} = [\text{Ag}^+] = [\text{Br}^-] = 7.082 \times 10^{-7} \text{ M}$$

Using the K_{sp} expression, we compute

$$K_{sp} = [\text{Ag}^+][\text{Br}^-] = (7.082 \times 10^{-7} \text{ M})^2 = 5.02 \times 10^{-13} \text{ M}^2$$

20-11   The solubility equilibrium is

$$\text{Mg(OH)}_2(s) \rightleftharpoons \text{Mg}^{2+}(aq) + 2\text{OH}^-(aq)$$

The solubility product constant expression is

$$K_{sp} = [\text{Mg}^{2+}][\text{OH}^-]^2 = 1.8 \times 10^{-11} \text{ M}^3$$

When $\text{Mg(OH)}_2(s)$ is in equilibrium with pure water, we have

$$[\text{OH}^-] = 2[\text{Mg}^{2+}]$$

The solubility of $\text{Mg(OH)}_2(s)$ in pure water is equal to $[\text{Mg}^{2+}]$ because each mole of Mg(OH)_2 that dissolves yields one mole of $\text{Mg}^{2+}(aq)$. The solubility of $\text{Mg(OH)}_2(s)$ is

$$s = [\text{Mg}^{2+}] = \frac{[\text{OH}^-]}{2}$$

or $[\text{OH}^-] = 2s$ where we have neglected the $[\text{OH}^-]$ arising from the dissociation of $\text{H}_2\text{O}(l)$.

Combining these results with the K_{sp} expression, we have

$$K_{sp} = (s)(2s)^2 = 4s^3 = 1.8 \times 10^{-11} \text{ M}^3$$
$$s = \left(\frac{1.8 \times 10^{-11} \text{ M}^3}{4} \right)^{1/3} = 1.65 \times 10^{-4} \text{ M}$$

The solubility in grams per liter is

$$s = (1.65 \times 10^{-4} \text{ mol} \cdot \text{L}^{-1}) \left(\frac{58.33 \text{ g Mg(OH)}_2}{1 \text{ mol Mg(OH)}_2} \right) = 9.6 \times 10^{-3} \text{ g} \cdot \text{L}^{-1}$$

20-13   The solubility quilibrium is

$$\text{KClO}_4(s) \rightleftharpoons \text{K}^+(aq) + \text{ClO}_4^-(aq)$$

The K_{sp} expression is

$$K_{sp} = [K^+][ClO_4^-]$$

From the reaction stoichiometry, at equilibrium $[K^+] = [ClO_4^-] =$ solubility of $KClO_4$. The solubility of $KClO_4$ is

$$s = \frac{0.75 \text{ g}}{0.100 \text{ L}} = (7.5 \text{ g} \cdot \text{L}^{-1}) \left(\frac{1 \text{ mol } KClO_4}{138.55 \text{ g } KClO_4} \right) = 0.054 \text{ M}$$

Substituting the values of $[K^+]$ and $[ClO_4^-]$ into the K_{sp} expression, we have

$$K_{sp} = (0.054 \text{ M})(0.054 \text{ M}) = 2.9 \times 10^{-3} \text{ M}^2$$

20-15 The equilibrium expression that describes the solubility of silver sulfate is

$$Ag_2SO_4(s) \rightleftharpoons 2Ag^+(aq) + SO_4^{2-}(aq)$$

and the solubility product expression is

$$K_{sp} = [Ag^+]^2[SO_4^{2-}] = 1.4 \times 10^{-5} \text{ M}^3 \qquad \text{(Table 20-1)}$$

The only source of $SO_4^{2-}(aq)$ is from the $Ag_2SO_4(s)$ that dissolves. If we let s be the solubility of $Ag_2SO_4(s)$ in 0.55 M $AgNO_3(aq)$, then

$$[SO_4^{2-}] = s$$

The $Ag^+(aq)$ is due to the 0.55 M $AgNO_3(aq)$ and the $Ag_2SO_4(s)$ that dissolves. Because each $Ag_2SO_4(s)$ that dissolves yields two $Ag^+(aq)$, we have at equilibrium

$$[Ag^+] = \underset{\substack{\text{from} \\ AgNO_3}}{0.55 \text{ M}} + \underset{\substack{\text{from} \\ Ag_2SO_4}}{2s}$$

If we substitute the expressions for $[Ag^+]$ and $[SO_4^{2-}]$ into the K_{sp} expression, then we obtain

$$K_{sp} = (0.55 \text{ M} + 2s)^2(s) = 1.4 \times 10^{-5} \text{ M}^3$$

Because Ag_2SO_4 is a slightly soluble salt, we expect the value of s to be small. Therefore, we neglect $2s$ compared to 0.55 M, and we write

$$(0.55 \text{ M})^2(s) \approx 1.4 \times 10^{-5} \text{ M}^3$$

$$s \approx \frac{1.4 \times 10^{-5} \text{ M}^3}{(0.55 \text{ M})^2} = 4.6 \times 10^{-5} \text{ M}$$

Note that s is small relative to 0.55 M, and therefore our approximation $0.55 \text{ M} + 2s \approx 0.55$ M is acceptable. The solubility of $Ag_2SO_4(s)$ in $g \cdot L^{-1}$ is

$$s = (4.63 \times 10^{-5} \text{ mol} \cdot L^{-1}) \left(\frac{311.9 \text{ g } Ag_2SO_4}{1 \text{ mol } Ag_2SO_4} \right) = 1.4 \times 10^{-2} \text{ g} \cdot L^{-1}$$

20-17   The equation that describes the solubility of $TlCl(s)$ is

$$TlCl(s) \rightleftharpoons Tl^+(aq) + Cl^-(aq)$$

The solubility product expression is

$$K_{sp} = [Tl^+][Cl^-] = 1.7 \times 10^{-4} \text{ M}^2$$

The only source of $Tl^+(aq)$ is from the $TlCl(s)$ that dissolves. If we let s be the solubility of $TlCl(s)$ in the solution, then

$$[Tl^+] = s$$

The $Cl^-(aq)$ in solution is due to 0.25 M $NaCl(aq)$ plus the dissolved TlCl.

$$[Cl^-] = 0.25 \text{ M} + s$$

Substituting these expressions into the K_{sp} expression yields

$$s(0.25 \text{ M} + s) = 1.7 \times 10^{-4} \text{ M}^2$$

Assuming $s \ll 0.25$ M because we expect s to be small yields

$$s \approx \frac{1.7 \times 10^{-4} \text{ M}^2}{0.25 \text{ M}} = 6.8 \times 10^{-4} \text{ M}$$

The solubility in grams per liter is

$$s = (6.8 \times 10^{-4} \text{ M}) \left(\frac{239.9 \text{ g } TlCl}{1 \text{ mol } TlCl} \right) = 0.16 \text{ g} \cdot L^{-1}$$

20-19   The equilibrium expression that describes the solubility of $AgI(s)$ is

$$AgI(s) \rightleftharpoons Ag^+(aq) + I^-(aq)$$

The solubility product expression is

$$K_{sp} = [Ag^+][I^-] = 8.3 \times 10^{-17} \text{ M}^2 \qquad \text{(Table 20-1)}$$

The only source of $Ag^+(aq)$ is $AgI(s)$; thus, the solubility s is equal to

$$s = [Ag^+]$$

The $I^-(aq)$ in solution is due to 0.20 M $CaI_2(aq)$ plus the dissolved AgI. Because CaI_2 dissociates completely, we have

$$[I^-] = (2 \times 0.20 \text{ M}) + s$$

Substitution of the above expressions for $[Ag^+]$ and $[I^-]$ into the K_{sp} expression yields

$$s(0.40 \text{ M} + s) = 8.3 \times 10^{-17} \text{ M}^2$$

Assuming that s is small compared to 0.40 M yields

$$s = \frac{8.3 \times 10^{-17} \text{ M}^2}{0.40 \text{ M}} = 2.08 \times 10^{-16} \text{ M}$$

The solubility in grams per liter is

$$s = (2.08 \times 10^{-16} \text{ M}) \left(\frac{234.8 \text{ g AgI}}{1 \text{ mol AgI}} \right) = 4.9 \times 10^{-14} \text{ g} \cdot \text{L}^{-1}$$

20-21   The equilibrium expression that describes the solubility of $Pb(IO_3)_2(s)$ is

$$Pb(IO_3)_2(s) \rightleftharpoons Pb^{2+}(aq) + 2IO_3^-(aq)$$

The solubility product expression is

$$K_{sp} = [Pb^{2+}][IO_3^-]^2 = 2.5 \times 10^{-13} \text{ M}^3 \qquad \text{(Table 20-1)}$$

The only source of $Pb^{2+}(aq)$ is $Pb(IO_3)_2(s)$; thus, the solubility s is equal to

$$s = [Pb^{2+}]$$

The $IO_3^-(aq)$ in solution is due to 0.10 M $KIO_3(aq)$ plus the dissolved $Pb(IO_3)_2(s)$. Because $KIO_3(aq)$ dissociates completely, we have

$$[IO_3^-] = 0.10 \text{ M} + 2s$$

Substitution of the above expressions for $[Pb^{2+}]$ and $[IO_3^-]$ into the K_{sp} expression yields

$$s(0.10 \text{ M} + 2s)^2 = 2.5 \times 10^{-13} \text{ M}^2$$

Assuming that $2s$ is small compared to 0.10 M yields

$$s = \frac{2.5 \times 10^{-13} \text{ M}^2}{(0.10 \text{ M})^2} = 2.5 \times 10^{-11} \text{ M}$$

The solubility in grams per liter is

$$s = (2.5 \times 10^{-11} \text{ M}) \left(\frac{557.0 \text{ g Pb(IO}_3)_2}{1 \text{ mol Pb(IO}_3)_2} \right) = 1.4 \times 10^{-8} \text{ g} \cdot \text{L}^{-1}$$

20-23 The equation for the solubility of $AgBr(s)$ in $NH_3(aq)$,

$$AgBr(s) + 2NH_3(aq) \rightleftharpoons Ag(NH_3)_2^+(aq) + Br^-(aq)$$

is the sum of the two equations

$$AgBr(s) \rightleftharpoons Ag^+(aq) + Br^-(aq) \qquad\qquad\qquad K_{sp} = 5.0 \times 10^{-13} \text{ M}^2$$
$$Ag^+(aq) + 2NH_3(aq) \rightleftharpoons Ag(NH_3)_2^+(aq) \qquad K_{comp} = 2.0 \times 10^7 \text{ M}^{-2}$$

The equilibrium constant for the solubility of $AgBr(s)$ in $NH_3(aq)$ is

$$K = K_{sp}K_{comp} = (5.0 \times 10^{-13} \text{ M}^2)(2.0 \times 10^7 \text{ M}^{-2}) = 1.0 \times 10^{-5}$$

We set up a concentration table.

	$AgBr(s) + 2NH_3(aq) \rightleftharpoons Ag(NH_3)_2^+(aq)$		$+ \;\; Br^-(aq)$
Initial concentration	0.35 M	0	≈ 0
Equilibrium concentration	0.35 M $- 2x$	x	x

The equilibrium-constant expression is

$$K = \frac{[Ag(NH_3)_2^+][Br^-]}{[NH_3]^2} = \frac{x^2}{(0.35 \text{ M} - 2x)^2} = 1.0 \times 10^{-5}$$

Taking the square root of both sides gives

$$\frac{x}{0.35 \text{ M} - 2x} = \pm 3.16 \times 10^{-3}$$

The negative root leads to a negative concentration, and the positive root gives

$$x = 1.10 \times 10^{-3} \text{ M}$$

Because the value of K_{sp} for $AgBr(s)$ is smaller than 1.0×10^{-5}, the solubility of

AgBr(s) in $NH_3(aq)$ is essentially equal to $[Ag(NH_3)_2^+]$. The solubility of AgBr(s) in grams per liter is

$$s = (1.10 \times 10^{-3} \text{ M}) \left(\frac{187.8 \text{ g AgBr}}{1 \text{ mol AgBr}} \right) = 0.21 \text{ g} \cdot \text{L}^{-1}$$

20-25   We set up the concentration table.

	$AgCl(s) + 2S_2O_3^{2-}(aq) \rightleftharpoons Ag(S_2O_3)_2^{3-}(aq) + Cl^-(aq)$		
Initial concentration	0.050 M	0	0
Equilibrium concentration	0.050 M $- 2x$ x		x

The equilibrium-constant expression is

$$K = \frac{[Ag(S_2O_3)_2^{3-}][Cl^-]}{[S_2O_3^{2-}]^2} = 5.20 \times 10^3$$

Substituting the values for $[Ag(S_2O_3)_2^{3-}]$, $[Cl^-]$, and $[S_2O_3^{2-}]$ from the concentration table into the equilibrium constant expression gives

$$\frac{x^2}{(0.050 \text{ M} - 2x)^2} = 5.20 \times 10^3$$

Taking the square root of both sides yields

$$\frac{x}{0.050 \text{ M} - 2x} = \pm 72.11$$

The negative root leads to a negative concentration, and the positive root gives

$$x = [Ag(S_2O_3)_2^{3-}] = 0.0248 \text{ M}$$

Because the value of K_{sp} for AgCl(s) is so small compared to 5.20×10^3, the solubility of AgCl(s) is given by $s = [Ag(S_2O_3)_2^{3-}] = 0.0248$ M. The solubility of AgCl(s) in grams per liter is

$$s = (0.0248 \text{ M}) \left(\frac{143.4 \text{ g AgCl}}{1 \text{ mol AgCl}} \right) = 3.6 \text{ g} \cdot \text{L}^{-1}$$

20-27   We set up a concentration table.

	$AgCl(s) + 2NH_3(aq) \rightleftharpoons Ag(NH_3)_2^+(aq) + Cl^-(aq)$		
Initial concentration	0.500 M	0	0.300 M
Equilibrium concentration	0.500 M − 2s	s	0.300 M + s

From Example 20-5, we have that the equilibrium-constant expression is

$$K = \frac{[Ag(NH_3)_2^+][Cl^-]}{[NH_3]^2} = 3.6 \times 10^{-3}\ M$$

Substituting in the expressions given in the concentration table yields

$$\frac{(s)(0.300\ M + s)}{(0.500\ M - 2s)^2} = 3.6 \times 10^{-3}\ M$$

As a first approximation, we neglect s with respect to 0.300 M and 0.500 M. This approximation yields $s = 3.0 \times 10^{-3}$ M. If we substitute this result into the 0.300 M + s and 0.500 M − s terms, then we obtain a second approximation, $s = 2.9 \times 10^{-3}$ M. Subsequent approximations yield the same value, so $s = 2.9 \times 10^{-3}$ M. The solubility of AgCl(s) in grams per liter is

$$s = (2.9 \times 10^{-3}\ M) \left(\frac{143.4\ \text{g AgCl}}{1\ \text{mol AgCl}} \right) = 0.42\ \text{g·L}^{-1}$$

20-29   The equilibrium expression is

$$PbF_2(s) \rightleftharpoons Pb^{2+}(aq) + 2F^-(aq)$$

Recall that HF is a weak acid, and a saturated PbF_2 solution would contain some HF(aq).

(a)   The solubility is increased; a decrease in pH is an increase in $[H_3O^+]$. The added $H_3O^+(aq)$ reacts with $F^-(aq)$ to form HF, thereby reducing the concentration of $F^-(aq)$ and causing a shift in the equilibrium from left to right.

(b)   The solubility is decreased, an increase in $[Pb^{2+}]$ shifts the equilibrium from right to left (common-ion effect).

20-31   The following compounds are more soluble at lower pH for the reasons stated:

(a) $CaCO_3$; $CO_3^{2-}(aq)$ is the conjugate base of the weak acid $HCO_3^-(aq)$.

(b) CaF_2; $F^-(aq)$ is the conjugate base of the weak acid HF(aq).

(c) $PbSO_3$; $SO_3^{2-}(aq)$ is the conjugate base of the weak acid $HSO_3^-(aq)$.
(e) $Fe(OH)_3$; $OH^-(aq)$ reacts with $H_3O^+(aq)$ so that $[OH^-]$ is decreased.
(f) ZnS; $S^{2-}(aq)$ is the conjugate base of the weak acid $HS^-(aq)$.

20-33   The equilibrium expression is

$$Mg(OH)_2(s) \rightleftharpoons Mg^{2+}(aq) + 2OH^-(aq)$$

The K_{sp} expression is

$$K_{sp} = [Mg^{2+}][OH^-]^2 = 1.8 \times 10^{-11} \text{ M}^3 \qquad \text{(Table 20-1)}$$

At pH = 8.50

$$[H_3O^+] = 10^{-8.50} = 3.16 \times 10^{-9} \text{ M}$$

Thus,

$$[OH^-] = \frac{K_w}{[H_3O^+]} = \frac{1.00 \times 10^{-14} \text{ M}^2}{3.16 \times 10^{-9} \text{ M}} = 3.16 \times 10^{-6} \text{ M}$$

Let s be the solubility of $Mg(OH)_2$. Then

$$[Mg^{2+}] = s$$
$$K_{sp} = (s)(3.16 \times 10^{-6} \text{ M})^2 = 1.8 \times 10^{-11} \text{ M}^3$$
$$s = \frac{1.8 \times 10^{-11} \text{ M}^3}{9.99 \times 10^{-12} \text{ M}^2} = 1.8 \text{ M}$$

20-35   The initial concentration of $Cl^-(aq)$ after mixing is

$$[Cl^-]_0 = \frac{(0.25 \text{ M}) (100 \text{ mL})}{(100 \text{ mL} + 5.0 \text{ mL})} = 0.24 \text{ M}$$

The initial concentration of $Ag^+(aq)$ after mixing is

$$[Ag^+]_0 = \frac{(0.10 \text{ M}) (5.0 \text{ mL})}{105 \text{ mL}} = 4.8 \times 10^{-3} \text{ M}$$

and NO_3^- is a spectator ion. The value of Q_{sp} is

$$Q_{sp} = [Ag^+]_0[Cl^-]_0 = (4.8 \times 10^{-3} \text{ M})(0.24 \text{ M}) = 1.2 \times 10^{-3} \text{ M}^2$$

The value of K_{sp} for $AgCl(s)$ is 1.8×10^{-10} M² (Table 20-1); thus,

$$\frac{Q_{sp}}{K_{sp}} = \frac{1.2 \times 10^{-3} \text{ M}^2}{1.8 \times 10^{-10} \text{ M}^2} = 6.7 \times 10^6 > 1$$

Thus, $AgCl(s)$ will precipitate from the solution.

20-37   The initial concentration of $Pb^{2+}(aq)$ after mixing is

$$[Pb^{2+}]_0 = \frac{(3.00\ \text{M})\ (40.0\ \text{mL})}{60.0\ \text{mL}} = 2.00\ \text{M}$$

The initial concentration of $I^-(aq)$ after mixing is

$$[I^-]_0 = \frac{(2.00 \times 10^{-3}\ \text{M})\ (20.0\ \text{mL})}{60.0\ \text{mL}} = 6.67 \times 10^{-4}\ \text{M}$$

The value of Q_{sp} is

$$Q_{sp} = [Pb^{2+}]_0[I^-]_0^2 = (2.00\ \text{M})(6.67 \times 10^{-4}\ \text{M})^2 = 8.90 \times 10^{-7}\ \text{M}^3$$

The value of K_{sp} is $7.1 \times 10^{-9}\ \text{M}^3$ (Table 20-1); thus,

$$\frac{Q_{sp}}{K_{sp}} = \frac{8.90 \times 10^{-7}\ \text{M}^3}{7.1 \times 10^{-9}\ \text{M}^3} = 130 > 1$$

Because $Q_{sp}/K_{sp} > 1$, $PbI_2(s)$ will precipitate from the solution. Because $[Pb^{2+}]_0 \gg [I^-]_0$, essentially all of the $I^-(aq)$ is precipitated as $PbI_2(s)$, and the final equilibrium value of $[Pb^{2+}]$ will still be 2.00 M. Thus, we have at equilibrium following the precipitation of $PbI_2(s)$

$$[Pb^{2+}]_0[I^-]^2 \simeq K_{sp} = 7.1 \times 10^{-9}\ \text{M}^3$$

Therefore,

$$[I^-] = \left(\frac{7.1 \times 10^{-9}\ \text{M}^3}{2.00\ \text{M}} \right)^{1/2} = 6.0 \times 10^{-5}\ \text{M}$$

The moles of I^- that precipitates is given by

$$\begin{pmatrix} \text{moles of } I^- \\ \text{precipitated} \end{pmatrix} = \begin{pmatrix} \text{initial} \\ \text{moles of } I^- \end{pmatrix} - \begin{pmatrix} \text{final moles} \\ \text{of } I^- \end{pmatrix}$$
$$= (6.67 \times 10^{-4}\ \text{M} - 6.0 \times 10^{-5}\ \text{M})(0.060\ \text{L})$$
$$= 3.64 \times 10^{-5}\ \text{mol}$$

The moles of $PbI_2(s)$ that precipitates is equal to one half the moles of I^- that precipitates because each mole of $PbI_2(s)$ contains 2 mol of I^-. Thus,

$$\text{moles of } PbI_2(s) = \tfrac{1}{2}(3.64 \times 10^{-5}\ \text{mol}) = 1.82 \times 10^{-5}\ \text{mol}$$

The equilibrium concentrations following the precipitation of $PbI_2(s)$ are

$$[Pb^{2+}] = 2.00\ \text{M} \qquad\qquad [NO_3^-] = 2[Pb^{2+}]_0 = 4.00\ \text{M}$$
$$[Na^+] = [I^-]_0 = 6.67 \times 10^{-4}\ \text{M} \qquad [I^-] = 6.0 \times 10^{-5}\ \text{M}$$

20-39 The initial concentration of $Cl^-(aq)$ after mixing is

$$[Cl^-]_0 = \frac{(2.00\ M)(0.100\ L)}{0.200\ L} = 1.00\ M$$

The initial concentration of $Ag^+(aq)$ after mixing is

$$[Ag^+]_0 = \frac{(0.020\ M)(0.100\ L)}{0.200\ L} = 0.010\ M$$

Thus,

$$Q_{sp} = [Ag^+]_0[Cl^-]_0 = (0.010\ M)(1.00\ M) = 0.010\ M^2$$

and

$$\frac{Q_{sp}}{K_{sp}} = \frac{0.010\ M^2}{1.8 \times 10^{-10}\ M^2} = 5.5 \times 10^7 > 1$$

Thus, $AgCl(s)$ will precipitate. Because $AgCl(s)$ is insoluble and $Cl^-(aq)$ is in excess, essentially all of the $Ag^+(aq)$ will precipitate from the solution. The number of moles of $AgCl(s)$ that precipitates is equal to the number of moles of $Ag^+(aq)$ initially present.

(a) moles of $AgCl(s)$ = moles of $Ag^+(aq)$ = (0.010 M)(0.200 L)
$$= 2.0 \times 10^{-3}\ mol$$

The number of grams of AgCl in 2.0×10^{-3} mol is

$$\text{mass of AgCl} = (2.0 \times 10^{-3}\ mol)\left(\frac{143.4\ g\ AgCl}{1\ mol\ AgCl}\right) = 0.29\ g$$

(b) The concentration of $Cl^-(aq)$ at equilibrium is

$$[Cl^-] = 1.00\ M - 0.010\ M = 0.99\ M$$

The concentration of Ag^+ at equilibrium can be found from the K_{sp} expression.

$$K_{sp} = [Ag^+][Cl^-] = [Ag^+][0.99\ M] = 1.8 \times 10^{-10}\ M^2$$
$$[Ag^+] = \frac{1.8 \times 10^{-10}\ M^2}{0.99\ M} = 1.8 \times 10^{-10}\ M$$

This confirms the statement that essentially all the $Ag^+(aq)$ is precipitated as $AgCl(s)$.

20-41 The equations for the two solubility equilibria are

$$PbI_2(s) \rightleftharpoons Pb^{2+}(aq) + 2I^-(aq) \qquad K_{sp} = 7.1 \times 10^{-9}\ M^3$$
$$Hg_2I_2(s) \rightleftharpoons Hg_2^{2+}(aq) + 2I^-(aq) \qquad K_{sp} = 4.5 \times 10^{-29}\ M^3$$

The expressions for the solubility quotients are

$$Q_{sp} = [Pb^{2+}]_0[I^-]_0^2$$
$$Q_{sp} = [Hg_2^{2+}]_0[I^-]_0^2$$

When $Q_{sp} \leq K_{sp}$, no precipitation occurs. We now calculate the concentration of $I^-(aq)$, at which $Q_{sp} = K_{sp}$ for $PbI_2(s)$.

$$Q_{sp} = (0.010 \text{ M})[I^-]^2 = 7.1 \times 10^{-9} \text{ M}^3$$

Solving for $[I^-]$, we have that

$$[I^-] = 8.4 \times 10^{-4} \text{ M}$$

At this concentration of $I^-(aq)$, Q_{sp} for Hg_2I_2 is

$$Q_{sp} = (0.010 \text{ M})(8.4 \times 10^{-4} \text{ M})^2 = 7.1 \times 10^{-9} \text{ M}^3$$

which is much greater than K_{sp}, and so precipitation of Hg_2I_2 will occur. The concentration of $Hg_2^{2+}(aq)$ remaining in solution with $[I^-]$ is given by

$$[Hg_2^{2+}](8.4 \times 10^{-4} \text{ M})^2 = 4.5 \times 10^{-29} \text{ M}^3$$
$$[Hg_2^{2+}] = 6.4 \times 10^{-23} \text{ M}$$

Thus, essentially all the $Hg_2^{2+}(aq)$ is precipitated, and so $Pb^{2+}(aq)$ may be separated from $Hg_2^{2+}(aq)$ by selective precipitation with $I^-(aq)$.

20-43   The equations for the two solubility equilibria are

$$CaSO_4(s) \rightleftharpoons Ca^{2+}(aq) + SO_4^{2-}(aq) \qquad K_{sp} = 9.1 \times 10^{-6} \text{ M}^2$$
$$Ag_2SO_4(s) \rightleftharpoons 2Ag^+(aq) + SO_4^{2-}(aq) \qquad K_{sp} = 1.4 \times 10^{-5} \text{ M}^3$$

The expressions for the solubility quotients are

$$Q_{sp} = [Ca^{2+}]_0[SO_4^{2-}]_0$$
$$Q_{sp} = [Ag^+]_0^2[SO_4^{2-}]_0$$

When $Q_{sp} \leq K_{sp}$, no precipitation occurs. We now calculate the concentration of $SO_4^{2-}(aq)$ at which $Q_{sp} = K_{sp}$ for the two equilibria.

$$Q_{sp} = (0.050 \text{ M})[SO_4^{2-}] = 9.1 \times 10^{-6} \text{ M}^2$$
$$[SO_4^{2-}] = 1.8 \times 10^{-4} \text{ M}$$
$$Q_{sp} = (0.025 \text{ M})^2[SO_4^{2-}] = 1.4 \times 10^{-5} \text{ M}^3$$
$$[SO_4^{2-}] = 0.022 \text{ M}$$

Thus, when the concentration of $SO_4^{2-}(aq)$ is 0.022 M, $Ag^+(aq)$ does not precipitate. The concentration of $Ca^{2+}(aq)$ at this concentration of $SO_4^{2-}(aq)$ is given by

$$[Ca^{2+}](0.022 \text{ M}) = 9.1 \times 10^{-6} \text{ M}^2$$

Solving for $[Ca^{2+}]$, we have

$$[Ca^{2+}] = 4.1 \times 10^{-4} \text{ M}$$

The percentage of $Ca^{2+}(aq)$ that precipitated is

$$\% \text{ precipitated} = \frac{0.050 \text{ M} - 4.1 \times 10^{-4} \text{ M}}{0.050 \text{ M}} \times 100 = 99\%$$

Thus, $Ca^{2+}(aq)$ can be separated from $Ag^+(aq)$ by selectively precipitating Ca^{2+} with $SO_4^{2-}(aq)$.

20-45 The solubility product expression of $Cr(OH)_3(s)$ is

$$K_{sp} = [Cr^{3+}][OH^-]^3 = 6.3 \times 10^{-31} \text{ M}^4 \qquad \text{(Table 20-1)}$$

$$\text{solubility of } Cr(OH)_3 = s = [Cr^{3+}] = \frac{6.3 \times 10^{-31} \text{ M}^4}{[OH^-]^3}$$

$$= \frac{6.3 \times 10^{-31} \text{ M}^4[H_3O^+]^3}{K_w^3}$$

At pH = 5.0

$$[H_3O^+] = 10^{-pH} = 10^{-5.0} = 1.0 \times 10^{-5} \text{ M}$$

Thus, the solubility of $Cr(OH)_3$ is

$$s = \frac{(6.3 \times 10^{-31} \text{ M}^4)(1.0 \times 10^{-5} \text{ M})^3}{(1.00 \times 10^{-14} \text{ M}^2)^3} = 6.3 \times 10^{-4} \text{ M}$$

The solubility product expression of $Ni(OH)_2$ is

$$K_{sp} = [Ni^{2+}][OH^-]^2 = 2.0 \times 10^{-15} \text{ M}^3$$

$$\text{solubility of } Ni(OH)_2 = s = [Ni^{2+}] = \frac{2.0 \times 10^{-15} \text{ M}^3}{[OH^-]^2}$$

$$= \frac{2.0 \times 10^{-15} \text{ M}^3[H_3O^+]^2}{K_w^2}$$

Thus, at pH = 5.0

$$s = \frac{(2.0 \times 10^{-15} \text{ M}^3)(1.0 \times 10^{-5} \text{ M})^2}{(1.00 \times 10^{-14} \text{ M}^2)^2} = 2.0 \times 10^3 \text{ M}$$

Of course, a solubility of 2000 M is physically unrealistic, but this result means that $Ni(OH)_2(s)$ is very soluble. Thus, at pH = 5.0, $Ni(OH)_2(s)$ is very soluble, while $Cr(OH)_3(s)$ is only slightly soluble; thus, a separation of $Ni^{2+}(aq)$ and $Cr^{3+}(aq)$ can be achieved by buffering the solution at pH = 5, where the $Cr(OH)_3(s)$ will precipitate and the $Ni^{2+}(aq)$ will remain in solution.

20-47 The K_{sp} expression for CuS(s) is

$$K_{sp} = [Cu^{2+}][S^{2-}] = 6.3 \times 10^{-36} \text{ M}^2 \qquad \text{(Table 20-1)}$$

The solubility of CuS is

$$s = [Cu^{2+}] = \frac{6.3 \times 10^{-36} \text{ M}^2}{[S^{2-}]}$$

From Equation (20-21)

$$[S^{2-}] = \frac{1.1 \times 10^{-21} \text{ M}^3}{[H_3O^+]^2}$$

At pH = 2.0

$$[H_3O^+] = 10^{-2.0} = 1.0 \times 10^{-2} \text{ M}$$

and thus,

$$[S^{2-}] = \frac{1.1 \times 10^{-21} \text{ M}^3}{(1.0 \times 10^{-2} \text{ M})^2} = 1.1 \times 10^{-17} \text{ M}$$

Therefore, the solubility of CuS is

$$s = \frac{6.3 \times 10^{-36} \text{ M}^2}{1.1 \times 10^{-17} \text{ M}} = 5.7 \times 10^{-19} \text{ M}$$

29-49 The solubility product expression for PbS is

$$K_{sp} = [Pb^{2+}][S^{2-}] = 8.0 \times 10^{-28} \text{ M}^2 \qquad \text{(Table 20-1)}$$

Thus, the solubility of PbS is

$$s = [Pb^{2+}] = \frac{8.0 \times 10^{-28} \text{ M}^2}{[S^{2-}]}$$

Substituting Equation (20-21) for $[S^{2-}]$ into the above equation for s, we obtain

$$s = \frac{8.0 \times 10^{-28} \text{ M}^2[H_3O^+]^2}{1.1 \times 10^{-21} \text{ M}^3} = 7.3 \times 10^{-7} \text{ M}^{-1}[H_3O^+]^2$$

Proceeding in an analogous manner, we find for the solubility of MnS as a function of $[H_3O^+]$

$$K_{sp} = [Mn^{2+}][S^{2-}] = 2.5 \times 10^{-13} \text{ M}^2 \qquad \text{(Table 20-1)}$$

$$s = [Mn^{2+}] = \frac{2.5 \times 10^{-13} \text{ M}^2}{[S^{2-}]} = \frac{2.5 \times 10^{-13} \text{ M}^2[H_3O^+]^2}{1.1 \times 10^{-21} \text{ M}^3}$$

$$= 2.3 \times 10^8 \text{ M}^{-1}[H_3O^+]^2$$

The $[H_3O^+]$ at which the solubility of PbS is 1×10^{-6} M is given by

$$1 \times 10^{-6} \text{ M} = 7.3 \times 10^{-7} \text{ M}^{-1}[H_3O^+]^2$$

Thus,

$$[H_3O^+]^2 = \frac{1 \times 10^{-6} \text{ M}}{7.3 \times 10^{-7} \text{ M}^{-1}} = 1.37 \text{ M}^2$$

$$[H_3O^+] = 1.2 \text{ M}$$

or

$$pH = -0.07$$

The solubility of MnS at $[H_3O^+] = 1.2$ M is

$$s = (2.3 \times 10^8 \text{ M}^{-1})(1.2 \text{ M})^2 = 3.3 \times 10^8 \text{ M}$$

which is much greater than 0.025 M. At a pH of -0.07, essentially all the $Pb^{2+}(aq)$ precipitates as PbS(s), and essentially all the $Mn^{2+}(aq)$ remains in solution.

20-51   The solution is basic and thus the equation is

$$Sn(OH)_2(s) + OH^-(aq) \rightleftharpoons Sn(OH)_3^-(aq)$$

From Table 20-4, the equilibrium-constant expression for this equation is

$$K = \frac{[Sn(OH)_3^-]}{[OH^-]} = 0.01 = \frac{s}{[OH^-]}$$

where we have used the fact that the solubility of $Sn(OH)_2$ is essentially equal to $[Sn(OH)_3^-]$. At pH = 13.0, $[H_3O^+] = 1.0 \times 10^{-13}$ M, and

$$[OH^-] = \frac{K_w}{[H_3O^+]} = \frac{1.00 \times 10^{-14} \text{ M}^2}{1.0 \times 10^{-13} \text{ M}} = 1.0 \times 10^{-1} \text{ M}$$

Thus,

$$s = (0.01)[OH^-] = (0.01)(0.10 \text{ M}) = 1 \times 10^{-3} \text{ M}$$

20-53   The $Pb^{2+}(aq)$ is removed from solution by the formation of insoluble $PbSO_4(s)$, which passes out of the body through the large intestine.

20-55   The equation for the precipitation reaction is

$$Pb(NO_3)_2(aq) + 2NaOH(aq) \rightleftharpoons Pb(OH)_2(s) + 2NaNO_3(aq)$$

The precipitate dissolves via the process

$$Pb(OH)_2(s) + OH^-(aq) \rightleftharpoons Pb(OH)_3^-(aq)$$

20-57   Calcium forms an insoluble oxalate, $CaC_2O_4(s)$, which is removed by vomiting. The excess $Ca^{2+}(aq)$ is removed by adding $MgSO_4(aq)$ to form $CaSO_4(s)$, which is insoluble in water and in stomach acid. Vomiting of the $CaC_2O_4(s)$ is necessary because the solubility of $CaC_2O_4(s)$ in stomach acid is sufficiently high to permit toxic levels of oxalic acid (a weak acid) to pass through the stomach walls into the bloodstream.

20-59   The equation for the precipitation reaction is

$$Zn(ClO_4)_2(aq) + 2KOH(aq) \rightleftharpoons Zn(OH)_2(s) + 2KClO_4(aq)$$

The precipitate dissolves via the formation of the $Zn(OH)_4^{2-}(aq)$ complex ion

$$Zn(OH)_2(s) + 2OH^-(aq) \rightleftharpoons Zn(OH)_4^{2-}(aq)$$

20-61   The fraction by mass of Cl in a sample of AgCl is equal to the ratio of the atomic mass of Cl to the formula mass of AgCl. Thus, for the 4.188-g sample we have

$$\text{mass of Cl} = (4.188\ g) \left(\frac{\text{atomic mass of Cl}}{\text{formula mass of AgCl}} \right)$$

$$= (4.188\ g) \left(\frac{35.45}{143.4} \right)$$

$$= 1.035\ g$$

Therefore the mass percentage of chloride in the original sample is

$$\text{mass \%} = \left(\frac{1.035\ g}{2.000\ g} \right) \times 100 = 51.75\%$$

20-63   We have for the value of K_{sp} (Table 20-1) for the equilibrium

(1) $AgCl(s) \rightleftharpoons Ag^+(aq) + Cl^-(aq)$ $K_{sp} = 1.8 \times 10^{-10}\ M^2$

Addition of the K_{sp} equilibrium equation to the complexation equation

(2) $Ag^+(aq) + 2NH_3(aq) \rightleftharpoons Ag(NH_3)_2^+(aq)$ $K_{comp} = 2.0 \times 10^7\ M^{-2}$

yields

(3) $AgCl(s) + 2NH_3(aq) \rightleftharpoons Ag(NH_3)_2^+(aq) + Cl^-(aq)$

for which we have

$$K_3 = K_{sp}K_{comp}$$
$$= (1.8 \times 10^{-10} \text{ M}^2)(2.0 \times 10^7 \text{ M}^{-2}) = 3.6 \times 10^{-3}$$

Because $K_3 \gg K_{sp}$ we have

$$[Ag(NH_3)_2^+] \gg [Ag^+]$$

and thus the solubility s is equal to

$$s = [Ag(NH_3)_2^+] = [Cl^-]$$

A concentration of 250 mg of AgCl in 100 mL of solution corresponds to a solubility of

$$s = \frac{(0.250 \text{ g})\left(\dfrac{1 \text{ mol AgCl}}{143.4 \text{ g AgCl}}\right)}{(0.100 \text{ L})} = 0.0174 \text{ M}$$

Substituting this value of s into the K_3 expression yields

$$K_3 = \frac{[Ag(NH_3)_2^+][Cl^-]}{[NH_3]^2} = \frac{s^2}{[NH_3]^2} = 3.6 \times 10^{-3}$$

or

$$[NH_3]^2 = \frac{(0.0174 \text{ M})^2}{3.6 \times 10^{-3}}$$

Thus,

$$[NH_3] = 0.29 \text{ M}$$

20-65 (a) We have from Table 20-1

(1) $Ag_2CrO_4(s) \rightleftharpoons 2Ag^+(aq) + CrO_4^{2-}(aq)$ $\qquad K_{sp1} = 1.1 \times 10^{-12} \text{ M}^3$

(2) $AgBr(s) \rightleftharpoons Ag^+(aq) + Br^-(aq)$ $\qquad K_{sp2} = 5.0 \times 10^{-13} \text{ M}^2$

If we reverse Equation (2), multiply through by 2, and then add the result to Equation (1), we obtain Equation (a):

(a) $Ag_2CrO_4(s) + 2Br^-(aq) \rightleftharpoons 2AgBr(s) + CrO_4^{2-}(aq)$

Thus,

$$K_{(a)} = \frac{K_{sp1}}{K_{sp2}^2}$$
$$= \frac{1.1 \times 10^{-12} \text{ M}^3}{(5.0 \times 10^{-13} \text{ M}^2)^2} = 4.4 \times 10^{12} \text{ M}^{-1}$$

(b)   We have from Table 20-1

(1) $PbCO_3(s) \rightleftharpoons Pb^{2+}(aq) + CO_3^{2-}(aq)$ $K_{sp1} = 7.4 \times 10^{-14}$ M^2

(2) $CaCO_3(s) \rightleftharpoons Ca^{2+}(aq) + CO_3^{2-}(aq)$ $K_{sp2} = 2.8 \times 10^{-9}$ M^2

If we reverse Equation (2), and add the result to Equation (1), then we obtain Equation (b):

(b) $PbCO_3(s) + Ca^{2+}(aq) \rightleftharpoons CaCO_3(s) + Pb^{2+}(aq)$

Thus,

$$K_{(b)} = \frac{K_{sp1}}{K_{sp2}}$$
$$= \frac{7.4 \times 10^{-14} \text{ M}^2}{2.8 \times 10^{-9} \text{ M}^2} = 2.6 \times 10^{-5}$$

20-67   The mass of $Pb(NO_3)_2$ present is given by

$$\begin{pmatrix} \text{mass of} \\ Pb(NO_3)_2 \end{pmatrix} = (12.79 \text{ g PbCl}_2) \left(\frac{1 \text{ mol PbCl}_2}{278.1 \text{ g PbCl}_2} \right)$$
$$\times \left(\frac{1 \text{ mol Pb(NO}_3)_2}{1 \text{ mol PbCl}_2} \right) \left(\frac{331.2 \text{ g Pb(NO}_3)_2}{1 \text{ mol Pb(NO}_3)_2} \right)$$
$$= 15.23 \text{ g}$$

The molarity of the solution is

$$M = \frac{(15.23 \text{ g}) \left(\dfrac{1 \text{ mol Pb(NO}_3)_2}{331.2 \text{ g Pb(NO}_3)_2} \right)}{(0.200 \text{ L})} = 0.230 \text{ M}$$

20-69   The equations for the two equilibria are

(1) $AgCH_3COO(s) \rightleftharpoons Ag^+(aq) + CH_3COO^-(aq)$ $K_{sp} = 4.0 \times 10^{-3}$ M^2

(2) $CH_3COO^-(aq) + H_3O^+(aq) \rightleftharpoons CH_3COOH(aq) + H_2O(l)$

$$K = \frac{1}{K_a} = \frac{1}{1.74 \times 10^{-5} \text{ M}} = 5.75 \times 10^4 \text{ M}^{-1}$$

The solubility s is given by

$$s = [Ag^+]$$

Because some of the $CH_3COO^-(aq)$ reacts with $H_3O^+(aq)$ to produce $CH_3COOH(aq)$, we have that

$$s = [CH_3COOH] + [CH_3COO^-]$$

From the second equilibrium, we have that

$$\frac{[CH_3COOH]}{[CH_3COO^-]} = [H_3O^+]K = [H_3O^+](5.75 \times 10^4 \ M^{-1})$$

At pH = 2.0, $[H_3O^+] = 1.0 \times 10^{-2}$ M.

$$\frac{[CH_3COOH]}{[CH_3COO^-]} = (1.0 \times 10^{-2} \ M)(5.75 \times 10^4 \ M^{-1}) = 575$$

$$[CH_3COOH] = 575 \ [CH_3COO^-]$$

$$s = 575 \ [CH_3COO^-] + [CH_3COO^-] = 576 \ [CH_3COO^-]$$

$$[CH_3COO^-] = \frac{s}{576}$$

$$K_{sp} = (s)\left(\frac{s}{576}\right) = 4.0 \times 10^{-3} \ M^2$$

$$s = 1.5 \ M$$

At pH = 4.0, $[H_3O^+] = 1.0 \times 10^{-4}$ M.

$$\frac{[CH_3COOH]}{[CH_3COO^-]} = (1.0 \times 10^{-4} \ M)(5.75 \times 10^4 \ M^{-1}) = 5.75$$

$$[CH_3COOH] = 5.75 \ [CH_3COO^-]$$

$$s = 5.75 \ [CH_3COO^-] + [CH_3COO^-] = 6.75 \ [CH_3COO^-]$$

$$[CH_3COO^-] = \frac{s}{6.75}$$

$$K_{sp} = (s)\left(\frac{s}{6.75}\right) = 4.0 \times 10^{-3} \ M^2$$

$$s = 0.16 \ M$$

At pH = 6.0, $[H_3O^+] = 1.0 \times 10^{-6}$ M.

$$\frac{[CH_3COOH]}{[CH_3COO^-]} = (1.0 \times 10^{-6} \ M)(5.75 \times 10^4 \ M^{-1}) = 0.058$$

$$[CH_3COOH] = 0.058 \ [CH_3COO^-]$$

$$s = 0.058 \ [CH_3COO^-] + [CH_3COO^-] = 1.058 \ [CH_3COO^-]$$

$$[CH_3COO^-] = \frac{s}{1.058}$$

$$K_{sp} = (s)\left(\frac{s}{1.058}\right) = 4.0 \times 10^{-3} \ M^2$$

$$s = 0.065 \ M$$

At pH above 7, the second equilibrium is not involved, and we have that the solubility is given by

$$s = [Ag^+] = [CH_3COO^-]$$
$$K_{sp} = s^2 = 4.0 \times 10^{-3} \text{ M}^2$$
$$s = 0.063 \text{ M}$$

20-71 We set up the equilibrium table.

	$Zn(OH)_2(s)$ + $2OH^-(aq)$	$\rightleftharpoons$ $Zn(OH)_4^{2-}(aq)$
Initial concentration	— 0.010 M	0
Equilibrium concentration	— 0.010 M − 2s	s

The equilibrium-constant expression is

$$K = \frac{[Zn(OH)_4^{2-}]}{[OH^-]^2} = \frac{s}{(0.010 \text{ M} - 2s)^2} = 0.05 \text{ M}^{-1}$$

The first approximation yields $s = 5.0 \times 10^{-6}$ M, which is confirmed by the second approximation.

20-73 (a) $Hg_2(ClO_4)_2(aq) + 2NaBr(aq) \rightarrow 2NaClO_4(aq) + Hg_2Br_2(s)$ (rule 3)

$$Hg_2^{2+}(aq) + 2Br^-(aq) \longrightarrow Hg_2Br_2(s)$$

(b) $Fe(ClO_4)_3(aq) + 3NaOH(aq) \rightarrow 3NaClO_4(aq) + Fe(OH)_3(s)$ (rule 5)

$$Fe^{3+}(aq) + 3OH^-(aq) \longrightarrow Fe(OH)_3(s)$$

(c) $Pb(NO_3)_2(aq) + 2LiIO_3(aq) \rightarrow 2LiNO_3(aq) + Pb(IO_3)_2(s)$ (rule 3)

$$Pb^{2+}(aq) + 2IO_3^-(aq) \longrightarrow Pb(IO_3)_2(s)$$

(d) $H_2SO_4(aq) + Pb(NO_3)_2(aq) \rightarrow 2HNO_3(aq) + PbSO_4(s)$ (rule 3)

$$Pb^{2+}(aq) + SO_4^{2-}(aq) \longrightarrow PbSO_4(s)$$

20-75 The equilibrium-constant expression is

$$K = \frac{[Ag(S_2O_3)_4^{3-}][Cl^-]}{[S_2O_3^{2-}]^2} = 5.20 \times 10^3$$

If we let s be the solubility of $AgCl(s)$ in $S_2O_3^{2-}(aq)$, then we have

$$[Cl^-] = s$$

The K_{sp} for $AgCl(s)$ is 1.8×10^{-10} M^2, which is very small compared to the K value above; thus, $[Ag^+] \ll [Ag(S_2O_3)_2^{3-}]$, and therefore

$$[Ag(S_2O_3)_2^{3-}] \simeq [Cl^-] = s$$

Substituting the expressions for $[Cl^-]$ and $[Ag(S_2O_3)_2^{3-}]$ into the K expression, we have

$$K = \frac{(s)(s)}{[S_2O_3^{2-}]^2} = \frac{s^2}{(0.010 \text{ M})^2} = 5.20 \times 10^3$$
$$s^2 = 0.52 \text{ M}^2$$

and

$$s = 0.72 \text{ M}$$

20-77 (a) Solubility is increased; an increase in the concentration of $S_2O_3^{2-}$ shifts the equilibrium from left to right.
   (b) Solubility remains unchanged; the amount of a solid reactant has no effect on the equilibrium concentrations.
   (c) Solubility is decreased; an increase in the concentration of $Br^-(aq)$ shifts the equilibrium from right to left.
   (d) Solubility remains unchanged; neither $Na^+(aq)$ nor $NO_3^-(aq)$ reacts with the species involved.

20-79 (a) soluble, rule 1   (b) insoluble, rule 3
   (c) insoluble, rule 3   (d) insoluble, rule 5

20-81 (a) soluble, rule 2   (b) insoluble, rule 3
   (c) soluble, rule 6   (d) insoluble, rule 5
   (e) soluble, rule 2

20-83 The equilibrium-constant expression is

$$K = \frac{[Al(OH)_4^-]}{[OH^-]} = 40$$

The solubility of $Al(OH)_3(s)$ in basic solution is

$$s = [Al(OH)_4^-] = 40[OH^-]$$

which is derived from the equilibrium expression. At $pH = 12.0$, $pOH = 2.0$, and

$$[OH^-] = 10^{-2.0} = 1.0 \times 10^{-2} \text{ M}$$

Thus,

$$s = (40)(1.0 \times 10^{-2} \text{ M}) = 0.40 \text{ M}$$

20-85   The solubility equilibrium is

$$Zn(OH)_2(s) \rightleftharpoons Zn^{2+}(aq) + 2OH^-(aq)$$

The K_{sp} expression is

$$K_{sp} = [Zn^{2+}][OH^-]^2 = 1.0 \times 10^{-15} \ M^3$$

From the reaction stoichiometry, at equilibrium we have

$$[OH^-] = 2[Zn^{2+}] \quad \text{or} \quad [Zn^{2+}] = \tfrac{1}{2}[OH^-]$$

Substituting $[Zn^{2+}] = \tfrac{1}{2}[OH^-]$ in the K_{sp} expression, we have

$$K_{sp} = \tfrac{1}{2}[OH^-][OH^-]^2 = \tfrac{1}{2}[OH^-]^3 = 1.0 \times 10^{-15} \ M^3$$
$$[OH^-] = (2.0 \times 10^{-15} \ M^3)^{1/3} = 1.26 \times 10^{-5} \ M$$

The pOH of the solution is

$$pOH = -\log [OH^-] = -\log (1.26 \times 10^{-5}) = 4.90$$

and the pH is

$$pH = 14.00 - pOH = 14.00 - 4.90 = 9.10$$

20-87   The equilibrium expression is

$$Cu(OH)_2(s) \rightleftharpoons Cu^{2+}(aq) + 2OH^-(aq)$$

The K_{sp} expression is

$$K_{sp} = [Cu^{2+}][OH^-]^2 = 2.2 \times 10^{-20} \ M^3 \qquad \text{(Table 20-1)}$$

At pH = 7.0

$$[H_2O^+] = 10^{-pH} = 10^{-7.0} = 1.0 \times 10^{-7} \ M$$

and thus,

$$[OH^-] = \frac{K_w}{[H_3O^+]} = \frac{1.00 \times 10^{-14} \ M^2}{1.0 \times 10^{-7} \ M} = 1.0 \times 10^{-7} \ M$$

Let s be the solubility of $Cu(OH)_2$. We have

$$[Cu^{2+}] = s$$

Therefore,

$$K_{sp} = (s)(1.0 \times 10^{-7} \ M)^2 = 2.2 \times 10^{-20} \ M^3$$

and

$$s = 2.2 \times 10^{-6} \text{ M}$$

20-89   (a) $ZnS(s) \rightleftharpoons Zn^{2+}(aq) + S^{2-}(aq)$
The solubility increases. The $H_3O^+(aq)$ from the added $HNO_3(aq)$ reacts with $S^{2-}(aq)$ to form $HS^-(aq)$, thereby decreasing the concentration of $S^{2-}(aq)$. A decrease in $[S^{2-}]$ shifts the equilibrium from left to right.

(b) $AgI(s) \rightleftharpoons Ag^+(aq) + I^-(aq)$
The solubility increases. Ammonia, $NH_3(aq)$, reacts with $Ag^+(aq)$ to form the soluble ion $Ag(NH_3)_2^+(aq)$, thereby reducing the amount of $Ag^+(aq)$. A decrease in $[Ag^+]$ shifts the equilibrium from left to right.

20-91   The equation for the solubility of mercury (I) chloride is

$$Hg_2Cl_2(s) \rightleftharpoons Hg_2^{2+}(aq) + 2Cl^-(aq)$$

The solubility product is

$$K_{sp} = [Hg_2^{2+}][Cl^-]^2 = 1.3 \times 10^{-18} \text{ M}^3 \qquad \text{(Table 20-1)}$$

(a)   The solubility s is equal to $[Hg_2^{2+}]$. Thus, we have that

$$(s)(1.5 \text{ M} + 2s)^2 = 1.3 \times 10^{-18} \text{ M}^3$$
$$s = 5.8 \times 10^{-19} \text{ M}$$

(b)   The solubility s is equal to $\frac{1}{2}[Cl^-]$. Thus, we have that

$$(1.5 \text{ M} + s)(2s)^2 = 1.3 \times 10^{-18} \text{ M}^3$$
$$s = 4.7 \times 10^{-10} \text{ M}$$

(c)   The solubility s is equal to $[Hg_2^{2+}]$. Thus, we have that

$$(s)(2s)^2 = 1.3 \times 10^{-18} \text{ M}^3$$
$$s = 6.9 \times 10^{-7} \text{ M}$$

20-93   After mixing, we have that

$$[Cl^-] = \left(\frac{50.0 \text{ mL}}{100 \text{ mL}}\right)(1.00 \text{ M}) = 0.500 \text{ M}$$

$$[Hg_2^{2+}] = \left(\frac{50.0 \text{ mL}}{100 \text{ mL}}\right)(1.00 \text{ M}) = 0.500 \text{ M}$$

The value of the reaction quotient is

$$Q_{sp} = [Hg_2^{2+}]_0[Cl^-]_0^2 = (0.500 \text{ M})(0.500 \text{ M})^2 = 0.125 \text{ M}^3$$

Because

$$Q_{sp} > K_{sp}$$

precipitation will occur. There is an excess of $Hg_2^{2+}(aq)$; thus, essentially all the $Cl^-(aq)$ precipitates. Because the equation for the precipitation reaction is

$$Hg_2^{2+}(aq) + 2Cl^-(aq) \longrightarrow Hg_2Cl_2(s)$$

then ½ mol of $Hg_2^{2+}(aq)$ will precipitate with the 1 mol of $Cl^-(aq)$, and so the equilibrium concentration of $Hg_2^{2+}(aq)$ is essentially 0.250(0.500 M/2) M. The solubility product expression is

$$K_{sp} = [Hg_2^{2+}][Cl^-]^2 = (0.250\ M)[Cl^-]^2 = 1.3 \times 10^{-18}\ M^3$$
$$[Cl^-] = 2.3 \times 10^{-9}\ M$$

20-95   After mixing, we have that

$$[Cl^-]_0 = \left(\frac{150\ mL}{250\ mL}\right)(0.300\ M) = 0.180\ M$$
$$[Hg_2^{2+}]_0 = \left(\frac{100\ mL}{250\ mL}\right)(0.200\ M) = 0.080\ M$$

There is an excess of $Cl^-(aq)$; thus, essentially all the $Hg_2^{2+}(aq)$ precipitates. The equilibrium concentration of $Cl^-(aq)$ is essentially 0.180 M − (2)(0.080 M) = 0.020 M. The solubility product expression is

$$K_{sp} = [Hg_2^{2+}][Cl^-]^2 = [Hg_2^{2+}](0.020\ M)^2 = 1.3 \times 10^{-18}\ M^3$$
$$[Hg_2^{2+}] = 3.3 \times 10^{-15}\ M$$

The fraction of $Hg_2^{2+}(aq)$ that is not precipitated is

$$\text{fraction} = \frac{3.3 \times 10^{-15}\ M}{0.080\ M} = 4.1 \times 10^{-14}$$

20-97   We set up the concentration table.

	$AgI(s) + 2NH_3(aq) \rightleftharpoons Ag(NH_3)^+(aq)$		$+\ I^-(aq)$
Initial concentration	14.0 M	0	0
Equilibrium concentration	14.0 M − $2x$	x	x

The value of the equilibrium constant for the reaction is

$$K = K_{sp}K_{comp} = (8.3 \times 10^{-17} \text{ M}^2)(2.0 \times 10^7 \text{ M}^{-2})$$
$$= 1.66 \times 10^{-9}$$

The equilibrium-constant expression is

$$K = \frac{[\text{Ag(NH}_3)_2^+][\text{I}^-]}{[\text{NH}_3]^2} = 1.66 \times 10^{-9}$$

Thus,

$$K = \frac{x^2}{(14.0 \text{ M} - 2x)^2} = 1.66 \times 10^{-9}$$

Taking the square root of both sides and solving for x yields

$$x = 5.7 \times 10^{-4} \text{ M}$$

Silver iodide is not soluble in the aqueous ammonia solution.

20-99   From Table 20-1, we have that

$$K_{sp} = [\text{Ag}^+]^2[\text{CrO}_4^{2-}] = 1.1 \times 10^{-12} \text{ M}^3$$
$$s = [\text{CrO}_4^{2-}]$$
$$(2s)^2(s) = 1.1 \times 10^{-12} \text{ M}^3$$
$$s = 6.5 \times 10^{-5} \text{ M}$$

and

$$K_{sp} = [\text{Ag}^+][\text{Cl}^-] = 1.8 \times 10^{-10} \text{ M}^2$$
$$s = [\text{Cl}^-]$$
$$(s)(s) = 1.8 \times 10^{-10} \text{ M}^2$$
$$s = 1.3 \times 10^{-5} \text{ M}$$

Ag_2CrO_4 is more soluble than AgCl in pure water.

E. ANSWERS TO THE SELF-TEST

1.   true

2.   solubility-product constant

3. $[Ca^{2+}][CO_3^{2-}]$

4.   true

5.   true

6.   false; each $Fe(OH)_2(s)$ that dissolves produces two $OH^-(aq)$

7. false

8. the decrease in solubility of a solid in an aqueous solution containing one of the ions that make up the solid

9. decreased

10. increased

11. $Ag^+(aq) + 2NH_3(aq) \rightleftharpoons$
$Ag(NH_3)_2^+(aq)$

12. $K_{sp}K_{comp}$

13. larger

14. true

15. lower

16. $C_6H_5COO^-(aq) + H_3O^+(aq) \rightleftharpoons$
$C_6H_5COOH(aq) + H_2O(l)$

17. true

18. $Cu(OH)_2(s)$ will

19. $Ag_2SO_4(s)$ will not

20. true

21. true

22. $(1.1 \times 10^{-21} \text{ M}^3)/[H_3O^+]^2$

23. true

24. false

25. true

26. true

27. a salt

28. false

29. $Al(OH)_3(s) \rightleftharpoons Al^{3+}(aq) +$
$3OH^-(aq)$

30. $Al(OH)_3(s) + 3H_3O^+(aq) \rightleftharpoons$
$Al^{3+}(aq) + 6H_2O(l)$

31. true

32. $Al(OH)_3(s) + OH^-(aq) \rightleftharpoons$
$Al(OH)_4^-(aq)$

33. identities of the species present

34. true

35. $Ag^+(aq) \ldots AgCl(s)$

36. true

37. false

38. true

21 OXIDATION-REDUCTION REACTIONS

A. OUTLINE OF CHAPTER 21

21-1   An oxidation state can be assigned to each atom in a chemical species.

Oxidation-reduction reactions are also called redox reactions.

The rules for the assignment of oxidation states to the elements in a chemical species are given in Section 4-1.

A more general method to assign oxidation states, based on Lewis formulas, is described on pages 753 and 754 of the text.

21-2   Oxidation-reduction reactions involve the transfer of electrons from one reactant to another.

A decrease in oxidation state is called reduction.

An increase in oxidation state is called oxidation.

The reactant that contains the element that is reduced is called the oxidizing agent.

The reactant that contains the element that is oxidized is called the reducing agent.

An oxidizing agent acts as an electron acceptor.

A reducing agent acts as an electron donor.

21-3   Electron-transfer reactions can be separated into two half reactions.

Electron-transfer reactions can be written as the sum of two half reactions.

The half reaction in which electrons appear on the right-hand side is the oxidation half reaction.

The half reaction in which electrons appear on the left-hand side is the reduction half reaction.

An electron-transfer reaction, in which the same species is oxidized and reduced, is called a disproportionation reaction.

21-4 Equations for oxidation-reduction reactions can be balanced by balancing each half-reaction separately.

The procedure for balancing equations by the method of half reactions is discussed in detail (pages 758 through 760 in the text).

The procedure also applies to balancing half reactions (Example 21-6).

21-5 Chemical equations for redox reactions occurring in basic solution are balanced using OH^- and H_2O.

The procedure for balancing equations that occur in basic solution differs in steps 3 and 4, as described on page 762 of the text.

21-6 Oxidation-reduction reactions are used in chemical analyses.

A quantitative reaction is a reaction for which the equilibrium constant is very large, and thus the reaction goes essentially to completion.

Redox reactions may be used to determine unknown concentrations of reducing and oxidizing agents.

21-7 Billions of dollars are spent each year to protect metals from corrosion.

Most metals when exposed to air develop an oxide film.

Humidity, acidity, and certain anions can destroy some metals.

Corrosion involves redox reactions between different sections of the same piece of metal or between two dissimilar metals in electrical contact with each other.

The corrosion of iron is shown in Figure 21-6.

The simplest method of corrosion prevention is to provide a protective layer of paint or of a corrosion-resistant metal.

Another method of corrosion prevention is to use a replaceable sacrificial metal, which is a piece of metal electrically connected to a less active metal (Figure 21-7).

Galvanization is an anticorrosion process using zinc as the sacrificial metal to prevent the corrosion of iron.

B. SELF-TEST

1. The oxidation state of a free element is _____ .

2. The alkali metals in compounds are always assigned an oxidation state of _____ .

3. The alkaline earth metals in compounds are always assigned an oxidation state of _ .

4. Fluorine in compounds is always assigned an oxidation state of _____ .

5. Oxygen in compounds usually is assigned an oxidation state of _____ .

6. Hydrogen in compounds usually is assigned an oxidation state of _____ .

7. The sum of the oxidation states of each atom in a chemical species is equal to the charge on the species. *True/False*

8. The Lewis formula of a species must be known when the simple rules for assigning oxidation states do not apply to the species. *True/False*

9. When using Lewis formulas to assign oxidation states, all the electrons in each bond are assigned to the (*more/less*) electronegative atom in the bond.

10. The oxidation state of an atom in a species is assigned according to the formula:

oxidation state = _____

_____ minus the total number of valence electrons assigned to the element.

11. The oxidation state of an element in a compound is always equal to the actual charge on the atom. *True/False*

12. Oxidation states can be assigned by analogy with another element in the periodic table. *True/False*

13. Oxidation-reduction reactions involve the transfer of _____ between reactants.

14. Oxidation is an (*increase/decrease*) in oxidation state.

15. An oxidizing agent is the reactant that contains _____.

16. An oxidizing agent acts as an electron (*donor/acceptor*).

17. Reduction is an (*increase/decrease*) in oxidation state.

18. A reducing agent is the reactant that contains _____.

19. A reducing agent acts as an electron (*donor/acceptor*).

20. The first step in balancing an oxidation-reduction equation in acidic solution is to separate the equation into the oxidation half reaction and the reduction half reaction. *True/False*

21. The second step is to balance the equation for each half reaction with respect to all the elements except _____.

22. The third step is to balance each half reaction with respect to oxygen by adding the appropriate number of _____ to the side deficient in oxygen.

23. The fourth step is to balance each half reaction with respect to hydrogen by adding the appropriate number of _____ to the side deficient in hydrogen.

24. The fifth step is to balance each half reaction with respect to charge by adding the appropriate number of _____ to the side with the excess (*positive/negative*) charge.

25. In a balanced oxidation-reduction equation, the number of electrons donated by the oxidation half reaction must equal _____

_____ .

26. The final balanced equation is obtained by _____

_____ .

27. Reactions occurring in basic solutions may contain $H^+(aq)$ in the final balanced equation. *True/False*

28. When the redox reaction takes place in basic solution, the third step is to balance each half reaction with respect to oxygen by adding the appropriate number of H_2O to the side with (*insufficient/excess*) oxygen.

29. When the redox reaction takes place in basic solution, the fourth step is to balance each half reaction with respect to hydrogen and oxygen by adding the appropriate number of _____ to the side opposite the added H_2O.

30. The appropriate number of OH^- in Question 29 is _____

_____ .

31. Oxidation-reduction reactions can be used in chemical analysis. *True/False*

32. The equilibrium constant for a quantitative reaction is very (*large/small*).

33. A quantitative reaction goes essentially to completion. *True/False*

34. The rusting of iron is an example of corrosion. *True/False*

35. Corrosion involves redox reactions. *True/False*

36. A layer of paint may provide corrosion protection to a metal. *True/False*

37. A replaceable sacrificial metal that prevents corrosion of another metal is (*more/less*) reactive than the metal being protected.

38. Galvanization is an anticorrosion process in which iron is protected from corrosion by the metal _____ .

C. CALCULATIONS YOU SHOULD KNOW HOW TO DO

1. Assign oxidation states to each atom in a molecule or an ion. This is done in one of the following ways:

    (a) Applying the rules given in Section 4-1.
    (b) Applying the rules given in Section 21-1 as in Example 21-1.
    (c) Analogy with elements in the same group in the periodic table as in Example 21-2.

    See Problems 21-1 through 21-10.

2. Use the method of half reactions to balance oxidation-reduction equations. See Examples 21-5 and 21-7 and Problems 21-19 through 21-30.

3. Balance half reactions. See Example 21-6 and Problems 21-31 through 21-36.

4. Carry out stoichiometric calculations involving oxidation-reduction reactions. See Example 21-8 and Problems 21-37 through 21-44.

D. ANSWERS TO THE ODD-NUMBERED PROBLEMS

21-1 We use the rules given in Section 4-2.

    (a) The oxidation state of oxygen is zero by rule 1.
    (b) The oxidation state of potassium is $+1$ by rule 3. The oxidation state of oxygen is $2x + (+1) = 0$ or $x = -1/2$ by rule 2.
    (c) The oxidation state of sodium is $+1$ by rule 3. The oxidation state of oxygen is $2x + 2(+1) = 0$ or $x = -1$ by rule 2.
    (d) The oxidation state of fluorine is -1 by rule 4. The oxidation state of oxygen is $x + 2(-1) = 0$ or $x = +2$ by rule 2.

21-3 We use the rules given in Section 4-2. The oxidation state of oxygen is -2 by rule 7.

    (a) The oxidation state of chlorine is $x + (-2) = -1$ or $x = +1$ by rule 2.
    (b) The oxidation state of chlorine is $x + (2)(-2) = -1$ or $x = +3$ by rule 2.
    (c) The oxidation state of chlorine is $x + (3)(-2) = -1$ or $x = +5$ by rule 2.
    (d) The oxidation state of chlorine is $x + (4)(-2) = -1$ or $x = +7$ by rule 2.

21-5 (a) The Lewis formula for HCHO is

$$\text{H}-\overset{\displaystyle |}{\underset{\displaystyle \text{H}}{\text{C}}}=\overset{\displaystyle ..}{\underset{\displaystyle ..}{\text{O}}}\!:$$

Because oxygen is more electronegative than carbon, we assign the electrons in the covalent bond to the oxygen atom, and because carbon is more electronegative than hydrogen, we assign the electrons in each covalent bond to the carbon atom.

$$\text{H} \quad :\overset{\bullet\bullet}{\underset{\bullet\bullet}{\text{C}}} \quad :\overset{\bullet\bullet}{\underset{\bullet\bullet}{\text{O}}}:$$
$$\text{H}$$

Therefore, the oxidation states are

$$\text{oxidation state of O in HCHO} = 6 - 8 = -2$$
$$\text{oxidation state of C in HCHO} = 4 - 4 = 0$$
$$\text{oxidation state of H in HCHO} = 1 - 0 = +1$$

(b)   The Lewis formula for CH_3OH is

$$\begin{array}{c}
\text{H} \\
| \\
\text{H}-\text{C}-\overset{\bullet\bullet}{\underset{\bullet\bullet}{\text{O}}}-\text{H} \\
| \\
\text{H}
\end{array}$$

Because oxygen is more electronegative than carbon and hydrogen, we assign the electrons in each covalent bond to the oxygen atom, and because carbon is more electronegative than hydrogen, we assign the electrons in each covalent bond to the carbon atom.

$$\begin{array}{c}
\text{H} \\
\text{H} \quad :\overset{\bullet\bullet}{\underset{\bullet\bullet}{\text{C}}} \quad :\overset{\bullet\bullet}{\underset{\bullet\bullet}{\text{O}}}: \quad \text{H} \\
\text{H}
\end{array}$$

Therefore, the oxidation states are

$$\text{oxidation state of O in } CH_3OH = 6 - 8 = -2$$
$$\text{oxidation state of C in } CH_3OH = 4 - 6 = -2$$
$$\text{oxidation state of H in } CH_3OH = 1 - 0 = +1$$

(c)   The Lewis formula for CH_3COCH_3 is

$$\begin{array}{ccc}
\text{H} & \overset{\bullet\bullet}{\text{O}} & \text{H} \\
| & \| & | \\
\text{H}-\text{C}-&\text{C}-&\text{C}-\text{H} \\
| & & | \\
\text{H} & & \text{H}
\end{array}$$

Because oxygen is more electronegative than carbon, we assign the electrons in the covalent bond to the oxygen atom, and because carbon is more electronegative than hydrogen, we assign the electrons in each covalent bond to the carbon atom. One electron is assigned to each carbon atom in the carbon-carbon covalent bond.

$$\begin{matrix} \text{H} & \overset{\cdot\cdot}{\text{:O:}} & \text{H} \\ & & \\ \text{H} \;\; \overset{\cdot\cdot}{\text{:C}}\cdot & \cdot\text{C}\cdot & \cdot\overset{\cdot\cdot}{\text{C:}} \;\; \text{H} \\ & & \\ \text{H} & & \text{H} \end{matrix}$$

Therefore, the oxidation states are

$$\text{oxidation state of O in CH}_3\text{COCH}_3 = 6 - 8 = -2$$

$$\text{oxidation state of each C in} -\text{CH}_3 \text{ in CH}_3\text{COCH}_3 = 4 - 7 = -3$$

$$\text{oxidation state of C in CO in CH}_3\text{COCH}_3 = 4 - 2 = +2$$

$$\text{oxidation state of each H in CH}_3\text{COCH}_3 = 1 - 0 = +1$$

(d)   The Lewis formula for CH_3COOH is

$$\begin{matrix} \text{H} & \overset{\cdot\cdot}{\text{O}} & \\ | & \parallel & \\ \text{H}-\text{C}-\text{C}-\overset{\cdot\cdot}{\underset{\cdot\cdot}{\text{O}}}-\text{H} \\ | & & \\ \text{H} & & \end{matrix}$$

Because oxygen is more electronegative than carbon, we assign the electrons in the covalent bond to the oxygen atom, and because carbon is more electronegative than hydrogen, we assign the electrons in each covalent bond to the carbon atom. One electron is assigned to each carbon atom in the covalent bond between them.

$$\begin{matrix} \text{H} & \overset{\cdot\cdot}{\text{:O:}} & \\ & & \\ \text{H} \;\; \overset{\cdot\cdot}{\text{:C}}\cdot & \cdot\text{C} \;\; \overset{\cdot\cdot}{\underset{\cdot\cdot}{\text{:O:}}} \;\; \text{H} \\ & & \\ \text{H} & & \end{matrix}$$

Therefore, the oxidation states are

$$\text{the oxidation state of each O in CH}_3\text{COOH} = 6 - 8 = -2$$

$$\text{the oxidation state of C in} -\text{CH}_3 \text{ in CH}_3\text{COOH} = 4 - 7 = -3$$

$$\text{the oxidation state of C in} -\text{COOH in CH}_3\text{COOH} = 4 - 1 = +3$$

$$\text{the oxidation state of each H in CH}_3\text{COOH} = 1 - 0 = +1$$

21-7   We use the rules given in Section 4-1.

(a)   The oxidation state of K is $+1$ by rule 3; the oxidation state of O is -2 by rule 7; and the oxidation state of Mn is $+1 + x + 4(-2) = 0$ or $x = +7$ by rule 2.

(b)   The oxidation state of O is -2 by rule 7, and the oxidation state of Mn is $x + 4(-2) = 0$ or $x = +6$ by rule 2.

(c)   The oxidation state of O is -2 by rule 7, and the oxidation state of Mn is $x + 2(-2) = 0$ or $x = +4$ by rule 2.

(d)   The oxidation state of O is -2 by rule 7; the oxidation state of Cl in ClO_4^- is $x + 4(-2) = -1$ or $x = +7$; and the oxidation state of Mn is $x + 3(-1) = 0$ or $x = +3$ by rule 2.

21-9   (a)   By analogy with oxygen, the oxidation state of Se is -2 by rule 7, and the oxidation state of Mo is $x + 2(-2) = 0$ or $x = +4$ by rule 2 (Section 4-1).

    (b)   Because carbon is more electronegative than silicon, the electrons in the covalent bonds are assigned to the carbon atom. Thus, we have

$$Si \quad :\overset{\cdot\cdot}{\underset{\cdot\cdot}{C}}:$$

The oxidation state of C is $4 - 8 = -4$, and the oxidation state of Si is $4 - 0 = +4$.

    (c)   By analogy with aluminium, the oxidation state of Ga is $+3$ and the oxidation state of As is $x + (+3) = 0$ or $x = -3$ by rule 2 (Section 4-1). We cannot use nitrogen or phosphorus as an analogy for arsenic because both nitrogen and phosphorus have various oxidation states.

    (d)   The oxidation state of K is $+1$ by rule 3; the oxidation state of O is -2 by rule 7; and the oxidation state of S is $2(+1) + 2x + 3(-2) = 0$ or $x = +2$ by rule 2 (Section 4-1).

21-11   The oxidation state of iodine decreases from 0 in I_2 to -1 in NaI. Thus, iodine is reduced, and I_2 acts as the oxidizing agent. The oxidation state of sulfur increases from $+2$ in $Na_2S_2O_3$ to $+\frac{10}{4}$ in $Na_2S_4O_6$. Thus, sulfur is oxidized, and $Na_2S_2O_3$ acts as a reducing agent.

21-13   The oxidation state of nitrogen decreases from $+5$ in $NaNO_3$ to $+3$ in $NaNO_2$. Thus, nitrogen is reduced; $NaNO_3$ is the oxidizing agent. The oxidation state of lead increases from 0 in Pb to $+2$ in PbO. Thus, lead is oxidized; Pb is the reducing agent.

21-15   (a)   The oxidation state of iodine increases from -1 in I^- to 0 in I_2. Thus, iodine is oxidized, and I^- acts as the reducing agent. The oxidation state of iron decreases from $+3$ in Fe^{3+} to $+2$ in Fe^{2+}. The iron is reduced, and Fe^{3+} is the oxidizing agent. The half reactions are

$$2I^-(aq) \longrightarrow I_2(s) + 2e^- \qquad \text{(oxidation half reaction)}$$
$$Fe^{3+}(aq) + e^- \longrightarrow Fe^{2+}(aq) \qquad \text{(reduction half reaction)}$$

    (b)   The oxidation state of titanium increases from $+2$ in Ti^{2+} to $+3$ in Ti^{3+}. Thus, Ti^{2+} is oxidized and acts as the reducing agent. The oxidation state of cobalt decreases from $+2$ in Co^{2+} to 0 in Co. Thus Co^{2+} is reduced and acts as the oxidizing agent. The half reactions are

$$Ti^{2+}(aq) \longrightarrow Ti^{3+}(aq) + e^- \qquad \text{(oxidation half reaction)}$$
$$Co^{2+}(aq) + 2e^- \longrightarrow Co(s) \qquad \text{(reduction half reaction)}$$

21-17   The oxygen in KO_2 is in an unusual oxidation state $(-\frac{1}{2})$. The common oxidation state of oxygen is -2, and so the oxygen in KO_2 is easily reduced from $-\frac{1}{2}$ to -2, thus making KO_2 a strong oxidizing agent.

21-19   (a)   The two half reactions are

$$MnO \longrightarrow MnO_4^- \quad \text{(oxidation)}$$
$$PbO_2 \longrightarrow Pb^{2+} \quad \text{(reduction)}$$

The various steps are

$$MnO + 3H_2O \longrightarrow MnO_4^- + 6H^+ \qquad \text{(oxidation)}$$
$$PbO_2 + 4H^+ \longrightarrow Pb^{2+} + 2H_2O \qquad \text{(reduction)}$$

$$MnO + 3H_2O \longrightarrow MnO_4^- + 6H^+ + 5e^- \qquad \text{(oxidation)}$$
$$PbO_2 + 4H^+ + 2e^- \longrightarrow Pb^{2+} + 2H_2O \qquad \text{(reduction)}$$

$$2MnO + 6H_2O \longrightarrow 2MnO_4^- + 12H^+ + 10e^- \qquad \text{(oxidation)}$$
$$5PbO_2 + 20H^+ + 10e^- \longrightarrow 5Pb^{2+} + 10H_2O \qquad \text{(reduction)}$$

Thus, the complete balanced equation is

$$2MnO(s) + 5PbO_2(s) + 8H^+(aq) \longrightarrow 2MnO_4^-(aq) + 5Pb^{2+}(aq) + 4H_2O(l)$$

electron donor	MnO
electron acceptor	PbO_2
oxidizing agent	PbO_2
reducing agent	MnO
species oxidized	Mn
species reduced	Pb

(b)   The oxidation state of arsenic does not change in this reaction. The two half reactions, balanced with respect to the elements other than oxygen and hydrogen are

$$As_2S_5 \longrightarrow 5HSO_4^- + 2H_3AsO_4 \quad \text{(oxidation)}$$
$$NO_3^- \longrightarrow NO_2 \quad \text{(reduction)}$$

The various steps are

$$As_2S_5 + 28H_2O \longrightarrow 5HSO_4^- + 2H_3AsO_4 + 45H^+ \qquad \text{(oxidation)}$$
$$NO_3^- + 2H^+ \longrightarrow NO_2 + H_2O \qquad \text{(reduction)}$$

$$As_2S_5 + 28H_2O \longrightarrow 5HSO_4^- + 2H_3AsO_4 + 45H^+ + 40e^- \qquad \text{(oxidation)}$$
$$NO_3^- + 2H^+ + e^- \longrightarrow NO_2 + H_2O \qquad \text{(reduction)}$$

$$As_2S_5 + 28H_2O \longrightarrow 5HSO_4^- + 2H_3AsO_4 + 45H^+ + 40e^- \qquad \text{(oxidation)}$$
$$40NO_3^- + 80H^+ + 40e^- \longrightarrow 40NO_2 + 40H_2O \qquad \text{(reduction)}$$

Thus, the complete balanced equation is

$$As_2S_5(s) + 40NO_3^-(aq) + 35H^+(aq) \longrightarrow$$
$$5HSO_4^-(aq) + 2H_3AsO_4(aq) + 40NO_2(g) + 12H_2O(l)$$

electron donor	As_2S_5
electron acceptor	NO_3^-
oxidizing agent	NO_3^-
reducing agent	As_2S_5
species oxidized	S
species reduced	N

21-21 (a) The two half reactions are

$$NH_4^+ \longrightarrow N_2O \qquad \text{(oxidation)}$$
$$NO_3^- \longrightarrow N_2O \qquad \text{(reduction)}$$

The various steps are

$$2NH_4^+ \longrightarrow N_2O \qquad\qquad\qquad\qquad \text{(oxidation)}$$
$$2NO_3^- \longrightarrow N_2O \qquad\qquad\qquad\qquad \text{(reduction)}$$

$$2NH_4^+ + H_2O \longrightarrow N_2O + 10H^+ \qquad\qquad \text{(oxidation)}$$
$$2NO_3^- + 10H^+ \longrightarrow N_2O + 5H_2O \qquad\qquad \text{(reduction)}$$

$$2NH_4^+ + H_2O \longrightarrow N_2O + 10H^+ + 8e^- \qquad \text{(oxidation)}$$
$$2NO_3^- + 10H^+ + 8e^- \longrightarrow N_2O + 5H_2O \qquad \text{(reduction)}$$

Thus, the complete balanced equation is

$$2NH_4^+(aq) + 2NO_3^-(aq) \longrightarrow 2N_2O(g) + 4H_2O(l)$$

or

$$NH_4^+(aq) + NO_3^-(aq) \longrightarrow N_2O(g) + 2H_2O(l)$$

(b) The notation $\cdot 3\ H_2O$ signifies three waters of hydration. We shall omit them and add them back after the half reaction is balanced. The two half reactions are

$$Fe \longrightarrow Fe_2O_3 \qquad \text{(oxidation)}$$
$$O_2 \longrightarrow H_2O \qquad \text{(reduction)}$$

The various steps are

$$2Fe \longrightarrow Fe_2O_3 \qquad \text{(oxidation)}$$
$$O_2 \longrightarrow 2H_2O \qquad \text{(reduction)}$$

$$2Fe \longrightarrow Fe_2O_3 + 3H_2O \qquad \text{(oxidation)}$$
$$O_2 + 2H_2O \longrightarrow 2H_2O + 2OH^- \qquad \text{(reduction)}$$

$$2Fe + 6OH^- \longrightarrow Fe_2O_3 + 3H_2O \qquad \text{(oxidation)}$$
$$O_2 + 4H_2O \longrightarrow 2H_2O + 4OH^- \qquad \text{(reduction)}$$

$$2Fe + 6OH^- \longrightarrow Fe_2O_3 + 3H_2O + 6e^-$$

Add the three waters of hydration.

$$2Fe + 6OH^- + 3H_2O \longrightarrow Fe_2O_3 \cdot 3H_2O + 3H_2O + 6e^-$$

$$2Fe + 6OH^- \longrightarrow Fe_2O_3 \cdot 3H_2O + 6e^- \qquad \text{(oxidation)}$$
$$O_2 + 2H_2O + 4e^- \longrightarrow 4OH^- \qquad \text{(reduction)}$$

$$4Fe + 12OH^- \longrightarrow 2Fe_2O_3 \cdot 3H_2O + 12e^- \qquad \text{(oxidation)}$$
$$3O_2 + 6H_2O + 12e^- \longrightarrow 12OH^- \qquad \text{(reduction)}$$

Thus, the complete balanced equation is

$$4Fe(s) + 3O_2(g) + 6H_2O(l) \longrightarrow 2Fe_2O_3 \cdot 3H_2O(s)$$

21-23 (a) The two half reactions are

$$Fe(OH)_2 \longrightarrow Fe(OH)_3 \qquad \text{(oxidation)}$$
$$O_2 \longrightarrow 2H_2O \qquad \text{(reduction)}$$

The various steps are (see Solution 21-21 (b))

$$Fe(OH)_2 + OH^- \longrightarrow Fe(OH)_3 \qquad \text{(oxidation)}$$
$$O_2 + 2H_2O \longrightarrow 4OH^- \qquad \text{(reduction)}$$

$$Fe(OH)_2 + OH^- \longrightarrow Fe(OH)_3 + e^- \qquad \text{(oxidation)}$$
$$O_2 + 2H_2O + 4e^- \longrightarrow 4OH^- \qquad \text{(reduction)}$$

$$4Fe(OH)_2 + 4OH^- \longrightarrow 4Fe(OH)_3 + 4e^- \qquad \text{(oxidation)}$$
$$O_2 + 2H_2O + 4e^- \longrightarrow 4OH^- \qquad \text{(reduction)}$$

The complete balanced equation is

$$4Fe(OH)_2(s) + O_2(g) + 2H_2O(l) \longrightarrow 4Fe(OH)_3(s)$$

(b)   The two half reactions are

$$Cu \longrightarrow Cu^{2+} \quad \text{(oxidation)}$$
$$NO_3^- \longrightarrow NO \quad \text{(reduction)}$$

The various steps are

$$Cu \longrightarrow Cu^{2+} \qquad\qquad\qquad \text{(oxidation)}$$
$$NO_3^- + 4H^+ \longrightarrow NO + 2H_2O \qquad \text{(reduction)}$$

$$Cu \longrightarrow Cu^{2+} + 2e^- \qquad\qquad \text{(oxidation)}$$
$$NO_3^- + 4H^+ + 3e^- \longrightarrow NO + 2H_2O \qquad \text{(reduction)}$$

$$3Cu \longrightarrow 3Cu^{2+} + 6e^- \qquad\qquad \text{(oxidation)}$$
$$2NO_3^- + 8H^+ + 6e^- \longrightarrow 2NO + 4H_2O \qquad \text{(reduction)}$$

The final balanced equation is

$$3Cu(s) + 2NO_3^-(aq) + 8H^+(aq) \longrightarrow 3Cu^{2+}(aq) + 2NO(g) + 4H_2O(l)$$

21-25   (a)   The two half reactions are

$$I^- \longrightarrow I_3^- \qquad \text{(oxidation)}$$
$$IO_4^- \longrightarrow IO_3^- \qquad \text{(reduction)}$$

The various steps are

$$3I^- \longrightarrow I_3^- \qquad\qquad\qquad \text{(oxidation)}$$
$$IO_4^- \longrightarrow IO_3^- \qquad\qquad\qquad \text{(reduction)}$$

$$3I^- \longrightarrow I_3^- \qquad\qquad\qquad \text{(oxidation)}$$
$$IO_4^- + H_2O \longrightarrow IO_3^- \qquad\qquad \text{(reduction)}$$

$$3I^- \longrightarrow I_3^- \qquad\qquad\qquad\qquad \text{(oxidation)}$$
$$IO_4^- + H_2O \longrightarrow IO_3^- + 2OH^- \qquad \text{(reduction)}$$

$$3I^- \longrightarrow I_3^- + 2e^- \qquad\qquad\qquad \text{(oxidation)}$$
$$IO_4^- + H_2O + 2e^- \longrightarrow IO_3^- + 2OH^- \qquad \text{(reduction)}$$

The final balanced equation is

$$IO_4^-(aq) + 3I^-(aq) + H_2O(l) \longrightarrow IO_3^-(aq) + I_3^-(aq) + 2OH^-(aq)$$

(b)   The two half reactions are

$$Cr^{2+} \longrightarrow Cr^{3+} \qquad \text{(oxidation)}$$
$$H_2MoO_4 \longrightarrow Mo \qquad \text{(reduction)}$$

The various steps are

$$Cr^{2+} \longrightarrow Cr^{3+} \qquad\qquad\qquad\qquad \text{(oxidation)}$$
$$H_2MoO_4 \longrightarrow Mo + 4H_2O \qquad\qquad \text{(reduction)}$$

$$Cr^{2+} \longrightarrow Cr^{3+} \qquad\qquad\qquad\qquad \text{(oxidation)}$$
$$H_2MoO_4 + 6H^+ \longrightarrow Mo + 4H_2O \qquad \text{(reduction)}$$

$$Cr^{2+} \longrightarrow Cr^{3+} + e^- \qquad\qquad\qquad \text{(oxidation)}$$
$$H_2MoO_4 + 6H^+ + 6e^- \longrightarrow Mo + 4H_2O \qquad \text{(reduction)}$$

$$6Cr^{2+} \longrightarrow 6Cr^{3+} + 6e^- \qquad\qquad\qquad \text{(oxidation)}$$
$$H_2MoO_4 + 6H^+ + 6e^- \longrightarrow Mo + 4H_2O \qquad \text{(reduction)}$$

The final balanced equation is

$$H_2MoO_4(aq) + 6Cr^{2+}(aq) + 6H^+(aq) \longrightarrow Mo(s) + 6Cr^{3+}(aq) + 4H_2O(l)$$

21-27 (a) The two half reactions are

$$Cl^- \longrightarrow ClO_2^- \qquad \text{(oxidation)}$$
$$CrO_4^{2-} \longrightarrow Cr^{3+} \qquad \text{(reduction)}$$

The various steps are

$$Cl^- + 2H_2O \longrightarrow ClO_2^- \qquad\qquad\qquad \text{(oxidation)}$$
$$CrO_4^{2-} \longrightarrow Cr^{3+} + 4H_2O \qquad\qquad\qquad \text{(reduction)}$$

$$Cl^- + 2H_2O \longrightarrow ClO_2^- + 4H^+ \qquad\qquad \text{(oxidation)}$$
$$CrO_4^{2-} + 8H^+ \longrightarrow Cr^{3+} + 4H_2O \qquad\qquad \text{(reduction)}$$

$$Cl^- + 2H_2O \longrightarrow ClO_2^- + 4H^+ + 4e^- \qquad\qquad \text{(oxidation)}$$
$$CrO_4^{2-} + 8H^+ + 3e^- \longrightarrow Cr^{3+} + 4H_2O \qquad\qquad \text{(reduction)}$$

$$3Cl^- + 6H_2O \longrightarrow 3ClO_2^- + 12H^+ + 12e^- \qquad \text{(oxidation)}$$
$$4CrO_4^{2-} + 32H^+ + 12e^- \longrightarrow 4Cr^{3+} + 16H_2O \qquad \text{(reduction)}$$

The final balanced equation is

$$4CrO_4^{2-}(aq) + 3Cl^-(aq) + 20H^+(aq) \longrightarrow$$
$$3ClO_2^-(aq) + 4Cr^{3+}(aq) + 10H_2O(l)$$

(b) The two half reactions are

$$S_2O_3^{2-} \longrightarrow S_4O_6^{2-} \qquad \text{(oxidation)}$$
$$Cu^{2+} \longrightarrow Cu^+ \qquad \text{(reduction)}$$

The various steps are

$$2S_2O_3^{2-} \longrightarrow S_4O_6^{2-} \qquad \text{(oxidation)}$$
$$Cu^{2+} \longrightarrow Cu^+ \qquad \text{(reduction)}$$

$$2S_2O_3^{2-} \longrightarrow S_4O_6^{2-} + 2e^- \qquad \text{(oxidation)}$$
$$Cu^{2+} + e^- \longrightarrow Cu^+ \qquad \text{(reduction)}$$

$$2S_2O_3^{2-} \longrightarrow S_4O_6^{2-} + 2e^- \qquad \text{(oxidation)}$$
$$2Cu^{2+} + 2e^- \longrightarrow 2Cu^+ \qquad \text{(reduction)}$$

Addition of the two half reaction equations yields

$$2S_2O_3^{2-}(aq) + 2Cu^{2+}(aq) \longrightarrow S_4O_6^{2-}(aq) + 2Cu^+(aq)$$

21-29   The two half reactions are

$$CrI_3 \longrightarrow CrO_4^{2-} + IO_4^- \qquad \text{(oxidation)}$$
$$Cl_2 \longrightarrow Cl^- \qquad \text{(reduction)}$$

Note that the oxidation state of chromium does not change. The various steps are

$$CrI_3 \longrightarrow CrO_4^{2-} + 3IO_4^- \qquad \text{(oxidation)}$$
$$Cl_2 \longrightarrow 2Cl^- \qquad \text{(oxidation)}$$

$$CrI_3 \longrightarrow CrO_4^{2-} + 3IO_4^- + 16H_2O \qquad \text{(oxidation)}$$
$$Cl_2 \longrightarrow 2Cl^- \qquad \text{(reduction)}$$

$$CrI_3 + 32OH^- \longrightarrow CrO_4^{2-} + 3IO_4^- + 16H_2O \qquad \text{(oxidation)}$$
$$Cl_2 \longrightarrow 2Cl^- \qquad \text{(reduction)}$$

$$CrI_3 + 32OH^- \longrightarrow CrO_4^{2-} + 3IO_4^- + 16H_2O + 27e^- \qquad \text{(oxidation)}$$
$$Cl_2 + 2e^- \longrightarrow 2Cl^- \qquad \text{(reduction)}$$

$$2CrI_3 + 64OH^- \longrightarrow 2CrO_4^{2-} + 6IO_4^- + 32H_2O + 54e^- \qquad \text{(oxidation)}$$
$$27Cl_2 + 54e^- \longrightarrow 54Cl^- \qquad \text{(reduction)}$$

The balanced equation is

$$2CrI_3(s) + 27Cl_2(g) + 64OH^-(aq) \longrightarrow$$
$$2CrO_4^{2-}(aq) + 6IO_4^-(aq) + 54Cl^-(aq) + 32H_2O(l)$$

21-31   We follow the first five steps given in Section 21-4.

(a) $Mo^{3+} \rightarrow MoO_2^{2+}$
      $Mo^{3+} + 2H_2O \rightarrow MoO_2^{2+}$
      $Mo^{3+} + 2H_2O \rightarrow MoO_2^{2+} + 4H^+$
      $Mo^{3+} + 2H_2O \rightarrow MoO_2^{2+} + 4H^+ + 3e^-$
      $Mo^{3+}(aq) + 2H_2O(l) \rightarrow MoO_2^{2+}(aq) + 4H^+(aq) + 3e^-$

(b) $P_4 \rightarrow H_3PO_4$
$P_4 \rightarrow 4H_3PO_4$
$P_4 + 16H_2O \rightarrow 4H_3PO_4$
$P_4 + 16H_2O \rightarrow 4H_3PO_4 + 20H^+$
$P_4 + 16H_2O \rightarrow 4H_3PO_4 + 20H^+ + 20e^-$
$P_4(s) + 16H_2O(l) \rightarrow 4H_3PO_4(aq) + 20H^+(aq) + 20e^-$

(c) $S_2O_8^{2-} \rightarrow HSO_4^-$
$S_2O_8^{2-} \rightarrow 2HSO_4^-$
$S_2O_8^{2-} + 2H^+ \rightarrow 2HSO_4^-$
$S_2O_8^{2-} + 2H^+ + 2e^- \rightarrow 2HSO_4^-$
$S_2O_8^{2-}(aq) + 2H^+(aq) + 2e^- \rightarrow 2HSO_4^-(aq)$

21-33 The steps are

(a) $2WO_3 \rightarrow W_2O_5$
$2WO_3 + 2H^+ \rightarrow W_2O_5 + H_2O$
$2WO_3(s) + 2H^+(aq) + 2e^- \rightarrow W_2O_5(s) + H_2O(l)$

(b) $U^{4+} + 2H_2O \rightarrow UO_2^+ + 4H^+$
$U^{4+}(aq) + 2H_2O(l) \rightarrow UO_2^+(aq) + 4H^+(aq) + e^-$

(c) $Zn + 4OH^- \rightarrow Zn(OH)_4^-$
$Zn(s) + 4OH^-(aq) \rightarrow Zn(OH)_4^{2-}(aq) + 2e^-$

21-35 The steps are

(a) $2SO_3^{2-} \rightarrow S_2O_4^{2-}$
$2SO_3^{2-} + 2H_2O \rightarrow S_2O_4^{2-} + 4OH^-$
$2SO_3^{2-}(aq) + 2H_2O(l) + 2e^- \rightarrow S_2O_4^{2-}(aq) + 4OH^-(aq)$

(b) $2Cu(OH)_2 \rightarrow Cu_2O$
$2Cu(OH)_2 \rightarrow Cu_2O + 4OH^-$
$2Cu(OH)_2 + 2OH^- \rightarrow Cu_2O + 4OH^- + H_2O$
$2Cu(OH)_2 \rightarrow Cu_2O + 2OH^- + H_2O$
$2Cu(OH)_2(s) + 2e^- \rightarrow Cu_2O(s) + 2OH^-(aq) + H_2O(l)$

(c) $2AgO \rightarrow Ag_2O$
$2AgO + H_2O \rightarrow Ag_2O + 2OH^-$
$2AgO(s) + H_2O(l) + 2e^- \rightarrow Ag_2O(s) + 2OH^-(aq)$

21-37 We must first balance the equation for the reaction. The steps are

$$Sb^{3+} \longrightarrow Sb^{5+} \qquad \text{(oxidation)}$$

$$BrO_3^- \longrightarrow Br^- \qquad \text{(reduction)}$$

$$Sb^{3+} \longrightarrow Sb^{5+} \qquad \text{(oxidation)}$$

$$BrO_3^- \longrightarrow Br^- + 3H_2O \qquad \text{(reduction)}$$

$$Sb^{3+} \longrightarrow Sb^{5+} \qquad \text{(oxidation)}$$

$$BrO_3^- + 6H^+ \longrightarrow Br^- + 3H_2O \qquad \text{(reduction)}$$

$$Sb^{3+} \longrightarrow Sb^{5+} + 2e^- \qquad \text{(oxidation)}$$

$$BrO_3^- + 6H^+ + 6e^- \longrightarrow Br^- + 3H_2O \qquad \text{(reduction)}$$

The complete balanced equation is

$$BrO_3^-(aq) + 6H^+(aq) + 3Sb^{3+}(aq) \longrightarrow Br^-(aq) + 3Sb^{5+}(aq) + 3H_2O(l)$$

The number of moles of BrO_3^- is

$$n = MV = (0.125 \text{ mol} \cdot L^{-1})(0.0437 \text{ L}) = 5.46 \times 10^{-3} \text{ mol}$$

The number of moles of antimony that reacts with 5.46×10^{-3} mol of BrO_3^- is

$$\text{moles of } Sb^{3+} = (5.46 \times 10^{-3} \text{ mol } BrO_3^-)\left(\frac{3 \text{ mol } Sb^{3+}}{1 \text{ mol } BrO_3^-}\right)$$

$$= 1.64 \times 10^{-2} \text{ mol}$$

The number of grams of antimony that reacts is

$$\text{mass of } Sb = (1.64 \times 10^{-2} \text{ mol})\left(\frac{121.8 \text{ g Sb}}{1 \text{ mol Sb}}\right) = 2.00 \text{ g}$$

$$\% \text{ antimony} = \frac{2.00 \text{ g Sb}}{9.62 \text{ g ore}} \times 100 = 20.8\%$$

21-39 We must first balance the equation for the reaction. The steps are

$$Sn^{2+} \longrightarrow Sn^{4+} \qquad \text{(oxidation)}$$
$$I_3^- \longrightarrow 3I^- \qquad \text{(reduction)}$$

$$Sn^{2+} \longrightarrow Sn^{4+} + 2e^- \qquad \text{(oxidation)}$$
$$I_3^- + 2e^- \longrightarrow 3I^- \qquad \text{(reduction)}$$

Thus,

$$I_3^-(aq) + Sn^{2+}(aq) \longrightarrow Sn^{4+}(aq) + 3I^-(aq)$$

The number of moles of $I_3^-(aq)$ used is

$$\text{moles of } I_3^- = (0.556 \text{ M})(0.0346 \text{ L}) = 0.0192 \text{ mol}$$

The number of moles of Sn^{2+} oxidized is

$$\text{moles } Sn^{2+} = (0.0192 \text{ mol } I_3^-)\left(\frac{1 \text{ mol } Sn^{2+}}{1 \text{ mol } I_3^-}\right) = 0.0192 \text{ mol}$$

The mass of tin is

$$\text{mass of tin} = (0.0192 \text{ mol Sn})\left(\frac{118.7 \text{ g Sn}}{1 \text{ mol Sn}}\right) = 2.28 \text{ g}$$

The percentage of tin in the sample is

$$\% \text{ tin} = \frac{2.28 \text{ g Sn}}{10.0 \text{ g sample}} \times 100 = 22.8\%$$

21-41   We must first balance the equation for the reaction. The steps are

$$2S_2O_3^{2-} \longrightarrow S_4O_6^{2-} \qquad \text{(oxidation)}$$
$$I_3^- \longrightarrow 3I^- \qquad \text{(reduction)}$$

$$2S_2O_3^{2-} \longrightarrow S_4O_6^{2-} + 2e^- \qquad \text{(oxidation)}$$
$$I_3^- + 2e^- \longrightarrow 3I^- \qquad \text{(reduction)}$$

Thus,

$$I_3^-(aq) + 2S_2O_3^{2-}(aq) \longrightarrow S_4O_6^{2-}(aq) + 3I^-(aq)$$

The number of millimoles of $Na_2S_2O_3$ used to titrate the I_3^- is

$$\text{millimoles of } Na_2S_2O_3 = (0.330 \text{ M})(36.4 \text{ mL}) = 12.0 \text{ mmol}$$

The number of millimoles of I_3^- is

$$\text{millimoles } I_3^- = (12.0 \text{ mmol } Na_2S_2O_3) \left(\frac{1 \text{ mmol } I_3^-}{2 \text{ mmol } Na_2S_2O_3} \right) = 6.00 \text{ mmol}$$

The concentration of I_3^- is

$$M = \frac{6.00 \text{ mmol}}{15.0 \text{ mL}} = 0.400 \text{ M}$$

21-43   The reactants and products are given as

$$P_4 + BaSO_4 \longrightarrow BaS + P_4O_{10}$$

The two half reactions are

$$P_4 \longrightarrow P_4O_{10}$$
$$BaSO_4 \longrightarrow BaS$$

The steps in balancing the first half reaction are

$$P_4 + 10H_2O \longrightarrow P_4O_{10}$$
$$P_4 + 10H_2O \longrightarrow P_4O_{10} + 20H^+$$
$$P_4 + 10H_2O \longrightarrow P_4O_{10} + 20H^+ + 20e^-$$

The steps in balancing the second half reaction are

$$BaSO_4 \longrightarrow BaS + 4H_2O$$
$$BaSO_4 + 8H^+ \longrightarrow BaS + 4H_2O$$
$$BaSO_4 + 8H^+ + 8e^- \longrightarrow BaS + 4H_2O$$

Multiply the first half reaction by 2 and the second by 5, then add the two. Add the phases to obtain

$$2P_4(s) + 5BaSO_4(s) \longrightarrow 2P_4O_{10}(s) + 5BaS(s)$$

The amount of phosphorus required to react with 2.16 g of $BaSO_4(s)$ is

$$\text{mass of P}_4 = (2.16 \text{ g BaSO}_4)\left(\frac{1 \text{ mol BaSO}_4}{233.4 \text{ g BaSO}_4}\right)\left(\frac{2 \text{ mol P}_4}{5 \text{ mol BaSO}_4}\right)\left(\frac{123.9 \text{ g P}_4}{1 \text{ mol P}_4}\right)$$
$$= 0.459 \text{ g}$$

21-45 The two half reactions are

$$MnO_4^- \longrightarrow MnO_2$$
$$H_2O \longrightarrow O_2$$

The steps in balancing the first half reaction are

$$MnO_4^- + 2H_2O \longrightarrow MnO_2$$
$$MnO_4^- + 2H_2O \longrightarrow MnO_2 + 4OH^-$$
$$MnO_4^- + 2H_2O + 3e^- \longrightarrow MnO_2 + 4OH^-$$

The steps in balancing the second half reaction are

$$2H_2O \longrightarrow O_2$$
$$2H_2O + 4OH^- \longrightarrow O_2 + 4H_2O$$
$$4OH^- \longrightarrow O_2 + 2H_2O + 4e^-$$

Multiply the first balanced half reaction by 4 and the second by 3 and add to obtain

$$4MnO_4^-(aq) + 2H_2O(l) \longrightarrow 4MnO_2(s) + 3O_2(g) + 4OH^-(aq)$$

21-47 (a)   The half reactions are

$$Cr_2O_7^{2-} \longrightarrow Cr^{3+}$$
$$I^- \longrightarrow I_3^-$$

The steps in balancing the first half reaction are

$$Cr_2O_7^{2-} \longrightarrow 2Cr^{3+}$$
$$Cr_2O_7^{2-} \longrightarrow 2Cr^{3+} + 7H_2O$$
$$Cr_2O_7^{2-} + 14H^+ \longrightarrow 2Cr^{3+} + 7H_2O$$
$$Cr_2O_7^{2-} + 14H^+ + 6e^- \longrightarrow 2Cr^{3+} + 7H_2O$$

and the steps in balancing the second half reaction are

$$3I^- \longrightarrow I_3^-$$
$$3I^- \longrightarrow I_3^- + 2e^-$$

We multiply the second half reaction by 3 and add the two half reactions to obtain

$$Cr_2O_7^{2-}(aq) + 9I^-(aq) + 14H^+(aq) \longrightarrow 2Cr^{3+}(aq) + 3I_3^-(aq) + 7H_2O(l)$$

(b) The two half reactions are

$$IO_4^- \longrightarrow I_3^-$$
$$I^- \longrightarrow I_3^-$$

The steps in balancing the first half reaction are

$$3IO_4^- \longrightarrow I_3^-$$
$$3IO_4^- \longrightarrow I_3^- + 12H_2O$$
$$3IO_4^- + 24H^+ \longrightarrow I_3^- + 12H_2O$$
$$3IO_4^- + 24H^+ + 22e^- \longrightarrow I_3^- + 12H_2O$$

and those for the second half reaction are

$$3I^- \longrightarrow I_3^-$$
$$3I^- \longrightarrow I_3^- + 2e^-$$

Multiply the second half reaction by 11 and add to obtain

$$3IO_4^-(aq) + 33I^-(aq) + 24H^+(aq) \longrightarrow 12I_3^-(aq) + 12H_2O(l)$$

21-49 The two half reactions are

$$Ag^{2+} \longrightarrow Ag^+$$
$$H_2O \longrightarrow O_2$$

The steps in balancing the first half reaction are

$$Ag^{2+} + e^- \longrightarrow Ag^+$$

and those for the second half reaction are

$$2H_2O \longrightarrow O_2$$
$$2H_2O \longrightarrow O_2 + 4H^+$$
$$2H_2O \longrightarrow O_2 + 4H^+ + 4e^-$$

Multiply the first half reaction by 4 and add the two half reactions to obtain

$$4Ag^{2+}(aq) + 2H_2O(l) \longrightarrow 4Ag^+(aq) + O_2(g) + 4H^+(aq)$$

21-51 (a) The two half reactions are

$$Cr_2O_7^{2-} \longrightarrow Cr^{3+}$$
$$H_2O_2 \longrightarrow O_2$$

The steps in balancing the first half reaction are

$$Cr_2O_7^{2-} \longrightarrow 2Cr^{3+}$$
$$Cr_2O_7^{2-} \longrightarrow 2Cr^{3+} + 7H_2O$$
$$Cr_2O_7^{2-} + 14H^+ \longrightarrow 2Cr^{3+} + 7H_2O$$
$$Cr_2O_7^{2-} + 14H^+ + 6e^- \longrightarrow 2Cr^{3+} + 7H_2O$$

and the steps in balancing the second half reaction are

$$H_2O_2 \longrightarrow O_2 + 2H^+$$
$$H_2O_2 \longrightarrow O_2 + 2H^+ + 2e^-$$

Multiply the second half reaction by 3 and add

$$Cr_2O_7^{2-}(aq) + 8H^+(aq) + 3H_2O_2(aq) \longrightarrow 2Cr^{3+}(aq) + 3O_2(g) + 7H_2O(l)$$

(b) The two half reactions are

$$Cr^{3+} \longrightarrow Cr^{2+}$$
$$Zn \longrightarrow Zn^{2+}$$

Both are balanced by just adding electrons

$$Cr^{3+} + e^- \longrightarrow Cr^{2+}$$
$$Zn \longrightarrow Zn^{2+} + 2e^-$$

and the complete, balanced equation is

$$2Cr^{3+}(aq) + Zn(s) \longrightarrow 2Cr^{2+}(aq) + Zn^{2+}(aq)$$

21-53 (a) If H_2O_2 is acting as an oxidizing agent, then the oxygen is reduced and the product is H_2O. The half reaction and the steps to balance it are

$$H_2O_2 \longrightarrow H_2O$$
$$H_2O_2 \longrightarrow 2H_2O$$
$$H_2O_2 + 2H^+ \longrightarrow 2H_2O$$
$$H_2O_2 + 2H^+ + 2e^- \longrightarrow 2H_2O$$
$$H_2O_2(aq) + 2H^+(aq) + 2e^- \longrightarrow 2H_2O(l)$$

(b) If H_2O_2 is acting as a reducing agent, then the oxygen is oxidized and the product is O_2. The half reaction and the steps to balance it are

$$H_2O_2 \longrightarrow O_2$$
$$H_2O_2 \longrightarrow O_2 + 2H^+$$
$$H_2O_2 \longrightarrow O_2 + 2H^+ + 2e^-$$
$$H_2O_2(aq) \longrightarrow O_2(g) + 2H^+(aq) + 2e^-$$

The reaction in which H_2O_2 oxidizes and reduces itself (disproportionation) is obtained by adding the two half reactions:

$$2H_2O_2(aq) \longrightarrow 2H_2O(l) + O_2(g)$$

21-55 The equations for the reactions that take place are

$$H_2S(g) + Cd^{2+}(aq) \longrightarrow CdS(s) + 2H^+(aq)$$
$$CdS(s) + I_2(aq) \longrightarrow CdI_2(aq) + S(s)$$
$$I_2(aq) + 2Na_2S_2O_3(aq) \longrightarrow Na_2S_4O_6(aq) + 2NaI(aq)$$

The excess of I_2 added is given by

$$\text{excess millimoles of } I_2 = (0.0750 \text{ M})(7.65 \text{ mL}) \left(\frac{1 \text{ mmol } I_2}{2 \text{ mmol } Na_2S_2O_3} \right)$$
$$= 0.287 \text{ mmol}$$

The total quantity of I_2 used is

$$\text{total millimoles of } I_2 = (0.0115 \text{ M})(30.00 \text{ mL}) = 0.345 \text{ mmol}$$

and the quantity of I_2 that reacts with the $CdS(s)$ is

$$\text{millimoles of } I_2 = 0.345 \text{ mmol} - 0.287 \text{ mmol} = 0.058 \text{ mmol}$$

According to the above chemical equations, this is equal to the number of milli-moles of $H_2S(g)$, and so the mass of H_2S in the air is

$$\text{mass of } H_2S = (0.058 \text{ mmol}) \left(\frac{10^{-3} \text{ mol}}{1 \text{ mmol}} \right) \left(\frac{34.08 \text{ g } H_2S}{1 \text{ mol } H_2S} \right)$$

$$= 1.98 \times 10^{-3} \text{ g}$$

and the mass percentage of H_2S in the air is

$$\% \ H_2S = \left(\frac{1.98 \times 10^{-3} \text{ g}}{10.75 \text{ g}} \right) \times 100 = 0.018\%$$

21-57   The quantity of H_2SO_4 neutralized is

$$\text{moles of } H_2SO_4 = (0.00250 \text{ M})(0.01850 \text{ L}) \left(\frac{1 \text{ mol } H_2SO_4}{2 \text{ mol NaOH}} \right)$$

$$= 2.31 \times 10^{-5} \text{ mol}$$

According to the equation for the reaction between $H_2O_2(aq)$ and $SO_2(g)$, this is also equal to the number of moles of SO_2. The mass of SO_2 is given by

$$\text{mass of } SO_2 = (2.31 \times 10^{-5} \text{ mol}) \left(\frac{64.06 \text{ g } SO_2}{1 \text{ mol } SO_2} \right) = 1.48 \times 10^{-3} \text{ g}$$

and the mass percentage of $SO_2(g)$ in the air is

$$\% \ SO_2 = \left(\frac{1.48 \times 10^{-3} \text{ g}}{812.1 \text{ g}} \right) \times 100 = 1.82 \times 10^{-4} \ \%$$

E. ANSWERS TO THE SELF-TEST

1. 0

2. $+1$

3. $+2$

4. -1

5. -2

6. $+1$

7. true

8. true

9. more

10. the group number of the element in the periodic table

11. false

12. true

13. electrons

14. increase

15. the element that is reduced

16. acceptor

17. decrease

18. the element that is oxidized

19. donor

20. true

21. oxygen and hydrogen

22. H_2O

23. H^+

24. electrons . . . positive

25. the number of electrons accepted by the reduction half reaction

26. adding the two balanced half reactions and canceling or combining any like terms

27. false

28. excess

29. OH^-

30. two times the number of added H_2O

31. true

32. large

33. true

34. true

35. true

36. true

37. more

38. zinc

22 ENTROPY, GIBBS FREE ENERGY, AND CHEMICAL REACTIVITY

A. OUTLINE OF CHAPTER 22

22-1 Not all spontaneous reactions evolve energy.

Exothermic reactions are energetically downhill. (Figure 22-1)

A spontaneous process takes place without the input of energy from an external source.

Spontaneous processes can occur with negative, zero, or positive values of the change in enthalpy.

$\Delta H_{rxn}^{\circ} < 0$ is not sufficient to guarantee reaction spontaneity.

22-2 The second law of thermodynamics places an additional restriction on energy transfers.

Spontaneous, unidirectional processes are referred to as irreversible processes.

The system is that part of the universe where the change of interest occurs and its surroundings is the rest of the universe.

The second law of thermodynamics arose from the analysis of the operation of heat engines by Sadi Carnot.

It is not possible to convert all the energy input as heat into energy output as work.

The ratio of the work output to the heat input depends upon the temperature difference between the system and its surroundings.

Entropy, denoted by S, is a thermodynamic state function that is closely associated with the transfer of energy as heat.

A statement of the second law of thermodynamics is that if energy is transferred to or from a system as heat q_{sys} at the absolute temperature T_{sys}, then the entropy change of the system ΔS_{sys} is greater than or equal to q_{sys}/T_{sys}.

A mathematical statement of the second law of thermodynamics is

$$\Delta S_{sys} \geq \frac{q_{sys}}{T_{sys}} \tag{22-1}$$

The equality sign applies if the process is reversible.

A reversible heat process is a process in which there is no heat generation due to friction and the heat is transferred at a uniform temperature.

The inequality sign applies if the process is irreversible.

The units of entropy are joules per kelvin, $J \cdot K^{-1}$.

The second law of thermodynamics also states that the total entropy change for spontaneous processes must always be positive.

The second law of thermodynamics tells us that the entropy of the universe always increases when a natural process occurs.

Entropy is not conserved in a natural process.

22-3 Entropy is a measure of the amount of disorder or randomness in a system.

Entropy arises from positional disorder and thermal disorder.

Positional disorder is the distribution of the particles in space.

Thermal disorder is the distribution of the available energy among the particles.

The entropy of a perfect crystal is zero at absolute zero, which is known as the third law of thermodynamics.

The entropy of a substance at fixed pressure increases as the temperature increases (Figure 22-6).

An increase in temperature leads to an increase in thermal disorder.

The greater the number of ways that the energy can be distributed among the energy levels of a substance, the greater is the entropy of the substance.

The entropy of a substance can also increase at a fixed temperature as a result of an increase in positional disorder.

22-4 There is an increase in entropy on melting and vaporization.

The entropy of the liquid phase of a substance is greater than the entropy of the solid phase at a given temperature and pressure.

The molar entropy change upon fusion is given by

$$\Delta S_{fus} = \frac{\Delta H_{fus}}{T_m} \tag{22-2}$$

where ΔH_{fus} is the molar enthalpy of fusion and T_m is the melting point in kelvins.

The entropy of the gaseous phase of a substance is greater than the entropy of the liquid phase at a given temperature and pressure.

The molar entropy change upon vaporization is given by

$$\Delta S_{vap} = \frac{\Delta H_{vap}}{T_b} \qquad (22\text{-}3)$$

where ΔH_{vap} is the molar enthalpy of vaporization and T_b is the boiling point in kelvins.

Gases are more disordered than liquids and liquids are more disordered than solids.

$$\Delta S_{vap} > \Delta S_{fus}$$

22-5 The molar entropy depends on molar mass and molecular structure.

The standard (1 atm) molar entropy is denoted by $S°$.

The units of $S°$ are $J \cdot K^{-1} \cdot mol^{-1}$.

The molar entropy of a gas or solute depends on concentration because of the change in positional disorder with concentration.

Values of $S°$ for some compounds are given in Table 22-1 and Appendix F.

Greater mass leads to a greater capacity to take up energy and thus to a higher entropy.

The more atoms of a given mass in a molecule, the higher is the entropy.

The more complex a molecule, the higher is the molar entropy.

22-6 $\Delta S°_{rxn}$ equals the entropy of the products minus the entropy of the reactants.

For the reaction represented by the chemical equation

$$aA + bB \longrightarrow yY + zZ$$

$$\Delta S°_{rxn} = yS°[Y] + zS°[Z] - aS°[A] - bS°[B] \qquad (22\text{-}6)$$

where $\Delta S°_{rxn}$ is the standard entropy change.

Standard entropy changes of reactions can be calculated from the values of the standard molar enthalpies of the products and reactants.

If there are more moles of gaseous products than gaseous reactants, then $\Delta S°_{rxn} > 0$.

22-7 Nature acts to minimize the energy and to maximize the entropy of all processes.

Spontaneous reactions with $\Delta S_{rxn} > 0$ are entropy-driven and are said to be entropy-favored.

Spontaneous reactions with $\Delta H_{rxn} < 0$ are energy (enthalpy)-favored.

If $\Delta H_{rxn} < 0$ and $\Delta S_{rxn} > 0$ for a reaction, then the reaction is spontaneous.

If $\Delta H_{rxn} > 0$ and $\Delta S_{rxn} < 0$ for a reaction, then the reaction is not spontaneous.

22-8 The sign of ΔG_{rxn} determines reaction spontaneity.

The Gibbs free energy change for a reaction that occurs at constant temperature is given by

$$\Delta G_{rxn} = \Delta H_{rxn} - T\Delta S_{rxn} \qquad (22\text{-}7)$$

Chemical reactions seek a compromise between energy minimization and entropy maximization.

The Gibbs criteria for reaction spontaneity are

If $\Delta G_{rxn} < 0$, then the reaction is spontaneous.

If $\Delta G_{rxn} > 0$, then the reaction is not spontaneous.

If $\Delta G_{rxn} = 0$, then the reaction is at equilibrium.

The Gibbs free energy is minimized in a spontaneous process.

The relationship between ΔG_{rxn}, the reaction quotient Q, and the equilibrium constant K is

$$\Delta G_{rxn} = RT \ln \left(\frac{Q}{K} \right) \tag{22-8}$$

If $Q/K > 1$, then $\Delta G_{rxn} > 0$.

If $Q/K < 1$, then $\Delta G_{rxn} < 0$.

If $Q/K = 1$, then $\Delta G_{rxn} = 0$.

Spontaneous is not synonymous with immediate.

A spontaneous reaction may not occur at a detectable rate.

For a reaction to occur, ΔG_{rxn} must be less than zero, but $\Delta G_{rxn} < 0$ does not guarantee that the reaction will occur at an appreciable rate.

The magnitude of ΔG_{rxn} is equal to the maximum energy that can be obtained from the reaction to perform work.

22-9 The values of ΔG_{rxn} and ΔG_{rxn}° are related.

The standard Gibbs free energy change, denoted by ΔG_{rxn}°, is equal to ΔG_{rxn} when the reactants and products are at standard conditions.

The standard Gibbs free energy change for a reaction is related to the equilibrium constant by

$$\Delta G_{rxn}^{\circ} = -RT \ln K \tag{22-10}$$

The sign of ΔG_{rxn}° determines reaction spontaneity only when all the products and reactants are at standard conditions.

The standard Gibbs free energy and the Gibbs free energy of a reaction are related by

$$\Delta G_{rxn} = \Delta G_{rxn}^{\circ} + RT \ln Q \tag{22-11}$$

where Q is the reaction quotient.

A reaction may be spontaneous at conditions other than standard conditions even when ΔG_{rxn}° is positive.

22-10 ΔG_{rxn}° values and equilibrium constants can be calculated from tabulated ΔG_f° values.

The standard Gibbs free energy of formation of a compound, denoted by ΔG_f°, is equal to the value of ΔG_{rxn}° for the reaction in which 1 mol of the compound at standard conditions is formed from its constituent elements at standard conditions.

For the reaction represented by the chemical equation

$$aA + bB \longrightarrow yY + zZ$$

$$\Delta G^\circ_{rxn} = y\,\Delta G^\circ_f[Y] + z\,\Delta G^\circ_f[Z] - a\,\Delta G^\circ_f[A] - b\,\Delta G^\circ_f[B] \quad (22\text{-}13)$$

Values of ΔG°_f for some compounds are given in Table 22-1 and Appendix F.

22-11 The van't Hoff equation governs the temperature dependence of equilibrium constants.

The van't Hoff equation is

$$\ln\left(\frac{K_2}{K_1}\right) = \frac{\Delta H^\circ_{rxn}}{R}\left(\frac{T_2 - T_1}{T_1 T_2}\right) \quad (22\text{-}15)$$

where K_2 is the equilibrium constant at temperature T_2 and K_1 is the equilibrium constant at T_1.

The value of ΔH°_{rxn} is assumed to remain constant over the temperature range T_1 to T_2.

The value of K increases with increasing T for an endothermic reaction.

The value of K decreases with increasing T for an exothermic reaction.

The Fischer-Tropsch synthesis involves the production of straight-chain hydrocarbons and alcohols from $H_2(g)$, $CO(g)$, and $CO_2(g)$.

22-12 The Clapeyron-Clausius equation predicts the equilibrium vapor-pressure curves of liquids and solids.

The Clapeyron-Clausius equation is given by

$$\ln\left(\frac{P_2}{P_1}\right) = \frac{\Delta H_{vap}}{R}\left(\frac{T_2 - T_1}{T_1 T_2}\right) \quad (22\text{-}17)$$

where P_1 is the equilibrium vapor pressure at the Kelvin temperature T_1, P_2 is the equilibrium vapor pressure at the Kelvin temperature T_2, R is the molar gas constant, and ΔH_{vap} is the molar enthalpy of vaporization.

The molar enthalpy ΔH_{vap} does not change appreciably with temperature.

22-13 The value of ΔH°_{rxn} is determined primarily by the difference in the bond enthalpies of the reactant and product molecules.

The molar bond enthalpy is equal to the energy input as heat required to break one mole of the covalent bond.

Molar bond enthalpy is often referred to as bond enthalpy or bond energy.

The value of the bond enthalpy for a chemical bond varies somewhat from compound to compound.

The enthalpy change for a reaction is given by

$$\Delta H^\circ_{rxn} \approx \begin{pmatrix} \text{heat energy input to} \\ \text{break all bonds in} \\ \text{reactants} \end{pmatrix} - \begin{pmatrix} \text{energy evolved as heat} \\ \text{on formation of all} \\ \text{bonds in products} \end{pmatrix} \quad (22\text{-}18)$$

$$\approx H(\text{bond})_R - H(\text{bond})_P$$

The values of average bond enthalpies for a variety of chemical bonds are given in Table 22-2.

The calculation of ΔH°_{rxn} values from bond enthalpies is restricted to gas-phase reactions.

The values of ΔH°_{rxn} calculated from average bond enthalpies are only approximations of the actual experimental values.

B. SELF-TEST

1. All spontaneous processes are exothermic processes. *True/False*

2. All naturally occurring processes are spontaneous. *True/False*

3. In thermodynamics, the system and its surroundings represent the universe. *True/False*

4. All naturally occurring processes are reversible. *True/False*

5. Entropy is closely associated with the transfer of energy as (*heat/work*).

6. The second law of thermodynamics states that if energy is tranferred to or from a system as heat q_{sys} at the absolute Kelvin temperature T_{sys}, then the entropy change of the system is (*less than or equal to/greater than or equal to*) q_{sys}/T_{sys}.

7. The mathematical statement of the second law of thermodynamics is _____ _____ where _____ _____.

8. The total entropy for a spontaneous process must always be (*positive/negative*).

9. Entropy is conserved in a spontaneous process. *True/False*

10. The units of entropy are _____.

11. The entropy of the universe is constantly (*increasing/decreasing*).

12. In any process, all the energy as heat can be used in the performance of work. *True/False*

13. Entropy arises from _____ disorder and _____ disorder.

14. The entropy of liquid water increases when the temperature is raised from 25°C to 50°C at constant pressure because _____ _____.

15. The entropy of water vapor (*increases/decreases*) when the volume of the gas is increased from 0.50 L to 1.0 L at 200°C.

16. The entropy of a substance increases upon melting because _____ _____ .

17. The entropy of a substance increases upon vaporization because _____ _____ .

18. The value of the molar entropy change of fusion is given by $\Delta S_{fus} =$ _____ .

19. The value of the molar entropy change of vaporization is given by $\Delta S_{vap} =$ _____ .

20. The molar entropy of $H_2O(l)$ is (*greater than/less than/the same as*) the molar entropy of $H_2O(g)$ at the same temperature and pressure.

21. The molar entropy (*decreases/increases*) with increasing molar mass.

22. The molar entropy depends upon the structure of the molecule. *True/False*

23. The standard entropy change for a reaction can be calculated from the _____ _____ of the reactants and the products.

24. The standard molar entropy of an element under standard conditions is equal to zero. *True/False*

25. All processes for which the change in entropy is positive are spontaneous. *True/False*

26. The sign of _____ determines whether a reaction is spontaneous.

27. Isothermal processes that lead to a decrease in the Gibbs free energy are spontaneous processes. *True/False*

28. If $\Delta G_{rxn} < 0$, then the reaction is (*spontaneous/not spontaneous*).

29. If $\Delta G_{rxn} > 0$, then the reaction is (*spontaneous/not spontaneous*).

30. If $\Delta G_{rxn} = 0$, then the reaction is _____ .

31. If a spontaneous reaction is endothermic, then the change in entropy for the reaction must be (*positive/negative*).

32. The value of the Gibbs free energy change of a reaction run at constant temperature is related to the changes in enthalpy and entropy of the reaction by the equation $\Delta G_{rxn} =$ _____ .

33. A spontaneous endothermic reaction is _____ driven.

34. A spontaneous reaction for which the entropy change is negative is ——————— driven.

35. If the value of ΔG_{rxn} is 123 kJ, then the reaction (*is/is not*) spontaneous.

36. All reactions with $\Delta H_{rxn} < 0$ and $\Delta S_{rxn} > 0$ are (*spontaneous/not spontaneous*).

37. All reactions with $\Delta H_{rxn} > 0$ and $\Delta S_{rxn} < 0$ are (*spontaneous/not spontaneous*).

38. The value of ΔG_{rxn} is independent of temperature. *True/False*

39. The maximum amount of work that can be obtained from a reaction is equal to

——————————————————— .

40. The relation between ΔG_{rxn} and Q/K is $\Delta G_{rxn} = $ ———————— .

41. A reaction for which $Q/K < 1$ (*is/is not*) spontaneous.

42. The standard Gibbs free energy change of a reaction is the Gibbs free energy change of the reaction when ———————————————— .

43. The standard Gibbs free energy change of a reaction is related to the equilibrium constant of the reaction by the equation $\Delta G_{rxn}^\circ = $ ———————————— .

44. Tables of standard molar Gibbs free energies of formation of compounds can be used to calculate values of ΔG_{rxn}°. *True/False*

45. The value of ΔG_{rxn}° for the dissociation of acetic acid is 27.1 kJ. The reaction is not spontaneous under any conditions. *True/False*

46. The value of ΔG_{rxn} for a reaction is -342 kJ. The reaction must proceed rapidly toward equilibrium. *True/False*

47. The value of ΔG_f° for $O_2(g)$ at 25°C is 0 kJ·mol^{-1}. *True/False*

48. The value of ΔH_f° for $O_2(g)$ at 25°C is 0 kJ·mol^{-1}. *True/False*

49. The value of S° for $O_2(g)$ at 25°C is 0 J·K^{-1}·mol^{-1}. *True/False*

50. The value of the equilibrium constant for a reaction does not change when the temperature increases. *True/False*

51. The van't Hoff equation governs the temperature dependence of ΔH_{rxn}°. *True/False*

52. The van't Hoff equation is

53. For an exothermic reaction, the value of the equilibrium constant (*increases/decreases*) with increasing temperature.

54. The Clapeyron-Clausius equation is given by

55. The value of the enthalpy of vaporization for a substance can be determined by measuring the equilibrium vapor pressure of the substance at two different temperatures. *True/False*

56. The bond enthalpy of a chemical bond can be determined experimentally. *True/False*

57. The bond enthalpy of a chemical bond is identical from compound to compound. *True/False*

58. The enthalpy change for a reaction can be estimated from the bond enthalpies of the chemical bonds in the product and reactant compounds. *True/False*

59. An input of energy is required to break a chemical bond. *True/False*

60. An input of energy is required to form a chemical bond. *True/False*

C. CALCULATIONS YOU SHOULD KNOW HOW TO DO

1. Calculate the values of ΔS_{fus} and ΔS_{vap} by using Equations (22-2) and (22-3). See Example 22-2 and Problems 22-1 through 22-5.

2. Calculate ΔS°_{rxn} from S° values of the reactants and products given in Table 22-1 and Appendix F. See Problems 22-19 through 22-22.

3. Calculate ΔG_{rxn} by using the equation $\Delta G_{rxn} = \Delta H_{rxn} - T \Delta S_{rxn}$. See Problems 22-25 through 22-28.

4. Calculate ΔG_{rxn} and ΔG°_{rxn} given the equilibrium constant by using the equations $\Delta G^\circ_{rxn} = -RT \ln K$ and $\Delta G_{rxn} = \Delta G^\circ_{rxn} + RT \ln Q$. See Examples 22-5 and 22-6 and Problems 22-29 through 22-40.

5. Calculate ΔG°_{rxn} by using tabulated ΔG°_f values of the reactants and products given in Table 22-1 and Appendix F. See Examples 22-7 and 22-8 and Problems 22-41 through 22-50.

6. Use the van't Hoff equation to calculate the value of the equilibrium constant at some other temperature. See Example 22-9 and Problems 22-55 and 22-56.

7. Use the van't Hoff equation to calculate the value of ΔH°_{rxn}. See Problems 22-51 through 22-54.

8. Use the Clapeyron-Clausius equation to calculate the value of the equilibrium vapor pressure at some other temperature. See Example 22-10 and Problems 22-57 through 22-60, 22-63, and 22-64.

9. Use the Clapeyron-Clausius equation to calculate the enthalpy of vaporization, ΔH_{vap}. See Problems 22-61 and 22-62.

10. Calculate the average bond enthalpy from the enthalpy change for a reaction. See Problems 22-65, 22-66, and 22-69 through 22-72.

11. Use the molar bond enthalpies given in Table 22-2 and Equation (22-18) to estimate ΔH°_{rxn} for a reaction. See Example 22-11 and Problems 22-67 through 22-70.

D. SOLUTIONS TO THE ODD-NUMBERED PROBLEMS

22-1 The values of ΔS_{fus} and ΔS_{vap} are given by

$$\Delta S_{fus} = \frac{\Delta H_{fus}}{T_m} \qquad \Delta S_{vap} = \frac{\Delta H_{vap}}{T_b}$$

The T_m and T_b are the melting and boiling points, respectively, on the Kelvin temperature scale. Thus, we have for CH_4

$$\Delta S_{fus} = \frac{937.0 \text{ J} \cdot \text{mol}^{-1}}{90.7 \text{ K}} = 10.3 \text{ J} \cdot \text{K}^{-1} \cdot \text{mol}^{-1}$$

$$\Delta S_{vap} = \frac{8907 \text{ J} \cdot \text{mol}^{-1}}{109.2 \text{ K}} = 81.57 \text{ J} \cdot \text{K}^{-1} \cdot \text{mol}^{-1}$$

For C_2H_6

$$\Delta S_{fus} = \frac{2859 \text{ J} \cdot \text{mol}^{-1}}{89.9 \text{ K}} = 31.8 \text{ J} \cdot \text{K}^{-1} \cdot \text{mol}^{-1}$$

$$\Delta S_{vap} = \frac{15.65 \times 10^3 \text{ J} \cdot \text{mol}^{-1}}{184.6 \text{ K}} = 84.78 \text{ J} \cdot \text{K}^{-1} \cdot \text{mol}^{-1}$$

For C_3H_8

$$\Delta S_{fus} = \frac{3525 \text{ J} \cdot \text{mol}^{-1}}{91.5 \text{ K}} = 38.5 \text{ J} \cdot \text{K}^{-1} \cdot \text{mol}^{-1}$$

$$\Delta S_{vap} = \frac{20.13 \times 10^3 \text{ J} \cdot \text{mol}^{-1}}{231.1 \text{ K}} = 87.11 \text{ J} \cdot \text{K}^{-1} \cdot \text{mol}^{-1}$$

22-3 We calculate ΔS_{fus} by using the relationship

$$\Delta S_{fus} = \frac{\Delta H_{fus}}{T_m}$$

Using the values of T_m and ΔH_{fus} given, we have

$$CH_3OH \qquad \Delta S_{fus} = \frac{3.177 \times 10^3 \text{ J}\cdot\text{mol}^{-1}}{175.4 \text{ K}} = 18.11 \text{ J}\cdot\text{K}^{-1}\cdot\text{mol}^{-1}$$

$$C_2H_5OH \qquad \Delta S_{fus} = \frac{5.021 \times 10^3 \text{ J}\cdot\text{mol}^{-1}}{158.7 \text{ K}} = 31.64 \text{ J}\cdot\text{K}^{-1}\cdot\text{mol}^{-1}$$

$$C_3H_7OH \qquad \Delta S_{fus} = \frac{5.195 \times 10^3 \text{ J}\cdot\text{mol}^{-1}}{147.1 \text{ K}} = 35.32 \text{ J}\cdot\text{K}^{-1}\cdot\text{mol}^{-1}$$

We calculate ΔS_{vap} by using the relationship

$$\Delta S_{vap} = \frac{\Delta H_{vap}}{T_b}$$

Using the values of T_b and ΔH_{fus} given, we have

$$CH_3OH \qquad \Delta S_{vap} = \frac{37.57 \times 10^3 \text{ J}\cdot\text{mol}^{-1}}{338.2 \text{ K}} = 111.1 \text{ J}\cdot\text{K}^{-1}\cdot\text{mol}^{-1}$$

$$C_2H_5OH \qquad \Delta S_{vap} = \frac{40.48 \times 10^3 \text{ J}\cdot\text{mol}^{-1}}{351.9 \text{ K}} = 115.0 \text{ J}\cdot\text{K}^{-1}\cdot\text{mol}^{-1}$$

$$C_3H_7OH \qquad \Delta S_{vap} = \frac{43.60 \times 10^3 \text{ J}\cdot\text{mol}^{-1}}{370.6 \text{ K}} = 117.6 \text{ J}\cdot\text{K}^{-1}\cdot\text{mol}^{-1}$$

22-5 $\qquad$ $H_2S \qquad \Delta S_{fus} = \dfrac{\Delta H_{fus}}{T_m} = \dfrac{2.38 \times 10^3 \text{ J}\cdot\text{mol}^{-1}}{187.6 \text{ K}} = 12.7 \text{ J}\cdot\text{K}^{-1}\cdot\text{mol}^{-1}$

$$\Delta S_{vap} = \frac{\Delta H_{vap}}{T_b} = \frac{18.7 \times 10^3 \text{ J}\cdot\text{mol}^{-1}}{212.5 \text{ K}} = 88.0 \text{ J}\cdot\text{K}^{-1}\cdot\text{mol}^{-1}$$

From Example 22-2 we have for H_2O

$$\Delta S_{fus} = 22.1 \text{ J}\cdot\text{K}^{-1}\cdot\text{mol}^{-1}$$
$$\Delta S_{vap} = 109 \text{ J}\cdot\text{K}^{-1}\cdot\text{mol}^{-1}$$

The larger values of ΔS_{fus} and ΔS_{vap} for H_2O are a result of the breaking of hydrogen bonds in the processes solid $\rightarrow$ liquid and liquid $\rightarrow$ gas. The hydrogen bonds make for a higher degree of order, and thus their breaking produces a greater increase in disorder.

22-7 $\quad$ (a) $\quad$ The mass of D_2O is greater than the mass of H_2O. Thus, we predict for the standard molar entropies of the gases

$$S°(H_2O) < S°(D_2O)$$

(b) Because of its ring structure, ethylene oxide has less freedom of movement than ethanol. Thus, we predict that

$$S°(\text{ethylene oxide}) < S°(\text{ethanol})$$

(c) Because of its ring structure, pyrrolidine has less freedom of movement than butyl amine. Thus, we predict that

$$S°(\text{pyrrolidine}) < S°(\text{butyl amine})$$

22-9 The molecular masses are, approximately, for CH_3Cl, 50.5; for CH_4, 16.0; and for CH_3OH, 32. The structures and numbers of atoms are about the same; thus, mass is the dominant factor, and we predict

$$S°(CH_4) < S°(CH_3OH) < S°(CH_3Cl)$$

22-11 The compound $Fe_3O_4(s)$ contains more atoms and has a greater mass than $Fe_2O_3(s)$. Thus, we would expect that

$$S°(Fe_2O_3) < S°(Fe_3O_4)$$

22-13 Bromine molecules in $Br_2(l)$ are much more restricted in movement than bromine molecules in $Br_2(g)$, which are free to move through the gas. The positional disorder in the gaseous state is greater than in the liquid state; thus, the entropy of bromine is greater in the gaseous state than in the liquid state at a given temperature and pressure.

22-15 (a) The argon atoms have a greater freedom of movement in the gaseous state. The entropy will increase.
(b) The oxygen atoms have a greater freedom of movement at lower pressure. The entropy will increase.
(c) Copper atoms have a greater thermal disorder at a higher temperature. The entropy will increase.
(d) The CO_2 molecules have a greater freedom of movement in the gaseous state. The entropy will decrease.

22-17 (a) We have the same number of moles of gaseous reactant and gaseous product ($\Delta n = 0$).
(b) We have 2 mol of gaseous reactants and no moles of gaseous products ($\Delta n = -2$).
(c) We have 4 mol of gaseous reactants and 2 mol of gaseous products ($\Delta n = -2$).
(d) We have 1 mol of gaseous reactant and 2 mol of gaseous products ($\Delta n = +1$).

The value of $\Delta S°_{rxn}$ increases as the net change in the number of moles of gas increases; thus,

$$\Delta S°_{rxn}(c) \approx \Delta S°_{rxn}(b) < \Delta S°_{rxn}(a) < \Delta S°_{rxn}(d)$$

22-19   The value of $\Delta S^\circ_{\text{rxn}}$ is given by

$$\Delta S^\circ_{\text{rxn}} = S^\circ(\text{products}) - S^\circ(\text{reactants})$$

(a) $\Delta S^\circ_{\text{rxn}} = 4S^\circ[\text{NO}_2(g)] + 6S^\circ[\text{H}_2\text{O}(g)] - 4S^\circ[\text{NH}_3(g)] - 7S^\circ[\text{O}_2(g)]$
$\Delta S^\circ_{\text{rxn}} = (4 \text{ mol})(240.4 \text{ J} \cdot \text{K}^{-1} \cdot \text{mol}^{-1}) + (6 \text{ mol})(188.7 \text{ J} \cdot \text{K}^{-1} \cdot \text{mol}^{-1})$
$\quad - (4 \text{ mol})(192.5 \text{ J} \cdot \text{K}^{-1} \cdot \text{mol}^{-1}) - (7 \text{ mol})(205.0 \text{ J} \cdot \text{K}^{-1} \cdot \text{mol}^{-1})$
$= -111.2 \text{ J} \cdot \text{K}^{-1}$

(b) $\Delta S^\circ_{\text{rxn}} = S^\circ[\text{CH}_3\text{OH}(l)] - S^\circ[\text{CO}(g)] - 2S^\circ[\text{H}_2(g)]$
$= (1 \text{ mol})(126.9 \text{ J} \cdot \text{K}^{-1} \cdot \text{mol}^{-1}) - (1 \text{ mol})(197.8 \text{ J} \cdot \text{K}^{-1} \cdot \text{mol}^{-1})$
$\quad - (2 \text{ mol})(130.6 \text{ J} \cdot \text{K}^{-1} \cdot \text{mol}^{-1})$
$= -332.1 \text{ J} \cdot \text{K}^{-1}$

(c) $\Delta S^\circ_{\text{rxn}} = S^\circ[\text{CO}(g)] + S^\circ[\text{H}_2(g)] - S^\circ[\text{C}(s, \text{graphite})] - S^\circ[\text{H}_2\text{O}(g)]$
$= (1 \text{ mol})(197.8 \text{ J} \cdot \text{K}^{-1} \cdot \text{mol}^{-1}) + (1 \text{ mol})(130.6 \text{ J} \cdot \text{K}^{-1} \cdot \text{mol}^{-1})$
$\quad - (1 \text{ mol})(5.74 \text{ J} \cdot \text{K}^{-1} \cdot \text{mol}^{-1}) - (1 \text{ mol})(188.7 \text{ J} \cdot \text{K}^{-1} \cdot \text{mol}^{-1})$
$= 134.0 \text{ J} \cdot \text{K}^{-1}$

(d) $\Delta S^\circ_{\text{rxn}} = 2S^\circ[\text{CO}_2(g)] - 2S^\circ[\text{CO}(g)] - S^\circ[\text{O}_2(g)]$
$= (2 \text{ mol})(213.6 \text{ J} \cdot \text{K}^{-1} \cdot \text{mol}^{-1}) - (2 \text{ mol})(197.8 \text{ J} \cdot \text{K}^{-1} \cdot \text{mol}^{-1})$
$\quad - (1 \text{ mol})(205.0 \text{ J} \cdot \text{K}^{-1} \cdot \text{mol}^{-1})$
$= -173.4 \text{ J} \cdot \text{K}^{-1}$

22-21   The value of $\Delta S^\circ_{\text{rxn}}$ is given by

$$\Delta S^\circ_{\text{rxn}} = S^\circ(\text{products}) - S^\circ(\text{reactants})$$

(a) $\Delta S^\circ_{\text{rxn}} = S^\circ[\text{CO}_2(g)] - S^\circ[\text{C}(s, \text{graphite})] - S^\circ[\text{O}_2(g)]$
$= (1 \text{ mol})(213.6 \text{ J} \cdot \text{K}^{-1} \cdot \text{mol}^{-1}) - (1 \text{ mol})(5.74 \text{ J} \cdot \text{K}^{-1} \cdot \text{mol}^{-1})$
$\quad - (1 \text{ mol})(205.0 \text{ J} \cdot \text{K}^{-1} \cdot \text{mol}^{-1})$
$= 2.9 \text{ J} \cdot \text{K}^{-1}$

(b) $\Delta S^\circ_{\text{rxn}} = 2S^\circ[\text{SO}_3(g)] - 2S^\circ[\text{SO}_2(g)] - S^\circ[\text{O}_2(g)]$
$= (2 \text{ mol})(256.3 \text{ J} \cdot \text{K}^{-1} \cdot \text{mol}^{-1}) - (2 \text{ mol})(248.4 \text{ J} \cdot \text{K}^{-1} \cdot \text{mol}^{-1})$
$\quad - (1 \text{ mol})(205.0 \text{ J} \cdot \text{K}^{-1} \cdot \text{mol}^{-1})$
$= -189.2 \text{ J} \cdot \text{K}^{-1}$

(c) $\Delta S^\circ_{\text{rxn}} = S^\circ[\text{CO}_2(g)] + 2S^\circ[\text{H}_2\text{O}(l)] - S^\circ[\text{CH}_4(g)] - 2S^\circ[\text{O}_2(g)]$
$= (1 \text{ mol})(213.6 \text{ J} \cdot \text{K}^{-1} \cdot \text{mol}^{-1}) + (2 \text{ mol})(69.9 \text{ J} \cdot \text{K}^{-1} \cdot \text{mol}^{-1})$
$\quad - (1 \text{ mol})(186.2 \text{ J} \cdot \text{K}^{-1} \cdot \text{mol}^{-1}) - (2 \text{ mol})(205.0 \text{ J} \cdot \text{K}^{-1} \cdot \text{mol}^{-1})$
$= -242.8 \text{ J} \cdot \text{K}^{-1}$

(d) $\Delta S^\circ_{\text{rxn}} = S^\circ[\text{C}_2\text{H}_4(g)] - S^\circ[\text{C}_2\text{H}_2(g)] - S^\circ[\text{H}_2(g)]$
$= (1 \text{ mol})(219.6 \text{ J} \cdot \text{K}^{-1} \cdot \text{mol}^{-1}) - (1 \text{ mol})(200.8 \text{ J} \cdot \text{K}^{-1} \cdot \text{mol}^{-1})$
$\quad - (1 \text{ mol})(130.6 \text{ J} \cdot \text{K}^{-1} \cdot \text{mol}^{-1})$
$= -111.8 \text{ J} \cdot \text{K}^{-1}$

22-23   The reaction is spontaneous because water spontaneously evaporates if the pressure of the water vapor is less than $P^\circ_{\text{H}_2\text{O}}$. Because the reaction is spontaneous, the sign of ΔG_{rxn} is negative. We learned in Chapter 8 that it requires energy to vaporize a liquid; thus, the sign of ΔH_{rxn} is positive. The sign of ΔS_{vap} is also positive because the entropy increases upon vaporization. Thus, vaporization is entropy-driven, because

$$\Delta H_{rxn} - T \Delta S_{rxn} < 0$$

and the negative $-T \Delta S_{rxn}$ term offsets the positive ΔH_{rxn} term.

22-25   The value of ΔS_{rxn}° is given by

$$
\begin{aligned}
\Delta S_{rxn}^{\circ} &= S^{\circ}(\text{products}) - S^{\circ}(\text{reactants}) = S^{\circ}[C_6H_6(l)] - 3S^{\circ}[C_2H_2(g)] \\
&= (1 \text{ mol})(172.8 \text{ J} \cdot \text{K}^{-1} \cdot \text{mol}^{-1}) - (3 \text{ mol})(200.8 \text{ J} \cdot \text{K}^{-1} \cdot \text{mol}^{-1}) \\
&= -429.6 \text{ J} \cdot \text{K}^{-1}
\end{aligned}
$$

We can calculate ΔG_{rxn}° by using the relationship

$$
\begin{aligned}
\Delta G_{rxn}^{\circ} &= \Delta H_{rxn}^{\circ} - T \Delta S_{rxn}^{\circ} \\
&= -631 \text{ kJ} - (298 \text{ K})(-429.6 \text{ J} \cdot \text{K}^{-1})\left(\frac{1 \text{ kJ}}{1000 \text{ J}}\right) = -503 \text{ kJ}
\end{aligned}
$$

The reaction is spontaneous in the direction

$$3C_2H_2(g) \longrightarrow C_6H_6(l)$$

when both $C_2H_2(g)$ and $C_6H_6(l)$ are at 1 atm and 25°C.

22-27 $\Delta S_{rxn}^{\circ} = 2S^{\circ}[CO_2(g)] + 2S^{\circ}[H_2O(g)] - S^{\circ}[C_2H_4(g)] - 3S^{\circ}[O_2(g)]$

$$
\begin{aligned}
&= (2 \text{ mol})(213.6 \text{ J} \cdot \text{K}^{-1} \cdot \text{mol}^{-1}) + (2 \text{ mol})(188.7 \text{ J} \cdot \text{K}^{-1} \cdot \text{mol}^{-1}) \\
&\quad - (1 \text{ mol})(219.6 \text{ J} \cdot \text{K}^{-1} \cdot \text{mol}^{-1}) - (3 \text{ mol})(205.0 \text{ J} \cdot \text{K}^{-1} \cdot \text{mol}^{-1}) \\
&= -30.0 \text{ J} \cdot \text{K}^{-1}
\end{aligned}
$$

We can calculate ΔG_{rxn}° by using

$$
\begin{aligned}
\Delta G_{rxn}^{\circ} &= \Delta H_{rxn}^{\circ} - T \Delta S_{rxn}^{\circ} \\
&= -1323 \text{ kJ} - (298 \text{ K})(-30.0 \text{ J} \cdot \text{K}^{-1})\left(\frac{1 \text{ kJ}}{1000 \text{ J}}\right) = -1314 \text{ kJ}
\end{aligned}
$$

Thus, when all the reactants and products are at standard conditions, the reaction is spontaneous in the direction

$$C_2H_4(g) + 3O_2(g) \longrightarrow 2CO_2(g) + 2H_2O(g)$$

The value of ΔG_{rxn} is given by

$$\Delta G_{rxn} = \Delta G_{rxn}^{\circ} + RT \ln Q$$

where

$$Q = \frac{P_{CO_2}^2 P_{H_2O}^2}{P_{C_2H_4} P_{O_2}^3}$$

Substituting in the values for the pressures yields

$$Q = \frac{(20 \text{ atm})^2 (0.010 \text{ atm})^2}{(0.010 \text{ atm})(0.020 \text{ atm})^2} = 5.0 \times 10^5$$

Thus,

$$\Delta G_{rxn} = -1314 \text{ kJ} + \frac{(8.314 \text{ J} \cdot \text{K}^{-1})(298 \text{ K})}{(1000 \text{ J} \cdot \text{kJ}^{-1})} \ln(5.0 \times 10^5)$$

$$= -1314 \text{ kJ} + 32.5 \text{ kJ} = -1281 \text{ kJ}$$

The value of ΔG_{rxn} is negative, and thus the reaction is spontaneous from left to right.

22-29 The relation between ΔG_{rxn} and ΔG_{rxn}° is

$$\Delta G_{rxn} = \Delta G_{rxn}^\circ + RT \ln Q$$

For the equation

$$\text{ATP}(aq) + \text{H}_2\text{O}(l) \rightleftharpoons \text{ADP}(aq) + \text{HPO}_4^{2-}(aq)$$

we have

$$Q = \frac{[\text{ADP}][\text{HPO}_4^{2-}]}{[\text{ATP}]} = \frac{(0.50 \times 10^{-3} \text{ M})(5.0 \times 10^{-3} \text{ M})}{(5.0 \times 10^{-3} \text{ M})}$$
$$= 0.50 \times 10^{-3} \text{ M}$$

Thus,

$$\Delta G_{rxn} = -30.5 \text{ kJ} + \frac{(8.314 \text{ J} \cdot \text{K}^{-1})(310 \text{ K})}{(1000 \text{ J} \cdot \text{kJ}^{-1})} \ln (0.50 \times 10^{-3})$$
$$= -30.5 \text{ kJ} - 19.6 \text{ kJ} = -50.1 \text{ kJ}$$

Because $\Delta G_{rxn} < 0$, the reaction is spontaneous from left to right at the stated conditions.

22-31 We have

$$\Delta G_{rxn}^\circ = -RT \ln K$$

At 250°C, we have

$$\Delta G_{rxn}^\circ = -(8.314 \text{ J} \cdot \text{K}^{-1})(523 \text{ K}) \ln (4.5 \times 10^3)$$
$$= -36.5 \times 10^3 \text{ J} = -36.5 \text{ kJ}$$

The reaction is spontaneous in the direction

$$\text{PCl}_5(g) \longrightarrow \text{PCl}_3(g) + \text{Cl}_2(g)$$

when PCl_5, PCl_3, and Cl_2 are at standard conditions. The value of ΔG_{rxn} at other conditions is given by

$$\Delta G_{rxn} = \Delta G^{\circ}_{rxn} + RT \ln Q$$

where

$$Q = \frac{P_{PCl_3} P_{Cl_2}}{P_{PCl_5}} = \frac{(0.20 \text{ atm})(0.80 \text{ atm})}{1.0 \times 10^{-6} \text{ atm}} = 1.6 \times 10^5 \text{ atm}$$

Thus, we have

$$\Delta G_{rxn} = -36.5 \text{ KJ} + \frac{(8.314 \text{ J} \cdot \text{K}^{-1})(523 \text{ K})}{1000 \text{ J} \cdot \text{kJ}^{-1}} \ln(1.6 \times 10^5)$$
$$= -36.5 \text{ kJ} + 52.1 \text{ kJ} = +15.6 \text{ kJ}$$

Because $\Delta G_{rxn} > 0$, the reaction to the right is not spontaneous under these conditions. The reaction from right to left is spontaneous.

22-33 The value of ΔG°_{rxn} is given by

$$\Delta G^{\circ}_{rxn} = -RT \ln K$$
$$= -(8.314 \text{ J} \cdot \text{K}^{-1})(298 \text{ K}) \ln (4.5 \times 10^{-4})$$
$$= 1.91 \times 10^4 \text{ J} = 19.1 \text{ kJ}$$

Because $\Delta G^{\circ}_{rxn} > 0$, nitrous acid will not dissociate spontaneously when $[NO_2^-] = [H^+] = [HNO_2] = 1.00$ M (standard conditions). The value of ΔG_{rxn} at any other conditions is given by

$$\Delta G_{rxn} = RT \ln \left(\frac{Q}{K}\right)$$

where,

$$Q = \frac{[NO_2^-][H^+]}{[HNO_2]} = \frac{(1.0 \times 10^{-5} \text{ M})(1.0 \times 10^{-5} \text{ M})}{1.0 \text{ M}} = 1.0 \times 10^{-10} \text{ M}$$

$$\Delta G_{rxn} = (8.314 \text{ J} \cdot \text{K}^{-1})(298 \text{ K}) \ln \left(\frac{1.0 \times 10^{-10} \text{ M}}{4.5 \times 10^{-4} \text{ M}}\right)$$

$$= -3.80 \times 10^4 \text{ J}$$

Because $\Delta G_{rxn} < 0$, nitrous acid will dissociate spontaneously under these conditions.

22-35 The value of ΔG°_{rxn} is given by

$$\Delta G^{\circ}_{rxn} = RT \ln K$$
$$= -(8.314 \text{ J} \cdot \text{K}^{-1})(298 \text{ K}) \ln (1.35 \times 10^{-3})$$
$$= 1.64 \times 10^4 \text{ J}$$

Chloroacetic acid will not dissociate spontaneously when $[CH_2ClCOO^-] =$ $[H^+] = [CH_2ClCOOH] = 1.0$ M. The value of ΔG_{rxn} at any other conditions is given by

$$\Delta G_{rxn} = RT \ln \left(\frac{Q}{K} \right)$$

where,

$$Q = \frac{[CH_2ClCOO^-][H^+]}{[CH_2ClCOOH]} = \frac{(0.0010 \text{ M})(1.0 \times 10^{-5} \text{ M})}{0.10 \text{ M}} = 1.0 \times 10^{-7} \text{ M}$$

$$\Delta G_{rxn} = (8.314 \text{ J} \cdot \text{K}^{-1})(298 \text{ K}) \ln \left(\frac{1.0 \times 10^{-7} \text{ M}}{1.35 \times 10^{-3} \text{ M}} \right)$$

$$= -2.36 \times 10^4 \text{ J}$$

Chloroacetic acid will dissociate spontaneously under these conditions.

22-37 The value of ΔG_{rxn}° is given by

$$\Delta G_{rxn}^\circ = -RT \ln K$$

Thus,

$$\Delta G_{rxn}^\circ = -(8.314 \text{ J} \cdot \text{K}^{-1})(298 \text{ K}) \ln (1.78 \times 10^{-10})$$
$$= 5.56 \times 10^4 \text{ J} = 55.6 \text{ kJ}$$

The reaction is spontaneous from right to left when $[Ag^+] = [Cl^-] = 1.0$ M (standard conditions). Therefore, it is not possible to prepare a solution that is 1.0 M in $Ag^+(aq)$ and $Cl^-(aq)$. Insoluble $AgCl(s)$ will precipitate out of the solution.

22-39 The value of ΔG_{rxn}° is given by

$$\Delta G_{rxn}^\circ = -RT \ln K$$

Thus,

$$\Delta G_{rxn}^\circ = -(8.314 \text{ J} \cdot \text{K}^{-1})(298 \text{ K}) \ln (2.5 \times 10^3)$$
$$= -1.94 \times 10^4 \text{ J}$$

The reaction is spontaneous from left to right when $Ag^+(aq)$, $NH_3(aq)$, and $Ag(NH_3)_2^+(aq)$ are at standard conditions. The value of ΔG_{rxn} at any other condition is given by

$$\Delta G_{rxn} = RT \ln \left(\frac{Q}{K} \right)$$

where

$$Q = \frac{[Ag(NH_3)_2^+]}{[Ag^+][NH_3]^2} = \frac{1.0 \times 10^{-3} \text{ M}}{(1.0 \times 10^{-3} \text{ M})(0.10 \text{ M})^2} = 100 \text{ M}^{-2}$$

Thus,

$$\Delta G_{rxn} = (8.314 \text{ J} \cdot \text{K}^{-1})(298 \text{ K}) \ln \left(\frac{100 \text{ M}^{-2}}{2.5 \times 10^3 \text{ M}^{-2}} \right)$$

$$= -7.97 \times 10^3 \text{ J}$$

The reaction is spontaneous left to right.

22-41 The value of ΔG_{rxn}° is calculated by using the data in Table 22-1 and the relationship

$$\Delta G_{rxn}^{\circ} = \Delta G_f^{\circ} (\text{products}) - \Delta G_f^{\circ} (\text{reactants})$$

(a) $\Delta G_{rxn}^{\circ} = \Delta G_f^{\circ} (\text{CH}_3\text{OH}(l)) - \Delta G_f^{\circ} [\text{CO}(g)] - 2 \Delta G_f^{\circ} [\text{H}_2(g)]$
$= (1 \text{ mol})(-166.3 \text{ kJ} \cdot \text{mol}^{-1}) - (1 \text{ mol})(-137.2 \text{ kJ} \cdot \text{mol}^{-1})$
$\quad -(2 \text{ mol})(0 \text{ kJ} \cdot \text{mol}^{-1})$
$= -29.1 \text{ kJ}$

Using the relationship of $\Delta G_{rxn}^{\circ} = -RT \ln K$, we have

$$\ln K = -\frac{\Delta G_{rxn}^{\circ}}{RT} = -\frac{(-29.1 \times 10^3 \text{ J})}{(8.314 \text{ J} \cdot \text{K}^{-1})(298 \text{ K})} = 11.75$$

Thus,

$$K = 1.26 \times 10^5 \text{ atm}^{-3}$$

(b) $\Delta G_{rxn}^{\circ} = \Delta G_f^{\circ} [\text{CO}(g)] + \Delta G_f^{\circ} [\text{H}_2(g)]$
$\quad - \Delta G_f^{\circ} [\text{C}(s, \text{ graphite})] - \Delta G_f^{\circ} [\text{H}_2\text{O}(g)]$
$= (1 \text{ mol})(-137.2 \text{ kJ} \cdot \text{mol}^{-1}) + (1 \text{ mol})(0 \text{ kJ} \cdot \text{mol}^{-1})$
$\quad -(1 \text{ mol})(0 \text{ kJ} \cdot \text{mol}^{-1}) - (1 \text{ mol})(-228.6 \text{ kJ} \cdot \text{mol}^{-1})$
$= 91.4 \text{ kJ}$

From the relationship of $\Delta G_{rxn}^{\circ} = -RT \ln K$, we have

$$\ln K = -\frac{\Delta G_{rxn}^{\circ}}{RT} = -\frac{(91.4 \times 10^3 \text{ J})}{(8.314 \text{ J} \cdot \text{K}^{-1})(298 \text{ K})} = -36.89$$

Thus,

$$K = 9.53 \times 10^{-17} \text{ atm}$$

(c) $\Delta G_{rxn}^{\circ} = \Delta G_f^{\circ} [\text{CH}_4(g)] + \Delta G_f^{\circ} [\text{H}_2\text{O}(g)] - \Delta G_f^{\circ} [\text{CO}(g)] - 3 \Delta G_f^{\circ} [\text{H}_2(g)]$
$= (1 \text{ mol})(-50.75 \text{ kJ} \cdot \text{mol}^{-1}) + (1 \text{ mol})(-228.6 \text{ kJ} \cdot \text{mol}^{-1})$
$\quad -(1 \text{ mol})(-137.2 \text{ kJ} \cdot \text{mol}^{-1}) - (3 \text{ mol})(0 \text{ kJ} \cdot \text{mol}^{-1})$
$= -142.2 \text{ kJ}$

From the relationship of $\Delta G_{rxn}^{\circ} = -RT \ln K$, we compute

$$\ln K = -\frac{\Delta G_{rxn}^{\circ}}{RT} = -\frac{(-142.2 \times 10^3 \text{ J})}{(8.314 \text{ J} \cdot \text{K}^{-1})(298 \text{ K})} = 57.39$$

$$K = 8.44 \times 10^{24} \text{ atm}^{-2}$$

22-43 The value of ΔG_{rxn}° is given by

$$\Delta G_{rxn}^{\circ} = \Delta G_f^{\circ}(\text{products}) - \Delta G_f^{\circ}(\text{reactants})$$

Thus,

$$\begin{aligned}
\Delta G_{rxn}^{\circ} &= 2 \Delta G_f^{\circ}[\text{HF}(g)] + \Delta G_f^{\circ}[\text{Cl}_2(g)] - 2 \Delta G_f^{\circ}[\text{HCl}(g)] - \Delta G_f^{\circ}[\text{F}_2(g)] \\
&= (2 \text{ mol})(-273 \text{ kJ} \cdot \text{mol}^{-1}) + (1 \text{ mol})(0 \text{ kJ} \cdot \text{mol}^{-1}) \\
&\quad -(2 \text{ mol})(-95.30 \text{ kJ} \cdot \text{mol}^{-1}) - (1 \text{ mol})(0 \text{ kJ} \cdot \text{mol}^{-1}) \\
&= -335 \text{ kJ}
\end{aligned}$$

The value of ΔH_{rxn}° is given by

$$\Delta H_{rxn}^{\circ} = \Delta H_f^{\circ}(\text{products}) - \Delta H_f^{\circ}(\text{reactants})$$

Thus,

$$\begin{aligned}
\Delta H_{rxn}^{\circ} &= 2 \Delta H_f^{\circ}[\text{HF}(g)] + \Delta H_f^{\circ}[\text{Cl}_2(g)] - 2 \Delta H_f^{\circ}[\text{HCl}(g)] - \Delta H_f^{\circ}[\text{F}_2(g)] \\
&= (2 \text{ mol})(-271.1 \text{ kJ} \cdot \text{mol}^{-1}) + (1 \text{ mol})(0 \text{ kJ} \cdot \text{mol}^{-1}) \\
&\quad -(2 \text{ mol})(-92.31 \text{ kJ} \cdot \text{mol}^{-1}) - (1 \text{ mol})(0 \text{ kJ} \cdot \text{mol}^{-1}) \\
&= -357.6 \text{ kJ}
\end{aligned}$$

From the relation $\Delta G_{rxn}^{\circ} = -RT \ln K$ we have

$$\ln K = -\frac{\Delta G_{rxn}^{\circ}}{RT} = -\frac{(-355 \times 10^3 \text{ J})}{(8.314 \text{ J} \cdot \text{K}^{-1})(298 \text{ K})} = 143.3$$

$$K = 1.7 \times 10^{62}$$

22-45 The value of ΔG_{rxn}° is given by

$$\begin{aligned}
\Delta G_{rxn}^{\circ} &= \Delta G_f^{\circ}(\text{products}) - \Delta G_f^{\circ}(\text{reactants}) \\
&= 2 \Delta G_f^{\circ}[\text{SO}_3(g)] - 2 \Delta G_f^{\circ}[\text{SO}_2(g)] - \Delta G_f^{\circ}[\text{O}_2(g)] \\
&= (2 \text{ mol})(-371.1 \text{ kJ} \cdot \text{mol}^{-1}) - (2 \text{ mol})(-300.2 \text{ kJ} \cdot \text{mol}^{-1}) \\
&\quad -(1 \text{ mol})(0 \text{ kJ} \cdot \text{mol}^{-1}) \\
&= -141.8 \text{ kJ}
\end{aligned}$$

The value of ΔH_{rxn}° is given by

$$\begin{aligned}
\Delta H_{rxn}^{\circ} &= \Delta H_f^{\circ}(\text{products}) - \Delta H_f^{\circ}(\text{reactants}) \\
&= 2 \Delta H_f^{\circ}[\text{SO}_3(g)] - 2 \Delta H_f^{\circ}[\text{SO}_2(g)] - \Delta H_f^{\circ}[\text{O}_2(g)] \\
&= (2 \text{ mol})(-395.7 \text{ kJ} \cdot \text{mol}^{-1}) - (2 \text{ mol})(-296.8 \text{ kJ} \cdot \text{mol}^{-1}) \\
&\quad -(1 \text{ mol})(0 \text{ kJ} \cdot \text{mol}^{-1}) \\
&= -197.8 \text{ kJ}
\end{aligned}$$

The value of $\ln K$ is given by

$$\ln K = -\frac{\Delta G^\circ_{\text{rxn}}}{RT} = -\frac{(-141.8 \times 10^3 \text{ J})}{(8.314 \text{ J} \cdot \text{K}^{-1})(298 \text{ K})} = 57.23$$

$$K = 7.18 \times 10^{24} \text{ atm}^{-1}$$

22-47   The value of $\Delta G^\circ_{\text{rxn}}$ is given by

$$\begin{aligned}
\Delta G^\circ_{\text{rxn}} &= \Delta G^\circ_f(\text{products}) - \Delta G^\circ_f(\text{reactants}) \\
&= \Delta G^\circ_f[\text{H}_2\text{O}(g)] + \Delta G^\circ_f[\text{CO}(g)] - \Delta G^\circ_f[\text{H}_2(g)] - \Delta G^\circ_f[\text{CO}_2(g)] \\
&= (1 \text{ mol})(-228.6 \text{ kJ} \cdot \text{mol}^{-1}) + (1 \text{ mol})(-137.2 \text{ kJ} \cdot \text{mol}^{-1}) \\
&\quad -(1 \text{ mol})(0 \text{ kJ} \cdot \text{mol}^{-1}) - (1 \text{ mol})(-394.4 \text{ kJ} \cdot \text{mol}^{-1}) \\
&= 28.6 \text{ kJ}
\end{aligned}$$

The value of $\Delta H^\circ_{\text{rxn}}$ is given by

$$\begin{aligned}
\Delta H^\circ_{\text{rxn}} &= \Delta H^\circ_f(\text{products}) - \Delta H^\circ_f(\text{reactants}) \\
&= \Delta H^\circ_f[\text{H}_2\text{O}(g)] + \Delta H^\circ_f[\text{CO}(g)] - \Delta H^\circ_f[\text{H}_2(g)] - \Delta H^\circ_f[\text{CO}_2(g)] \\
&= (1 \text{ mol})(-241.8 \text{ kJ} \cdot \text{mol}^{-1}) + (1 \text{ mol})(-110.5 \text{ kJ} \cdot \text{mol}^{-1}) \\
&\quad - (1 \text{ mol})(0 \text{ kJ} \cdot \text{mol}^{-1}) - (1 \text{ mol})(-393.5 \text{ kJ} \cdot \text{mol}^{-1}) \\
&= 41.2 \text{ kJ}
\end{aligned}$$

The value of $\Delta S^\circ_{\text{rxn}}$ is given by

$$\begin{aligned}
\Delta S^\circ_{\text{rxn}} &= S^\circ(\text{products}) - S^\circ(\text{reactants}) \\
&= S^\circ[\text{H}_2\text{O}(g)] + S^\circ[\text{CO}(g)] - S^\circ[\text{H}_2(g)] - S^\circ[\text{CO}_2(g)] \\
&= (1 \text{ mol})(188.7 \text{ J} \cdot \text{K}^{-1} \cdot \text{mol}^{-1}) + (1 \text{ mol})(197.8 \text{ J} \cdot \text{K}^{-1} \cdot \text{mol}^{-1}) \\
&\quad - (1 \text{ mol})(130.6 \text{ J} \cdot \text{K}^{-1} \cdot \text{mol}^{-1}) - (1 \text{ mol})(213.6 \text{ J} \cdot \text{K}^{-1} \cdot \text{mol}^{-1}) \\
&= 42.3 \text{ J} \cdot \text{K}^{-1}
\end{aligned}$$

Because $\Delta G^\circ_{\text{rxn}} = +28.6$ kJ, the reaction is spontaneous right to left when all the reactants and all the products are at standard conditions. The reaction is enthalpy-driven to the left.

22-49   The equation for the combustion of ethane is

$$\text{C}_2\text{H}_6(g) + \tfrac{7}{2}\text{O}_2(g) \longrightarrow 2\text{CO}_2(g) + 3\text{H}_2\text{O}(l)$$

The value of $\Delta G^\circ_{\text{rxn}}$ is given by

$$\begin{aligned}
\Delta G^\circ_{\text{rxn}} &= \Delta G^\circ_f(\text{products}) - \Delta G^\circ_f(\text{reactants}) \\
&= 2\Delta G^\circ_f[\text{CO}_2(g)] + 3\Delta G^\circ_f[\text{H}_2\text{O}(l)] - \Delta G^\circ_f[\text{C}_2\text{H}_6(g)] - \tfrac{7}{2}\Delta G^\circ_f[\text{O}_2(g)] \\
&= (2 \text{ mol})(-394.4 \text{ kJ} \cdot \text{mol}^{-1}) + (3 \text{ mol})(-237.2 \text{ kJ} \cdot \text{mol}^{-1}) \\
&\quad - (1 \text{ mol})(-32.89 \text{ kJ} \cdot \text{mol}^{-1}) - (\tfrac{7}{2} \text{ mol})(0 \text{ kJ} \cdot \text{mol}^{-1}) \\
&= -1468 \text{ kJ}
\end{aligned}$$

The maximum amount of work that can be obtained from the combustion of

1 mol of ethane when $CO_2(g)$, $H_2O(l)$, $C_2H_6(g)$, and $O_2(g)$ are at standard conditions is 1468 kJ.

22-51 The van't Hoff equation is

$$\ln\left(\frac{K_2}{K_1}\right) = \frac{\Delta H^\circ_{rxn}}{R}\left(\frac{T_2 - T_1}{T_1 T_2}\right)$$

Substituting the values of the second and fourth sets of data, for example, we have

$$\ln\left(\frac{16.9 \times 10^{-4}}{6.86 \times 10^{-4}}\right) = \frac{\Delta H^\circ_{rxn}(2300\ K - 2100\ K)}{(8.314\ J\cdot K^{-1})(2300\ K)(2100\ K)}$$

$$0.9016 = (4.980 \times 10^{-6}\ J^{-1})\Delta H^\circ_{rxn}$$

$$\Delta H^\circ_{rxn} = 1.81 \times 10^5\ J = 181\ kJ$$

22-53 The van't Hoff equation is

$$\ln\left(\frac{K_2}{K_1}\right) = \frac{\Delta H^\circ_{rxn}}{R}\left(\frac{T_2 - T_1}{T_1 T_2}\right)$$

Substituting the values of the last two sets of data, for example, we have

$$\ln\left(\frac{1.77}{1.34}\right) = \frac{\Delta H^\circ_{rxn}(1273\ K - 1173\ K)}{(8.314\ J\cdot K^{-1})(1273\ K)(1173\ K)}$$

$$0.2783 = (8.055 \times 10^{-6}\ J^{-1})\Delta H^\circ_{rxn}$$

Thus,

$$\Delta H^\circ_{rxn} = 3.46 \times 10^4\ J = 34.6\ kJ$$

22-55 The value of ΔH°_{rxn} is given by

$$\begin{aligned}\Delta H^\circ_{rxn} &= \Delta H^\circ_f[PCl_5(g)] - \Delta H^\circ_f[PCl_3(g)] - \Delta H^\circ_f[Cl_2(g)]\\ &= (1\ mol)(-375.0\ kJ\cdot mol^{-1}) - (1\ mol)(-306.4\ kJ\cdot mol^{-1})\\ &\quad - (1\ mol)(0\ kJ\cdot mol^{-1})\\ &= -68.6\ kJ\end{aligned}$$

Using the van't Hoff equation, we have

$$\ln\left(\frac{K_p}{0.562\ atm^{-1}}\right) = \frac{(-68.6 \times 10^3\ J)(673\ K - 523\ K)}{(8.314\ J\cdot K^{-1})(673\ K)(523\ K)} = -3.516$$

Thus, taking antilogarithms of both sides yields

$$\frac{K_p}{0.562\ atm^{-1}} = 2.971 \times 10^{-2}$$

and

$$K_p = 0.0167 \text{ atm}^{-1} \text{ at } 400°C$$

22-57 The Clapeyron-Clausius equation is

$$\ln\left(\frac{P_2}{P_1}\right) = \frac{\Delta H_{vap}}{R}\left(\frac{T_2 - T_1}{T_1 T_2}\right)$$

At the normal boiling point, $P = 760$ torr. Let these values be T_2 and P_2 and let $T_1 = 293.2$ K. Then

$$\ln\left(\frac{760 \text{ torr}}{P_1}\right) = \frac{(31.97 \times 10^3 \text{ J} \cdot \text{mol}^{-1})}{(8.314 \text{ J} \cdot \text{K}^{-1} \cdot \text{mol}^{-1})}\left[\frac{36.2 \text{ K}}{(293.2 \text{ K})(329.4 \text{ K})}\right]$$
$$= 1.441$$

or

$$\frac{760 \text{ torr}}{P_1} = e^{1.441} = 4.226$$

or

$$P_1 = \frac{760 \text{ torr}}{4.226} = 180 \text{ torr}$$

22-59 Let $P_1 = 387$ torr, $T_1 = 333.2$ K, $P_2 = 760$ torr, and write the Clapeyron-Clausius equation as

$$\ln\left(\frac{760 \text{ torr}}{387 \text{ torr}}\right) = \frac{(32.3 \times 10^3 \text{ J} \cdot \text{mol}^{-1})}{(8.314 \text{ J} \cdot \text{K}^{-1} \cdot \text{mol}^{-1})}\left[\frac{T_2 - 333.2 \text{ K}}{(333.2 \text{ K})(T_2)}\right]$$

or

$$\frac{T_2 - 333.2 \text{ K}}{T_2} = \frac{(8.314 \text{ J} \cdot \text{mol}^{-1} \cdot \text{K}^{-1})(333.2 \text{ K})}{(32.3 \times 10^3 \text{ J} \cdot \text{mol}^{-1})}\ln\left(\frac{760}{387}\right) = 0.05789$$

Thus,

$$T_2 - 333.2 \text{ K} = 0.05789 T_2$$
$$T_2 = \frac{333.2 \text{ K}}{0.9421} = 353.7 \text{ K} = 80.5°C$$

22-61 Let $P_1 = 92.68$ torr, $T_1 = 296.65$ K, $P_2 = 221.6$ torr, $T_2 = 318.15$ K, and write the Clapeyron-Clausius equation as

$$\ln\left(\frac{221.6 \text{ torr}}{92.68 \text{ torr}}\right) = \frac{\Delta H_{vap}}{(8.314 \text{ J} \cdot \text{K}^{-1} \cdot \text{mol}^{-1})}\left[\frac{318.15 \text{ K} - 296.65 \text{ K}}{(296.65 \text{ K})(318.15 \text{ K})}\right]$$

$$0.8717 = (2.740 \times 10^{-5} \text{ J}^{-1} \cdot \text{mol}) \Delta H_{vap}$$

Solving for ΔH_{vap} gives

$$\Delta H_{vap} = 3.18 \times 10^4 \text{ J} \cdot \text{mol}^{-1} = 31.8 \text{ kJ} \cdot \text{mol}^{-1}$$

22-63 Let P_2 be the vapor pressure of lead at $1300°C$ and P_1 be that at $500°C$. Then the Clapeyron-Clausius equation is

$$\ln\left(\frac{P_2}{P_1}\right) = \frac{(178 \times 10^3 \text{ J} \cdot \text{mol}^{-1})}{(8.314 \text{ J} \cdot \text{K}^{-1} \cdot \text{mol}^{-1})}\left[\frac{800 \text{ K}}{(773 \text{ K})(1573 \text{ K})}\right]$$
$$= 14.09$$

$$\frac{P_2}{P_1} = e^{14.09} = 1.31 \times 10^6$$

22-65 The reaction involves breaking three Cl—F bonds; thus,

$$\Delta H°_{rxn} \approx 3H(\text{Cl—F})$$

Given the value of $\Delta H°_{rxn}$, we can calculate $H(\text{Cl—F})$.

$$514 \text{ kJ} \approx (3 \text{ mol})H(\text{Cl—F})$$

$$H(\text{Cl—F}) \approx \frac{514 \text{ kJ}}{3 \text{ mol}} = 171 \text{ kJ} \cdot \text{mol}^{-1}$$

22-67 In the reaction

$$\text{CCl}_4(g) + 2\text{F}_2(g) \longrightarrow \text{CF}_4(g) + 2\text{Cl}_2(g)$$

we break four Cl—Cl bonds and two F—F bonds; we make four C—F bonds and two Cl—Cl bonds. The enthalpy required to break the bonds of the reactant molecules is given by

$$\begin{aligned} H(\text{bond})_R &= 4H(\text{C—Cl}) + 2H(\text{F—F}) \\ &= (4 \text{ mol})(331 \text{ kJ} \cdot \text{mol}^{-1}) + (2 \text{ mol})(155 \text{ kJ} \cdot \text{mol}^{-1}) \\ &= 1634 \text{ kJ} \end{aligned}$$

The enthalpy released upon the formation of the bonds in the product molecules is

$$\begin{aligned} H(\text{bond})_P &= 4H(\text{C—F}) + 2H(\text{Cl—Cl}) \\ &= (4 \text{ mol})(439 \text{ kJ} \cdot \text{mol}^{-1}) + (2 \text{ mol})(243 \text{ kJ} \cdot \text{mol}^{-1}) \\ &= 2242 \text{ kJ} \end{aligned}$$

The enthalpy change of the reaction is given by

$$\Delta H°_{rxn} \approx H(\text{bond})_R - H(\text{bond})_P = 1634 \text{ kJ} - 2242 \text{ kJ} = -608 \text{ kJ}$$

22-69 $\Delta H°_{rxn} \approx 2H(\text{H—H}) + H(\text{O—O}) - 4H(\text{O—H})$

The value of ΔH°_{rxn} is

$$\Delta H^\circ_{rxn} = 2\Delta H^\circ_f[H_2O(g)] = (2 \text{ mol})(-241.8 \text{ kJ}\cdot\text{mol}^{-1}) = -483.6 \text{ kJ}$$

Thus, we can write

$$\begin{aligned}
-483.6 \text{ kJ} &\approx 2H(\text{H}-\text{H}) + H(\text{O}-\text{O}) - 4H(\text{O}-\text{H}) \\
&= (2 \text{ mol})(435 \text{ kJ}\cdot\text{mol}^{-1}) + (1 \text{ mol})H(\text{O}-\text{O}) \\
&\quad - (4 \text{ mol})(464 \text{ kJ}\cdot\text{mol}^{-1}) \\
&= (1 \text{ mol})H(\text{O}-\text{O}) - 986 \text{ kJ}
\end{aligned}$$

$$H(\text{O}-\text{O}) \approx +502 \text{ kJ}\cdot\text{mol}^{-1}$$

22-71 The relevant equation is

$$CH_4(g) \longrightarrow C(g) + 4H(g)$$

We have that $\Delta H^\circ_{rxn} \approx 4H(\text{C}-\text{H})$ and

$$\begin{aligned}
\Delta H^\circ_{rxn} &= \Delta H^\circ_f[C(g)] + 4\Delta H^\circ_f[H(g)] - \Delta H^\circ_f[CH_4(g)] \\
&= (1 \text{ mol})(709 \text{ kJ}\cdot\text{mol}^{-1}) + (4 \text{ mol})(218 \text{ kJ}\cdot\text{mol}^{-1}) \\
&\quad - (1 \text{ mol})(-74.86 \text{ kJ}\cdot\text{mol}^{-1}) \\
&= 1656 \text{ kJ}
\end{aligned}$$

Therefore, we have that $4H(\text{C}-\text{H}) = 1656 \text{ kJ}$, or

$$H(\text{C}-\text{H}) = \frac{1656 \text{ kJ}}{4 \text{ mol}} = 414 \text{ kJ}\cdot\text{mol}^{-1}$$

22-73 We calculate ΔS_{fus} by using the relationship

$$\Delta S_{fus} = \frac{\Delta H_{fus}}{T_m}$$

Using the value of T_m and ΔH_{fus} given, we have

$$\text{Li} \quad \Delta S_{fus} = \frac{2.99 \times 10^3 \text{ J}\cdot\text{mol}^{-1}}{454 \text{ K}} = 6.59 \text{ J}\cdot\text{K}^{-1}\cdot\text{mol}^{-1}$$

$$\text{Na} \quad \Delta S_{fus} = \frac{2.60 \times 10^3 \text{ J}\cdot\text{mol}^{-1}}{371 \text{ K}} = 7.01 \text{ J}\cdot\text{K}^{-1}\cdot\text{mol}^{-1}$$

$$\text{K} \quad \Delta S_{fus} = \frac{2.33 \times 10^3 \text{ J}\cdot\text{mol}^{-1}}{336 \text{ K}} = 6.93 \text{ J}\cdot\text{K}^{-1}\cdot\text{mol}^{-1}$$

$$\text{Rb} \quad \Delta S_{fus} = \frac{2.34 \times 10^3 \text{ J}\cdot\text{mol}^{-1}}{312 \text{ K}} = 7.50 \text{ J}\cdot\text{K}^{-1}\cdot\text{mol}^{-1}$$

$$\text{Cs} \quad \Delta S_{fus} = \frac{2.10 \times 10^3 \text{ J}\cdot\text{mol}^{-1}}{302 \text{ K}} = 6.95 \text{ J}\cdot\text{K}^{-1}\cdot\text{mol}^{-1}$$

We calculate ΔS_{vap} by using the relationship

$$\Delta S_{vap} = \frac{\Delta H_{vap}}{T_b}$$

Using the values of T_b and ΔH_{vap} given, we have

$$\text{Li} \quad \Delta S_{vap} = \frac{134.7 \times 10^3 \text{ J} \cdot \text{mol}^{-1}}{1615 \text{ K}} = 83.41 \text{ J} \cdot \text{K}^{-1} \cdot \text{mol}^{-1}$$

$$\text{Na} \quad \Delta S_{vap} = \frac{89.6 \times 10^3 \text{ J} \cdot \text{mol}^{-1}}{1156 \text{ K}} = 77.5 \text{ J} \cdot \text{K}^{-1} \cdot \text{mol}^{-1}$$

$$\text{K} \quad \Delta S_{vap} = \frac{77.1 \times 10^3 \text{ J} \cdot \text{mol}^{-1}}{1033 \text{ K}} = 74.6 \text{ J} \cdot \text{K}^{-1} \cdot \text{mol}^{-1}$$

$$\text{Rb} \quad \Delta S_{vap} = \frac{69 \times 10^3 \text{ J} \cdot \text{mol}^{-1}}{956 \text{ K}} = 72 \text{ J} \cdot \text{K}^{-1} \cdot \text{mol}^{-1}$$

$$\text{Cs} \quad \Delta S_{vap} = \frac{66 \times 10^3 \text{ J} \cdot \text{mol}^{-1}}{942 \text{ K}} = 70 \text{ J} \cdot \text{K}^{-1} \cdot \text{mol}^{-1}$$

22-75 No, because if K is infinite, then ΔG_{rxn}° is negative infinite, and thus an infinite amount of work could be obtained from the reaction. If K is zero, then ΔG_{rxn}° is positive infinite, and thus an infinite amount of energy would be necessary to make the reaction take place.

22-77 The values of ΔG_{rxn}° for the equation

(1) $$H_2(g) + O_2(g) \rightleftharpoons H_2O_2(l)$$

is given by

$$\begin{aligned}\Delta G_{rxn}^{\circ} &= \Delta G_f^{\circ}[H_2O_2(l)] - \Delta G_f^{\circ}[H_2(g)] - \Delta G_f^{\circ}[O_2(g)] \\ &= (1 \text{ mol})(-120.4 \text{ kJ} \cdot \text{mol}^{-1}) - (1 \text{ mol})(0 \text{ kJ} \cdot \text{mol}^{-1}) \\ &\quad - (1 \text{ mol})(0 \text{ kJ} \cdot \text{mol}^{-1}) \\ &= -120.4 \text{ kJ}\end{aligned}$$

We must calculate ΔG_{rxn}° for the production of 1 mol of $H_2O_2(l)$. The equation is

(2) $$H_2O(l) + \tfrac{1}{2}O_2(g) \rightleftharpoons H_2O_2(l)$$

and ΔG_{rxn}° is given by

$$\begin{aligned}\Delta G_{rxn}^{\circ} &= \Delta G_f^{\circ}[H_2O_2(l)] - \Delta G_f^{\circ}[H_2O(l)] - \tfrac{1}{2}\Delta G_f^{\circ}[O_2(g)] \\ &= (1 \text{ mol})(-120.4 \text{ kJ} \cdot \text{mol}^{-1}) - (1 \text{ mol})(-237.2 \text{ kJ} \cdot \text{mol}^{-1}) \\ &\quad - (\tfrac{1}{2} \text{ mol})(0 \text{ kJ} \cdot \text{mol}^{-1}) \\ &= +116.8 \text{ kJ}\end{aligned}$$

The reaction between hydrogen and oxygen is the more energy-efficient because $\Delta G_{rxn}^{\circ}(1) < \Delta G_{rxn}^{\circ}(2)$.

22-79   Using the data in Problem 22-51, we have

$(1/T)/10^{-4}$ K^{-1}	ln K_p
5.000	-7.804
4.762	-7.285
4.545	-6.812
4.348	-6.383
4.167	-5.987

A plot of ln K_p versus $1/T$ is a straight line (see plot). The equation for the straight line is

$$\ln K = -\left(\frac{\Delta H^{\circ}_{\text{rxn}}}{R}\right)\left(\frac{1}{T}\right) + b$$

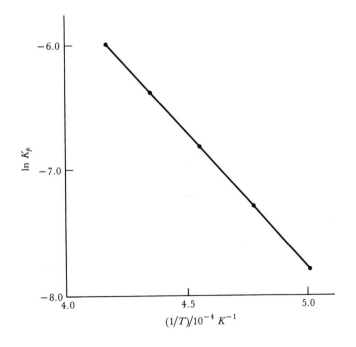

The slope of the line is $-\Delta H_{rxn}^{\circ}/R$. The slope can be calculated by using any two points on the line; thus,

$$\text{slope} = \frac{\ln K_2 - \ln K_1}{\left(\dfrac{1}{T_2} - \dfrac{1}{T_1}\right)}$$

$$= \frac{-7.285 - (-7.804)}{(4.762 - 5.000)(10^{-4}\ K^{-1})} = -2.181 \times 10^4\ K$$

$$\text{slope} = -\frac{\Delta H_{rxn}^{\circ}}{R} = -2.181 \times 10^4\ K$$

$$\Delta H_{rxn}^{\circ} = -(-2.181 \times 10^4\ K)(8.314\ J \cdot K^{-1}) = 181 \times 10^5\ J = 181\ kJ$$

22-81 Starting with $\Delta G_{rxn}^{\circ} = -RT \ln K$, we have at 25°C

$$\Delta G_{rxn}^{\circ} = -(8.314\ J \cdot K^{-1})(298\ K)\ln K$$
$$= -2.478\ kJ\ \ln K$$

The difference in the values of ΔG_{rxn}° when there is a tenfold increase in K at 25°C is calculated as follows:

$$\Delta G_{rxn_2}^{\circ} - \Delta G_{rxn_1}^{\circ} = -(2.478\ kJ)\ln K_2 + (2.478\ kJ)\ln K_1$$
$$= (2.478\ kJ)\ln\left(\frac{K_1}{K_2}\right)$$

where we have used the property of logarithms

$$\ln a - \ln b = \ln\left(\frac{a}{b}\right)$$

We are given that $K_2 = 10K_1$; thus,

$$\Delta G_{rxn_2}^{\circ} - \Delta G_{rxn_1}^{\circ} = (2.478\ kJ)\ln\left(\frac{K_1}{10K_1}\right) = (2.478\ kJ)\ln\left(\frac{1}{10}\right)$$
$$= -5.71\ kJ$$

A tenfold increase in K corresponds to a change of -5.71 kJ in ΔG_{rxn}°.

22-83 The gas solubility equilibrium is

$$X(g) \rightleftharpoons X(soln)$$

Thus,

$$K = \frac{[X]_{soln}}{P_X}$$

Because the solubility of gases decreases with increasing temperature, K decreases with increasing temperature. In other words, when $T_2 > T_1$, $K_2 < K_1$. From the van't Hoff equation

$$\ln\left(\frac{K_2}{K_1}\right) = \frac{\Delta H^\circ_{soln}}{R}\left(\frac{T_2 - T_1}{T_1 T_2}\right)$$

we see that $\ln (K_2/K_1)$ has a negative value. The right-hand side of the equation must also have a negative value. Because R, T_1, T_2, and $T_2 - T_1$ all have a positive value, ΔH°_{soln} must have a negative value.

22-85   The value of ΔH°_{rxn} for the equation

$$CO_2(aq) \rightleftharpoons CO_2(g)$$

can be calculated by using the data given and the van't Hoff equation, because the Henry's law constants given are equilibrium constants for the equation above. Thus,

$$\ln\left(\frac{K_2}{K_1}\right) = \frac{\Delta H^\circ_{rxn}}{R}\left(\frac{T_2 - T_1}{T_1 T_2}\right)$$

and

$$\ln\left(\frac{29.4}{13.2}\right) = \left(\frac{\Delta H^\circ_{rxn}}{8.314\ J\cdot K^{-1}}\right)\left[\frac{25\ K}{(298\ K)(273\ K)}\right]$$

from which we obtain

$$\Delta H^\circ_{rxn} = +21.7\ kJ$$

22-87   The equation is

$$Mg(s) + 2HCl(aq) \rightleftharpoons H_2(g) + MgCl_2(aq)$$

The value of ΔH°_{rxn} is

$$\begin{aligned}
\Delta H^\circ_{rxn} &= \Delta H^\circ_f[MgCl_2(aq)] + \Delta H^\circ_f[H_2(g)] - \Delta H^\circ_f[Mg(s)] - 2\,\Delta H^\circ_f[HCl(aq)]\\
&= (1\ mol)(-801.2\ kJ\cdot mol^{-1}) + 0 - 0 - (2\ mol)(-167.2\ kJ\cdot mol^{-1})\\
&= -466.8\ kJ
\end{aligned}$$

The value of ΔS°_{rxn} is

$$\begin{aligned}
\Delta S^\circ_{rxn} &= S^\circ[MgCl_2(aq)] + S^\circ[H_2(g)] - S^\circ[Mg(s)] - 2S^\circ[HCl(aq)]\\
&= (1\ mol)(-25.1\ J\cdot K^{-1}\cdot mol^{-1}) + (1\ mol)(130.6\ J\cdot K^{-1}\cdot mol^{-1})\\
&\quad -(1\ mol)(32.6\ J\cdot K^{-1}\cdot mol^{-1}) - (2\ mol)(56.5\ J\cdot K^{-1}\cdot mol^{-1})\\
&= -40.1\ J\cdot K^{-1}
\end{aligned}$$

The value of ΔG°_{rxn} is given by

$$\Delta G^\circ_{rxn} = \Delta H^\circ_{rxn} - T\,\Delta S^\circ_{rxn}$$

Thus,

$$\Delta G^\circ_{rxn} = -466.8 \text{ kJ} - \frac{(298 \text{ K})(-40.1 \text{ J}\cdot\text{K}^{-1})}{(1000 \text{ J}\cdot\text{kJ}^{-1})} = -454.9 \text{ kJ}$$

22-89   In the reaction

We are breaking one C—N bond, one C—H bond, and one C—C bond, and we are forming one N—H bond and one C=C bond. The enthalpy change of the reaction is given by

$$\Delta H^\circ_{rxn} \approx H(\text{C—N}) + H(\text{C—H}) + H(\text{C—C}) - H(\text{N—H}) - H(\text{C}=\text{C})$$

Using the data given in Table 22-2, we write

$$\begin{aligned}\Delta H^\circ_{rxn} &= 293 \text{ kJ} + 414 \text{ kJ} + 347 \text{ kJ} - 390 \text{ kJ} - 615 \text{ kJ}\\ &= 49 \text{ kJ}\end{aligned}$$

22-91 $\Delta H^\circ_{rxn} = (1 \text{ mol})\,\Delta H^\circ_f[\text{I}^-(aq)] + (1 \text{ mol})\,\Delta H^\circ_f[\text{Ag}^+(aq)] - (1 \text{ mol})\,\Delta H^\circ_f[\text{Ag}(s)]$
$\qquad = (1 \text{ mol})(-55.9 \text{ kJ}\cdot\text{mol}^{-1}) + (1 \text{ mol})(105.9 \text{ kJ}\cdot\text{mol}^{-1})$
$\qquad\quad - (1 \text{ mol})(-62.4 \text{ kJ}\cdot\text{mol}^{-1})$
$\qquad = 112.4 \text{ kJ}$

We use the van't Hoff equation to calculate K_{sp} at 35°C.

$$\ln\left(\frac{K_{sp}}{(8.3 \times 10^{-17} \text{ M}^2)}\right) = \frac{(112.4 \times 10^3 \text{ J})(308 \text{ K} - 298 \text{ K})}{(8.314 \text{ J}\cdot\text{K}^{-1})(308 \text{ K})(298 \text{ K})}$$
$$= 1.47$$

$$\frac{K_{sp}}{8.3 \times 10^{-17} \text{ M}^2} = e^{1.47} = 4.35$$

$$K_{sp} = 3.6 \times 10^{-16} \text{ M}^2$$

22-93   (a)   We first must balance the equation.

$$2\text{V}_2\text{O}_3 + 4\text{H}^+ + 4e^- \longrightarrow 4\text{VO} + 2\text{H}_2\text{O} \qquad \text{(reduction)}$$
$$\text{V}_2\text{O}_3 + 2\text{H}_2\text{O} \longrightarrow \text{V}_2\text{O}_5 + 4\text{H}^+ + 4e^- \qquad \text{(oxidation)}$$
$$3\text{V}_2\text{O}_3(s) \longrightarrow 4\text{VO}(s) + \text{V}_2\text{O}_5(s)$$

The above equation is the sum of the following equations:

$$3V_2O_3(s) \rightleftharpoons 6V(s) + \tfrac{9}{2}O_2(g) \qquad \Delta G^\circ_{rxn} = -(3)(-272.3 \text{ kJ}) = 816.9 \text{ kJ}$$

$$4V(s) + \tfrac{4}{2}O_2(g) \rightleftharpoons 4VO(s) \qquad \Delta G^\circ_{rxn} = (4)(-96.6 \text{ kJ}) = -386.4 \text{ kJ}$$

$$2V(s) + \tfrac{5}{2}O_2(g) \rightleftharpoons V_2O_5(s) \qquad \Delta G^\circ_{rxn} = (1)(-339.2 \text{ kJ}) = -339.2 \text{ kJ}$$

$$3V_2O_3 \rightleftharpoons 4VO(s) + V_2O_5(s)$$
$$\Delta G^\circ_{rxn} = 816.9 \text{ kJ} + (-386.4 \text{ kJ}) + (-339.2 \text{ kJ}) = 91.3 \text{ kJ}$$

(b) We use the equation

$$\Delta S^\circ_{rxn} = \frac{\Delta H^\circ_{rxn} - \Delta G^\circ_{rxn}}{T}$$

(1) $\qquad \Delta S^\circ_{rxn} = \dfrac{-103.2 \text{ kJ} - (-96.6 \text{ kJ})}{298 \text{ K}} = -0.022 \text{ kJ} \cdot \text{K}^{-1}$

(2) $\qquad \Delta S^\circ_{rxn} = \dfrac{-171.5 \text{ kJ} - 12.3 \text{ kJ}}{298 \text{ K}} = -0.617 \text{ kJ} \cdot \text{K}^{-1}$

(3) $\qquad \Delta S^\circ_{rxn} = \dfrac{-291.3 \text{ kJ} - (-272.3 \text{ kJ})}{298 \text{ K}} = -0.064 \text{ kJ} \cdot \text{K}^{-1}$

(4) $\qquad \Delta S^\circ_{rxn} = \dfrac{-370.6 \text{ kJ} - (-339.3 \text{ kJ})}{298 \text{ K}} = -0.105 \text{ kJ} \cdot \text{K}^{-1}$

Reaction (2) takes place with the greatest decrease in entropy.

22-95 (a) $\qquad \Delta G^\circ_{rxn} = (1 \text{ mol}) \Delta G^\circ_f[\text{HCl}(g)] + (1 \text{ mol}) \Delta G^\circ_f[\text{NH}_3(g)]$
$\qquad\qquad\qquad - (1 \text{ mol}) \Delta G^\circ_f[\text{NH}_4\text{Cl}(s)]$

$\qquad\qquad\qquad = (1 \text{ mol})(-95.3 \text{ kJ} \cdot \text{mol}^{-1}) + (1 \text{ mol})(-16.64 \text{ kJ} \cdot \text{mol}^{-1})$
$\qquad\qquad\qquad - (1 \text{ mol})(-202.97 \text{ kJ} \cdot \text{mol}^{-1})$

$\qquad\qquad\qquad = 91.0 \text{ kJ}$

(b) $\qquad \ln K = -\dfrac{\Delta G^\circ_{rxn}}{RT} = -\dfrac{91.0 \times 10^3 \text{ J}}{(8.314 \text{ J} \cdot \text{K}^{-1})(298 \text{ K})} = -36.74$

$\qquad\qquad K_c = 1.10 \times 10^{-16} \text{ M}^2$

(c) $\quad K_p = K_c(RT)^2 = (1.10 \times 10^{-16} \text{ M}^2)(0.0821 \text{ L} \cdot \text{atm} \cdot \text{mol}^{-1} \cdot \text{K}^{-1})^2(298 \text{ K})^2$
$\qquad\qquad = 6.58 \times 10^{-14} \text{ atm}^2$

$$K_p = P_{\text{NH}_3} P_{\text{HCl}} = 6.58 \times 10^{-14} \text{ atm}^2$$

We have that $P_{\text{NH}_3} = P_{\text{HCl}}$; thus,

$$P^2_{\text{HCl}} = 6.58 \times 10^{-14} \text{ atm}^2$$

$$P_{\text{HCl}} = 2.57 \times 10^{-7} \text{ atm}$$

$$P_{\text{NH}_3} = 2.57 \times 10^{-7} \text{ atm}$$

$$P_{tot} = P_{\text{HCl}} + P_{\text{NH}_3} = 5.14 \times 10^{-7} \text{ atm}$$

22-97   The equation for the reaction is

$$\tfrac{1}{2}H_2(g) + \tfrac{1}{2}Cl_2(g) \longrightarrow HCl(g)$$

The value of the equilibrium constant is given by

$$\ln K = -\frac{\Delta G^{\circ}_{rxn}}{RT} = \frac{95.3 \times 10^3\,J}{(8.314\,J\cdot K^{-1})(298\,K)} = 38.47$$

$$K = 5.07 \times 10^{16}$$

The value of the concentration quotient is

$$Q = \frac{P_{HCl}}{P_{H_2}^{1/2} P_{Cl_2}^{1/2}} = \frac{0.31\,atm}{(3.5\,atm)^{1/2}(1.5\,atm)^{1/2}} = 0.135$$

The value of ΔG_{rxn} for the reaction is

$$\Delta G_{rxn} = RT \ln\left(\frac{Q}{K}\right) = (8.314\,J\cdot K^{-1})\,(298\,K) \ln\left(\frac{0.135}{5.07 \times 10^{16}}\right) = -100\,kJ$$

The reaction is more favorable under the conditions given.

E. ANSWERS TO THE SELF-TEST

1. false
2. true
3. true
4. false
5. heat
6. greater than or equal to
7. $\Delta S_{sys} \geq q_{sys}/T_{sys}$; ΔS_{sys} is the entropy change of the system, q_{sys} is the heat transferred, and T_{sys} is the Kelvin temperature
8. positive
9. false
10. $J \cdot K^{-1}$
11. increasing
12. false
13. positional . . . thermal

14. of the increased energy that is distributed among the molecules (increase in thermal disorder)
15. increases
16. of the increased positional disorder of the molecules
17. of the increased positional disorder of the molecules
18. $\Delta H_{fus}/T_m$
19. $\Delta H_{vap}/T_b$
20. less than
21. increases
22. true
23. standard molar entropies, S°
24. false
25. false

26. ΔG_{rxn}

27. true

28. spontaneous

29. not spontaneous

30. at equilibrium

31. positive

32. $\Delta H_{\text{rxn}} - T \Delta S_{\text{rxn}}$

33. entropy

34. energy (enthalpy)

35. is not

36. spontaneous

37. not spontaneous

38. false

39. the magnitude of ΔG_{rxn}

40. $RT \ln (Q/K)$

41. is

42. the reactants and products are at standard conditions

43. $-RT \ln K$

44. true

45. false

46. false

47. true

48. true

49. false

50. false

51. false

52. $\ln \left(\dfrac{K_2}{K_1} \right) = \dfrac{\Delta H^{\circ}_{\text{rxn}}}{R} \left(\dfrac{T_2 - T_1}{T_1 T_2} \right)$

53. decreases

54. $\ln \left(\dfrac{P_2}{P_1} \right) = \dfrac{\Delta H^{\circ}_{\text{rxn}}}{R} \left(\dfrac{T_2 - T_1}{T_1 T_2} \right)$

55. true

56. true

57. false (They differ slightly.)

58. true

59. true

60. false

23 ELECTROCHEMISTRY

A. OUTLINE OF CHAPTER 23

23-1 Chemical reactions can occur as a result of the passage of an electric current through a solution.

> A voltaic pile can produce an electric current as a result of a chemical reaction (Figure 23-1).
>
> Voltage is the electrical energy per unit of charge.
>
> The SI unit of voltage is the volt: one volt is equal to one joule per coulomb.
>
> The SI unit of electric current is the ampere.
>
> One ampere is a flow of one coulomb of charge per second.
>
> The charge Z in coulombs is given by

$$Z = It \qquad (23-2)$$

> where I is the current in amperes and t is the time in seconds.

23-2 An electrochemical cell produces electricity directly from a chemical reaction.

> An electrochemical cell enables us to obtain an electric current from an oxidation-reduction reaction (Figure 23-3).
>
> An electrochemical cell involves an oxidation half reaction and a reduction half reaction.
>
> Electrodes are used to enable electrons to enter or leave an electrochemical cell.
>
> The electron transfer processes occur at the electrodes.
>
> The reduction half reaction takes place at the cathode; the oxidation half reaction takes place at the anode.
>
> Salt bridges are used to make an electrical connection between two different electrolyte solutions in a cell.

Current in the form of moving ions can pass through the salt bridge from one electrolyte solution to the other.

Current within the cell electrolyte(s) is carried by ions.

A cell must be designed so that the reducing agent and the oxidizing agent are physically separated.

Discharge denotes that current is drawn from the cell.

Electric current flows spontaneously through metallic conductors from a region of negative electrical potential to a region of positive electrical potential.

23-3 A cell diagram is used to represent an electrochemical cell.

By convention, the half reaction of the left electrode of a cell is written as an oxidation half reaction, and the half reaction for the right electrode of a cell is written as a reduction half reaction (oxidation at the left electrode, reduction at the right electrode).

The net cell reaction is given by the sum of the two electrode half reactions adjusted so that the number of electrons is the same in both half reactions.

23-4 The cell voltage depends on the concentrations of the reactants and products of the cell reaction.

The cell voltage is a quantitative measure of the driving force of the cell reaction.

The effect of a change in the concentration of a reactant or product on the cell voltage can be predicted by applying Le Châtelier's principle.

The Nernst equation expresses the quantitative relationship between the cell voltage E_{cell} and the value of ratio of the reaction quotient Q to the equilibrium constant K.

A relationship between the Gibbs free energy and the cell voltage is

$$\Delta G_{rxn} = -nFE_{rxn} \tag{23-5}$$

where n is the number of moles of electrons transferred from the reducing agent to the oxidizing agent and F is Faraday's constant.

Faraday's constant is equal to the charge in coulombs on one mole of electrons or $F = 9.65 \times 10^4 \ C \cdot mol^{-1}$

The Nernst equation is given by

$$E_{rxn} = -\left(\frac{RT}{nF}\right) \ln \left(\frac{Q}{K}\right) \tag{23-7}$$

At 25°C, the Nernst equation in terms of base 10 logarithms is

$$E_{rxn} = -\left(\frac{0.0592 \ V}{n}\right) \log \left(\frac{Q}{K}\right) \tag{23-9}$$

If $E_{rxn} > 0$, then the cell reaction is spontaneous from left to right as written.
If $E_{rxn} < 0$, then the cell reaction is spontaneous from right to left as written.

If $E_{rxn} = 0$, then the cell reaction is at equilibrium.

The standard cell voltage E_{rxn}° is the voltage of the cell when all the species in the cell reaction are in their standard state; or $Q = 1$.

The standard cell voltage can be designated by E_{rxn}° or E_{cell}°.

The relation between the standard cell voltage and the equilibrium constant for the cell reaction at 25°C is given by

$$E_{rxn}^{\circ} = \left(\frac{0.0592 \text{ V}}{n}\right) \log K \tag{23-11}$$

Another form of the Nernst equation is

$$E_{rxn} = E_{rxn}^{\circ} - \left(\frac{0.0592 \text{ V}}{n}\right) \log Q \tag{23-13}$$

A plot of E_{rxn} versus log Q is shown in Figure 23-7.

23-5 E_{rxn}° values can be assigned to half reaction equations.

A standard cell voltage is the difference in standard reduction voltages between the two cell electrodes:

$$E_{rxn}^{\circ} = E_{red}^{\circ} + E_{ox}^{\circ} \tag{23-15}$$

For a particular half reaction,

$$E_{ox}^{\circ} = -E_{red}^{\circ} \tag{23-16}$$

By convention, for the half reaction

$$2H^+(aq, 1 \text{ M}) + 2e^- \longrightarrow H_2(g, 1 \text{ atm}) \qquad E_{red}^{\circ} = 0$$

Some standard reduction voltages are given in Table 23-1.

The standard reduction voltages are used to calculate standard cell voltages.

The more positive the value of E_{red}° for a half reaction, the stronger is the oxidizing agent in the half reaction.

The more negative the value of E_{red}° value for a half reaction, the stronger is the reducing agent in the half reaction.

The arrangement of half reactions in order of the standard reduction voltages is called the electromotive series (emf series).

The E_{red}° values are used to predict many of the reactions that can occur between oxidizing and reducing agents.

23-6 Electrochemical cells are used to determine concentrations of ions.

If E_{rxn} is measured and if E_{rxn}° and all the concentration terms in Q but one are known, then the unknown concentration can be calculated.

The pH of a solution can be determined by electrochemical cell measurements.

23-7 The electrical energy released from an electrochemical cell can do useful work.

The Gibbs free energy change for an oxidation-reduction reaction is related to the cell voltage by Equation (23-5), $\Delta G_{rxn} = -nFE_{rxn}$.

If ΔG_{rxn} is negative, then the magnitude of ΔG_{rxn} is equal to the maximum amount of work that can be obtained from the reaction.

If ΔG_{rxn} is positive, then ΔG_{rxn} is equal to the minimum amount of work that must be done to make the reaction occur.

23-8 Electrolysis is described quantitatively by Faraday's laws.

Electrolysis is the process by which a chemical reaction is made to occur by the passage of an electric current through the solution.

The metal ions of many salts are deposited as the metal when an electric current is passed through an aqueous solution of their salts.

Faraday's laws of electrolysis are given on page 836 of the text.

Faraday's laws are summarized as

$$\left(\begin{array}{c} \text{mass deposited as} \\ \text{metal or evolved as gas} \end{array}\right) = m = \left(\frac{It}{F}\right)\left(\frac{M}{n}\right) \qquad (23\text{-}23)$$

where I is the current in amperes, t is the time in seconds, F is Faraday's constant, M is the molar mass, and n is the number of electrons involved in the equation for the reaction.

23-9 Many chemicals are produced on an industrial scale by electrolysis.

The chlor-alkali process involves the electrolysis of $NaCl(aq)$ to produce $Cl_2(g)$ and $NaOH(aq)$ (Figures 23-10 and 23-11).

The Hall process involves the electrolysis of $Al_2O_3(s)$ dissolved in cryolite, Na_3AlF_6, to produce aluminum metal (Figure 23-12).

The production of a layer of protective metal by electrochemical deposition is called electroplating.

B. SELF-TEST

1. An electric current can be produced by a chemical reaction. *True/False*

2. (*Voltage/Current*) is a measure of how strongly an electric current is driven through a wire.

3. (*Voltage/Current*) is the flow of electric current.

4. The SI unit of current is the ——————————————————.

5. The SI unit of electric charge is the ——————————————.

6. An electric current can be obtained from an oxidation-reduction reaction. *True/False*

7. In an electrochemical cell, the _____ reaction is separated from the _____ reaction.

8. The function of a salt bridge in an electrochemical cell is _____

_____.

9. The current through the solution of an electrochemical cell is carried by _____.

10. The electrode at which the reduction half reaction occurs is the _____.

11. The electrode at which the oxidation half reaction occurs is the _____.

12. When current flows through an electrochemical cell, the cations in the solution move toward the (*anode/cathode*).

13. In the external circuit of an electrochemical cell the current of electrons flows through the metal conductors in the external circuit from the (*positive, negative*) electrode to the (*positive, negative*) electrode.

14. Consider the electrochemical cell whose cell diagram is

$$Cd(s)|Cd(NO_3)_2(aq)||Pb(NO_3)_2(aq)|Pb(s)$$

(a) Oxidation occurs at the _____ electrode.

(b) The oxidation half reaction is _____.

(c) The reduction half reaction is _____.

15. Consider the cell diagram

$$Zn(s)|ZnSO_4(aq)|Hg_2SO_4(s)|Hg(l)$$

(a) The electrolyte solution is _____.

(b) The substances in contact with $ZnSO_4(aq)$ are _____.

16. The function of platinum in a hydrogen gas electrode is _____

_____.

17. The measured voltage of an electrochemical cell depends on the size of the electrodes. *True/False*

18. Consider the cell whose cell diagram is

$$Zn(s)|ZnSO_4(aq)||CuSO_4(aq)|Cu(s)$$

(a) The measured voltage (*increases/decreases/is unchanged*) when the concentration of $ZnSO_4(aq)$ is increased.

(b) The measured voltage (*increases/decreases/is unchanged*) when the concentration of $CuSO_4(aq)$ is increased.

(c) The measured voltage (*increases/decreases/is unchanged*) when the amount of $Zn(s)$ is increased.

(d) The measured voltage (*increases/decreases/is unchanged*) when the salt bridge is removed.

19. The sign of the cell voltage is $(+/-)$ when the cell reaction is spontaneous left to right.

20. The Gibbs free energy change for an oxidation-reduction reaction is related to the corresponding cell voltage. *True/False*

21. The Gibbs free energy change for an oxidation-reduction reaction is related to the corresponding cell voltage by the relationship $\Delta G_{rxn} =$ _____.

22. The amount of charge contained in one mole of electrons is called _____

_____.

23. Faraday's constant is equal to _____.

24. The standard cell voltage is the measured cell voltage when _____

_____.

25. The Nernst equation, which relates the cell voltage E_{rxn} and the reaction concentration-quotient Q at $25°C$ is given by $E_{rxn} =$ _____.

26. If the value of the cell voltage E_{rxn} is greater than zero, then the cell reaction is spontaneous from (*left/right*) to (*left/right*).

27. If the value of the cell voltage E_{rxn} is less than zero, then the cell reaction is spontaneous from (*left/right*) to (*left/right*).

28. The measured cell voltage when the cell reaction is at equilibrium is _____.

29. The standard cell voltage E_{rxn}° is the cell voltage when all the species are in their standard states. *True/False*

30. The measured cell voltage is always equal to E_{rxn}°. *True/False*

31. The equation for the relationship between the standard cell voltage E_{rxn}° and the equilibrium constant K of a reaction at $25°C$ is given by $E_{rxn}^{\circ} =$ _____.

32. The standard voltage of a half reaction is a directly measured quantity. *True/False*

33. The standard cell voltage is always a positive quantity. *True/False*

34. The standard reduction voltage of the hydrogen electrode is set equal to _____.

35. Standard electrode voltages are tabulated in the text for the (*oxidation/reduction*) half reactions.

36. A large, positive standard reduction voltage for a half reaction indicates a strong oxidizing agent in the half reaction. *True/False*

37. The stronger the reducing agent, the more (*negative, positive*) is the half reaction E°_{red} value.

38. The standard cell voltage E°_{rxn} for a reaction can be calculated from the standard electrode voltages of the half reactions of the reaction. *True/False*

39. The standard oxidation voltage for an oxidation half reaction is related to the standard reduction voltage of the reverse half-reaction. *True/False*

40. Measurement of the cell voltage can be used to calculate the concentration of a species in solution by using the _____ equation.

41. The pH of a solution is related to the _____ of the hydrogen electrode.

42. The maximum amount of work that can be obtained from a reaction is equal to

_____ .

43. The value of ΔG_{rxn} for the combustion of a certain fuel is -2330 kJ. The maximum amount of work that can be obtained from the utilization of the fuel in a cell is

_____ .

44. Electrical energy can be used to drive a reaction for which ΔG_{rxn} is positive. *True/False*

45. During electrolysis of a solution of $AgNO_3(aq)$, silver metal is deposited from solution. *True/False*

46. During electrolysis of a solution of $AgNO_3(aq)$, silver metal is deposited from solution at the (*anode/cathode*).

47. The amount of silver deposited by the electrolysis of a $AgNO_3(aq)$ solution depends on the amount of charge that flows through the solution. *True/False*

48. The amount of charge that flows is related to the current by the relationship: charge = _____ .

49. Faraday's laws are expressed quantitatively by the equation $m =$ _____

where _____

_____.

50. Faraday's laws can be used to calculate the amount of metal deposited by passing a current through an aqueous solution of one of its salts. *True/False*

51. The chlor-alkali process is used industrially to prepare _____

_____ by the electrolysis of _____.

52. The Hall process is used industrially to prepare _____ by the electrol-

ysis of _____.

C. CALCULATIONS YOU SHOULD KNOW HOW TO DO

1. Write cell reactions and cell diagrams for electrochemical cells. See Example 23-3 and Section 23-3 and Problems 23-5 through 23-10.

2. Use the Nernst equation in the form of Equation (23-11) or (23-12) to calculate the value of E_{rxn}° or the value of K, the equilibrium constant. See Examples 23-5 and 23-7 and Problems 23-19 through 23-24.

3. Use the Nernst equation in the form of Equation (23-13) to calculate the value of E_{rxn}° from the measured cell voltage for known concentrations of products and reactants. See Example 23-6.

4. Calculate the standard voltage of a cell using Equation (23-15) and the standard reduction voltages given in Table 23-1 and Appendix I. See Example 23-7 and Problems 23-31 and 23-32.

5. Use E_{rxn}° values calculated from standard reduction voltages and the Nernst equation to determine whether a reaction will occur. See Examples 23-8 and 23-9 and Problems 23-35 and 23-36, and 23-43 through 23-46.

6. Use the Nernst equation to determine the concentration of an ion in solution. See Example 23-10 and Problems 23-27 through 23-30.

7. Calculate ΔG_{rxn} for reactions used in electrochemical cells by using Equation (23-5). See Example 23-11 and Problems 23-47 through 23-54.

8. Calculate the current required to deposit a given amount of a metal or the amount of metal deposited during the passage of a given amount of current by using Faraday's laws in the form of Equation (23-23). See Problems 23-55 through 23-64.

D. *SOLUTIONS TO THE ODD-NUMBERED PROBLEMS*

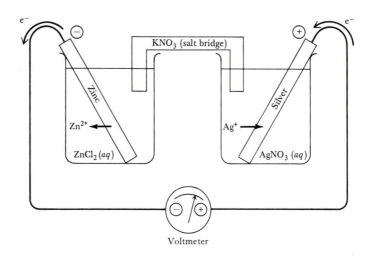

Voltmeter

23-1   Electrons flow from the negative electrode to the positive electrode in the external circuit. The reaction at the negative electrode is

$$Zn(s) \longrightarrow Zn^{2+}(aq) + 2e^-$$

Positive ions, $Zn^{2+}(aq)$, are produced at the negative electrode. The reaction at the positive electrode is

$$2Ag^+(aq) + 2e^- \longrightarrow 2Ag(s)$$

Positive ions, $Ag^+(aq)$, are consumed at the positive electrode. We write the cell diagram with the oxidation half reaction occurring at the left electrode and the reduction half reaction occurring at the right electrode. Thus, the cell diagram is

$$Zn(s)|ZnCl_2(aq)\|AgNO_3(aq)|Ag(s)$$

23-3   The reaction at the negative electrode is

$$V(s) \longrightarrow V^{2+}(aq) + 2e^-$$

The reaction at the positive electrode is

$$Cu^{2+}(aq) + 2e^- \longrightarrow Cu(s)$$

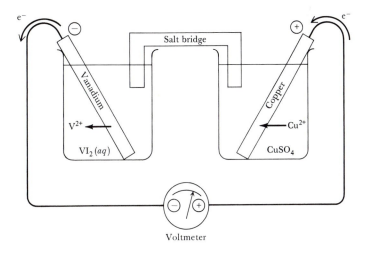

The cell diagram is

$$V(s)|VI_2(aq)||CuSO_4(aq)|Cu(s)$$

23-5 Oxidation takes place at the left electrode. Thus the half reaction at the left electrode is

$$Pb(s) + 2I^-(aq) \longrightarrow PbI_2(s) + 2e^-$$

The reaction is not $Pb(s) \rightarrow Pb^{2+}(aq) + 2e^-$ because the oxidized lead is in the form of $PbI_2(s)$. The half reaction at the right electrode is

$$2H^+(aq) + 2e^- \longrightarrow H_2(g)$$

The net cell reaction is

$$Pb(s) + 2HI(aq) \longrightarrow PbI_2(s) + H_2(g)$$

23-7 Oxidation takes place at the left electrode; thus,

$$In(s) \longrightarrow In^{3+}(aq) + 3e^-$$

The reaction that occurs at the right electrode is

$$Cd^{2+}(aq) + 2e^- \longrightarrow Cd(s)$$

The net cell reaction is

$$2In(s) + 3Cd^{2+}(aq) \longrightarrow 2In^{3+}(aq) + 3Cd(s)$$

A sketch of the cell is as follows.

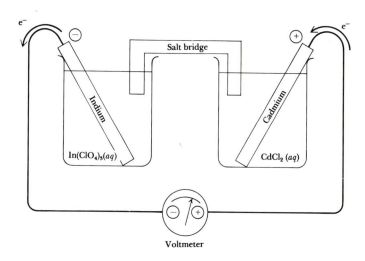

23-9 The half reaction that occurs at the left electrode is

$$H_2(g) \longrightarrow 2H^+(aq) + 2e^- \qquad \text{(oxidation)}$$

Platinum is an inert electrode. The half reaction that occurs at the right electrode is

$$Hg_2Cl_2(s) + 2e^- \longrightarrow 2Hg(l) + 2Cl^-(aq) \qquad \text{(reduction)}$$

The net cell reaction is

$$H_2(g) + Hg_2Cl_2(s) \longrightarrow 2Hg(l) + 2H^+(aq) + 2Cl^-(aq)$$

23-11 (a) A change in the amount of Ca(s) has no effect on the cell voltage.
   (b) An increase in P_{H_2} corresponds to an increase in the concentration of $H_2(g)$, which drives the reaction from right to left. The cell voltage decreases.
   (c) An increase in [HCl] drives the reaction from left to right. The cell voltage increases.
   (d) Dissolution of $Ca(NO_3)_2$ in the $CaCl_2(aq)$ solution leads to an increase in $[Ca^{2+}]$. An increase in $[Ca^{2+}]$ drives the reaction from right to left. The cell voltage decreases.

23-13 (a) A change in the amount of $PbCl_2(s)$ has no effect on the cell voltage.
   (b) Dilution of the cell solution decreases the value of [HCl]. A decrease in [HCl] increases the driving force of the reaction from left to right and thus increases the cell voltage.

(c) Addition of NaOH(s) to the cell solution decreases [H$^+$] and thus increases the reaction driving force from left to right. The cell voltage increases.

(d) Addition of HClO$_4$ increases the concentration of H$^+$(aq) and thus decreases the driving force of the reaction from left to right. The cell voltage decreases.

(e) A decrease in P_{H_2} decreases the driving force of the reaction from left to right and thus decreases the cell voltage.

(f) Because $\Delta H_{rxn}^\circ > 0$, an increase in temperature increases the driving force of the reaction from left to right. Thus, the cell voltage increases.

23-15 (a) An increase in [HClO] drives the reaction from left to right. The cell voltage increases.

(b) The size of the electrode has no effect on the cell voltage.

(c) An increase in the pH corresponds to a decrease in [H$^+$]. A decrease in [H$^+$] drives the reaction from right to left. The cell voltage decreases.

(d) Dissolving KCl(s) in the cell solution containing Cl$^-$(aq) increases [Cl$^-$]. An increase in [Cl$^-$] drives the reaction from right to left. The cell voltage decreases.

23-17 (a) Taking the oxygen half reaction we have

$$2O_2(g) + 8H^+(aq) + 8e^- \longrightarrow 4H_2O(l)$$

Thus $n = 8$.

(b) Taking the zinc half reaction we have

$$8OH^-(aq) + 2Zn(s) \longrightarrow 2Zn(OH)_4^{2-}(aq) + 4e^-$$

Thus $n = 4$.

23-19 We shall use the Nernst equation in the form of Equation (23-12)

$$\log K = \frac{nE_{rxn}^\circ}{0.0592 \text{ V}}$$

The value of n for the reaction is 2, and thus,

$$\log K = \frac{(2)(0.47 \text{ V})}{(0.0592 \text{ V})} = 15.9$$

Solving for K yields

$$K = 10^{15.9} = 7.6 \times 10^{15}$$

23-21 We first calculate E_{rxn}° by using the Nernst equation:

$$E_{rxn} = E_{rxn}^\circ - \left(\frac{0.0592 \text{ V}}{n}\right) \log Q$$

Thus, using the given equation for the reaction,

$$E_{rxn} = E_{rxn}^\circ - \left(\frac{0.0592\ V}{n}\right) \log\left(\frac{[Cd^{2+}]}{[Pb^{2+}]}\right)$$

For this reaction $n = 2$. Thus, using the given concentrations, we have

$$0.293\ V = E_{rxn}^\circ - \left(\frac{0.0592\ V}{2}\right) \log\left(\frac{0.0250\ M}{0.150\ M}\right)$$

$$0.293\ V = E_{rxn}^\circ + 0.023\ V$$

$$E_{rxn}^\circ = 0.293\ V - 0.023\ V = 0.270\ V$$

Using Equation (23-12),

$$\log K = \frac{nE_{rxn}^\circ}{0.0592\ V}$$

and using the above value of E_{rxn}°, we compute

$$\log K = \frac{(2)(0.270\ V)}{(0.0592\ V)} = 9.12$$

Thus,

$$K = 10^{9.12} = 1.3 \times 10^9$$

23-23   Application of the Nernst equation, Equation (23-13), to the cell reaction yields

$$E_{rxn} = E_{rxn}^\circ - \left(\frac{0.0592\ V}{n}\right) \log\left(\frac{[Al^{3+}]}{[Fe^{3+}]}\right)$$

Substituting in the given values with $n = 3$ yields

$$1.59\ V = E_{rxn}^\circ - \left(\frac{0.0592\ V}{3}\right) \log\left(\frac{0.250\ M}{0.0050\ M}\right)$$

$$1.59\ V = E_{rxn}^\circ - 0.0335\ V$$

and

$$E_{rxn}^\circ = 1.59\ V + 0.034\ V = 1.62\ V$$

We shall use Equation (23-12).

$$\log K = \frac{nE_{rxn}^\circ}{0.0592\ V}$$

Thus,

$$\log K = \frac{(3)(1.62 \text{ V})}{0.0592 \text{ V}} = 82.09$$

$$K = 10^{82.09} = 1.2 \times 10^{82}$$

23-25   The oxidation half reaction is

$$Zn(s) \longrightarrow Zn^{2+}(aq) + 2e^-$$

The reduction half reaction is

$$Hg_2Cl_2(s) + 2e^- \longrightarrow 2Hg(l) + 2Cl^-(aq)$$

The cell reaction is

$$Zn(s) + Hg_2Cl_2(s) \longrightarrow 2Hg(l) + Zn^{2+}(aq) + 2Cl^-(aq)$$

Application of the Nernst equation to the cell reaction ($n = 2$) yields

$$E_{rxn} = E_{rxn}^\circ - \left(\frac{0.0592 \text{ V}}{2}\right) \log ([Zn^{2+}][Cl^-]^2)$$

Thus,

$$E_{rxn} = 1.03 \text{ V} - \left(\frac{0.0592 \text{ V}}{2}\right) \log [(0.040)(0.080)^2]$$

$$= 1.03 \text{ V} + 0.11 \text{ V} = 1.14 \text{ V}$$

23-27   Application of the Nernst equation to the cell reaction yields

$$E_{rxn} = E_{rxn}^\circ - \left(\frac{0.0592 \text{ V}}{n}\right) \log ([Zn^{2+}][Cl^-]^2)$$

The oxidation half reaction is

$$Zn(s) \longrightarrow Zn^{2+}(aq) + 2e^-$$

and thus $n = 2$. The value of the cell voltage when $Q = 1.00$ is E_{rxn}°; thus, $E_{rxn}^\circ = 1.03$ V. Substitution of the data given into the Nernst equation yields

$$1.21 \text{ V} = 1.03 \text{ V} - \left(\frac{0.0592 \text{ V}}{2}\right) \log ([Zn^{2+}](0.10)^2)$$

or

$$\log ([Zn^{2+}](0.10)^2) = -\frac{2(1.21 \text{ V} - 1.03 \text{ V})}{0.0592 \text{ V}} = -6.08$$

Therefore

$$[Zn^{2+}](0.10)^2 = 10^{-6.08} = 8.3 \times 10^{-7}$$

and

$$[Zn^{2+}] = 8.3 \times 10^{-5} \text{ M}$$

23-29   Oxidation occurs at the left electrode.

$$\tfrac{1}{2}H_2(g) \longrightarrow H^+(aq) + e^-$$

Reduction occurs at the right electrode.

$$Ag^+(aq) + e^- \longrightarrow Ag(s)$$

The net cell reaction is

$$\tfrac{1}{2}H_2(g) + Ag^+(aq) \rightleftharpoons H^+(aq) + Ag(s) \qquad n = 1$$

Application of the Nernst equation to the cell reaction yields

$$E_{rxn} = E_{rxn}^{\circ} - \left(\frac{0.0592 \text{ V}}{1}\right) \log\left(\frac{[H^+]}{P_{H_2}^{1/2}[Ag^+]}\right)$$

Thus, using the data given, we have

$$0.900 \text{ V} = 0.800 \text{ V} - (0.0592 \text{ V}) \log\left[\frac{[H^+]}{(1.00 \text{ atm})^{1/2}(1.00 \text{ M})}\right]$$
$$= 0.800 \text{ V} - (0.0592 \text{ V}) \log[H^+]$$

and solving for $-\log[H^+] = pH$, we find

$$pH = -\log[H^+] = \frac{(0.900 \text{ V} - 0.800 \text{ V})}{(0.0592 \text{ V})} = 1.69$$

23-31   (a)   The reduction half reaction is

$$Fe(CN)_6^{3-}(aq) + e^- \longrightarrow Fe(CN)_6^{4-}(aq) \qquad E_{red}^{\circ} = +0.36 \text{ V}$$

The oxidation half reaction is

$$Cu(s) \longrightarrow Cu^+(aq) + e^- \qquad E_{ox}^{\circ} = -E_{red}^{\circ} = -0.52 \text{ V}$$

where the E_{red}° values for the half reactions are obtained from Table 23-1 and Appendix I. The E_{rxn}° value for the equation is

$$E_{rxn}^{\circ} = E_{red}^{\circ}[Fe(CN)_6^{3-}/Fe(CN)_6^{4-}] + E_{ox}^{\circ}[Cu/Cu^+]$$
$$= 0.36 \text{ V} - 0.52 \text{ V} = -0.16 \text{ V}$$

(b)　The reduction half reaction is

$$Fe^{3+}(aq) + e^- \longrightarrow Fe^{2+}(aq) \qquad E^\circ_{red} = +0.77 \text{ V}$$

The oxidation half reaction is

$$Ag(s) \longrightarrow Ag^+(aq) + e^- \qquad E^\circ_{ox} = -E^\circ_{red} = -0.80 \text{ V}$$

The E°_{rxn} value for the equation is

$$E^\circ_{rxn} = E^\circ_{red}[Fe^{3+}/Fe^{2+}] + E^\circ_{ox}[Ag/Ag^+] = 0.77 \text{ V} - 0.80 \text{ V} = -0.03 \text{ V}$$

(c)　The reduction half reaction is

$$F_2(g) + 2e^- \longrightarrow 2F^-(aq) \qquad E^\circ_{red} = +2.87 \text{ V}$$

The oxidation half reaction is

$$Zn(s) \longrightarrow Zn^{2+}(aq) + 2e^- \qquad E^\circ_{ox} = -E^\circ_{red} = +0.76 \text{ V}$$

The E°_{rxn} value for the equation is

$$E^\circ_{rxn} = E^\circ_{red}[2F^-/F_2] + E^\circ_{ox}[Zn/Zn^{2+}] = 2.87 \text{ V} + 0.76 \text{ V} = 3.63 \text{ V}$$

23-33　The oxidation half reaction is

$$Cr^{2+}(aq) \longrightarrow Cr^{3+}(aq) + e^-$$

and the reduction half reaction is

$$HClO(aq) + H^+(aq) + 2e^- \longrightarrow Cl^-(aq) + H_2O(l)$$

The E°_{rxn} value for the complete cell is given by

$$E^\circ_{rxn} = E^\circ_{red}[HClO/Cl^-] + E^\circ_{ox}[Cr^{2+}/Cr^{3+}]$$

From Appendix I we obtain

$$E^\circ_{ox}[Cr^{2+}/Cr^{3+}] = -E^\circ_{red}[Cr^{3+}/Cr^{2+}] = -(-0.41 \text{ V}) = 0.41 \text{ V}$$

and

$$E^\circ_{rxn} = 1.80 \text{ V} = E^\circ_{red}[HClO/Cl^-] + 0.41 \text{ V}$$
$$E^\circ_{red}[HClO/Cl^-] = 1.39 \text{ V}$$

23-35　The oxidation half reaction is

$$Zn(s) \longrightarrow Zn^{2+}(aq) + 2e^-$$

and the reduction half reaction is

$$Cd^{2+}(aq) + 2e^- \longrightarrow Cd(s)$$

The value of E°_{rxn} for the complete cell reaction is given by

$$E^\circ_{rxn} = E^\circ_{red}[Cd^{2+}/Cd] + E^\circ_{ox}[Zn/Zn^{2+}]$$

From Appendix I we obtain

$$E^\circ_{rxn} = -0.40 \text{ V} + 0.76 \text{ V} = +0.36 \text{ V}$$

Thus, the net reaction

$$Cd^{2+}(aq) + Zn(s) \longrightarrow Zn^{2+}(aq) + Cd(s)$$

occurs spontaneously at 1.00-M concentrations. Application of the Nernst equation to the cell reaction yields

$$E_{rxn} = 0.36 \text{ V} - \left(\frac{0.0592 \text{ V}}{2}\right) \log\left(\frac{[Zn^{2+}]}{[Cd^{2+}]}\right)$$

Thus, at $[Zn^{2+}] = 1.00$ M and $[Cd^{2+}] = 0.0010$ M we have

$$E_{rxn} = 0.36 \text{ V} - \left(\frac{0.0592 \text{ V}}{2}\right) \log\left(\frac{1.00 \text{ M}}{0.0010 \text{ M}}\right)$$
$$= 0.36 \text{ V} - 0.09 \text{ V}$$
$$= 0.27 \text{ V} > 0$$

Because $E_{rxn} > 0$, the reaction is spontaneous under these conditions.

23-37 The reduction half reaction is

$$H^+(aq) + e^- \longrightarrow \tfrac{1}{2}H_2(g)$$

and the oxidation half reaction is

$$V^{2+}(aq) \longrightarrow V^{3+}(aq) + e^-$$

Thus, we have for E°_{rxn}

$$E^\circ_{rxn} = E^\circ_{red}[H^+/H_2] + E^\circ_{ox}[V^{2+}/V^{3+}]$$

From Table 23-1 and the data given we have

$$E^\circ_{rxn} = 0 + 0.24 \text{ V} = +0.24 \text{ V}$$

From the relation between E_{rxn}° and K we have

$$\log K = \frac{nE_{rxn}^{\circ}}{0.0592 \text{ V}} = \frac{(1)(0.24 \text{ V})}{0.0592 \text{ V}} = 4.05$$

$$K = 10^{4.05} = 1.1 \times 10^4 \text{ atm}^{1/2} \text{ M}^{-1}$$

Application of the Nernst equation to the cell reaction yields

$$E_{rxn} = 0.24 \text{ V} - (0.0592 \text{ V}) \log \left(\frac{[V^{3+}]P_{H_2}^{1/2}}{[V^{2+}][H^+]} \right)$$

Thus, using the concentrations given, we obtain

$$E_{rxn} = 0.24 \text{ V} - (0.0592 \text{ V}) \log \left[\frac{(1.00 \times 10^{-4})(1.00)^{1/2}}{(1.0)(1.0)} \right]$$

$$= 0.48 \text{ V} > 0$$

Hence, the reaction is spontaneous and $V^{2+}(aq)$ can liberate $H_2(g)$ under the given conditions.

23-39 The value of E_{rxn}° is given by

$$E_{rxn}^{\circ} = E_{red}^{\circ}[O_2/H_2O] + E_{ox}^{\circ}[Zn/Zn^{2+}]$$

From Table 23-1 we obtain

$$E_{rxn}^{\circ} = 1.23 \text{ V} + 0.76 \text{ V} = 1.99 \text{ V}$$

Application of the Nernst equation to the cell reaction yields ($n = 4$)

$$E_{rxn} = E_{rxn}^{\circ} - \left(\frac{0.0592 \text{ V}}{4} \right) \log \left(\frac{[Zn^{2+}]^2}{P_{O_2}[H^+]^4} \right)$$

Thus,

$$E_{rxn} = 1.99 \text{ V} - \left(\frac{0.0592 \text{ V}}{4} \right) \log \left[\frac{(0.0010)^2}{(0.20)(0.20)^4} \right]$$

$$= 1.99 \text{ V} + 0.037 \text{ V} = 2.03 \text{ V}$$

23-41 The oxidation half reaction in basic solution is

$$S_2O_3^{2-}(aq) + 6OH^-(aq) \longrightarrow 2SO_3^{2-}(aq) + 3H_2O(l) + 4e^-$$

The reduction half reaction in basic solution is

$$O_2(g) + 2H_2O(l) + 4e^- \longrightarrow 4OH^-(aq)$$

From Table 23-1 we obtain

$$E_{red}^{\circ}[O_2/OH^-] = 0.40 \text{ V}$$

We can obtain the value of $E_{red}^{\circ}[SO_3^{2-}/S_2O_3^{2-}]$ from the value of E_{rxn}°:

$$E_{rxn}^{\circ} = E_{red}^{\circ}[O_2/OH^-] + E_{ox}^{\circ}[S_2O_3^{2-}/SO_3^{2-}]$$
$$= E_{red}^{\circ}[O_2/OH^-] - E_{red}^{\circ}[SO_3^{2-}/S_2O_3^{2-}]$$

Thus,

$$0.98 \text{ V} = 0.40 \text{ V} - E_{red}^{\circ}[SO_3^{2-}/S_2O_3^{2-}]$$
$$E_{red}^{\circ}[SO_3^{2-}/S_2O_3^{2-}] = -0.98 \text{ V} + 0.40 \text{ V} = -0.58 \text{ V}$$

23-43   The reduction half reaction is

$$Co^{3+}(aq) + e^- \longrightarrow Co^{2+}(aq) \qquad E_{red}^{\circ} = +1.81 \text{ V}$$

The oxidation half reaction is

$$2H_2O(l) \longrightarrow O_2(g) + 4H^+(aq) + 4e^- \qquad \begin{aligned} E_{ox}^{\circ} &= -(+1.23 \text{ V}) \\ &= -1.23 \text{ V} \end{aligned}$$

The equation for the complete reaction is

$$4Co^{3+}(aq) + 2H_2O(l) \longrightarrow 4Co^{2+}(aq) + O_2(g) + 4H^+(aq)$$

and

$$E_{rxn}^{\circ} = 1.81 \text{ V} - 1.23 \text{ V} = 0.58 \text{ V}$$

Application of the Nernst equation to the reaction yields ($n = 4$)

$$E_{rxn} = 0.58 \text{ V} - \left(\frac{0.0592 \text{ V}}{4}\right) \log \left(\frac{[Co^{2+}]^4 P_{O_2}[H^+]^4}{[Co^{3+}]^4}\right)$$

Substitution of the values given yields

$$E_{rxn} = 0.58 \text{ V} - \left(\frac{0.0592 \text{ V}}{4}\right) \log \left[\frac{(1.0 \times 10^{-4})^4(0.20)(0.30)^4}{(0.20)^4}\right]$$
$$= 0.58 \text{ V} + 0.24 \text{ V} = 0.82 \text{ V} > 0$$

The positive value of E_{rxn} means that the reaction is spontaneous under the stated conditions.

23-45   The reduction half reaction is

$$Co^{3+}(aq) + e^- \longrightarrow Co^{2+}(aq) \qquad E_{red}^{\circ} = 1.81 \text{ V}$$

and the oxidation half reaction is

$$2H_2O(l) \longrightarrow O_2(g) + 4H^+(aq) + 4e^- \qquad \begin{aligned} E°_{ox} &= -(+1.23 \text{ V}) \\ &= 1.23 \text{ V} \end{aligned}$$

The equation for the complete reaction is

$$4Co^{3+}(aq) + 2H_2O(l) \longrightarrow 4Co^{2+}(aq) + O_2(g) + 4H^+(aq)$$

and

$$E°_{rxn} = 1.81 \text{ V} - 1.23 \text{ V} = 0.58 \text{ V}$$

Application of the Nernst equation to the cell reaction yields ($n = 4$)

$$E_{rxn} = 0.58 \text{ V} - \left(\frac{0.0592 \text{ V}}{4}\right) \log \left(\frac{[Co^{2+}]^4 P_{O_2}[H^+]^4}{[Co^{3+}]^4}\right)$$

If the solution is open to the atmosphere, the pressure of O_2 must be 0.20 atm for O_2 gas to be evolved from the solution. Thus we have, with $[Co^{3+}] = 0.010$ M, $[H^+] = 10^{-pH} = 10^{-1.0} = 0.10$ M,

$$\begin{aligned} E_{rxn} &= 0.58 \text{ V} - \left(\frac{0.0592 \text{ V}}{4}\right) \log \left\{\frac{[Co^{2+}]^4(0.20)(0.10)^4}{(0.010)^4}\right\} \\ &= 0.58 \text{ V} - \left(\frac{0.0592 \text{ V}}{4}\right) \log \left\{\frac{(0.20)(0.10)^4}{(0.010)^4}\right\} - \left(\frac{0.0592 \text{ V}}{4}\right) \log [Co^{2+}]^4 \\ &= 0.58 \text{ V} - 0.05 \text{ V} - (0.0592 \text{ V}) \log [Co^{2+}] \\ &= 0.53 \text{ V} - (0.0592 \text{ V}) \log [Co^{2+}] \end{aligned}$$

Thus, the value of E_{rxn} will be positive for any possible value of $[Co^{2+}]$ when $[Co^{3+}] = 0.010$ M, and thus the oxidation of water by $Co^{3+}(aq)$ is a spontaneous process.

23-47   The value of ΔG_{rxn} is given by

$$\Delta G_{rxn} = -nFE_{rxn}$$

Two moles of electrons are transferred in this reaction; thus, $n = 2$.

$$\begin{aligned} \Delta G_{rxn} &= -(2 \text{ mol})(96{,}500 \text{ C} \cdot \text{mol}^{-1})(1.05 \text{ V}) \\ &= -203{,}000 \text{ J} = -203 \text{ kJ} \end{aligned}$$

23-49   The value of $\Delta G°_{rxn}$ is given by

$$\Delta G°_{rxn} = -nFE°_{rxn}$$

In the reaction 1 mol of copper is oxidized from an oxidization state of zero to an oxidation state of $+2$. Thus, the reaction requires 2 mol of electrons, and so $n = 2$.

$$\Delta G^\circ_{rxn} = -(2 \text{ mol})(96,500 \text{ C} \cdot \text{mol}^{-1})(0.65 \text{ V})$$
$$= -130,000 \text{ J} = -130 \text{ kJ}$$

23-51 (a)   The oxidation half reaction is

$$2Ag(s) \longrightarrow 2Ag^+(aq) + 2e^-$$

and the reduction half reaction is

$$F_2(g) + 2e^- \longrightarrow 2F^-(aq)$$

The standard cell voltage E°_{rxn} is given by

$$E^\circ_{rxn} = E^\circ_{red}[F_2/F^-] + E^\circ_{ox}[Ag/Ag^+]$$
$$= 2.87 \text{ V} + 0.80 \text{ V}$$
$$= 2.07 \text{ V}$$

The value of ΔG°_{rxn} is given by

$$\Delta G^\circ_{rxn} = -nFE^\circ_{rxn}$$

Two electrons are transferred in this reaction; thus,

$$\Delta G^\circ_{rxn} = -(2 \text{ mol})(96,500 \text{ C} \cdot \text{mol}^{-1})(2.07 \text{ V})$$
$$= -400,000 \text{ J} = -400 \text{ kJ}$$

(b)   The oxidation half reaction is

$$\tfrac{1}{2}H_2(g) \longrightarrow H^+(aq) + e^-$$

and the reduction half reaction is

$$Fe^{3+}(aq) + e^- \longrightarrow Fe^{2+}(aq)$$

The standard cell voltage E°_{rxn} is

$$E^\circ_{rxn} = E^\circ_{red}[Fe^{3+}/Fe^{2+}] + E^\circ_{ox}[H_2/H^+]$$
$$= 0.77 \text{ V} + 0 \text{ V} = 0.77 \text{ V}$$

The value of ΔG°_{rxn} is given by

$$\Delta G^\circ_{rxn} = -nFE^\circ_{rxn}$$

One electron is transferred in this reaction; thus, $n = 1$.

$$\Delta G^\circ_{rxn} = -(1 \text{ mol})(96,500 \text{ C} \cdot \text{mol}^{-1})(0.77 \text{ V})$$
$$= -74,000 \text{ J} = -74 \text{ kJ}$$

23-53   The equation for the cell reaction is

$$Zn(s) + Cd^{2+}(aq) \longrightarrow Zn^{2+}(aq) + Cd(s)$$

The value of the standard cell voltage is

$$E^\circ_{rxn} = E^\circ_{red}[Cd^{2+}/Cd] + E^\circ_{ox}[Zn/Zn^{2+}]$$
$$= -0.40\ V - (-0.76\ V) = 0.36\ V$$

The value of ΔG°_{rxn} is

$$\Delta G^\circ_{rxn} = -nFE^\circ_{rxn}$$

In the reaction 2 mol of electrons are transferred; thus, $n = 2$.

$$\Delta G^\circ_{rxn} = -(2\ mol)(96,500\ C \cdot mol^{-1})(0.36\ V)$$
$$= -69,000\ J = -69\ kJ$$

The value of ΔG_{rxn} is given by

$$\Delta G_{rxn} = \Delta G^\circ_{rxn} + RT \ln Q$$
$$= \Delta G^\circ_{rxn} + RT \ln \left(\frac{[Zn^{2+}]}{[Cd^{2+}]} \right)$$
$$= -69\ kJ + (8.314\ J \cdot K^{-1})(298\ K) \ln \left(\frac{0.010}{0.050} \right)$$
$$= -69\ kJ - 3.99\ kJ = -73\ kJ$$

The value of E_{rxn} for the cell under the conditions given is

$$E_{rxn} = -\frac{\Delta G_{rxn}}{nF}$$
$$= -\frac{-73 \times 10^3\ J}{(2\ mol)(96,500\ C \cdot mol^{-1})}$$
$$= 0.38\ V$$

23-55   We shall use Equation (23-23):

$$m = \left(\frac{It}{F} \right) \left(\frac{M}{n} \right)$$

The number of moles of copper in the solution is

$$n = MV = (0.150\ M)(0.500\ L) = 0.0750\ mol$$

Thus, the mass of the copper is

$$m = (0.0750\ mol)(63.55\ g \cdot mol^{-1}) = 4.77\ g$$

The half reaction for the deposition of Cu(s) is

$$Cu^{2+}(aq) + 2e^- \longrightarrow Cu(s)$$

and therefore $n = 2$. The time required to deposit 4.77 g of Cu(s) using a 1.25-A current is

$$t = \frac{mFn}{IM}$$
$$= \frac{(4.77\text{ g})(96{,}500\text{ C}\cdot\text{mol}^{-1})(2)}{(1.25\text{ A})(63.55\text{ g}\cdot\text{mol}^{-1})}$$
$$= 1.16 \times 10^4\text{ s}$$

(In canceling units in the expression for t, recall that $1\text{ A} = 1\text{ C}\cdot\text{s}^{-1}$.)

23-57   The mass deposited is calculated by using Equation (23-23):

$$m = \left(\frac{It}{F}\right)\left(\frac{M}{n}\right)$$

The electrode reaction for the deposition of Cs(s) is

$$Cs^+(aq) + e^- \longrightarrow Cs(s)$$

and thus $n = 1$. Substituting the data given into Equation (23-23) yields

$$m = \frac{(0.500\text{ A})(30\text{ min})\left(\dfrac{60\text{ s}}{1\text{ min}}\right)(132.9\text{ g}\cdot\text{mol}^{-1})}{(9.65 \times 10^4\text{ C}\cdot\text{mol}^{-1})(1)}$$
$$= 1.24\text{ g}$$

23-59   The mass of gas evolved is calculated by using Equation (23-23):

$$m = \left(\frac{It}{F}\right)\left(\frac{M}{n}\right)$$

The electrode reaction for the evolution of $F_2(g)$ is

$$2F^-(KF) \longrightarrow F_2(g) + 2e^-$$

and thus, $n = 2$. Substitution of the data given into Equation (23-23) yields

$$m = \frac{(1500\text{ A})(24\text{ h})\left(\dfrac{3600\text{ s}}{1\text{ h}}\right)\left(\dfrac{38.00\text{ g F}_2}{1\text{ mol F}_2}\right)}{(9.65 \times 10^4\text{ C}\cdot\text{mol}^{-1})(2)}$$
$$= 2.55 \times 10^4\text{ g}$$

Liquid HF is not used in the commercial electrolysis because HF is a covalent compound and thus a poor conductor of electric current. Liquid HF is also very corrosive and very toxic.

23-61 The mass deposited is given by Equation (23-23):

$$m = \left(\frac{It}{F}\right)\left(\frac{M}{n}\right)$$

The electrode reaction for the deposition of Ga(s) is

$$Ga^{3+}(aq) + 3e^- \longrightarrow Ga(s)$$

and thus, $n = 3$. Substitution of the data given into Equation (23-23) yields

$$m = \frac{(0.50\ A)(30\ min)\left(\dfrac{60\ s}{1\ min}\right)\left(\dfrac{69.72\ g\ Ga}{1\ mol\ Ga}\right)}{(9.65 \times 10^4\ C \cdot mol^{-1})(3)}$$

$$= 0.22\ g$$

23-63 The mass of gas evolved is calculated by using Equation (23-23):

$$m = \left(\frac{It}{F}\right)\left(\frac{M}{n}\right)$$

The electrode reaction for the evolution of $O_2(g)$ is

$$2H_2O(l) \longrightarrow O_2(g) + 4H^+(aq) + 4e^-$$

and thus, $n = 4$. Substitution of the data given into Equation (23-23) yields

$$m = \frac{(30.35\ A)(2.00\ h)\left(\dfrac{3600\ s}{1\ h}\right)\left(\dfrac{32.00\ g\ O_2}{1\ mol\ O_2}\right)}{(9.65 \times 10^4\ C \cdot mol^{-1})(4)}$$

$$= 18.1\ g$$

The number of moles of $O_2(g)$ produced is

$$n = (18.1\ g)\left(\frac{1\ mol\ O_2}{32.00\ g\ O_2}\right) = 0.566\ mol$$

The volume of $O_2(g)$ produced is calculated by using the ideal-gas equation:

$$V = \frac{nRT}{P}$$

$$= \frac{(0.566\ mol)(0.0821\ L \cdot atm \cdot K^{-1} \cdot mol^{-1})(298\ K)}{(1.00\ atm)}$$

$$= 13.8\ L$$

23-65   Oxidation occurs at the left electrode; thus,

$$\text{Mn}^{2+}(aq) \longrightarrow \text{MnO}_4^-(aq)$$

The balanced half reaction is

$$4\text{H}_2\text{O}(l) + \text{Mn}^{2+}(aq) \longrightarrow \text{MnO}_4^-(aq) + 8\text{H}^+(aq) + 5e^-$$

Reduction occurs at the right electrode; thus

$$\text{IO}_3^-(aq) \longrightarrow \text{I}^-(aq)$$

The balanced half reaction is

$$6e^- + 6\text{H}^+(aq) + \text{IO}_3^-(aq) \longrightarrow \text{I}^-(aq) + 3\text{H}_2\text{O}(l)$$

The equation for the complete cell reaction is ($n = 6 \times 5 = 30$)

$$9\text{H}_2\text{O}(l) + 6\text{Mn}^{2+}(aq) + 5\text{IO}_3^-(aq) \longrightarrow 18\text{H}^+(aq) + 6\text{MnO}_4^-(aq) + 5\text{I}^-(aq)$$

23-67   The equation for the anode half-reaction is

$$\text{Fe}(s) + 2\text{OH}^-(aq) \longrightarrow \text{Fe(OH)}_2(s) + 2e^- \qquad (1)$$

The equation for the cathode half-reaction is

$$\text{NiOOH}(s) + \text{H}_2\text{O}(l) + e^- \longrightarrow \text{Ni(OH)}_2(s) + \text{OH}^-(aq) \qquad (2)$$

Multiplication of Equation (2) by 2 and addition to Equation (1) yields the net cell equation.

$$\text{Fe}(s) + 2\,\text{NiOOH}(s) + 2\text{H}_2\text{O}(l) \longrightarrow 2\text{Ni(OH)}_2(s) + \text{Fe(OH)}_2(s)$$

23-69   The number of moles in one metric ton of copper is

$$\text{moles of Cu} = (1 \text{ metric ton}) \left(\frac{1000 \text{ kg}}{1 \text{ metric ton}} \right) \left(\frac{1000 \text{ g}}{1 \text{ kg}} \right) \left(\frac{1 \text{ mol Cu}}{63.55 \text{ g Cu}} \right)$$
$$= 1.574 \times 10^4 \text{ mol}$$

The amount of electrons required to deposit 1.574×10^4 mol of copper is

$$\text{moles of electrons} = (1.574 \times 10^4 \text{ mol Cu}) \left(\frac{2 \text{ mol } e^-}{1 \text{ mol Cu}} \right)$$
$$= 3.148 \times 10^4 \text{ mol}$$

The charge that corresponds to 3.148×10^4 moles of electrons is

$$\text{charge} = (3.148 \times 10^4 \text{ mol})(96{,}500 \text{ C} \cdot \text{mol}^{-1}) = 3.038 \times 10^9 \text{ C}$$

One ampere-hour is a current of one ampere that flows for one hour. Thus, the number of coulombs in 1 A·h is

$$1 \text{ A·h} = (1 \text{ C·s}^{-1})(1 \text{ h})\left(\frac{3600 \text{ s}}{1 \text{ h}}\right) = 3.600 \times 10^3 \text{ C}$$

The number of ampere-hours in 3.038×10^9 C is

$$\text{ampere-hour} = (3.038 \times 10^9 \text{ C})\left(\frac{1 \text{ A·h}}{3.600 \times 10^3 \text{ C}}\right) = 8.44 \times 10^5 \text{ A·h}$$

23-71 The mass of silver deposited is given by

$$m = \left(\frac{It}{F}\right)\left(\frac{M}{n}\right)$$

The half reaction for the deposition of Ag(s) is

$$\text{Ag}^+(aq) + \text{e}^- \longrightarrow \text{Ag}(s)$$

and thus, $n = 1$. Thus, we have for the silver deposition

$$\frac{It}{F} = \frac{nm}{M} = \frac{(1)(0.876 \text{ g})}{(107.9 \text{ g·mol}^{-1})} = 8.12 \times 10^{-3} \text{ mol}$$

The half reaction for the deposition of Cd(s) is

$$\text{Cd}^{2+}(aq) + 2\text{e}^- \longrightarrow \text{Cd}(s)$$

and thus, $n = 2$. Using the above value of It/F, we obtain for the mass of Cd(s) deposited

$$m = \left(\frac{It}{F}\right)\left(\frac{M}{n}\right) = (8.12 \times 10^{-3} \text{ mol})\left(\frac{112.4 \text{ g·mol}^{-1}}{2}\right)$$
$$= 0.456 \text{ g}$$

23-73 Oxidation takes place at the left electrode.

$$\text{Ag}(s) + \text{Br}^-(aq) \longrightarrow \text{AgBr}(s) + \text{e}^-$$

Reduction occurs at the right electrode.

$$\text{Ag}^+(aq) + \text{e}^- \longrightarrow \text{Ag}(s)$$

The net cell reaction is

$$\text{Ag}^+(aq) + \text{Br}^-(aq) \rightleftharpoons \text{AgBr}(s)$$

and thus, $n = 1$. We shall use Equation (23-12)

$$\log K = \frac{nE^\circ_{rxn}}{0.0592 \text{ V}}$$

where

$$K = \frac{1}{[Ag^+][Br^-]} = \frac{1}{K_{sp}}$$

$$\log K = \frac{(1)(0.728 \text{ V})}{0.0592 \text{ V}} = 12.30$$

$$K = 10^{12.30} = 1.98 \times 10^{12} \text{ M}^{-2}$$

$$K_{sp} = \frac{1}{K} = \frac{1}{1.98 \times 10^{12} \text{ M}^{-2}} = 5.0 \times 10^{-13} \text{ M}^2$$

23-75   The voltage measured between the $Zn(s)$ and $Cu(s)$ rods will be zero because the cell is short-circuited. The reducing agent $Zn(s)$ must be physically separated from the oxidizing agent $CuSO_4(aq)$. In this case, the $Zn(s)$ rod is in the $Cu^{2+}(aq)$ solution.

23-77   We shall use the Nernst equation in the form

$$E_{rxn} = -\left(\frac{2.303RT}{nF}\right) \log \left(\frac{Q}{K}\right)$$

The cell reaction will take place until $Q = K$, which occurs when equilibrium has been reached:

$$E_{rxn} = -\left(\frac{2.303RT}{nF}\right) \log (1) = 0$$

23-79   We shall use Equation (23-23) to calculate the molar mass.

$$m = \left(\frac{It}{F}\right)\left(\frac{M}{n}\right)$$

$$M = \frac{mnF}{It}$$

$$= \frac{(2.42 \text{ g})(1)(9.65 \times 10^4 \text{ C} \cdot \text{mol}^{-1})}{(0.600 \text{ A})(3600 \text{ s})}$$

$$= 108 \text{ g} \cdot \text{mol}^{-1}$$

The atomic mass of the metal is 108; thus the metal is silver.

E. ANSWERS TO THE SELF-TEST

1. true

2. voltage

3. current

4. ampere

5. coulomb

6. true

7. oxidation . . . reduction

8. to allow the flow of current carry-
ing ions from one solution to the
other

9. ions

10. cathode

11. anode

12. cathode

13. negative . . . positive

14. (a) left or Cd(s)
    (b) $Cd(s) \longrightarrow Cd^{2+}(aq) + 2e^-$
    (c) $Pb^{2+}(aq) + 2e^- \longrightarrow Pb(s)$

15. (a) $ZnSO_4(aq)$
    (b) $Zn(s)$ and $Hg_2SO_4(s)$

16. source of electrons

17. false

18. (a) decreases
    (b) increases
    (c) is unchanged
    (d) decreases (It becomes zero.)

19. $+$

20. true

21. $-nFE_{rxn}$

22. Faraday's constant

23. $9.65 \times 10^4 \ C \cdot mol^{-1}$

24. all the reactants and products are at
standard conditions

25. $-(0.0592 \ V/n) \log (Q/K)$

26. left . . . right

27. right . . . left

28. zero

29. true

30. false

31. $(0.0592 \ V/n) \log K$

32. false

33. false

34. zero

35. reduction

36. true

37. negative

38. true

39. true

40. Nernst

41. voltage

42. Gibbs free energy change

43. 2330 kJ

44. true

45. true

46. cathode

47. true

48. current $\times$ time (in seconds)

49. $(It/F)(M/n)$ . . . m is the mass deposited as a metal or evolved as a
gas, I is the current in amperes, t is
the time in seconds, F is Faraday's
constant, n is the number of electrons transferred in the reaction,
and M is the molar mass.

50. true

51. $Cl_2(g)$ and $NaOH(aq)$ . . .
$NaCl(aq)$

52. aluminum . . . cryolite, Na_3AlF_6

24 NUCLEAR CHEMISTRY AND RADIOCHEMISTRY

A. OUTLINE OF CHAPTER 24

24-1   Many nuclei spontaneously emit small particles.

        Nuclei consist of protons and neutrons, which are collectively called nucleons.

        The number of protons is the atomic number Z.

        The number of nucleons is the mass number A.

        Atoms that contain the same number of protons but different numbers of neutrons are called isotopes.

        An isotope is denoted by a symbol $^{A}_{Z}X$.

        The spontaneous disintegration of a nucleus is called radioactivity; such a nucleus is radioactive.

        A radioactive isotope is called a radioisotope.

        A nuclear equation describes a reaction involving nuclei.

        Nuclear equations must be balanced with respect to charge and the number of nucleons.

24-2   There are several types of emissions from radioactive nuclei.

        The types of particles that are emitted in radioactive processes are listed in Table 24-1.

        Electron capture is a type of nuclear transformation in which one of the innermost electrons of an atom is absorbed by the nucleus.

        Radioactive decay is the process in which a nucleus emits a particle and transforms to another nucleus.

24-3   Emissions from radioactive substances can be detected by several means.

        A film badge detects emissions by the action of the radioactive emissions on photographic film, which darkens to an extent that is a measure of the amount of radiation that strikes the film.

A Geiger counter detects and measures α and β particles by measuring the current produced by the ionization of a gas by the particles emitted by the radioactive source.

A scintillation counter detects and measures radioactive emissions by the fluorescence of certain materials by the interaction with the emissions from a radioactive source.

24-4 The proton-to-neutron ratio is a useful indicator of nuclear stability.

The stable nuclei form a band of stability in a plot of stable nuclei as a function of their number of protons Z versus their number of neutrons N (Figure 24-6).

For elements with an atomic number lower than 21, the proton-neutron ratio is unity or slightly less.

As the atomic number increases, the proton-to-neutron ratio decreases to less than unity.

Nuclei that lie below the band of stability emit β-particles.

Nuclei that lie above the band of stability emit positrons or α-particles or undergo electron capture.

All nuclei with $Z > 83$ are unstable.

Only a few stable nuclei have odd numbers of both protons and neutrons.

The number of stable nuclei with even and odd numbers of protons and neutrons are given in Table 24-3.

Nuclei that contain 2, 8, 28, 50, or 82 protons or neutrons (called magic numbers) are particularly stable.

24-5 The rate of decay of a radioactive isotope is a first-order process.

Different radioisotopes have different half-lives.

The half-lives of some radioisotopes are listed in Table 24-4.

The number of nuclei remaining N after a decay time t is given by

$$\ln\left(\frac{N_0}{N}\right) = \frac{0.693\,t}{t_{1/2}} \tag{24-3}$$

where N_0 is the initial number of nuclei and $t_{1/2}$ is the half-life of the radioisotope.

24-6 Radioactivity can be used to determine the age of rocks.

The age of a rock can be determined by its relative content of uranium-238 and lead-206.

24-7 Carbon-14 can be used to date certain archaeological objects.

Radiocarbon dating is used to date carbon-containing objects derived from formerly living materials.

Radiocarbon dating is useful for objects less than about 30,000 years old.

Living organisms contain a fixed percentage of carbon-14.

The radiation due to carbon-14 in living organisms is 15.3 disintegrations per minute per gram of total carbon.

The age of a carbon-containing object is given by

$$t = (8.27 \times 10^3 \text{ years}) \ln \left(\frac{15.3}{R} \right) \qquad (24\text{-}6)$$

where R is the number of disintegrations per minute per gram of carbon in the object.

24-8   Radioisotopes can be produced in the laboratory.

A nuclear reaction in which one element is converted to another is called transmutation.

Radioisotopes that are not found in nature but are produced in the laboratory are called artificial radioisotopes.

Elements beyond uranium in the periodic table are produced in the laboratory.

The elements beyond uranium in the periodic table are called the transuranium elements.

24-9   Nuclear chemistry can be used to detect extremely small quantities of the elements.

In neutron activation analysis, the sample is irradiated by a beam of neutrons.

The frequencies of the γ-rays emitted are characteristic of the isotopes present in the sample.

In a PIXE (*proton-induced X-ray emission*) analysis, the sample is irradiated by a beam of protons.

The energies of the X-rays emitted are characteristic of the elements present in the sample.

24-10   Enormous amounts of energy accompany nuclear reactions.

The mass lost in a nuclear reaction is converted into energy.

The relation between mass lost and energy produced is given by

$$\Delta E = c^2 \, \Delta m \qquad (24\text{-}8)$$

where c is the speed of light, $3.00 \times 10^8 \text{ m} \cdot \text{s}^{-1}$.

Nuclear reactions are about a million times more energetic than chemical reactions.

The energy required to break up a nucleus into its constituent protons and neutrons is called the binding energy.

The binding energy is calculated by calculating Δm for the formation of the nucleus from its constituent protons and neutrons and then using the equation $\Delta E = c^2 \, \Delta m$.

The graph of the binding energy per nucleon versus the mass number, called a binding energy curve, is shown in Figure 24-13.

24-11   Some nuclei fragment in nuclear reactions.

A nuclear reaction in which a nucleus splits into two smaller fragments is called a fission reaction.

Energy is released in the fission reaction of uranium-235.

24-12   The fission of uranium-235 can initiate a chain reaction.

After absorbing a neutron, uranium-235 undergoes a fission reaction releasing several neutrons that can lead to further fission reactions.

The fission reactions of uranium-235 and neutrons lead to a chain reaction (Figure 24-15).

The smallest quantity of fissionable material that can support a chain reaction is called the critical mass.

24-13   A nuclear reactor utilizes a controlled chain nuclear reaction.

The chain reaction in a nuclear reactor is controlled by control rods, which are made out of substances that readily absorb neutrons.

The heat produced by a nuclear reaction in the core of the reactor is used to generate electricity (Figure 24-18).

24-14   A nuclear breeder reactor produces more fissionable material than it consumes.

Uranium-238 is converted to the fissionable material plutonium-239 in a breeder reactor.

24-15   Fusion reactions release more energy than fission reactions.

A nuclear reaction in which small nuclei join to form larger nuclei is called a fusion reaction.

The sun's energy is due to a nuclear reaction involving the fusion of four protons into a helium-4 nucleus and two positrons.

The hydrogen bomb utilizes a fusion reaction.

24-16   Exposure to radiation damages cells, tissues, and genes.

A measure of the activity of a radioactive substance is its specific activity.

The specific activity of a radioisotope is inversely proportional to its half-life $t_{1/2}$.

The specific activity of a radioisotope is given by

$$\text{specific activity} = \frac{4.17 \times 10^{23} \text{ disintegrations} \cdot \text{g}^{-1}}{M t_{1/2}} \qquad (24\text{-}10)$$

where M is the atomic mass of the isotope and $t_{1/2}$ is the half-life in seconds.

The quantity 3.7×10^{10} disintegrations per second is called a curie and is designated by Ci.

The specific activities of some important radioisotopes are listed in Table 24-5.

The becquerel is the SI unit of the rate of radioactive decay.

One becquerel is defined as one disintegration per second.

The extent of the damage produced by radiation depends on the energy and type of radiation.

24-17   Radon is a major health hazard.

Radon is a naturally occurring radioactive noble gas formed in the radioactive decay of radium-226, which arises from the radioactive decay of uranium ores.

Three radon isotopes are produced in the decay of uranium ores.

Only radon-222 has a significant half-life, which is 3.8 days.

B. SELF-TEST

1. A radioisotope is _____.

2. Alpha particles are _____.

3. Beta particles are _____.

4. Gamma rays are _____.

5. Nuclei contain electrons as well as protons and neutrons. *True/False*

6. A nuclear equation must be balanced with respect to _____

and _____.

7. Nuclei that lie above the band of stability emit _____.

8. Nuclei that lie below the band of stability emit _____.

9. Elements with an odd number of protons usually have only one or two stable isotopes. *True/False*

10. A nucleus containing an even number of both protons and neutrons is likely to be unstable. *True/False*

11. Nuclei that contain a magic number of protons or neutrons are usually stable. *True/False*

12. All nuclei with Z _____ are radioactive.

13. All radioactive nuclei decay at the same rate. *True/False*

14. The rate of radioactive decay varies with temperature. *True/False*

15. Radioactive decay is a (*first/second*)-order rate process.

16. The half-life of a radioactive sample is the time _____

_____.

17. The age of a uranium-containing rock can be determined by measuring the masses of _____ and _____ in the rock sample.

18. Carbon-14 dating is used to determine the age of once-living material. *True/False*

19. All living organisms contain carbon-14. *True/False*

20. The rate of β-decay in living organisms is _____ disintegrations per minute per gram of carbon.

21. Objects of any age can be dated by the carbon-14 method. *True/False*

22. The atmospheric source of carbon-14 in living organisms is _____.

23. Radioisotopes that are made in the laboratory are called _____.

24. Neutron activation analysis is used to measure trace quantities of elements. *True/False*

25. To carry out a neutron activation analysis, the sample is irradiated by a beam of _____.

26. In the PIXE analytical method, the sample is irradiated by a beam of _____.

27. The transuranium elements are laboratory made. *True/False*

28. The mass lost in a nuclear reaction is converted into _____.

29. The relation between mass and energy is given by _____.

30. The value of ΔE for a typical nuclear reaction is similar to that for a typical chemical reaction. *True/False*

31. Energy is (*absorbed/evolved*) when a nucleus is broken up into its constituent protons and neutrons.

32. The binding energy per nucleon is the same for all nuclei. *True/False*

33. The binding energy curve is a plot of _____ versus _____.

34. In the uranium-235 fission reaction, when uranium-235 absorbs a neutron, the uranium-235 nucleus _____.

35. The uranium-235 fission reaction can support a chain reaction because _____ _____.

36. A quantity of uranium-235 less than the critical mass will not support a chain reaction. *True/False*

37. In a reactor the rate of the uranium-235 fission reactions can be controlled by inserting material that absorbs neutrons. *True/False*

38. The uranium-235 fission reaction can be used in a nuclear reactor to produce energy. *True/False*

39. A breeder reactor produces more fissionable material than is consumed. *True/False*

40. The major source of the sun's energy is the fusion of _____
into _____ .

41. The specific activity of a radioactive substance is _____

_____ .

42. The specific activity of a radioisotope is related to the half-life and the atomic mass of the isotope by the equation: specific activity = _____ .

43. One curie is equal to _____ .

44. One bequerel is equal to _____ .

45. Radon is a naturally occurring radioactive noble gas. *True/False*

C. CALCULATIONS YOU SHOULD KNOW HOW TO DO

1. Complete and balance nuclear equations. See Examples 24-1 and 24-2 and Problems 24-1 through 24-4.

2. Calculate the quantity of a radioisotope that remains after a certain time. See Example 24-4 and Problems 24-13 through 24-17 and 24-19 and 24-22.

3. Calculate the time required for a radioisotope to decay to a certain level using Equation (24-3). See Problems 24-18, 24-20, and 24-21.

4. Determine the ages of rocks by uranium-lead dating. See Problems 24-29 and 24-30.

5. Use Equation (24-6) to determine the ages of objects by radiocarbon dating. See Example 24-6 and Problems 24-23 through 24-28.

6. Calculate the binding energy and the binding energy per nucleon of a given nucleus. See Example 24-8 and Problems 24-31 through 24-34.

7. Calculate the energy released in nuclear reactions using Equation (24-8). See Problems 24-35 through 24-37 and 24-39 through 24-42.

8. Calculate the specific activity of a radioisotope from its half-life using Equation (24-10). See Section 24-16 and Problems 24-43 through 24-48.

D. SOLUTIONS TO THE ODD-NUMBERED PROBLEMS

24-1 (a) $^{72}_{30}Zn \rightarrow {}^{0}_{-1}e + {}^{72}_{31}Ga$

(b) $^{230}_{92}U \rightarrow {}^{4}_{2}He + {}^{226}_{90}Th$

(c) $^{136}_{57}La \rightarrow {}^{0}_{+1}e + {}^{136}_{56}Ba$

(d) $^{14}_{7}N + {}^{1}_{0}n \rightarrow {}^{1}_{1}H + {}^{14}_{6}C$

24-3 (a) $^{25}_{12}Mg + {}^{4}_{2}He \rightarrow {}^{28}_{13}Al + {}^{1}_{1}H$

(b) $^{27}_{13}Al + {}^{1}_{0}n \rightarrow {}^{4}_{2}He + {}^{24}_{11}Na$

(c) $^{17}_{8}O + {}^{1}_{1}H \rightarrow {}^{4}_{2}He + {}^{14}_{7}N$

(d) $^{63}_{29}Cu + {}^{1}_{1}H \rightarrow {}^{63}_{30}Zn + {}^{1}_{0}n$

24-5 The proton-to-neutron ratio in calcium-38 is $\left(\dfrac{20}{18}\right) = 1.11$, which lies above the band of stability. Thus we predict that calcium-38 is a positron emitter.

24-7 (a) The proton-to-neutron ratio in rubidium-76 is $\left(\dfrac{37}{39}\right) = 0.95$, which lies above the band of stability. Thus rubidium-76 is a positron emitter.

(b) The proton-to-neutron ratio in germanium-80 is $\left(\dfrac{32}{48}\right) = 0.67$, which lies below the band of stability. Thus germanium-80 is a β-emitter.

(c) The proton-to-neutron ratio in chlorine-32 is $\left(\dfrac{17}{15}\right) = 1.13$, which lies above the band of stability. Thus chlorine-32 is a positron emitter.

(d) The proton-to-neutron ratio in iron-62 is $\left(\dfrac{26}{36}\right) = 0.72$, which lies below the band of stability. Thus iron-62 is a β-emitter.

24-9 (a) The proton-to-neutron ratio in argon-35 is $\left(\dfrac{18}{17}\right) = 1.06$, which lies above the band of stability. Thus argon-35 decays by positron emission or electron capture.

(b) Magnesium-24 has an even number of both protons and neutrons, and thus we predict that it is not radioactive.

(c)   Calcium-40 has a magic number of both protons and neutrons, and thus we predict that it is not radioactive.

(d)   Neon-20 has an even number of both protons and neutrons, and thus we predict that it is not radioactive.

24-11 (a)   Sodium-24 has an odd number of both protons and neutrons. In addition, the proton-to-neutron ratio of sodium-24 is $\left(\dfrac{11}{13}\right) = 0.85$, which lies below the band of stability. Thus we predict that sodium-24 is a β-emitter.

(b)   All elements with $Z > 83$ are radioactive.

(c)   Tin-118 has a magic number of protons and an even number of neutrons. Thus we predict that tin-118 is not radioactive.

(d)   All elements with $Z > 83$ are radioactive.

24-13   If we use Equation (24-3), then we have

$$\ln\left(\frac{N_0}{N}\right) = \frac{0.693\,t}{t_{1/2}}$$

$$= \frac{(0.693)(1\text{ day})\left(\dfrac{24\text{ h}}{1\text{ day}}\right)\left(\dfrac{60\text{ min}}{1\text{ h}}\right)}{110\text{ min}} = 9.072$$

$$\frac{N_0}{N} = e^{9.072} = 8710$$

Inverting both sides gives the fraction remaining

$$\frac{N}{N_0} = 1.15 \times 10^{-4}$$

24-15   According to Example 24-5, we can use Equation (24-3) with mass m replacing number of nuclei (N):

$$\ln\left(\frac{m_0}{m}\right) = \ln\left(\frac{0.200\text{ mg}}{m}\right) = \frac{0.693\,t}{t_{1/2}} = \frac{(0.693)(182\text{ d})}{(46.9\text{ d})} = 2.689$$

$$\frac{0.200\text{ mg}}{m} = e^{2.689} = 14.72$$

$$m = \frac{0.200\text{ mg}}{14.72} = 0.0136\text{ mg}$$

24-17   We use Equation (24-3).

$$\ln\left(\frac{N_0}{N}\right) = \frac{0.693\,t}{t_{1/2}} = \frac{(0.693)(30\text{ d})}{45\text{ d}} = 0.462$$

$$\frac{N_0}{N} = e^{0.462} = 1.6$$

$$\frac{N}{N_0} = 0.63$$

24-19 We use Equation (24-3).

$$\ln\left(\frac{N_0}{N}\right) = \frac{0.693\,t}{t_{1/2}} = \frac{(0.693)(14\text{ d})}{(14.3\text{ d})} = 0.678$$

$$\frac{N_0}{N} = e^{0.678} = 1.97$$

$$\frac{N}{N_0} = 0.51$$

24-21 We use Equation (24-3) with $N = 0.0010N_0$ (0.10% of the initial value). Thus we write

$$\ln\left(\frac{N_0}{N}\right) = \ln\left(\frac{1}{0.0010}\right) = \frac{0.693\,t}{t_{1/2}} = \frac{0.693\,t}{12.8\text{ h}}$$

Solving for t gives

$$t = \left(\frac{12.8\text{ h}}{0.693}\right)\ln(1.0 \times 10^3) = 128\text{ h} = 5.3\text{ d}$$

24-23 We use Equation (24-6).

$$t = (8.27 \times 10^3\text{ years})\ln\left(\frac{15.3}{R}\right)$$

$$= (8.27 \times 10^3\text{ years})\ln\left(\frac{15.3}{5.37}\right) = 8660\text{ years}$$

24-25 We use Equation (24-6) with $R = (0.145)(15.3$ disintegrations $\cdot$ min$^{-1} \cdot$ g^{-1} of carbon). Thus,

$$t = (8.27 \times 10^3\text{ years})\ln\left(\frac{15.3}{R}\right)$$

$$= (8.27 \times 10^3\text{ years})\ln\left(\frac{1}{0.145}\right) = 16{,}000\text{ years}$$

24-27 We use Equation (24-6), but first we must convert the decay rate to disintegrations $\cdot$ min$^{-1} \cdot$ g^{-1}.

$$R = (498\text{ disintegrations}\cdot\text{h}^{-1}\cdot\text{g}^{-1})\left(\frac{1\text{ h}}{60\text{ min}}\right)$$

$$= 8.30\text{ disintegrations}\cdot\text{min}^{-1}\cdot\text{g}^{-1}$$

Thus,

$$t = (8.27 \times 10^3\text{ years})\ln\left(\frac{15.3}{R}\right)$$

$$= (8.27 \times 10^3\text{ years})\ln\left(\frac{15.3}{8.30}\right) = 5060\text{ years}$$

24-29   The age of the uranite is given by

$$t = \frac{t_{1/2}}{0.693} \ln\left(\frac{N_0}{N}\right) = \frac{4.51 \times 10^9 \text{ years}}{0.693} \ln\left(\frac{N_0}{N}\right)$$

where

$$\frac{N_0}{N} = \frac{\text{initial mass of } {}^{238}_{92}\text{U}}{\text{present mass of } {}^{238}_{92}\text{U}}$$

The initial mass of ${}^{238}_{92}\text{U}$ is the sum of the present mass and the mass that has decayed:

$$\frac{N_0}{N} = \frac{\text{mass of } {}^{238}_{92}\text{U that has decayed} + \text{present mass of } {}^{238}_{92}\text{U}}{\text{present mass of } {}^{238}_{92}\text{U}}$$

$$= \frac{\text{mass of } {}^{238}_{92}\text{U that has decayed}}{\text{present mass of } {}^{238}_{92}\text{U}} + 1$$

The mass of lead resulting from the decay of uranium is

$$\text{mass of } {}^{206}_{82}\text{Pb} = (\text{mass of } {}^{238}_{92}\text{U that has decayed})\left(\frac{1 \text{ mol } {}^{238}_{92}\text{U}}{238 \text{ g } {}^{238}_{92}\text{U}}\right)$$

$$\times \left(\frac{1 \text{ mol } {}^{206}_{82}\text{Pb}}{1 \text{ mol } {}^{238}_{92}\text{U}}\right)\left(\frac{206 \text{ g } {}^{206}_{82}\text{Pb}}{1 \text{ mol } {}^{206}_{82}\text{Pb}}\right)$$

$$= \left(\frac{206}{238}\right)(\text{mass of } {}^{238}_{92}\text{U that has decayed})$$

The mass ratio ${}^{206}_{82}\text{Pb}/{}^{238}_{92}\text{U}$ is given as

$$\text{mass ratio} = \frac{\text{mass of } {}^{206}_{82}\text{Pb}}{\text{present mass of } {}^{238}_{92}\text{U}} = 0.395$$

$$= \frac{\left(\frac{206}{238}\right)(\text{mass of } {}^{238}_{92}\text{U that has decayed})}{\text{present mass of } {}^{238}_{92}\text{U}}$$

Therefore,

$$\frac{\text{mass of } {}^{238}_{92}\text{U that has decayed}}{\text{present mass of } {}^{238}_{92}\text{U}} = \left(\frac{238}{206}\right)(0.395)$$

and

$$\frac{N_0}{N} = \left(\frac{238}{206}\right)(0.395) + 1 = 1.456$$

and the age of the uranite is

$$t = \frac{4.51 \times 10^9 \text{ years}}{0.693} \ln 1.456$$
$$= 2.44 \times 10^9 \text{ years}$$

24-31 The mass difference between $^{206}_{82}Pb$ and its constituent nucleons (82 protons and 124 neutrons) is

$$\Delta m = (82 \times 1.0078 \text{ amu}) + (124 \times 1.0087 \text{ amu}) - 205.97446 \text{ amu}$$
$$= 1.7439 \text{ amu}$$

which corresponds to an energy of

$$\Delta E = c^2 \, \Delta m = (3.00 \times 10^8 \text{ m} \cdot \text{s}^{-1})^2 (1.7439 \text{ amu})(1.66 \times 10^{-27} \text{ kg} \cdot \text{amu}^{-1})$$
$$= 2.61 \times 10^{-10} \text{ J}$$

The binding energy per nucleon is obtained by dividing this result by 206:

$$\text{binding energy per nucleon} = \frac{2.61 \times 10^{-10} \text{ J}}{206 \text{ nucleon}}$$
$$= 1.26 \times 10^{-12} \text{ J} \cdot \text{nucleon}^{-1}$$

24-33 The mass difference between $^{35}_{17}Cl$ and its constituent particles is

$$\Delta m = (17 \times 1.0078 \text{ amu}) + (18 \times 1.0087 \text{ amu}) - 34.9689 \text{ amu}$$
$$= 0.3203 \text{ amu}$$

which corresponds to an energy

$$\Delta E = c^2 \, \Delta m = (9.00 \times 10^{16} \text{ m}^2 \cdot \text{s}^{-2})(0.3203 \text{ amu})(1.66 \times 10^{-27} \text{ kg} \cdot \text{amu}^{-1})$$
$$= 4.79 \times 10^{-11} \text{ J}$$

The binding energy per nucleon is

$$\frac{4.79 \times 10^{-11} \text{ J}}{35 \text{ nucleons}} = 1.37 \times 10^{-12} \text{ J} \cdot \text{nucleon}^{-1}$$

24-35 The difference in mass between products and reactants is

$$\Delta m = 140.9137 \text{ amu} + 87.9142 \text{ amu} + (7 \times 1.0087 \text{ amu}) - 235.0439 \text{ amu}$$
$$- 1.0087 \text{ amu}$$
$$= -0.1638 \text{ amu}$$

which corresponds to an energy of

$$\Delta E = c^2 \, \Delta m = (9.00 \times 10^{16} \text{ m}^2 \cdot \text{s}^{-2})(-0.1638 \text{ amu})(1.66 \times 10^{-27} \text{ kg} \cdot \text{amu}^{-1})$$
$$= -2.447 \times 10^{-11} \text{ J} \cdot \text{atom}^{-1} \text{ of } ^{235}_{92}U$$

The energy *released* per gram of $^{235}_{92}U$ is

$$\Delta E = (2.447 \times 10^{-11} \text{ J} \cdot \text{atom}^{-1})(6.022 \times 10^{23} \text{ atom} \cdot \text{mol}^{-1})$$
$$\times \left(\frac{1 \text{ mol } ^{235}_{92}U}{235.0439 \text{ g } ^{235}_{92}U} \right)$$
$$= 6.27 \times 10^{10} \text{ J} \cdot \text{g}^{-1}$$

The energy released in a 50-kiloton bomb is

$$\Delta E = (50 \text{ kton}) \left(\frac{10^3 \text{ ton}}{1 \text{ kton}} \right) \left(\frac{10^3 \text{ kg}}{1 \text{ ton}} \right) (2.5 \times 10^6 \text{ J} \cdot \text{kg}^{-1})$$
$$= 1.25 \times 10^{14} \text{ J}$$

and

$$\text{mass of } ^{235}_{92}U \text{ consumed} = \frac{1.25 \times 10^{14} \text{ J}}{6.27 \times 10^{10} \text{ J} \cdot \text{g}^{-1}} = 1.99 \times 10^3 \text{ g}$$
$$= 1.99 \text{ kg}$$

24-37 The mass of an electron or a positron is 9.10953×10^{-31} kg. The loss of mass is

$$\Delta m = 2 \times 9.10953 \times 10^{-31} \text{ kg} = 1.821906 \times 10^{-30} \text{ kg}$$

The energy produced by the reaction is

$$\Delta E = c^2 \, \Delta m$$
$$= (9.00 \times 10^{16} \text{ m}^2 \cdot \text{s}^{-2})(1.821906 \times 10^{-30} \text{ kg})$$
$$= 1.64 \times 10^{-13} \text{ J}$$

The energy of each γ-ray is

$$E = \left(\frac{1}{2} \right) (1.64 \times 10^{-13} \text{ J}) = 8.20 \times 10^{-14} \text{ J}$$

Recall that the energy and frequency of electromagnetic radiation are related by [Equation (9-3)]

$$E = h\nu$$

and so the frequency of each γ-ray is

$$\nu = \frac{E}{h} = \frac{8.20 \times 10^{-14} \text{ J}}{6.626 \times 10^{-34} \text{ J} \cdot \text{s}} = 1.24 \times 10^{20} \text{ Hz}$$

24-39 The mass of the products is

$$m_{\text{prod}} = 133.8969 \text{ amu} + 94.9125 \text{ amu} + 4.0026 \text{ amu} + 3(1.0087 \text{ amu})$$
$$= 235.8381 \text{ amu}$$

The mass of the reactants is

$$m_{\text{react}} = 235.0439 \text{ amu} + 1.0087 \text{ amu} = 236.0526 \text{ amu}$$

and the difference is

$$\Delta m = m_{\text{prod}} - m_{\text{react}} = -0.2145 \text{ amu}$$

This corresponds to an energy of

$$\begin{aligned}
\Delta E = c^2 \, \Delta m &= (3.00 \times 10^8 \text{ m} \cdot \text{s}^{-1})^2(-0.2145 \text{ amu})(1.66 \times 10^{-27} \text{ kg} \cdot \text{amu}^{-1}) \\
&= -3.20 \times 10^{-11} \text{ J} \cdot \text{molecule}^{-1} \\
&= (-3.20 \times 10^{-11} \text{ J} \cdot \text{molecule}^{-1})(6.022 \times 10^{23} \text{ molecule} \cdot \text{mol}^{-1}) \\
&= -1.93 \times 10^{13} \text{ J} \cdot \text{mol}^{-1}
\end{aligned}$$

24-41 The number of moles of uranium-235 required to produce 3×10^{17} kJ is

$$\text{moles of } {}^{235}_{92}\text{U} = \frac{3 \times 10^{20} \text{ J}}{2 \times 10^{13} \text{ J} \cdot \text{mol}^{-1}} = 1.5 \times 10^7 \text{ mol}$$

$$\begin{aligned}
\text{mass of } {}^{235}_{92}\text{U} &= (1.5 \times 10^7 \text{ mol})\left(\frac{235 \text{ g } {}^{235}_{92}\text{U}}{1 \text{ mol } {}^{235}_{92}\text{U}}\right)\left(\frac{1 \text{ kg}}{1000 \text{ g}}\right)\left(\frac{1 \text{ ton}}{1000 \text{ kg}}\right) \\
&= 3.5 \times 10^3 \text{ metric tons}
\end{aligned}$$

The mass of naturally occurring uranium needed to produce 3.5×10^3 metric tons of ^{235}U is

$$\text{mass of uranium} = \frac{3.5 \times 10^3 \text{ metric tons}}{0.007} = 5 \times 10^5 \text{ metric tons}$$

The time that the world supply would last is

$$\text{time} = \frac{10^6 \text{ metric tons}}{5 \times 10^5 \text{ metric tons} \cdot \text{y}^{-1}} = 2 \text{ y}$$

24-43 We use Equation (24-10).

$$\begin{aligned}
\text{specific activity} &= \frac{4.17 \times 10^{23} \text{ disintegrations} \cdot \text{g}^{-1}}{M t_{1/2}} \\
&= \frac{4.17 \times 10^{23} \text{ disintegrations} \cdot \text{g}^{-1}}{(18.00)(110 \text{ min})\left(\dfrac{60 \text{ s}}{1 \text{ min}}\right)} \\
&= 3.51 \times 10^{18} \text{ disintegrations} \cdot \text{s}^{-1} \cdot \text{g}^{-1}
\end{aligned}$$

In units of curies we have

$$\text{specific activity} = (3.51 \times 10^{18} \text{ disintegrations} \cdot \text{s}^{-1} \cdot \text{g}^{-1})$$
$$\times \left(\frac{1 \text{ curie}}{3.7 \times 10^{10} \text{ disintegrations} \cdot \text{s}^{-1}} \right)$$
$$= 9.5 \times 10^{7} \text{ Ci} \cdot \text{g}^{-1}$$

24-45 We first must convert the units of half-life to seconds:

$$t_{1/2} = (78 \text{ h}) \left(\frac{60 \text{ min}}{1 \text{ h}} \right) \left(\frac{60 \text{ s}}{1 \text{ min}} \right) = 2.81 \times 10^{5} \text{ s}$$

The specific activity of gallium-67 is

$$\text{specific activity} = \frac{4.17 \times 10^{23} \text{ disintegrations} \cdot \text{g}^{-1}}{(67.00)(2.81 \times 10^{5} \text{ s})}$$
$$= 2.21 \times 10^{16} \text{ disintegrations} \cdot \text{s}^{-1} \cdot \text{g}^{-1}$$

The activity of the sample is

$$\text{activity} = (350 \text{ mCi}) \left(\frac{1 \text{ Ci}}{1000 \text{ mCi}} \right) (3.7 \times 10^{10} \text{ disintegrations} \cdot \text{s}^{-1} \cdot \text{Ci}^{-1})$$
$$= 1.30 \times 10^{10} \text{ disintegrations} \cdot \text{s}^{-1}$$

The number of grams of gallium-67 required to produce this activity is

$$\text{mass of } {}^{67}_{31}\text{Ga} = \frac{1.30 \times 10^{10} \text{ disintegrations} \cdot \text{s}^{-1}}{2.21 \times 10^{16} \text{ disintegrations} \cdot \text{s}^{-1} \cdot \text{g}^{-1}}$$
$$= 5.88 \times 10^{-7} \text{ g}$$

The mass of ${}^{67}_{31}\text{GaCl}_3$ is

$$\text{mass} = (5.88 \times 10^{-7} \text{ g } {}^{67}_{31}\text{Ga}) \left(\frac{1 \text{ mol } {}^{67}_{31}\text{Ga}}{67.00 \text{ g } {}^{67}_{31}\text{Ga}} \right) \left(\frac{1 \text{ mol } {}^{67}_{31}\text{GaCl}_3}{1 \text{ mol } {}^{67}_{31}\text{Ga}} \right) \left(\frac{173 \text{ g } {}^{67}_{31}\text{GaCl}_3}{1 \text{ mol } {}^{67}_{31}\text{GaCl}_3} \right)$$
$$= 1.5 \times 10^{-6} \text{ g} = 1.5 \text{ } \mu\text{g}$$

24-47 The activity of the sample in μCi is

$$\text{activity} = (6.5 \times 10^{4} \text{ disintegrations} \cdot \text{min}^{-1}) \left(\frac{1 \text{ min}}{60 \text{ s}} \right) = 1.08 \times 10^{3} \text{ disintegrations} \cdot \text{s}^{-1}$$
$$= (1.08 \times 10^{3} \text{ disintegrations} \cdot \text{s}^{-1}) \left(\frac{1 \text{ Ci}}{3.7 \times 10^{10} \text{ disintegrations} \cdot \text{s}^{-1}} \right) \left(\frac{10^{6} \text{ } \mu\text{Ci}}{1 \text{ Ci}} \right)$$
$$= 0.029 \text{ } \mu\text{Ci}$$

The specific activity of uranium-234 is given by Equation (24-10):

$$\text{specific activity} = \frac{4.17 \times 10^{23} \text{ disintegrations} \cdot \text{g}^{-1}}{Mt_{1/2}}$$

$$= \frac{4.17 \times 10^{23} \text{ disintegrations} \cdot \text{g}^{-1}}{(234)(2.44 \times 10^5 \text{ y})\left(\dfrac{365 \text{ d}}{1 \text{ y}}\right)\left(\dfrac{24 \text{ h}}{1 \text{ d}}\right)\left(\dfrac{3600 \text{ s}}{1 \text{ h}}\right)}$$

$$= (2.32 \times 10^8 \text{ disintegrations} \cdot \text{s}^{-1} \cdot \text{g}^{-1})$$

$$\times \left(\frac{1 \text{ Ci}}{3.7 \times 10^{10} \text{ disintegrations} \cdot \text{s}^{-1}}\right)\left(\frac{10^6 \, \mu\text{Ci}}{1 \text{ Ci}}\right)$$

$$= 6.26 \times 10^3 \, \mu\text{Ci} \cdot \text{g}^{-1}$$

The mass of uranium-234 is obtained by dividing its activity by its specific activity

$$\text{mass of } {}^{234}_{92}\text{U} = \frac{0.029 \, \mu\text{Ci}}{6.27 \times 10^3 \, \mu\text{Ci} \cdot \text{g}^{-1}} = 4.6 \times 10^{-6} \text{ g} = 4.6 \, \mu\text{g}$$

24-49 (a) $\quad {}^{12}_{6}\text{C} + {}^{1}_{1}\text{H} \rightarrow {}^{13}_{7}\text{N} + \gamma$

(b) $\quad {}^{13}_{7}\text{N} \rightarrow {}^{13}_{6}\text{C} + {}^{0}_{+1}\text{e}$

(c) $\quad {}^{13}_{6}\text{C} + {}^{1}_{1}\text{H} \rightarrow {}^{14}_{7}\text{N} + \gamma$

(d) $\quad {}^{14}_{7}\text{N} + {}^{1}_{1}\text{H} \rightarrow {}^{15}_{8}\text{O} + \gamma$

(e) $\quad {}^{15}_{8}\text{O} \rightarrow {}^{15}_{7}\text{N} + {}^{0}_{+1}\text{e}$

(f) $\quad {}^{15}_{7}\text{N} + {}^{1}_{1}\text{H} \rightarrow {}^{12}_{6}\text{C} + {}^{4}_{2}\text{He}$

The net equation is

$$4{}^{1}_{1}\text{H} \rightarrow {}^{4}_{2}\text{He} + 2 \, {}^{0}_{+1}\text{e} + 3\gamma$$

24-51 We have

$$\ln\left(\frac{N_0}{N}\right) = \frac{0.693 \, t}{t_{1/2}} = \frac{(0.693)(50.0 \text{ y})}{(12.3 \text{ y})} = 2.817$$

$$\frac{N_0}{N} = e^{2.817} = 16.73$$

$$\frac{N}{N_0} = 0.0598$$

The fraction remaining is 5.98×10^{-2}.

24-53   We have that

$$\ln\left(\frac{N_0}{N}\right) = \frac{0.693\ t}{t_{1/2}} = \frac{(0.693)(6\ \text{h})\left(\dfrac{60\ \text{min}}{1\ \text{h}}\right)}{25.0\ \text{min}} = 9.979$$

$$\frac{N_0}{N} = e^{9.979} = 21{,}570$$

$$\frac{N}{N_0} = 4.64 \times 10^{-5}$$

Recall that the rate of decay is proportional to the number of nuclei present; thus,

$$\frac{N}{N_0} = \frac{\text{rate}}{\text{rate}_0}$$

and

$$\text{rate at 2 P.M.} = (4.63 \times 10^{-5})(10{,}000\ \text{disintegrations} \cdot \text{min}^{-1})$$
$$= 0.46\ \text{disintegrations} \cdot \text{min}^{-1}$$

or about one disintegration every two minutes.

24-55   The difference in mass between products and reactants is

$$\Delta m = 4.0026\ \text{amu} - (2 \times 2.0141\ \text{amu}) = -0.0256\ \text{amu}$$

The loss of mass corresponds to a *release* of energy of

$$\Delta E = c^2\ \Delta m = (9.00 \times 10^{16}\ \text{m}^2 \cdot \text{s}^{-2})(0.0256\ \text{amu})(1.66 \times 10^{-27}\ \text{kg} \cdot \text{amu}^{-1})$$
$$= 3.825 \times 10^{-12}\ \text{J per atom of helium}$$

The energy released per mole of helium is

$$\Delta E = (3.825 \times 10^{-12}\ \text{J} \cdot \text{atom}^{-1})(6.022 \times 10^{23}\ \text{atom} \cdot \text{mol}^{-1})$$
$$= 2.303 \times 10^{12}\ \text{J} \cdot \text{mol}^{-1}$$

The energy released per gram of helium is

$$\Delta E = (2.303 \times 10^{12}\ \text{J} \cdot \text{mol}^{-1})\left(\frac{1\ \text{mol}\ {}_2^4\text{He}}{4.0026\ \text{g}\ {}_2^4\text{He}}\right) = 5.75 \times 10^{11}\ \text{J} \cdot \text{g}^{-1}$$

The number of moles of octane that must be burned to produce this amount of energy is

$$\text{moles of octane} = \frac{5.75 \times 10^{11}\ \text{J}}{5.45 \times 10^6\ \text{J} \cdot \text{mol}^{-1}} = 1.06 \times 10^5\ \text{mol}$$

The mass of octane is

$$\text{mass of octane} = (1.06 \times 10^5 \text{ mol}) \left(\frac{114.22 \text{ g C}_8\text{H}_{18}}{1 \text{ mol C}_8\text{H}_{18}} \right)$$
$$= 1.21 \times 10^7 \text{ g} = 1.21 \times 10^4 \text{ kg} = 12.1 \text{ metric ton}$$

24-57 The mass of $^{235}_{92}\text{U}$ in 1 kg of the uranium fuel is

$$\text{mass of } ^{235}_{92}\text{U} = (0.03)(1 \text{ kg}) = 30 \text{ g}$$

The mass of $^{235}_{92}\text{U}$ that can be used is

$$\text{mass of } ^{235}_{92}\text{U} \text{ that reacts} = \left(\frac{1}{3} \right)(30 \text{ g}) = 10 \text{ g}$$

The number of $^{235}_{92}\text{U}$ atoms in 10 g is

$$\text{number of } ^{235}_{92}\text{U} = (10 \text{ g}) \left(\frac{1 \text{ mol } ^{235}_{92}\text{U}}{235 \text{ g } ^{235}_{92}\text{U}} \right)(6.022 \times 10^{23} \text{ atom} \cdot \text{mol}^{-1})$$
$$= 2.56 \times 10^{22} \text{ atoms}$$

The energy released by the fission of 2.56×10^{22} atoms of $^{235}_{92}\text{U}$ is

$$\Delta E = (2.9 \times 10^{-14} \text{ kJ} \cdot \text{atom}^{-1})(2.56 \times 10^{22} \text{ atom})$$
$$= 7.43 \times 10^8 \text{ kJ} = 7.43 \times 10^{11} \text{ J}$$

The available energy per kilogram of the uranium fuel is

$$\Delta E = (0.30)(7.43 \times 10^{11} \text{ J}) = 2.23 \times 10^{11} \text{ J}$$

The energy produced by the power plant in 1 year is

$$\Delta E = (1000 \text{ MW}) \left(\frac{10^6 \text{ W}}{1 \text{ MW}} \right) \left(\frac{1 \text{ J} \cdot \text{s}^{-1}}{1 \text{ W}} \right) \left(\frac{60 \text{ s}}{1 \text{ min}} \right) \left(\frac{60 \text{ min}}{1 \text{ h}} \right)$$
$$\times \left(\frac{24 \text{ h}}{1 \text{ d}} \right) \left(\frac{365 \text{ d}}{1 \text{ y}} \right) (1 \text{ y})$$
$$= 3.15 \times 10^{16} \text{ J}$$

The amount of the uranium fuel required is

$$\text{mass of uranium} = \frac{3.15 \times 10^{16} \text{ J}}{2.23 \times 10^{11} \text{ J} \cdot \text{kg}^{-1}}$$
$$= 1.4 \times 10^5 \text{ kg}$$

24-59 We use 6000 min^{-1} for R_0 and calculate $t_{1/2}$ for each subsequent time.

$$t_{1/2} = \frac{0.693}{\ln(6000/R)}$$

t/day	R/min^{-1}	$t_{1/2}$/day
10.0	4680	27.9
20.0	3650	27.9
30.0	2840	27.8
40.0	2200	27.6
50.0	1725	27.8
	average	27.8

The half-life of chromium-51 is 27.8 days.

24-61   The source of the water is

$$CH_3-C(\!\!=\!\!O)(O\text{-}H) + H\text{-}\overset{*}{O}\text{-}CH_3 \longrightarrow CH_3-C(\!\!=\!\!O)(\overset{*}{O}\text{-}CH_3) + H_2O$$

Thus, if the oxygen atom in methanol is labeled with oxygen-18, all the oxygen-18 will appear in the ester. If acetic acid were labeled with oxygen-18, then oxygen-18 would appear in the H$_2$O *and* the ester because of the acid-dissociation reaction of acetic acid:

$$CH_3-C(\!\!=\!\!O)(\overset{*}{O}\text{-}H) + H_2O \longrightarrow CH_3-C(\!\!=\!\!O)(\overset{*}{O}{}^{\ominus}) + H_3O^+(aq)$$

$$CH_3-C(\!\!=\!\!\overset{*}{O})(O\text{-}H) \longleftarrow H_3O^+(aq) + CH_3-C(\!\!=\!\!\overset{*}{O})(O^{\ominus})$$

$$CH_3-C(\!\!=\!\!O)(\overset{*}{O}\text{-}H) + H\text{-}O\text{-}CH_3 \longrightarrow CH_3-C(\!\!=\!\!O)(O\text{-}CH_3) + H_2O^*$$

$$CH_3-C(\!\!=\!\!\overset{*}{O})(O\text{-}H) + H\text{-}O\text{-}CH_3 \longrightarrow CH_3-C(\!\!=\!\!\overset{*}{O})(O\text{-}CH_3) + H_2O$$

Thus, the labeled oxygen atoms appear in water and in the ester.

24-63 The mass difference is

$$\Delta m = 4.0026 \text{ amu} + 1.0087 \text{ amu} - 2.0141 \text{ amu} - 3.0161 \text{ amu}$$
$$= -0.0189 \text{ amu}$$

and the energy *released* is

$$\Delta E = c^2 \, \Delta m = (3.00 \times 10^8 \text{ m} \cdot \text{s}^{-1})^2(0.0189 \text{ amu})(1.66 \times 10^{-27} \text{ kg} \cdot \text{amu}^{-1})$$
$$= 2.82 \times 10^{-12} \text{ J}$$
$$= (2.85 \times 10^{-12} \text{ J})(6.022 \times 10^{23} \text{ atom} \cdot \text{mol}^{-1})$$
$$= 1.70 \times 10^{12} \text{ J} \cdot \text{mol}^{-1}$$

24-65 The mass difference is

$$\Delta m = 12.0000 \text{ amu} - 3(4.0026 \text{ amu}) = -0.0078 \text{ amu}$$

and the energy *released* is

$$\Delta E = c^2 \, \Delta m = (3.00 \times 10^8 \text{ m} \cdot \text{s}^{-1})^2(0.0078 \text{ amu})(1.66 \times 10^{-27} \text{ kg} \cdot \text{amu}^{-1})$$
$$= 1.2 \times 10^{-12} \text{ J} = 7.0 \times 10^{11} \text{ J} \cdot \text{mol}^{-1}$$

24-67 The number of water molecules in 1.00 kg is

$$\text{molecules H}_2\text{O} = (1000 \text{ g}) \left(\frac{1 \text{ mol H}_2\text{O}}{18.02 \text{ g H}_2\text{O}} \right) \left(\frac{6.022 \times 10^{23} \text{ molecule}}{1 \text{ mol}} \right)$$
$$= 3.34 \times 10^{25} \text{ molecules}$$

The number of hydrogen atoms is twice this result

$$\text{H atoms} = 6.68 \times 10^{25} \text{ atoms}$$

The number of deuterium atoms is given by

$$\text{D atoms} = (6.68 \times 10^{25} \text{ H atoms}) \left(\frac{1 \text{ D atom}}{6700 \text{ H atoms}} \right)$$
$$= 9.97 \times 10^{21} \text{ D atoms}$$

The reaction for the fusion of two deuterium atoms into a helium atom is

$$2 \, {}^{2}_{1}\text{H} \longrightarrow {}^{4}_{2}\text{He}$$

and the associated energy is

$$\Delta E = c^2 \, \Delta m$$
$$= (3.00 \times 10^8 \text{ m} \cdot \text{s}^{-1})^2[4.0026 \text{ amu} - 2(2.0141 \text{ amu})]$$
$$\times (1.66 \times 10^{-27} \text{ kg} \cdot \text{amu}^{-1})$$
$$= -3.82 \times 10^{-12} \text{ J}$$

The number of such reactions is equal to one half of the total number of deuterium atoms in one kilogram of water, and so

$$\Delta E = (-3.82 \times 10^{-12} \text{ J}) \left(\frac{9.97 \times 10^{21}}{2} \right)$$
$$= -1.91 \times 10^{10} \text{ J per kilogram water}$$

24-69 It requires large energies to overcome the coulombic repulsion involved in a fusion nuclear reaction. We can estimate the energies involved by using Coulomb's law (Equation 10-1).

$$E = \frac{(2.31 \times 10^{-16} \text{ J} \cdot \text{pm}) Z_1 Z_2}{d}$$

For two deuterium nuclei, $Z_1 = Z_2 = 1$, and d is the diameter of a deuterium nucleus, which is about 10^{-15} m, or 10^{-3} pm. Thus the coulombic repulsion involved in bringing two deuterium nuclei together is about

$$E = 2 \times 10^{-13} \text{ J}$$

For one mole of deuterium nuclei pairs we have

$$E = (2 \times 10^{-13} \text{ J})(6.022 \times 10^{23} \text{ mol}^{-1}) = 1 \times 10^{11} \text{ J} \cdot \text{mol}^{-1}$$

If we assume that the deuterium nuclei are in thermal equilibrium, then their average kinetic energy is [Equation (7-26)]

$$E = \tfrac{3}{2}RT$$

The value of the temperature required to achieve a kinetic energy of $1 \times 10^{11} \text{ J} \cdot \text{mol}^{-1}$ is

$$T = \frac{2E}{3R} = \frac{2 \times 10^{11} \text{ J} \cdot \text{mol}^{-1}}{3 \times 8.314 \text{ J} \cdot \text{mol}^{-1} \cdot \text{K}^{-1}} = 1 \times 10^{10} \text{ K}$$

24-71 Assuming that the activity due to phosphorus-32 remains constant during the time of the experiment, we write

$$(\text{activity})_{\text{start}} = (\text{activity})_{\text{later}}$$

or

$$50,000 \text{ disintegrations} \cdot \text{min}^{-1} = (10.0 \text{ disintegrations} \cdot \text{min}^{-1} \cdot \text{mL}^{-1})$$
$$\times (\text{volume of blood})$$

Solving for the volume of blood, we find that

$$\text{volume of blood} = \frac{50,000 \text{ disintegrations} \cdot \text{min}^{-1}}{10.0 \text{ disintegrations} \cdot \text{min}^{-1} \cdot \text{mL}^{-1}}$$
$$= 5000 \text{ mL} = 5.00 \text{ L}$$

24-73   The number of grams of barium in the precipitate is

$$\text{mass of Ba}^{2+} = \frac{3270 \text{ disintegrations} \cdot \text{min}^{-1}}{7.6 \times 10^7 \text{ disintegrations} \cdot \text{min}^{-1} \cdot \text{g}^{-1}}$$
$$= 4.30 \times 10^{-5} \text{ g}$$

The number of moles of barium-131 is

$$\text{moles of } {}^{131}_{56}\text{Ba} = (4.30 \times 10^{-5} \text{ g}) \left(\frac{1 \text{ mol Ba-131}}{131 \text{ g Ba-131}} \right) = 3.28 \times 10^{-7} \text{ mol}$$

The number of moles of $SO_4^{2-}(aq)$ is

$$\text{moles of } SO_4^{2-}(aq) = \text{moles of } BaSO_4(s) = \text{moles of } {}^{131}_{56}\text{Ba}$$
$$= 3.28 \times 10^{-7} \text{ mol}$$

Assuming that essentially all the sulfate is precipitated (barium is in excess), we calculate the concentration of sulfate ion as

$$[SO_4^{2-}] = \frac{3.28 \times 10^{-7} \text{ mol}}{0.010 \text{ L}} = 3.3 \times 10^{-5} \text{ M}$$

24-75   The total activity of sulfur-35 is

$$\text{activity} = (14{,}000 \text{ disintegrations} \cdot \text{min}^{-1} \cdot \text{mL}^{-1})(75 \text{ mL})$$
$$= 1.05 \times 10^6 \text{ disintegrations} \cdot \text{min}^{-1}$$

The total number of moles of $SO_4^{2-}(aq)$ is

$$\text{moles of } SO_4^{2-}(aq) = (0.010 \text{ M})(0.075 \text{ L}) = 7.50 \times 10^{-4} \text{ mol}$$

The molar activity of sulfur-35 is

$$\text{molar activity} = \frac{1.05 \times 10^6 \text{ disintegrations} \cdot \text{min}^{-1}}{7.50 \times 10^{-4} \text{ mol}}$$
$$= 1.40 \times 10^9 \text{ disintegrations} \cdot \text{min}^{-1} \cdot \text{mol}^{-1}$$

The total activity of sulfur-35 remaining in solution after the two solutions are mixed is

$$\text{activity} = (183 \text{ disintegrations} \cdot \text{min}^{-1} \cdot \text{mL}^{-1})(150 \text{ mL})$$
$$= 2.745 \times 10^4 \text{ disintegrations} \cdot \text{min}^{-1}$$

The number of moles of $SO_4^{2-}(aq)$ in the solution is

$$\text{moles of } SO_4^{2-}(aq) = \frac{2.745 \times 10^4 \text{ disintegrations} \cdot \text{min}^{-1}}{1.40 \times 10^9 \text{ disintegrations} \cdot \text{min}^{-1} \cdot \text{mol}^{-1}}$$
$$= 1.96 \times 10^{-5} \text{ mol}$$

The concentration of sulfate ion is

$$[SO_4^{2-}] = \frac{1.96 \times 10^{-5} \text{ mol}}{0.150 \text{ L}} = 1.31 \times 10^{-4} \text{ M}$$

The solutions of $Pb(NO_3)_2(aq)$ and $Na_2SO_4(aq)$ are mixed on an equimolar basis, and so the concentration of $Pb^{2+}(aq)$ is

$$[Pb^{2+}] = [SO_4^{2-}] = 1.31 \times 10^{-4} \text{ M}$$

The solubility product of $PbSO_4$ is given by

$$\begin{aligned} K_{sp} &= [Pb^{2+}][SO_4^{2-}] \\ &= (1.31 \times 10^{-4} \text{ M})(1.31 \times 10^{-4} \text{ M}) \\ &= 1.7 \times 10^{-8} \text{ M}^2 \end{aligned}$$

24-77   The activity of the radiation in the lungs is

$$\begin{aligned} \text{activity} &= (0.2 \text{ pCi} \cdot \text{L}^{-1}) \left(\frac{1 \text{ Ci}}{10^{12} \text{ pCi}} \right) \\ &\times \left(\frac{3.7 \times 10^{10} \text{ disintegrations} \cdot \text{s}^{-1}}{1 \text{ Ci}} \right) (6.0 \text{ L}) \\ &= 0.044 \text{ disintegration} \cdot \text{s}^{-1} \end{aligned}$$

The number of disintegrations per day is given by

$$\begin{aligned} \text{disintegrations} &= (0.044 \text{ disintegration} \cdot \text{s}^{-1}) \left(\frac{60 \text{ s}}{1 \text{ min}} \right)\left(\frac{60 \text{ min}}{1 \text{ h}} \right)\left(\frac{24 \text{ h}}{1 \text{ day}} \right) \\ &= 4000 \text{ disintegrations} \end{aligned}$$

24-79   We use Equation (24-3):

$$\ln \left(\frac{N_0}{N} \right) = \frac{0.693 \, t}{t_{1/2}}$$

where $t_{1/2}$ is equal to 1.3×10^9 y, N_0 is the original number of potassium-40 nuclei, and N is the present number of potassium-40 nuclei. The original number of potassium-40 nuclei is equal to the present number of potassium-40 nuclei plus the number of potassium nuclei that decayed. We are given that 10.7 percent of the potassium-40 nuclei that decay produce argon-40 nuclei. Thus, the number of potassium nuclei that decayed is given by

$$\text{number of } {}^{40}\text{K that decayed} = \frac{\text{present number of } {}^{40}\text{Ar}}{0.107}$$

The original number of potassium nuclei is given by

$$N_0 = \text{present number of } {}^{40}K + \frac{\text{present number of } {}^{40}Ar}{0.107}$$

$$= N + \frac{\text{present number of } {}^{40}Ar}{0.107}$$

However, we are given that

$$\frac{\text{present number of } {}^{40}Ar}{N} = 0.0102$$

Thus,

$$\text{present number of } {}^{40}Ar = (0.0102)N$$

and so

$$N_0 = N + \frac{(0.0102)N}{0.107}$$

Therefore,

$$\frac{N_0}{N} = \frac{N + \dfrac{(0.0102)N}{0.107}}{N} = 1 + \frac{0.0102}{0.107} = 1.095$$

Thus,

$$\ln 1.095 = 0.09105 = \frac{0.693\,t}{1.3 \times 10^9 \text{ y}}$$

Solving for t, we have that

$$t = 1.7 \times 10^8 \text{ y}$$

The sedimentary rocks are 1.7×10^8 or 170 million years old.

24-81 We are given that

$$\text{specific activity} = \frac{kN}{m}$$

We shall rewrite the above equation as

$$\text{specific activity} = \frac{k}{m/N}$$

The ratio m/N is equal to the mass of one atom. If we multiply m/N by Avogadro's number N_0, then we have the mass of 1 mol of atoms, which is the atomic mass of the isotope.

$$\frac{mN_0}{N} = M$$

Thus, if we multiply the numerator and denominator by Avogadro's number N_0, then we obtain

$$\text{specific activity} = \frac{kN_0}{mN_0/N} = \frac{kN_0}{M}$$

From Equation (16-14), we have that

$$k = \frac{0.693}{t_{1/2}}$$

Substituting the above expression for k into the equation for specific activity, we obtain

$$\text{specific activity} = \frac{0.693\,N_0}{Mt_{1/2}}$$

Substituting in the value of Avogadro's number, we obtain

$$\text{specific activity} = \frac{4.17 \times 10^{23}}{Mt_{1/2}}$$

E. ANSWERS TO THE SELF-TEST

1. radioactive isotope
2. helium-4 nuclei
3. electrons
4. high-energy electromagnetic radiation
5. false
6. charge and number of nuclei
7. positrons or α-particles or undergo electron capture
8. β-particles
9. true

10. false
11. true
12. >83
13. false
14. false
15. first
16. that it takes one half of the radioactive nuclei to decay
17. uranium-238 . . . lead-206
18. true

19. true

20. 15.3

21. false

22. carbon dioxide, CO_2

23. artificial radioisotopes

24. true

25. neutrons

26. protons

27. true

28. energy

29. $E = mc^2$

30. false

31. evolved

32. false

33. the binding energy per nucleon . . . the mass number

34. splits into two smaller fragments and several neutrons

35. several neutrons are emitted which can be absorbed by further uranium-235 nuclei

36. true

37. true

38. true

39. true

40. four protons . . . a helium-4 nucleus and two positrons

41. a measure of the radioactivity of a radioisotope

42. $(4.17 \times 10^{23} \text{ disintegrations} \cdot g^{-1})/(Mt_{1/2})$

43. 3.7×10^{10} disintegrations per second

44. one disintegration per second

45. true

25 THE CHEMISTRY OF THE MAIN-GROUP ELEMENTS I

A. OUTLINE OF CHAPTER 25

25-1 Chemical properties show trends within the periodic table.

    25-1A The metallic character of the elements increases moving down a column or right to left across a row.

        A comparison of some physical properties of metals, semimetals, and nonmetals is given in Table 25-1.

        The chlorides of the reactive metals are ionic compounds and the chlorides of the nonmetals are covalent compounds.

        Oxides become more basic moving down a column and moving right to left across a row.

        An amphoteric oxide reacts with both acids and bases.

    25-1B Chemical properties show periodic trends within each group.

        The maximum oxidation state found in a main group is numerically equal to the group number.

        Not all elements in a group may occur in the maximum oxidation state.

        The common oxidation states of the p-block elements vary in steps of two down from the maximum oxidation state.

        The chemical properties of the first member of a group differ in a number of ways from those of the remaining members of the group.

        The elements in the third and higher rows have d orbitals available for bonding.

25-2 The chemical properties of hydrogen.

    Hydrogen is a diatomic gas under ordinary conditions.

    The oxidation states of hydrogen are $+1$, 0, and -1.

25-2A   There are three hydrogen isotopes.

The three isotopes of hydrogen have mass numbers 1, 2, and 3.

The isotope of mass number 2 is called deuterium and is given the symbol D.

Some physical properties of the isotopic molecules H_2 and D_2 are given in Table 25-6.

The isotope of mass number 3 is called tritium and is given the symbol T.

Tritium is radioactive and emits β-particles.

25-2B   Hydrogen is the most abundant element in the universe.

Very little elemental hydrogen occurs in the earth's crust or atmosphere.

Most of the earth's hydrogen is in the form of water.

Most of the hydrogen in the universe is atomic hydrogen.

25-2C   Hydrogen can be prepared in the laboratory by a variety of reactions.

Hydrogen may be prepared by reacting a metal whose standard reduction voltage is less than zero volts with an aqueous acid.

Hydrogen is a product of the reaction between a reactive metal (Groups 1 and 2) and water.

Hydrogen is prepared commercially and in the laboratory by the electrolysis of water.

Major industrial methods for the preparation of hydrogen are the steam re-forming of natural gas, the water-gas reaction, and the reaction between carbon monoxide and steam at 400°C over an iron oxide catalyst.

The major industrial use of hydrogen is in the synthesis of ammonia.

Large quantities of hydrogen are used in the hydrogenation of unsaturated oils.

Hydrogen is used in the oxyhydrogen torch.

25-3   The chemical properties of the alkali metals.

The Group 1 elements do not occur as the free metal in nature because of their high reactivity.

The chemistry of the Group 1 metals involves the metals and the $+1$ ions.

Lithium differs in some of its properties from the other members of the group.

25-3A   The properties of the Group 1 metals depend on the sizes of the atoms.

The Group 1 metals are also called the alkali metals because their hydroxides are all soluble, strong bases in water.

Some properties of the alkali metals are given in Table 25-8.

The alkali metals can be obtained by electrolysis of the molten chlorides.

Potassium, rubidium, and cesium can also be prepared by the reaction of their chlorides with gaseous sodium metal.

25-3B Group 1 compounds are generally ionic, water-soluble salts.

The alkali metals react directly with all the nonmetals except the noble gases.

The increasing reactivity of the alkali metals with increasing atomic number is demonstrated by their reaction with water to produce hydrogen and the metal hydroxide.

The alkali metals react with alcohols to produce alkoxides and hydrogen.

Molten lithium is one of the most reactive substances known.

The alkali metals react directly with oxygen, forming lithium oxide, Li_2O; sodium peroxide, Na_2O_2; potassium superoxide, KO_2; rubidium superoxide, RbO_2; and cesium superoxide, CeO_2.

The alkali metals react with hydrogen to form hydrides, which are ionic compounds that contain the hydride ion H^-.

Lithium is the only element that reacts directly with nitrogen at room temperature.

Some lithium salts are insoluble in water.

The reactions of the Group 1 metals are summarized in Figure 25-10.

Some important compounds of the Group 1 metals are given in Table 25-9.

25-4 The chemical properties of the alkaline earth metals.

The Group 2 metals are called the alkaline earth metals.

The Group 2 metals are not as reactive as the Group 1 metals, but they are too reactive to be found free in nature.

The chemistry of the alkaline earth metals involves the metals and the $+2$ ions.

Some properties of the Group 2 metals are given in Table 25-11.

25-4A The small size of Be^{2+} makes the chemistry of beryllium different from that of the other Group 2 metals.

All Be(II) compounds involve appreciable covalent bonding and there are no crystalline compounds or solutions involving Be^{2+}.

Beryllium metal is prepared by electrolysis of the halides and by reduction of BeF_2 with magnesium.

Beryllium metal is fairly unreactive at room temperature.

Some reactions of beryllium are given in Figure 25-13.

25-4B Magnesium, calcium, strontium, and barium form ionic compounds involving M^{2+} ions.

The alkaline earth metals are prepared by high-temperature electrolysis of the molten chloride.

The alkaline earth metals readily form M^{2+} ions.

The alkaline earth metal ions are neutral in aqueous solution.

The alkaline earth metals react readily with water, but at slower rates than the alkali metals.

The alkaline earth metals burn in oxygen to form oxides having the formula MO.

Strontium and barium also form the peroxides, SrO_2 and BaO_2.

Except for beryllium, the alkaline metals react vigorously with dilute acids.

The Group 2 metals, except for Be, react with most of the nonmetals to form ionic binary compounds.

Some of the reactions of the alkaline earth metals are given in Figure 25-14.

25-4C Many alkaline earth metal compounds are important commercially.

Suspensions of magnesium hydroxide are sold as the antacid Milk of Magnesia.

Calcium oxide, also known as quicklime, is made by heating limestone, which is mainly calcium carbonate.

Calcium hydroxide, also known as slaked lime, is prepared by dissolving calcium oxide in water.

Some commercially important compounds of the Group 2 metals are listed in Table 25-12.

25-5 The chemical properties of the Group 3 elements.

Boron is a semimetal and the other members of the Group 3 elements are metals.

The common oxidation states of the Group 3 elements are 0 and $+3$.

Some properties of the Group 3 elements are given in Table 25-13.

25-5A The bonding in boron compounds is covalent.

Boron forms no simple cations of the type B^{3+}.

The boron trihalides react with water to form boric acid, $B(OH)_3$ and the corresponding hydrohalic acid.

Boric acid is usually made by adding hydrochloric acid to borax, $NaB_4O_7 \cdot 10H_2O$.

Boric acid acts as a Lewis acid by accepting a hydroxide ion to form the ion $B(OH)_4^-$ and a hydrogen ion.

The principal oxide of boron, B_2O_3, is prepared by heating boric acid.

25-5B Boron hydrides involve multicenter bonds.

Boron forms a number of hydrides.

Diborane, B_2H_6, can be prepared by reacting sodium borohydride, $NaBH_4$, with sulfuric acid.

Pyrolysis of diborane, in some cases in the presence of hydrogen, is used to prepare more complex boron hydrides.

Boron hydrides contain B—H—B bonds, which are three-center bonds.

The structures of some boron hydrides are shown in Figures 25-18 and 25-19.

The bonding in diborane is illustrated in Figure 25-20.

25-5C Aluminum is the most abundant metal in the earth's crust.

The chief source of aluminum is bauxite, AlO(OH).

Aluminum metal is obtained from cryolite, Na_3AlF_6, by the Hall process.

The other Group 3 metals are obtained by the electrolysis of the appropriate molten halide salt or by electrolysis of aqueous solutions of their salts.

25-5D The Group 3 oxides become increasingly basic on descending the group.

The stability of the M(III) state relative to the M(I) state decreases upon descending the group.

The metallic character for identical oxidation states increases upon descending the group.

The increase in metallic character upon descending the group is illustrated by the increasing basicity of the metal oxides.

Aluminum and gallium are amphoteric.

The oxides and hydroxides of aluminum and gallium are amphoteric.

Compounds of the Group 3 metals exhibit both ionic and covalent bonding; ionic bonding is somewhat favored.

The Group 3 metals react with halogens to form compounds with the empirical formula, MX_3.

The Group 3 fluorides are ionic compounds.

The other Group 3 halides are covalent compounds that are dimeric in the vapor state (Figure 25-23).

The hydrides $LiAlH_4$ and $LiGaH_4$ are useful reducing agents but are violently decomposed by water.

Some commercially important compounds of the Group 3 metals are given in Table 25-15.

25-6 The chemical properties of the Group 4 elements.

The Group 4 elements are among the most abundant in nature.

Some properties of the Group 4 elements are given in Table 25-17.

Carbon differs markedly from the other members of the group.

25-6A Diamond and graphite are allotropic forms of carbon.

Diamond has an extended, covalently bonded tetrahedral structure.

Graphite has a layered structure with weak interactions between the layers.

Graphite is the stable form of carbon at ordinary temperatures and pressures.

Diamond can be produced from graphite by applying high temperatures and pressures that lie above the graphite-diamond line in the graphite-diamond phase diagram shown in Figure 25-26.

Activated carbon is an amorphous form of carbon that is characterized by its ability to absorb large quantities of gases.

25-6B Carbon ranks second among the elements in the number of compounds formed.

Hydrogen forms the greatest number of compounds.

Most carbon compounds are organic compounds.

Carbides are binary compounds in which carbon is combined with less electronegative elements.

Calcium carbide, one of the most important carbides, is prepared by the reaction of lime, CaO, and coke, C.

The reaction between calcium carbide and water produces acetylene, C_2H_2.

Silicon carbide, also known as carborundum, has a structure similar to that of diamond and is one of the hardest-known substances.

Carbon has two stable oxides, $CO(g)$ and $CO_2(g)$.

When carbon is burned in a limited amount of oxygen, carbon monoxide predominates; when an excess of oxygen is used, carbon dioxide predominates.

Carbon disulfide is the only stable sulfide of carbon at room temperature.

Carbon forms several important nitrogen-containing compounds, for example, hydrogen cyanide, HCN.

25-6C Silicon is a semimetal.

Elemental silicon is made by the high-temperature reduction of silicon dioxide with carbon.

Ultrapure silicon is prepared by a method of recrystallization called zone refining (Figure 25-29).

Silicon must be ultrapure (99.9999%) for use in transistor and semiconductor manufacture.

Semiconductors have electrical conductivities intermediate between those of conductors and insulators.

In an insulator, the energy separation of the valence band and conduction band is large and there are essentially no electrons in the conduction band.

In a conductor, the energy separation of the valence band and conduction band is zero and all the valence electrons are conduction electrons.

In a semiconductor, the energy separation of the valence band and conduction band is comparable to thermal energies, and thus some of the valence electrons can occupy the conduction band.

An insulator like silicon can be converted to a semiconductor by the addition of selected impurity atoms.

An *n*-type silicon semiconductor is produced when trace amounts of atoms with five valence electrons (phosphorus or antimony) are added to silicon.

The excess electrons on the impurity atoms in an *n*-type semiconductor become the current carrier in the crystal.

A *p*-type silicon semiconductor is produced when trace amounts of atoms with three valence electrons (boron or indium) are added to silicon.

The deficiency of valence electrons in a *p*-type semiconductor functions as "holes" by means of which electrons can "hop" through the crystal.

25-6D   The most common and important compounds of silicon involve oxygen.

Silicon is fairly unreactive.

Silicon reacts with halides to form tetrahalides.

Silanes, the silicon hydrides analogous to the hydrocarbons, are much less stable than the corresponding hydrocarbons.

Silicon reacts with dilute alkalis to give a number of silicates, which are found widely in nature.

The simplest silicate anion is the tetrahedral orthosilicate ion SiO_4^{4-}, whose structure is shown in Figure 25-32.

Silicates can form a number of polymeric anions.

Some commercially important compounds of silicon are given in Table 25-18.

25-6E   Most glasses are silicates.

Quartz is a crystalline material with the composition SiO_2 (Figure 25-38).

When crystalline quartz is melted and cooled quickly, quartz glass results.

A wide variety of glass properties can be produced by the addition of other components to the quartz sand.

Photochromic eyeglasses have a small amount of silver chloride dispersed throughout the glass.

25-6F   Germanium is a semimetal; tin and lead are metals.

Germanium is prepared in a manner similar to that of silicon.

Tin is produced by heating SnO_2, the form present in its ores, with carbon.

Lead is produced by roasting its ore, PbS, to produce PbO, which is reduced to the metal with carbon.

The elements show a trend to metallic properties and to increasing stability of the $+2$ oxidation state descending the group.

The salts of germanium, tin, and lead show increasing ionic character upon descending the group.

Some commercially important compounds of germanium, tin, and lead are given in Table 25-19.

B. SELF-TEST

1. The most metallic element is ――――――――――――――.

2. The most nonmetallic element is ――――――――――――――.

3. The metallic character of the elements increases moving down a column in the periodic table. *True/False*

4. The metallic character of the elements increases moving from left to right across a row in the periodic table. *True/False*

5. The oxides become more (*acidic/basic*) moving down a column in the periodic table.

6. The oxides become more (*acidic/basic*) moving across a row in the periodic table.

7. An amphoteric oxide reacts with both acids and bases. *True/False*

8. The maximum oxidation state found in a main group element is numerically equal to its ――――――――――――――.

9. All the elements in a group may occur in the maximum oxidation state. *True/False*

10. The common oxidation states of the *p*-block elements in a group vary in steps of ―――――――― down from the maximum oxidation state.

11. The first member of a group differs somewhat in chemical properties from those of the other members of the group. *True/False*

12. Describe some of the physical properties of hydrogen.

――

13. The possible oxidation states of hydrogen are ――――――――――――.

14. The three isotopes of hydrogen are ――――――――――― , ――――――――――― , and ――――――――――.

15. Heavy water is composed of ――――――――――― and ――――――――――.

16. Hydrogen may be prepared by reacting zinc with ――――――――――.

17. The equation for the reaction between calcium and water is _____

_____ .

18. The reaction between a Group 1 metal and water is not used to produce hydrogen

because _____ .

19. The equation for the electrolysis of a dilute aqueous solution of sulfuric acid is ___

_____ .

20. The charge on the hydride ion is _____ .

21. Most compounds containing hydrogen are covalent compounds. *True/False*

22. The Group 1 metals are very reactive metals. *True/False*

23. Why must the Group 1 metals be stored under kerosene?

_____ .

24. The Group 1 metals are also called the _____ metals.

25. Sodium is prepared commercially by _____ .

26. The Group 1 metals generally form ionic compounds. *True/False*

27. Most alkali metal compounds are soluble in water. *True/False*

28. The alkali metals react with water to form _____ and

_____ .

29. Sodium reacts with alcohols to produce _____ and _____

_____ .

30. The alkali metals react with oxygen to form oxides with the general formula M_2O only. *True/False*

31. The only element to react directly with nitrogen at room temperature is _____ .

32. All lithium salts are soluble. *True/False*

33. The Group 2 metals are also called the _____ metals.

34. The chemistry of beryllium is similar to that of the other Group 2 metals. *True/False*

35. The Be(II) compounds always involve the beryllium ion Be^{2+}. *True/False*

36. The alkaline earth metals are prepared by _____ .

37. The charge of a Group 2 metal ion is _____ .

38. The equation for the reaction of an alkaline earth metal M with water is _____

_____.

39. The Group 2 metals react with most nonmetals to form binary (*covalent/ionic*) compounds.

40. The Group 3 elements are all metals. *True/False*

41. Boron (III) compounds are (*ionic/covalent*) compounds.

42. The equation for the reaction between boric acid, $B(OH)_3$, and water is _____

_____.

43. Boric acid is a (*Lewis/Arhennius*) acid.

44. The bonding in diborane, B_2H_6, is analogous to that in ethane, C_2H_6. *True/False*

45. The B—H—B bond in diborane is an example of a _____.

46. Aluminum is prepared from cryolite by the _____ process.

47. Gallium and indium are prepared by the electrolysis of _____.

48. The Group 3 oxides become (*increasingly/decreasingly*) basic upon descending the group.

49. The stability of the M(III) state relative to the M(I) state (*increases/decreases*) upon descending the group.

50. The Group 3 metals form only ionic compounds. *True/False*

51. The hydride $LiAlH_4$ is a strong (*oxidizing/reducing*) agent.

52. Two allotropic forms of carbon are _____ and _____.

53. Diamond is a (*covalent/ionic*) compound.

54. The stable form of carbon at ordinary temperatures and pressures is _____.

55. Graphite can be converted into diamond. *True/False*

56. The element that forms the greatest number of compounds is _____.

57. Most carbon-containing compounds are (*covalent/ionic*) compounds.

58. A binary compound in which carbon is combined with a less electronegative element is called a _____.

59. Calcium carbide reacts with water to form _____ and _____.

60. The two stable oxides of carbon are _____ and _____.

61. Elemental silicon is made by the high-temperature reduction of _____
_____ with carbon.

62. Zone refining is used to purify _____.

63. In an insulator, the energy separation of the valence band and conduction band is large. *True/False*

64. In a semiconductor, the energy separation of the valence band and the conduction band is comparable to _____ energy.

65. An *n*-type semiconductor is produced when a trace amount of an element with _____ valence electrons is added to silicon.

66. The addition of trace amounts of boron to silicon produces an ____-type semiconductor.

67. Polysilicate anions contain the elements _____ and _____.

68. Naturally occurring silicates have a variety of structures. *True/False*

69. Quartz has the composition _____.

70. The Group 4 metals are _____.

71. The _____ oxidation state is the predominant state of lead.

72. The salts of germanium, tin, and lead show a(n) (*increasing/decreasing*) ionic character on descending the group.

73. The oxides of germanium, tin, and lead become increasingly (*acidic/basic*) upon descending the group.

C. CALCULATIONS YOU SHOULD KNOW HOW TO DO

There are no new types of calculations in this chapter.

D. SOLUTIONS TO THE ODD-NUMBERED PROBLEMS

25-1   Consult Figure 9-35 in the text.

25-3   The inner transition metal series are so-called because the electrons in the *f* orbitals lie closer to the nucleus than do the *d* electrons in the transition metals.

25-5 (a) metal (b) nonmetal (c) semimetal (d) metal
(e) metal (f) metal (g) semimetal (h) semimetal

25-7 (a) s-block (b) d-block (c) p-block (d) p-block
(e) p-block (f) p-block (g) f-block (h) f-block

25-9 The compounds may be covalent molecules or ionic compounds; therefore, X may be in Group 3 or Group 5.

25-11 (a) decreases (b) increases (c) decreases
(d) increases (e) increases

25-13 The elements in the third row can use the $3d$ orbitals in addition to the $3s$ and $3p$ orbitals in forming bonds, and thus the total number of bonds may exceed four. The $3d$ orbitals of the elements in the second row are of too high energy compared to the $2s$ and $2p$ orbitals to be involved in bonding.

25-15 The atomic mass of naturally occurring hydrogen is given by

$$\text{atomic mass} = \left(\frac{99.985}{100}\right)(1.0078) + \left(\frac{0.0148}{100}\right)(2.0141) + \text{trace}$$
$$= 1.0079$$

25-17 We shall use Equation (24-10) to calculate the specific activity of the sample. We must first calculate the half-life of tritium in seconds.

$$t_{1/2} = (12.4 \text{ y})\left(\frac{365 \text{ d}}{1 \text{ y}}\right)\left(\frac{24 \text{ h}}{1 \text{ d}}\right)\left(\frac{60 \text{ min}}{1 \text{ h}}\right)\left(\frac{60 \text{ s}}{1 \text{ min}}\right)$$
$$= 3.91 \times 10^8 \text{ s}$$

Using Equation (24-10), we have

$$\text{specific activity} = \frac{4.17 \times 10^{23} \text{ disintegration} \cdot \text{g}^{-1}}{M t_{1/2}}$$
$$= \frac{4.17 \times 10^{23} \text{ disintegration} \cdot \text{g}^{-1}}{(3.0160)(3.91 \times 10^8 \text{ s})}$$
$$= 3.536 \times 10^{14} \text{ disintegration} \cdot \text{s}^{-1} \cdot \text{g}^{-1}$$

The number of disintegrations per second per one micromole is given by

$$\text{activity} = (3.536 \times 10^{14} \text{ disintegration} \cdot \text{s}^{-1} \cdot \text{g}^{-1})\left(\frac{3.0160 \text{ g}}{1 \text{ mol}}\right)\left(\frac{1 \text{ mol}}{10^6 \, \mu \text{ mol}}\right)$$
$$= 1.07 \times 10^9 \text{ disintegration} \cdot \text{s}^{-1} \cdot \mu \text{ mol}^{-1}$$

25-19 (a) $Fe_2O_3(s) + 3H_2(g) \xrightarrow{\text{high T}} 2Fe(s) + 3H_2O(g)$
(b) $LiH(s) + H_2O(l) \rightarrow LiOH(aq) + H_2(g)$
(c) $Mg(s) + H_2(g) \rightarrow MgH_2(s)$
(d) $2K(s) + H_2(g) \rightarrow 2KH(s)$

25-21 $2Li(s) + 2CH_3CH_2OH(l) \rightarrow 2LiCH_3CH_2O(alc) + H_2(g)$

25-23   The equation for the reaction is

$$Zn(s) + 2HCl(aq) \longrightarrow ZnCl_2(aq) + H_2(g)$$

The number of moles of hydrogen generated is found by using the ideal-gas equation:

$$\text{moles} = \frac{PV}{RT} = \frac{(740 \text{ torr})\left(\dfrac{1 \text{ atm}}{760 \text{ torr}}\right)(0.500 \text{ L})}{(0.0821 \text{ L} \cdot \text{atm} \cdot \text{K}^{-1} \cdot \text{mol}^{-1})(293 \text{ K})} = 0.02024 \text{ mol}$$

The number of grams of zinc required is

$$\text{mass} = (0.02024 \text{ mol } H_2)\left(\frac{1 \text{ mol Zn}}{1 \text{ mol } H_2}\right)\left(\frac{65.38 \text{ g Zn}}{1 \text{ mol Zn}}\right) = 1.32 \text{ g}$$

25-25   The equation for the reaction is

$$WO_3(s) + 3H_2(g) \longrightarrow W(s) + 3H_2O(g)$$

The number of moles of hydrogen required is

$$\begin{aligned}
\text{moles} = {}&(2.50 \text{ metric tons } WO_3)\left(\frac{1000 \text{ kg}}{1 \text{ metric ton}}\right)\left(\frac{10^3 \text{ g}}{1 \text{ kg}}\right) \\
&\times \left(\frac{1 \text{ mol } WO_3}{231.9 \text{ g } WO_3}\right)\left(\frac{3 \text{ mol } H_2}{1 \text{ mol } WO_3}\right) \\
= {}&3.234 \times 10^4 \text{ mol}
\end{aligned}$$

The volume of hydrogen required is given by the ideal-gas equation:

$$\begin{aligned}
V = \frac{nRT}{P} &= \frac{(3.234 \times 10^4 \text{ mol})(0.0821 \text{ L} \cdot \text{atm} \cdot \text{K}^{-1} \cdot \text{mol}^{-1})(523 \text{ K})}{10.0 \text{ atm}} \\
&= 1.39 \times 10^5 \text{ L}
\end{aligned}$$

25-27   We use Equation (23-5) to calculate the maximum voltage that can be obtained:

$$\Delta G^\circ_{\text{rxn}} = -nFE^\circ_{\text{rxn}}$$

For the reaction, two electrons are transferred and thus $n = 2$. Therefore,

$$E^\circ_{\text{rxn}} = -\frac{\Delta G^\circ_{\text{rxn}}}{nF} = -\frac{-191 \times 10^3 \text{ J}}{(2 \text{ mol})(96,500 \text{ C} \cdot \text{mol}^{-1})} = 0.990 \text{ V}$$

25-29   The alkali metals must be stored under kerosene because they react with oxygen and water vapor in the air but not with kerosene.

25-31   Sodium metal is produced commercially by the electrolysis of molten sodium chloride.

25-33 (a) $4Li(s) + O_2(g) \rightarrow 2Li_2O(s)$

(b) $2Na(s) + O_2(g) \rightarrow Na_2O_2(s)$

(c) $K(s) + O_2(g) \rightarrow KO_2(s)$

(d) $Cs(s) + O_2(g) \rightarrow CsO_2(s)$

25-35 Any sodium metal produced by the electrolysis of an aqueous solution of sodium chloride will react with the water to produce sodium hydroxide and hydrogen gas.

25-37 Potassium, rubidium, and cesium form superoxides in addition to only oxides when oxidized.

25-39 The number of grams of NaH required is given by

$$\text{mass} = (1 \text{ kg TiCl}_4) \left(\frac{10^3 \text{ g}}{1 \text{ kg}} \right) \left(\frac{1 \text{ mol TiCl}_4}{189.70 \text{ g TiCl}_4} \right) \left(\frac{4 \text{ mol NaH}}{1 \text{ mol TiCl}_4} \right)$$
$$\times \left(\frac{24.00 \text{ g NaH}}{1 \text{ mol NaH}} \right)$$
$$= 506 \text{ g}$$

25-41 The number of moles of oxygen produced is given by

$$\text{moles} = (454 \text{ g Li}_2O_2) \left(\frac{1 \text{ mol Li}_2O_2}{45.88 \text{ g Li}_2O_2} \right) \left(\frac{1 \text{ mol O}_2}{2 \text{ mol Li}_2O_2} \right)$$
$$= 4.948 \text{ mol}$$

The volume of oxygen produced is given by the ideal-gas equation:

$$V = \frac{nRT}{P} = \frac{(4.948 \text{ mol})(0.0821 \text{ L} \cdot \text{atm} \cdot \text{K}^{-1} \cdot \text{mol}^{-1})(310 \text{ K})}{1.0 \text{ atm}}$$
$$= 126 \text{ L}$$

25-43 (a) $Ca(s) + H_2(g) \xrightarrow{500°C} CaH_2(s)$

(b) $3Mg(s) + N_2(g) \xrightarrow{500°C} Mg_3N_2(s)$

(c) $Sr(s) + S(s) \xrightarrow{500°C} SrS(s)$

(d) $Ba(s) + O_2(g) \xrightarrow{500°C} BaO_2(s)$

25-45 (a) $Be(s) + 2HCl(aq) \rightarrow BeCl_2(aq) + H_2(g)$

(b) $Be(s) + NaOH(aq) + 2H_2O(l) \rightarrow Na_2Be(OH)_4(aq) + H_2(g)$

(c) $3Be(s) + N_2(g) \xrightarrow{500°C} Be_3N_2(s)$

(d) $2Be(s) + O_2(g) \xrightarrow{300°C} 2BeO(s)$

25-47 (a) $2HCl(aq) + Mg(OH)_2(s) \rightarrow MgCl_2(aq) + 2H_2O(l)$

(b)   The number of milligrams of $Mg(OH)_2(s)$ required is given by

$$\text{mass} = (0.10 \text{ mmol HCl} \cdot mL^{-1})(1.0 \text{ mL}) \left(\frac{1 \text{ mmol Mg(OH)}_2}{2 \text{ mmol HCl}} \right)$$
$$\times \left(\frac{58.33 \text{ mg Mg(OH)}_2}{1 \text{ mmol Mg(OH)}_2} \right)$$
$$= 2.9 \text{ mg}$$

25-49 Magnesium chloride may be prepared from magnesium carbonate by reacting the magnesium carbonate with an aqueous solution of hydrochloric acid followed by evaporation of the water and the $HCl(g)$. The equation for the reaction is

$$MgCO_3(s) + 2HCl(aq) \longrightarrow MgCl_2(aq) + CO_2(g) + H_2O(g)$$

25-51 (a)   The Lewis formula for beryllium chloride is

$$:\overset{\cdot\cdot}{Cl}-Be-\overset{\cdot\cdot}{Cl}:$$

Beryllium chloride is an AX_2 class molecule and thus is linear.

(b)   The two bonds in beryllium chloride are formed by combining each of the two sp hybrid orbitals on the beryllium atom with one $3p$ orbital on each of the chlorine atoms.

25-53 The equation for the solubilities of the alkaline earth hydroxides in water is

$$M(OH)_2(s) \longrightarrow M^{2+}(aq) + 2OH^-(aq)$$

where M stands for the alkaline earth metal ion. The solubility s for each hydroxide is given by

$$s = [M^{2+}]$$

and therefore,

$$[OH^-] = 2s$$

For $Mg(OH)_2$, we have that

$$[OH^-] = (2) \left(\frac{9 \times 10^{-4} \text{ g}}{0.10 \text{ L}} \right) \left(\frac{1 \text{ mol Mg(OH)}_2}{58.33 \text{ g Mg(OH)}_2} \right) = 3.1 \times 10^{-4} \text{ M}$$

The pOH of the solution is given by

$$\text{pOH} = -\log [OH^-] = -\log (3.1 \times 10^{-4}) = 3.5$$

The pH of the solution is given by

$$\text{pH} = 14.00 - \text{pOH} = 14.00 - 3.5 = 10.5$$

For $Ca(OH)_2$, we have that

$$[OH^-] = (2)\left(\frac{0.18 \text{ g}}{0.10 \text{ L}}\right)\left(\frac{1 \text{ mol Ca(OH)}_2}{74.10 \text{ g Ca(OH)}_2}\right) = 0.0486 \text{ M}$$

$$pOH = -\log(0.0486) = 1.31$$

$$pH = 14.00 - 1.31 = 12.69$$

For $Sr(OH)_2$, we have that

$$[OH^-] = (2)\left(\frac{0.93 \text{ g}}{0.10 \text{ L}}\right)\left(\frac{1 \text{ mol Sr(OH)}_2}{121.6 \text{ g Sr(OH)}_2}\right) = 0.153 \text{ M}$$

$$pOH = -\log(0.153) = 0.815$$

$$pH = 14.00 - 0.815 = 13.18$$

For $Ba(OH)_2$, we have that

$$[OH^-] = (2)\left(\frac{5.8 \text{ g}}{0.10 \text{ L}}\right)\left(\frac{1 \text{ mol Ba(OH)}_2}{171.3 \text{ g Ba(OH)}_2}\right) = 0.677 \text{ M}$$

$$pOH = -\log(0.677) = 0.17$$

$$pH = 14.00 - 0.17 = 13.83$$

25-55 The acidities of the oxides of the Group 3 elements decrease upon descending the group.

25-57 Boric acid is an electron-pair acceptor. The equation for the reaction with water is

$$B(OH)_3(aq) + 2H_2O(l) \longrightarrow B(OH)_4^-(aq) + H_3O^+(aq)$$

25-59 (a) $(3 \times 3) + (8 \times 1) + 1 = 18$ valence electrons

(b) $(10 \times 3) + (10 \times 1) + 2 = 42$ valence electrons

25-61 (a) :Cl—Ga—Cl: AX_3 trigonal planar
            |
          :Cl:

(b) :F—Ga—F: AX_2 linear   (Ga has $\oplus$ charge)

(c) :Br—Ga—Br: AX_4 tetrahedral   (Ga has $\ominus$ charge, with :Br: above and :Br: below)

25-63 See the discussion of the bonding in diborane given in Section 25-5B.

25-65 The equation for the dissociation reaction for each of the Group 3 M^{3+} ions is given by (see Section 18-11)

$$M(H_2O)_6^{3+}(aq) + H_2O(l) \longrightarrow M(OH)(H_2O)_5^{2+}(aq) + H_3O^+(aq)$$

The equilibrium-constant expression for each equation is given by

$$K_a = \frac{[M(OH)(H_2O)_5^{2+}][H_3O^+]}{[M(H_2O)_6^{3+}]}$$

We also have the equilibrium concentration expressions

$$[M(OH)(H_2O)_5^{2+}] = [H_3O^+]$$
$$[M(H_2O)_6^{3+}] = 0.10 \text{ M} - [H_3O^+]$$

For $Al^{3+}(aq)$, the value of the equilibrium constant is

$$K_a = 10^{-pK_a} = 10^{-4.96} = 1.096 \times 10^{-5} \text{ M}$$

The equilibrium-constant expression is given by

$$K_a = \frac{[H_3O^+]^2}{0.10 \text{ M} - [H_3O^+]} = 1.096 \times 10^{-5} \text{ M}$$

The method of successive approximation yields

$$[H_3O^+] = 1.04 \times 10^{-3} \text{ M}$$

The pH of the solution is

$$pH = -\log(1.04 \times 10^{-3}) = 2.98$$

For $Ga^{3+}(aq)$, we have that

$$K_a = 10^{-2.60} = 2.51 \times 10^{-3} \text{ M}$$
$$K_a = \frac{[H_3O^+]^2}{0.10 \text{ M} - [H_3O^+]} = 2.51 \times 10^{-3} \text{ M}$$

The method of successive approximation yields 0.0146 M. Thus we have that

$$pH = -\log(0.0146) = 1.83$$

For $In^{3+}(aq)$, we have that

$$K_a = 10^{-2.66} = 2.19 \times 10^{-3} \text{ M}$$
$$K_a \frac{[H_3O^+]^2}{0.10 \text{ M} - [H_3O^+]} = 2.19 \times 10^{-3} \text{ M}$$

The method of successive approximation yields 0.0137 M. Thus we have that

$$pH = -\log (0.0137) = 1.86$$

For $Tl^{3+}(aq)$, we have that

$$K_a = 10^{-1.15} = 0.0708$$

$$K_a = \frac{[H_3O^+]^2}{0.10 \text{ M} - [H_3O^+]} = 0.0708 \text{ M}$$

We shall use the quadratic equation to solve for $[H_3O^+]$ because the value of K_a is large.

$$[H_3O^+]^2 + 0.0708 \text{ M} [H_3O^+] - 0.00708 \text{ M}^2 = 0$$

$$[H_3O^+] = \frac{-0.0708 \text{ M} \pm \sqrt{5.013 \times 10^{-3} \text{ M}^2 - (4)(1)(-0.00708 \text{ M}^2)}}{2}$$

$$= \frac{-0.0708 \text{ M} \pm 0.183 \text{ M}}{2} = 0.0559 \text{ M}$$

$$pH = -\log (0.0559) = 1.25$$

25-67 (a) $Al_2O_3(s) + 3C(s) \xrightarrow[\text{furnace}]{\text{electric}} 2Al(s) + 3CO(g)$

(b) $CaC_2(s) + 2D_2O(l) \xrightarrow{\text{heat}} Ca(OD)_2(s) + C_2D_2(g)$

(c) $2PbS(s) + 3O_2(g) \rightarrow 2PbO(s) + 2SO_2(g)$

25-69 When pressure is applied, the system will shift toward the phase with the smaller molar volume in order to relieve the stress. Diamond is the stable form at high pressure, and so has the smaller molar volume and therefore the greater density.

25-71 Diamond is essentially one large covalently bonded molecule and thus has no electrons available to conduct an electric current. Graphite has many delocalized electrons in the covalently bonded layers. The delocalized electrons are mobile and can carry an electric current.

25-73 An element with five valence electrons is added to silicon to make an n-type semiconductor; thus we would add an element from Group 5 such as antimony or bismuth. An element with three valence electrons is added to silicon to make a p-type semiconductor; thus we would add an element from Group 3 such as aluminum or gallium.

25-75 See Section 4-2 for the rules on assigning oxidation states to elements.

(a) The oxidation state of O is -2; the oxidation state of Pb is $x + 2(-2) = 0$ or $+4$.

(b) The oxidation state of O is -2; the oxidation state of Pb is $3x + 4(-2) = 0$ or $+8/3$.

(c) The oxidation state of Al is $+3$; the oxidation state of C is $3x + 4(+3) = 0$ or -4.

(d) The oxidation state of Ca is $+2$; the oxidation state of C is $2x + (+2) = 0$ or -1.

25-77 $SiO_2(s) + 6HF(aq) \rightarrow H_2SiF_6(aq) + 2H_2O(l)$
   glass

25-79   The Lewis formula of SiH_4 is

$$
\begin{array}{c}
\text{H} \\
| \\
\text{H}-\text{Si}-\text{H} \\
| \\
\text{H}
\end{array}
$$

According to VSEPR theory, SiH_4 is tetrahedral, and so it is appropriate to use sp^3 hybrid orbitals on the silicon atom and $1s$ atomic orbitals on the hydrogen atoms.
   The Lewis formula of Si_2H_6 is

$$
\begin{array}{cc}
\text{H} & \text{H} \\
| & | \\
\text{H}-\text{Si}-\text{Si}-\text{H} \\
| & | \\
\text{H} & \text{H}
\end{array}
$$

Again, it is appropriate to use sp^3 hybrid orbitals on the silicon atom and $1s$ atomic orbitals on the hydrogen atoms. The Si—Si bond can be described by combining an sp^3 hybrid orbital on each of the silicon atoms.

25-81   See Section 25-6E.

E. ANSWERS TO THE SELF-TEST

1. cesium
2. fluorine
3. true
4. false
5. basic
6. acidic
7. true
8. group number
9. false
10. two
11. true
12. colorless, odorless gas
13. $+1, 0, -1$

14. hydrogen-1, deuterium, and tritium
15. deuterium and oxygen
16. an aqueous dilute acid
17. $Ca(s) + 2H_2O(l) \rightarrow$
    $\qquad\qquad Ca(OH)_2(aq) + H_2(g)$
18. The reaction is too vigorous.
19. $2H_2O(l) \rightarrow 2H_2(g) + O_2(g)$
20. -1
21. true
22. true
23. The metals react with oxygen and water vapor in the air
24. alkali

25. electrolysis of molten sodium chloride

26. true

27. true

28. the metal hydroxide and hydrogen

29. sodium alkoxides and hydrogen

30. false

31. lithium

32. false

33. alkaline earth

34. false

35. false

36. electrolysis of the molten chlorides

37. $+2$

38. $M(s) + 2H_2O(l) \rightarrow$
    $$M(OH)_2(aq) + H_2(g)$$

39. ionic

40. false

41. covalent

42. $B(OH)_3(aq) + 2H_2O(l) \rightarrow$
    $$B(OH)_4^-(aq) + H_3O^+(aq)$$

43. Lewis

44. false

45. three-center bond

46. Hall

47. the halide salts

48. increasingly

49. decreases

50. false

51. reducing

52. graphite and diamond

53. covalent

54. graphite

55. true

56. hydrogen

57. covalent

58. carbide

59. acetylene and calcium hydroxide

60. CO and CO_2

61. silicon dioxide

62. silicon and germanium

63. true

64. thermal

65. five

66. p

67. silicon and oxygen

68. true

69. SiO_2

70. tin and lead

71. $+2$

72. increasing

73. basic

26 THE CHEMISTRY OF THE MAIN-GROUP ELEMENTS II

A. OUTLINE OF CHAPTER 26

26-1 The chemical properties of the Group 5 elements.

Some properties of the Group 5 elements are given in Table 26-2.

26-1A Nitrogen, N_2, has a very strong triple bond.

Nitrogen is very unreactive.

The conversion of nitrogen from the free element to nitrogen compounds is called nitrogen fixation.

The nitrogen-fixing bacterium, *Rhizobium bacterium,* invades the roots of leguminous plants.

26-1B Most nitrogen is converted to ammonia by the Haber process.

Ammonia, NH_3, is prepared from nitrogen and hydrogen at high pressure and temperature.

Ammonia is very soluble in water, forming a basic solution.

Ammonia reacts with acids to form ammonium compounds, many of which are important commercially, especially as fertilizers.

26-1C Nitric acid is produced by the Ostwald process.

The three steps in the Ostwald process are

1. The reaction of ammonia and oxygen in the presence of a platinum catalyst to produce nitrogen oxide and water.
2. The reaction between nitrogen oxide and oxygen to produce nitrogen dioxide.
3. The dissolution of nitrogen dioxide in water to produce an aqueous solution of nitric acid and nitrogen oxide gas.

Nitric acid, HNO_3, has a number of industrial uses.

26-1D Nitrogen forms several important compounds with hydrogen and oxygen.

Hydrazine, N_2H_4, is produced by the Raschig synthesis, in which ammonia is reacted with hypochlorite ion $ClO^-(aq)$ in basic solution.

The reaction between hydrazine and oxygen releases a large amount of energy.

The principal oxides of nitrogen are given in Table 26-3.

The preparations of the various nitrogen oxides are discussed on pages 933 and 934.

Nitrous acid, HNO_2, is a weak acid, whose salts are called nitrites.

Hydrazoic acid, HN_3, is produced by the reaction between nitrous acid and hydrazine.

Some salts of hydrazoic acid, called azides, are explosive.

Some commercially important compounds of nitrogen are given in Table 26-4.

26-1E There are two principal allotropes of solid phosphorus.

White phosphorus consists of tetrahedral P_4 molecules, and is extremely reactive.

Red phosphorus is an amorphous substance, which is much less reactive than white phosphorus.

Red phosphorus is produced by heating white phosphorus above $250°C$ in the absence of air.

Phosphorus is not found free in nature.

The principal source of phosphorus is phosphate rock.

Phosphate rock is insoluble and cannot be used directly as a fertilizer.

Treatment of phosphate rock with sulfuric acid produces a water-soluble product, $Ca(H_2PO_4)_2$, called superphosphate.

26-1F The oxides of phosphorus are acid anhydrides.

White phosphorus reacts directly with oxygen to produce the oxides P_4O_6, phosphorus trioxide, and P_4O_{10}, phosphorus pentoxide.

The structures of the two oxides are shown in Figure 26-10.

The phosphorus oxides P_4O_6 and P_4O_{10} react with water to give the acids H_3PO_3, phosphorous acid, and H_3PO_4, phosphoric acid, respectively.

Phosphorus pentoxide is a powerful dehydrating agent.

Hypophosphorous acid, H_3PO_2, is a monoprotic oxyacid of phosphorus.

The pK_a values of the oxyacids of phosphorus are given in Table 26-5.

The structures of the three oxyacids of phosphorus are shown in Figure 26-11.

26-1G   Phosphorus forms a number of binary compounds.

Phosphorus reacts directly with reactive metals to form phosphides.

Phosphine, PH_3, can be prepared by the reaction between white phosphorus and a strong base.

Phosphorus reacts directly with the halogens to form the halides PX_3 and PX_5.

Phosphorus reacts with sulfur to form the crystalline compound P_4S_3.

26-1H   Many phosphorus compounds are important biologically.

The energy requirements for many biochemical reactions are supplied by adenosine triphosphate, ATP.

Many organic phosphates are potent insecticides.

Some commercially important compounds of phosphorus are given in Table 26-6.

26-1I   Arsenic and antimony are semimetals and bismuth is the only Group 5 metal.

Both semimetals are produced by first roasting their ores, As_2S_3 and Sb_2S_3, in oxygen and then reducing the oxide with carbon or hydrogen to the free element.

Arsenic tends to favor an oxidation state (III) two less than the maximum for the group.

The oxides tend from acidic to basic with increasing atomic number.

Bismuth is a pink-white metal; its common source is the sulfide ore Bi_2S_3.

The principal compounds of bismuth contain Bi(III), an oxidation state two less than the group number.

Some commercially important compounds of arsenic, antimony, and bismuth are given in Table 26-7.

26-2 The chemical properties of the Group 6 elements.

Oxygen is the most abundant element on earth.

Some properties of the Group 6 elements are given in Table 26-9.

26-2A   Over 35 billion pounds of oxygen are produced annually in the United States.

Oxygen is produced industrially by the fractional distillation of liquid air.

Oxygen can be prepared in the laboratory by heating potassium chlorate in the presence of the catalyst MnO_2 or by adding sodium peroxide to water.

26-2B   Oxygen in the earth's atmosphere is produced by photosynthesis.

Photosynthesis is the process by which green plants combine carbon dioxide and water into carbohydrates and oxygen under the influence of visible light.

26-2C   Oxygen reacts directly with most other elements.

Oxygen forms oxides with many elements.

Hydrocarbons burn in oxygen to give carbon dioxide and water.

26-2D   Some metals react with oxygen to yield peroxides.

Peroxides are compounds in which the negative ion is the peroxide ion O_2^{2-}.

Hydrogen peroxide, H_2O_2, is a strong oxidizing agent and has a number of uses.

26-2E   Ozone is a potent oxidizing agent.

Ozone, O_3, is produced when a spark is passed through oxygen.

Ozone is extremely reactive.

Ozone is produced in the stratosphere by the action of radiation on oxygen.

Ozone screens out ultraviolet light in the stratosphere.

26-2F   Sulfur exists as rings of eight sulfur atoms.

Sulfur does not react with dilute acids or bases.

Sulfur reacts with many metals at elevated temperatures to form metal sulfides.

Sulfur is extracted from underground deposits by the Frasch process (Figure 26-21).

When sulfur is heated from room temperature to above its boiling point, it undergoes a number of physical and chemical changes (Figure 26-23).

Sulfur exists as rings of S_8 below 150°C; the rings begin to break apart above this temperature.

Sulfur in the vapor state exists as S_8 rings.

26-2G   Sulfuric acid is the leading industrial chemical.

Most sulfuric acid is made by the contact process.

In the contact process, sulfur is burned in oxygen to produce sulfur dioxide, which is converted to sulfur trioxide by reaction with oxygen in the presence of a vanadium pentoxide catalyst. The sulfur trioxide is absorbed into nearly pure liquid sulfuric acid to form fuming sulfuric acid, which is added to water or aqueous sulfuric acid to produce the desired concentration of sulfuric acid.

Sulfuric acid is a strong dehydrating agent.

Dilute solutions of sulfuric acid react with reactive metals to produce the metal salts and hydrogen.

Hot, concentrated sulfuric acid acts as an oxidizing agent.

26-2H   Sulfur forms several widely used compounds.

Sulfur undergoes combination reactions with most metals and nonmetals.

Sulfur forms sulfides with most metals, most of which are insoluble.

The reaction of a solution of a strong acid with a metal sulfide yields hydrogen sulfide.

Mercaptans are foul-smelling organic compounds that contain an —SH group.

Sulfur dioxide is very soluble in water.

Salts of sulfurous acid are called sulfites.

The thiosulfate ion is produced when the sulfite ion is boiled in the presence of sulfur.

The thiosulfate ion is a mild reducing agent that is used in the determination of iodine in analytical chemistry.

Some commercially important compounds of sulfur are given in Table 26-11.

26-2I   Selenium and tellurium behave like sulfur.

The major source of the two elements is as a by-product from the electrolytic refining of copper.

Selenium and tellurium form covalently bonded compounds.

Some important compounds of selenium and tellurium are given in Table 26-12.

26-3   The chemical properties of the halogens.

The halogens are the Group 7 elements.

The halogens are all nonmetals.

Some properties of the Group 7 elements are given in Table 26-14.

26-3A   Fluorine is the most reactive element.

Fluorine reacts directly, and in most cases vigorously, with all the elements except helium, neon, and argon.

Fluorine reacts spontaneously with the low molecular mass hydrocarbons.

Elemental fluorine is obtained by electrolysis of hydrogen fluoride in molten potassium fluoride.

Hydrogen fluoride reacts with glass.

Hydrogen fluoride is a nonelectrolyte consisting of $(HF)_6$ hydrogen-bonded rings.

In aqueous solution, hydrofluoric acid is a weak acid that reacts with a base to form both the fluoride salt and the acid fluoride salt, HF_2^-.

26-3B   Chlorine is obtained from chlorides by electrolysis.

Chlorine may be produced in the laboratory by heating a mixture of hydrochloric acid and manganese dioxide.

Chlorine combines directly with most other elements except carbon, nitrogen, oxygen, and the noble gases.

Most metals react with chlorine to form ionic chlorides.

Nonmetals react with chlorine to form covalent chlorides.

Chlorine is a strong oxidizing agent.

Many chlorinated hydrocarbons present a serious health hazard to humans.

26-3C Bromine and iodine are obtained by oxidation of bromides and iodides with chlorine.

Bromine is used to prepare a variety of metal bromides and organo-bromide compounds.

Iodine is the only halogen to occur in nature in a positive oxidation state as the iodate ion, IO_3^-.

Iodine can be prepared from the iodate ion by reduction with sodium hydrogen sulfite.

Iodine is not very soluble in water, but it is soluble in $KI(aq)$ due to the formation of the ion $I_3^-(aq)$.

Some important compounds of the halogens are given in Table 26-15.

26-3D The hydrogen halides are prepared from halogen salts.

The hydrogen halides, except for HI, can be prepared by the reaction between the corresponding alkali metal halide and concentrated sulfuric acid.

Hydrogen chloride is a by-product in the chlorination of hydrocarbons and can also be prepared by the reaction between phosphorus pentachloride and water.

Hydrogen bromide is produced industrially by the reaction between hydrogen and bromine.

Hydrogen iodide can be prepared by the reaction between iodine and hydrazine, N_2H_4.

26-3E The halogens form a family of oxyacids.

The halogens form a series of oxyacids in which the oxidation state of the halogen atom can be $+1$, $+3$, $+5$, or $+7$.

Some of the halogen oxyacids are listed in Table 26-16.

There are no oxyacids of fluorine; the only oxidation state of fluorine is -1.

The strength of an oxyacid depends strongly upon the number of oxygen atoms without attached hydrogen atoms that are attached to the central atom.

The value of the pK_a of the oxyacid is related to the number of oxygen atoms without attached hydrogen atoms, n_o, by Equation (26-1).

The strength of an oxyacid increases by a factor of about 10^5 for a unit increase in n_o.

The values of pK_a for some oxyacids are given in Table 26-17.

26-4 The chemical properties of the noble gases.

The noble gases are the Group 8 elements.

Some properties of the Group 8 elements are given in Table 26-18.

26-4A   The noble gases were not discovered until 1893.

The noble gases were discovered in air.

The noble gases are relatively inert.

26-4B   Xenon forms compounds with fluorine and oxygen.

Prior to 1962, it was thought that the noble gases did not form chemical compounds.

In 1962, Neil Bartlett reacted platinum hexafluoride with xenon to prepare the first noble-gas compound.

Three xenon fluorides can be prepared by direct combination of xenon and fluorine in a nickel vessel.

Xenon tetrafluoride, XeF_4, hydrolyzes to form the oxide XeO_3.

The only known compound of krypton is KrF_2.

There are no reported compounds containing the remaining noble gases.

B. SELF-TEST

1. Nitrogen is a reactive molecule. *True/False*

2. The bond in the nitrogen molecule is (*strong/weak*).

3. Nitrogen fixation is the conversion of nitrogen to _____.

4. Nitrogen is obtained commercially from _____.

5. Most nitrogen is used to produce _____.

6. Briefly describe the Haber process. _____

_____

7. Solutions of ammonia are (*acidic/neutral/basic*).

8. Ammonia reacts with acids to form _____ compounds.

9. An important use of ammonium compounds is fertilizers. *True/False*

10. The Ostwald process is used to produce _____.

11. The Raschig process is used to produce _____.

12. Nitrogen forms only one compound with hydrogen. *True/False*

13. Nitrogen forms only one compound with oxygen. *True/False*

14. Nitrous acid is a (*strong/weak*) acid.

15. The more reactive allotrope of phosphorus is (*white/red*) phosphorus.

16. Red phosphorus is prepared by heating white phosphorus in the absence of air. *True/False*

17. Phosphorus is obtained from _____.

18. Phosphate rock can be used directly as a fertilizer. *True/False*

19. The molecular formula of phosphorus pentoxide is _____.

20. Phosphorus trioxide is the acid anhydride of the acid _____.

21. All the phosphorus oxyacids are triprotic acids. *True/False*

22. Phosphorus reacts directly with reactive metals to form _____.

23. Phosphine is a base in water. *True/False*

24. Phosphorus forms two halides of the form _____ and _____.

25. Arsenic and antimony are metals. *True/False*

26. Arsenic tends to favor the $+3$ oxidation state. *True/False*

27. The oxides of arsenic are (*acidic/amphoteric/basic*).

28. The oxides of antimony are (*acidic/amphoteric/basic*).

29. Bismuth is a metal. *True/False*

30. Oxygen is produced commercially from _____.

31. In the process of photosynthesis, green plants combine _____ and _____ to form carbohydrates and oxygen.

32. Oxygen can be prepared in the laboratory by the thermal decomposition of _____.

33. Oxygen reacts with most elements. *True/False*

34. Oxygen reacts with many metals to form _____.

35. The alkali metals react with oxygen to form only oxides of the form X_2O. *True/False*

36. Hydrogen peroxide is a strong oxidizing agent. *True/False*

37. Ozone is extremely reactive. *True/False*

38. Ozone in the stratosphere screens out (*infrared/visable/ultraviolet*) radiation.

39. Sulfur occurs as the free element in nature. *True/False*

40. The molecular units of monoclinic sulfur are _____.

41. The molecular units of plastic sulfur are _____.

42. The molecular units of gaseous sulfur are _____.

43. The contact process is used to produce _____.

44. Hot, concentrated sulfuric acid is an (*oxidizing/reducing*) agent.

45. Sulfur reacts with many metals to form _____.

46. Most sulfides are (*soluble/insoluble*) in water.

47. Organic compounds that contain an —SH group are called _____.

48. Aqueous solutions of sulfur dioxide are (*acidic/neutral/basic*).

49. Sulfurous acid is a strong acid. *True/False*

50. The thiosulfate ion is used in analytical chemistry to determine the amount of iodine present in a sample. *True/False*

51. Selenium and tellurium are metals. *True/False*

52. Selenium and tellurium have similar chemical properties to sulfur. *True/False*

53. Fluorine reacts with almost all the elements. *True/False*

54. The only oxidation state of fluorine is _____.

55. Elemental fluorine is produced by the electrolysis of _____

_____.

56. Hydrofluoric acid is a weak acid. *True/False*

57. Hydrofluoric acid reacts with sodium hydroxide to produce water, _____,

and _____.

58. Chlorine is obtained by _____.

59. Chlorine reacts with most other elements. *True/False*

60. Chlorine is a strong (*oxidizing/reducing*) agent.

61. Bromine is obtained by _____.

62. _____ is the only halogen to occur naturally in a positive oxidation state.

63. Iodine is soluble in aqueous KI solution because of the formation of _____ .

64. Chlorine, bromine, and iodine have similar chemical properties. *True/False*

65. The hydrogen halides, HCl, HBr, and HI, act as (*strong/weak*) acids in aqueous solution.

66. The halogens can have a maximum oxidation state of _____ in their oxyacids.

67. The names of the chlorine oxyacids are based on $HClO_3$, which is called _____ .

68. The strength of an oxyacid depends upon the number of _____ _____ _____ .

69. Hypochlorous acid, HClO, is a (*stronger/weaker*) acid than chlorous acid, $HClO_2$.

70. The most notable chemical property of the noble gases is their _____ .

71. The noble gases were discovered in the atmosphere. *True/False*

72. Xenon forms compounds with the elements _____ and _____ .

73. Xenon reacts directly with fluorine. *True/False*

74. All the noble gases react with fluorine. *True/False*

C. CALCULATIONS YOU SHOULD KNOW HOW TO DO

There are no new types of calculations presented in this chapter.

D. SOLUTIONS TO THE ODD-NUMBERED PROBLEMS

26-1 (a) $P_4(s) + 5O_2(g) \rightarrow P_4O_{10}(s)$
            excess

(b) $P_4O_6(s) + 6H_2O(l) \rightarrow 4H_3PO_3(aq)$

(c) $P_4O_{10}(s) + 6H_2O(l) \rightarrow 4H_3PO_4(aq)$

26-3 We recognize that D is deuterium, an isotope of hydrogen, and so by analogy with the reaction of lithium nitride and water, we can use the reaction

$$Li_3N(s) + 3D_2O(l) \longrightarrow 3LiOD(aq) + ND_3(g)$$

26-5 Nitrogen can form only four covalent bonds using a total of eight electrons.

26-7 (a)

$$
\begin{array}{c}
\text{H} \\
| \\
\text{H}-\ddot{\text{N}}-\ddot{\text{O}}-\text{H}
\end{array}
$$

(b)

$$
\begin{array}{c}
^{\ominus}\ddot{\text{O}} \\
\diagdown \\
\overset{\oplus}{\text{N}}=\ddot{\text{O}}\ddot{} \\
\diagup \\
_{\ominus}\ddot{\text{O}}\ddot{}
\end{array}
\qquad \text{plus other resonance forms}
$$

(c) $\quad ^{\ominus}\ddot{\text{O}}-\text{N}=\ddot{\text{O}}\ddot{} \longleftrightarrow \ddot{\text{O}}=\text{N}-\ddot{\text{O}}\ddot{}^{\ominus}$

(d) $\quad ^{\ominus}\ddot{\text{N}}=\overset{\oplus}{\text{N}}=\ddot{\text{O}}\ddot{} \longleftrightarrow \ddot{}\text{N}\equiv\overset{\oplus}{\text{N}}-\ddot{\text{O}}\ddot{}^{\ominus}$

(e) $\quad ^{\ominus}\ddot{\text{O}}-\overset{\oplus}{\text{N}}=\ddot{\text{O}}\ddot{} \longleftrightarrow \ddot{\text{O}}=\overset{\oplus}{\text{N}}-\ddot{\text{O}}\ddot{}^{\ominus}$

26-9 $\quad P_4O_{10}(s) + 6H_2SO_4(l) \rightarrow 4H_3PO_4(l) + 6SO_3(g)$

$ P_4O_{10}(s) + 12HNO_3(l) \rightarrow 4H_3PO_4(l) + 6N_2O_5(s)$

26-11

Lewis Formula	Class of Molecule	Shape
(a) $\ddot{\text{Cl}}-\overset{\overset{\ddot{\text{O}}}{\|}}{\underset{\underset{\ddot{\text{Cl}}}{\|}}{\text{P}}}-\ddot{\text{Cl}}$	AX_4	Tetrahedral
(b) $^{\ominus}\ddot{\text{O}}-\overset{\overset{\ddot{\text{O}}}{\|}}{\underset{\underset{\ddot{\text{O}}^{\ominus}}{\|}}{\text{P}}}-\ddot{\text{O}}^{\ominus}$	AX_4	Tetrahedral
(c) $\ddot{\text{Cl}}-\text{P}-\ddot{\text{Cl}}$ (with Cl above and below, $\ominus$)	AX_6	Octahedral

26-13 $\quad As_4O_6(s) + 12NaOH(aq) \longrightarrow 4Na_3AsO_3(aq) + 6H_2O(l)$

$ As_4O_{10}(s) + 20NaOH(aq) \longrightarrow 4Na_3AsO_4(aq) + 10H_2O(l)$

$ Sb_2O_3(s) + 6NaOH(aq) \longrightarrow 2Na_3SbO_3(aq) + 3H_2O(l)$

$ Sb_2O_3(s) + 6HCl(aq) \longrightarrow 2SbCl_3(aq) + 3H_2O(l)$

$ Bi_2O_3(s) + 6HCl(aq) \longrightarrow 2BiCl_3(aq) + 3H_2O(l)$

26-15 The source of most of the oxygen in the earth's atmosphere is from photosynthesis in green plants, the process of using carbon dioxide, water, and energy from the sun to produce oxygen and carbohydrates.

26-17 $\quad 2C_2H_2(g) + 5O_2(g) \longrightarrow 4CO_2(g) + 2H_2O(g)$

26-19   Consult Section 26-2F.

26-21   The loss of gas in a tennis ball is due to effusion of the gas. The molecular mass of $SF_6(g)$ is much greater than the molecular masses of the molecules in air, mainly $N_2(g)$ and $O_2(g)$; therefore, the rate of effusion of $SF_6(g)$ is much slower than that of air. Recall that Graham's law states that the rate of effusion is inversely proportional to the square root of the molecular mass of the gas.

26-23

$$H-\overset{..}{\underset{..}{O}}-\overset{\overset{..}{\overset{\displaystyle O}{\|}}}{\underset{\underset{..}{\underset{\displaystyle O}{\|}}}{S}}-\overset{..}{\underset{..}{O}}-\overset{\overset{..}{\overset{\displaystyle O}{\|}}}{\underset{\underset{..}{\underset{\displaystyle O}{\|}}}{S}}-\overset{..}{\underset{..}{O}}-H \qquad \text{plus other resonance forms}$$

26-25   The Lewis formulas of the two molecules are

$$\text{SF}_4 \qquad \text{SF}_6$$

The structure of SF_4 is square pyramidal, and the structure of SF_6 is octahedral. One reason for the relative lack of reactivity of SF_6 is that the sulfur atom in SF_6 is effectively hidden by six fluorine atoms, and thus the water molecule has difficulty in reacting with the sulfur atom. The sulfur atom in SF_4 is more open to reaction with a water molecule, and hence the reaction takes place more rapidly. Another factor is that the lone pair of electrons in SF_4 can act as a reactive center.

26-27   The equation for the reaction is

$$2S_2O_3^{2-}(aq) + I_3^-(aq) \longrightarrow 3I^-(aq) + S_4O_6^{2-}(aq)$$

The number of millimoles of $I_3^-(aq)$ that reacted is

$$\text{millimoles of } I_3^- = (0.150 \text{ mmol } S_2O_3^{2-}\cdot\text{mL}^{-1})(28.5 \text{ mL})\left(\frac{1 \text{ mmol } I_3^-}{2 \text{ mmol } S_2O_3^{2-}}\right)$$

$$= 2.138 \text{ mmol}$$

The concentration of $I_3^-(aq)$ is

$$M = \frac{n}{V} = \frac{2.138 \text{ mmol}}{35.0 \text{ mL}} = 0.0611 \text{ M}$$

26-29   (a)   The equations for the balanced half reactions are

$$2Cl^- \longrightarrow Cl_2 + 2e^- \qquad\qquad \text{(oxidation)}$$

$$MnO_2 + 4H^+ + 2e^- \longrightarrow Mn^{2+} + 2H_2O \qquad \text{(reduction)}$$

Add the above two equations to obtain

$$2Cl^- + MnO_2 + 4H^+ \longrightarrow Mn^{2+} + Cl_2 + 2H_2O$$

Add the ions needed to form neutral compounds and the phases to obtain

$$4NaCl(aq) + MnO_2(s) + 2H_2SO_4(aq) \longrightarrow$$
$$MnCl_2(aq) + Cl_2(g) + 2H_2O(l) + 2Na_2SO_4(aq)$$

Note that an additional 2 mol of NaCl are required to balance the ionic charges.

(b) The equations for the balanced half reactions are

$$2IO_3^- + 12H^+ + 10e^- \longrightarrow I_2 + 6H_2O \qquad \text{(reduction)}$$
$$5HSO_3^- + 5H_2O \longrightarrow 5SO_4^{2-} + 15H^+ + 10e^- \qquad \text{(oxidation)}$$

Add the above two equations to obtain

$$2IO_3^- + 5HSO_3^- \longrightarrow I_2 + 5SO_4^{2-} + 3H^+ + H_2O$$

We shall need an even number of H^+ ions to form H_2SO_4, and so we must multiply the equation by 2. Add the ions to form neutral compounds and the phases to obtain

$$4NaIO_3(aq) + 10NaHSO_3(aq) \longrightarrow$$
$$2I_2(s) + 3H_2SO_4(aq) + 7Na_2SO_4(aq) + 2H_2O(l)$$

(c) The equations for the balanced half reactions (in basic solution) are

$$5Br_2 + 10e^- \longrightarrow 10Br^- \qquad \text{(reduction)}$$
$$Br_2 + 12OH^- \longrightarrow 2BrO_3^- + 6H_2O + 10e^- \qquad \text{(oxidation)}$$

Note that bromine is acting as both an oxidizing agent and a reducing agent. Add the above two equations to obtain

$$6Br_2 + 12OH^- \longrightarrow 2BrO_3^- + 10Br^- + 6H_2O$$

Add sodium ions to form neutral compounds and add the phases to obtain

$$6Br_2(l) + 12NaOH(aq) \longrightarrow 2NaBrO_3(aq) + 10NaBr(aq) + 6H_2O(l)$$

26-31 The equations for the balanced half reactions are

$$3I_2 + 18H_2O \longrightarrow 6IO_3^- + 36H^+ + 30e^- \qquad \text{(oxidation)}$$
$$10NO_3^- + 40H^+ + 30e^- \longrightarrow 10NO + 20H_2O \qquad \text{(reduction)}$$

Add the above two equations to obtain

$$3I_2 + 10NO_3^- + 4H^+ \longrightarrow 6IO_3^- + 10NO + 2H_2O$$

Add six H^+ ions to form the neutral acids and add the phases to obtain

$$3I_2(s) + 10HNO_3(aq) \longrightarrow 6HIO_3(aq) + 10NO(g) + 2H_2O(l)$$

26-33   The names of the acids are

(a)   nitrous acid   (b)   sulfurous acid
(c)   hypophosphorous acid   (d)   phosphorous acid
(e)   hyponitrous acid

The names of the salts are

(a)   potassium sulfite   (b)   calcium nitrite
(c)   potassium iodite   (d)   magnesium hypobromite

26-35   The electronegativity of nitrogen is greater than that of hydrogen, and so the nitrogen atom in NH_3 has a partial negative charge. Therefore, the lone pair on the nitrogen atom is held relatively loosely and can be easily donated. In NF_3, on the other hand, the nitrogen has a partial positive charge because fluorine is more electronegative than nitrogen. Therefore, the single lone pair on the nitrogen atom is held relatively tightly and cannot be easily donated.

26-37   The Lewis formula of Cl_2O_7 is

plus other resonance forms

According to VSEPR theory, the bonding around each chlorine atom is tetrahedral and the geometry around the central oxygen atom is bent.

26-39   (a)   The oxidation state of fluorine is -1 and that of iodine is $+5$.
(b)   The oxidation state of chlorine is $+1$.
(c)   The oxidation state of bromine is $+5$.
(d)   The oxidation state of fluorine is -1 and that of chlorine is $+1$.
(e)   The oxidation state of iodine is $+5$.

26-41   The value of ΔG°_{rxn} is given by

$$\Delta G^\circ_{rxn} = (1 \text{ mol})\, \Delta G^\circ_f[I_3^-(aq)] - (1 \text{ mol})\, \Delta G^\circ_f[I_2(aq)] - (1 \text{ mol})\, \Delta G^\circ_f[I^-(aq)]$$
$$= (1 \text{ mol})(-51.40 \text{ kJ} \cdot \text{mol}^{-1}) - (1 \text{ mol})(16.40 \text{ kJ} \cdot \text{mol}^{-1})$$
$$- (1 \text{ mol})(-51.57 \text{ kJ} \cdot \text{mol}^{-1})$$
$$= -16.23 \text{ kJ}$$

The relationship between ΔG°_{rxn} and the equilibrium constant is given by Equation (22-10).

$$\ln K = -\frac{\Delta G^\circ_{rxn}}{RT} = -\frac{-16.23 \times 10^3 \text{ J}}{(8.314 \text{ J} \cdot \text{K}^{-1})(298 \text{ K})} = +6.551$$

$$K = e^{6.551} = 700 \text{ M}^{-1}$$

26-43 The number of moles of $I_3^-(aq)$ that reacted is

$$\text{moles of } I_3^- = (0.0350 \text{ mol } S_2O_3^{2-} \cdot L^{-1})(0.010 \text{ L}) \left(\frac{1 \text{ mol } I_3^-}{2 \text{ mol } S_2O_3^{2-}} \right)$$

$$= 1.75 \times 10^{-4} \text{ mol}$$

The number of moles of iodine is given by

$$\text{moles of } I_2 = \text{moles of } I_3^- = 1.75 \times 10^{-4} \text{ mol}$$

The number of moles of carbon monoxide required is

$$\text{moles of CO} = (1.75 \times 10^{-4} \text{ mol } I_2) \left(\frac{5 \text{ mol CO}}{1 \text{ mol } I_2} \right) = 8.75 \times 10^{-4} \text{ mol}$$

26-45 See Section 26-4A.

26-47 See Figure 26-36. The remaining nitrogen may be removed by liquification or by the reaction with lithium metal.

26-49 He reasoned that the electronegativity of O_2 was about the same value as the electronegativity of Xe and that Xe might replace O_2 in $O_2^+PtF_6^-$.

26-51 The van der Waals constant b is proportional to the volume of the gas molecule, and the volume of a noble gas molecule increases with increasing atomic number. The van der Waals constant a is related to the attraction between molecules. The London forces, or van der Waals forces, between molecules increase as the size of the molecules increases, and so the value of a increases with increasing atomic number.

E. ANSWERS TO THE SELF-TEST

1. false

2. strong

3. nitrogen compounds

4. air

5. ammonia

6. Hydrogen and nitrogen are reacted at high temperature and pressure in the presence of a catalyst to produce ammonia.

7. basic

8. ammonium

9. true

10. nitric acid

11. hydrazine, N_2H_4

12. false

13. false

14. weak

15. white

16. true

17. phosphate rock

18. false

19. P_4O_{10}

20. phosphorous acid, H_3PO_3

21. false

22. phosphides

23. false

24. PX_3 and PX_5

25. false

26. true

27. basic

28. amphoteric

29. true

30. air

31. carbon dioxide, $CO_2(g)$, and water, $H_2O(g)$

32. potassium chlorate

33. true

34. metal oxides

35. false

36. true

37. true

38. ultraviolet

39. true

40. rings of S_8

41. entangled chains

42. rings of S_8

43. sulfuric acid

44. oxidizing

45. metal sulfides

46. insoluble

47. mercaptans

48. acidic

49. false

50. true

51. false

52. true

53. true

54. -1

55. hydrogen fluoride in molten potassium fluoride

56. true

57. the fluoride ion, $F^-(aq)$, and acid fluoride ion, $HF_2^-(aq)$

58. electrolysis of a chloride, such as NaCl

59. true

60. oxidizing

61. reaction of a bromide with chlorine

62. iodine

63. $I_3^-(aq)$

64. true

65. strong

66. $+7$

67. chloric acid

68. the number of oxygen atoms without attached hydrogen atoms that are attached to the central atom

69. weaker

70. lack of chemical reactivity

71. true

72. fluorine and oxygen

73. true

74. false

27

THE CHEMISTRY OF THE TRANSITION METALS

A. OUTLINE OF CHAPTER 27

27-1 The maximum oxidation states of scandium through manganese are equal to the total number of $4s$ and $3d$ electrons.

Several oxidation states are available to the transition metals.

The trends in the chemistry of the transition metals are given on pages 976 and 977 of the text.

Scandium

Scandium is somewhat similar to aluminum in its chemistry.

Scandium has a $+3$ oxidation state in almost all its compounds.

Titanium

The most common oxidation state is $+4$; these compounds are covalently bonded.

The pure metal is prepared from its ore TiO_2 by first converting it to $TiCl_4$, and then reacting the $TiCl_4$ with magnesium to give $Ti(s)$ and $MgCl_2(s)$.

Some important titanium compounds are given in Table 27-3.

Vanadium

The maximum oxidation state of vanadium is $+5$; the $+2$, $+3$, and $+4$ states are also common.

Except for V_2O_5, the compounds of vanadium have limited commercial importance.

Some important compounds of titanium and vanadium are given in Table 27-3.

27-2 The $+6$ oxidation state of chromium and the $+7$ oxidation state of manganese are strongly oxidizing.

Chromium

The maximum oxidation state of chromium is $+6$.

The $+2$ and $+3$ states are the common oxidation states of chromium.

Chromium (VI), as in the dichromate ion $Cr_2O_7^{2-}$, is strongly oxidizing.

The chromium(II) ion is a fairly strong reducing agent.

A chromous bubbler is used to remove traces of oxygen from gases.

Manganese

The maximum oxidation state of manganese is $+7$.

The reagent $KMnO_4$ is an important strong oxidizing agent.

Manganese (II) forms soluble salts with most anions.

The most important compounds for the $+3$ and $+4$ oxidation states are Mn_2O_3 and MnO_2.

Some important compounds of chromium and manganese are given in Table 27-3.

27-3 Iron is produced in a blast furnace.

The maximum oxidation state of iron is $+6$.

Only the $+2$ and $+3$ oxidation states of iron are common.

Iron is the most abundant transition metal.

Iron is an essential element in mammalian cells.

Iron is produced by the reduction of Fe_2O_3 with coke, as diagrammed in Figure 27-9.

Steel is an alloy of iron and small, but definite amounts of other metals, and between 0.1 and 1.5 percent carbon.

Steel can be produced from pig iron, the product from a blast furnace, by the basic oxygen process (Figure 27-10).

The various types of steel are described.

27-4 The $+2$ oxidation state is the most important oxidation state for cobalt, nickel, copper, and zinc.

Cobalt

Cobalt is a relatively unreactive metal.

Most simple, cobalt salts involve Co(II).

The species $Co^{3+}(aq)$ is a strong oxidizing agent.

Nickel

Nickel is produced by roasting its sulfide ore to form NiO, which is then reduced to the metal by hydrogen or carbon.

Nickel is more reactive than cobalt.

The aqueous chemistry of nickel involves the species $Ni^{2+}(aq)$.

Some important compounds of cobalt and nickel are given in Table 27-5.

Copper

Copper generally exists in the form of sulfides in nature but may exist as the free metal.

Copper is an excellent conductor of electricity.

Brass and bronze are alloys of copper and zinc and of copper and tin, respectively.

Copper is fairly unreactive; it does not replace hydrogen from dilute acids but does react with oxidizing acids.

Most compounds of copper involve Cu(II).

When $NH_3(aq)$ is added to $Cu^{2+}(aq)$, a copper-ammonia complex ion, $[Cu(NH_3)_4(H_2O)_2]^{2+}$, is formed.

The $Cu^+(aq)$ ion is unstable and disproportionates to $Cu(s)$ and $Cu^{2+}(aq)$.

Zinc

Zinc is produced by roasting its ore to form ZnO, which is reduced to the metal with carbon.

Zinc behaves more like a Group 2 metal than a transition metal because its $3d$ subshell is filled.

Zinc metal is a strong reducing agent.

The only important oxidation state of zinc is $+2$.

Some important compounds of copper and zinc are given in Table 27-6.

27-5   Gold, silver, and mercury have been known since ancient times.

Gold

Gold is found in nature as the free metal.

Gold is very unreactive.

The amount of gold in its alloys is expressed in karats.

Pure gold is 24 karat.

Gold is extracted from ores by a reaction with NaCN and oxygen to form $Au(CN)_2^-(aq)$, from which gold is produced by reduction with zinc or by electrolysis.

Silver

Most silver is produced as a by-product of the production of other metals.

Silver has the highest electrical conductivity of all metals.

The silver halides are insoluble in water except for AgF, which is very soluble in water.

Mercury

Mercury is the only metal that is a liquid at $25°C$.

Mercury is produced from its ore, HgS, by reaction with oxygen.

Mercury is not very reactive; on being heated it reacts with oxygen, sulfur, and the halogens.

The oxide HgO decomposes to the metal and oxygen when heated.

Mercury does not replace hydrogen from dilute acids, but does react with oxidizing acids.

In its compounds mercury occurs as Hg(I) or Hg(II).

Hg(I) salts are generally insoluble.

Hg(I) salts consist of the diatomic ion Hg_2^{2+}.

Many of the salts containing mercury ions are covalently bonded.

Mercury compounds are very poisonous.

Some important compounds of gold, silver, and mercury are given in Table 27-7.

B. SELF-TEST

1. For scandium through manganese, the highest oxidation state is equal to the total number of _____.

2. Except for scandium and titanium, all the $3d$ transition metals form divalent ions in aqueous solution. *True/False*

3. For a given metal, the oxides become (*more/less*) acidic with increasing oxidation state.

4. For a given metal, the halides become more (*ionic/covalent*) with increasing oxidation state.

5. Scandium behaves chemically similarly to the element _____.

6. Scandium has a _____ oxidation state in almost all its compounds.

7. The most common oxidation state of titanium is _____.

8. Titanium metal is produced by the reaction of titanium dioxide with carbon. *True/False*

9. Most titanium is used in the production of _____.

10. Vanadium can exist in several oxidation states. *True/False*

11. The chemical formula of vanadium pentoxide is _____.

12. Compounds of Cr(VI) are strong (*oxidizing/reducing*) agents.

13. Compounds of Cr(II) are strong (*oxidizing/reducing*) agents.

14. The most stable oxidation state of chromium is _____.

15. A chromous bubbler is a device used to remove _____.

16. Potassium permanganate, $KMnO_4$, is a strong (*oxidizing/reducing*) agent.

17. Manganese (II) forms soluble salts with most common anions. *True/False*

18. The common oxidation states of iron are _____ and _____.

19. Iron is produced in the blast furnace by the reaction of _____ with _____.

20. The actual reducing agent in the blast furnace is hot _____.

21. Steel is made from pig iron by removing the impurities with _____.

22. All steels contain carbon. *True/False*

23. Different metals give different properties to steel. *True/False*

24. The most common oxidation state of cobalt is the _____ oxidation state.

25. The species $Co^{3+}(aq)$ is a strong (*oxidizing/reducing*) agent.

26. Cobalt is a reactive metal. *True/False*

27. Nickel reacts with dilute acids. *True/False*

28. The aqueous chemistry of nickel involves the species _____(aq).

29. Copper is _____ in color.

30. Copper is a reactive metal. *True/False*

31. Brass is an alloy of copper and _____.

32. Bronze is an alloy of copper and _____.

33. Copper replaces hydrogen from dilute acid. *True/False*

34. Most compounds of copper involve copper in the _____ oxidation state.

35. Many copper (II) salts are _____ in color.

36. The $Cu^+(aq)$ ion disproportionates to _____ and _____.

37. Zinc behaves more like a _____ metal than a transition metal.

38. Zinc has a completely filled $3d$ subshell. *True/False*

39. Zinc is a reactive metal. *True/False*

40. Zinc replaces hydrogen from dilute acids. *True/False*

41. The only important oxidation state of zinc is _____.

42. Gold is a reactive metal. *True/False*

43. Gold occurs as the free metal in nature. *True/False*

44. Gold is an excellent conductor of electricity. *True/False*

45. Gold is extracted from its ores by reaction with _____ and _____.

46. The amount of gold in its alloys is expressed in _____.

47. Silver is an excellent conductor of electricity. *True/False*

48. All the silver halides are insoluble. *True/False*

49. Mercury is a reactive metal. *True/False*

50. Mercury is produced from its ores by reaction with _____.

51. Mercury replaces hydrogen from dilute acids. *True/False*

52. Most Hg(I) salts are insoluble. *True/False*

53. The Hg(I) salts consist of the _____ ion.

54. All Hg(II) salts are ionic compounds. *True/False*

55. Mercury compounds are poisonous. *True/False*

C. CALCULATIONS YOU SHOULD KNOW HOW TO DO

There are no new calculations in this chapter.

D. ANSWERS TO THE ODD-NUMBERED PROBLEMS

27-1   Tungsten has the highest melting point.

27-3   Iron is the most abundant transition metal.

27-5 $TiO_2(s) + 2Cl_2(g) \longrightarrow TiCl_4(l) + O_2(g)$
$TiCl_4(l) + Mg(l) \longrightarrow Ti(s) + MgCl_2(l)$

27-7   Vanadium(V) oxide, V_2O_5, is a catalyst in the production of sulfuric acid by the contact process.

27-9   The equations for the two half reactions are

$$Cr_2O_7^{2-} + 8H^+ + 6e^- \longrightarrow Cr_2O_3 + 4H_2O \quad \text{(reduction)}$$
$$2H_2O \longrightarrow O_2 + 4H^+ + 4e^- \quad \text{(oxidation)}$$

Multiply the first equation by 2 and the second by 3. Add the two resulting equations and cancel like terms. Adding the phases yields

$$2Cr_2O_7^{2-}(aq) + 4H^+(aq) \longrightarrow 2Cr_2O_3(s) + 2H_2O(l) + 3O_2(g)$$

27-11   The $KMnO_4$ is decomposed to $MnO_2(s)$ in the presence of light.

27-13

Lewis Formula		Class of Molecule	Structure
(a)	$:\overset{..}{\underset{..}{Cl}}:$ $\|$ $:\overset{..}{\underset{..}{Cl}}-Ti-\overset{..}{\underset{..}{Cl}}:$ $\|$ $:\overset{..}{\underset{..}{Cl}}:$	AX_4	Tetrahedral
(b)	$:\overset{..}{\underset{..}{F}}:$ $\|$ $:\overset{..}{F}-V-\overset{..}{F}:$ $\diagup \ \diagdown$ $F \quad F$	AX_5	Bipyramidal
(c)	$:\overset{..}{\underset{}{O}}:^{\ominus}$ $\|$ $\overset{.}{.}O=Cr=O\overset{.}{.}$ $\|$ $:\overset{}{\underset{..}{O}}:_{\ominus}$ plus other resonance forms	AX_4	Tetrahedral
(d)	$\overset{.}{\underset{}{O}}\overset{.}{.}$ $\|$ $\overset{.}{.}O=Mn=O\overset{.}{.}$ $\|$ $:\overset{}{\underset{..}{O}}:_{\ominus}$ plus other resonance forms	AX_4	Tetrahedral

27-15   See Section 27-3 of the text.

27-17   The cobalt(III) ion is unstable and is readily reduced to cobalt(II). The equation for the reaction in aqueous solution is

$$4Co^{3+}(aq) + 2H_2O(l) \longrightarrow 4Co^{2+}(aq) + O_2(g) + 4H^+(aq)$$

27-19   The number of moles of copper produced is

$$\text{moles of Cu} = (1 \text{ metric ton})\left(\frac{10^3 \text{ kg}}{1 \text{ metric ton}}\right)\left(\frac{10^3 \text{ g}}{1 \text{ kg}}\right)\left(\frac{1 \text{ mol Cu}}{63.55 \text{ g Cu}}\right)$$
$$= 1.574 \times 10^4 \text{ mol}$$

All the copper in chalcopyrite, $CuFeS_2$, ends up as copper metal. Thus, the number of metric tons of $CuFeS_2$ required is

$$\text{mass of CuFeS}_2 = (1.574 \times 10^4 \text{ mol Cu})\left(\frac{1 \text{ mol CuFeS}_2}{1 \text{ mol Cu}}\right)\left(\frac{183.52 \text{ g CuFeS}_2}{1 \text{ mol CuFeS}_2}\right)$$
$$= 2.888 \times 10^6 \text{ g} = 2.888 \times 10^3 \text{ kg} = 2.888 \text{ metric ton}$$

The number of metric tons of ore required is given by

$$\% \ CuFeS_2 = \frac{mass \ of \ CuFeS_2}{mass \ of \ ore} \times 100$$

$$mass \ of \ ore = \frac{mass \ CuFeS_2 \times 100}{\% \ CuFeS_2} = \frac{2.888 \ metric \ ton \times 100}{2.65\%}$$

$$= 109 \ metric \ ton$$

27-21 Twenty-four-karat gold is 100 percent gold. We have the following relationship between 14-karat gold and 24-karat gold:

$$\frac{14\text{-karat gold}}{24\text{-karat gold}} = \frac{x}{100\%}$$

Solving for x, we have that

$$x = 58\%$$

Thus 14-karat gold is 58 percent gold.

27-23 The extraction of metallic gold from its ores is described by the equation

$$4Au(s) + 8CN^-(aq) + O_2(g) + 2H_2O(l) \longrightarrow 4[Au(CN)_2]^-(aq) + 4OH^-(aq)$$

The recovery of the metallic gold is described by the equation

$$2[Au(CN)_2]^-(aq) + Zn(s) \longrightarrow [Zn(CN)_4]^{2-}(aq) + 2Au(s)$$

27-25 (a)   The mass percentage of cobalt in vitamin B-12 is given by

$$mass \ \% \ Co = \frac{molecular \ mass \ Co}{molecular \ mass \ vitamin \ B\text{-}12} \times 100$$

$$= \frac{58.93}{1161.25} \times 100 = 5.075\%$$

(b)   The mass of cobalt in 6.0 μg of vitamin B-12 is given by

$$mass \ of \ Co = (6.0 \times 10^{-6} \ g \ B\text{-}12) \left(\frac{1 \ mol \ B\text{-}12}{1161.25 \ g \ B\text{-}12} \right)$$

$$\times \left(\frac{1 \ mol \ Co}{1 \ mol \ B\text{-}12} \right) \left(\frac{58.93 \ g \ Co}{1 \ mol \ Co} \right)$$

$$= 3.0 \times 10^{-7} \ g = 0.30 \ \mu g$$

27-27 The mass of ore required is given by

$$\% \text{ Cu} = \frac{\text{mass Cu}}{\text{mass ore}} \times 100$$

or

$$\text{mass of ore} = \frac{\text{mass Cu} \times 100}{\% \text{ Cu}} = \frac{91 \text{ metric ton} \times 100}{0.25\%}$$
$$= 36,400 \text{ metric tons}$$

E. ANSWERS TO THE SELF-TEST

1.	$4s$ and $3d$ electrons	20.	carbon monoxide, CO
2.	true	21.	hot, pure O_2
3.	more	22.	true
4.	covalent	23.	true
5.	aluminum	24.	+2
6.	+3	25.	oxidizing
7.	+4	26.	false
8.	false	27.	false
9.	titanium steels	28.	Ni^{2+}
10.	true	29.	reddish-gold
11.	V_2O_5	30.	false
12.	oxidizing	31.	zinc
13.	reducing	32.	tin
14.	+3	33.	false
15.	traces of oxygen from gases	34.	+2
16.	oxidizing	35.	blue
17.	true	36.	$Cu(s)$ . . . $Cu^{2+}(aq)$
18.	+2 . . . +3	37.	Group 2
19.	Fe_2O_3 . . . coke (C)	38.	true
		39.	true

40. true

41. $+2$

42. false

43. true

44. true

45. NaCN . . . O_2

46. karats

47. true

48. false; AgF is soluble.

49. false

50. O_2

51. false

52. true

53. Hg_2^{2+}

54. false

55. true

28

TRANSITION-METAL COMPLEXES

A. OUTLINE OF CHAPTER 28

28-1   Each d transition-metal ion has a characteristic number of d electrons.

There are five d orbitals for each value of the principal quantum number, $n \geqslant 3$.

For d orbitals, $l = 2$ and $m_l = 2, 1, 0, -1$, or -2.

Sc to Zn form the $3d$ transition-metal series.

Y to Cd form the $4d$ transition-metal series.

Lu to Hg form the $5d$ transition-metal series.

The shapes and relative orientations of the five d orbitals are shown in Figure 28-2.

The electron configurations of the M(II) transition-metal ions follow the arithmetic sequence: the $n = 1$ level is filled first, then the $n = 2$ level is filled, then the $n = 3$ level is filled, and so on.

Transition-metal ions with x electrons in the outer d orbitals are called d^x ions.

The number of d electrons in the M(II) ion is the same as the position of the element within the d transition-metal series (d^1 is first, d^2 is second, and so on).

The M(III) ions of the $3d$ series have one fewer electron than the M(II) ion.

The second digit of the atomic number of the M(II) $3d$ transition metals is the same as x in d^x [except for Zn(II)].

28-2   Transition-metal complexes consist of central metal atoms or ions that are bonded to ligands.

A ligand is an anion or neutral molecule that is attached to the central atom or ion in a complex.

The coordination number is the number of ligands that are attached to the central atom or ion in a complex.

Transition-metal complexes are often called coordination complexes.

A ligand-substitution reaction is a reaction involving a change in the ligands attached to the central metal ion in a complex ion.

A change in ligands in a transition-metal complex may change the color of the complex.

The complex ion $[Fe(CN)_6]^{4-}$ has an octahedral structure with six cyanide ions bonded to the central iron atom (Figure 28-3).

The complex ion $[Ni(CN)_4]^{2-}$ has a square-planar structure with four cyanide ions bonded to the central nickel atom (Figure 28-6).

The complex ion $[CoCl_4]^{2-}$ has a tetrahedral structure with four chloride ions bonded to the central cobalt atom (Figure 28-7).

The complex ion $[Ag(NH_3)_2]^+$ has a linear structure with two nitrogen atoms bonded to the central silver atom (Figure 28-8).

Some examples of transition-metal complexes are listed in Table 28-1.

28-3   Transition-metal complexes have a systematic nomenclature.

The rules for naming transition-metal complexes are given on page 1000 of the text.

The names of some common ligands are given in Table 28-2.

The chemical formula of a complex ion can be written when its name is given (see page 1001 of the text and Example 28-4).

28-4   Polydentate ligands bind to more than one coordination position around the metal ion.

Ligands that attach to a metal ion at more than one coordination position are called polydentate ligands or chelating agents.

The atoms of a polydentate ligand that attach to the metal ion are called ligating atoms.

The oxalate ion (ox) and ethylenediamine (en) are examples of bidentate ligands.

Ethylenediaminetetraacetate ($EDTA^{4-}$) is a hexadentate ligand.

The prefixes *bis-* for two ligands and *tris-* for three ligands are used in the nomenclature of complexes that contain chelating agents.

28-5   Some octahedral and square-planar transition-metal complexes can exist in isomeric forms.

Cis-trans isomers of square-planar and octahedral complexes exist (Figures 28-10 and 28-11).

The prefix *cis-* designates the structure in which the identical ligands are adjacent to each other.

The prefix *trans-* designates the structure in which identical ligands are opposite to each other.

28-6   The five *d* orbitals of a transition-metal ion in an octahedral complex are split into two groups by the ligands.

The five *d* orbitals are split into the two sets:

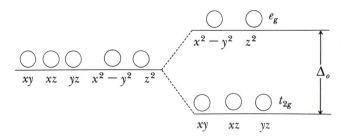

The magnitude of Δ_o depends on the central metal ion, its charge, and the ligands.

This splitting pattern arises from the orientation of the *d* orbitals relative to the ligands, as shown in Figure 28-15.

To determine the ground-state *d*-orbital configuration of a transition metal in an octahedral complex, electrons are placed in the t_{2g} and e_g orbitals according to the Pauli exclusion principle and Hund's rule.

The colors of transition-metal octahedral complexes arise from electronic transitions within the metal *d* orbitals.

The energy difference Δ_o often corresponds to the visible region of the spectrum.

The visible absorption spectrum of the complex ion $[Ti(H_2O)_6]^{3+}(aq)$ is shown in Figure 28-18.

The colors of many gemstones are due to electronic excitations of transition-metal ions.

28-7 *d*-Orbital electron configuration is the key to understanding many properties of the *d* transition-metal ions.

Molecules with no unpaired electrons cannot be magnetized by an external magnetic field and are called diamagnetic.

Molecules with unpaired electrons can be magnetized by an external magnetic field and are called paramagnetic.

Paramagnetic substances are drawn into a magnetic field (Figure 28-19).

Electrons are placed into the sets of *d* orbitals according to Hund's rule and the value of Δ_o.

Certain octahedral complexes can be either high spin or low spin.

The value of Δ_o and the ligands determine whether a *d* electron configuration will be low spin or high spin.

If Δ_o is small, then the *d* electrons will occupy the e_g orbitals before they pair up in the t_{2g} orbitals (high spin).

If Δ_o is large, then the d electrons will fill the t_{2g} orbitals completely before occupying the higher-energy e_g orbitals (low spin).

The various possible ground-state d electron configurations for octahedral ions are given in Figure 28-21.

The high-spin configuration has the maximum number of unpaired d electrons.

The low-spin configuration has the minimum number of unpaired d electrons.

28-8 Ligands can be ordered according to their ability to split the transition-metal d orbitals.

The spectrochemical series orders ligands according to the magnitude of the splitting of the d orbitals that they cause (Figure 28-23).

The spectrochemical series can be used to predict whether a given complex will be low spin or high spin.

The value of Δ_o increases as the oxidation state of the metal increases.

The value of Δ_o increases as we move down a column in the periodic table for a given oxidation state.

With the exception of the metal halide complexes, all M(III) and higher oxidation state complex ions of the $4d$ and $5d$ transition-metal series are low spin.

28-9 The d-orbital splitting patterns in square-planar and tetrahedral complexes are different from those in octahedral complexes.

The splitting patterns for the d orbitals in square-planar and tetrahedral complexes are

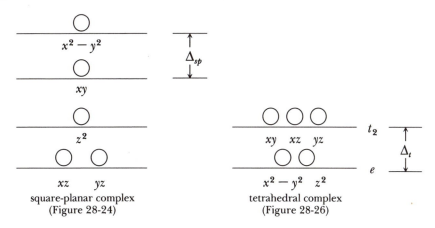

square-planar complex
(Figure 28-24)

tetrahedral complex
(Figure 28-26)

All d^8 square-planar complexes are low spin because Δ_{sp} is relatively large.

There are no low-spin tetrahedral complexes because Δ_t is relatively small.

Magnetic studies can determine whether a complex is square-planar or tetrahedral.

28-10 Transition-metal complexes are classified as either inert or labile.

A labile complex undergoes ligand substitution reactions relatively rapidly.

An inert complex undergoes ligand substitution reactions relatively slowly or not at all.

Taube's rules are used to predict whether a complex is inert or labile.

Taube's rules say that $t_{2g}^3 e_g^0$, $t_{2g}^4 e_g^0$, $t_{2g}^5 e_g^0$, and $t_{2g}^6 e_g^0$ octahedral complexes are inert, whereas all other octahedral complexes are labile.

B. SELF-TEST

1. There are a total of _____ metals in each d transition-metal series.

2. Iron(II) is a d^6 ion; iron (II) has _____ d electrons.

3. Chromium(II), whose atomic number is 24, is a $d^{(\ )}$ ion.

4. Iron (III), whose atomic number is 26, is a $d^{(\ )}$ ion.

5. Transition metals often have more than one possible oxidation state. *True/False*

6. In the complex ion $[Fe(CN)_6]^{4-}$, the metal ion is _____ and the ligands are

_____ .

7. When the compound $K_4[Fe(CN)_6]$ is dissolved in water, the major species present are $K^+(aq)$, $Fe^{2+}(aq)$, and $CN^-(aq)$. *True/False*

8. Most transition-metal complexes are colored due to the electronic transitions of

_____ .

9. All nickel(II) complexes are the same color. *True/False*

10. The ligands NH_3 in $[Ni(NH_3)_6]^{2+}$ are arranged around the nickel atom in a(an)

_____ structure.

11. The ligands CN^- in $[Pt(CN)_4]^{2-}$ are arranged around the platinum atom in a(an)

_____ structure.

12. In any complex ion or complex molecule, name the _____

first and then the _____ .

13. The name of the ligand NH_3 is _____ .

14. The name of the ligand CO is _____ .

15. The number of ligands of a particular type in a complex ion or complex molecule is

denoted by _____ .

16. The name tetrachlorocobaltate(II) indicates that the complex is an anion. *True/False*

17. The name tetrachlorocobaltate(II) indicates that there are _____ Cl^- ligands attached to the cobalt ion.

18. The oxidation state of the metal atom in a complex is denoted by _____ _____ .

19. *cis*-Diamminedichloroplatinum(II) is identical in physical properties to *trans*-diamminedichloroplatinum(II). *True/False*

20. The two Cl^- ligands in $[Pt(NH_3)_2Cl_2]$ in the *cis* isomer are located (*adjacent/opposite*) to one another in a square-planar structure.

21. There are (1/2/3/4) geometrical isomers in a $[MX_4Y_2]$ octahedral complex.

22. A chelating ligand is a ligand that is a _____ .

23. Ethylenediamine is an example of a bidentate ligand. *True/False*

24. The number of chelating ligands in a complex is indicated by _____ _____ .

25. The five *d* orbitals in a metal ion without any attached ligands have the same energy. *True/False*

26. The five *d* orbitals in a metal ion with attached ligands have the same energy. *True/False*

27. The *d* orbitals of a transition-metal ion in an octahedral complex are split into a lower set, called _____ , and an upper set, called _____ .

28. The t_{2g} set of orbitals can accommodate a maximum of _____ electrons.

29. The difference in energy between the t_{2g} and e_g orbitals is the same in all octahedral complexes. *True/False*

30. The difference in energy between the t_{2g} and e_g orbitals in an octahedral complex is denoted by _____ .

31. If the value of Δ_o is small compared to the pairing energy, then the *d* electrons will pair up in the t_{2g} orbitals before occupying the e_g orbitals. *True/False*

32. The *d* electron configuration of an octahedral complex is high spin when _____ _____ .

33. If the value of Δ_o is small compared to the pairing energy, then the d electron configuration is (*low/high*) spin.

34. The value of Δ_o depends on the nature of the ligands in an octahedral complex. *True/False*

35. In the _____ series, the ligands are ordered in increasing ability to split the metal d orbitals.

36. The value of Δ_o for the ligand Cl^- is (*greater than/less than*) that for the ligand CN^-.

37. The splitting pattern of the d orbitals in a square-planar complex is (*different from/the same as*) that in a tetrahedral complex.

38. The lower set of d orbitals in a tetrahedral complex is called the _____ orbitals and the upper set is called the _____ orbitals.

39. The d-orbital splitting energy for a tetrahedral complex is designated by _____.

40. All tetrahedral complexes are (*low/high*) spin.

41. A paramagnetic molecule behaves like a magnet in an externally applied magnetic field. *True/False*

42. A paramagnetic molecule contains (*paired/unpaired*) electrons.

43. A high-spin d^6 complex is (*paramagnetic/diamagnetic*).

44. The rate at which an octahedral complex exchanges its ligands depends on the d electron configuration of the transition-metal ion. *True/False*

45. A (*labile/inert*) complex undergoes rapid ligand substitution reactions.

46. A $t_{2g}^5 e_g^0$ octahedral complex is (*inert/labile*).

47. All octahedral complexes are inert. *True/False*

C. CALCULATIONS YOU SHOULD KNOW HOW TO DO

Although there are no numerical calculations in Chapter 28, here are some things that you should be able to do.

1. Determine electron configurations of transition-metal ions. See Example 28-1 and Problems 28-1 through 28-4.

2. Determine the oxidation states of the metal atoms in transition-metal complexes. See Example 28-2 and Problems 28-7 through 28-10.

3. Write the name of a transition-metal complex from its chemical formula. See Example 28-3 and Problems 28-15 through 28-18, 28-23, and 28-24.

4. Write the chemical formula of a transition-metal complex from its name. See Examples 28-4 and 28-5 and Problems 28-19 through 28-22, 28-25, and 28-26.

5. Draw structures of the geometrical isomers of complexes. See Example 28-6 and Problems 28-27 through 28-30.

6. Write d-orbital electron configurations of metals in octahedral complexes. See Examples 28-7 and 28-8 and Problems 28-31, 28-32, 28-35, and 28-36.

7. Use the spectrochemical series to predict whether a given complex is high spin or low spin. See Example 28-9 and Problems 28-33 through 28-40.

8. Write the d electron configuration for square-planar complexes and tetrahedral complexes. See Examples 28-10 and 28-11 and Problems 28-35 through 28-42.

9. Use Taube's rules, to predict whether a given complex is inert or labile. See Example 28-12 and Problems 28-43 through 28-46.

D. SOLUTIONS TO THE ODD-NUMBERED PROBLEMS

28-1
(a) $1s^2 2s^2 2p^6 3s^2 3p^6 3d^5$ or $[Ar]3d^5$
(b) $1s^2 2s^2 2p^6 3s^2 3p^6 3d^2$ or $[Ar]3d^2$
(c) $1s^2 2s^2 2p^6 3s^2 3p^6 3d^{10} 4s^2 4p^6 4d^6$ or $[Kr]4d^6$
(d) $1s^2 2s^2 2p^6 3s^2 3p^6 3d^{10} 4s^2 4p^6 4d^{10} 4f^{14} 5s^2 5p^6 5d^6$ or $[Xe]4f^{14}5d^6$

28-3
(a) Ag(I) has one more electron than Ag(II), and so it has ten $4d$ electrons.
(b) Pd(IV) has two less electrons than Pd(II), and so it has six $4d$ electrons.
(c) Ir(III) has one less electron than Ir(II), and so it has six $5d$ electrons.
(d) Co(II) has seven $3d$ electrons.

28-5
(a) The d^6 ions with a III oxidation state are those ions which are d^7 ions in a II oxidation state. Thus the answer is Co(III), Rh(III), and Ir(III).
(b) The d^4 ions with a IV oxidation state are those ions which are d^6 ions in a II oxidation state. Thus the answer is Fe(IV), Ru(IV), and Os(IV).
(c) The d^{10} ions with a (I) oxidation state are those ions which are d^9 ions in a II oxidation state. Thus the answer is Cu(I), Ag(I), and Au(I).

28-7
(a) $[Os(NH_3)_4Cl_2]^+$. The charge on the NH_3 ligand is 0 and that on the Cl^- is -1. The overall charge of the complex ion is $+1$, and so if x is the charge on Os, then

$$x + 4(0) + 2(-1) = +1$$
$$x = +3$$

(b) $[CoCl_6]^{3-}$. The charge on each Cl^- ligand is -1. The overall charge of the complex ion is -3, and so if x is the charge on Co, then

$$x + 6(-1) = -3$$
$$x = +3$$

(c) $[Fe(CN)_6]^{4-}$. The charge on each CN^- ligand is -1. The overall charge of the complex ion is -4, and so if x is the charge on Fe, then

$$x + 6(-1) = -4$$
$$x = +2$$

(d) $[Nb(NO_2)_6]^{3-}$. The charge on each NO_2^- ligand is -1. The overall charge of the complex ion is -3, and so if x is the charge on Nb, then

$$x + 6(-1) = -3$$
$$x = +3$$

28-9 (a) $[Cd(CN)_4]^{2-}$. The charge on each CN^- ligand is -1. The overall charge of the complex ion is -2, and so if x is the charge on Cd, then

$$x + 4(-1) = -2$$
$$x = +2$$

(b) $[Pt(NH_3)_6]^{2+}$. The charge on an NH_3 ligand is 0. The overall charge of the complex ion is $+2$, and so if x is the charge on Pt, then

$$x + 6(0) = +2$$
$$x = +2$$

(c) $[Pt(NH_3)_4Cl_2]$. The charge on an NH_3 ligand is 0 and that on a Cl^- ligand is -1. The overall charge of the complex ion is 0, and so if x is the charge on Pt, then

$$x + 4(0) + 2(-1) = 0$$
$$x = +2$$

(d) $[RhBr_6]^{3-}$. The charge on each Br^- ligand is -1. The overall charge of the complex ion is -3, and so if x is the charge on Rh, then

$$x + 6(-1) = -3$$
$$x = +3$$

28-11 (a) three moles of $K^+(aq)$ and one mole of $[Fe(CN)_6]^{3-}(aq)$
The name of the complex ion is hexacyanoferrate(III).

(b) one mole of $[Ir(NH_3)_6]^{3+}(aq)$ and three moles of $NO_3^-(aq)$
The name of the complex ion is hexammineiridium(III).

(c) one mole of $[Pt(NH_3)_4Cl_2]^{2+}(aq)$ and two moles of $Cl^-(aq)$
The name of the complex ion is tetramminedichloroplatinum(IV).

(d) one mole of $[Ru(NH_3)_6]^{3+}(aq)$ and three moles of $Br^-(aq)$
The name of the complex ion is hexammineruthenium(III).

28-13 The key point is that only the chloride ions that exist in solution as $Cl^-(aq)$, and not the chloride ions that are complexed with the platinum ions, are precipitated by $Ag^+(aq)$ as $AgCl(s)$. Let's look at each case in turn.

$PtCl_4 \cdot 6NH_3$ Because all four chloride ions per formula unit are precipitated by $Ag^+(aq)$, all four chloride ions must exist in solution as $Cl^-(aq)$. The chemical formula of the complex salt must be $[Pt(NH_3)_6]Cl_4$.

$PtCl_4 \cdot 5NH_3$ One of the four chloride ions must be complexed to the platinum ion because it is not precipitated by $Ag^+(aq)$. The chemical formula of the complex salt must be $[Pt(NH_3)_5Cl]Cl_3$.

$PtCl_4 \cdot 4NH_3$ Two of the four chloride ions must be complexed to the platinum ion because they are not precipitated by $Ag^+(aq)$. The chemical formula of the complex salt must be $[Pt(NH_3)_4Cl_2]Cl_2$.

$PtCl_4 \cdot 3NH_3$ Three of the four chloride ions must be complexed to the platinum ion because they are not precipitated by $Ag^+(aq)$. The chemical formula of the complex salt must be $[Pt(NH_3)_3Cl_3]Cl$.

$PtCl_4 \cdot 2NH_3$ All four of the chloride ions must be complexed to the platinum ion because none are precipitated by $Ag^+(aq)$. The chemical formula of the complex must be $[Pt(NH_3)_2Cl_4]$.

28-15   (a)   The complex ion is $[Cr(CN)_6]^{3-}$. If the oxidation state of Cr is denoted by x, then $x + 6(-1) = -3$, or $x = +3$. The name of the compound is potassium hexacyanochromate (III).

     (b)   The complex ion is $[Cr(H_2O)_5Cl]^{2+}$. If x is the oxidation state of Cr, then $x + 5(0) + (-1) = +2$, or $x = +3$. The name of the compound is pentaaquachlorochromium(III) perchlorate.

     (c)   The complex ion is $[Co(CO)_4Cl_2]^+$. If x is the oxidation state of Co, then $x + 4(0) + 2(-1) = +1$, or $x = +3$. The name of the compound is tetracarbonyldichlorocobalt(III) perchlorate.

     (d)   The complex ion is $[Pt(NH_3)_4Br_2]^{2+}$. If x is the oxidation state of Pt, then $x + 4(0) + 2(-1) = +2$, or $x = +4$. The name of the compound is tetraamminedibromoplatinum(IV) chloride.

28-17   (a)   The complex ion is $[Co(NO_2)_6]^{3-}$. Denoting the oxidation state of cobalt by x, we have $x + 6(-1) = -3$, or $x = +3$. The compound is called ammonium hexanitrocobaltate(III).

     (b)   The complex ion is $[Ir(NH_3)_4Br_2]^+$. Denoting the oxidation state of iridium by x, we have $x + 4(0) + 2(-1) = +1$, or $x = +3$. The compound is called tetraamminedibromoiridium(III) bromide.

     (c)   The complex ion is $[CuCl_4]^{2-}$. Denoting the oxidation state of copper by x, we have $x + 4(-1) = -2$, or $x = +2$. The compound is called potassium tetrachlorocuprate(II).

     (d)   The complex is $[Ru(CO)_5]$. Denoting the oxidation state of ruthenium by x, we have $x + 5(0) = 0$, or $x = 0$. The molecule is called pentacarbonylruthenium(0).

28-19   (a)   The complex consists of a central iron atom with five cyanide ions, CN^-, and one carbon monoxide, CO, as ligands. The oxidation state of the iron is $+2$, and so the charge on the complex ion is $+2 + 5(-1) + 0 = -3$. The formula of the compound is $Na_3[Fe(CN)_5CO]$.

     (b)   The complex ion consists of a central gold atom with two chloride ions and two iodide ions as ligands. The oxidation state of the gold is $+3$, and so the charge on the complex ion is $+3 + 2(-1) + 2(-1) = -1$. The formula of the compound is $trans$-$NH_4[AuCl_2I_2]$.

(c)   The complex ion consists of a central cobalt atom with six cyanide ions as ligands. The oxidation state of the cobalt is $+3$, and so the charge on the complex is $+3 + 6(-1) = -3$. The formula of the compound is $K_3[Co(CN)_6]$.

(d)   The complex ion consists of a central cobalt atom with six NO_2^- ions as ligands. The oxidation state of the cobalt is $+3$, and so the charge on the complex ion is $+3 + 6(-1) = -3$. The formula of the compound is $Ca_3[Co(NO_2)_6]_2$.

28-21 (a)   The complex ion consists of a central platinum atom with a chloride ion and three ammonia molecules as ligands. The oxidation state of the platinum is $+2$, and so the charge on the complex ion is $+2 + (-1) + 3(0) = +1$. The formula of the compound is $[Pt(NH_3)_3Cl]NO_3$.

(b)   The complex ion consists of a central copper atom with four fluoride ions as ligands. The oxidation state of the copper is $+2$, and so the charge on the complex ion is $+2 + 4(-1) = -2$. The formula of the compound is $Na_2[CuF_4]$.

(c)   The complex ion consists of a central cobalt atom with six nitrite ions as ligands. The oxidation state of the cobalt is $+2$, and so the charge on the complex ion is $+2 + 6(-1) = -4$. The formula of the compound is $Li_4[Co(NO_2)_6]$.

(d)   The complex ion consists of a central iron atom with six cyanide ions as ligands. The oxidation state of the iron is $+2$, and so the charge on the complex ion is $+2 + 6(-1) = -4$. The formula of the compound is $Ba_2[Fe(CN)_6]$.

28-23 (a)   The structure of the complex is octahedral. The possible arrangements of the ligands around the central cobalt ion are

(b)

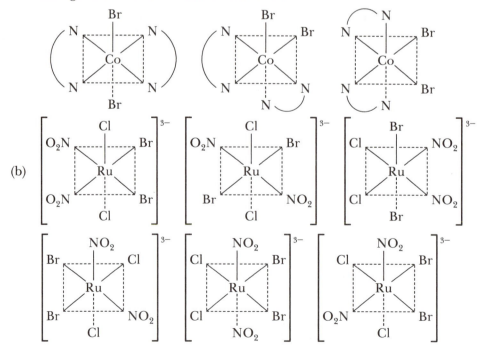

These structures are easier to visualize if you make a model of an octahedron as given in Appendix C of the text.

28-25 (a)   The complex is square planar:

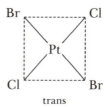

trans

(b)   The complex ion is square planar:

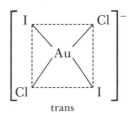

trans

(c)   The complex is octahedral:

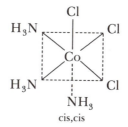

cis,cis

(d)   The complex ion is octahedral:

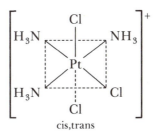

cis,trans

28-27 (a)   Niobium(III) is a d^2 ion. The d-orbital electron configuration of Nb^{3+} is

$$
\begin{array}{c}
\underline{\;\;\bigcirc\qquad\bigcirc\;\;}\;\;e_g^0 \\
{\scriptstyle x^2-y^2\quad z^2} \\[4pt]
\underline{\;\;\uparrow\quad\uparrow\quad\bigcirc\;\;}\;\;t_{2g}^2 \\
{\scriptstyle xy\quad xz\quad yz}
\end{array}
$$

or simply t_{2g}^2.

(b) Molybdenum(II) is a d^4 ion. If Δ_o is greater than the energy that is required to pair electrons, then the d-orbital electron configuration of Mo^{2+} is

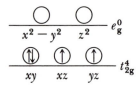

or t_{2g}^4.

(c) Manganese(II) is a d^5 ion. If Δ_o is less than the energy that is required to pair electrons, then the d-orbital electron configuration of Mn^{2+} is

or $t_{2g}^3 e_g^2$.

(d) Gold(I) is a d^{10} ion. The d-orbital electron configuration of Au^+ is

or $t_{2g}^6 e_g^4$.

(e) Iridium(III) is a d^6 ion. If Δ_o is greater than the energy that is required to pair electrons, then the d-orbital electron configuration of Ir^{3+} is

or $t_{2g}^6 e_g^0$.

28-29 We set up the following table:

Ion	Oxidation State of Central Ion	x in d^x	Low-Spin d-Orbital Electron Configuration (Number of Unpaired Electrons)	High-Spin d-Orbital Electron Configuration (Number of Unpaired Electrons)
(a) $[Fe(CN)_6]^{4-}$	Fe(II)	6	$t_{2g}^6 e_g^0$ (0)	$t_{2g}^4 e_g^2$ (4)
(b) $[Fe(CN)_6]^{3-}$	Fe(III)	5	$t_{2g}^5 e_g^0$ (1)	$t_{2g}^3 e_g^2$ (5)
(c) $[Co(NH_3)_6]^{2+}$	Co(II)	7	$t_{2g}^6 e_g^1$ (1)	$t_{2g}^5 e_g^2$ (3)
(d) $[CoF_6]^{3-}$	Co(III)	6	$t_{2g}^6 e_g^0$ (0)	$t_{2g}^4 e_g^2$ (4)
(e) $[Mn(H_2O)_6]^{2+}$	Mn(II)	5	$t_{2g}^5 e_g^0$ (1)	$t_{2g}^3 e_g^2$ (5)

Thus we see that $[Fe(CN)_6]^{4-}$ is low spin, $[Fe(CN)_6]^{3-}$ is low spin, $[Co(NH_3)_6]^{2+}$ is high spin, $[CoF_6]^{3-}$ is high spin, and $[Mn(H_2O)_6]^{2+}$ is high spin.

28-31 (a) The oxidation state of the vanadium atom in $[VCl_6]^{3-}$ is $+3$. Vanadium(III) is a d^2 ion, and so there are two unpaired electrons in $[VCl_6]^{3-}$.

(b) The oxidation state of the cobalt atom in $[CoCl_4]^{2-}$ is $+2$. Cobalt(II) is a d^7 ion. The d electron configuration is $e^4 t_2^3$, and so there are three unpaired electrons in $[CoCl_4]^{2-}$.

(c) The oxidation state of the chromium atom in $[Cr(CO)_6]$ is 0. Chromium(0) is d^6 and CO gives rise to low-spin complexes. Thus the d electron configuration is $t_{2g}^6 e_g^0$, and so there are no unpaired electrons in $[Cr(CO)_6]$.

(d) The oxidation state x of the chromium ion in $[Cr(CN)_6]^{4-}$ is $+2$. Chromium(II) is a d^4 ion and CN^- gives rise to low-spin complexes. Thus the d electron configuration is $t_{2g}^4 e_g^0$, and so there are two unpaired electrons in the complex ion $[Cr(CN)_6]^{4-}$.

28-33 (a) Cobalt(III) is a d^6 ion. The d electron configuration of low-spin Co(III) is $t_{2g}^6 e_g^0$. There are no unpaired electrons; thus $[Co(en)_3]^{3+}$ is diamagnetic.

(b) Iron(II) is a d^6 ion. The CN^- ligand is a low-spin ligand, and the d electron configuration of low-spin Fe(II) is $t_{2g}^6 e_g^0$. There are no unpaired electrons; thus $[Fe(CN)_6]^{4-}$ is diamagnetic.

(c) Nickel(II) is a d^8 ion. The d electron configuration is $e^4 t_2^4$. There are two unpaired electrons; thus $[NiF_4]^{2-}$ is paramagnetic.

(d) Cobalt(II) is a d^7 ion. The d electron configuration is $e^4 t_2^3$. There are three upaired electrons; thus, $[CoBr_4]^{2-}$ is paramagnetic.

28-35 Each complex ion could be either square planar or tetrahedral. Nickel(II) is a d^8 ion, and the two possible d electron configurations of $[NiF_4]^{2-}$ are

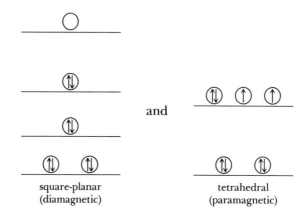

square-planar   tetrahedral
(diamagnetic)   (paramagnetic)

Because $[NiF_4]^{2-}$ is paramagnetic, we conclude that $[NiF_4]^{2-}$ is tetrahedral. The two possibilities for $[Ni(CN)_4]^{2-}$ are the same as shown above. Because $[Ni(CN)_4]^{2-}$ is diamagnetic, we predict that it is square planar.

28-37 (a) The oxidation state of the titanium atom in $[Ti(H_2O)_6]^{3+}$ is $+3$ Titanium(III) is a d^1 ion. The d electron configuration is $t_{2g}^1 e_g^0$, and so $[Ti(H_2O)_6]^{3+}$ is labile.

(b) The oxidation state of the vanadium atom in $[VF_6]^{3-}$ is $+3$. Vanadium(III) is a d^2 ion, and so the d electron configuration is $t_{2g}^2 e_g^0$ and $[VF_6]^{3-}$ is labile.

(c) The oxidation state of the chromium ion in $[Cr(NO_2)_6]^{3-}$ is $+3$. Chromium(III) is a d^3 ion with a $t_{2g}^3 e_g^0$ d electron configuration. Therefore $[Cr(NO_2)_6]^{3-}$ is inert.

(d) The oxidation state of the copper atom in $[CuCl_6]^{4-}$ is $+2$. Copper(II) is a d^9 ion. The d electron configuration is $t_{2g}^6 e_g^3$ and $[CuCl_6]^{4-}$ is labile.

28-39 (a) The charge on the complex ion is given by $+3 + 6(-1) = -3$. The formula is $[Co(NO_2)_6]^{3-}$.

(b) The charge on the complex ion is given by $+4 + 2(-1) + 2(0) = +2$. The formula is *trans*-$[PtCl_2(en)_2]^{2+}$.

(c) The charge on the complex ion is given by $+2 + 5(-1) + 0 = -3$. The formula is $[Fe(CN)_5CO]^{3-}$.

(d) The charge on the complex ion is given by $+3 + 2(-1) + 2(-1) = -1$. The formula is *trans*-$[AuCl_2I_2]^-$.

28-41 (a) The data indicate the number of chloride ions that are not bonded as ligands to the cobalt atom (see Solution 28-13). Thus we have

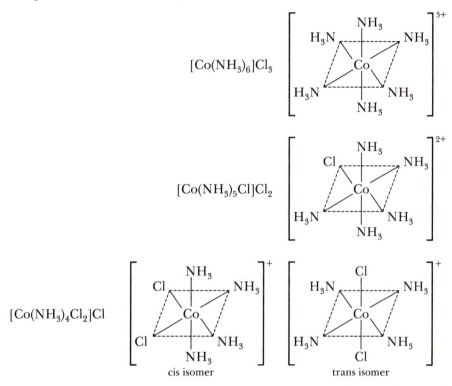

$[Co(NH_3)_6]Cl_3$

$[Co(NH_3)_5Cl]Cl_2$

$[Co(NH_3)_4Cl_2]Cl$

cis isomer

trans isomer

(b) There are cis and trans isomers of $[Co(NH_3)_4Cl_2]^+$ (as shown above) and hence two colors.

28-43 (a) The oxidation state of the cobalt atom in $[Co(CN)_6]^{3-}$ is $+3$, and in $[CoCl_6]^{4-}$ it is $+2$. Thus $[CoCl_6]^{4-}$ has more d electrons, and the statement is false.

(b) False [see part (a)].

(c) From part (a), we see that the cobalt atom in $[Co(CN)_6]^{3-}$ is d^6, while in $[CoF_6]^{4-}$ it is d^7. The CN^- causes the $[Co(CN)_6]^{3-}$ to be low spin, with a $t_{2g}^6 e_g^0$ d electron configuration. The F^- causes the $[CoF_6]^{4-}$ to be high spin, with a $t_{2g}^5 e_g^2$ d electron configuration. Thus $[CoF_6]^{4-}$ is paramagnetic and $[Co(CN)_6]^{3-}$ is diamagnetic. Thus the statement is false.

(d) True [see part (c)].

28-45 (a) The oxidation state of the nickel atom in $[NiCl_4]^{2-}$ is $+2$. Nickel(II) is a d^8 ion, and the d electron configuration is $e^4 t_2^4$. Thus there are two unpaired electrons in $[NiCl_4]^{2-}$.

(b) The oxidation state of the cobalt atom in $[CoCl_4]^{2-}$ is $+2$. Cobalt(II) is a d^7 ion, and the d electron configuration is $e^4 t_2^3$. Thus there are three unpaired electrons in $[CoCl_4]^{2-}$.

(c)   The oxidation state of the cobalt atom in $[Co(CO)_6]^{3+}$ is $+3$. Cobalt(III) is a d^6 ion, and the complex is low spin because CO is a low-spin ligand. Thus the d electron configuration is $t_{2g}^6 e_g^0$ and there are no unpaired electrons.

(d)   The oxidation state of the iron atom in $[Fe(CN)_6]^{3-}$ is $+3$. Iron(III) is a d^5 ion. The complex is low spin because CN^- is a low spin ligand and the d electron configuration is $t_{2g}^5 e_g^0$. Thus there is one unpaired electron in $[Fe(CN)_6]^{3-}$.

28-47   The equation for the reaction is

(1) $Pb_2[Fe(CN)_6](s) + 4I^-(aq) \rightleftharpoons 2PbI_2(s) + [Fe(CN)_6]^{4-}(aq)$

and its equilibrium constant is

$$K_1 = \frac{[[Fe(CN)_6]^{4-}]}{[I^-]^4} = \frac{(0.11\ M)}{(0.57\ M)^4} = 1.04\ M^{-3}$$

The equations for the dissolution of both $Pb_2[Fe(CN)_6](s)$ and $PbI_2(s)$ are

(2) $Pb_2[Fe(CN)_6](s) \rightleftharpoons 2Pb^{2+}(aq) + [Fe(CN)_6]^{4-}(aq)$

(3) $PbI_2(s) \rightleftharpoons Pb^{2+}(aq) + 2I^-(aq)$

with

$$K_{sp(2)} = [Pb^{2+}]^2[[Fe(CN)_6]^{4-}]$$
$$K_{sp(3)} = [Pb^{2+}][I^-]^2$$

If we multiply equation (3) by 2, then reverse it, and add equation (2), then we obtain equation (1). Thus,

$$K_1 = \frac{K_{sp(2)}}{K_{sp(3)}^2}$$

and so

$$K_{sp(2)} = K_1 K_{sp(3)}^2 = (1.04\ M^{-3})(7.1 \times 10^{-9}\ M^3)^2 = 5.2 \times 10^{-17}\ M^3$$

28-49   (a)   The oxidation state of the chromium atom in $[Cr(CN)_6]^{4-}$ is given by $x + 6(-1) = -4$, or $x = 2$. Chromium(II) is a d^4 ion and CN^- is a strong ligand, and so the d-electron configuration is t_{2g}^4. Thus $[Cr(CN)_6]^{4-}$ is inert.

(b)   The oxidation state of the platinum atom in $[Pt(NH_3)_6]^{4+}$ is given by $x + 6(0) = +4$, or $x = +4$. Platinum(IV) is a low-spin d^6 ion, and so $[Pt(NH_3)_6]^{4+}$ is inert.

(c)   The oxidation state of the copper atom in $[Cu(NH_3)_6]^{2+}$ is given by $x + 6(0) = +2$, or $x = +2$. Copper(II) is a d^9 ion and so $[Cu(NH_3)_6]^{2+}$ is labile.

(d)   The oxidation of the tungsten atom in $[W(NH_3)_6]^{2+}$ is given by $x + 6(0) = +2$, or $x = +2$. Tungsten(II) is a low-spin d^4 ion, and so $[W(NH_3)_6]^{2+}$ is inert.

E. ANSWERS TO THE SELF-TEST

1. ten
2. six
3. 4
4. 5
5. true
6. iron . . . cyanide ions
7. false
8. the d electrons of the metal ion
9. false
10. octahedral
11. square-planar
12. ligands . . . metal
13. ammine
14. carbonyl
15. a Greek prefix
16. true
17. four
18. a Roman numeral in parentheses
19. false
20. adjacent
21. 2
22. polydentate ligand
23. true
24. a prefix such as *bis* for two ligands or *tris* for three ligands

25. true
26. false
27. t_{2g} . . . e_g
28. six
29. false
30. Δ_o
31. false
32. the value of Δ_o is less than the energy that is required to pair electrons.
33. high
34. true
35. spectrochemical
36. less than
37. different from
38. e . . . t_2
39. Δ_t
40. high
41. true
42. unpaired
43. paramagnetic
44. true
45. labile
46. inert
47. false

29 REACTIONS OF ORGANIC COMPOUNDS

A. OUTLINE OF CHAPTER 29

29-1   Alkanes are hydrocarbons that contain only single bonds.

Hydrocarbons consist of only hydrogen and carbon.

Alkanes have the general formula C_nH_{2n+2} where $n = 1, 2, 3, \ldots$

Condensed structural formulas may be considered to be abbreviations of Lewis formulas.

Additional hydrogen atoms cannot be bonded to carbon atoms in alkanes.

Molecules that have the same chemical formula but different structures are called structural isomers.

Structural isomers have different chemical and physical properties.

Alkanes are saturated hydrocarbons.

The bonding in alkanes can be described in terms of sp^3 hybrid orbitals on the carbon atoms.

All the carbon-carbon bonds in alkanes are σ bonds.

Rotation can occur about carbon-carbon single bonds.

The number of possible structural isomers increases with the number of carbon atoms in an alkane.

29-2   Alkanes are not very reactive.

Alkanes react with oxygen in combustion reactions, which are highly exothermic.

The substitution reaction between an alkane and a halogen in the presence of ultraviolet light yields alkyl halides, or haloalkanes.

The chlorination of methane is a free-radical reaction; the ultraviolet radiation produces free chlorine atoms.

Free-radical reactions lead to more than a single product.

Atoms or groups of atoms that replace a hydrogen atom bonded to a carbon atom are called substituents.

29-3 Alkanes and substituted alkanes can be named systematically according to IUPAC rules.

The IUPAC rules for naming alkanes and their derivatives are given on pages 1031 and 1032 in the text.

The names of some common groups are listed in Table 29-3.

The formula of a compound can be written from the IUPAC name of the compound.

29-4 Hydrocarbons that contain double bonds are called alkenes.

In an unsaturated hydrocarbon, not all the carbon atoms are bonded to four other atoms.

Alkenes are unsaturated hydrocarbons that contain one or more double bonds.

The bonding in ethene can be described by sp^2 orbitals on each carbon atom.

The double bond in alkenes consists of a σ bond and a π bond.

The double bond in an alkene forces the alkene to have a planar region about the double bond.

The π orbital prevents rotation around the double bond.

Cis-trans isomers of alkenes may exist.

The IUPAC nomenclature for alkenes and their derivatives is given on page 1035 in the text.

29-5 Alkenes undergo addition reactions as well as combustion reactions and substitution reactions.

Alkenes undergo these addition reactions:

(a) Addition of hydrogen in the presence of a catalyst and high temperature and pressure.

(b) Addition of chlorine or bromine.

(c) Addition of hydrogen chloride.

(d) Addition of water in the presence of acid.

Markovnikov's rule states that when HX adds to an alkene, the hydrogen atom becomes bonded to the carbon atom in the double bond already bearing the larger number of hydrogen atoms.

Markovnikov's rule is used to predict the primary products in addition reactions.

29-6 Hydrocarbons that contain a triple bond are called alkynes.

The IUPAC name for an alkyne is formed by replacing the *-ane* ending of the parent alkane with *-yne*.

Alkynes undergo the same addition reactions, substitution reactions, and combustion reactions as alkenes.

29-7   Benzene belongs to a class of hydrocarbons called aromatic hydrocarbons.

Benzene is represented by the structure [⬡] .

The benzene ring is stable due to the π-electron delocalization.

Benzene undergoes few reactions.

Benzene undergoes substitution reactions with the halogens, HNO_3, H_2SO_4, and haloalkanes.

A common way of naming disubstituted benzenes is the *ortho-*, *meta-*, and *para-* system (see page 1042 of the text).

29-8   Alcohols are organic compounds that contain an —OH group.

The IUPAC nomenclature for alcohols is given on page 1043 in the text.

Low-molecular-mass alcohols form hydrogen bonds and are completely miscible with water.

High-molecular-mass alcohols have low solubilities in water.

Alcohols undergo some reactions analogous to water to form an alkoxide ion RO^-.

Alcohols can be converted to alkenes by acid-catalyzed dehydration.

A primary alcohol has one carbon atom bonded to the carbon atom bearing the —OH group, a secondary alcohol has two such carbon atoms, and a tertiary alcohol has three such carbon atoms.

The reactions of alcohols depend on whether the alcohol is a primary, secondary, or tertiary alcohol.

29-9   Aldehydes and ketones contain a carbon-oxygen double bond.

Aldehydes have the general formula RCHO.

The aldehyde group —CHO is planar.

Aldehydes are obtained from the oxidation of primary alcohols.

Aldehydes are readily oxidized to carboxylic acids.

Secondary alcohols are oxidized to ketones.

A ketone has the general formula
$$\begin{matrix} R' \\ \diagdown \\ C=O \\ \diagup \\ R \end{matrix}$$

Tertiary alcohols cannot be oxidized to a molecule containing a carbon-oxygen double bond.

29-10   Amines are organic derivatives of ammonia.

Amines are derivatives of ammonia in which one or more of the hydrogen atoms in NH_3 are replaced by hydrocarbon groups.

Amines can be named by combining the names of the attached alkyl groups with the word amine.

A primary amine has one attached group, a secondary amine has two attached groups, and a tertiary amine has three attached groups.

Amines are soluble in water due to the formation of hydrogen bonds with water molecules.

Amines are weak bases.

29-11   The reaction of a carboxylic acid with an alcohol produces an ester.

A carboxylic acid can be obtained by the oxidation of an aldehyde or a primary alcohol.

Carboxylic acids react with alcohols in the presence of an acid catalyst to form esters.

$$\text{The general formula of an ester is} \quad \begin{array}{c} R \\ \diagdown \\ C{=}O \\ \diagup \\ R'O \end{array}$$

where the R' group comes from the alcohol.

Esters are named by first naming the alkyl group from the alcohol and then designating the acid with the *-ic* ending changed to *-ate*.

The yields of esterification reactions are often low.

29-12   The empirical formulas of many organic compounds can be determined by combustion analysis.

Combustion analysis is the analysis of the products obtained when an organic compound is burned in an excess of oxygen.

The masses of CO_2 and H_2O produced in the combustion reaction are converted to the stoichiometrically equivalent masses of carbon and hydrogen by the relations

$$\text{mass of C} = (\text{mass of } CO_2 \text{ formed}) \left(\frac{\text{atomic mass of C}}{\text{formula mass of } CO_2} \right)$$

$$\text{mass of H} = (\text{mass of } H_2O \text{ formed}) \left(\frac{2 \times \text{atomic mass of H}}{\text{formula mass of } H_2O} \right)$$

The mass percentage of carbon and hydrogen in the original sample are given by

$$\text{mass \% of C} = \left(\frac{\text{mass of C}}{\text{mass of sample}} \right) \times 100$$

$$\text{mass \% of H} = \left(\frac{\text{mass of H}}{\text{mass of sample}} \right) \times 100$$

The mass percentage of oxygen in organic compounds containing carbon, hydrogen, and oxygen is given by

$$\text{mass \% of O} = 100\% - \text{mass \% of C} - \text{mass \% of H.}$$

The molecular formula does not indicate the structural formula of the compound.

B. SELF-TEST

1. The bonding in alkanes can be described in terms of _____ hybrid orbitals on the carbon atoms.

2. Structural isomers have the (*same/different*) molecular formulas.

3. Structural isomers have the (*same/different*) structural formulas.

4. Structural isomers have identical physical properties. *True/False*

5. The number of structural isomers of pentane is (*greater than/less than/the same as*) the number of structural isomers of heptane.

6. Ethane, CH_3CH_3, is a planar molecule. *True/False*

7. Alkanes undergo a great variety of reactions. *True/False*

8. Alkanes react with oxygen in a combustion reaction to form _____ and _____.

9. Hydrocarbons are used as fuels because _____ _____.

10. The reaction between chlorine and an alkane is an example of a _____.

11. The reaction between chlorine and an alkane requires _____ for the reaction to occur.

12. The reaction between chlorine and an alkane produces _____.

13. In the IUPAC system for naming saturated hydrocarbons, the carbons in the main chain are numbered starting at the end that leads to assigning the (*lowest/highest*) numbers to the attached groups.

14. If two methyl groups are attached to the same carbon atom in an alkane, then only one number is necessary to designate the location of the methyl groups in the IUPAC name. *True/False*

15. Lewis formulas can be written if the IUPAC name is known. *True/False*

16. Alkenes are saturated hydrocarbons. *True/False*

17. Alkenes contain one or more carbon-carbon double bonds. *True/False*

18. The double bond in an alkene can be described in terms of a _____ bond and a _____ bond.

19. The region around the double bond in an alkene has a _____ geometry.

20. In the cis isomer of 1,2-dichloroethene, the two chlorine atoms lie _____

_____ .

21. Describe how alkenes are named. _____

_____

22. The reaction between hydrogen and an alkene is an example of a(an) _____ reaction.

23. The reaction between hydrogen and ethene in the presence of a catalyst produces

_____ .

24. The reaction between chlorine and ethene produces _____ .

25. The reaction between hydrogen chloride and ethene produces _____ .

26. The reaction between water and ethene in the presence of acid produces _____ .

27. Alkynes contain one or more carbon-carbon (*double/triple*) bonds.

28. Alkynes undergo reactions similar to alkenes. *True/False*

29. Describe how alkynes are named. _____

_____

30. A compact way of writing the benzene structure is _____ .

31. Benzene contains three double bonds. *True/False*

32. Benzene undergoes the same reactions as an alkene. *True/False*

33. Derivatives of benzene are named by placing one of the substituents at the number 1 position of the benzene ring and then numbering the carbon atoms in benzene sequentially. *True/False*

34. The designation *para-* in *p*-dichlorobenzene indicates that the two chlorine atoms are in the _____ and _____ positions on the benzene ring.

35. Aromatic hydrocarbons are stabilized by _____ .

36. An alcohol contains the _____ group.

37. Describe how alcohols are named. _____

38. The position of the —OH group must be designated by a number in the IUPAC name for an alcohol. *True/False*

39. (*Low/High*) molecular mass alcohols are miscible in water.

40. Some alcohols form hydrogen bonds. *True/False*

41. Alcohols react with alkali metals to form _____ and _____ .

42. Alcohols can be converted to _____ by acid-catalyzed dehydration.

43. A secondary alcohol has (*1/2/3*) carbon atoms bonded to the carbon atom bearing the —OH group.

44. Aldehydes have the general formula _____ .

45. Ketones have the general formula _____ .

46. Aldehydes are obtained from the oxidation of _____ .

47. Ketones are obtained from the oxidation of _____ .

48. Carboxylic acids can be obtained from the oxidation of _____ or _____ .

49. The reaction between a carboxylic acid and an alcohol yields a(an) _____ .

50. An ester has the general formula _____ where _____ .

51. An esterification reaction is a quantitative reaction. *True/False*

52. The (*molecular/empirical*) formula of many organic compounds can be determined by combustion analysis.

53. When an organic compound that consists of carbon, hydrogen, and water is burned in excess oxygen, all the carbon ends up in _____ and all the hydrogen ends up in _____ .

54. The mass of carbon in an organic compound can be determined from the mass of carbon dioxide produced in the combustion of the compound. *True/False*

55. The mass percentage of oxygen in an organic compound cannot be determined in a combustion analysis of the compound. *True/False*

56. The molecular formula of a compound indicates how the atoms are bonded together in the compound. *True/False*

C. CALCULATIONS YOU SHOULD KNOW HOW TO DO

There are no new types of calculations in this chapter. You should know how to determine the empirical formula of an organic compound from combustion analysis. See Example 29-13 and Problems 29-61 through 29-66. You should also know how to do the following:

1. Name organic compounds according to IUPAC rules.

2. Write formulas from the IUPAC names of organic compounds.

3. Write chemical equations for the reactions involving organic compounds.

D. SOLUTIONS TO THE ODD-NUMBERED PROBLEMS

29-1 (a) $C_5H_{12}(g) + 8O_2(g) \rightarrow 5CO_2(g) + 6H_2O(l)$

(b) $C_2H_6(g) + Cl_2(g) \xrightarrow{\text{dark}} \text{N.R.}$

(c) $C_4H_{10}(g) + H_2SO_4(aq) \rightarrow \text{N.R.}$

(d) $CH_4(g) + Cl_2(g) \xrightarrow{\text{UV}} CH_3Cl(g) + HCl(g)$
plus other chloromethanes such as $CH_2Cl_2(g)$

29-3 (a) The molecules are identical: One can be rotated 180° to superimpose upon the other.
(b) The molecules are identical. The chlorine atom is attached to the second carbon atom in each molecule.

29-5 n-Hexane is

Different chloro isomers are obtained by attaching a chlorine atom to the first, second, or third carbon atoms in the chain. Therefore, you get three isomers of chlorohexane. Their IUPAC names are 1-chlorohexane, 2-chlorohexane, and 3-chlorohexane.

29-7 The equation for the reaction is

$$C_4H_{10}(g) + \tfrac{13}{2}O_2(g) \longrightarrow 4CO_2(g) + 5H_2O(l)$$

The value of ΔH_{rxn}° for the combustion of butane is given by

$$\Delta H_{rxn}^\circ = 5\,\Delta H_f^\circ[H_2O(l)] + 4\,\Delta H_f^\circ[CO_2(g)] - \Delta H_f^\circ[C_4H_{10}(g)] - \tfrac{13}{2}\,\Delta H_f^\circ[O_2(g)]$$
$$= (5\text{ mol})(-285.8\text{ kJ}\cdot\text{mol}^{-1}) + (4\text{ mol})(-393.5\text{ kJ}\cdot\text{mol}^{-1})$$
$$- (1\text{ mol})(-126.1\text{ kJ}\cdot\text{mol}^{-1}) - (\tfrac{13}{2}\text{ mol})(0)$$
$$= -2876.9\text{ kJ}$$

The heat of combustion per gram of butane is given by

$$\text{heat per gram} = (-2876.9 \text{ kJ} \cdot \text{mol}^{-1}) \left(\frac{1 \text{ mol C}_4\text{H}_{10}}{58.12 \text{ g C}_4\text{H}_{10}} \right)$$
$$= -49.50 \text{ kJ} \cdot \text{g}^{-1}$$

29-9   The Lewis formulas for cyclopropane and cyclobutane are

cyclopropane   cyclobutane

29-11   (a)   The longest chain consists of four carbon atoms. The IUPAC name is 2-bromo-3-chlorobutane or 2-chloro-3-bromobutane.

(b)   The longest chain consists of three carbon atoms. The IUPAC name is 2,2-dimethylpropane.

29-13   (a)   The name violates rule 3. The chain was not numbered to give the lowest number to the carbon atom that has an attached group. The correct IUPAC name is 2-methylpentane.

(b)   The formula for 2-ethylbutane is

$$\underset{\displaystyle \underset{\displaystyle CH_3}{|} \atop \displaystyle \underset{CH_2}{|}}{CH_3CHCH_2CH_3}$$

This name violates rule 2. The correct IUPAC name is 3-methylpentane.

(c)   The formula for 2-propylhexane is

$$CH_3-\underset{\displaystyle \underset{\displaystyle CH_3}{|} \atop \displaystyle \underset{\displaystyle CH_2}{|} \atop \displaystyle \underset{CH_2}{|}}{CH}-CH_2-CH_2-CH_2-CH_3$$

This name violates rule 2. The correct IUPAC name is 4-methyloctane.

(d)   One of the methyl groups has not been numbered (rule 6). The correct IUPAC name is 2,2-dimethylpropane.

29-15   (a) $CH_3-\underset{\displaystyle \underset{CH_3}{|}}{CH}-CH_2-CH_3$ 2-methylbutane

(b) $Cl-CH_2-CH_2-Br$ 1-bromo-2-chloroethane
2-bromo-1-chloroethane

(c)
$$CH_3-\underset{\underset{Cl}{|}}{\overset{\overset{Cl}{|}}{C}}-\underset{\underset{Cl}{|}}{\overset{\overset{Cl}{|}}{C}}-Cl$$ 1,1,1,2,2-pentachloropropane

(d)
$$CH_3-\underset{\underset{CH_3}{|}}{\overset{\overset{CH_3}{|}}{C}}-CH_3$$ 2,2-dimethylpropane

29-17 (a) The parent alkane is butane. The name indicates that a methyl group is bonded to the second and third carbon atoms. Thus the structural formula is

$$CH_3-\underset{\underset{CH_3}{|}}{CH}-\underset{\underset{CH_3}{|}}{CH}-CH_3$$

(b) The parent alkane is butane. The name indicates that two methyl groups are bonded to the second carbon atom and one methyl group is bonded to the third carbon atom. The structural formula is

$$CH_3-\underset{\underset{CH_3}{|}}{\overset{\overset{CH_3}{|}}{C}}\!-\!-\!-\!\underset{\underset{CH_3}{|}}{CH}-CH_3$$

(c) The parent alkane is hexane. The name indicates that two methyl groups are bonded to the third carbon atom and an ethyl group is bonded to the fourth carbon atom. The structural formula is

$$CH_3-CH_2-\underset{\underset{CH_3}{|}}{\overset{\overset{CH_3}{|}}{C}}\!-\!-\!-\!\underset{\underset{\underset{\underset{CH_3}{|}}{CH_2}}{|}}{CH}-CH_2-CH_3$$

(d) The parent alkane is octane. The name indicates that an isopropyl group is bonded to the fourth carbon atom. The structural formula is

$$CH_3-CH_2-CH_2-\underset{\underset{CH_3-CH-CH_3}{|}}{CH}-CH_2-CH_2-CH_2-CH_3$$

29-19 (a) The longest chain consists of four carbon atoms, and so we write

$$
\begin{array}{cc}
CH_3 & CH_3 \\
| & | \\
H_3C-C\!\!-\!\!\!-\!\!\!-\!\!C-CH_3 \\
| & | \\
H & H
\end{array}
$$

(b) The longest chain consists of four carbon atoms, and so we write

$$
\begin{array}{cc}
NH_2 & CH_3 \\
| & | \\
H_3C-C\!\!-\!\!\!-\!\!\!-\!\!C-CH_3 \\
| & | \\
H & H
\end{array}
$$

(c) The longest chain consists of five carbon atoms, and so we write

$$
\begin{array}{ccc}
H & Cl & H \\
| & | & | \\
H_3C-C-C\!\!-\!\!\!-\!\!\!-\!\!C-CH_3 \\
| & | & | \\
H & CH_2 & H \\
 & | & \\
 & CH_3 &
\end{array}
$$

(d) The longest chain consists of three carbon atoms, and so we write

$$
\begin{array}{ccc}
Cl & Cl & Cl \\
| & | & | \\
H-C-C-C-H \\
| & | & | \\
H & H & H
\end{array}
$$

29-21 (a)

$$
\begin{array}{cc}
H & CH_3 \\
\!\!\searrow & \!\!\swarrow \\
C\!=\!C \\
\!\!\nearrow & \!\!\nwarrow \\
H & H
\end{array}
\;+\; HCl \longrightarrow
\begin{array}{c}
Cl \\
| \\
H_3C-C-CH_3 \\
| \\
H
\end{array}
$$

2-chloropropane

(b)

$$
\begin{array}{cc}
H_3C & CH_3 \\
\!\!\searrow & \!\!\swarrow \\
C\!=\!C \\
\!\!\nearrow & \!\!\nwarrow \\
H & H
\end{array}
\;+\; HBr \longrightarrow
\begin{array}{cc}
H & Br \\
| & | \\
H_3C-C-C-CH_3 \\
| & | \\
H & H
\end{array}
$$

2-bromobutane

(c)

$$
\begin{array}{cc}
H & CH_2CH_3 \\
\!\!\searrow & \!\!\swarrow \\
C\!=\!C \\
\!\!\nearrow & \!\!\nwarrow \\
H & H
\end{array}
\;+\; HCl \longrightarrow
\begin{array}{cccc}
H & Cl & H & H \\
| & | & | & | \\
H-C-C-C-C-H \\
| & | & | & | \\
H & H & H & H
\end{array}
$$

2-chlorobutane

(d)

$$H_3C, \quad CH_2CH_3$$
$$\begin{array}{c} H_3C \\ C=C \\ H \end{array} \quad \begin{array}{c} CH_2CH_3 \\ H \end{array} + HBr \longrightarrow H_3C-\overset{\overset{\displaystyle H}{|}}{\underset{\underset{\displaystyle H}{|}}{C}}-\overset{\overset{\displaystyle Br}{|}}{\underset{\underset{\displaystyle H}{|}}{C}}-CH_2CH_3$$

3-bromopentane

and $H_3C-\overset{\overset{\displaystyle Br}{|}}{\underset{\underset{\displaystyle H}{|}}{C}}-\overset{\overset{\displaystyle H}{|}}{\underset{\underset{\displaystyle H}{|}}{C}}-CH_2CH_3$

2-bromopentane

29-23 (a)

$$\begin{array}{c} H \\ C=C \\ H \end{array} \begin{array}{c} CH_2CH_3 \\ H \end{array} + HCl \longrightarrow CH_3-\overset{\overset{\displaystyle Cl}{|}}{\underset{\underset{\displaystyle H}{|}}{C}}-CH_2CH_3$$

2-chlorobutane

(b)

$$\begin{array}{c} H \\ C=C \\ H \end{array} \begin{array}{c} CH_3 \\ H \end{array} + Cl_2 \longrightarrow H-\overset{\overset{\displaystyle Cl}{|}}{\underset{\underset{\displaystyle H}{|}}{C}}-\overset{\overset{\displaystyle Cl}{|}}{\underset{\underset{\displaystyle H}{|}}{C}}-CH_3$$

1,2-dichloropropane

29-25 (a) We shall react H_2O in the presence of an acid with an alkene in accord with Markovnikov's rule to obtain the desired alcohol. We use the alkenes

$$\begin{array}{c} H_3C \\ C=C \\ H_3C \end{array} \begin{array}{c} H \\ CH_3 \end{array} \quad \text{or} \quad CH_2=C \begin{array}{c} CH_3 \\ CH_2CH_3 \end{array}$$

(b) We react H_2O in the presence of an acid with an alkene in accord with Markovnikov's rule to obtain the desired alcohol. We use the alkene $CH_2=CHCH_3$.

(c) We react one mole of H_2 with the corresponding alkyne $CH_3C\equiv CCH_3$.

29-27 (a) propyne (b) 2-butyne (c) 4,4-dimethyl-2-hexyne

29-29 We can break the reaction down into two steps. We shall use Markovnikov's rule to predict the product of each step. The first step is

$$CH_3C\equiv CH(g) + HBr(g) \longrightarrow CH_3\underset{\underset{\displaystyle Br}{|}}{C}=CH_2(g)$$

The second step is

$$CH_3C{=}CH_2(g) + HBr(g) \longrightarrow CH_3\overset{\displaystyle Br}{\underset{\displaystyle Br}{C}}CH_3(l)$$

with Br below the first carbon on the left.

The product is 2,2-dibromopropane.

29-31 (a) This reaction can be broken down into two steps. We shall use Markovnikov's rule to predict the product in each step. The first step is

$$CH_3C{\equiv}CH(g) + HCl(g) \longrightarrow CH_3\underset{\displaystyle Cl}{C}{=}CH_2(g)$$

The second step is

$$CH_3\underset{\displaystyle Cl}{C}{=}CH_2(g) + HCl(g) \longrightarrow CH_3\overset{\displaystyle Cl}{\underset{\displaystyle Cl}{C}}CH_3(l)$$

(b) $$CH_3C{\equiv}CH(g) + 2Br_2(l) \rightarrow CH_3\overset{\displaystyle Br}{\underset{\displaystyle Br}{C}}CHBr_2(l)$$

29-33 (a)   (b)   (c)   (d)

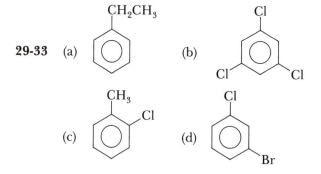

29-35 (a) The longest chain consists of three carbon atoms and the —OH group is attached to the second carbon atom, 2-propanol.
  (b) The longest chain consists of four carbon atoms and the —OH group is attached to the first carbon atom, 2,2-dimethyl-1-butanol.
  (c) The longest chain consists of three carbon atoms and the —OH group is attached to the second carbon atom, 1,3-dichloro-2-propanol.
  (d) The longest chain consists of four carbon atoms and the —OH group is attached to the second carbon atom, 2-butanol.

29-37 (a) a primary alcohol   (b) a secondary alcohol
       (c) a secondary alcohol   (d) a tertiary alcohol

29-39 (a) $\qquad 2CH_3CH_2OH(l) + 2Na(s) \longrightarrow 2Na^+CH_3CH_2O^-(s) + H_2(g)$

     (b) $\qquad 2CH_3CH_2CH_2OH(l) + 2Na(s) \longrightarrow 2Na^+CH_3CH_2CH_2O^-(s) + H_2(g)$

29-41 (a) $\qquad CH_3CHOHCH_2CH_3(l) \xrightarrow{Cr_2O_7^{2-}} CH_3-\overset{\displaystyle O}{\overset{\|}{C}}-CH_2CH_3(l)$

     (b) $\qquad CH_3CH_2OH(l) + NaH(s) \longrightarrow NaOCH_2CH_3(alc) + H_2(g)$

29-43 (a) butanal   (b) 3-methylbutanal
       (c) propanal   (d) 3,4-dimethylpentanal

29-45 (a) The formula of diethyl ketone is

$$CH_3CH_2 \diagdown \underset{\underset{O}{\|}}{C} \diagup CH_2CH_3$$

        The alcohol to use to prepare diethyl ketone is 3-pentanol.
     (b) The formula of methyl propyl ketone is

$$H_3C \diagdown \underset{\underset{O}{\|}}{C} \diagup CH_2CH_2CH_3$$

        The alcohol to use to prepare methyl propyl ketone is 2-pentanol.
     (c) The formula of ethyl propyl ketone is

$$CH_3CH_2 \diagdown \underset{\underset{O}{\|}}{C} \diagup CH_2CH_2CH_3$$

        The alcohol to use to prepare ethyl propyl ketone is 3-hexanol.

29-47 (a) propanone (number not needed)
       (b) 3-pentanone
       (c) butanone (number not needed)
       (d) butanone (number not needed)

29-49 (a) *N*-ethyl-1-amino-2-methylpropane
       (b) *N*-methyl-2-amino-3-methylpentane

29-51 (a) $C_2H_5NH_2(aq) + HBr(aq) \rightarrow C_2H_5NH_3Br(aq)$

     (b) $2(CH_3)_2NH(aq) + H_2SO_4(aq) \rightarrow [(CH_3)_2NH_2]_2SO_4(aq)$

(c)   [benzene ring]–NH$_2$ (aq) + HCl(aq) $\longrightarrow$ [benzene ring]–NH$_3$Cl (aq)

29-53   (a)   CH$_3$CH$_2$COOH   or   CH$_3$CH$_2$–$\overset{\overset{\displaystyle O}{\|}}{C}$–OH

(b)   CH$_3\underset{\underset{\displaystyle CH_3}{|}}{C}$HCOOH   or   CH$_3\underset{\underset{\displaystyle CH_3}{|}}{C}$H–$\overset{\overset{\displaystyle O}{\|}}{C}$–OH

(c)   CH$_3\underset{\underset{\displaystyle CH_3}{|}}{\overset{\overset{\displaystyle CH_3}{|}}{C}}CH_2$COOH   or   CH$_3\underset{\underset{\displaystyle CH_3}{|}}{\overset{\overset{\displaystyle CH_3}{|}}{C}}CH_2$–$\overset{\overset{\displaystyle O}{\|}}{C}$–OH

(d)   CH$_3$CH$_2\underset{\underset{\displaystyle CH_3}{|}}{C}HCH_2$COOH   or   CH$_3CH_2\underset{\underset{\displaystyle CH_3}{|}}{C}HCH_2$–$\overset{\overset{\displaystyle O}{\|}}{C}$–OH

29-55   (a)   This is a neutralization reaction. The balanced equation is

$$HCOOH(aq) + NaOH(aq) \longrightarrow NaHCOO(aq) + H_2O(l)$$

(b)   The reaction between an acid and an alcohol yields an ester. The balanced equation is

$$HCOOH(aq) + CH_3OH(aq) \xrightarrow{H^+(aq)} \begin{array}{c} H \\ \diagdown \\ \end{array} C{=}O(aq) + H_2O(l) \quad \begin{array}{c} \diagup \\ CH_3O \end{array}$$

(c)   This is a neutralization reaction. The balanced equation is

$$2HCOOH(aq) + Ca(OH)_2(aq) \longrightarrow Ca(HCOO)_2(aq) + 2H_2O(l)$$

29-57   (a)   CH$_3$CH$_2$COOH(aq) + KOH(aq) $\rightarrow$ KCH$_3$CH$_2$COO(aq) + H$_2$O(l)
                         potassium propanoate

(b)   CH$_3\underset{\underset{\displaystyle CH_3}{|}}{C}$HCOOH$(aq)$ + KOH(aq) $\rightarrow$ KCH$_3\underset{\underset{\displaystyle CH_3}{|}}{C}$HCOO$(aq)$ + H$_2$O(l)
                         potassium 2-methylpropanoate

(c)   2Cl$_2$CHCOOH(aq) + Ca(OH)$_2$$(aq)$ $\rightarrow$ Ca(Cl$_2$CHCOO)$_2$$(aq)$ + 2H$_2$O(l)
                         calcium dichloroethanoate

29-59   Each of these reactions is the reaction between a carboxylic acid and an alcohol to yield an ester.

(a)
O
C
OH
$HOCH_2CH_3$
$H^+(aq)$
O
C
OCH_2CH_3
$+ H_2O(l)$
benzoic acid
ethanol
ethyl benzoate

(b)
$HO-C(=O)-C(=O)-OH + 2CH_3CH_2CH_2OH$
$H^+(aq)$
oxalic acid
1-propanol
$CH_3CH_2CH_2O-C(=O)-C(=O)-OCH_2CH_2CH_3 + 2H_2O$
propyl oxalate

(c)
CH_3-C
O
OH
$+ CH_3-CH-CH_3$
OH
$H^+(aq)$
CH_3-C
O
O—CH
CH_3
CH_3
$+ H_2O$
acetic acid
2-propanol
(isopropyl
alcohol)
isopropyl
acetate

29-61 The masses of carbon and hydrogen in the original sample are

$$\text{mass of C} = (1.500 \text{ g } CO_2)\left(\frac{12.01 \text{ g C}}{44.01 \text{ g } CO_2}\right) = 0.4093 \text{ g C}$$

$$\text{mass of H} = (0.409 \text{ g } H_2O)\left(\frac{2 \times 1.008 \text{ g H}}{18.02 \text{ g } H_2O}\right) = 0.04576 \text{ g H}$$

The mass percentages of C and H are

$$\text{mass \% of C} = \frac{0.4093 \text{ g}}{1.000 \text{ g}} \times 100 = 40.93\%$$

$$\text{mass \% of H} = \frac{0.04576 \text{ g}}{1.000 \text{ g}} \times 100 = 4.576\%$$

The mass percentage of oxygen is obtained by difference

$$\text{mass \% of O} = 100.00 - 40.93 - 4.58 = 54.49$$

Take a 100-g sample and write

$$40.93 \text{ g C} \simeq 4.576 \text{ g H} \simeq 54.49 \text{ g O}$$

$$3.408 \text{ mol C} \simeq 4.540 \text{ mol H} \simeq 3.406 \text{ mol O}$$

$$1.000 \text{ mol C} \simeq 1.333 \text{ mol H} \simeq 1.000 \text{ mol O}$$

or

$$3.00 \text{ mol C} \simeq 4.00 \text{ mol H} \simeq 3.00 \text{ mol O}$$

The empirical formula is $C_3H_4O_3$.

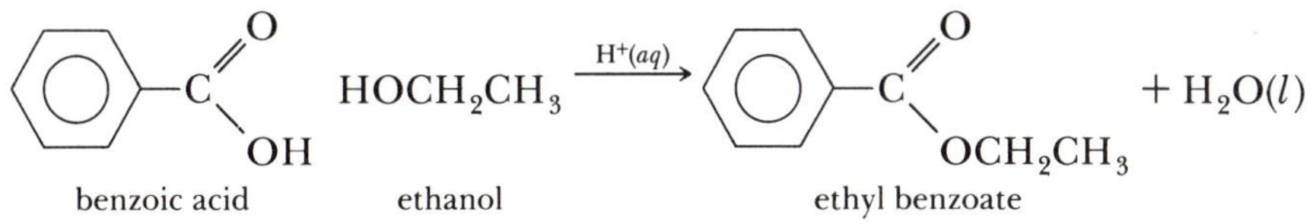

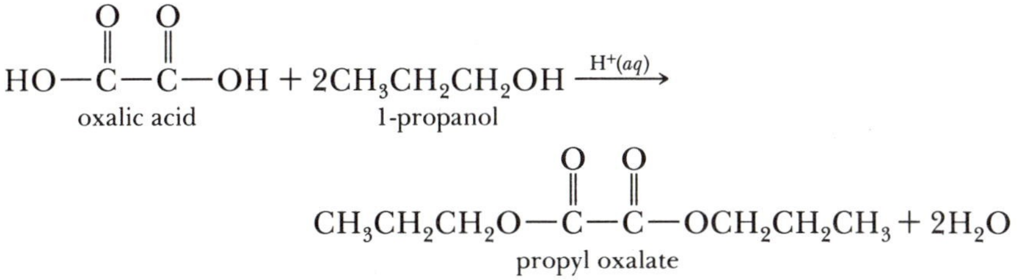

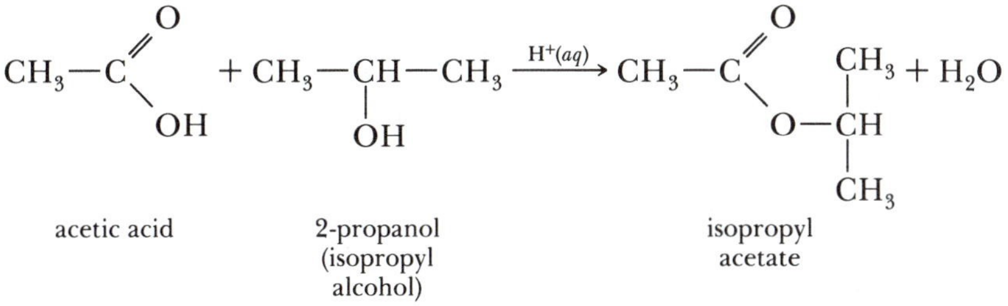

29-63 The masses of carbon and hydrogen in the original sample are

$$\text{mass of C} = (2.92 \text{ g CO}_2)\left(\frac{12.01 \text{ g C}}{44.01 \text{ g CO}_2}\right) = 0.7968 \text{ g C}$$

$$\text{mass of H} = (1.49 \text{ g H}_2\text{O})\left(\frac{2 \times 1.008 \text{ g H}}{18.02 \text{ g H}_2\text{O}}\right) = 0.1667 \text{ g H}$$

The mass percentages of C and H are

$$\text{mass \% of C} = \frac{0.7968 \text{ g}}{1.23 \text{ g}} \times 100 = 64.78\%$$

$$\text{mass \% of H} = \frac{0.1667 \text{ g}}{1.23 \text{ g}} \times 100 = 13.55\%$$

The mass percentage of oxygen is obtained by difference.

$$\text{mass \% of O} = 100.00 - 64.78 - 13.55 = 21.67\%$$

Take a 100-g sample and write

$$64.78 \text{ g C} \backsimeq 13.55 \text{ g H} \backsimeq 21.67 \text{ g O}$$
$$5.394 \text{ mol C} \backsimeq 13.44 \text{ mol H} \backsimeq 1.354 \text{ mol O}$$
$$3.98 \text{ mol C} \backsimeq 9.93 \text{ mol H} \backsimeq 1.00 \text{ mol O}$$

The empirical formula is $C_4H_{10}O$.

29-65 The mass percentages of carbon, hydrogen, and nitrogen are

$$\text{mass \% of C} = \frac{(1.518 \text{ g CO}_2)\left(\dfrac{12.01 \text{ g C}}{44.01 \text{ g CO}_2}\right)}{0.546 \text{ g}} \times 100 = 75.87\%$$

$$\text{mass \% of H} = \frac{(0.311 \text{ g H}_2\text{O})\left(\dfrac{2 \times 1.008 \text{ g H}}{18.02 \text{ g H}_2\text{O}}\right)}{0.546 \text{ g}} \times 100 = 6.372\%$$

$$\text{mass \% of N} = 100.00 - 75.87 - 6.372 - 17.76\%$$

Take a 100-g sample and write

$$75.87 \text{ g C} \backsimeq 6.372 \text{ g H} \backsimeq 17.76 \text{ g N}$$
$$6.317 \text{ mol C} \backsimeq 6.321 \text{ mol H} \backsimeq 1.268 \text{ mol N}$$
$$4.98 \text{ mol C} \backsimeq 4.99 \text{ mol H} \backsimeq 1.00 \text{ mol N}$$

The empirical formula is C_5H_5N.

29-67 1-chloropropane and "3-chloropropane" are the same molecule. One formula can be obtained from the other by just rotating it 180°.

29-69 The IUPAC names of the alkanes are

1:	2,3-dimethylpentane	(a heptane)
2:	2,3,3-trimethylpentane	(an octane)
3:	3,4-dimethylhexane	(an octane)
4:	2,3-dimethylpentane	(a heptane)

(a) 1 and 4 are the same compound.
(b) 2 and 3 are isomers of octane.
(c) 3 is a derivative of hexane.
(d) 2 has the most methyl groups.

29-71 There is hydrogen bonding in 1-butanol, but none in diethyl ether or pentane.

29-73 The gas must be propene.

$$H_2C=CH-CH_3 + HCl \longrightarrow CH_3-CHCl-CH_3$$

propene   2-chloropropane

29-75 A tank of gasoline (75 L) corresponds to

$$\text{mass} = (75 \times 10^3 \text{ mL})(0.80 \text{ g} \cdot \text{mL}^{-1}) = 6.0 \times 10^4 \text{ g}$$

Using octane as the gasoline, we have

$$\text{moles of octane} = (6.0 \times 10^4 \text{ g}) \left(\frac{1 \text{ mol C}_8\text{H}_{18}}{114.2 \text{ g C}_8\text{H}_{18}} \right) = 525 \text{ mol}$$

The equation for the combustion of octane is

$$C_8H_{18}(l) + \frac{25}{2} O_2(g) \longrightarrow 8CO_2(g) + 9H_2O(l)$$

and so we have

$$\text{moles of O}_2 = (525 \text{ mol C}_8\text{H}_{18}) \left(\frac{\frac{25}{2} \text{ mol O}_2}{1 \text{ mol C}_8\text{H}_{18}} \right) = 6.57 \times 10^3 \text{ mol}$$

The volume of $O_2(g)$ is

$$V = \frac{nRT}{P} = \frac{(6.57 \times 10^3 \text{ mol})(0.0821 \text{ L} \cdot \text{atm} \cdot \text{K}^{-1} \cdot \text{mol}^{-1})(293 \text{ K})}{1.0 \text{ atm}}$$
$$= 1.58 \times 10^5 \text{ L}$$

and the volume of air is

$$V_{air} = \frac{V_{O_2}}{0.21} = \frac{1.58 \times 10^5 \text{ L}}{0.21} = 7.5 \times 10^5 \text{ L}$$

which is about 10,000 times the volume of gasoline (75 L) burned.
    The energy produced is given by

$$\text{energy produced} = (6.0 \times 10^4 \text{ g C}_8\text{H}_{18}) \left(\frac{48 \text{ kJ}}{1 \text{ g C}_8\text{H}_{18}} \right) = 2.9 \times 10^6 \text{ kJ}$$

29-77 Take a 100-g sample and write

$$81.71 \text{ g C} \eqsim 18.29 \text{ g H}$$

$$6.803 \text{ mol C} \eqsim 18.14 \text{ mol H}$$

$$1.000 \text{ mol C} \eqsim 2.667 \text{ mol H}$$

Thus the empirical formula of the hydrocarbon is C_3H_8. The formula mass of the hydrocarbon is determined from the gas data:

$$n = \frac{PV}{RT} = \frac{\left(\frac{750}{760} \text{ atm} \right)(0.386 \text{ L})}{(0.0821 \text{ L} \cdot \text{atm} \cdot \text{mol}^{-1} \cdot \text{K}^{-1})(273 \text{ K})} = 0.0170 \text{ mol}$$

and so we have

$$0.0170 \text{ mol} \eqsim 0.75 \text{ g}$$

or

$$1.00 \text{ mol} \eqsim 44.1 \text{ g}$$

Thus the formula mass is 44.1, which corresponds closely with that of C_3H_8. The gas is propane.

29-79 The Lewis formula for 1-butene is

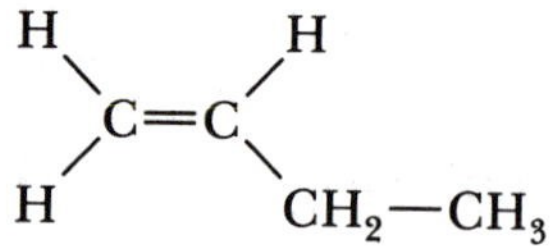

(a)   1-butene + $Cl_2(g) \longrightarrow$ Cl—CH_2—$\underset{\underset{Cl}{|}}{CH}$—$CH_2$—$CH_3$

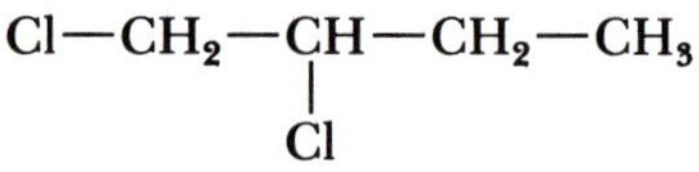

1,2-dichlorobutane

(b) We must use Markovnikov's rule in this case:

$$\text{1-butene} + \text{HCl}(g) \longrightarrow \underset{\underset{\displaystyle \text{2-chlorobutane}}{|}}{\underset{\underset{\displaystyle \text{Cl}}{|}}{CH_3\text{—}CH\text{—}CH_2\text{—}CH_3}}$$

(c) We must use Markovnikov's rule in this case:

$$\text{1-butene} + H_2O(l) \xrightarrow{H^{+(aq)}} \underset{\underset{\displaystyle \text{2-butanol}}{OH}}{CH_3\text{—}CH\text{—}CH_2\text{—}CH_3}$$

(d) $\text{1-butene} + H_2(g) \xrightarrow{\text{Pt}} \underset{\text{butane}}{CH_3\text{—}CH_2\text{—}CH_2\text{—}CH_3}$

29-81 We shall add HCl to each alkene according to Markovnikov's rule.

(a) $\underset{\underset{\displaystyle Cl}{|}}{CH_2\text{=}CHCHCH_3} + HCl \rightarrow \underset{\underset{\displaystyle Cl\ Cl}{|\ \ |}}{CH_3CHCHCH_3}$

(b) $BrCH_2CH\text{=}CHCH_3 + HCl \rightarrow \underset{\underset{\displaystyle Cl}{|}}{BrCH_2CHCH_2CH_3}$ and $\underset{\underset{\displaystyle Cl}{|}}{BrCH_2CH_2CHCH_3}$

(c) $\underset{\underset{\displaystyle CH_3}{|}}{CH_2\text{=}CCH_3} + HCl \rightarrow \underset{\underset{\displaystyle CH_3}{|}}{\overset{\overset{\displaystyle Cl}{|}}{CH_3CCH_3}}$

(d) $ClCH\text{=}CHCH_3 + HCl \rightarrow \underset{\underset{\displaystyle Cl}{|}}{ClCH_2CHCH_3}$ and $Cl_2CHCH_2CH_3$

29-83 (a) The alcohol is $CH_3CH_2CH_2OH$. We would use propanal

$$CH_3CH_2C\overset{\displaystyle \diagup H}{\underset{\displaystyle \diagdown\!\!\diagdown O}{}}$$

(b) The alcohol is $\underset{\underset{\displaystyle CH_3}{|}}{CH_3CHCH_2OH}$. We would use 2-methylpropanal

$$\underset{\underset{\displaystyle CH_3}{|}}{CH_3CHC}\overset{\displaystyle \diagup H}{\underset{\displaystyle \diagdown\!\!\diagdown O}{}}$$

(c)   The alcohol is CH₃CH₂ĊCH₂OH. We would use 2,2-dimethylbutanal

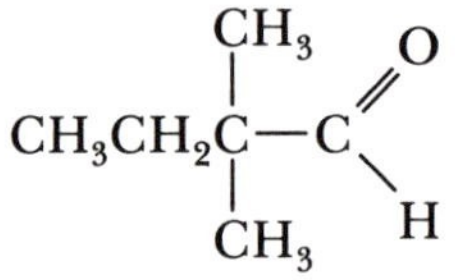

$$CH_3CH_2\underset{\displaystyle CH_3}{\overset{\displaystyle CH_3}{C}} - \overset{\displaystyle O}{\underset{\displaystyle H}{C}}$$

29-85   The primary amines are

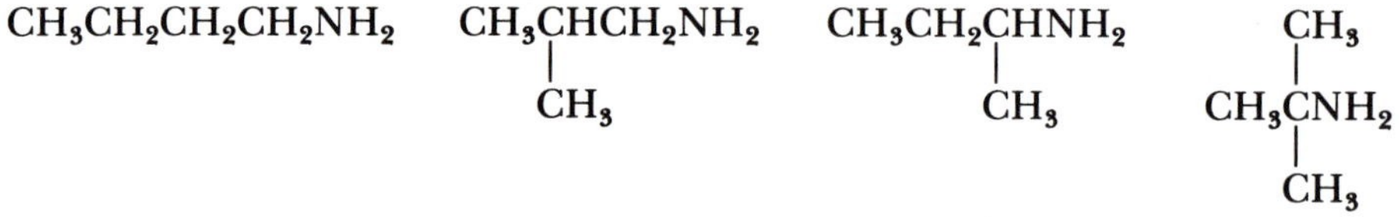

$$CH_3CH_2CH_2CH_2NH_2 \quad CH_3\underset{CH_3}{CH}CH_2NH_2 \quad CH_3CH_2\underset{CH_3}{CH}NH_2 \quad \underset{CH_3}{\overset{CH_3}{CH_3CNH_2}}$$

The secondary amines are

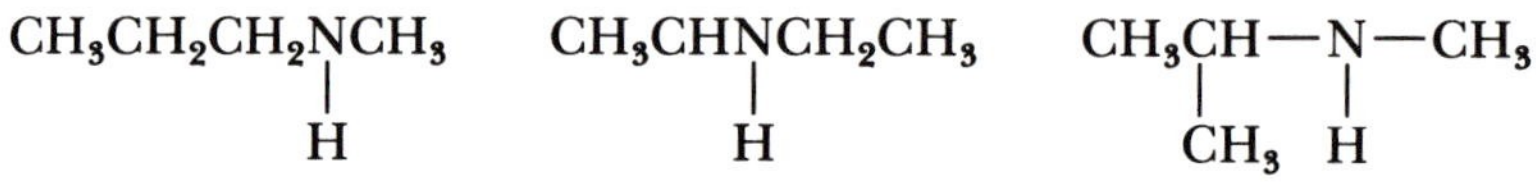

$$CH_3CH_2CH_2\underset{H}{N}CH_3 \quad CH_3\underset{H}{CH}NCH_2CH_3 \quad CH_3\underset{CH_3}{CH}-\underset{H}{N}-CH_3$$

The tertiary amine is

$$CH_3CH_2\underset{CH_3}{N}CH_3$$

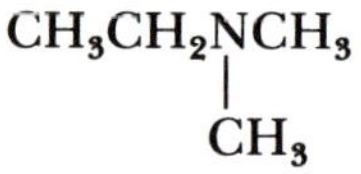

E. ANSWERS TO THE SELF-TEST

1. sp^3
2.   same
3.   different
4.   false
5.   less than
6.   false
7.   false
8. $CO_2 \ldots H_2O$
9.   The combustion reaction of hydrocarbons is highly exothermic; they produce a large amount of heat when burned.
10.   substitution
11.   UV light
12.   a haloalkane, or an alkyl halide
13.   lowest
14.   false
15.   true
16.   false
17.   true

18. $\sigma \ldots \pi$

19. planar

20. on the same side of the double bond

21. See Section 29-4 of the text.

22. addition

23. ethane

24. 1,2-dichloroethane

25. chloroethane

26. ethanol (CH_3CH_2OH)

27. triple

28. true

29. See Section 29-6 of the text.

30.

31. false

32. false

33. true

34. 1 . . . 4

35. π electron delocalization

36. —OH

37. See Section 29-8 of the text.

38. true

39. low

40. true

41. alkoxides and hydrogen

42. alkenes

43. 2

44. RCHO

45.

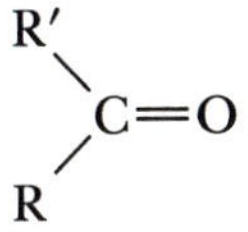

46. primary alcohols

47. secondary alcohols

48. primary alcohols or aldehydes

49. ester

50.

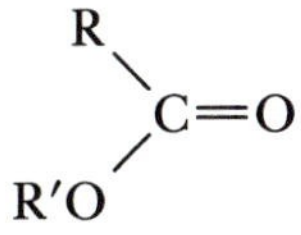

where R is from the acid and R' is from the alcohol

51. false

52. empirical

53. $CO_2 \ldots H_2O$

54. true

55. false

56. false

30 SYNTHETIC AND NATURAL POLYMERS

A. OUTLINE OF CHAPTER 30

30-1   Polymers are composed of many molecular subunits joined end to end.

Polymerization is the repeated addition of small molecules to form a long, continuous chain called a polymer.

The small molecules of units from which polymers are synthesized are called monomers.

The polymerization of ethylene, $H_2C{=}CH_2$, can be initiated by a free radical.

The monomers of polyethylene are ethylene.

Polyethylene can be written $-(CH_2CH_2-)_n$ where n is large.

Polymer synthesis produces polymer molecules with a range of n values.

Some common polymers are listed in Table 30-1.

30-2   Nylon and Dacron are made by condensation reactions.

In a condensation reaction, a small molecule, such as H_2O, is split out as each monomer is added to the polymer chain.

A polymer made from more than one type of monomer is called a copolymer.

The condensation reaction to form nylon proceeds by the formation of an amide linkage between a carboxyl group $-COOH$ and an amine group $-NH_2$.

The condensation reaction to form Dacron proceeds by the formation of an ester linkage between a carboxyl group $-COOH$ and an alcohol group $-OH$.

Polyesters are copolymers linked by an ester linkage.

30-3   Polymers with cross-linked chains are elastic.

At a given temperature, the longer the average length of the polymer chains, the less liquidlike the polymer is.

The polymer chains in a sample are coiled and intertangled with each other.

The relative movement of polymer chains that occurs when a sample is stretched can be decreased by connecting the chains through chemical bonds called cross-links (Figure 30-4).

When a cross-linked network is stretched, the coiled chains become elongated, but when the stress is released, the polymer network returns to its original coiled state.

A cross-linked polymer that exhibits elastic behavior is called an elastomer.

Natural rubber is cross-linked by the formation of —S—S— bonds between the polyisoprene chains (Figure 30-5).

30-4   Amino acids are the monomer units of polymers called proteins.

Proteins are naturally occurring polymers.

Amino acids have an amino group and a carboxyl group attached to a central carbon atom.

Amino acids have the general formula

$$H_2N-\underset{\underset{G}{|}}{\overset{\overset{H}{|}}{C}}-COOH$$

where G is called the side group.

There are a total of 20 different amino acids commonly found in proteins (Table 30-3).

The four bonds about the central atom in an amino acid are tetrahedrally oriented.

The amino acids have optical isomers.

Optical isomers are nonsuperimposable mirror-image molecules (Figure 30-6).

The two optical isomers are distinguished by the letters d and l.

Optical isomers ordinarily display the same chemical properties.

The l isomers of amino acids occur in proteins.

Biochemical reactions are stereospecific; they are dependent upon the shape of the reactants.

30-5   Proteins are formed by condensation reactions of amino acids.

Proteins are formed by condensation reactions between the carboxyl group —COOH on one amino acid and the amino group —NH$_2$ on another amino acid.

The amide bond in proteins is called a peptide bond.

Polypeptides are composed of long chains of amino acids joined together by peptide bonds.

The chain to which the amino acid side groups are attached is called the polypeptide backbone.

Proteins are naturally occurring polypeptides.

It is necessary to specify the order of the amino acids in a polypeptide.

The amino acids are given three-letter symbols (Table 30-3).

The order of the amino acid units in a polypeptide is called the primary structure of the polypeptide.

The primary structure uniquely characterizes a protein.

30-6   The shape of a protein molecule is called its tertiary structure.

The α-helix shape of certain proteins results from the formation of hydrogen bonds between peptide linkages in the peptide chain (Figure 30-9).

The structure within regions of a protein is called secondary structure (for example, α-helical structure).

Tertiary structure denotes the three-dimensional shape of a protein and is determined by X-ray analysis.

The shape of a protein depends on the nature and the order of the amino acid units making up the protein chain.

Tertiary structure plays a major role in protein function.

30-7   DNA is a double helix.

Genetic information is stored in DNA.

DNA and RNA are polynucleotides.

Nucleotides, the monomers of DNA and RNA, consist of a sugar portion, a phosphate group, and a nitrogen-containing ring compound called a base.

Deoxyribose and ribose are the sugars in the nucleotides of DNA and RNA, respectively.

DNA contains the four bases adenine (A), guanine (G), cytosine (C), and thymine (T).

RNA contains the four bases A, G, C, and uracil (U).

Nucleotides are joined by a condensation reaction between the phosphate group of one nucleotide and the 3-hydroxyl group on another nucleotide.

The DNA double helix consists of two polynucleotide chains intertwined in a helical fashion (Figure 30-12).

The two polynucleotide chains in the DNA double helix are held together by hydrogen bonding between base pairs: A and T bond together and G and C bond together (Figure 30-13).

The bases lie in the interior of the double helix; the sugar-phosphate backbone lies on the outside.

The two strands of double helical DNA are complementary to one another.

30-8   DNA can duplicate itself.

Each strand of the DNA double helix can act as a template for building a complementary strand.

The two new double helices are identical to the original double helix.

The genetic information that calls for the production of all the proteins characteristic of an organism is stored in DNA.

Each consecutive series of three bases along a DNA segment is a code for a particular amino acid.

The genetic code is a triplet code.

A gene is a segment along a DNA molecule that codes the synthesis of one polypeptide.

DNA can have a molecular mass of over 100 million.

B. SELF-TEST

1. Define polymerization.

_____

_____

2. The monomers of polyethylene are _____.

3. The polymer polyethylene can be written schematically as _____.

4. All the polymer molecules are of the same length. *True/False*

5. The polymerization of nylon is an example of an (*addition/condensation*) polymerization reaction.

6. The formation of Dacron is an example of a(an) (*addition/condensation*) polymerization reaction.

7. In the condensation polymerization reaction to form nylon, water is split out during the reaction. *True/False*

8. An amide bond in a polymer is formed by the reaction between an amino group and an acid group. *True/False*

9. An ester bond in a polymer is formed in the reaction between an _____ and an _____.

10. A copolymer is made from one type of monomer. *True/False*

11. The polymer chains in a sample are normally elongated to their maximum length. *True/False*

12. A cross-linked polymer can exhibit elastic behavior. *True/False*

13. Natural rubber is cross-linked by vulcanizing with sulfur. *True/False*

14. Amino acids are monomers from which _____ are built.

15. The general formula for an amino acid is

16. Amino acids can occur as (*structural/optical*) isomers.

17. The two isomers of an amino acid are designated by the letters _____ and _____ .

18. Both isomers of an amino acid occur in biological systems. *True/False*

19. The amino acids in a polypeptide are linked by (*amide/ester*) bonds.

20. Two amino acids form a dipeptide when the _____ group of one amino acid reacts with the _____ group of the other amino acid to form a peptide bond.

21. The polypeptide backbone is the same for all tripeptides. *True/False*

22. The order of attachment of the amino acids to the polypeptide backbone is the same for all tripeptides. *True/False*

23. The primary structure of a protein is unique to that protein. *True/False*

24. All proteins contain the same number of amino acid units. *True/False*

25. A polypeptide may have a helical shape because of _____ bonds between _____ .

26. The α-helix is an example of (*primary/secondary/tertiary*) structure.

27. The three-dimensional shape of a protein is its (*primary/secondary/tertiary*) structure.

28. Nucleotides are the monomers from which the polymers _____ and _____ are built up.

29. Nucleotides consist of a _____ , _____ _____ , and a _____ .

30. The sugar in DNA is _____ .

31. The sugar in RNA is _____ .

32. Uracil is found in RNA but not in DNA. *True/False*

33. Two nucleotides can be joined by a condensation reaction between the _____ _____ of one nucleotide and the _____ on another nucleotide.

34. The sugar-phosphate backbone is the same for all polynucleotides. *True/False*

35. The order of the attachment of the bases to the sugar-phosphate backbone is the same for all polynucleotides. *True/False*

36. The two polynucleotide chains in DNA are arranged as a _____ .

37. In DNA, the amount of guanine is equal to the amount of _____ .

38. In DNA, adenine on one chain is always paired with _____ on the other chain.

39. The two chains in DNA are joined together by _____ bonds.

40. Each strand of DNA acts as a template for reproducing itself. *True/False*

41. Genetic information for the production of _____ is stored in DNA.

42. Each amino acid is coded by a sequence of _____ base pairs on the DNA segment.

C. CALCULATIONS YOU SHOULD KNOW HOW TO DO

There are no calculations in this chapter.

D. SOLUTIONS TO THE ODD-NUMBERED PROBLEMS

30-1   In addition polymerization, monomers are joined to each other directly, without the formation of any small molecules as additional products. In condensation polymerization, monomers are joined together with the formation of small molecules as joint products. Usually condensation polymerization involves more than one kind of monomer, such as a diacid and a dialcohol. Teflon is an example of an addition polymer, and Dacron is an example of a condensation polymer.

30-3   Both the dicarboxylic acid and diamine monomers of nylon 66 contain six carbon atoms (see page 1065).

30-5 (a) No optical isomers. (The four substituents must be different.)

(b) 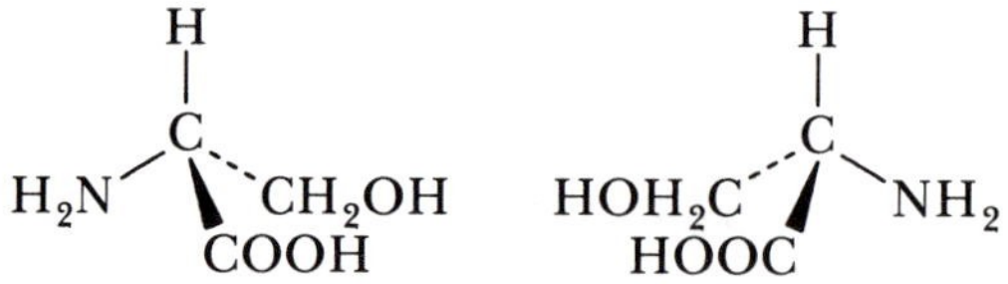

(c) 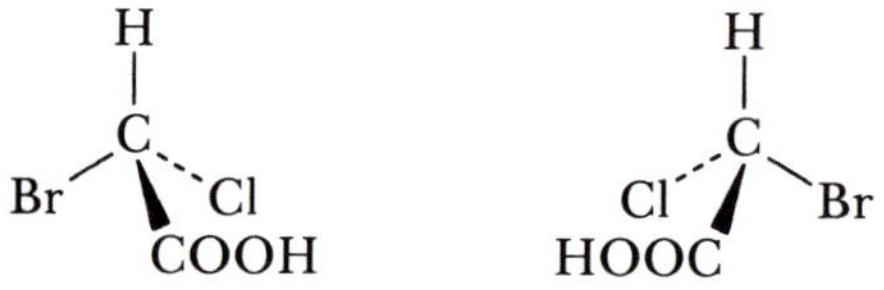

(d) 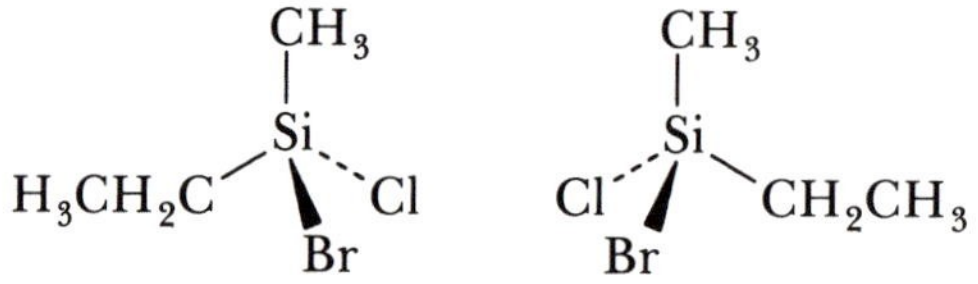

30-7

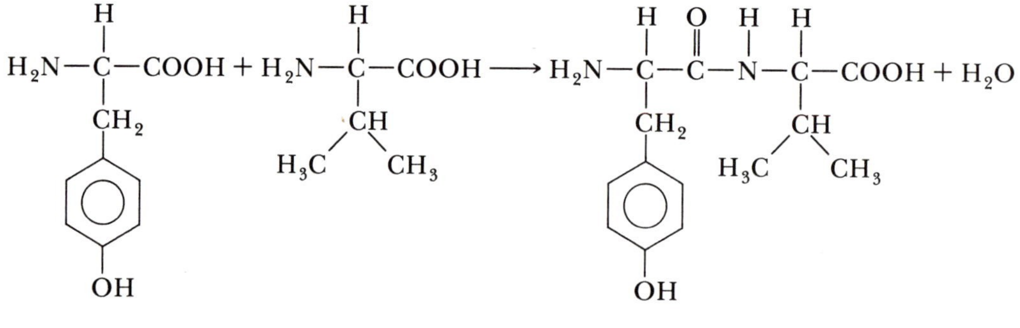

or

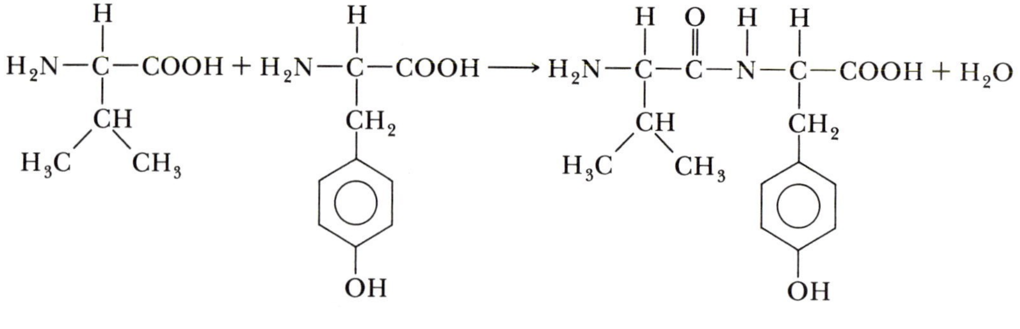

30-9 Two different dipeptides result because we can form the peptide bond in two ways depending on which carboxyl and amino groups are linked:

$$
\begin{array}{c}
\text{H} \quad \text{O} \qquad\quad \text{H} \\
| \qquad || \qquad\quad | \\
\text{H}_2\text{N}-\text{C}-\text{C}-\text{N}-\text{C}-\text{COOH} \\
| \qquad\qquad | \quad | \\
\text{H} \qquad\quad \text{H} \;\; \text{CH}_3 \\
\text{gly} \qquad\qquad \text{ala}
\end{array}
\qquad \text{or} \qquad
\begin{array}{c}
\text{H} \quad \text{O} \qquad\quad \text{H} \\
| \qquad || \qquad\quad | \\
\text{H}_2\text{N}-\text{C}-\text{C}-\text{N}-\text{C}-\text{COOH} \\
| \qquad\qquad | \quad | \\
\text{CH}_3 \qquad \text{H} \;\; \text{H} \\
\text{ala} \qquad\qquad \text{gly}
\end{array}
$$

30-11 We can form six different tripeptides from two different amino acids. If we represent the side groups of the two amino acids by G_1 and G_2, then the tripeptides are

$$
\begin{array}{c}
\text{H} \quad \text{O} \qquad \text{H} \quad \text{O} \qquad \text{H} \\
| \qquad || \qquad | \qquad || \qquad | \\
\text{H}_2\text{N}-\text{C}-\text{C}-\text{N}-\text{C}-\text{C}-\text{N}-\text{C}-\text{COOH} \\
| \qquad\quad | \quad | \qquad\quad | \quad | \\
\text{G}_1 \qquad \text{H} \;\; \text{G}_1 \qquad \text{H} \;\; \text{G}_2
\end{array}
$$

$$
\begin{array}{c}
\text{H} \quad \text{O} \qquad \text{H} \quad \text{O} \qquad \text{H} \\
| \qquad || \qquad | \qquad || \qquad | \\
\text{H}_2\text{N}-\text{C}-\text{C}-\text{N}-\text{C}-\text{C}-\text{N}-\text{C}-\text{COOH} \\
| \qquad\quad | \quad | \qquad\quad | \quad | \\
\text{G}_1 \qquad \text{H} \;\; \text{G}_2 \qquad \text{H} \;\; \text{G}_1
\end{array}
$$

$$
\begin{array}{c}
\text{H} \quad \text{O} \qquad \text{H} \quad \text{O} \qquad \text{H} \\
| \qquad || \qquad | \qquad || \qquad | \\
\text{H}_2\text{N}-\text{C}-\text{C}-\text{N}-\text{C}-\text{C}-\text{N}-\text{C}-\text{COOH} \\
| \qquad\quad | \quad | \qquad\quad | \quad | \\
\text{G}_2 \qquad \text{H} \;\; \text{G}_1 \qquad \text{H} \;\; \text{G}_1
\end{array}
$$

$$
\begin{array}{c}
\text{H} \quad \text{O} \qquad \text{H} \quad \text{O} \qquad \text{H} \\
| \qquad || \qquad | \qquad || \qquad | \\
\text{H}_2\text{N}-\text{C}-\text{C}-\text{N}-\text{C}-\text{C}-\text{N}-\text{C}-\text{COOH} \\
| \qquad\quad | \quad | \qquad\quad | \quad | \\
\text{G}_2 \qquad \text{H} \;\; \text{G}_2 \qquad \text{H} \;\; \text{G}_1
\end{array}
$$

$$
\begin{array}{c}
\text{H} \quad \text{O} \qquad \text{H} \quad \text{O} \qquad \text{H} \\
| \qquad || \qquad | \qquad || \qquad | \\
\text{H}_2\text{N}-\text{C}-\text{C}-\text{N}-\text{C}-\text{C}-\text{N}-\text{C}-\text{COOH} \\
| \qquad\quad | \quad | \qquad\quad | \quad | \\
\text{G}_2 \qquad \text{H} \;\; \text{G}_1 \qquad \text{H} \;\; \text{G}_2
\end{array}
$$

$$
\begin{array}{c}
\text{H} \quad \text{O} \qquad \text{H} \quad \text{O} \qquad \text{H} \\
| \qquad || \qquad | \qquad || \qquad | \\
\text{H}_2\text{N}-\text{C}-\text{C}-\text{N}-\text{C}-\text{C}-\text{N}-\text{C}-\text{COOH} \\
| \qquad\quad | \quad | \qquad\quad | \quad | \\
\text{G}_1 \qquad \text{H} \;\; \text{G}_2 \qquad \text{H} \;\; \text{G}_2
\end{array}
$$

30-13

$$H_2N-\underset{\underset{\underset{\underset{COOH}{|}}{CH_2}}{\underset{|}{CH_2}}}{\overset{\overset{H}{|}}{C}}-\underset{}{\overset{\overset{O}{\|}}{C}}-\underset{}{\overset{\overset{H}{|}}{N}}-\underset{\underset{\underset{COOH}{|}}{CH_2}}{\overset{\overset{H}{|}}{C}}-\underset{}{\overset{\overset{O}{\|}}{C}}-\underset{}{\overset{\overset{H}{|}}{N}}-\underset{\underset{CH_2}{|}}{\overset{\overset{H}{|}}{C}}-COOH$$

30-15 The sugar in DNA polynucleotides is deoxyribose. The DNA triplet is deoxy-guanosine-deoxyadenosine-deoxythymidine. The structures of the nucleotides are given in Section 30-7. The structural formula for the DNA triplet GAT is

30-17 The sugar in RNA is ribose. The RNA triplet UCU is uridine-cytidine-uridine. The structural formula for the RNA triplet UCU is

(structural formula of the RNA triplet UCU)

30-19 The two sequences must be complementary to each other: A and T must be opposite to each other, and G and C must be opposite to each other. The other sequence must have the base sequence TTCAGAGCT.

30-21 We must have T and A opposite each other and G and C opposite each other. The complementary base sequence is

G A T C A A T

30-23 The two strands come apart to give

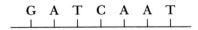 ①

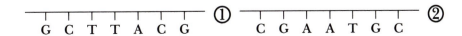 ②

The complements to the two strands are

$$\underline{\text{G C T T A C G}} \quad ① \quad \underline{\text{C G A A T G C}} \quad ②$$
$$\underline{\text{C G A A T G C}} \quad\quad\quad \underline{\text{G C T T A C G}}$$

30-25 We learned in Chapter 22 that the maximum amount of work that can be obtained from a reaction that is run under standard conditions is equal to the value of ΔG°_{rxn}. Thus,

$$\text{work} = \Delta G^\circ_{rxn} = (2.87 \times 10^3 \text{ kJ} \cdot \text{mol}^{-1}) \left(\frac{1 \text{ mol glucose}}{180.16 \text{ g glucose}} \right) (1.0 \text{ g}) = 15.9 \text{ kJ}$$

30-27 We can obtain the equation for the combustion of sucrose from the three equations given:

$$\text{sucrose}(aq) + H_2O(l) \longrightarrow \text{glucose}(aq) + \text{fructose}(aq)$$
$$\Delta G^\circ_{rxn} = -29.3 \text{ kJ}$$
$$\text{fructose}(aq) \longrightarrow \text{glucose}(aq)$$
$$\Delta G^\circ_{rxn} = -1.6 \text{ kJ}$$
$$2 \text{ glucose}(aq) + 12O_2(g) \longrightarrow 12CO_2(g) + 12H_2O(l)$$
$$\Delta G^\circ_{rxn} = (2)(-2.87 \times 10^3 \text{ kJ}) = -5.74 \times 10^3 \text{ kJ}$$

If we add these three equations, then we have

$$\text{sucrose}(aq) + 12O_2(g) \longrightarrow 12CO_2(g) + 11H_2O(l)$$
$$\Delta G^\circ_{rxn} = -29.3 \text{ kJ} - 1.6 \text{ kJ} - 5.74 \times 10^3 \text{ kJ}$$
$$= -5.77 \times 10^3 \text{ kJ}$$

The value of ΔG°_{rxn} for 1 mol of sucrose is -5.77×10^3 kJ. The value of ΔG°_{rxn} for 1.0 g of sucrose is

$$\Delta G^\circ_{rxn} = (-5.77 \times 10^3 \text{ kJ} \cdot \text{mol}^{-1}) \left(\frac{1 \text{ mol sucrose}}{342.3 \text{ g sucrose}} \right) = -16.9 \text{ kJ} \cdot \text{g}^{-1}$$

The maximum amount of work that can be obtained from the process is $16.9 \text{ kJ} \cdot \text{g}^{-1}$.

30-29 The value of the equilibrium constant is given by

$$\ln K = -\frac{\Delta G^\circ_{rxn}}{RT}$$
$$= -\frac{-31 \times 10^3 \text{ J} \cdot \text{mol}^{-1}}{(8.314 \text{ J} \cdot \text{mol}^{-1} \cdot \text{K}^{-1})(310 \text{ K})} = 12.0$$
$$K = 10^{12.0} = 1.7 \times 10^5 \text{ M}$$

30-31   The value of the equilibrium constant is given by

$$\ln K = -\frac{\Delta G^{\circ}_{\text{rxn}}}{RT}$$

$$= -\frac{-200 \times 10^3 \text{ J} \cdot \text{mol}^{-1}}{(8.314 \text{ J} \cdot \text{mol}^{-1} \cdot \text{K}^{-1})(298 \text{ K})}$$

$$= 80.72$$

$$K = 10^{80.72} = 1.1 \times 10^{35} \text{ M}$$

E. ANSWERS TO THE SELF-TEST

1. Polymerization is the repeated addition of small molecules to form a long continous chain.

2. ethene

3. $+CH_2CH_2\!\!+_n$

4. false

5. condensation

6. condensation

7. true

8. true

9. acid . . . alcohol

10. false

11. false

12. true

13. true

14. proteins

15.
$$H_2N-\overset{\overset{\displaystyle H}{|}}{\underset{\underset{\displaystyle G}{|}}{C}}-COOH$$

16. optical

17. d . . . l

18. false

19. amide

20. NH_2 . . . COOH

21. true

22. false

23. true

24. false

25. hydrogen . . . peptide bonds

26. secondary

27. tertiary

28. DNA . . . RNA

29. sugar, a phosphate group, and a nitrogen-containing ring compound called a base

30. deoxyribose

31. ribose

32. true

33. phosphate group . . . 3-hydroxyl group

34. true

35. false

36. double helix

37. cytosine

38. thymine

39. hydrogen

40. true

41. proteins

42. three

GLOSSARY

absolute alcohol (538): pure ethanol.

absolute temperature, T (11): the temperature given in the unit kelvin.

absolute temperature scale (201): the fundamental temperature scale. The temperature on the absolute temperature scale is found by adding $273.15°C$ to the temperature on the Celsius scale. The unit of absolute temperature is the kelvin, K. The absolute temperature scale is also called the Kelvin scale.

absolute zero (202): the lowest possible temperature on the Kelvin scale.

absorption spectrum (314): the spectrum obtained when atoms or molecules absorb electromagnetic radiation and are raised to excited states.

acceleration of gravity (244): the proportionality constant, g, that appears in the equation for the potential energy, E_p, of a mass, m, that is a height, h, above the ground: $E_p = gmh$. A value of g is 9.81 $m \cdot s^{-1}$.

accuracy (14): the degree to which a measurement is close to the actual value.

acid (114): a compound that yields hydrogen ions when it is dissolved in water.

acid-base titration (182, 681): the exact neutralization of an acid by a base of known concentration, or of a base by an acid of known concentration, carried out using a buret.

acid-dissociation constant, K_a (646): the equilibrium constant for the proton transfer reaction between an acid and water. For the general equation,

$$HB(aq) + H_2O(l) \rightleftharpoons$$
$$B^-(aq) + H_3O^+(aq),$$

$$K_a = \frac{[B^-][H_3O^+]}{[HB]}$$

acid-dissociation reaction (646): the proton-transfer reaction between an acid and water. The general equation is

$$HB(aq) + H_2O(l) \rightleftharpoons$$
$$B^-(aq) + H_3O^+(aq).$$

acid rain (241): rain that is more acidic than normal rain.

acidic anhydride (114): an oxide that yields an acid when it is dissolved in water.

acidic anion (660): an anion that reacts with water to produce $H_3O^+(aq)$ in aqueous solution.

acidic cation (661): a cation that reacts with water to form $H_3O^+(aq)$ in aqueous solution.

acidic hydrogen atom (115): a hydrogen atom in a species that dissociates in solution to become $H_3O^+(aq)$.

acidic proton (115): a dissociable hydrogen atom in a species.

acidic solution (126, 637): an aqueous solution in which $[H_3O^+] > [OH^-]$.

acidity (640): the concentration of $H_3O^+(aq)$ in an aqueous solution.

actinide series (82, 339): the inner transition series that is headed by actinium. The actinide series contains the elements actinium ($Z = 89$) through nobelium ($Z = 102$).

activated carbon (914): an amorphous form of carbon that is characterized by its ability to absorb large quantities of gases.

activation energy diagram (575): a diagram that shows the energies involved in the reaction of A and B to give C and D.

activation energy, E_a (574): the minimum energy necessary to cause a reaction between the colliding reactant molecules.

actual yield (161): the experimentally determined mass of a product that results from a reaction.

addition polymerization reaction (1065): a reaction that involves a direct addition of monomer molecules to form a polymer chain.

addition reaction (1036): a reaction in which atoms or molecules are added to a molecule.

alcohol (1043): an organic compound that contains an —OH group attached to a hydrocarbon chain.

aldehyde (1046): a class of organic compounds that have the general formula RCHO.

aldehyde group (1046): the —CHO group.

alkali metal (76): any of the elements (lithium, sodium, potassium, rubidium, cesium, and francium) that constitute Group 1 of the periodic table.

alkaline (75): having the property of producing a basic solution.

alkaline earth metal (76): any of the elements (beryllium, magnesium, calcium, strontium, barium, and radium) that make up Group 2 of the periodic table.

alkane (1025): a hydrocarbon that contains only carbon-carbon single bonds. (See **saturated hydrocarbon.**)

alkene (1034): a hydrocarbon that contains one or more carbon-carbon double bonds.

alkoxide (896): the ion, RO^-, that results from the loss of a proton from an alcohol.

alkyl group (638): a group that is derived from an alkane by removing a hydrogen atom.

alkyl halide (1030): a substance that is derived from an alkane by removing one or more hydrogen atoms and replacing them with halogen atoms. (See **haloalkane.**)

alkynes (1038): the class of hydrocarbons that contain one or more carbon-carbon triple bonds.

allotrope (912, 934): a substance that can exist in different modifications in the same physical state; for example, graphite and diamond are allotropes of solid carbon.

allotropy (912, 934): a property wherein a substance has more than one possible molecular form.

α-emitter (851): a radioisotope that decays by emitting α particles.

α-helix (1076): the helical shape of a polypeptide that results from the formation of hydrogen bonds between oxygen and hydrogen atoms in peptide bonds that are separated by three peptide units along the chain.

α-particle (45): a helium-4 nucleus emitted in certain types of nuclear decay.

amines (653): a class of organic compounds that contain nitrogen. The general formulas for amines are RNH_2,

$$R\!-\!\overset{..}{N}\!-\!H, \text{ and } R\!-\!\overset{..}{N}\!-\!R''.$$
$$\qquad\underset{R'}{|} \qquad\qquad \underset{R'}{|}$$

amino acid (1070): a monomer from which proteins are built. The general formula for an amino acid is

$$\overset{\displaystyle H}{\underset{\displaystyle G}{H_2N\!-\!\overset{|}{\underset{|}{C}}\!-\!COOH}}$$

where G represents a side group.

amino acid side group (1070): the organic group represented by —G that is attached to amino acids.

amorphous solid (507): characteristic of a solid that does not have a definite crystal structure (e.g., glass).

ampere (814): the SI unit of current. One ampere is equal to the flow of one coulomb of charge per second.

amphoteric (889): soluble in both acids and bases.

amphoteric metal hydroxide (736): a metal hydroxide that is soluble in both acidic and basic solutions, but insoluble in neutral solutions.

analytical chemistry (1054): the area of chemistry that deals with the chemical composition of substances.

anion (52, 359): a negatively charged ion.

anode (816): the electrode at which oxidation occurs. During current flow, anions in solution move toward the anode.

antibonding orbital (442): a molecular orbital that has one or more nodes in the region between two nuclei.

antifreeze (528): a substance that is used to lower the freezing point of water.

aqueous solution (108): a solution in which a solute is dissolved in water. An aqueous solution is designated by (*aq*).

aromatic hydrocarbon (1040): a class of hydrocarbons that have rings, which are stabilized by π electron delocalization.

Arrhenius acid (635): a substance that produces $H^+(aq)$ in aqueous solution.

Arrhenius base (635): a substance that produces $OH^-(aq)$ in aqueous solution.

Arrhenius equation (578): the equation that describes the temperature dependence of a rate constant. The Arrhenius equation is

$$\ln(k_2/k_1) = (E_a/R)(T_2 - T_1)/T_1 T_2$$

where k_1 and k_2 are the rate constants at the absolute temperatures T_1 and T_2, respectively; E_a is the activation energy; and R is the gas constant.

artificial radioisotopes (864): radioactive isotopes that are produced by nuclear reactions in the laboratory and are not found in nature.

asbestos (919): a mineral containing polymeric, cyclic polysilicate chains.

atom (33): a basic component of matter.

atomic absorption spectra (302): the spectra that are produced when atoms absorb electromagnetic energy.

atomic crystal (503): a crystal whose constituent particles are atoms.

atomic emission spectra (299): the spectra that result when atoms in excited states return to lower energy states.

atomic mass (37): the mass of an atom relative to the mass of an atom of carbon-12, which is assigned an atomic mass of exactly 12.

atomic mass ratio (37): the atomic mass.

atomic mass unit (37): a unit based on the mass of carbon-12, which contains exactly 12 atomic mass units. The symbol for atomic mass unit is amu.

atomic number (47): the number of protons in the nucleus of an atom. The atomic number of an atom is designated by the symbol Z.

atomic radius (340): the distance from the nucleus of an atom to the point where the electronic charge density is very small. Atoms do not have well-defined radii.

atomic spectroscopy (302): the study of the spectra of atoms.

atomic spectrum (302): a line spectrum resulting from the radiation emitted by gaseous, excited atoms or the radiation absorbed by gaseous atoms.

atomic substance (140): a substance that is composed of single atoms.

atomic theory (Dalton's) (36): a theory that postulates that (1) matter is composed of small, indivisible particles called atoms; (2) the atoms of a given element all have the same mass and are identical in all respects, including chemical behavior; (3) the atoms of different elements differ in mass and in chemical behavior; (4) chemical compounds are composed of two or more different types of atoms joined together in simple fixed ratios. The particle that results when two or more atoms join together is called a molecule; (5) in a chemical reaction, the atoms involved are rearranged to form different molecules; no atoms are created or destroyed.

ATP (938): adenosine triphosphate, which supplies the energy requirements of many biochemical reactions.

average (220): the quotient obtained by dividing the sum of a set of numbers by the number of numbers in the set.

average kinetic energy, $\bar{E}_k$ (220): the average kinetic energy of one mole of a gas, is given by $\bar{E}_k = (3/2)RT$, where T is the Kelvin temperature and R is the gas constant.

Avogadro's law (205): a law stating that equal volumes of gases at the same pressure and temperature contain equal numbers of molecules.

Avogadro's number (142): the number of formula units contained in one mole of a substance. One mole of a substance contains 6.022×10^{23} formula units.

AX_mE_n **(419):** a general representation of a molecule or ion where A represents the central atom, X_m represents m ligands bonded to the central atom, and E_n represents n lone electron pairs (denoted by E) on the central atom.

axial vertex (415): one of the two vertices that lie above or below the equilateral-triangle midplane of a trigonal bipyramid. The two axial vertices are equivalent.

azeotrope (538): a solution that distills without change in composition.

azide (933): a compound that contains the azide ion, N_3^-.

azimuthal quantum number (320): the quantum number, l, that specifies the shape of an atomic orbital. The allowed values of l are $0, 1, 2, \ldots, n-1$.

B

baking powder (937): a mixture of $Ca(H_2PO_4)_2(s)$ and $NaHCO_3(s)$ that is used to cause cakes and breads to rise.

balancing by inspection (72): a method of balancing chemical equations by trial and error.

balanced chemical equation (72): a chemical equation that has the same number of each type of atom on its reactant and product sides.

balancing coefficient (72): the number placed in front of the chemical formula of a reactant or product in a chemical equation so that the chemical equation is balanced.

ball-and-stick molecular model (411): a model of a molecule that is constructed to display the angles between the bonds in the molecule.

Balmer series (313): the series of lines in the hydrogen atomic spectrum emitted during transitions from higher states ($n > 2$) to the $n = 2$ state.

band gap (506): the energy difference between the top of the valence band and the bottom of the conduction band.

band of stability (856): the region of stable nuclei in a plot of the number of protons versus the number of neutrons of all known stable nuclei.

barometer (198): a device used to measure the pressure of the atmosphere.

barometric pressure (198): the pressure exerted by the atmosphere.

base (113): a compound that yields hydroxide ions, $OH^-(aq)$, when it is dissolved in water. A base is a proton acceptor.

base (1077): a nitrogen-containing ring compound that comprises a nucleotide. The five bases found in DNA or RNA are adenine, guanine, cytosine, uracil, and thymine.

base protonation constant, K_b (655): the equilibrium constant for the proton transfer reaction between a base and water. For the general equation

$$B^-(aq) + H_2O(l) \rightleftharpoons$$
$$BH(aq) + OH^-(aq)$$
$$K_b = \frac{[BH][OH^-]}{[B^-]}$$

basic anhydride (113): an oxide that yields a base when it is dissolved in water.

basic anion (660): an anion that reacts with water to produce $OH^-(aq)$ in aqueous solution.

basic oxygen process (984): a process for making steel from pig iron in which hot, pure O_2 is blown through molten pig iron.

basic SI units (7): the SI units from which all other units may be derived.

basic solution (126, 637): an aqueous solution in which $[OH^-] > [H_3O^+]$.

battery (846): an electrochemical cell or group of cells designed for use as a power source.

becquerel, Bq, (877): the SI unit of the rate of radioactive decay; one becquerel is defined as one disintegration per second.

bent (418): the shape of a triatomic species in which the bond angle is less than $180°$; the shape of an AX_2E and an AX_2E_2 molecule or ion.

beryl (900): a green mineral that is the chief source of beryllium and is used as a gem.

β-emitter (857): a radioisotope that emits electrons.

β-particle (45): an electron emitted in certain types of nuclear decay.

bidentate (1002): a chelating ligand that attaches to two metal coordination positions.

bifluoride ion (958): the $HF_2^-(aq)$ ion that is formed when hydrofluoric acid is dissolved in water.

$$2\,HF(aq) \longrightarrow H^+(aq) + HF_2^-(aq)$$

binary acid (115): an acid that consists of hydrogen and one other element.

binary compound (39): a compound that consists of two different elements.

binding energy (nuclear) (868): the energy required to separate the nucleons in a nucleus.

binding energy curve (870): a plot of the binding energy per nucleon versus the mass number for nuclei.

biopolymer (1077): a naturally occurring polymer.

bismuthinite (928): a sulfide ore (Bi_2S_3) that is the most common source of bismuth.

blackbody radiation (302): the emission of light by a solid body which has been heated to a high temperature.

blast furnace (983): a furnace in which Fe_2O_3 is reacted with coke at high temperatures to produce iron commercially.

body-centered cubic (500): the geometry of a unit cell in which the components of a crystal are located at the corners and in the center of a cube.

boiler scale (713): a deposit of metal carbonates from hard water.

boiling-point elevation (525): the increase in the boiling point of a solution containing nonvolatile solutes over the boiling point of the pure solvent. The boiling-point elevation is given by the equation $T_b - T_b^\circ = K_b m_c$, where T_b is the boiling point of the solution; T_b° is the boiling point of the pure solvent; m_c is the colligative molality of the solution; and K_b is the proportionality constant, called the boiling-point elevation constant, for the solvent.

boiling-point elevation constant (526): the proportionality constant between the boiling-point elevation and the colligative molality of a solution. The value depends only on the solvent. The symbol is K_b and the units are $K \cdot mc^{-1}$.

bomb calorimeter (273): a sealed reaction vessel that is used to measure the heat of combustion of a substance.

bond energy (800): bond enthalpy.

bond enthalpy (800): the energy as heat required to break one mole of bonds between atom X and atom Y at constant pressure. The symbol for bond enthalpy is $H(X - Y)$.

bond length (377): the average distance between the nuclei of the two atoms that are joined by a covalent bond.

bond order (443): one half of the net number of bonding electrons in a diatomic molecule [see Equation (13-1)].

bond polarity (401): the unequal sharing of the electrons in a covalent bond due to the difference in the electronegativities of the two atoms.

bonding orbital (442): a molecular orbital that is concentrated in a region between two nuclei.

borax (905): $NaB_4O_7 \cdot 10\ H_2O(s)$.

Born-Haber cycle (370): a closed sequence of thermodynamic steps used in calculating lattice energies of ionic compounds.

Boyle's law (201): a law stating that the volume of a fixed mass of gas at constant temperature is inversely proportional to the pressure of the gas.

brass (987): an alloy of copper and zinc.

breeder reactor (875): a nuclear reactor that produces more fissionable material than is consumed.

Brønsted-Lowry acid (635): a proton donor.

Brønsted-Lowry base (635): a proton acceptor.

bronze (987): an alloy of copper and tin.

buffer (675): a solution that is resistant to changes in pH upon the addition of an acid or a base.

buret (183): a precision-made piece of glassware that is used to measure accurately the volume of a solution that is added to another container.

C

calorie (275): the amount of energy as heat required to raise the temperature of one gram of water at $14.5\,^\circ C$ by one Celsius degree; 1 calorie $= 4.184$ J.

calorimeter (272): a device used to measure the amount of heat evolved or absorbed in a process.

capillary action (489): the rise of a liquid in a thin tube.

carbohydrate (275): a compound composed of carbon, hydrogen, and oxygen with the general formula $C_x(H_2O)_y$.

carbon-14 dating (862): the determination of the age of formerly living materials from the rate of disintegration of carbon-14 in the sample.

carbonyl group (1047): the $\diagdown C = O$ group

carborundum (915): silicon carbide, SiC.

carboxyl group (638): the —COOH group.

carboxylate ion (640): the anion that results from the dissociation or neutralization of a carboxylic acid. A carboxylate ion has the general formula $RCOO^-$.

carboxylic acid (638): a class of organic compounds that contain the —COOH group.

carrier gas (58): the inert gas used to carry a mixture of species through a gas chromatography column.

cassiterite (911): the mineral SnO_2, which is the primary ore of tin.

catalyst (579): a substance that increases the rate of a reaction but is not a reactant. A catalyst acts by providing a different and faster reaction pathway than that which would prevail in its absence.

cathode (816): the electrode at which reduction occurs. During electrolysis cations in solution move toward the cathode.

cation (52, 359): a positively charged ion.

caustic soda (898): sodium hydroxide, NaOH.

cell diagram (819): a notation used to represent an electrochemical cell. By convention, oxidation occurs at the left electrode and reduction occurs at the right electrode.

cell voltage (822): the voltage of an electrochemical cell.

Celsius temperature (11): the SI temperature scale in which the boiling point of water is 100 degrees and the freezing point of water is 0 degrees. The unit of Celsius temperature is °C.

centi- (9): a prefix meaning one hundredth.

centimeter (8): one hundredth of a meter.

centrifuge (740): an instrument that hastens the settling of a precipitate by centrifugation.

ceramics (516): a complex oxide of several metals.

chain reaction (872): a self-sustaining reaction in which the products initiate further reaction. This term is commonly applied to certain nuclear fission reactions in which neutrons that are produced by the fission process cause additional nuclei to fission, hence propagating the reaction.

charge delocalization (467): the distribution of the electronic charge over more than one atom in a polyatomic species.

Charles's law (202): a law stating that the volume of a fixed mass of gas at constant pressure is directly proportional to the absolute temperature of the gas.

chelate (1002): a complex that contains a metal ion and at least one chelating ligand.

chelating ligand (1002): a ligand that attaches to a metal ion at more than one coordination position.

chemical equation (72): the representation of a chemical reaction using the chemical formulas of the reactants and the products, separated by an arrow.

chemical equilibrium (595): chemical equilibrium is attained when the rate of the forward reaction is equal to the rate of the reverse reaction. A true chemical equilibrium is approachable from either the reactant or the product side of the reaction. At equilibrium the reactant and product concentrations do not change with time.

chemical formula (40): the chemical notation for a compound. The relative number of atoms of each element in the compound are indicated by subscripts.

chemical nomenclature (39): the system of naming chemical compounds.

chemical reaction (36): the formation of new substances from other substances by a rearrangement of the constituent atoms.

chemical symbol (32): the abbreviation that is used to designate an element.

chlor-alkali process (837): a method of preparing chlorine, Cl_2, and sodium hydroxide, NaOH(aq), by the electrolysis of an aqueous sodium chloride, NaCl(aq), solution.

chromatogram (58): a recording of the results of a chromatographic separation.

chromatography (57): the separation of the components of a solution by their different tendencies of adsorption on a condensed phase. The solution phase flows over the condensed (stationary) phase.

chromous bubbler (980): a device used to remove traces of oxygen from gases. It is prepared by reacting zinc metal with chromium(III) nitrate to form chromium(II).

***cis*- (462):** a prefix meaning on the same side.

cisplatin (1004): *cis*-diamminedichloroplatinum(II), an anticancer drug.

cis-trans isomerism (462, 1004): a form of

geometric isomerism. The designation *cis* indicates that two identical ligands are adjacent to each other in some sense, while the designation *trans* indicates that two identical ligands are directly opposite each other in some sense.

Clapeyron-Clausius equation (798): the equation that gives the temperature dependence of the equilibrium vapor pressure. The equation is

$$\ln(P_2/P_1) = (\Delta H_{vap}/R)(T_2 - T_1)/T_1 T_2$$

where P_2 is the equilibrium vapor pressure at the Kelvin temperature T_2, P_1 is the equilibrium vapor pressure at the Kelvin temperature T_1, ΔH_{vap} is the molar enthalpy of vaporization, and R is the molar gas constant, 8.314 $J \cdot K^- \cdot mol^{-1}$.

classical physics (302): physics as it was developed during the nineteenth century.

Claus process (948): the production of sulfur from hydrogen sulfide that occurs in some deposits of natural gas. Hydrogen sulfide is burned in air to produce sulfur dioxide, which reacts with hydrogen sulfide to produce sulfur.

coal gasification (253): a process for producing combustible gaseous products from coal.

colligative molality (522): a concentration scale for solute particles in a solution. The colligative molality, m_c, is defined as m_c = moles of solute particles/kilograms of solvent. The units of colligative molality are $mol \cdot kg^{-1}$.

colligative molarity (533): a concentration scale for solute particles in a solution. The colligative molarity, M_c, is defined as M_c = moles of solute particles/liters of solution. The units of colligative molarity are $mol \cdot L^{-1}$.

colligative properties (519): the properties of a solution that depend on the ratio of the number of solute particles to the number of solvent particles. The major colligative properties are vapor pressure lowering, boiling-point elevation, freezing-point depression, and osmotic pressure.

collision frequency, z (227): the number of collisions that a molecule undergoes per second. Collision frequency is given by $z = v_{rms}/l$, where l is the mean free path

and v_{rms} is the root-mean-square speed of the molecules.

collision theory (573): the postulate that two molecules must collide with sufficient energy in order to react.

combination reaction (108): a reaction between two different substances in which a single product is formed.

combustion (1052): the burning of a substance in oxygen.

combustion analysis (1052): a determination of the mass percent composition of a compound from the amounts of CO_2 and H_2O formed by combustion of the compound.

combustion reaction (109): a reaction in which a substance is burned in oxygen.

common ion (721): the same ionic constituent of two or more salts, usually in aqueous solution.

common-ion effect (721): the decrease in the solubility of an ionic solid caused by the presence in the solution of one of its constituent ions due to another salt.

common logarithm, log (A-3): the power to which 10 must be raised to obtain a certain number. The common logarithm of a number a is log $a = x$ such that $a = 10^x$.

complementary base pairs (1080): the base pairs adenine and thymine (A—T) and guanine and cytosine (G—C), which are always opposite each other on the two polynucleotide chains in a DNA double helix.

complex ion (772, 997): a charged chemical species containing a metal ion with one or more attached ligands.

complexation reaction (722): a reaction in which a complex ion is formed.

components (171): the species of a solution.

compound (31): a pure substance that can be broken down into simpler substances. A compound is composed of two or more different kinds of atoms.

compound unit (12): a unit of measurement that is expressed in terms of two or more units.

compressibility (197): the extent to which a substance changes its volume with increasing pressure.

concentration (172): the quantity of solute dissolved in a given quantity of solvent or solution.

concentration quotient, Q_{sp}, (728): an expression that has the same algebraic form as the K_{sp} expression (see **solubility product constant**) but into which arbitrary (nonequilibrium) values of the ionic concentrations can be substituted.

concentration table (605): a table of initial concentrations and equilibrium concentrations of the reactants and products of a reaction.

condensation polymerization reaction (1065): the formation of a polymer from two different monomer molecules by splitting out a small molecule, such as water.

condensed structural formula (1026): an abbreviation of a Lewis formula.

condenser (56): a component of a distillation apparatus in which vapors are cooled and thereby converted to liquid.

conduction band (506): a densely packed set of orbitals that extend throughout a crystal and that are analogous to antibonding orbitals in a molecule. Electrons move through a metal by means of the conduction band.

conduction electrons (916): electrons in the conduction band.

conjugate acid (656): an acid that is formed when a base accepts a proton from another species.

conjugate acid-base pair (656): an acid and a base that can be interconverted by transfer of a proton.

conjugate base (656): a base that is formed when an acid transfers a proton to another species.

conservation of energy (243): the law describing the concept that the total energy never changes. Energy according to their law cannot be created or destroyed, only transferred.

conservation of mass (37): the law stating that in a chemical reaction, the mass of the products is equal to the mass of the reactants that produced the products. In a chemical reaction, mass is neither created nor destroyed.

conservation of orbitals (452): the principle stating that when atomic orbitals are combined, the total number of new hybrid atomic orbitals is equal to the number of orbitals used to make them.

contact process (949): a method for the production of sulfuric acid, H_2SO_4, in which sulfur is oxidized to sulfur trioxide, which is then combined with water.

continuous spectrum (299): electromagnetic radiation that contains radiation of all the wavelengths in some region.

control rod (873): one of the cadmium or boron rods used to control the rate of the reaction in a nuclear reactor by absorbing neutrons.

coordinate-covalent bond (394): a covalent bond that is formed when one species contributes both electrons to the bond.

coordination number (997): the number of positions on the central atom or ion to which ligands are attached in a complex.

coordination position (1002): the position on a transition metal at which ligands attach to form a transition-metal complex.

copolymer (1066): a polymer that is made from more than one type of monomer.

coulomb (22): the SI unit of charge.

Coulomb's law (366): a law that gives the energy change involved when two infinitely separated ions are brought to a separation distance d as in the following:

$$E = (2.31 \times 10^{-16} \, \text{J} \cdot \text{pm})(Z_1 Z_2 / d),$$

where Z_1 and Z_2 are the ionic charges of the two ions. When d is expressed in picometers, the energy is given in joules.

covalent bond (377): the bond formed between two atoms by a shared electron pair.

covalent network crystal (505): a crystal in which the constituent particles are held together by covalent bonds.

criteria of reaction spontaneity (787): the value of a quantity associated with a reaction, such as ΔG_{rxn}° or Q/K, that can be used to predict whether a reaction is spontaneous or not.

critical mass (872): the minimum mass that will support a nuclear chain reaction.

critical point (494): the point in the phase diagram of a substance at which the liquid-gas curve abruptly terminates.

critical temperature (494): the minimum temperature above which a gas cannot be liquefied, no matter what its pressure; the temperature at the critical point.

crystal lattice (498): the ordered arrangement of the constituent molecules or ions of a crystalline solid.

crystallographic radius (340, 500): an atomic radius that has been determined from X-ray analysis of a crystal containing the given atoms.

cubic centimeter (9): a unit of volume. One cubic centimeter is equal to one milliliter.

curie (877): a measure of the radioactivity of a sample; $1 \text{ Ci} = 3.7 \times 10^{10}$ disintegrations per second.

cylindrically symmetric (323): a characteristic of a function that depends only on the distance from an axis. A cylindrically symmetric function has a circular cross section.

D

d-**block element (339):** a transition metal.

d **orbital (320):** an orbital for which $l = 2$. There are five *d* orbitals for each value of $n \geq 3$.

d-**orbital splitting pattern (1006):** the splitting of the five *d* orbitals of a transition-metal ion in a complex into groups of orbitals of differing energies, by the ligands in the complex.

d **transition-metal series (85):** the *d* transition-metal series are as follows: $3d$ series: $Z = 21$ to $Z = 30$ (Sc to Zn); $4d$ series: $Z = 39$ to $Z = 48$ (Y to Cd); $5d$ series: $Z = 71$ to $Z = 80$ (Lu to Hg).

*dsp*³ (**sp³d**) **(458):** one of the five hybrid atomic orbitals obtained by combining a $3s$ orbital, three $3p$ orbitals, and one $3d$ orbital on the same atom. The dsp^3 orbitals have trigonal bipyramidal symmetry.

*d*² *sp*³ (**sp³d²**) **(458):** one of the six equivalent hybrid atomic orbitals obtained by combining a $3s$ orbital, three $3p$ orbitals, and two $3d$ orbitals on the same atom. The $d^2 sp^3$ orbitals point to the vertices of a regular octahedron.

*d*ˣ **ion (996):** a transition-metal ion that has *x* electrons in its outer *d* orbitals.

$d_{xy}, d_{xz}, d_{yz}, d_{x^2-y^2}, d_{z^2}$ **orbitals (995):** the set of five *d* orbitals.

Dalton's law of partial pressures (214): a law stating that the total pressure exerted by a mixture of gases is equal to the sum of the partial pressures of each of the gases. For a mixture of two gases, $P_{total} = P_1 + P_2$. Each gaseous component exerts a pressure independent of the other gases.

deBroglie wavelength (306): the wavelength associated with a moving particle. The wavelength is given by $\lambda = h / mv$, where h is Planck's constant, m is the mass of the particle, and v is the speed of the particle.

decomposition reaction (116): a reaction in which a substance is broken up into two or more simpler substances.

degree of dissociation (357): the extent to which a compound dissociates into ions in solution.

deionize (713): to remove the ions in water by means of ion-exchange resins.

deionized water (713): water from which all cations and anions have been removed.

delocalized electron (467): an electron that occupies a delocalized orbital.

delocalized orbital (467): π orbitals in a molecule that are not associated with a particular pair of atoms.

Δ_o **(1008):** the magnitude of the splitting of the two sets of *d* orbitals, t_{2g} and e_g, on a metal ion in an octahedral complex.

Δ_{sp} **(1013):** the magnitude of the splitting of the two highest-energy *d* orbitals, d_{xy} and $d_{x^2-y^2}$ on a metal ion in a square-planar complex.

Δ_t **(1014):** the magnitude of the splitting of the two sets of *d* orbitals, e and t_2, on a metal ion in a tetrahedral complex.

ΔG_f^o **(793):** the standard molar Gibbs free energy of the formation of a substance from its constituent elements

ΔG_{rxn} **(787):** the Gibbs free energy change of a reaction. The value of ΔG_{rxn} for a reaction run at a constant temperature is given by $\Delta G_{rxn} = \Delta H_{rxn} - T \Delta S_{rxn}$, where ΔH_{rxn} is the enthalpy change of the reaction, ΔS_{rxn} is the entropy change of the reaction, and T is the temperature in kelvins. The value of ΔG_{rxn} is the maximum amount of work that can be obtained from the reaction under the stated conditions.

ΔG_{rxn}^o **(790):** the standard Gibbs free energy change for a reaction, which is the Gibbs free energy change for the reaction run under standard conditions.

ΔH_f^o **(261):** the standard molar enthalpy of formation of a compound from its constituent elements.

ΔH_{rxn} **(251):** the heat, q_P, absorbed or evolved by a reaction when the reaction occurs at constant pressure.

ΔS_{fus}^o **(781):** the standard molar entropy change upon fusion. The value of ΔS_{fus}^o is given by $\Delta S_{fus}^o = \Delta H_{fus}^o / T_m$, where ΔH_{fus}^o

is the molar enthalpy of fusion and T_m is the melting point in kelvins at one atmosphere.

ΔS_{rxn} **(778):** the entropy change of a reaction at conditions other than standard conditions.

ΔS°_{rxn} **(785):** the entropy change of a reaction when the reactants and products are at standard conditions. The value of ΔS°_{rxn} is given by $\Delta S^{\circ}_{rxn} = S^{\circ}_{products} - S^{\circ}_{reactants}$, where $S^{\circ}_{products}$ is the total entropy of all the product species and $S^{\circ}_{reactants}$ is the total entropy of all the reactant species.

ΔS°_{vap} **(781):** the standard molar entropy change upon vaporization. The value of ΔS°_{vap} is given by $\Delta S^{\circ}_{vap} = \Delta H^{\circ}_{vap}/T_b$, where ΔH°_{vap} is the standard molar enthalpy of vaporization and T_b is the boiling point in kelvins at one atmosphere.

ΔU_{rxn} **(250):** the heat, q_v, absorbed or evolved during a reaction that occurs at constant volume.

density (12): the mass per unit volume of a substance.

derived SI units (7): units that are defined in terms of the basic SI units.

desalination (712): the process of removing dissolved solids from seawater.

deuterium (48): the isotope of hydrogen that has a mass number of 2.

developer (971): a mild reducing agent that preferentially reduces the silver ion in the grains of an emulsion that have been exposed in order to intensify the latent image.

deviation from ideality (228): the behavior of a gas such that for one mole of the gas, the ratio PV/RT is not equal to 1.

dew point (494): the air temperature at which the relative humidity is 100 percent; this depends on the partial pressure of water vapor in the air.

Dewar flask (272): an insulated vessel that heat cannot readily leave or enter. A thermos bottle is an example of a Dewar flask.

diagonal relationship (88): the similarities in chemical reactivity between the first element of one group and the second element in the following group.

diamagnetic (444): not magnetized by an external magnetic field. Diamagnetic molecules contain no unpaired electrons.

diatomic molecule (34): a molecule that is composed of two atoms.

dielectric constant (490): a constant that relates the coulombic energy between two separated charges in a particular medium to that in a vacuum. The value of the dielectric constant depends on the medium.

dilution (175): a decrease in the concentration of a solution, obtained by adding solvent.

dimensional analysis (17): the treatment of the units of the various quantities involved in calculations as quantities that follow the rules of algebra.

dimensions (12): the units of a quantity.

dimer (1065): a pair of identical molecules bonded together into a single unit.

dipeptide (1074): a molecule composed of two amino acids joined by a peptide bond.

dipole-dipole attraction (482): the attraction between polar molecules.

dipole moment (403): a measure of the polarity of a molecule. The dipole moment has both magnitude and direction. The direction of the dipole moment is represented by an arrow ($\leftrightarrow$) pointing from the positive charge to the negative charge.

diprotic acid (699): an acid that has two dissociable protons.

direction of reaction spontaneity (621): the direction (left to right or right to left) in which a reaction proceeds toward equilibrium.

discharge (817): the production of an electric current from an electrochemical cell.

disproportionation (757): a chemical reaction in which a single species acts as both oxidizing and reducing agent.

dissociate (356): to break up into ions.

dissolution (55): the process of one substance dissolving in another substance.

distillation (55): a process by which a substance is separated from a liquid phase by volatilization upon heating. The vapor is then condensed to liquid by cooling.

disubstituted (1030): replace two hydrogen atoms in a compound with two other atoms or groups of atoms.

disulfide bond (1069): the bond formed between two cysteine side groups on the same or neighboring polypeptide chains. The disulfide bond is of the type $-CH_2-S-S-CH_2-$.

DNA (1077): deoxyribonucleic acid, the substance that contains the genetic information in cells. DNA is a polynucleotide composed of the sugar deoxyribose, the phosphate group, and the four bases adenine, thymine, guanine, and cytosine.

donor-acceptor complex (394): the product of the formation of a coordinate covalent bond between two species.

doping (916): the addition of selected impurity atoms to a pure semiconductor.

double bond (387): the bond formed between two atoms by two shared electron pairs.

double helix (1079): the shape of DNA: two polynucleotide chains intertwined in a helical fashion.

double replacement reaction (123): a reaction of the type $AB + CB \rightarrow AD + BC$ in which the cations in each compound exchange anionic partners.

driving force (123): the formation of insoluble, gaseous, or un-ionized products that results in a chemical reaction.

dry cell (846): a primary battery utilizing the cell

$$\ominus Zn(s)|ZnCl_2 \cdot 2NH_3(s)|NH_4Cl(aq)|$$
$$MnO_2(s), Mn_2O_3 \cdot H_2O(s)|C(s)\oplus$$

ductile (32): able to be drawn into a wire.

dynamic equilibrium (491, 595): a state of balance between forward and reverse processes such that no net change in the system takes place.

E

e_g (1006): the set of d orbitals of higher energy on a metal ion in an octahedral complex.

e orbitals (1014): the set of d orbitals of lower energy on a metal ion in a tetrahedral complex. The e orbitals consist of the $d_{x^2-y^2}$ and d_{z^2} orbitals.

effusion (224): the process whereby a gas exits through a very small hole in a container.

Einstein (unit) (304): the energy of a mole of photons at a given frequency.

18-electron outer configuration (360): the relatively stable outer electron configuration $ns^2np^6nd^{10}$.

18-electron rule (360): the unusual stability of the 18-electron outer configuration.

elastomer (1069): a polymeric substance that can be stretched and will return to its original shape when the stretching force is released.

electrical conductance (357): the property that measures the ability to conduct an electric current.

electrochemical cell (815): an experimental setup that enables an electric current to be obtained from a chemical reaction.

electrochemistry (813): the study of the chemical processes involved when an electric current is passed through different materials.

electrode (816): a metal conductor used to establish electrical contact with an electrolyte solution.

electrolysis (834): a chemical reaction that occurs when an electric current is passed through a solution.

electrolyte (356): a substance that dissolves in water to produce solutions that conduct an electric current.

electromagnetic spectrum (298): the range of wavelengths or frequencies of electromagnetic radiation.

electromotive force (emf) (813): the standard reduction voltage.

electromotive force series (830): the arrangement of half reactions in decreasing order of their standard reduction voltages.

electron (44): a subatomic particle that has a negative charge and a mass that is 1/1837 that of a hydrogen atom.

electron acceptor (107): the reactant that gains electrons in an electron transfer reaction. The electron acceptor is reduced in an oxidation-reduction reaction.

electron affinity, E_A (365): the energy released in the process of adding an electron to an atom. The equation for this process is

$$atom(g) + electron \rightarrow ion(g) + E_A$$

electron arrangement (89, 291): the placement of electrons in the orbitals of an atom or the electronic structure of an atom.

electron capture (854): a type of nuclear transformation in which one of the innermost electrons of an atom is absorbed by the nucleus.

electron configuration (331): the assignment of electrons to orbitals according to the Pauli exclusion principle.

electron deficiency or electron-deficient compound (393): a condition or a compound in which one or more of the atoms other than hydrogen has less than eight valence electrons; such a compound violates the octet rule.

electron diffraction (308): the scattering of a beam of electrons in a characteristic pattern by a substance.

electron donor (107): the reactant that loses electrons in an electron transfer reaction. An electron donor is oxidized in an oxidation-reduction reaction.

electron microscope (308): an instrument that uses the wavelike property of electrons to investigate subcellular and molecular structures.

electron-pair acceptor (665): an electron-deficient species that can act as a Lewis acid.

electron-pair donor (665): a species with a lone pair of electrons that can act as a Lewis base.

electron-pairing energy (1010): the energy required to pair up two electrons in the d orbitals of a metal in a complex.

electron-transfer reaction (100): a reaction in which one species is oxidized and another species is reduced.

electronegativity (399): a measure of the force with which an atom attracts the electrons that it is sharing in a covalent bond.

electronic structure (291): the arrangement of the electrons within an atom.

electroplating (839): the production of a layer of protective metal by electrochemical deposition.

electrostatic force (358): the coulombic attraction between oppositely charged ions.

element (31): a substance that contains only one kind of atom. There are over 100 known elements.

elementary process (570): a chemical reaction that occurs in a single step—the reactants go directly to the products without the involvement of intermediates.

emf series (830): electromotive force series.

emission spectrum (314): the spectrum obtained when gaseous excited atoms or molecules return to their ground state and emit light of characteristic wavelengths.

empirical formula (145): the simplest chemical formula of a substance.

emulsion (photographic) (971): a dispersion of silver halide crystals in a gelatinous substance.

end point (684): the point in a titration at which the indicator changes color.

endothermic reaction (252): a reaction that absorbs energy as heat ($\Delta H_{rxn} > 0$).

energy change (250): difference in energy of a system before and after a process has occurred. The change in energy of a reaction is given the symbol ΔU_{rxn}, where $\Delta U_{rxn} = U_{products} - U_{reactants}$ denotes the change in energy of a reaction.

energy-favored reaction (786): a reaction for which $\Delta U_{rxn} < 0$.

energy state (310): one of the discrete set of energies that an atom or molecule can have.

energy transfer function (250): a function whose value depends on how a process is carried out.

enthalpy, H (251): a defined quantity given by the equation $H = U + PV$, where U is the energy, P is the pressure, and V is the volume of a system.

enthalpy change, ΔH_{rxn} (251): the energy evolved or absorbed as heat during a reaction when the reaction takes place at constant pressure. The enthalpy change is given as $\Delta H_{rxn} = H_{products} - H_{reactants}$ for a reaction.

enthalpy of formation (261): the enthalpy change for a reaction in which one mole of a compound is formed from its elements.

entropy, s (778): a quantitative measure of the amount of disorder in a substance. The SI unit of entropy is joules per kelvin, $J \cdot K^{-1}$.

entropy change, ΔS_{rxn} (778): the difference between entropy of the reaction's products and the reactants.

entropy-driven reaction (786): a spontaneous reaction for which $\Delta S_{rxn} > 0$ and $\Delta H_{rxn} > 0$.

entropy-favored reaction (786): a reaction for which $\Delta S_{rxn} > 0$.

enzyme (583): a protein that catalyzes a chemical reaction in biological systems.

equatorial vertex (415): one of the three vertices of the equilateral triangle that forms the shared base of a trigonal bipyramid. The three equatorial vertices are equivalent.

equilibrium (491): a state characterized by the equality of forward and reverse rates for a process.

equilibrium concentration (597): the value of the concentration of a reactant or of a product when the reaction has attained equilibrium.

equilibrium constant (600): the algebraic relationship between reactant and product concentrations that exists at equilibrium. The form of the equilibrium-constant expression for a chemical reaction is obtained by applying the law of concentration action to the balanced chemical equation.

equilibrium-constant expression (601): the expression for the equilibrium constant for a chemical reaction, given by the law of concentration action. For the balanced chemical equation

$$aA(g) + bB(soln) + cC(s) \rightleftharpoons xX(g) + yY(soln) + zZ(l)$$

the equilibrium-constant expression is given by

$$K_c = \frac{[X]^x[Y]^y}{[A]^a[B]^b}$$

equilibrium shift (611): the response of an equilibrium chemical reaction to a displacement from equilibrium produced by a change in conditions that affect the reaction equilibrium.

equilibrium state (595): a state in which the forward reaction rate is equal to the reverse reaction rate.

equilibrium vapor pressure (491): the pressure of a vapor in equilibrium with its liquid.

equivalence point (681): the pH or point in a titration at which all the acid or base initially present is just neutralized.

escaping tendency (532): the tendency of a substance to leave a solution.

ester (1050): a class of organic compounds that results from the reaction between a carboxylic acid and an alcohol.

esterification reaction (1051): a reaction between an organic acid and an alcohol, which produces an ester.

eutrophication (751): the depletion of the oxygen in a body of water caused by decaying organisms such as algae.

evaporation (55): the removal of water from a solution as a result of heat.

excess reactant (159): a reactant that is present in larger quantity than is necessary to react completely with the other reactants in a reaction.

excited state (311): an energy state that is higher than the ground state.

exothermic reaction (252): a reaction that evolves energy as heat ($\Delta H_{rxn} < 0$).

expanded valence shell (394): the idea that elements beyond the second row of the periodic table can accommodate more than eight electrons by using d orbitals. Such elements need not obey the octet rule.

extensive property (13): a property of a substance that depends on the amount of the substance.

external circuit (816): the part of an electrical circuit involving an electrochemical cell that does not include the electrochemical cell.

extraction (57): the separation of one substance from a mixture or solution.

F

f-block elements (339): the inner transition metals.

f orbital (320): an orbital for which $l = 3$. There are seven f orbitals for each value of $n \geq 4$.

face-centered cubic unit cell (500): the unit cell in which the components of a crystal are located at the corners of the cube and in the centers of the six faces of the cube.

Fahrenheit temperature (11): the temperature scale in which the boiling point of water is 212 degrees and the freezing point of water is 32 degrees. The unit of Fahrenheit temperature is °F.

faraday (823): the value of Faraday's constant, 9.65×10^4 C·mol^{-1}.

Faraday's constant (823): the charge on one mole of electrons. Faraday's constant, F, is equal to 96,500 C·mol^{-1}.

Faraday's laws (836): the first law states that the extent of an electrochemical reaction depends solely on the quantity of electricity that is passed through a solution. The second law states that the mass of a substance that is deposited as a metal or evolved as a gas by the passage of a given quantity of electricity is directly proportional to the molar mass of the substance divided by the number of electrons consumed or produced per formula unit. Faraday's laws are expressed quantitatively by the equation

$$\left(\begin{array}{c}\text{mass deposited as metal}\\\text{or evolved as gas}\end{array}\right) = m = \left(\frac{It}{F}\right)\left(\frac{M}{n}\right)$$

where I is the current in amperes, t is the time in seconds, F is Faraday's constant, M is the molar mass of the metal or gas, and n is the number of electrons required to produce one formula unit of metal or gas.

fatty acid (1051): a carboxylic acid whose R group is a long-chain hydrocarbon.

film badge (855): a device that detects radioactive emissions by their action on photographic film.

filtration (55): the process of separating a liquid phase from a solid phase by the use of a material, such as special papers, through which the liquid phase can pass.

first electron affinity, E_{A1}, (365): the energy associated with the process of adding an electron to an isolated atom.

first excited state (311): the first energy state above the ground state.

first ionization energy (291): the minimum energy required to remove an electron from a neutral atom, A, to produce the A^+ ion.

first law of thermodynamics (248): the law of conservation of energy.

first-order rate law (554): a law in which the reaction rate is proportional to the first power of the concentration of a reactant: rate = $k[A]$.

first-order reaction (554): a reaction whose rate law is a first-order rate law.

Fisher-Tropsch process (synthesis) (799): a syntheses involving reactions of hydrogen with carbon monoxide to produce straight-chain hydrocarbons and alcohols with up to 10 carbon atoms.

fission (871): a nuclear reaction in which a nucleus splits into two smaller fragments of roughly equal size.

fixed (971): the removal of all the unexposed silver halide from an exposed emulsion.

flame test (314): the identification of alkali metals in a sample using the color of the flame produced by the sample.

flammability limits (288): the composition limits of the ratio of fuel to oxidizer between which the propagation of a flame can occur.

flash point (289): the minimum temperature at which the vapor pressure of a liquid fuel reaches the lower flammability limit.

flocculent (736): fluffy; a precipitate that does not settle readily.

flotation (57): a technique used to separate two or more solids based on their different densities.

fluorescence (856): the emission of visible light.

fool's gold (982): iron pyrite, FeS_2.

formal charge (383): the assignment of a charge to an atom in a molecule or ion. The formal charge of an atom is found by the relationship, formal charge = (total number of valence electrons in the free atom) − (total number of lone-pair electrons) − $\frac{1}{2}$ (total number of shared electrons)

formula mass (42): the relative mass of a formula unit. The formula mass is the sum of the atomic masses of all the atoms that make up the formula unit.

formula unit (91, 140): the simplest component of a substance. The formula unit may be an atom, a molecule, or a group of ions. A formula unit is defined by the chemical formula.

forward rate (596): the rate of the forward reaction.

forward reaction (595): the formation of the reaction products from the reactants, or the reaction that proceeds left to right for the equation as written.

forward reaction rate (596): the rate for the reaction that takes place left to right as written.

fossil fuel (278): oil, coal, or natural gas.

fountain effect (930): a demonstration of the solubility of ammonia in water.

4d transition-metal series (338): the second transition-metal series, which involves the sequential filling of the 4d orbitals.

fractional distillation (537): a distillation that involves a long distillation column in which the vapor is continuously condensed and revaporized.

Frasch process (947): a process for extracting sulfur from underground deposits by using hot (180°C) high-pressure water to melt the sulfur.

free radical (392): a species with one or more unpaired electrons. All species with an odd number of electrons are free radicals.

free radical reaction (1029): a reaction involving free radicals.

freezing-point depression (528): the decrease in the freezing point of a solution below the freezing point of the pure solvent. The freezing-point depression is given by the equation $T_f^o - T_f = K_f m_c$, where T_f^o is the freezing point of the pure solvent, T_f is the freezing point of the solution, m_c is the colligative molality of the solution, and K_f is the freezing-point-depression constant of the solvent.

freezing-point-depression constant (528): the proportionality constant between the freezing-point depression and the colligative molality of a solution. Its value depends only on the solvent. The symbol for this constant is K_f and the units are $K \cdot m_c^{-1}$.

frequency (297): the number of maxima or minima of a wave that pass a given point per second. The symbol of frequency is ν and the units of frequency are cycles per second, s^{-1}, or hertz, Hz (1 Hz = 1 cycle·s^{-1}).

fuel (109): a substance that can be used in a chemical reaction to provide energy for the performance of tasks.

fuel rod (873): one of the rods in a nuclear reactor that contain the isotope that undergoes fission.

fuming sulfuric acid (949): $H_2S_2O_7$ (35% in H_2SO_4), which is obtained by absorbing SO_3 into nearly pure sulfuric acid.

fundamental equation of the kinetic theory of gases (222): the equation,

$$PV = (1/3)N_o M \overline{v^2}$$

where P is the pressure, V is the volume, N_o is Avogadro's number, m is the mass of one molecule, and $\overline{v^2}$ is the average of the squares of the speeds of the molecules.

fusion (875): the process by which a nucleus is produced from smaller nuclei.

G

galena (911): the ore PbS.

γ-ray (45): a high-energy electromagnetic wave emitted in certain types of nuclear decay.

gas (196): the physical state of matter such that the substance occupies the entire volume and assumes the shape of its container, and has a large compressibility.

gas constant (207): the constant, R, in the ideal-gas equation. Its value depends on the units of P, V, and T:

$$R = 0.0821 \text{ L} \cdot \text{atm} \cdot \text{mol}^{-1} \cdot \text{K}^{-1}$$

and

$$8.314 \text{ J} \cdot \text{mol}^{-1} \cdot \text{K}^{-1}.$$

gas electrode (820): an electrode involving a gaseous species.

gas-liquid chromatography (58): an analytical technique used to separate mixtures of gases or volatile liquids by flowing the vapor mixture over a liquid phase in which the vapor components have different solubilities.

gasohol (278): a mixture of gasoline and ethyl alcohol.

gas thermometer (204): a thermometer that uses the volume of a fixed mass of a gas to measure temperature.

Gay-Lussac's law of combining volumes (205): a law stating that the volumes of gases that combine to form reaction products are in the ratio of small whole numbers. These volumes must be measured at the same temperature and pressure.

Geiger counter (855): a device that detects and measures α- and β-particles by measuring the current produced by the ionization of a gas by the particles emitted by the radioactive source.

gene (1081): a segment along a DNA molecule that codes the synthesis of a particular polypeptide.

geometric isomers (410): molecules that

have the same chemical formula but different geometric arrangements of the atoms. Geometric isomers have different chemical and physical properties.

Gibbs criteria (of reaction spontaneity) (787): criteria used to predict whether a process is spontaneous or not. The Gibbs criteria are

1. $\Delta G_{rxn} < 0$; the reaction is spontaneous
2. $\Delta G_{rxn} > 0$; the reaction is not spontaneous — product formation requires energy input
3. $\Delta G_{rxn} = 0$; the reaction is at equilibrium

Gibbs free energy (787): a quantity that serves as a compromise function between enthalpy and entropy. For a reaction run at constant temperature, the Gibbs free energy change is given by $\Delta G_{rxn} = \Delta H_{rxn} - T\,\Delta S_{rxn}$.

Gibbs free energy change, ΔG_{rxn} (787): the change in Gibbs free energy for a reaction. ΔG_{rxn} is given by

$$\Delta G_{rxn} = G(\text{products}) - G(\text{reactants})$$

Glauber's salt (280): the chemical

$$Na_2SO_4 \cdot 10H_2O(s).$$

Graham's law of effusion (225): the relation between the rates of effusion of two gases, which is given by $\text{rate}_A/\text{rate}_B = (M_B/M_A)^{1/2}$, where M_A and M_B are the molecular masses of gas A and of gas B, respectively.

gram (10): a unit of mass that is equal to $1/1000$ of a kilogram.

greenhouse effect (240): the increase in temperature of the troposphere resulting from absorption of infrared radiation by atmospheric gases, primarily carbon dioxide and water.

ground electronic state (331): the state of lowest energy. The ground electronic state is obtained by filling up the atomic orbitals of lowest energy according to the Pauli exclusion principle and Hund's rule.

ground state (311): the lowest-possible energy state of an atom or molecule.

ground-state wave function (439): the wave function for the lowest energy state of a species.

group (80): the collection of elements in the same column in the periodic table. A group of elements is also called a family of elements.

Group 1 metals (80): the elements that appear in the first column of the periodic table. The Group 1 metals are called the alkali metals.

Group 2 metals (80): the elements that appear in the second column of the periodic table. The Group 2 metals are called the alkaline earth metals.

Guggenheim notation (21): a notation used to label table headings and graph axes that yields dimensionless numbers for tabulation and graphing.

H

Haber process (894): the method by which ammonia is produced commercially from nitrogen and hydrogen. The reaction

$$N_2(g) + 3\,H_2(g) \rightarrow 2\,NH_3(g)$$

is carried out on a commercial scale at 500°C and 300 atm, with the aid of a catalyst.

half-life, $t_{1/2}$ (561, 859): the time it takes for one half of a sample to undergo reaction.

half reaction (757): one part of an electron transfer reaction representing either the loss of electrons or the gain of electrons by a reactant.

halide (77): a binary compound consisting of a halogen and another element.

Hall process (838): the industrial method of preparing aluminum by the electrolysis of a solution of aluminum oxide, Al_2O_3, dissolved in molten cryolite, $Na_3AlF_6(l)$.

haloalkane (1030): a substance derived from an alkane in which one or more hydrogen atoms of the alkane are replaced by halogen atoms (also called alkyl halide).

halogen (77): a member of the group of elements — fluorine, chlorine, bromine, iodine, and astatine — which are the Group 7 elements in the periodic table.

hard water (713): water that contains appreciable amounts of divalent cations, such as Ca^{2+}, Mg^{2+}, and Fe^{2+}, together with either the anion HCO_3^- or SO_4^{2-} the anion, or both.

heat, q (249): a mode of energy transfer that occurs as a result of a temperature difference.

heat capacity, c_P (266): the heat required to raise the temperature of a substance by one Kelvin degree. The SI units of heat capacity are joules per kelvin, $J \cdot K^{-1}$.

heat of combustion (252): the heat evolved in a combustion reaction.

heat of formation (261): the molar enthalpy of formation.

heat of reaction (252): the amount of energy evolved or absorbed as heat when a chemical reaction occurs.

heating curve (476): a plot of how the temperature of a substance varies with time if it is heated at a constant rate.

heavy water (48): water that is composed of deuterium and oxygen, D_2O.

Heisenberg uncertainty principle (316): a principle stating that it is not possible to measure accurately the position and the momentum of a particle simultaneously.

Henderson-Hasselbalch equation (677): an equation that relates the pH of a buffer solution to the stoichiometric concentrations of a conjugate acid-base pair.

Henry's law (539): a law stating that the solubility of a gas in a liquid is directly proportional to the equilibrium partial pressure of the gas over the solution. The equation is $P_{gas} = k_h M_{gas}$, where P_{gas} is the equilibrium gas pressure over the solution, M_{gas} is the concentration of the gas in the solution, and k_h is Henry's law constant for the gas.

Henry's law constant (539): the proportionality constant between the solubility of a gas in a liquid and the equilibrium pressure of the gas over the solution. The symbol is k_h and the units are $atm \cdot M^{-1}$.

hertz, Hz (298): the SI unit of frequency. One hertz is equal to one cycle per second.

Hess's law (256): A law stating that if two or more chemical equations are added together, then the value of ΔH_{rxn} for the resulting equation is equal to the sum of the ΔH_{rxn} values for the separate equations.

heterogeneous (55): not uniform in chemical composition; for example, a mixture of salt and sugar.

heterogeneous catalysis (582): catalysis in which the catalyst is in a different phase from the reactants.

heterogeneous catalyst (582): a solid catalyst that catalyzes a gas- or solution-phase reaction.

heterogeneous mixture (55): a mixture in which the components are not uniform from point to point.

heteronuclear diatomic molecule (400): a molecule that consists of two different nuclei.

high-spin configuration (1010): a d electron configuration of a complex in which the d electrons occupy a higher-energy set of d orbitals before they pair up in the lower-energy set of d orbitals.

homogeneous (55, 71): having uniform properties throughout.

homonuclear diatomic molecule (400): a molecule that consists of two atoms of the same element.

Hund's rule (333): a rule stating that for any set of orbitals of the same energy, the ground-state electron configuration is obtained by placing the electrons in different orbitals of this set with parallel spins until each of the orbitals has one electron, before pairing up any of the electrons.

hybrid atomic orbital (448): an orbital that is the result of combining different atomic orbitals on the same atom according to a procedure of quantum mechanics.

hydrated salt (hydrate) (136): A salt that forms crystals containing a definite percentage by mass of water; for example, $CuSO_4 \cdot 5H_2O$.

hydrocarbon (149, 1025): a class of organic compounds that consist of only hydrogen and carbon.

hydrogen bonding (482): a special type of dipole-dipole attraction that involves the electrostatic interaction of a hydrogen atom in one species with an electronegative atom in another species.

hydrogen electrode (827): an electrode consisting of a platinum electrode and hydrogen gas, $H_2(g)$, at 1 atm pressure. By convention, the standard reduction voltage of the hydrogen electrode is zero.

hydrogen ion, $H^+(aq)$ (114): a proton that occurs when a hydrogen atom has lost its electron. In aqueous solution the hydrogen ion exists primarily as the $H_3O^+(aq)$ ion.

hydrogen molecular ion (440): the species H_2^+, which consists of two protons and one electron.

hydrogenation (1036): addition of hydrogen to a molecule.

hydronium ion (635, 114): the species, $H_3O^+(aq)$, that is the dominant form of the hydrogen ion in aqueous solution.

hypo- **(129):** the prefix of an oxyacid indicating that one less oxygen atom is attached to the central atom than that of the *-ous* acid.

hypothesis (4): a proposition put forth as a possible explanation for the occurrence of a phenomenon, which serves as a guide to further investigation.

hydroxide ion (113): the species OH^-.

I

ideal gas (207): a gas that obeys the ideal-gas law.

ideal-gas equation (207): the equation $PV = nRT$, where P is the pressure, V is the volume, T is the kelvin temperature, and n is the number of moles of the gas; R is the gas constant.

ideal-gas law (207): the combination of Charles's law, Boyle's law, and Avogadro's law. The ideal-gas law equation is $PV = nRT$.

ideal solution (536): a solution in which the solute molecules and the solvent molecules are randomly distributed throughout the solution.

ignition source (288): the means by which a fire is started.

ignition temperature (288): the minimum temperature necessary to initiate combustion.

indicator (683): a substance that is used to signal, by a color change, the end point in a titration.

inert complex (1015): a complex that exchanges its ligands slowly with other available ligands.

inert gas (81): one of the noble gases.

infrared spectrum (296): a plot of the infrared radiation energy absorbed versus the wavelength of the radiation for a species.

inner transition metals (86, 338): the two 14-member series of metals that involve the sequential filling of the f orbitals. These series are called the lanthanides and the actinides, respectively.

intensive property (13): a property of a substance that is independent of the amount of the substance.

intercept (202): the value of y at which a straight line crosses the y axis. The intercept is b in the equation for a straight line, $y = mx + b$.

interhalogen compound (423): a compound in which a central halogen atom is bonded to one or more atoms of a more electronegative halogen.

intermediate species (571): a species that is formed from the reactants and is involved in the conversion of reactants to products, but that does not appear as a reactant or product in the overall reaction.

internuclear axis (442): the line joining two nuclei.

intrinsic electron spin (325): a characteristic property of an electron due to the spin of the electron around its axis in one of two directions.

ion (52): a species that has either a deficiency of electrons (in which case the ion is positively charged) or an excess of electrons (in which case the ion is negatively charged).

ion-exchange resin (713): an organic polymer containing acidic and/or basic groups that can remove cations (via the acid groups), or anions (via the basic groups), by means of an ion-exchange reaction.

ion pair (367): a positive and negative ion held together by electrostatic attraction.

ion-product constant of water (636): the equilibrium constant for the equation

$$H_2O(l) + H_2O(l) \rightleftharpoons H_3O^+(aq) + OH^-(aq)$$

The ion product constant is given by $K_w = [H_3O^+][HO^-]$. The value of K_w is 1.00×10^{-14} M^2 at 25°C.

ionic bond (358): the electrostatic attraction that holds oppositely charged ions together.

ionic charge (92): the positive or negative charge of an ion.

ionic compound (91, 359): a compound that is composed of positive and negative ions. An ionic compound has no net charge.

ionic crystal (376): an ordered array of negatively and positively charged ions.

ionic equation (108): a chemical equation that shows directly only the ions involved in a reaction.

ionic radius (362): the radius of an ion that is obtained from X-ray crystallographic measurements.

ionization energy (291): the minimum energy that is required to remove an electron completely from a gaseous atom or ion.

ionosphere (239): the outermost region of the atmosphere that contains ions and electrons produced by solar radiation.

isoelectronic (53): possessing the same number of electrons as another species (*iso* means the same).

isotope (48): an atom of an element that has a particular mass number. Isotopes of an element have the same number of protons but different numbers of neutrons.

IUPAC nomenclature (1031): the system of naming organic compounds that has been recommended by the International Union of Pure and Applied Chemistry.

J

joule (218): the SI unit of energy: $1\,J = 1\ kg \cdot m^2 \cdot s^{-2}$.

K

karat (989): a unit for expressing the amount of gold in alloys. Pure gold is 24 karat.

K_c (601): the equilibrium constant expressed in terms of concentrations of the products and reactants.

kelvin, K (11): the SI unit of temperature defined on the Kelvin temperature scale.

Kelvin temperature (11): the temperature scale that defines the SI unit of temperature. Kelvin temperature is related to Celsius temperature by the equation: $T(K) = t(°C) + 273.15$.

ketone (1047): a class of organic compounds that has the general formula:

$$R-\underset{\underset{O}{\|}}{C}-R'$$

key oxyacid (128): the oxyacid on which the names of the oxyacids of a particular element are based.

kilo- (9): a prefix meaning one thousand.

kilogram (9): the basic SI unit of mass.

kilometer (8): an SI unit of length. One kilometer is equal to 1000 meters.

kilowatt (247): a unit of power. One kilowatt is 1000 joules per second.

kilowatt-hour (247): a unit of energy. One kilowatt-hour is the energy produced by a one-kilowatt power source operating for one hour.

kinetic energy (218): the energy of a body due to its motion. The SI unit of kinetic energy is the joule, J. The formula relating kinetic energy to the speed of the body is $E = (1/2)mv^2$, where m is the mass of the body and v is its speed.

kinetic theory of gases (218): a molecular theory of gases. A gas is considered mostly empty space. Gas molecules are viewed as tiny noninteracting spheres in constant motion. The molecules are traveling about at high speeds and are continually colliding with each other and with the walls of the container. The pressure of a gas is due to the collisions of the molecules of the gas with the walls of the container.

K_p (603): an equilibrium constant expressed in terms of equilibrium partial pressures of products and/or reactants.

L

labile complex (1015): a complex that exchanges its ligands rapidly with other available ligands.

lanthanide series (82, 338): the inner transition-metal series headed by lanthanum. The lanthanide series contains the elements lanthanum ($Z = 57$) through ytterbium ($Z = 70$).

laser (290): a device that harnesses and channels stimulated radiation. Laser is an acronym for *l*ight *a*mplification by *s*timulated *e*mission of *r*adiation.

latent image (971): the distribution of silver clusters over the surface of an emulsion according to the pattern and intensity of the exposure.

lattice energy (369): the energy released when isolated negative ions and isolated positive ions combine to form an ionic crystal.

laughing gas (933): nitrous oxide, $N_2O(g)$.

law of concentration action (600): law stating that the equilibrium-constant expression for an equation is given by the ratio of product equilibrium concentrations to reactant equilibrium concentrations, with each concentration factor raised to a power equal to the stoichiometric coefficient of that species in the balanced equation. Pure liquids and solids, whose concentrations cannot be varied, do not appear in the equilibrium-constant expression.

law of conservation of energy (243): the law stating that the total energy of a system never changes.

law of conservation of mass (6): the law stating that in an ordinary chemical reaction, the total mass of the reacting substances is equal to the total mass of the products formed.

law of constant composition (34): the law stating that the mass percentage of each element in a compound is always the same, regardless of the source of the compound or how it is prepared.

LeChâtelier's principle (611): principle stating that if a chemical reaction at equilibrium is subjected to a change in conditions that displaces the reaction from equilibrium, then the direction in which the reaction proceeds toward a new equilibrium state will be such as to at least partially offset this change in conditions.

lead storage battery (846): a group of the following cells arranged in series

$$\ominus Pb(s)|PbSO_4(s)|H_2SO_4(aq)|$$
$$PbO_2(s), PbSO_4(s)|Pb(s)\oplus$$

Lewis acid (665): an electron-pair acceptor.

Lewis base (665): an electron-pair donor.

Lewis electron-dot formula (295): a pictorial representation of an atom. The nucleus and inner-core electrons are indicated by the chemical symbol of the atom, and the outer electrons are indicated by dots placed around the chemical symbol.

Lewis formula (377): the formula for a molecule or other species in which covalent bonds are indicated by lines and lone pairs by dots.

ligand (419, 997): an anion or neutral molecule that binds to metal ions to form a complex ion. Also, a ligand is an atom that is bonded to a central atom in a molecule or ion.

ligand-substitution reaction (997): a reaction involving a change in ligands attached to the central metal ion in a complex.

ligating atoms (1002): the atoms of a ligand that attach to a metal ion in a complex species.

limiting reactant (159): the reactant that is consumed completely in a reaction in which nonstoichiometric amounts of reactants are allowed to react.

line spectrum (299): the resolution of the components of electromagnetic radiation that contains radiation of only a few discrete wavelengths. The spectrum consists of a few lines corresponding to these wavelengths.

linear molecule (404): a molecule in which all the atoms lie on a straight line.

liquid (196): the physical state of matter having the properties of fixed volume, assumption of the shape of its container, and very small compressibility.

liquid-solid chromatography (57): a means of separating compounds in solution by passing the solution through a column packed with a pulverized solid compound.

liter, L (9): an SI unit of volume.

liter atmosphere, L·atm (249): a unit of energy. $1 \text{ L·atm} = 101.3 \text{ J}.$

lithium battery (846): a primary battery whose cell diagram is

$$Li(s)|LiCl(SOCl_2)|MnO_2(s),$$
$$Mn_2O_3(s)|Fe(s)$$

litmus paper (126): a paper impregnated with litmus, a vegetable substance that is red in acidic solutions and blue in basic solutions.

localized bond orbital (447): the orbital that describes the bonding electrons in a covalent bond between two atoms. The bonding electrons are concentrated primarily in the region between the two atoms joined by the covalent bond.

localized covalent bond (447): a bond that results when two electrons occupy a localized bond orbital.

lock-and-key theory (583): the postulate that an enzyme acts as a specific template to one of the reactants to catalyze a chemical reaction.

logarithm (A-3): the power to which a number must be raised to equal a given number.

London force (484): the attractive force between nonpolar molecules and atoms arising from the correlation of their electron distributions.

lone electron pair, or lone pair (377): a pair of valence electrons that is not shared between two atoms in a molecule.

lone pair (377): a lone electron pair.

low-spin configuration (1010): a d electron configuration in a complex in which the d electrons pair up in the lower-energy d orbitals before they occupy the higher-energy set of d orbitals.

LOX (279): liquid oxygen.

Lyman series (312): the series of lines in the hydrogen atomic spectrum due to transitions from higher states ($n > 1$) to the ground state ($n = 1$).

M

macromolecules (1063): molecules that contain thousands of atoms. Polymers are macromolecules.

magic numbers (858): the numbers, 2, 8, 20, 28, 50, and 82. Nuclei that contain a magic number of protons or neutrons are particularly stable and abundant in nature.

magnetic quantum number (323): the quantum number, m_1, that determines the spatial orientation of an orbital. The allowed values of m_1 are $-l, \ldots, -1, 0, +1, \ldots, +l$ or $-l \leq m_1 \leq l$.

main-group elements (81): the elements in groups headed by the numbers 1 through 8 in the periodic table.

malleable (32): able to be rolled into thin sheets.

manganese nodules (710): porous chunks of metallic oxides that form spontaneously in the vicinity of vent holes in the ocean floor.

manometer (197): a device used to measure the pressure of a gas.

Markovnikov's rule (1037): a rule stating that when HX is added to an unsaturated hydrocarbon, the hydrogen atom bonds to the carbon atom in the double or triple bond that already bears the larger number of hydrogen atoms.

mass, m (9): the inherent amount of material in an object.

mass number (47): the total number of protons and neutrons in an atom. The mass number is designated by the symbol A.

mass percentage (34): the percent by mass of an element in a chemical compound. The mass percentage is given by the equation, mass percentage = (mass of element/mass of compound) $\times$ 100.

mass spectrometer (52): an instrument used to measure the relative masses and amounts of atoms and molecules present in a sample.

mass spectrometry (52): the study of the fragmentation patterns of molecular ions produced in a mass spectrometer. Mass spectra are often used to identify compounds.

mass spectrum (52): a plot of the relative numbers of ions of various masses versus the mass of the ions; obtained from a mass spectrometer.

material balance condition (647): the condition stating that in an aqueous solution, an anion derived from an acid must be in the form of the undissociated acid or the anion.

Maxwell-Boltzmann distribution (220): the distribution of molecular speeds in a gas.

mean free path, l (226): the average distance that a gas molecule travels between collisions. The mean free path depends on the pressure, the temperature, and the size of the molecule.

mega-, M (9): prefix meaning 10^6.

melting point curve (495): a plot of the pressure at which the solid and liquid phases of a substance are in equilibrium versus the temperature.

meniscus (489): the surface of a liquid in a capillary.

mercaptans (952): organic compounds that contain the —SH group.

mercury battery (821): a battery using the cell

$$\ominus steel|Zn(s)|ZnO(s)$$
$$|KOH(aq, 40\%)|HgO(s)|Hg(l)|steel\oplus$$

mesosphere (239): the region above the stratosphere wherein the temperature decreases with altitude.

meta- (1042): designation for disubstituted benzenes with substituents at the 1 and 3 positions.

metal (32): a substance that has the following properties: it has a characteristic luster; it can be rolled into sheets; it can be drawn into wires; it can be melted and cast into various shapes; and it is a good conductor of electricity and heat.

metallic crystal (506): the crystalline form of a metal, in which the lattice sites are occupied by ions and the valence electrons are delocalized.

metalloid (84): a semimetal.

meter, m (7): the basic SI unit of length.

method of half reactions (758): a system of balancing oxidation-reduction equations in which the oxidation half reaction and the reduction half reaction are balanced separately.

method of initial rates (553): a means of obtaining the rate law from a determination of the reaction rate at the start of the reaction, during which time the concentrations of the reactants do not change appreciably.

method of successive approximations (651): a method of solving a quadratic equation.

metric system (7): a system of scientific units of measurement based on the meter, the kilogram, and the second as the base units of length, mass, and time, respectively.

metric ton (156): a mass of 1000 kg or 2205 lb.

mica (919): a mineral containing two-dimensional polymeric sheets with the composition $Si_2O_5^{2-}$.

micelles (748): small droplets that are soluble in water as a result of polar groups on the surface.

micro-, μ (9): a prefix meaning 10^{-6}.

midpoint (691): the point on the titration curve that is halfway between the starting point and the equivalence point.

milli-, m (9): a prefix meaning one thousandth.

milliliter, mL (9): one thousandth of a liter.

millimole, mmol (183): one thousandth of a mole.

mixture (31): a composition of two or more substances that may be separated into various pure components.

mol (140): the symbol for the unit mole.

molality (521): a concentration scale for a solute in a solution. The molality, m, is defined as $m = $ moles of solute/kilograms of solvent. The units of molality are $mol \cdot kg^{-1}$.

molar (173): pertaining to one mole.

molar bond enthalpy, H(bond) (800): the

enthalpy change associated with the dissociation of one mole of a given type of bond. The units are joules per mole, $J \cdot mol^{-1}$, or kilojoules per mole, $kJ \cdot mol^{-1}$.

molar conductance (357): the electrical conductance of a one-molar solution.

molar enthalpy of fusion, ΔH_{fus} (476): the energy that is required to melt one mole of a substance. The SI units are $kJ \cdot mol^{-1}$.

molar enthalpy of sublimation, ΔH_{sub} (480): the energy that is required to sublime one mole of a substance. The SI units are $kJ \cdot mol^{-1}$.

molar enthalpy of vaporization, ΔH_{vap} (476): the energy that is required to vaporize one mole of a substance. The SI units are $kJ \cdot mol^{-1}$.

molar entropy of fusion, ΔS_{fus}^{o} (781): the entropy change that occurs upon melting one mole of a substance. The value of ΔS_{fus}^{o} is given by $\Delta S_{fus}^{o}/T_m$, where ΔH_{fus}^{o} is the molar enthalpy of fusion and T_m is the melting point in kelvins.

molar entropy of vaporization, ΔS_{vap}^{o} (781): the entropy change upon vaporization of one mole of a substance. The value of ΔS_{vap}^{o} is given by $\Delta S_{vap}^{o} = \Delta H_{vap}^{o}/T_b$, where ΔH_{vap}^{o} is the molar enthalpy of vaporization and T_b is the boiling point in kelvins.

molar gas constant (207): the constant, R, in the ideal-gas equation.

molar heat capacity, C_p (267): the heat capacity per mole of a substance. The SI unit of molar heat capacity is $J \cdot K^{-1} \cdot mol^{-1}$.

molar heat of formation, ΔH_f^{o} (261): the enthalpy change at 25°C for the reaction in which one mole of a substance at 1 atm is formed from the elements at 1 atm.

molar heat of fusion (476): molar enthalpy of fusion.

molar heat of sublimation (480): molar enthalpy of sublimation.

molar heat of vaporization (476): molar enthalpy of vaporization.

molar mass (140, 213): the mass in grams of one mole of a substance. The units of molar mass are $g \cdot mol^{-1}$.

molar volume (207): the volume occupied by one mole of a substance. At 0°C and 1.00 atm the molar volume of an ideal gas is equal to 22.4 L.

molarity, M (172): the concentration of a solution, expressed as the number of moles of solute per liter of solution. The units of molarity are $mol \cdot L^{-1}$ and the symbol of molarity is M.

mole (140): the quantity of a substance that is equal to its formula mass in grams. The official SI definition of a mole is the amount of substance of a system that contains as many elementary entities as there are atoms in exactly 0.012 kg of carbon-12. The symbol for a mole is mol.

mole fraction (214, 520): an expression for the concentration of a solution. In a solution containing n_1 moles of solvent and n_2 moles of solute, the mole fraction of the solvent is defined as $X_1 = n_1/(n_1 + n_2)$. The mole fraction is a unitless quantity.

molecular compound (109): a compound composed of molecules; such compounds generally have low melting and low boiling points.

molecular crystal (377, 504): a three-dimensional ordered array of molecules.

molecular diameter, σ (227): the experimentally determined diameter of a molecule.

molecular formula (148): the chemical formula of a molecular compound, which gives the number of atoms of each element making up one formula unit of the compound.

molecular mass (42): the mass of a molecule relative to the atomic mass of carbon-12. The molecular mass is the sum of the atomic masses of the atoms that make up the molecule.

molecular orbital (440): a wave function that describes an electron in a molecule.

molecular orbital theory (440): a theory of bonding based on orbitals that extend over two or more atoms.

molecular substance (140): a substance that is composed of molecules.

molecule (34): an entity in which two or more atoms are joined together.

momentum, p (306): the product of the mass and the speed of an object.

monoclinic (948): in the text, a crystalline form of sulfur.

monomers (1063): small molecules that are joined together to form a polymer

monopolymers (1066): a polymer that is made of one type of monomer.

muriatic acid (959): hydrochloric acid, $HCl(aq)$.

N

nano-, n (9): a prefix meaning 10^{-9}.

natural abundance (50): the percentage of an isotope of an element that is present in the naturally occurring element.

natural law (4): a concise summary of experimental observations regarding some aspects of the behavior of matter in nature.

natural logarithm, ln (A-6): the power to which the number e is raised to obtain a given number. The natural logarithm of a number a is $\ln a = x$, such that $a = e^x$.

negative deviation (Raoult's law) (538): a deviation from Raoult's law by a nonideal solution whose vapor pressure is less than that calculated by Raoult's law.

Nernst equation (882): the quantitative relationship between the cell voltage, E_{rxn}, and the value of the concentration quotient Q. At 25°C the Nernst equation is given by

$$E_{rxn} = E^{\circ}_{rxn} - (0.0592 \text{ V}/n)\log Q \quad \text{or}$$
$$E_{rxn} = (0.0592 \text{ V}/n)\log (Q/K)$$

where E°_{rxn} is the standard cell voltage, n is the number of moles of electrons transferred in the cell reaction as written, and K is the equilibrium constant.

net ionic equation (123): a ionic equation which omits the spectator ions.

neutral anion (660): an anion that does not react with water in aqueous solution to produce either $H_3O^+(aq)$ or $OH^-(aq)$.

neutral atom (47): an atom that has zero charge. In a neutral atom, the number of electrons is equal to the number of protons.

neutral cation (660): a cation that does not react with water in aqueous solution to produce either $H_3O^+(aq)$ or $OH^-(aq)$.

neutral solution (637): an aqueous solution in which $[H_3O^+] = [OH^-]$.

neutralized (125): the condition attained upon complete reaction of an acid with a base or vice versa.

neutralization reaction (125): a reaction between an acid and a base.

neutron (47): a subatomic particle that has almost the same mass as a proton and has no charge.

neutron activation analysis (866): an analytical method used to measure trace quantities of elements. The sample is irradiated by a beam of neutrons, and the nuclear products emit γ-rays of energies that are characteristic of each isotope.

nickel-cadmium battery (848): a battery utilizing the cell

$$\ominus steel|Cd(s)|Cd(OH)_2(s)|LiOH(aq)|$$
$$NiOOH(s), Ni(OH)_2(s)|steel\oplus$$

nitride (929): a compound that contains the nitride ion, N^{3-}.

nitrogen fixation (930): a process whereby $N_2(g)$ is converted into nitrogen-containing compounds.

noble-gas outer electron configuration (88): the outer electron configuration ns^2np^6.

NMR spectrum (434): the spectrum of a molecule obtained from a nuclear magnetic resonance spectrometer.

noble gases (81): the Group 8 elements. The noble gases are helium, neon, argon, krypton, xenon, and radon.

noble-gas electron arrangement (89): the electron configuration of one of the noble gases.

nodal plane (322): a plane on which the value of an orbital is zero.

nodal surface (322): a surface over which the value of an orbital is zero.

nonbonding electron (469): an electron that occupies a nonbonding orbital.

nonbonding orbital (468): a molecular orbital whose electrons neither contribute to nor detract from bonding.

nonelectrolyte (356): a substance that dissolves in water to produce solutions that do not conduct an electric current.

nonmetal (32): a substance that does not have the properties of a metal. The nonmetals are not uniform in their physical appearance or chemical properties.

nonpolar bond (401): a pure covalent bond.

nonpolar molecule (409, 428): a molecule that has no net dipole moment.

normal boiling point (493): the temperature at which the equilibrium vapor pressure of a liquid equals exactly 1 atm.

normal melting point (495): the melting point of a solid at 1 atm.

normal sublimation point (497): the temperature at which the equilibrium vapor pressure of a solid is exactly 1 atm.

n-type semiconductor (916): a semiconductor produced when atoms with five valence electrons are added in minute amounts to silicon or germanium; n stands for negative.

nuclear equation (852): an equation representing a nuclear reaction.

nuclear magnetic resonance (NMR) (434): absorption of radio-wave region electromagnetic energy by a sample in a magnetic field.

nuclear reactor (873): a device using a controlled chain reaction to produce thermal energy to power a heat engine.

nucleon (851): a proton or a neutron in a nucleus.

nucleotide (1077): a monomer of DNA or RNA that consists of a sugar, a phosphate group, and a nitrogen-containing ring compound called a base.

nucleus (46): the central part of an atom in which essentially all the mass and all the positive charge of the atom are concentrated.

number density (222): the number of particles per unit volume.

O

octahedral (416): shaped like an octahedron; the shape of an AX_6 molecule.

octahedron (416): a regular solid body that has six vertices and eight faces, each of which is an identical equilateral triangle. All six vertices are equivalent.

octet rule (378): states that many elements form covalent bonds in order to end up with eight electrons in their outer shells. The octet rule is particularly useful for compounds that contain carbon, nitrogen, oxygen, and fluorine.

oleum (949): fuming sulfuric acid, $H_2S_2O_7$ (35% in H_2SO_4).

opposite spins (330): spins in opposite directions. Two electrons that have opposite spins have different values of m_s.

optical isomer (1071): nonsuperimposable isomers that are mirror images of each other.

orbital (318): a one-electron wave function.

organic acid (638): an organic compound that contains dissociable protons.

organic compound (1025): a compound that contains carbon atoms.

ortho- **(1042):** the designation for disubstituted benzenes with substituents at the 1 and 2 positions.

orthosilicate ion (918): the simplest silicate anion, SiO_4^{4-}

osmosis (532): the spontaneous passage of a solvent (usually water) from a dilute solution to a more concentrated solution through a semipermeable membrane.

osmotic pressure (532): the hydrostatic pressure produced when a solvent passes through a rigid semipermeable membrane from a dilute solution to a more concentrated solution. The osmotic pressure, π, is given by the equation $\pi = RTM_c$, where R is the gas constant, T is the absolute temperature, and M_c is the colligative molarity of the solution.

Ostwald process (931): the conversion of ammonia to nitric acid involving the oxidation of NH_3 to NO_2 and the dissolution of NO_2 in water.

outer electron (295): an electron in the shell with the highest value of the principal quantum number, n.

overlap (440): the combining of an orbital of one atom with an orbital of another atom.

oxidation (104): a process that involves an increase in the oxidation state of an atom. oxidation involves a loss of electrons.

oxidation half reaction (105): the half reaction in which electrons appear on the right-hand side.

oxidation number (101): the oxidation state of an atom.

oxidation-reduction reaction (100): a chemical reaction involving a transfer of electrons from one species to another.

oxidation state (101): a number assigned to an atom in a chemical species by a set of rules based on the number of electrons and the electronegativities of the various atoms in the species.

oxidation-state method of balancing equations (105): the method of balancing oxidation reduction on the basis of changes of oxidation states coupled with the conservation of electrons.

oxidized (104): having transferred electrons or having an increased oxidation state.

oxidizer (109): the reactant in a combustion reaction that acts as the oxidizing agent.

oxidizing agent (104): the reactant that contains the atom that is reduced in an electron transfer reaction.

oxyacid (115): an inorganic acid that contains oxygen atoms.

oxyanion (128): the anion derived from an oxyacid.

ozone layer (630): the region between 15 and 30 km in altitude in the atmosphere that contains ozone, O_3, produced photochemically by the action of solar ultraviolet light on oxygen.

P

p-**block element (339):** an element in Groups 3 through 8 in the periodic table.

p **orbital (320):** an orbital for which $l = 1$. All p orbitals are cylindrically symmetric. There are three p orbitals for each value of $n \geq 2$.

paper chromatography (59): an analytical technique for separating compounds in solution using treated porous paper that is dipped in a liquid which is allowed to wick along the paper, thus separating the compounds.

para- **(1042):** a designation for disubstituted benzenes with substituents at the 1 and 4 positions.

parallel spins (332): spins in the same direction. Electrons that have parallel spins have the same value of m_s.

paramagnetic (444): magnetized by an external magnetic field and consequently attracted to the region between the poles of a magnet. Paramagnetic molecules contain at least one unpaired electron.

partial ionic character (402): the character of a covalent bond in which the electrons are not shared equally by the two atoms comprising the bond, resulting in a partial ionic charge on each atom.

partial pressure (214): the pressure exerted by one component in a mixture of gases.

pascal (199): the SI unit of pressure. A pascal, Pa, is equal to 1 newton per square meter, $1 \ N \cdot m^{-2}$; 1 kilopascal corresponds to about $1/100$ of an atmosphere.

Pauli exclusion principle (329): the principle that no two electrons in the same atom can have the same set of four quantum numbers.

peptide bond (1073): the bond formed when

the amino group on one amino acid reacts with the carboxylic acid group on another amino acid. The peptide bond is

$$-\overset{\overset{\displaystyle ||}{O}}{C}-\overset{\overset{\displaystyle ..}{N}}{\underset{|}{}}-$$
$$\qquad\quad H$$

per- **(129):** the prefix of the name of an oxyacid indicating that the oxyacid has one more oxygen attached to the central atom than the key oxyacid.

percent dissociation (520, 645): the percentage of an acid in aqueous solution that has transferred a proton to water.

percentage yield (161): the ratio of the actual yield of a reaction to the theoretical yield, multiplied by 100.

period (78): a horizontal row of the periodic table.

periodic table of the elements (78): an arrangement of the elements according to increasing atomic number, such that elements that have similar properties appear in the same column of the table.

periodicity (78): the regular repetitive pattern of some property of the elements.

peroxide (897): a compound with an oxygen-oxygen single bond ($-O-O-$). The peroxide ion is O_2^{2-}.

pH (640): a measure of the acidity of an aqueous solution. The pH is defined as $pH = -\log[H_3O^+]$.

pH meter (642): an instrument used to determine pH by means of electrochemical measurements.

pH transition range of an indicator (685): the pH region in which the acid form and the base form of an indicator are present simultaneously in similar amounts. The pH range is equal to $pK_{a1} \pm 1$, where K_{a1} is the acid-dissociation constant of the indicator.

phase diagram (494): a simultaneous plot of the equilibrium vapor pressure curve, the equilibrium sublimation pressure curve, and the solid-liquid equilibrium curve (melting point curve) of a substance.

phosphate rock (928): ores containing the phosphate group PO_4^{3-}.

photochromic glass (920): glass that contains small dispersed amounts of silver chloride.

photodissociation (240): the dissociation of a molecule produced by light absorption.

photodissociation reaction (240): a reaction that occurs as a result of the absorption of radiation.

photoelectric effect (303): the phenomenon in which electrons are ejected from the surface of a metal if the surface is irradiated with certain ultraviolet radiation.

photon (303): a packet of energy that constitutes electromagnetic energy. The energy of one photon is given by $E = h\nu$.

photosynthesis (944): the conversion of CO_2 and H_2O into carbohydrates and O_2 in plants, a process that is driven by the energy of absorbed sunlight.

π bond (460): the result of two electrons occupying a π orbital.

π orbital (442): a localized bond orbital that is the result of combining p atomic orbitals from different atoms. The cross section of a π orbital is similar to that of an atomic p orbital.

π^* orbital (442): the designation for an antibonding π orbital.

pico-, p (9): a prefix meaning 10^{-12}.

pig iron (983): iron obtained directly from a blast furnace.

PIXE (particle-induced X-ray emission) (866): an analytical method that is used to measure quantities of elements in very small samples. A sample that is irradiated with a beam of protons yields nuclear products that emit X-rays of energies characteristic of each isotope.

pK_a (650): a measure of the strength of an acid. The value of pK_a is given by $pK_a = -\log K_a$.

pK_b (655): a measure of the strength of a base. The value of pK_b is given by $pK_b = -\log K_b$.

planar (410): two-dimensional, flat.

planar molecule (392): a molecule in which all the atoms lie in a plane.

planar node (323): a plane in an atomic orbital on which the electron probability density is zero. The number of planar nodes in an orbital is equal to the angular quantum number, l.

Planck's constant, h (302): the proportionality constant that relates the energy, E, and frequency, ν, of electromagnetic radiation. The value of h is 6.626×10^{-34} J·s.

plastic sulfur (949): a rubbery form of sulfur that occurs when liquid sulfur is quenched.

pOH (642): by definition $pOH = -\log [OH^-]$.

polar bond (401): a covalent bond in which the electron pair is not shared equally by each atom, but is more likely to be found near the more electronegative atom.

polar molecule (404, 429): a molecule that has a dipole moment.

polarity (818): a characteristic of an electrochemical cell, such that one of the electrodes is negative and the other is positive with respect to the direction of current flow in the external circuit.

polyatomic ion (111): an ion that consists of more than one atom.

polyatomic molecule (439): a molecule composed of three or more atoms.

polydentate ligand (1002): a ligand that attaches to a metal ion at more than one coordination position.

polymer (918): a long chainlike molecule that is formed by joining together many small molecules called monomers.

polymerization (1063): a reaction in which monomers combine to form a polymer.

polynucleotide (1063): a polymer made up of nucleotides.

polypeptide (1074): a molecule composed of a chain of amino acids joined together by peptide bonds.

polypeptide backbone (1074): the chain in a polypeptide to which the amino acid side groups are attached.

polyprotic acid (699): an acid that can donate more than one proton per acid molecule.

population inversion (351): the situation in which there are many more molecules in the upper energy level than in the lower energy level.

positional disorder (779): the distribution of the particles of a substance over positions in space.

positive deviation (Raoult's law) (539): deviation from Raoult's law by a nonideal solution whose vapor pressure is greater than that calculated by Raoult's law.

positron (853): a particle that has the same mass as an electron, but is positively charged.

potential energy (244): the energy that a mass has as a result of its location.

power (247): the rate of production or utilization of energy.

precipitate (123): an insoluble product of a reaction that occurs in solution.

precipitation reaction (123): a double replacement reaction involving the formation of a precipitate.

precision (14): the agreement of repeated measurements with one another.

pressure (197): force per unit area. Gas pressure is the force exerted by a gas per unit area on the wall of its container. Common units of pressure are the torr and the standard atmosphere, atm. The SI unit of pressure is the pascal, Pa (1 atm = 760 torr = 101.3 kPa = 14.7 psi).

primary alcohol (1045): an alcohol in which the $-OH$ group is attached to a carbon atom that is attached to only one other carbon atom, as in RCH_2OH.

primary battery (846): a nonrechargeable battery.

primary structure (1075): the order of the amino acid units in a polypeptide.

principal quantum number (319): the integer, n, that specifies the energy of the electron in the hydrogen atom. The principal quantum number can take on the values 1, 2, 3, ...

principal species (692): the species of significant stoichiometric concentration in a solution.

probability density (318): the probability that an electron will be found in a small volume, ΔV, surrounding the point (x, y, z). The probability density is given by $\psi^2 \Delta V$, where ψ^2 is the square of the wave function.

product (72): a substance that is formed in a chemical reaction.

proof (alcohol) (538): twice the percentage by volume of alcohol in an alcohol-water solution.

protein (1070): a naturally occurring polypeptide.

proton (46): a subatomic particle that has a positive charge equal in magnitude but opposite in sign to that of an electron, and a mass almost equal to that of a hydrogen atom.

proton acceptor (635): a species capable of accepting a proton from an acid.

proton donor (635): a species capable of donating a proton to a base.

proton magnetic resonance, PMR

(434): nuclear magnetic resonance spectroscopy due to protons only.

proton magnetic resonance spectrum (435): the record of the absorption of radiation at a fixed frequency by a sample.

proton-transfer reaction (636): a reaction in which a proton is transmitted from one species to another. It is also called a protonation reaction.

protonation reaction (636): a reaction involving the transfer of a proton from one species to another. It is also called a proton transfer reaction.

p-type semiconductor (916): a semiconductor produced when atoms with three valence electrons are added in minute amounts to silicon or germanium; p stands for positive.

pure covalent bond (401): a bond in which the electron pair is shared equally by each atom joined by the covalent bond.

pure ionic bond (401): an electrostatic bond formed when an electron from one atom is transferred completely to another atom.

pyrolysis (906): the transformation of one substance into another by heat without oxidation.

Q

Q_c (620): the reaction quotient, expressed in terms of the concentrations of gaseous or dissolved reactants and products.

Q_p (620): the reaction quotient, expressed in terms of the pressures of gaseous reactants and products.

Q/K (621): the value of the ratio of the reaction quotient to the equilibrium constant of a chemical reaction. The numerical value of Q/K indicates the direction in which a nonequilibrium reaction system spontaneously proceeds toward equilibrium.

quadratic equation (606): an algebraic equation that can be put in the form $ax^2 + bx + c = 0$, where a, b, and c are known and x is unknown.

quadratic formula (606): the solution to a quadratic equation. The two roots of a quadratic equation are given by

$$x = \frac{-b \pm \sqrt{b^2 - 4ac}}{2a}$$

qualitative analysis (731): the determination of the species present in a sample.

qualitative data (3): descriptive observations.

qualitative observation (6): expression of the results of an observation or experiment as general (nonnumerical) characteristics.

quanta (302): discrete units of electromagnetic energy.

quantitative analysis (731): the determination of the amount of each species in a sample.

quantitative data (3): numbers obtained by measurements.

quantitative measurement (6): expression of the result of an observation or experiment as a number.

quantitative reaction (764): a chemical reaction whose equilibrium constant is very large.

quantization of energy (306): the restriction of energy to certain discrete values.

quantized (310): restricted to certain fixed values.

quantum condition (310): the relationship between the radius of the orbit of an electron around a hydrogen nucleus and the wavelength of the electron. In an equation, $2\pi r = n\lambda$; $n = 1, 2, 3, \ldots$.

quantum number (318): an integer or half-integer that, in sets of four, characterizes the energy states of atoms.

quantum theory (306): the theory that predicts the quantized energies of particles.

quenching (949): quickly placing a heated substance into cold water.

R

radioactive (45): decomposed spontaneously by the emission of a small particle such as an α-particle or a β-particle.

radioactive decay (854): the process by which a radioactive nucleus emits a particle and is transformed into another nucleus.

radioactivity (45): a property of certain nuclei that spontaneously emit small particles such as α particles or β particles.

radiocarbon dating (862): carbon-14 dating.

radioisotope (852): a radioactive isotope.

Raoult's law (524): a law stating that the equilibrium vapor pressure of a solvent over a solution is proportional to the mole fraction of the solvent. The equation is $P_1 = X_1 P_1^o$, where P_1 is the equilibrium vapor pressure of the solution, P_1^o is the equilibrium vapor pressure of the pure solvent at the same temperature as the solution, and X_1 is the mole fraction of the solvent.

rare earth element (86): any member of the lanthanides, or the series of elements lanthanum ($Z = 57$) through ytterbium ($Z = 70$). The rare earths occur because of the sequential filling of the $4f$ orbitals.

Raschig synthesis (932): the formation of hydrazine, N_2H_4, via the reaction of ammonia, NH_3, with sodium hypochlorite, $NaClO$.

rate constant (553): the proportionality constant between the reaction rate and the concentrations of the reactants that appear in the rate law.

rate-determining step (572): the step in a reaction mechanism that controls the overall reaction rate. The rate-determining step is much slower than any other step in the reaction mechanism.

rate law (553): the rate of a reaction, expressed in terms of the concentrations of the species that affect the reaction rate.

reactant (72): a substance consumed in a chemical reaction; it appears on the left side of the chemical equation.

reaction mechanism (570): the sequence of elementary processes by which reactants are converted to products.

reaction quotient, Q (620): the ratio of arbitrary or initial product concentrations to arbitrary or initial reactant concentrations, with each concentration factor raised to a power equal to the stoichiometric coefficient of that species in the balanced equation. Pure liquids and solids do not appear in the reaction quotient expression. For the balanced chemical equation

$$aA(g) + bB(soln) + cC(s) \rightleftharpoons xX(g) + yY(soln) + zZ(l)$$

the reaction quotient is given by

$$Q = \frac{[X]_o^x [Y]_o^y}{[A]_o^a [B]_o^b}$$

reaction rate (549): the rate at which a reactant, A, is consumed or a product, P, is produced. The rate is defined as rate $= -\Delta[A]/t = \Delta[P]/t$. Units of the reaction rate are $M \cdot s^{-1}$ (moles per liter per second).

reactivity series (120): an ordering of the metals according to their chemical reactivity.

red phosphorus (934): an allotrope of phosphorus, which is less reactive than the white form.

redox reaction (100): an oxidation-reduction or electron-transfer reaction.

reduced (104): the decreased oxidation state of a species; having gained electrons.

reducing agent (104): the reactant that contains the atom that is oxidized in an electron transfer reaction.

reduction (104): a process that involves a decrease in the oxidation state of an atom. Reduction involves a gain of electrons.

reduction half reaction (757): the half reaction in which electrons appear on the left-hand side.

relative humidity (493): the ratio of the partial pressure of the water vapor in the atmosphere to the equilibrium vapor pressure of water at the same temperature, multiplied by 100

relative humidity $= (P_{H_2O}/P_{H_2O}^o) \times 100$.

resonance (389): the procedure of superimposing each of the possible Lewis formulas for a molecule or ion to obtain a more accurate picture of the electron distribution.

resonance (434): the absorption of radiation by a molecule when the magnetic field strength is just enough to cause protons to make transitions.

resonance form (389): one of the possible Lewis formulas that can be written for a molecule or an ion without altering the positions of the nuclei.

resonance hybrid (389): a superimposed formula of a species for which it is possible to write more than one satisfactory Lewis formula.

resonance stabilization (392): the stability of a species, ascribed to the fact that the energy of the species, represented by a superposition of Lewis formulas, is lower than the energy of any of its individual Lewis formulas.

reverse osmosis (533): the process in which the solvent (usually water) passes through a rigid semipermeable membrane from a solution to the pure solvent, as a result of applying to the solution a pressure in excess of the osmotic pressure.

reverse rate (596): the rate of the reverse (right to left) reaction.

reverse reaction (595): the formation of the reactants from the reaction products, or the reaction that proceeds right to left for the equation as written.

reverse reaction rate (596): the rate of the reaction that occurs right to left as the equation is written.

rhombic (949): in the text, a crystalline form of sulfur.

RNA (1077): ribonucleic acid. RNA is a polynucleotide composed of the sugar ribose, the phosphate group, and the four bases adenine, uracil, guanine, and cytosine.

root-mean-square (223): the square root of the average or mean of the square of some property.

root-mean-square speed (223): the square root of the average or mean of the square of the molecular speeds.

rule of Dulong and Petit (271): the experimental observation that for solid metals the value of the molar heat capacity, C_p, is approximately equal to $3R$, where R is the molar gas constant.

Rydberg constant (301): the constant that appears in the Rydberg-Balmer equation. The value of the Rydberg constant is 1.0974×10^7 m^{-1}.

Rydberg-Balmer equation (300): the empirical equation for the wavelengths of lines in the visible spectrum of hydrogen

$$1/\lambda = (1.0974 \times 10^7 \text{ m}^{-1})(1/4 - 1/n^2)$$
$$\text{for } n = 3, 4, 5, \ldots .$$

S

s-block element (339): Groups 1 and 2 in the periodic table.

s orbital (320): an orbital for which $l = 0$. All s orbitals are spherically symmetric.

sacrificial anode (767): a reactive piece of metal that is electrically connected to a less active metal and is preferentially oxi-

dized, thereby protecting the less active metal against corrosion.

salt (125): an ionic compound formed in a neutralization reaction or in the reaction of a metal and a nonmetal.

salt bridge (816): a concentrated electrolyte solution suspended in a gel that is used to make electrical contact between two different electrolyte solutions that are part of an electrochemical cell.

saturated hydrocarbon (1026): a hydrocarbon in which the bonding about each carbon atom is tetrahedral, and no more hydrogen atoms can be added to any carbon atom. (See **alkane**.)

saturated solution (172): a solution in which no more solute can be dissolved.

Schrödinger equation (318): the central equation of quantum theory, which takes into account the wave nature of electrons and yields the discrete energy levels of atoms and molecules.

scientific law (4): a concise summary of a large body of experimental results.

scientific method (3): the use of carefully controlled experiments to answer certain questions.

scientific theory (4): an explanation of a scientific law and an aid in making predictions that lead to new knowledge.

scintillation counter (856): a device that detects and measures radioactive emissions by analyzing the fluorescence of certain materials, using their interaction with the emissions from a radioactive source.

second electron affinity, E_{A2} (366): the energy associated with the process of adding an electron to a singly charged negative ion according to

$$X^-(g) + e^- \rightarrow X^{2-}(g).$$

second excited state (311): the second available energy state above the ground state.

second ionization energy (291): the minimum energy required to remove an electron from an A^+ ion to produce the A^{2+} ion.

second law of thermodynamics (777): a law stating that the total entropy change for a spontaneous process must always be positive. The entropy of the universe can only increase.

second-order rate law (556): a rate law in which the reaction rate is proportional to the second power of the concentration of a reactant. The rate law is of the form rate = $k[A]^2$ or rate = $k[A][B]$.

second-order reaction (556): a reaction whose rate law is a second-order rate law.

secondary alcohol (1045): an alcohol in which the —OH group is attached to a carbon atom that is attached to two other carbon atoms.

secondary battery (846): a battery that is rechargeable.

secondary structure (1077): the coiled helical portion in different regions of a protein chain.

seesaw (422): the shape of a AX_4E class molecule or ion.

selective precipitation (730): the separation of two ions by the formation of salts with different solubilities.

semiconductor (84, 506): a semimetal. A semimetal conducts electricity and heat less well than metals but better than nonmetals.

semimetal (79): an element that has properties intermediate to metals and nonmetals.

semipermeable membrane (532): a membrane that allows the passage of only certain species, such as water molecules.

shell (294, 330): the energy level designated by the principal quantum number, n. The $n = 1$ shell is called the K shell; the $n = 2$ shell, the L shell; the $n = 3$ shell, the M shell; and so forth.

SI units (7): the internationally accepted set of units in the metric system. SI stands for Système International.

side group (1070): the group represented by —G in the general formula for an amino acid.

σ bond (450): a bond that occurs when a σ orbital is occupied by one electron or two electrons of opposite spin.

σ-bond framework (460): all the σ bonds that are formed in a molecule or an ion.

σ orbital (442): a molecular orbital that is cylindrically symmetric when viewed along a line drawn between the two nuclei joined by the covalent bond.

σ^* orbital (442): designation for an antibonding σ orbital.

significant figures (14): the precision of a measured quantity as indicated by the number of digits used to express the result.

silicate polyanion (polysilicate ion) (919): an anionic polymer composed of silicon and oxygen.

silicones (199): polymers that consist of silicon and hydrocarbon groups.

simple cubic unit cell (500): the unit cell in which the components of a crystal are located at the corners of a cube.

simplest formula (145): the formula of a substance derived from an analysis of the composition of the compound. It gives the relative number of atoms in the formula unit, using the smallest possible integers.

single replacement reaction (118): a reaction in which one element in a compound is replaced by another element.

slag (983): a molten calcium silicate by-product of the production of iron in a blast furnace.

slope (202): the ratio of the change in the vertical coordinate to the change in the horizontal coordinate, $\Delta y / \Delta x$.

soda ash (898): sodium carbonate, Na_2CO_3.

softened water (713): water from which divalent cations have been removed.

solid (196): the physical state of matter having the properties of fixed volume, fixed shape, and very small compressibility.

solubility (172, 715): the maximum quantity of solute that can be dissolved in a given quantity of solvent in ordinary circumstances.

solubility product constant, K_{sp} (715): the equilibrium-constant expression obtained by applying the law of concentration action to the equilibrium between an ionic solid and its constitutent ions in solution.

solubility rules (181): a set of guidelines that can be used to predict whether an ionic compound is soluble or insoluble in water.

solute (172): a substance that is dissolved in another substance to form a solution.

solution (55, 171): a mixture of two or more substances that is uniform and homogeneous at the molecular level.

solvated (108, 519): dissolved in water.

solvent (172): the substance in which a solute is dissolved to form a solution. The solvent is generally present in greater quantity than the solute.

sp orbital (448): one of the two equivalent hybrid orbitals obtained by the combination of an ns orbital and one np orbital on the same atom. The two sp orbitals point 180° from each other.

sp^2 **orbital (451):** one of the three equivalent hybrid orbitals obtained by the combination of an ns orbital and two np orbitals on the same atom. The three sp^2 orbitals point to the vertices of an equilateral triangle.

sp^3 **orbital (452):** one of the four equivalent hybrid atomic orbitals obtained by the combination of an ns orbital and the three np orbitals on the same atom. The four sp^3 orbitals point to the vertices of a tetrahedron.

space-filling molecular model (411): a model of a molecule that is constructed to represent the relative sizes of the atoms in the molecule and the angles between bonds.

specific activity (877): a measure of the activity of a radioactive substance; the number of nuclei that disintegrate per second per gram of radioactive isotope.

specific heat, c_{sp} (268): the heat capacity per gram of substance. The SI unit of specific heat is joules per kelvin per gram, $J \cdot K^{-1} \cdot g^{-1}$.

spectator ion (123): an ion present in a solution that does not participate directly in a reaction.

spectrochemical series (1012): an arrangement of ligands in order of the increasing ability of the ligands to split the metal d orbitals: $Cl^- < F^- < H_2O < NH_3 < NO_2^- < CN^- < CO$.

spectroscopic method (434): any of a number of analytical methods that employ spectra to identify or measure the amounts of compounds.

spectrum (299): a plot of the electromagnetic energy absorbed versus the wavelength of the energy for a species.

speed of light, c (297): light travels with a speed of 3.00×10^8 m·s^{-1}.

spherical node (323): a spherical surface in an atomic orbital over which the electron probability density is zero. The number of spherical nodes in an orbital is equal to $n - 1 - l$, where n is the principal quantum number and l is the angular quantum number.

spherically symmetric (319): characteristic of a function that depends only on its distance from a center and not on its direction in space.

spin down (330): the state in which the spin quantum number, m_s, is $-1/2$.

spin-paired electrons (330): two electrons that occupy the same orbital.

spin quantum number (325): the quantum number, m_s, that designates the spin state of an electron. The allowed values of m_s are $+1/2$ or $-1/2$.

spin-unpaired electron (330): a single electron in an orbital.

spin up (330): the state in which the spin quantum number, m_s, is $+1/2$.

spontaneous (776): proceeding without external input.

spontaneous emission (350): the process of the spontaneous return of an excited molecule to the ground state with the emission of a photon.

square planar (425): the shape of an AX_4E_2 molecule.

square pyramidal (424): shaped like a pyramid with a square base; the shape of an AX_5E molecule.

standard atmosphere (atm) (199): a unit of pressure. One atmosphere is equal to 760 torr.

standard cell voltage, E_{rxn}° (825): the voltage of an electrochemical cell when Q for the cell reaction is equal to 1. The standard cell voltage, is related to the equilibrium constant for the reaction by the equation $E^{\circ} = (0.0592 \text{ V}/n) \log K$. at 25°C, where n is the number of moles of electrons transferred in the equation as written.

standard enthalpy change, ΔH_{rxn}° (253): the amount of energy absorbed or evolved as heat by a reaction when all gases are at 1 atm and all solution species are at 1 M.

standard enthalpy of formation, ΔH_f°, (260): the enthalpy of formation of a substance at standard conditions from its elements at standard conditions.

standard entropy, S° (783): the entropy of a substance at 1 atm. If the substance is a solute, then its concentration is 1 M.

standard entropy change, ΔS_{rxn}° (783): the value of the entropy change for a reaction when all gases, solids, and liquids are at 1 atm and all solutes are at 1 M.

standard Gibbs free energy change, ΔG_{rxn}° (790): the value of ΔG_{rxn} at 25°C when all gases are at 1 atm and all solution species are at one molar.

standard molar enthalpy of formation, ΔH_f° (261): the enthalpy change for the reaction in which one mole of substance at 1 atm and 25°C is formed from the elements at 1 atm and 25°C.

standard molar Gibbs free energy of formation, ΔG_f° (793): the Gibbs free energy change for the formation of a substance at 25°C from its constituent elements, all at standard conditions.

standard reduction voltage, E_{red}° (827: the standard voltage for a half reaction based on the assignment of $E_{red}^\circ = 0$ for the electrode reaction

$$2\,H^+(aq,\,1\,M) + 2\,e^- \rightarrow H_2(g,\,1\,atm)$$

The standard cell voltage, E_{cell}°, is given by $E_{cell}^\circ = E_{red}^\circ + E_{ox}^\circ$, where E_{red}° is the standard reduction voltage of the right electrode in an electrochemical cell and E_{ox}° $(-E_{red}^\circ)$ is the standard oxidation voltage of the left electrode in an electrochemical cell.

state function (250): a function that depends only on the state of a system and not on how that state is achieved.

stationary state (311): one of the allowed energy states of an atom or a molecule.

steam reforming reaction (893): a large-scale process of producing hydrogen from natural gas and steam according to the reaction:

$$CH_4(g) + H_2O(g) \rightarrow CO(g) + 3\,H_2(g).$$

steel (984): an alloy composed primarily of iron with variable amounts of carbon and other substances added to produce special properties.

stereoisomers (462): molecules with the same atom-to-atom bonding but different spatial arrangements of the atoms.

stereospecific reaction (1073): a reaction in which the relative orientations of the groups attached to a specific atom are unchanged.

stimulated emission (351): the process in which an excited molecule encounters photons whose energies just match the energy difference between the excited state and a lower state, which causes the molecule to emit light.

stoichiometric coefficient (149): a numerical coefficient of a reactant or product species in a balanced chemical equation; a balancing coefficient.

stoichiometric unit conversion factor (150): the relationship between two species in a balanced chemical equation.

stoichiometrically equivalent to $\rightleftharpoons$ (145): a symbol that denotes a stoichiometric correspondence between two quantities.

stoichiometry (145): the procedures for the calculations of the quantities of elements or compounds involved in chemical reactions.

stratosphere (232): the region of the atmosphere from 10 to 50 km in altitude, which lies above the troposphere.

strong acid (638): an acid that is completely dissociated in aqueous solution.

strong base (638): a base that produces a concentration of $OH^-(aq)$ in aqueous solution equal to the concentration of the base.

strong electrolyte (356): an electrolyte that is 100 percent or close to 100 percent dissociated into ions in solutions.

structural chemistry (411): the area of chemistry in which the shapes and sizes of molecules are studied.

structural formula (1026): a formula that indicates the various attachments of the atoms in a molecule.

structural isomers (1027): compounds that have the same molecular formula but different arrangements of the atoms.

subatomic particle (44): one of the particles of which atoms are composed; a proton, neutron, or electron.

subcritical mass (872): a mass of a spontaneously fissionable substance that is insufficient to sustain a nuclear chain reaction.

sublimation (480): the process whereby a solid is converted directly to a gas.

sublimation pressure curve (495): a plot of the pressure of a vapor in equilibrium with its solid versus the temperature.

subshell (330): the group of orbitals designated by a particular l value within a shell.

substituent (1030): atoms or groups of atoms that replace a hydrogen atom bonded to a carbon atom.

substitution reaction (118, 1029): a single replacement reaction.

substrate (583): the reactant that binds to the enzyme in an enzyme-catalyzed reaction.

sugar-phosphate backbone (1079): the chain in a polynucleotide to which the bases are attached. The nucleotides are joined by bonds with the oxygen atom at the five-carbon position on one nucleotide, through a phosphate group, to the three-carbon position on a second nucleotide.

superconductor (514): a conductor that has zero electrical resistance.

supercritical mass (872): a mass of a spontaneously fissionable substance that exceeds the mass necessary to sustain a nuclear chain reaction.

supernatant (740): the solution that remains after a substance has been precipitated from the original solution.

superoxide (897): a substance that contains the ion O_2^-, which is called the superoxide ion.

superphosphate (936): water-soluble compound $Ca(H_2PO_4)_2$, which is an important fertilizer.

surface tension (488): the net attractive force toward the interior of a liquid that a molecule at the surface of the liquid experiences.

surfactant (489): a substance that lowers the surface tension of a liquid.

systematic name (93): the nomenclature of ionic compounds that indicates the charge on a metallic ion that may have more than one charge by using a Roman numeral.

T

t_2 orbitals (1014): the set of d orbitals of higher energy on a metal ion in a tetrahedral complex. The t_2 orbitals consist of d_{xy}, d_{xz}, and d_{yz}.

t_{2g} (1006): the set of d orbitals of lower energy on a metal ion in an octahedral complex. The t_{2g} orbitals consist of $d_{x^2-y^2}$ and d_{z^2}.

Taube's rules (1015): the rules that predict whether a complex is inert or labile.

temperature (10): a quantitative measure of the relative tendency of heat to escape from a body.

temperature scale (11): measurement of temperature established by assigning temperatures to two reference systems.

termination reaction (1064): a reaction that stops the growth of a polymer chain.

tertiary alcohol (1045): an alcohol in which the —OH group is attached to a carbon atom that is, in turn, attached to three other carbon atoms.

tertiary structure (1077): the three-dimensional shape of a protein.

tetrahedral bond angle (411): the angle between two bonds formed by an atom whose bonds are directed toward the vertices of a tetrahedron. The tetrahedral bond angle is 109.5°.

tetrahedron (410): a regular solid body that has four equivalent vertices and four equivalent faces, each of which is an equilateral triangle.

tetravalent (411): bonded to four other atoms.

theoretical yield (161): the mass of a particular product that is calculated to result if all the limiting reactant in a balanced chemical equation is converted to products.

theory (4): an explanation of a scientific law.

thermal decomposition (116): a decomposition reaction that occurs when the temperature of a substance is increased.

thermal disorder (779): the distribution of available energy among the particles of a substance.

thermite reaction (160): a reaction between aluminum metal and a metal oxide that yields Al_2O_3 and the free metal.

thermochemistry (252): the study of the heat evolved or absorbed in chemical reactions.

thermodynamics (248): the study of energy transfers.

thermometer (11): a device used to measure temperature.

thin-layer chromatography (59): see **paper chromatography.**

thio- (953): a prefix that designates the replacement of an oxygen atom with a sulfur atom.

three-center bond (906): a bond that results when electrons occupy a three-center bond orbital.

three-center bond orbital (906): a localized bond orbital that is spread over three atoms and can be described as a combination of one atomic orbital from each of the three atoms.

3d transition-metal series (337): the series of ten elements with atomic numbers 21 through 30.

threshold frequency (303): the minimum frequency of ultraviolet radiation necessary for electrons to be ejected from a particular metal. The threshold frequency is denoted by v_0.

tin disease (920): the conversion of white tin to gray tin.

tincture of iodine (962): an alcohol solution of iodine, formerly used as an antiseptic.

titrant (182): the solution of acid (base) that is added to a solution of base (acid) during a titration.

titration (182, 681): the nuetralization of a given volume of a basic (acidic) solution by slowly adding an acidic (basic) solution of known concentration until the base (acid) has been completely neutralized.

titration curve (681): a plot of the pH of the solution that results when an acid solution is titrated with a base solution (or when a base solution is titrated with an acid solution) as a function of the volume of the added solution.

torr (198): a convenient unit of pressure. The pressure of a gas is expressed as the height of a column of mercury that is supported by the pressure of the gas. One torr is one millimeter of mercury.

trans- **(462):** across from.

trans-cis isomerism (462, 1004): geometric isomerism.

transitional metal (85): an element in the groups that are not headed by a number in the periodic table. They serve as a transition between the reactive Group 1 and Group 2 metals and the nonmetals.

translational motion (474): movement through space from one point to another.

transmutation (864): a nuclear reaction in which one element is converted to another.

transuranium element (339): an element of atomic number greater than that of uranium ($Z > 92$).

tridentate (1002): a chelating ligand that attaches to three metal coordination positions.

trigonal bipyramid (415): a solid body that has five vertices and six equilateral triangular faces. A trigonal bipyramid has the shape of two triangular pyramids that share an equilateral triangular base.

trigonal planar (414): the shape of a molecule in which the three atoms bonded to a central atom lie at the vertices of an equilateral triangle, with the central atom in the center of the triangle. All four atoms lie in the same two-dimensional surface.

trigonal pyramid (417): a solid body that has four vertices and four triangular faces. Three of the four faces are identical. The unique face serves as a triangular base for the pyramid.

triple bond (387): the bond formed between two atoms by three shared electron pairs.

triple point (495): a point on a phase diagram at which three phases of a substance coexist in equilibrium.

triplet code (1081): the sequence of three bases or nucleotides along a DNA segment that codes for a particular amino acid.

triprotic acid (699): an acid that has three dissociable protons.

tritium (891): the isotope of hydrogen containing one proton and two neutrons.

troposphere (238): the lowest region of the atmosphere, from 0 to 10 km above sea level, which accounts for 80 percent of the mass and contains essentially all the earth's weather.

T-shaped (422): the shape of an AX_3E_2 molecule.

U

uncertainty principle (316): the fundamental principle of nature that states that it is not possible to determine exactly the momentum and position of a particle simultaneously.

unit cell (499): the smallest subunit of a crystal lattice that contains all the structural information about the crystal.

unit conversion factor (18): an expression that is used to convert a physical quantity from one unit to another.

unsaturated hydrocarbon (1034): a hydrocarbon that contains carbon atoms which are bonded to each other by double or triple bonds. Hydrogen atoms may be added to unsaturated hydrocarbons.

unsaturated solution (172): a solution in which more of the same solute can be dissolved.

uranium-lead dating (862): the determination of the age of a rock sample from the masses of uranium-238 and lead-206 in the sample.

V

valence bond (506): the set of bonding orbitals that extend throughout a crystal and that are analogous to bonding orbitals in a molecule.

valence-bond theory (447): a theory of chemical bonding that involves localized bond orbitals and possibly delocalized bond orbitals.

valence electron (295, 376): an electron in the outer shell of an atom.

valence-shell electron-pair repulsion (VSEPR) theory (413): the theory that the shape of a molecule is determined by the mutual repulsion of the electron pairs in the valence shell of the central atom.

valence-shell electrons (295, 334): the electrons that are located in the outermost occupied shell of an atom.

van der Waals constants (228): the two constants, a, and b, that appear in the van der Waals equation. The values of a and b depend on the particular gas.

van der Waals equation (228): an equation that describes the behavior of a nonideal gas. The equation is

$$(P + n^2a/V^2)(V - nb) = nRT$$

where a and b are van der Waals constants, whose values depend on the particular gas.

van der Waals forces (484): a general term for the attractive forces between molecules; van der Waals forces include dipole-dipole forces and London forces.

van't Hoff equation (796): the equation that governs the temperature dependence of the equilibrium constant. The van't Hoff equation is

$$\ln(K_2/K_1) = \Delta H^o_{rxn}/R)(T_2 - T_1/T_1T_2)$$

where ΔH^o_{rxn} is the standard enthalpy of reaction and R is the gas constant.

van't Hoff factor, i (529): the number of solute particles produced per dissolving formula unit.

vapor pressure (492): the pressure of a gas above its liquid.

vapor pressure curve (492): a plot of the equilibrium vapor pressure of a liquid verses its temperature.

vapor pressure lowering (524): the amount by which the equilibrium vapor pressure of a solution is less than the equilibrium vapor pressure of the pure solvent.

vector (403): a quantity that has both magnitude and direction.

viscosity (486): the resistence of a liquid to flow.

visible region (298): the region of the electromagnetic spectrum from 400 to 700 nm.

volt (814): the SI unit of voltage.

volatile (57): readily converted from a liquid to a gas; a substance with a relatively high equilibrium vapor pressure.

voltaic cell (815): electrochemical cell.

voltaic pile (813): a device, consisting of alternate discs of dissimilar metals, each separated by an electrolyte layer, that is used to produce a voltage.

volumetric flask (173): a precision-made piece of glassware that can be used to prepare a precise liquid volume.

vulcanization (1069): the formation of —S—S— cross-links between polyisoprene chains by reaction with sulfur.

W

water-gas reaction (253): the reaction in which steam is passed over hot carbon to produce a mixture of carbon monoxide and hydrogen:

$$C(s) + H_2O(g) \rightarrow CO(g) + H_2(g)$$

watt (247): the SI unit of power. One watt is one joule per second.

wave function (318): a function that describes the positions of the electrons in an atom or a molecule. A wave function is denoted by ψ, and has the physical interpretation that $\psi^2\Delta V$ is the probability that an electron will be located in the little volume ΔV.

wave-particle duality (306): the concept that both light and matter appear to be particlelike under certain conditions and to be wavelike under other conditions.

wavelength, λ (297): the distance between successive maxima or minima of a wave. A unit of wavelength is the meter.

weak acid (638): an acid that is incompletely dissociated in aqueous solution or that reacts only partially with water.

weak base (638): a base that reacts only partially with water.

weak electrolyte (356): a substance that dissolves in water to produce solutions that conduct an electric current poorly.

weight (10): the force of attraction of an object to a large body, such as the earth.

white phosphorus (934): an allotrope of

phosphorus, which is very reactive chemically, consisting of tetrahedral P_4 molecules.

work, *w*, (248): energy transferred when a force acts to cause a displacement of the boundaries of a system.

work function, Φ (304): the minimum energy required to remove an electron from the surface of a metal.

X

X-ray crystallography (504): the determination of the positions of atoms in the unit cell of a crystal by use of X-ray diffraction studies.

X-ray diffraction (307): the scattering of X-rays in a definite pattern by a crystalline substance.

X-ray diffraction pattern (498): the array of spots on an X-ray film that results when X-rays pass through a crystal.

Y

yield (161): the percent conversion of reactants to products in a chemical reaction.

Z

zero-order reaction (557): a reaction whose reaction rate does not depend on the concentration of any species.

zone refining (916): a recrystallization method involving a moving melted zone that is used to prepare extremely pure crystals.

Physical Constants

Constant	Symbol	Value
atomic mass unit	amu	1.66056×10^{-27} kg
Avogadro's number	N	6.02205×10^{23} mol^{-1}
Bohr radius	a_0	5.292×10^{-11} m
Boltzmann constant	k	1.38066×10^{-23} J$\cdot$K^{-1}
charge of a proton	e	1.60219×10^{-19} C
Faraday constant	F	96,485 C$\cdot$mol^{-1}
gas constant	R	8.31441 J$\cdot$K$^{-1}\cdot$mol^{-1}
		0.08206 L$\cdot$atm$\cdot$K$^{-1}\cdot$mol^{-1}
mass of an electron	m_e	9.10953×10^{-31} kg
		5.48580×10^{-4} amu
mass of a neutron	m_n	1.67495×10^{-27} kg
		1.00866 amu
mass of a proton	m_p	1.67265×10^{-27} kg
		1.00728 amu
Planck's constant	h	6.62618×10^{-34} J$\cdot$s
speed of light	c	2.997925×10^8 m$\cdot$s^{-1}

SI Prefixes

Prefix	Multiple	Symbol	Prefix	Multiple	Symbol
tera	10^{12}	T	deci	10^{-1}	d
giga	10^{9}	G	centi	10^{-2}	c
mega	10^{6}	M	milli	10^{-3}	m
kilo	10^{3}	k	micro	10^{-6}	μ
			nano	10^{-9}	n
			pico	10^{-12}	p
			femto	10^{-15}	f
			atto	10^{-18}	a